Advances in Biological Solid-State NMR
Proteins and Membrane-Active Peptides

New Developments in NMR

Editor-in-Chief:
Professor William S. Price, *University of Western Sydney, Australia*

Series Editors:
Professor Bruce Balcom, *University of New Brunswick, Canada*
Professor István Furó, *Industrial NMR Centre at KTH, Sweden*
Professor Masatsune Kainosho, *Nagoya University, Japan*
Professor Maili Liu, *Chinese Academy of Sciences, Wuhan, China*

Titles in the Series:

1: Contemporary Computer-Assisted Approaches to Molecular Structure Elucidation
2: New Applications of NMR in Drug Discovery and Development
3: Advances in Biological Solid-State NMR: Proteins and Membrane-Active Peptides

How to obtain future titles on publication:
A standing order plan is available for this series. A standing order will bring delivery of each new volume immediately on publication.

For further information please contact:
Book Sales Department, Royal Society of Chemistry, Thomas Graham House, Science Park, Milton Road, Cambridge, CB4 0WF, UK
Telephone: +44 (0)1223 420066, Fax: +44 (0)1223 420247
Email: booksales@rsc.org
Visit our website at www.rsc.org/books

Advances in Biological Solid-State NMR
Proteins and Membrane-Active Peptides

Edited by

Frances Separovic
School of Chemistry, Bio21 Institute, University of Melbourne, VIC 3010, Australia
Email: fs@unimelb.edu.au

and

Akira Naito
Graduate School of Engineering, Yokohama National University, Yokohama 240-8501, Japan
Email: naito@ynu.ac.jp

ROYAL SOCIETY
OF CHEMISTRY

New Developments in NMR No. 3

ISBN: 978-1-84973-910-8
ISSN: 2044-253X

A catalogue record for this book is available from the British Library

Published by The Royal Society of Chemistry,
Thomas Graham House, Science Park, Milton Road,
Cambridge CB4 0WF, UK

Registered Charity Number 207890

For further information see our web site at www.rsc.org

Preface

This book describes the methodology and applications of solid-state NMR spectroscopy to studies of proteins, membrane-active peptides and model biological membranes. As well as structural studies, membrane interactions and molecular motions are described. Experts who are leaders in the development and application of biological solid-state NMR have contributed an exciting range of topics that cover recent advances in high-resolution structure determination of membrane-associated proteins and peptides by multidimensional solid-state NMR spectroscopy; studies of orientation, conformation, dynamics and interaction of proteins and peptides in magnetically aligned membrane systems; developments of novel solid-state NMR techniques, including resolution and sensitivity enhancement; and computational and dynamic information of biological molecules.

Solid-state NMR is a unique technique in structural biology, since local motions can be studied without the complexity of overall tumbling of biological complexes over a wide dynamic range in relation to biological function. The molecular arrangement of supramolecular complexes and fibril formation in relation to molecular folding, misfolding and aggregation can also be addressed. We are grateful to our contributors who have showcased developments in biological solid-state NMR, which have been arranged under the following sections: (i) introduction to biological solid-state NMR (Chapters 1–5); (ii) applications to biological membranes (Chapters 6–10); (iii) membrane-active peptides (Chapters 11–16); (iv) membrane proteins (Chapters 17–23); and (v) protein assemblies and in-cell NMR (Chapters 24–28). We envisage these topics will prove useful to both novices and seasoned practitioners of NMR and signal a resurgence in solid-state applications to biological systems.

New Developments in NMR No. 3
Advances in Biological Solid-State NMR: Proteins and Membrane-Active Peptides
Edited by Frances Separovic and Akira Naito

Published by the Royal Society of Chemistry, www.rsc.org

Finally, we would like to thank Professor William S. Price for inviting us to edit this book as part of the *New Developments in NMR* series. We also thank Ms Rosalind Searle of the Royal Society of Chemistry for her encouragement and assistance in putting the book together.

Frances Separovic
Akira Naito

Contents

New Developments in NMR No. 3
Advances in Biological Solid-State NMR: Proteins and Membrane-Active Peptides
Edited by Frances Separovic and Akira Naito

Published by the Royal Society of Chemistry, www.rsc.org

CHAPTER 1

Introduction to Biological Solid-State NMR

M. WEINGARTH AND M. BALDUS*

Bijvoet Center for Biomolecular Research, Faculty of Science,
Utrecht University, Padualaan 8, 3584 CH Utrecht, The Netherlands
*Email: M.Baldus@uu.nl

1.1 Preface

Solid-state nuclear magnetic resonance has emerged as an established spectroscopic technique to provide atomic-scale information in complex biological systems. Recent years, with the appearance of the first *de novo* structures of membrane proteins[1] and solid-state NMR spectroscopists unraveling the details of native cellular components at atomic resolution,[2] indeed seem to mark a watershed moment for solid-state NMR becoming a leading technique to study highly disordered or heterogeneous biological systems.

These stunning advances, however, are hardwon. What catches one's eye, especially for eyes used to the spectral resolution of liquid-state NMR, is the broadness of non-modulated solid-state NMR signals. For liquid-state spectroscopists, such data may appear as a featureless blob, from which structural parameters are hardly deducible, let alone at atomic resolution (Figure 1.1). The broadness of the signals, which easily exceeds dozens of kHz or even several MHz, results from the presence of anisotropic interactions in solid-state NMR spectra, and a large part of solid-state NMR

New Developments in NMR No. 3
Advances in Biological Solid-State NMR: Proteins and Membrane-Active Peptides
Edited by Frances Separovic and Akira Naito

Published by the Royal Society of Chemistry, www.rsc.org

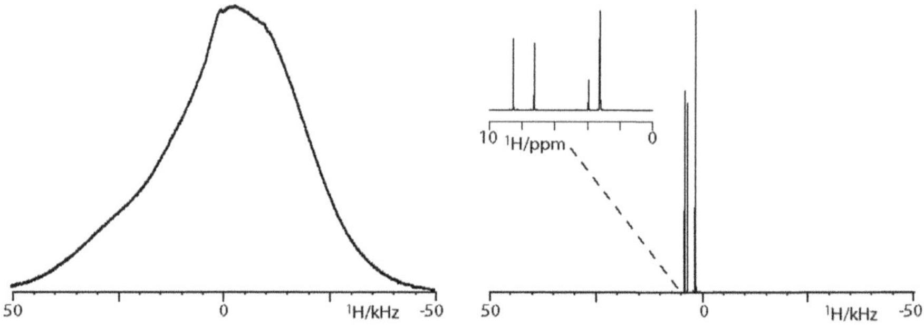

Figure 1.1 Comparison of a conventional proton spectrum of natural abundance histidine measured in solution (*right*) and as a static solid powder (*left*). While anisotropic interactions broaden the lines in solid samples, nature solves this issue in liquid samples by rapid molecular tumbling, which is usually orders of magnitude larger than the anisotropic interactions.

methodology refers to the manipulation of the Hamiltonian to dissect the anisotropic interactions or to suppress their influence on NMR spectra in a controlled manner.

In the following, we will first introduce the reader to the most relevant NMR interactions in solid-like biological systems and afterwards briefly discuss the basic components of solid-state NMR pulse sequences for biological applications.

1.2 Interactions in Biological Solid-State NMR

Interactions which affect the spin system and its associated Hamiltonian can be classified as external and internal (eqn 1.1). The external Hamiltonian is directly under the control of the NMR operator and consists of the Zeeman interaction and the rf (radio-frequency) fields, while the internal Hamiltonian includes the interaction of the spins with the local electronic environment (chemical shift), with each other (dipolar and scalar couplings) and with electric field gradients (quadrupolar coupling). The internal interactions may be modulated by the spectroscopist by means of rf pulses or magic angle spinning.

$$H_{\text{total}} = H_{\text{external}} + H_{\text{internal}} \tag{1.1}$$

1.2.1 General and Rotational Properties of NMR Interactions

Before we summarize the most relevant interactions in biological solid-state NMR, we will briefly discuss their general appearance and properties.

Whether external or internal interactions, their Hamiltonians can be depicted as:

$$H = I \cdot \tilde{A} \cdot S = (I_x, I_y, I_z) \begin{pmatrix} A_{xx} & A_{xy} & A_{xz} \\ A_{yx} & A_{yy} & A_{yz} \\ A_{zx} & A_{zy} & A_{zz} \end{pmatrix} \begin{pmatrix} S_x \\ S_y \\ S_z \end{pmatrix} \tag{1.2}$$

with I and S as row and column vectors, respectively, and $\tilde{A}$ as a second-rank Cartesian tensor as a means to represent the anisotropy of an interaction, corresponding to a 3×3 square matrix. In this equation, I is a spin operator, while the other vector S represents either a second spin (dipolar or scalar coupling), or a magnetic (chemical shift) or an electric field (quadrupolar coupling). Subjected to rapid isotropic molecular motion, $\tilde{A}$ is averaged to its trace, which is the sum of the diagonal elements:

$$\frac{1}{3}\mathrm{Tr}\{A_{\alpha\beta}\} = \frac{1}{3}\sum_{\alpha} A_{\alpha\alpha} \tag{1.3}$$

Besides the Cartesian representation, irreducible spherical tensor operators are a particularly appropriate basis to describe transformations in solid-state NMR:

$$H = \sum_{k=0}^{2} \sum_{m=-k}^{+k} (-1) A_{k-m} T_{km} \tag{1.4}$$

In this representation, it is easily visible that any NMR interaction can be broken down into a spatial part, represented by spherical tensor A, and a spin part, represented by spherical tensor operators T. Indices k indicate the rank of a tensor, which has $2m+1$ orders (which are irreducible spherical tensors). Component A_{00} is the isotropic, A_{1-m} is the antisymmetric and A_{2-m} is the anisotropic part of an interaction. Since the antisymmetric part can usually be ignored, and since the static magnetic field B_0 is usually orders of magnitude larger than the internal interactions (with the exception of certain quadrupolar interactions), the summation can be reduced in the so-called secular approximation to:

$$H = A_{00} T_{00} + A_{20} T_{20} \tag{1.5}$$

neglecting all terms that vanish over time and retaining only those that commute with B_0. Rotations of spherical tensors are carried out by Wigner rotation matrices D of the same rank, with the Euler angles α, β and γ (eqn 1.6). Details of all these operations can be found in standard references.[3]

$$A_{2m} = \sum_{m'=-2}^{+2} A_{2m'} D^{2}_{m'm}(\alpha, \beta, \gamma) \tag{1.6}$$

NMR interactions have their simplest form in their principal axis system (PAS), in which all off-diagonal elements are zero. Yet, NMR signals

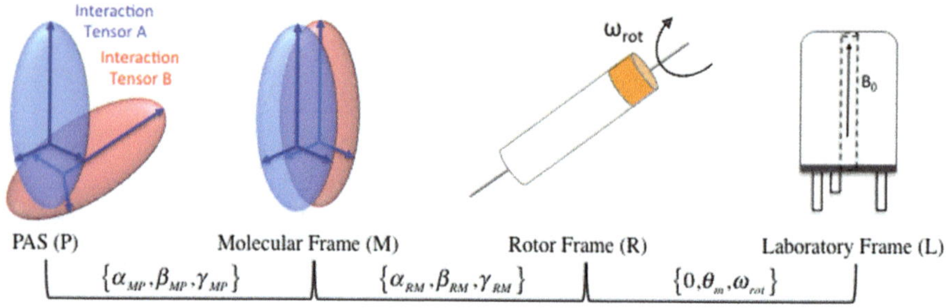

Figure 1.2 Pictorial representation of the usual transformations in solid-state NMR.

are detected in the laboratory frame, into which the interactions thus have to be transformed to analyze their effect on the spectrum. Transformations between different frames obey the following relation, which is called the addition theorem of Wigner rotation matrices:

$$D^2_{n'n}(\alpha, \beta, \gamma)_{ac} = \sum_{m=-2}^{+2} D^2_{n'm}(\alpha, \beta, \gamma)_{ab} D^2_{mn}(\alpha, \beta, \gamma)_{bc} \tag{1.7}$$

According to these rules, NMR interactions can be transformed from one frame into another. Typically, in solid-state NMR, transformations include a transformation from the PAS to a molecular frame, which is fixed to the molecular structure and in which all NMR interactions share a common orientation, followed by a transformation to a rotor frame to take into account the powder averaging, and finally into the laboratory frame (Figure 1.2).

1.2.2 Chemical Shift Anisotropy

Nuclei possessing a spin I have an associated nuclear magnetic dipole moment $\mu = \gamma \hbar I$, with γ as the gyromagnetic ratio. For an isolated spin-$\frac{1}{2}$ nucleus, subjected to a static magnetic field B_0 along the z-axis, this gives rise to energy levels separated by $\Delta E = -\gamma \hbar B_0 = \hbar \omega_0$, with ω_0 denoting the Larmor frequency. The resonance frequencies observed in NMR spectra, however, usually slightly differ from the Larmor frequency. This difference is called the chemical shift, which is caused by variations in the electron distribution around nuclei. The electron distribution around a given nucleus is rarely spherically symmetric, and the resulting anisotropic local field is termed the chemical shift anisotropy (CSA; see Figure 1.3). The CSA may be described as a Cartesian second-rank tensor, the trace of which is the isotropic chemical shift:

$$\sigma_{cs} = \begin{pmatrix} \sigma_{xx} & \sigma_{xy} & \sigma_{xz} \\ \sigma_{yx} & \sigma_{yy} & \sigma_{yz} \\ \sigma_{zx} & \sigma_{zy} & \sigma_{zz} \end{pmatrix}, \ \mathrm{Tr}\{\sigma_{cs}\} = \sigma_{iso} \tag{1.8}$$

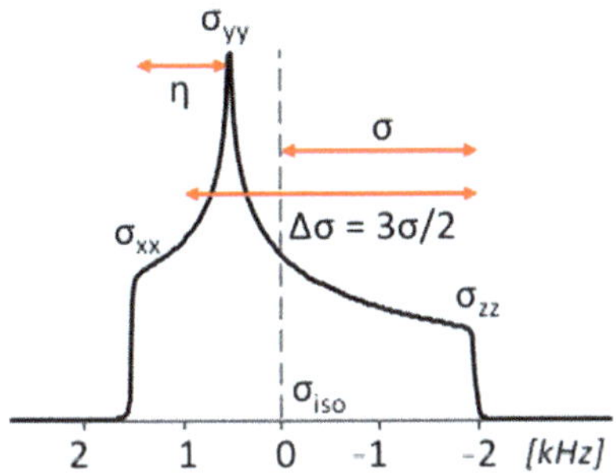

Figure 1.3 A chemical shift anisotropy powder pattern with a reduced anisotropy $\sigma = 2$ kHz and an asymmetry $\eta = 0.5$. The pattern reflects that, in solid powders, molecules adopt all possible orientations. The orientation dependence of the chemical shift is deducible from eqn (1.10).

The chemical shift tensor can be characterized (note that there are different nomenclature conventions; here we use the Haeberlen convention with $|\sigma_{zz} - \sigma_{\text{iso}}| \geq |\sigma_{xx} - \sigma_{\text{iso}}| \geq |\sigma_{yy} - \sigma_{\text{iso}}|$) by its reduced anisotropy σ and anisotropy $\Delta\sigma = 3\sigma/2$, which reflect the deviation from cubic symmetry, and its asymmetry η, which reflects the deviation from axial symmetry (Figure 1.3):

$$\sigma = \sigma_{zz} - \sigma_{\text{iso}},\ \eta = (\sigma_{yy} - \sigma_{xx})/\sigma \tag{1.9}$$

For an axially symmetric CSA tensor, with θ defining the direction of B_0 in the PAS of the CSA tensor, the resonance frequency can be expressed as:

$$\omega = \omega_0 \sigma_{zz}(3\cos^2\theta - 1) \tag{1.10}$$

The non-zero elements of the irreducible spherical tensor operators in the PAS of the chemical shift are

$$A_{00}^{\text{PAS}} = \sigma_{\text{iso}},\ A_{20}^{\text{PAS}} = \sqrt{3/2}\sigma,\ A_{2\pm2}^{\text{PAS}} = (1/2)\eta\sigma \tag{1.11}$$

1.2.3 Dipolar Coupling

The local magnetic moments $\vec{\mu}$ of spins do not only interact with B_0, but also with each other. For example, in the context of the dipolar coupling, the magnetic moments interact directly through space. The interaction energy between two magnetic moments $\vec{\mu}$ with a distance r_{12} (assuming that both dipoles have the same orientation to r_{12}; see Figure 1.4) is given by:

$$E = (\mu_0/4\pi)\left(\vec{\mu_1} \cdot \vec{\mu_2}\right) r_{12}^{-3}\left[1 - 3\cos^2\theta\right] \tag{1.12}$$

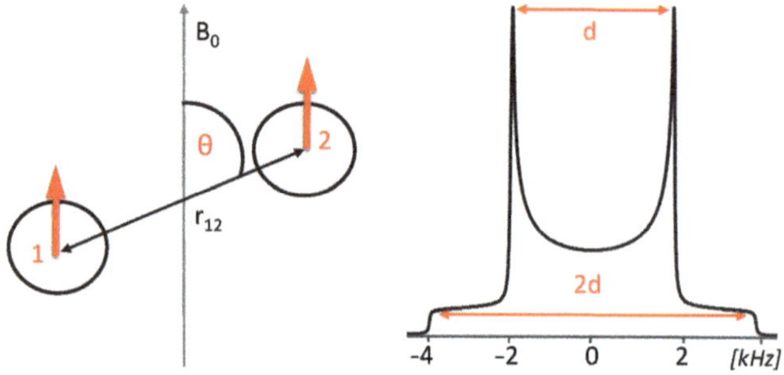

Figure 1.4 *Left*: The strength of the dipolar coupling depends on both the internuclear distance between spins 1 and 2 and the angle θ of the internuclear vector with the static magnetic field B_0. In a powder sample, the presence of all possible molecular orientations results in a typical two-horned Pake pattern, with a maximum splitting of $3d$ in the homonuclear and $2d$ in the heteronuclear case. *Right*: Simulation of a ^{13}C dipolar powder pattern of a ^{13}C–^{1}H spin pair with a dipolar coupling constant $d_{CH} = 4$ kHz ($r_{CH} = 1.96$ Å).

which after replacement of $\vec{\mu} = \gamma\hbar I$ and reformulation in polar coordinates yields the dipolar Hamiltonian with the so-called dipolar alphabet:

$$H_D = \frac{\mu_0 \gamma_1 \gamma_2 h^2}{16\pi^3 r^3}[A + B + C + D + E + F] \tag{1.13}$$

where

$$A = I_{1z} I_{2z}(3\cos^2\theta - 1) \quad \text{and} \quad B = (1/4)\left[I_1^+ I_2^- + I_1^- I_2^+\right](3\cos^2\theta - 1) \tag{1.14}$$

The terms C–F contain single-quantum $I_1^{\pm} I_{2z}$ and double-quantum operators $I_1^{\pm} I_2^{\pm}$ and can therefore be safely neglected in the secular approximation. Note that if $\gamma_1 \neq \gamma_2$ the B term is also non-secular and can be omitted, which is the reason for the different expression for heteronuclear H^{IS} and homonuclear H^{II} dipolar Hamiltonians (eqn 1.15). Theta (θ) is defined by the direction of the internuclear vector in a coordinate system in which the B_0 field is in the direction of the z-axis.

$$\begin{aligned} H_D^{IS} &= d(3\cos^2\theta - 1) I_{1z} I_{2z} \quad \text{and} \\ H_D^{II} &= (1/2) d(3\cos^2\theta - 1)([3I_{1z} I_{2z} - I_1 \cdot I_2]) \end{aligned} \tag{1.15}$$

with $d = \dfrac{\mu_0 \gamma_1 \gamma_2 h^2}{16\pi^3 r^3}$ as the dipolar coupling constant. The dipolar coupling tensor is axially symmetric and traceless; hence A_{20}^{PAS} is its only PAS component in the secular approximation.

1.2.4 Scalar Coupling

Scalar couplings are indirect interactions mediated through electrons, which usually act *via* covalent bonds. The term "scalar" already indicates this interaction to be usually rotational-invariant, *i.e.* isotropic with the scalar coupling tensor J_{ij} reduced to its trace. Scalar couplings are often small in comparison to other interactions in solid-state NMR and may be obscured by a broad powder pattern. However, they can of course be exploited in the solid state, too, especially in combination with magic angle spinning.[4] The homo- and heteronuclear scalar coupling interactions are respectively described by:

$$H = \hbar J_{12} I_1 I_2 \quad \text{and} \quad H = \hbar J_{12} I_{1z} S_{2z} \tag{1.16}$$

1.2.5 Quadrupolar Coupling

Nuclei possessing a spin $I > \frac{1}{2}$ have an electric quadrupole moment arising from a non-spherical distribution of the electric charge around the nucleus. This moment interacts with the electric field gradient, which arises by virtue of the distribution of other nuclei and electrons in the vicinity of the nucleus. Since quadrupolar nuclei are, except for deuterium, non-standard nuclei for biological solid-state NMR, we will not dwell further on this NMR interaction and refer the interested reader elsewhere.[3]

1.2.6 Magic Angle Spinning

Solid-state NMR experiments of biological samples are usually subjected to magic angle spinning (MAS),[5] which is a technique to average the spatial part of second-rank interactions by making the interactions time-dependent. MAS can greatly enhance spectral resolution and sensitivity. It implies spinning a rotor containing the sample at a certain angle with respect to the static B_0 field (Figure 1.5), at which the spatial anisotropic part of the interactions, describable as second-rank tensors like the dipolar coupling or the CSA, are averaged out. The orientation dependence of these interactions is proportional to $3(\cos^2\theta - 1)/2$, which is the second Legendre polynomial $P2(\cos\,\theta)$. This expression becomes zero at $\theta = 54.74°$ which is, therefore, also called the magic angle θ_m (Figure 1.5). For axial symmetric tensors, it can be shown that MAS yields the average orientation function:

$$\langle 3\cos^2\theta_{av} - 1\rangle = \frac{1}{2}(3\cos^2\theta_r - 1)(3\cos^2\theta - 1) \tag{1.17}$$

For asymmetric CSA tensors, one obtains a term depending on the orientation of all three principal axes to the spinning axis. Since this term is also multiplied by $P2(\cos\,\theta)$ due to MAS, both axial symmetric and asymmetric CSA tensors are averaged out by fast spinning. If the MAS frequency is smaller than the magnitude of the CSA, a spinning sideband pattern is observed in the NMR spectrum, which at higher spinning frequencies collapses into one resolved peak (Figure 1.5).

For an interaction that commutes with itself over time (referred to as inhomogeneous by Maricq and Waugh[6]), MAS achieves complete averaging.

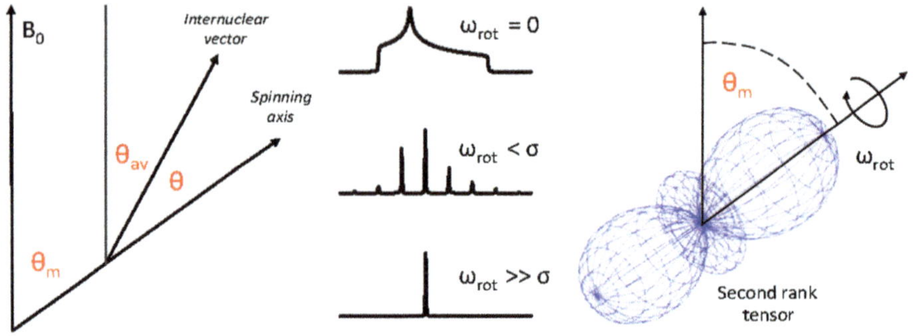

Figure 1.5 *Left*: MAS means rotating the sample rapidly around the spinning axis at the magic angle θ_m. If the spinning is fast enough, the average orientation of the sample with the static magnetic field is $\theta_{av} = \theta_m$, or any other angle to which the spinning axis is set with respect to B_0. Note that for axially symmetric CSA tensors, the internuclear vector corresponds to the orientation of the principal *z*-axis of the shielding tensor with the spinning axis. *Middle*: At intermediate spinning frequencies the powder pattern collapses into spinning sidebands, which are separated from the isotropic frequency by multiples of the spinning frequency. At spinning frequencies exceeding the anisotropy of (heterogeneous) interactions, the signal intensity is centered at the isotropic frequency (signals are normalized).

This is, however, not the case for time-dependent homogeneous interactions (which do not commute with themselves over time), like strong homonuclear dipolar proton couplings. We will refer to a hand-waving explanation to illustrate the latter. As discussed above (see Section 1.2.3), unlike H_D^{IS}, H_D^{II} exhibits a secular *B*-term which contains zero-quantum operators inducing flip-flop transitions among spins *I* at a ratio proportional to the strengths of H_D^{II}. This means that the spin states and thus the local dipolar fields are not constant over time (they are mixed by the *B*-term), while MAS only works efficiently if the interaction is static for at least one rotor period. It is important to keep in mind that the presence of one homogeneous interaction suffices to render all internal interactions homogeneous and thus difficult to be spun out completely. The reason for this behaviour is that NMR interactions are entangled with the homonuclear couplings in higher order cross-terms (see below and elsewhere[3,6,7] for further reading). Interestingly, the efficiency of flip-flop transitions decreases with increasing chemical shift differences, *i.e.* with increasing B_0, reducing the homogenous character of interactions like dipolar proton–proton couplings. These dependencies become, for example, apparent using average Hamiltonian or Floquet theory.

Note that MAS is limited to second-rank tensors, which does not suffice to suppress interactions like quadrupolar couplings, which in addition to a $P2(\cos\theta)$ orientation dependence also exhibit a $P4(\cos\theta)$ dependence. An elegant means to suppress these interactions further is the double rotation (DOR)[8] technique, which requires spinning the sample at two angles simultaneously.

1.3 Cross Polarization

Cross polarization (CP)[9] is a method to enhance the sensitivity of heteronuclear NMR experiments, in which polarization is transferred from the abundant proton bath across the heteronuclear dipolar couplings to the heteronuclei. The technique, analogously to the INEPT[10] experiment in solution, exploits the much higher equilibrium Boltzmann polarization of the protons.

Cross polarization requires simultaneous rf irradiation of the protons (I) and the heteronuclei (S), spin-locking both species. The irradiation has to fulfill the Hartmann–Hahn matching condition $\omega_1^S = \omega_1^I$ in the static case, and $\omega_1^S = \omega_1^I \pm n\omega_{\mathrm{rot}}$ for spinning samples (with $n = 1,2$).[11]

Phenomenologically (Figure 1.6), the resonance frequencies of both spins match (while the resonance frequencies are very different when subjected to the Zeeman field alone) at the Hartmann–Hahn condition so that polarization transfer can occur. In a quantum mechanical description, it can be shown that at the Hartmann–Hahn condition[12] (in the so-called doubly rotating frame), one obtains an effective Hamiltonian (see Section 1.5), which contains operators of the form I_xS_x and I_yS_y. Reformulating these operators in terms of raising and lowering operators $I_x = (1/2)(I^+ + I^-)$ and $I_y = (1/2i)(I^+ + I^-)$, it can be shown that, at static conditions and slow to medium spinning frequencies, CP transfer can be induced *via* zero-quantum flip-flop transitions $I^{\pm}S^{\mp}$. Note that in the spin-lock field, homonuclear dipolar couplings are still active, merely scaled by a factor of -0.5. Another benefit of the CP technique is that experiments can be accumulated faster, because they depend on ^{1}H-T_1, which is usually much shorter than ^{13}C-T_1 or ^{15}N-T_1.

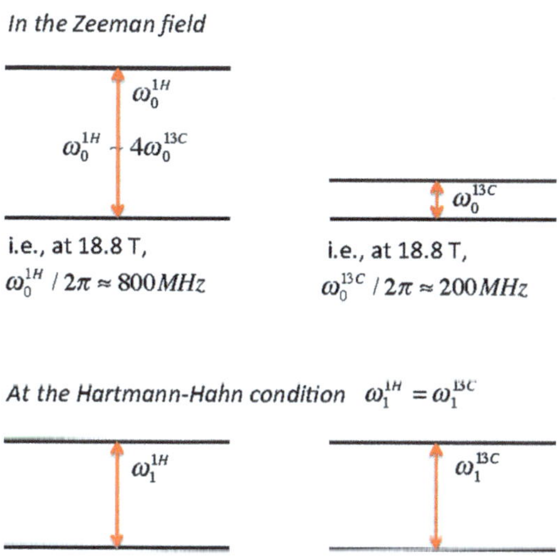

Figure 1.6 Illustration of the cross-polarization technique.

1.4 Dipolar Decoupling

As delineated in Section 1.2, internal NMR interactions can be decomposed into two factors, which are the space and the spin parts. Fast MAS averages the space part, yet not to completeness, predominately because of the homogeneous character of the homonuclear dipolar proton interactions, rendering the whole internal Hamiltonian homogeneous. To further narrow the lines and enhance sensitivity, one thus has to average simultaneously the spin part by rf pulses, which is called decoupling.

Heteronuclear dipolar decoupling in biological solid-state NMR relates to the decoupling of heteronuclear spins from the proton spins. The mode of action of strong rf pulses applied to the proton spins may be imagined as to cause a fast precession of the proton spins around the effective field, of a frequency higher than the heteronuclear dipolar couplings. In fact, the frequency also has to be larger than the homonuclear dipolar proton couplings. One should keep in mind that such a simplified description does not provide much insight into the effect of different decoupling sequences and we refer interested readers to detailed reviews on this subject.[13] In its simplest form, heteronuclear decoupling refers to the application of high-power continuous wave (CW) rf irradiation on the proton channel. The performance of CW decoupling, however, can be compromised by interference effects between heteronuclear dipolar couplings and the proton CSA.[13a] This interference is significantly reduced in the case of phase-modulated sequences like two-pulse phase modulation (TPPM)[14] and its phase-cycled version, small phase incremental alternation (SPINAL).[15] Both are to date the standard decoupling methods at low to medium spinning frequencies. Note that there exists a large number of modifications of these two basic decoupling sequences.[16] With increasing spinning frequency, decoupling sequences like XiX[17] and PISSARRO[18] usually perform better than TPPM or SPINAL. One reason is the residual sensitivity of TPPM and SPINAL to interferences between heteronuclear dipolar couplings and the proton CSA, which weighs more with increasing spinning frequency (at lower spinning frequency, this interference is partly suppressed by the self-decoupling effect induced by strong proton–proton dipolar couplings[13a]). On the other hand, both XiX and PISSARRO are impeded by the effect of the proton–proton dipolar couplings,[19] which become evidently reduced with increasing spinning frequency, so that these sequences unfold their potential more at fast MAS frequencies. Besides high-power irradiation, all these heteronuclear decoupling sequences can be applied as low-power variants at very fast spinning frequencies.[20] Low-power decoupling prevents excessive sample heating, which may be of particular importance for biological samples, although slight losses in performance usually have to be accepted. Generally, the subtle differences among these decoupling pulse sequences all manifest in the higher order terms of their effective Hamiltonians (see Section 1.5), which is the reason why a description (and analysis) of these sequences is not straightforward.

Homonuclear dipolar decoupling in biological solid-state NMR usually relates to proton–proton decoupling. A powerful way to achieve additional spin space averaging of these couplings is the Lee–Goldburg (LG) irradiation technique.[21] It comprises tilting the effective field to the magic angle by means of offset irradiation on the proton channel, which suppresses dipolar proton–proton couplings to the lowest order, leaving the chemical shift Hamiltonian scaled by $1/\sqrt{3}$. The LG technique is nowadays often applied in its frequency- and phase-switched versions,[22] which average out the proton–proton couplings to higher order (see further reading[23]). Besides the LG technique, sequences using continuously phase-modulated rf pulses[24] and sequences based on symmetry principles[25] also perform well to decouple proton–proton interactions.

1.5 Recoupling

Recoupling means the reintroduction of anisotropic NMR interactions, which are otherwise averaged out by MAS. Since these interactions, as delineated in Section 1.2, are orientation and/or distance dependent, their controlled reintroduction can yield structural information. Today, a plethora of recoupling pulse sequences exists and there are numerous options to reintroduce the desired interactions.[26] Here we will sketch some options to assign proteins and to yield structural information. Recoupling methods (and generally solid-state NMR methods) are often analyzed in terms of average Hamiltonian theory (AHT),[7] in which a time-dependent Hamiltonian is expressed as a time-independent effective Hamiltonian (see further reading[3,7]):

$$\bar{H}_{\text{eff}} = \bar{H}^{(0)} + \bar{H}^{(1)} + \bar{H}^{(2)} + \dots \tag{1.18}$$

where $\bar{H}^{(0)}$ is the lowest order term (the actual "average Hamiltonian"), $\bar{H}^{(1)}$ is the first order correction term, and so on. These terms can be expressed in terms of a Magnus expansion, which shows that, while $\bar{H}^{(0)}$ is independent of the spinning frequency ν_{rot}, higher order terms of order k scale with $1/\nu_{\text{rot}}^{k-1}$, *i.e.* $\bar{H}^{(1)}$ is inversely proportional to ν_{rot}. It is advisable to understand which terms of $\bar{H}_{\text{eff}}$ are reintroduced by a recoupling sequence to fully exploit its potential.

1.5.1 Protein Backbone Assignment

The protein backbone reads as $C\alpha_{(i)}$–$N_{(i)}$–$CO_{(i+1)}$–$C\alpha_{(i+1)}$. Hence, to obtain sequential assignments, correlations $C\alpha_{(i)} \rightarrow N_{(i)} \rightarrow CO_{(i+1)}$ (or *vice versa*) need to be established, while at the same time the transfer should be restrained to this pathway to avoid misleading correlations over more than one amino acid. This is usually achieved using specific $^{13}C \leftrightarrow {}^{15}N$ cross polarization techniques,[27] which exploit a frequency-selective

Hartmann–Hahn condition to reintroduce the heteronuclear dipolar ^{13}C–^{15}N couplings to lowest order, which implies that this method also works at very fast spinning frequencies (Figure 1.7). In the specific CP experiment, chemical shift-dependent transfer characteristics are introduced by using a controlled frequency offset in combination with relatively weak rf fields on the heteronuclei. Alternatively, $C\alpha_{(i)} \rightarrow N_{(i)} \rightarrow CO_{(i+1)}$ correlations may be brought about with the PAIN sequence,[28] which is a third-spin assisted recoupling (TSAR)-based method (see below).[29]

The single amino acids can be identified by means of their characteristic chemical shifts and connectivity patterns, for example among carbon resonances. As mentioned above, methods to reintroduce homonuclear dipolar ^{13}C–^{13}C couplings are widespread. A popular class of experiments relies on the proton-mediated reintroduction of the carbon homonuclear dipolar couplings, which is usually referred to as a longitudinal

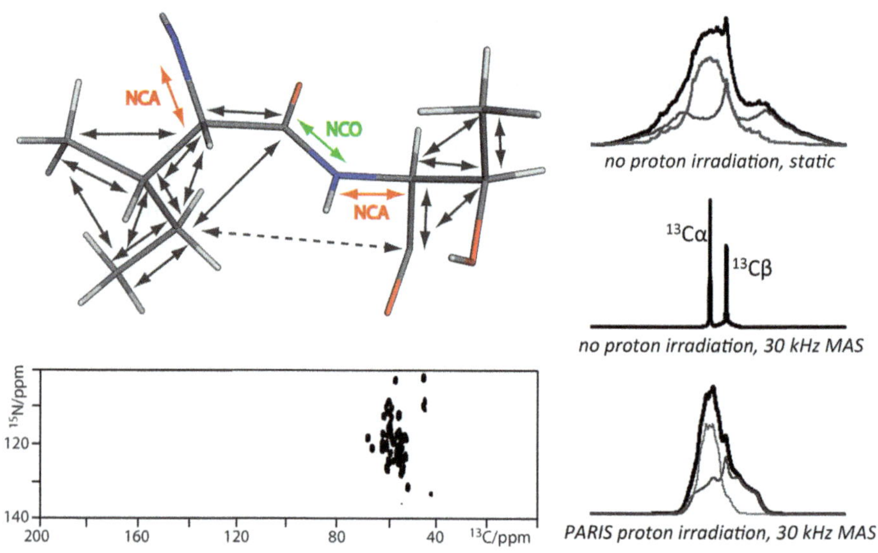

Figure 1.7 *Left, top*: Illustration of experimental options to assign protein backbone and side-chain resonances. Protein backbone and sequential assignments can be obtained *via* NCA[27] (*red arrows*) and NCO[27] (*green*) experiments, carbon side-chain assignments *via* spin diffusion-based experiments (*gray*), and long-range and sequential assignments/restraints (*dashed, gray*) *via* TSAR,[29] rotational resonance at the weak coupling condition[48] and spin diffusion-based experiments.[30–32,49] *Left, bottom*: A two-dimensional NCA spectrum of a protein sample. *Right*: Spin diffusion magnetization transfer requires spectral overlap, which can be achieved by switching off the ^{1}H decoupling during the mixing time. At higher MAS frequencies, heteronuclear dipolar couplings are significantly reduced by MAS, which curtails the spin diffusion efficiency. With active recoupling, here shown by the example of PARIS[31] recoupling, the spectral broadening is independent of the MAS frequency (signals are normalized).

magnetization transfer by spin diffusion facilitating zero-quantum flip-flop transitions among ^{13}C nuclei. In its basic form, known as proton driven spin diffusion (PDSD), it suffices to switch off the decoupling field on the proton channel (while ^{13}C magnetization is stored along the *z*-axis) to enhance spin diffusion. The reason why proton decoupling is switched off during the spin diffusion mixing time is to reintroduce heteronuclear dipolar couplings, making the influence of flip-flop transitions more efficient. This reintroduction occurs, however, only to higher order, *i.e.* scales inversely with the spinning frequency. Other methods like DARR,[30] PARIS[31] or MIRROR,[32] therefore, outperform PDSD with increasing spinning frequency, because these methods reintroduce the heteronuclear dipolar couplings independently of the MAS frequency (Figure 1.7). However, since the flip-flop transitions appear in the correction terms of the effective Hamiltonian, spin diffusion efficiency invariably decreases with increasing spinning frequency, although the transfer can still be efficient at very fast MAS.[33]

1.5.2 Protein Geometry Restraints

As for solid-state NMR experiments to assign protein resonances, numerous possibilities exist to collect information on protein geometry. For example, the rotational resonance (R^2) experiment,[34] which reintroduces homonuclear dipolar couplings between two nuclei to lowest order if the resonance condition $n\nu_{\mathrm{rot}} = \Delta CS$, with ΔCS being the chemical shift difference of the two nuclei, is matched. This recoupling technique is very efficient in sparsely labeled samples and, in particular, allows exact measurements of internuclear distances. However, lowest order recoupling sequences are sensitive to the so-called dipolar truncation effect, which means that in the presence of a strong homonuclear dipolar coupling, transfer S across a small coupling is significantly reduced. It has been shown[35] that if

$$\frac{S_{12}}{S_{23}} \approx \left(\frac{d_{12}}{d_{23}}\right)^2 \tag{1.19}$$

i.e. if two pairs of homonuclear dipolar couplings d differ by one order of magnitude, the transfer across the small coupling is quenched by two orders of magnitude. The latter situation, with weak couplings in the presence of strong couplings, is, however, the common situation in uniformly labeled biomolecules, so that long-range distances are hardly collectable with conventional lowest order recoupling techniques. A remedy to this issue is to use frequency-selective lowest order recoupling[36] to reintroduce only homonuclear interactions between certain spin pairs, to selectively label the protein sample, or to resort to higher order recoupling methods like spin diffusion,[30–32] TSAR[29] (which is based on higher order cross-terms of two heteronuclear couplings) or CHHC[37] (which reads out ^1H–^1H spin diffusion on ^{13}C nuclei), which are much less sensitive to dipolar truncation. Note that there are many other methods to obtain information on protein geometry, *e.g.* by analyzing the relative orientation of chemical shift tensors.[38]

1.5.3 Intermolecular Geometry Restraints

The measurement of distance restraints across biological molecular interfaces by solid-state NMR usually involves mixtures of differentially labeled samples (involving nucleic species X and Y in Figure 1.8a), because the direct detection of intermolecular contacts at the protein–protein interface in uniformly labeled samples is usually prohibited by spectral crowding and dipolar truncation in the presence of short-ranged intramolecular interactions. A frequently applied strategy refers to the use of equimolar mixtures of ^{13}C and ^{15}N labeled proteins in combination with ^{15}N–^{13}C transfer schemes.[39] In general, polarization transfer across the molecular interface can be brought about either *via* the relatively small dipolar ^{15}N–^{13}C couplings by REDOR- or TEDOR-based transfer schemes[40] (Figure 1.8), or by involving protons in the context of NHHC[41] and PAIN[28] experiments (Figure 1.8).

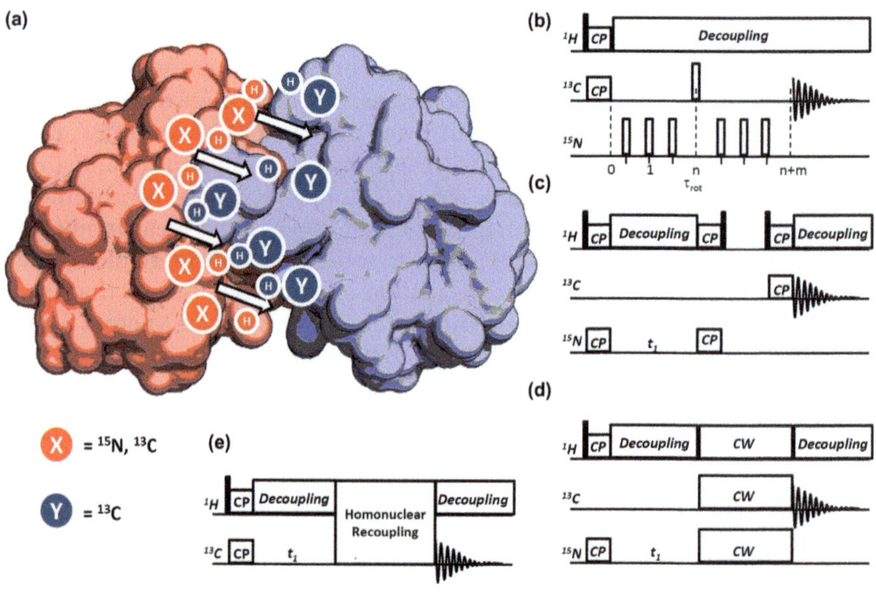

Figure 1.8 (a) Identification of protein–protein binding interfaces by using mixtures labeled with species X and Y and dedicated solid-state NMR schemes. For X = ^{15}N and Y = ^{13}C, polarization transfer can be brought about by heteronuclear transfer sequences like: (b) REDOR,[50] (c) NHHC,[37,41] or (d) PAIN.[28] Also (e) homonuclear recoupling sequences[30–32] can be used for X–Y transfer (such as X = labeled on the basis of [1-^{13}C]glucose and Y = labeled using [2-^{13}C]glucose). *Filled* and *unfilled* rectangles represent 90° and 180° pulses, respectively, if not indicated otherwise.
(Reproduced from Weingarth and Baldus,[39] with permission from the American Chemical Society.)

In principle any combination of spin-$\frac{1}{2}$ species X and Y can be studied by the schemes shown in Figure 1.8. For example, in complexes involving nucleic acids, ^{31}P nuclei are a convenient source to establish X–Y heteronuclear transfers and can yield unambiguous intermolecular information.[42] Using mixtures of proteins labeled on the basis of [1-^{13}C]- and [2-^{13}C]glucose, in combination with homonuclear ^{13}C–^{13}C recoupling sequences, has emerged as another strategy to probe intermolecular contacts of larger proteins.[43] Note that next to distance restraints, chemical shift perturbation studies, *i.e.* comparing chemical shifts before and after complex formation, are another convenient means to obtain protein binding interfaces.[44] For further information on the investigation of biomolecular supramolecular structure, we refer the interested reader elsewhere.[39]

1.5.4 Protein Dynamics

A particular advantage of solid-state NMR (and NMR in general) spectroscopy is its potential to provide information on the dynamics of biomolecules in close reference to structure and function. Rigid and mobile protein segments can readily be site-specifically distinguished[44,45] by comparing spectra employing dipolar (like spin diffusion[30–32]) and scalar (like INEPT[10]) transfer. Dedicated solid-state NMR experiments principally allow characterization of the dynamics of biomolecules from the nano- to the millisecond timescale and beyond.[46] For example, the nano- to millisecond timescale can readily be explored by measuring the spin–lattice relaxation in the laboratory (R_1) or the rotating frame ($R_{1\rho}$), while real-time solid-state NMR allows the course of protein refolding or of enzymatic reactions to be followed.[47]

References

1. (a) S. A. Shahid, B. Bardiaux, W. T. Franks, L. Krabben, M. Habeck, B. J. van Rossum and D. Linke, *Nat. Methods*, 2012, **9**, 1212–1217; (b) S. H. Park, B. B. Das, F. Casagrande, Y. Tian, H. J. Nothnagel, M. Chu, H. Kiefer, K. Maier, A. A. De Angelis, F. M. Marassi and S. J. Opella, *Nature*, 2012, **491**, 779–783.
2. (a) M. Renault, R. Tommassen-van Boxtel, M. P. Bos, J. A. Post, J. Tommassen and M. Baldus, *Proc. Natl. Acad. Sci. U. S. A.*, 2012, **109**, 4863–4868; (b) R. Fu, X. Wang, C. Li, A. N. Santiago-Miranda, G. J. Pielak and F. Tian, *J. Am. Chem. Soc.*, 2011, **133**, 12370–12373.
3. (a) M. Mehring, *Principles of High Resolution NMR in Solids*, Springer Verlag, Berlin, 1983; (b) K. S. Schmidt-Rohr and H. W. Spiess, *Multi-dimensional Solid State NMR and Polymers*, Academic Press, London, 1994; (c) M. Duer, *Introduction to Solid-State NMR Spectroscopy*, Wiley-Blackwell, Oxford, 2004.
4. M. Baldus and B. H. Meier, *J. Magn. Reson., Ser. A*, 1996, **121**, 65–69.

5. (a) E. R. Andrew, A. Bradbury and R. G. Eades, *Nature*, 1958, **182**, 1659–1659; (b) I. J. Lowe, *Phys. Rev. Lett.*, 1959, **2**, 285–287.
6. M. M. Maricq and J. S. Waugh, *J. Chem. Phys.*, 1979, **70**, 3300–3316.
7. U. Haeberlen and J. S. Waugh, *Phys. Rev.*, 1968, **175**, 453–467.
8. A. Samoson, E. Lippmaa and A. Pines, *Mol. Phys.*, 1988, **65**, 1013–1018.
9. A. Pines, M. G. Gibby and J. S. Waugh, *J. Chem. Phys.*, 1973, **59**, 569–590.
10. G. A. Morris and R. Freeman, *J. Am. Chem. Soc.*, 1979, **101**, 760–762.
11. J. Schaefer and E. O. Stejskal, *J. Am. Chem. Soc.*, 1976, **98**, 1031–1032.
12. S. R. Hartmann and E. L. Hahn, *Phys. Rev.*, 1962, **128**, 2042–2053.
13. (a) M. Ernst, *J. Magn. Reson.*, 2003, **162**, 1–34; (b) P. Hodgkinson, *Prog. Nucl. Magn. Reson. Spectrosc.*, 2005, **46**, 197–222.
14. A. E. Bennett, C. M. Rienstra, M. Auger, K. V. Lakshmi and R. G. Griffin, *J. Chem. Phys.*, 1995, **103**, 6951–6958.
15. B. M. Fung, A. K. Khitrin and K. Ermolaev, *J. Magn. Reson.*, 2000, **142**, 97–101.
16. R. S. Thakur, N. D. Kurur and P. K. Madhu, *Chem. Phys. Lett.*, 2006, **426**, 459–463.
17. (a) P. Tekely, P. Palmas and D. Canet, *J. Magn. Reson., Ser A*, 1994, **107**, 129–133; (b) A. Detken, E. H. Hardy, M. Ernst and B. H. Meier, *Chem. Phys. Lett.*, 2002, **356**, 298–304.
18. M. Weingarth, P. Tekely and G. Bodenhausen, *Chem. Phys. Lett.*, 2008, **466**, 247–251.
19. (a) M. Ernst, H. Geen and B. H. Meier, *Solid State Nucl. Magn. Reson.*, 2006, **29**, 2–21; (b) M. Weingarth, G. Bodenhausen and P. Tekely, *Chem. Phys. Lett.*, 2011, **502**, 259–265.
20. (a) M. Ernst, A. Samoson and B. H. Meier, *Chem. Phys. Lett.*, 2001, **348**, 293–302; (b) M. Ernst, A. Samoson and B. H. Meier, *J. Magn. Reson.*, 2003, **163**, 332–339; (c) M. Weingarth, G. Bodenhausen and P. Tekely, *J. Magn. Reson.*, 2009, **199**, 238–241.
21. M. Lee and W. I. Goldburg, *Phys. Rev.*, 1965, **140**, 1261–1271.
22. (a) A. Bielecki, A. C. Kolbert and M. H. Levitt, *Chem. Phys. Lett.*, 1989, **155**, 341–346; (b) E. Vinogradov, P. K. Madhu and S. Vega, *Chem. Phys. Lett.*, 1999, **314**, 443–450.
23. (a) M. E. Halse and L. Emsley, *Phys. Chem. Chem. Phys.*, 2012, **14**, 9121–9130; (b) P. K. Madhu, *Solid State Nucl. Magn. Reson.*, 2009, **35**, 2–11.
24. D. Sakellariou, A. Lesage, P. Hodgkinson and L. Emsley, *Chem. Phys. Lett.*, 2000, **319**, 253–260.
25. O. Lafon, Q. Wang, B. Hu, J. Trebosc, F. Deng and J. P. Amoureux, *J. Chem. Phys.*, 2009, **130**, 014504.
26. M. Baldus, *J. Biomol. NMR*, 2007, **39**, 73–86.
27. M. Baldus, A. T. Petkova, J. Herzfeld and R. G. Griffin, *Mol. Phys.*, 1998, **95**, 1197–1207.
28. J. R. Lewandowski, G. De Paepe and R. G. Griffin, *J. Am. Chem. Soc.*, 2007, **129**, 728–729.

29. G. De Paepe, J. R. Lewandowski, A. Loquet, A. Bockmann and R. G. Griffin, *J. Chem. Phys.*, 2008, **129**, 245101.
30. K. Takegoshi, S. Nakamura and T. Terao, *Chem. Phys. Lett.*, 2001, **344**, 631–637.
31. M. Weingarth, D. E. Demco, G. Bodenhausen and P. Tekely, *Chem. Phys. Lett.*, 2009, **469**, 342–348.
32. I. Scholz, M. Huber, T. Manolikas, B. H. Meier and M. Ernst, *Chem. Phys. Lett.*, 2008, **460**, 278–283.
33. M. Weingarth, Y. Masuda, K. Takegoshi, G. Bodenhausen and P. Tekely, *J. Biomol. NMR*, 2011, **50**, 129–136.
34. (a) D. P. Raleigh, M. H. Levitt and R. G. Griffin, *Chem. Phys. Lett.*, 1988, **146**, 71–76; (b) M. G. Colombo, B. H. Meier and R. R. Ernst, *Chem. Phys. Lett.*, 1988, **146**, 189–196.
35. M. J. Bayro, M. Huber, R. Ramachandran, T. C. Davenport, B. H. Meier, M. Ernst and R. G. Griffin, *J. Chem. Phys.*, 2009, **130**, 114506.
36. M. J. Bayro, T. Maly, N. R. Birkett, C. M. Dobson and R. G. Griffin, *Angew. Chem. Int. Ed.*, 2009, **48**, 5708–5710.
37. A. Lange, K. Seidel, L. Verdier, S. Luca and M. Baldus, *J. Am. Chem. Soc.*, 2003, **125**, 12640–12648.
38. B. J. Wylie, L. J. Sperling, A. J. Nieuwkoop, W. T. Franks, E. Oldfield and C. M. Rienstra, *Proc. Natl. Acad. Sci. U. S. A.*, 2011, **108**, 16974–16979.
39. M. Weingarth and M. Baldus, *Acc. Chem. Res.*, 2013, **46**, 2037–2046.
40. (a) A. T. Petkova and R. Tycko, *J. Magn. Reson.*, 2004, **168**, 137–146; (b) A. J. Nieuwkoop and C. M. Rienstra, *J. Am. Chem. Soc.*, 2010, **132**, 7570–7571.
41. M. Etzkorn, A. Bockmann, A. Lange and M. Baldus, *J. Am. Chem. Soc.*, 2004, **126**, 14746–14751.
42. S. Jehle, M. Falb, J. P. Kirkpatrick, H. Oschkinat, B. J. van Rossum, G. Althoff and T. Carlomagno, *J. Am. Chem. Soc.*, 2010, **132**, 3842–3846.
43. (a) S. Jehle, P. Rajagopal, B. Bardiaux, S. Markovic, R. Kuhne, J. R. Stout, V. A. Higman, R. E. Klevit, B. J. van Rossum and H. Oschkinat, *Nat. Struct. Mol. Biol.*, 2010, **17**, 1037–1042; (b) A. Loquet, N. G. Sgourakis, R. Gupta, K. Giller, D. Riedel, C. Goosmann, C. Griesinger, M. Kolbe, D. Baker, S. Becker and A. Lange, *Nature*, 2012, **486**, 276–279.
44. M. Etzkorn, K. Seidel, L. Li, S. Martell, M. Geyer, M. Engelhard and M. Baldus, *Structure*, 2010, **18**, 293–300.
45. M. Renault, M. P. Bos, J. Tommassen and M. Baldus, *J. Am. Chem. Soc.*, 2011, **133**, 4175–4177.
46. J. R. Lewandowski, *Acc. Chem. Res.*, 2013, **46**, 2018–2027.
47. S. J. Ullrich, U. A. Hellmich, S. Ullrich and C. Glaubitz, *Nat. Chem. Biol.*, 2011, 7, 263–270.
48. K. Seidel, A. Lange, S. Becker, C. E. Hughes, H. Heise and M. Baldus, *Phys. Chem. Chem. Phys.*, 2004, **6**, 5090–5093.
49. M. Weingarth, G. Bodenhausen and P. Tekely, *J. Am. Chem. Soc.*, 2009, **131**, 13937–13939.
50. T. Gullion and J. Schaefer, *J. Magn. Reson.*, 1989, **81**, 196–200.

CHAPTER 2

Combining NMR Spectroscopic Measurements and Molecular Dynamics Simulations to Determine the Orientation of Amphipathic Peptides in Lipid Bilayers

B. SCOTT PERRIN JR.,[a] RICHARD W. PASTOR[a] AND MYRIAM COTTEN*[b]

[a] Laboratory of Computational Biology, National Heart, Lung and Blood Institute, National Institutes of Health, Bethesda, MD 20892, USA;
[b] Department of Chemistry, Hamilton College, Clinton, NY 13323, USA
*Email: mcotten@hamilton.edu

2.1 Introduction

The combination of solid-state NMR (ssNMR) and molecular dynamics (MD) simulations provides a rich context for the structural, topological, and dynamic studies of peptides in lipid bilayers and the establishment of important structure–function relationships.[1–12] At minimum, they corroborate measured and theoretical properties. Good agreement between methods supports properties from MD that may not have an experimental analogue, or might otherwise be difficult to obtain. If there is disagreement

New Developments in NMR No. 3
Advances in Biological Solid-State NMR: Proteins and Membrane-Active Peptides
Edited by Frances Separovic and Akira Naito

Published by the Royal Society of Chemistry, www.rsc.org

between methods, analysis of the MD trajectory or auxiliary calculations guide convergence and reveal weaknesses in one or both methods.

Oriented sample (OS) ssNMR takes advantage of orientational restraints such as the dipolar coupling (DC), chemical shifts (CS), or ^{2}H quadrupolar interactions to determine the orientation of peptides aligned mechanically (*e.g.* glass plates) or magnetically (*e.g.* bicelles) in lipid bilayers.[13–16] Two-dimensional separated local field experiments, such as polarization inversion spin exchange at the magic angle (PISEMA) and ^{15}N–^{1}H heteronuclear correlation (HETCOR) spectroscopy applied to multiply ^{15}N-labeled peptides, yield 2D spectra that simultaneously provide DC and CS values for each labeled residue. Two-dimensional spectra of α-helices yield ellipsoidal resonance patterns called PISA (polar index slant angle) wheels. Since the tilt (τ) defines the pattern, shape, and size of the PISA wheels, determining the tilt does not require the assignment of the individual signals on the wheel. The rotation (ρ) is provided by indexing and, therefore, assigning the resonances on the wheel.[17–20] Alternatively, the DC and CS values from the 2D spectra can be plotted as a function of residue number and fitted with sinusoidal waves to obtain the orientation of the peptide (Figures 2.1A and 2.1B). The amplitude of the wave is related to τ and the phase is related to ρ. Since not only the orientation but also the periodicity of the α-helix is reflected in dipolar waves, dipolar wave-fit analysis can be used to reveal deviations from the ideal α-helix, such as kinks and bends.[21–25] When an α-helix is allowed to explore its full conformational space, its dipolar wave, CS wave, and PISA wheel are inherently related to each other, as demonstrated in Figure 2.1D. The gray 3D coil in the center of Figure 2.1D is determined by the orientations of the dipolar and chemical shift tensors with respect to the static magnetic field.

The orientation of membrane-bound peptides, as well as the DC and CS values of their individual peptide planes, can be independently and directly determined from MD simulations. At the beginning of this chapter, the calculation of DC and CS values from simulations and the derivation of the dipolar and CS waves are presented. Next, it is demonstrated that dipolar and CS waves may give inaccurate orientations or even multiple solutions when typical uncertainties in the experimental measurements are taken into account. Comparing orientations deduced from the dipolar wave to those sampled in MD can reduce this ambiguity. Since the NMR observables can be affected by peptide dynamics, the effect of dynamics on the order parameter of the NH bond and the implications for the rigorous determination of peptide structures are also discussed. Specifically, the bias in orientation due to peptide dynamics, such as fluctuations about the helical axis, is investigated. For a peptide with a fluctuation in $\tau > 20°$, wave fitting may result in a τ value that is not equal to the average τ obtainable by MD. Bringing into agreement the analyses of the NMR and MD data as presented here instills confidence in properties directly obtainable by MD but not NMR, such as the depth of insertion of the

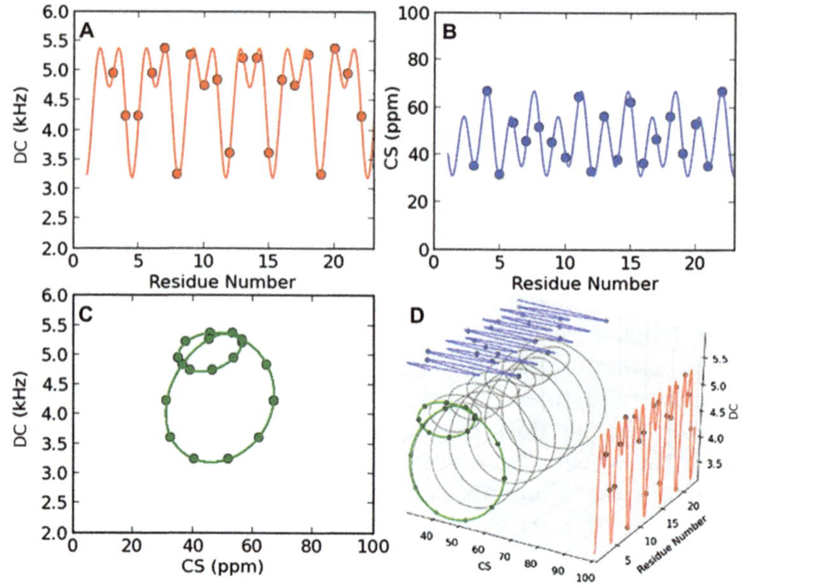

Figure 2.1 Examples of (A) a dipolar wave, (B) a chemical shift (CS) wave, and (C) a PISA wheel for an ideal α-helix ($\phi/\psi = -61°/-45°$) with $\tau = 85°$ and $\rho = 260°$. (D) 3D plot demonstrating the relationship between plots (A–C). The tilt, τ, determines the amplitude of the dipolar or CS waves and size of the PISA wheel, while the rotation, ρ, determines the phase of the waves. Intertwined wheels are observed in (C). Waves were generated using average D_a and principal elements for the ^{15}N chemical shift tensor interaction.

peptide in the bilayer. The use of dipolar waves and MD simulations to assist in the assignment of DC signals from multiply labeled peptides is also presented.

In previous work,[26] the orientation of two antimicrobial amphipathic cationic peptides, piscidin 1 (p1, FFHHIFRGIVHVGKTIHRLVTG) and piscidin 3 (p3, FIHHIFRGIVHAGRSIGRFLTG), bound to oriented lipid bilayers that mimic bacterial cell membranes was studied with dipolar waves, CS waves, NMR structure refinement by Xplor-NIH,[27] and MD. Experimentally, these structural studies are particularly challenging because the orientation of amphipathic helices, almost parallel to the membrane surface, translates into crowded NMR spectra where the ^{15}N chemical shifts and ^{15}N–^{1}H dipolar couplings are smaller than 80 ppm and 11 kHz, respectively (Figure 2.2). Some of the results from the investigation of p1 and p3 are elaborated here to illustrate nuances of the aforementioned methods. While these examples more directly pertain to interfacial peptides, much is also applicable to TM peptides and proteins.

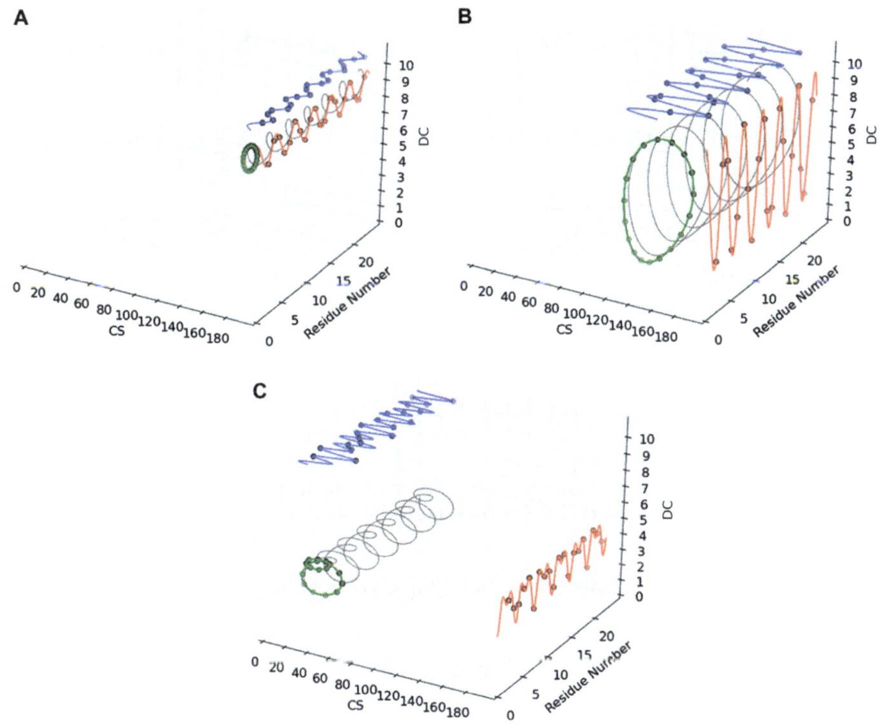

Figure 2.2 CS, DC, and residue number for an ideal 22-residue α-helix with (A) $\tau = 5°$, (B) $\tau = 30°$, and (C) $\tau = 85°$; $\rho = 260°$ for all plots.

2.2 Orientation of an α-Helix from ^1H–^{15}N Dipolar Couplings and ^{15}N Chemical Shifts

Since both the DC and CS interaction tensors are fixed with respect to the molecular frame but depend on their orientation relative to the static magnetic field, it is straightforward to express the orientation of the interaction tensors with respect to the helical axis and magnetic field. DC is calculated as:

$$DC = D_a \left\langle \frac{3\cos^2(\theta) - 1}{2} \right\rangle \tag{2.1}$$

where D_a is the static internuclear dipolar coupling magnitude ($D_a = \mu_0 \hbar \gamma_H \gamma_N / 4\pi^2 r_{NH}{}^3$), and θ is the angle of the NH vector with respect to the bilayer normal, $\vec{n}$ (Figure 2.3).[22] The ^{15}N chemical shift is calculated as:[18,20]

$$CS = \sigma_{11}\lambda_1^2 + \sigma_{22}\lambda_2^2 + \sigma_{33}\lambda_3^2 \tag{2.2}$$

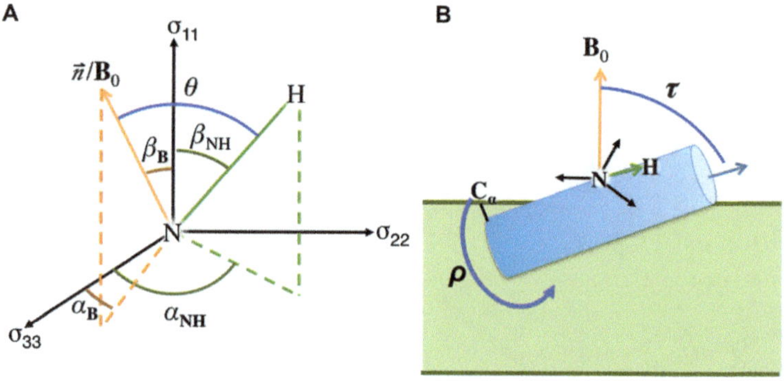

Figure 2.3 (A) Geometric relationship of the angles relating the ^{15}N–^{1}H tensor and the principal axis frame of the ^{15}N chemical shift tensor to the static magnetic field. (B) Polar angles τ and ρ that define the orientation of an α-helix with respect to the helical axis and the static magnetic field. The N-terminus of the peptide is on the left-hand side. By convention, B_0 is along the z-axis of the orthogonal (x,y,z) coordinate system.

where σ_{11}, σ_{22}, and σ_{33} are the principal elements of the ^{15}N chemical shift tensor interaction, and

$$\begin{aligned}\lambda_1 &= \sin(\beta_B)\cos(\alpha_B)\\ \lambda_2 &= \sin(\beta_B)\cos(\alpha_B)\\ \lambda_3 &= \cos(\beta_B)\end{aligned} \tag{2.3}$$

where α_B is the angle between σ_{33} and the projection of $\vec{n}$ onto the $\sigma_{22}\sigma_{33}$ plane, and β_B is the angle between σ_{11} and $\vec{n}$. This is illustrated in Figure 2.3A.

For an α-helix, DC is fit to a sinusoidal wave with a periodicity of 3.6 residues per turn, which restricts the values of θ.[22] θ oscillates between $\theta+\delta$ and $\theta-\delta$, where δ is the angle of the tensor principal axes for ^{15}N CS or ^{1}H–^{15}N DC interactions in the molecular frame. θ is propagated along the length of the helical axis as:

$$\theta = \theta - \delta\cos(x\Omega) \tag{2.4}$$

where x is the residue number and Ω is the rotation angle per residue. DC becomes:

$$\mathrm{DC} = D_a\left(\frac{3\cos^2(\theta - \delta\cos(x\Omega)) - 1}{2}\right) \tag{2.5}$$

Equation (2.1) can also be derived for the orientation of the ^{1}H–^{15}N vector, ***NH***, relative to the static magnetic field, $\boldsymbol{B}_0$:

$$\mathrm{DC} = D_a\left[\frac{3(\boldsymbol{B}_0\cdot\boldsymbol{NH})^2 - 1}{2}\right] \tag{2.6}$$

where

$$\boldsymbol{B}_0 = \begin{bmatrix} \sin(\beta_B)\cos(\alpha_B) \\ \sin(\beta_B)\sin(\alpha_B) \\ \cos(\beta_B) \end{bmatrix}, \boldsymbol{NH} = \begin{bmatrix} \sin(\beta_{NH})\cos(\alpha_{NH}) \\ \sin(\beta_{NH})\sin(\alpha_{NH}) \\ \cos(\beta_{NH}) \end{bmatrix} \tag{2.7}$$

In the helical axis frame (HAF), $\alpha_{NH} = 0°$, $\beta_{NH} = \angle \text{CNH-}\beta_D = \delta$, $\angle \text{CNH} = 122°$, and $\beta_D = 105°$,[28,29] resulting in an ***NH*** that is related to the tilt of a peptide in a bilayer aligned such that the normal, $\vec{n}$, is parallel to the magnetic field (Figure 2.3). Therefore, Eqn (2.7) written in terms of $\tau(0 < \tau < 90)$ and $\rho(0 < \rho < 180)$ becomes:

$$\boldsymbol{B}_0^{\text{HAF}} = \begin{bmatrix} \sin(\tau)\cos(\rho - x\Omega) \\ \sin(\tau)\sin(\rho - x\Omega) \\ \cos(\tau) \end{bmatrix}, \boldsymbol{NH} = \begin{bmatrix} \sin\delta \\ 0 \\ \cos\delta \end{bmatrix} \tag{2.8}$$

where δ is 17° and Ω is 100°;[15] however, δ and Ω can change significantly with ϕ and ψ. The preceding values for δ and Ω were for $\phi = -61°$ and $\psi = -45°$, as found for α-helices in most proteins. Equation (2.2) is solved relative to τ and ρ when:

$$\begin{aligned} \lambda_1 &= \sin(\tau)\cos(\rho - x\Omega) \\ \lambda_2 &= \sin(\tau)\sin(\rho - x\Omega) \\ \lambda_3 &= \cos(\tau) \end{aligned} \tag{2.9}$$

Dipolar and CS waves are continuous plots in the HAF of DC or CS as a function of residue number along the α-helix. The wave is best fit to a set of DC or CS data by adjusting τ and ρ. DC and CS for the MD structures presented here were calculated using CHARMM 36b2.[30] DC was determined from Eqn (2.1) with $D_a = 10.735$ kHz; and CS was calculated using the NMR facility with ^{15}N chemical shift tensor principal elements of $\sigma_{11} = 32$ ppm, $\sigma_{22} = 54$ ppm, and $\sigma_{33} = 202$ ppm referenced with respect to an aqueous solution of ^{15}N-labeled ammonium sulfate.[19,20]

The symmetry of nuclear spin interactions results in ambiguities when solving for the orientation of helices by OS ssNMR.[18] Indeed, angles τ and $(180 - \tau)$ give rise to the same NMR resonances in uniaxially oriented samples. In addition, the sign of DC is uncertain when it is smaller than half the maximum value (10.735 kHz), as is the case in membrane-bound piscidin.[31] However, simultaneously considering the DC and ^{15}N CS values of piscidin indicates that its DC values are negative.[31] When this sign is known, the number of ambiguities can be reduced to two orientations ($180-\tau, \rho$ and $\tau, 180+\rho$) of the helical axis, which are magnetically equivalent and consistent with the dipolar waves. However, only one combination yields a peptide orientation such that the hydrophobic side chains are pointing toward the hydrophobic bilayer. The other combination, which directs hydrophilic side chains toward the core of the bilayer, is energetically unfavorable. Consequently, only the former orientation is considered in the rest of this chapter.

Motions affect the NMR observables, including the dipolar couplings and chemical shifts.[14,16,32,33] Therefore, the derivation of reliable structures from these variables requires investigating motional averaging in terms of rate, axis, and amplitude. Considering the amphipathic helix of piscidin, molecular motions include the diffusion of the peptide about the bilayer normal,[34] which is aligned parallel to the static magnetic field, and the rocking of the peptide about its helical axis. In the aligned samples considered in this chapter, the helical axis is approximately perpendicular to the static magnetic field and, therefore, signal averaging arises mostly from the rocking motion as well as local motions around the torsion angles. As demonstrated previously through the comparison of NMR and MD data,[26] and expanded upon in this chapter, the NMR observables collected for piscidin are reliable restraints for structural determination at high resolution.

2.3 Crosschecking ^{1}H–^{15}N Dipolar Coupling Assignments of Multiply Labeled Peptides

The amide ^{15}N–^{1}H dipolar couplings for multiply labeled peptides are not straightforward to assign if the ^{15}N-amide sites have similar signals or the residue has multiple signals due to ^{15}N-containing side chains. Dipolar waves can help guide the selection of signals; however, the ability to fit the experimental data with a wave for short helical segments may become difficult because the experimental data points may not span a full period of the wave. Furthermore, kinks and bends in the helix distort the dipolar wave, and fraying at the termini may result in smaller than expected DC due to increased dynamics. In Figure 2.4, dipolar waves are fit to a set of DC for the 22-residue peptide p1 in 1 : 1 phosphatidylethanolamine/phosphatidylglycerol (PE/PG). A glycine at position 13 adds flexibility and a kink to the peptide, necessitating a fit of residues 3–10 using (τ_N, ρ_N) separately from residues 14 to 20 that are fitted using (τ_C, ρ_C). Two residues at each terminus do not belong to the best-fit wave due to helical fraying.

In the process of fitting the experimental data of p1 in PE/PG to a kinked helix, Arg18 was initially assigned a DC of 1.9 kHz. MD simulations of the same peptide in the same bilayers predicted a DC of 4.8 kHz. This was a clear outlier as the other sites had experimental and simulated DC values that agreed within 1 kHz. Further inspection of the experimental data revealed the presence of another DC splitting at 5.1 kHz. Thanks to the MD predictions, the larger splitting could be confidently assigned to the backbone of Arg18.

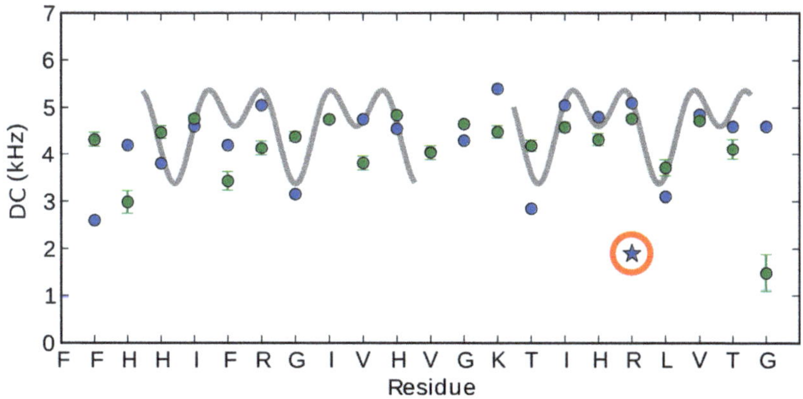

Figure 2.4 DC for piscidin 1 in PE/PG from ssNMR (*blue*) and MD (*green*). DC for the side chain of Arg18 is circled in *red.* Dipolar wave fits to the experimentally observed DC are shown in *gray* for an ideal α-helix, assuming dihedral angles ϕ of $-61°$ and $\psi = -45°$. A kink at position 13 is identified by the need to use different τ and ρ values to fit the experimental data at the N- and C-ends of the peptide. τ,ρ values of 86°, 251° and 83°, 248° were used at the N- and C-ends, respectively.

2.4 Inaccurate Orientations from Dipolar and Chemical Shift Waves

In some cases, the orientations determined from best-fit dipolar or CS waves are not accurate. This can be due to the wave fit yielding multiple (τ, ρ) solutions within the uncertainty of the measured DC or CS. The remainder of this section considers fitting of the dipolar and CS waves within experimental error. Two examples are provided: one for p1 in 3 : 1 phosphatidylcholine/phosphatidylglycerol (PC/PG) with a single expected solution; and one for p3 in 1 : 1 PE/PG with multiple solutions. Finally, the use of MD simulations to resolve the more likely value is presented.

Based on experimental linewidths, errors in the experimental DC (0.2 kHz) and CSA (3.0 ppm) were propagated into the dipolar and chemical shift waves from an ensemble of synthetic data. Specifically, 1000 sets of DC for the two examples were generated with the DC for each residue calculated as a random number from a Gaussian distribution with a mean equal to the measured values and a standard deviation (σ) of 0.2 kHz for DC and 3.0 ppm for CS. Dipolar or CS waves were fit to the DC or CS values for residues 3–10 or residues 14–20 by a Levenberg–Marquardt minimization.[35,36] For p3 in 1 : 1 PE/PG, which does not have an experimental DC for residue 3, the dipolar wave was fit to residues 4–10. Statistically, instances where multiple solutions exist within the experimental error will be characterized by high standard deviations in the solution ($\sigma[\tau]$ or $\sigma[\rho]$).

2.4.1 Single-Solution Example: Orientation of Piscidin 1 in PC/PG

In the example of p1 in PC/PG, the 1000 waves for DC are tightly fit to one solution, as seen by the significant overlap (Figure 2.5A). Figure 2.5B shows a narrow band of solutions for τ_N and τ_C at 88° and 82°, respectively. The narrow range of solutions for τ_N and τ_C demonstrates the sensitivity of the dipolar waves to a change in τ and corroborates results on TM helices.[19,22] On the other hand, the solutions for ρ_N and ρ_C cover broad ranges of solutions with averages at 259° and 249°, respectively. This insensitivity to variation in ρ demonstrates the small effect that rotation around the helical axis has on the orientation of the amide bond, which is parallel to the helical axis and perpendicular to the magnetic field for interfacial helices in mechanically aligned samples. The NMR-derived values are comparable to those obtained by MD, specifically τ_N, ρ_N of 91°, 265° and τ_C, ρ_C of 87°, 245°. Therefore, this first example is characterized by an experimentally determined orientation that is accurate.

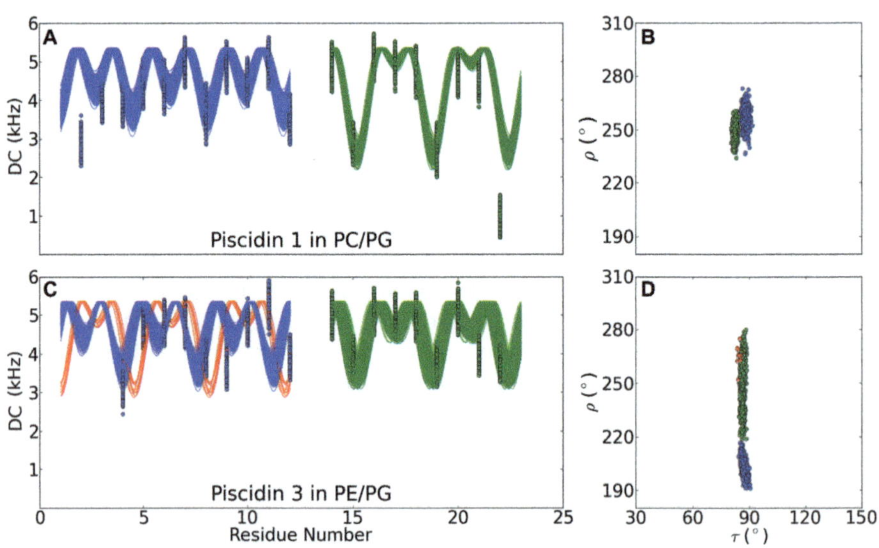

Figure 2.5 Dipolar wave fits to the 1000 sets of DC randomly selected from a Gaussian distribution with a mean equal to the experimental DC and $\sigma = 0.2$ kHz. Simulated data for piscidin 1 in PC/PG and piscidin 3 in PE/PG are shown in (A) and (C), respectively. Corresponding τ and ρ values for the wave fits are plotted in (B) and (D). Orientations for residues 3 to 10 (4 to 10 for p3) are in *blue* and those for 14 to 20 are in *green.* Secondary solutions for residues 4 to 10 of piscidin 3 in PE/PG are shown in *red.*

2.4.2 Multiple-Solution Example: Orientation of Piscidin 3 in PE/PG

In the case of p3 in PE/PG, a straightforward solution for ρ could not be identified using wave fits. Figure 2.5C shows a narrow band of solutions for τ; however, when ρ_N is considered, there are multiple solutions to the dipolar wave fit. One of these solutions has a ρ_N near 200° (blue in Figures 2.5C and 2.5D), while the second is closer to 260° degrees (red). In the simulated system, ρ_N is 241° with a root-mean squared fluctuation of 11°, while ρ_N is 250° for the NMR-derived wave and refined structure. This indicates that $\rho_N = 260°$ rather than 200° is the likely solution.

Since the percent error in CS is higher than that in DC for in-plane helices, multiple solutions are more common for CS than DC wave-fit analysis. In Figure 2.6, which shows CS wave-fit analysis, multiple solutions for τ_N and ρ_N are obtained for both p1 in PC/PG and p3 in PE/PG. Moreover, the solutions are clustered with τ,ρ of 82°, 240° or 100°, 260° for p1 and τ,ρ of 80°, 260° or 100°, 280° for p3. In both systems, the smaller solution for τ_N is less than the values obtained by MD (τ_N,ρ_N of 93°, 241°), while the larger solution is significantly too large to be in agreement with the dipolar waves and MD values. The discrepancy between solutions and difficulty in relying on experiments to obtain an accurate solution are likely due to the percent

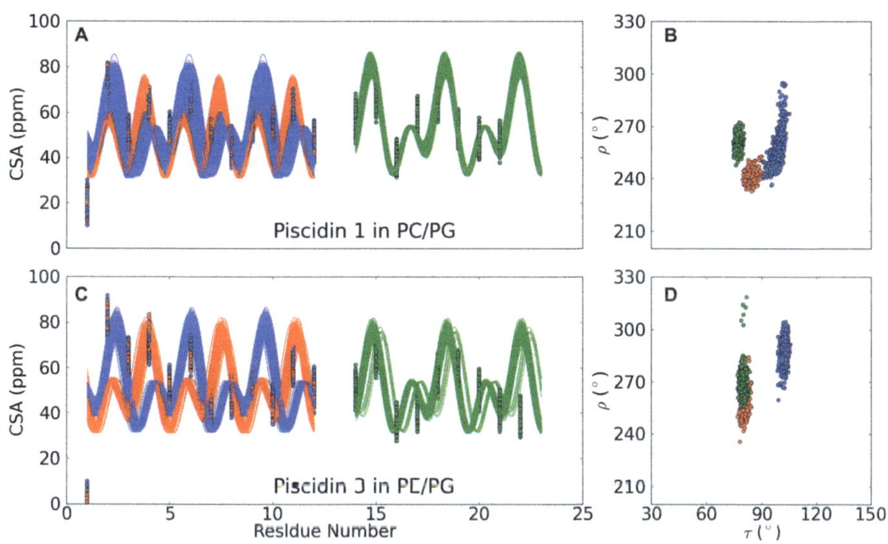

Figure 2.6 CS wave fits to the 1000 sets of CS randomly selected from a Gaussian distribution with a mean equal to the experimental CS and $\sigma = 3.0$ ppm. Simulated data are shown for piscidin 1 in PC/PG and piscidin 3 in PE/PG in (A) and (C), respectively. Corresponding τ and ρ for the wave fits are plotted in (B) and (D). Orientations for residues 3 to 10 (4 to 10 for p3) are in *blue* and those for 14 to 20 are in *green*. Secondary solutions for residues 4 to 10 are shown in *red*.

error in the experimental CS and/or the use of a common ^{15}N chemical shift tensor for all residues in the CS wave fits.

2.4.3 Resolving the Accurate Orientation with MD or Structural Refinement

MD simulations can help resolve the problem of multiple solutions in two ways. First, the sampling of wave-fits to points within the experimental error *via* a best-fit algorithm is the primary method for unbiased generation of solutions. In the aforementioned case of p3 in PE/PG, for which multiple solutions exist and differ in ρ by 50°, the difference between solutions is well within the range that can be resolved by a MD simulation of the peptide in the bilayer. While the experimental data alone do not yield an accurate solution in this example, combining NMR refinement and MD simulations resolves this issue and helps identify a τ,ρ combination in the vicinity of 90°, 250°.

2.4.4 Change to the Static Internuclear Dipolar Coupling and Anisotropic Order Parameter due to Peptide Dynamics

Proteins and peptides bound to bilayers are not rigid, and a rigorous determination of protein structure should include the effect of dynamics on the NMR variables.[37,38] Specifically, when refining the structure of a peptide through simulated annealing the experimental order parameters (which are used as restraints) are commonly scaled to take into account experimental error, mosaic spread (residual sample disorder), and fast librations of the peptide. While experimental error and mosaic spread are not likely to be predicted from an atomistic simulation, reduction of order parameters by fast motions of the amides can be examined by MD.

Here, estimating this reduction of order parameters is based on $C_{20}(t)$, the correlation function of Y_{20}, the second-order spherical harmonic:

$$C_{20}(t) = \langle \mathrm{Y}_{20}(0)\mathrm{Y}_{20}(t)\rangle = \left\langle \left(\frac{3z^2(0)-1}{2}\right)\left(\frac{3z^2(t)-1}{2}\right)\right\rangle \tag{2.10}$$

where $z(t)$ is the z-component of the unit vector along the N–H bond. The long-time or plateau value of $C_{20}(t)$ is:

$$\underset{t\to\infty}{C_{20}(t)} = \left\langle \frac{3z^2-1}{2}\right\rangle^2 = \left(\frac{\mathrm{DC}}{D_a}\right)^2 \tag{2.11}$$

The second equality follows from Eqn (2.1) and the definition $z=\cos\theta$.

Figure 2.7A shows plots of $C_{20}(t)$ for the NH vectors of residues 2 (toward the N-terminus), 8 (center of the helix), and 22 (the C-terminus) of p1 based on MD simulations in PC/PG;[26] Figure 2.7B shows the normalized

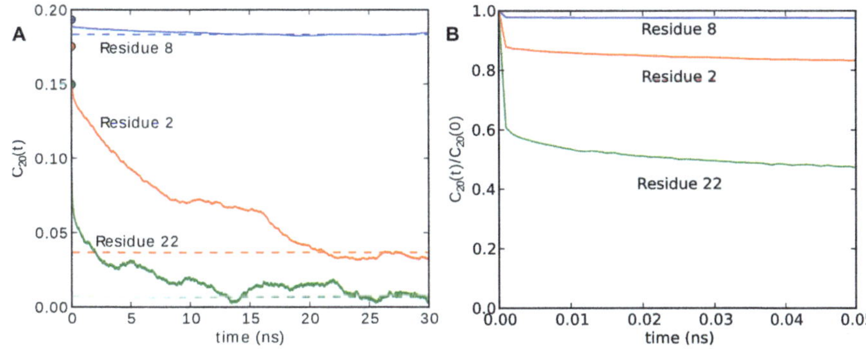

Figure 2.7 (A) Correlation function for the *z*-component of the amide bond for residues 2, 8, and 22; the *dashed lines* show $(DC/D_a)^2$ obtained as an average from the trajectory. (B) Normalized correlation functions from (A) at short times.

correlation function from 0 to 50 ps. It is clear that there are at least two exponential decays: a very rapid one on the ps timescale, and a slower one of approximately 5–10 ns. The former arises from fast internal motion, primarily libration in the torsional minima; the slower one arises from tilting motions of the helix. The dashed lines in Figure 2.7A show $(DC/D_a)^2$ obtained as an average from the trajectory, as opposed to from the correlation function. The confluence of order parameters calculated in these two ways is an important consistency check, and indicates that MD directly simulates the motions that modulate DC.

Returning to the fast motions, it is useful to write $C_{20}(t)$ as:

$$C_j(t) = S_j^2 S_s^2 + S_j^2(1 - S_s^2)\exp(-t/\tau_s) + (1 - S_j)^2 \exp(-t/\tau_j) \qquad (2.12)$$

where the subscripts *j* and *S* denote correlation times and order parameters associated with fast motions (in principle different for each vector) and slow motions (common to all vectors), respectively.[39] S_j^2 is analogous to the "generalized order parameter" in the Lipari–Szabo treatment of solution NMR;[40] the observed order parameter is proportional to S_jS_S. S_j^2 can be obtained from fitting the simulated correlation function to Eqn (2.12) or, for the current example, simply read from Figure 2.7B: it is 0.98, 0.88, and 0.61 for residues 8, 2, and 22, respectively. As may be inferred from the large variation in the preceding S_j^2, scaling observed dipolar couplings by a single constant could introduce errors in structural refinements.

2.5 Error in Measured Orientation due to Large Fluctuations in τ

Orientations of TM peptides obtained by NMR and MD have previously been shown to disagree. For WALP23, NMR gives tilt values of ~5° while the MD simulations predict a much larger tilt of >30°. In the process of reproducing the NMR observables from simulations of a rigid peptide fluctuating in τ, the PISA spectra were demonstrated to result in a bias of the average τ when fluctuations in τ were >20°.[10] This bias has been observed in experimental data consisting of CS, DC, and ^{2}H quadrupolar splittings.[10,41] In this regard, specific approaches that have been scrutinized include the PISEMA experiment, which provides ^{15}N CS and ^{15}N–^{1}H DC, and the geometric analysis of labeled alanine (GALA) method, which relies on ^{2}H quadrupolar splittings. This section demonstrates the bias in the context of the dipolar waves, CS waves, and PISA wheels.

The effect of fluctuations in τ and ρ can be incorporated into the DC and CS wave fits. To illustrate this effect, 1000 τ angles were randomly sampled from a Gaussian distribution of angles with a specified mean ("reference") τ and σ. The DC for each residue was averaged over each of the 1000 τ values. Next, each DC was averaged over the 1000 sets and dipolar waves were fit to the average DC values to give a fitted τ. The sampling was repeated for ρ and to fit CS, rather than DC. The mean range of τ values chosen here were between 0° and 90° with σ of 0°, 10°, and 20°, while the mean range of ρ values was between 180° and 270° with σ of 0°, 10°, 20°, and 40°.

Figure 2.8 shows the ensemble average orientation versus input ("reference τ,ρ"). It is clear that a peptide with fluctuations in $\tau \leq 10°$ or $\rho \leq 20°$ will have

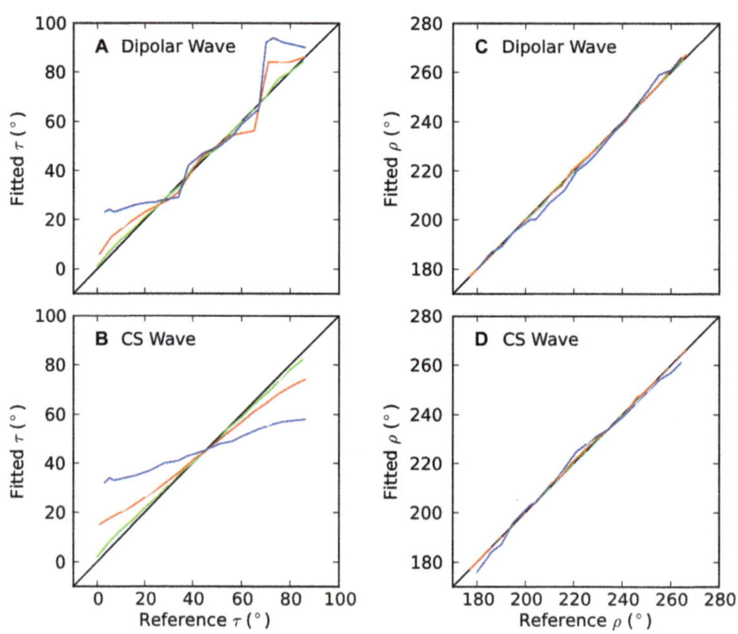

Figure 2.8 Fitted *vs.* reference (specified mean) values of τ and ρ with fluctuations of 10° (*green*), 20° (*red*), and 40° (*blue*). Fluctuations of 0° are equivalent to the *black diagonal line*, where fitted values are equal to the reference values. The further away the fitted values are from the diagonal, the less satisfactory is the fit.

an observed average similar to that of the ensemble average. However, fluctuations $\geq 20^\circ$ result in a significant difference between ensemble and observed averages when τ is near 0° or 90°. In the case of piscidin, fluctuations $<10^\circ$ were observed and, therefore, the NMR structure is accurate.[26] It is likely that other amphipathic α-helical peptides in the monomeric state will have fluctuations similar to those of piscidin. These results are in agreement with those obtained by Esteban-Martín *et al.*,[10] who calculated orientations by fitting a single reference peak with PISA wheels, which were simulated for different fluctuations around a mean tilt. The significance of the simulations presented here is that the individual bias in DC and CS can be characterized and that all amide bonds rather than a single reference peak were considered. Advantageously, using individual DC and CS values to plot PISA wheels, as done here, results in the expected two intertwined ellipses, as seen in Figure 2.1C, rather than the single ellipse shown in Esteban-Martín *et al.*[10]

2.6 Characterizing Atomistic Details not Directly Measurable by Solid-State NMR

The limitations of MD simulations include approximations in the force fields[42–44] and lack of temporal and spatial sampling. Hence, it is essential to compare simulation and experimental results before drawing conclusions from a trajectory. Where there is agreement between experiment and theory, the simulation can be used to investigate properties impossible or not easy to apprehend experimentally. For ssNMR, such properties include the depth of insertion of the peptide, membrane thinning, solvent accessibility to side chains, or structural deviations from ideality. The latter is important in peptides that include a kink or bend. As demonstrated here and alluded in previous work on piscidin,[26] dipolar and CS waves are unable to precisely determine small changes in τ and ρ. Therefore, small characteristics, such as the angle of a kink, may not be accurately determined; however MD simulations are very helpful in this regard. On another topic, the depth of insertion can be determined from either a long, equilibrated simulation or umbrella sampling at several specified depths of insertion.

MD simulations can also include restraints to determine the effect of different sample conditions. For instance, a peptide with ionizable residues can be simulated with different protonation states to examine possible effects on the peptide structure or orientation. Specifically, p1 contains four histidine residues and simulations of different charge states did not show significant differences in orientation with protonation of His3, 4, or 11. However, protonation of His17 in p1 caused a significant change in τ.[26] This information may be used to predict peptide behavior as a function of pH.

2.7 Summary and Perspective

OS ssNMR provides crucial orientational restraints to elucidate the structure, orientation, and dynamics of peptides and proteins in lipid bilayers.

In this chapter, the focus was on in-plane α-helices that have not been as widely studied as TM helices. While ^{15}N CS and ^{15}N–^{1}H dipolar couplings were used here to investigate piscidin, the ^{2}H quadrupolar splitting represents another orientational restraint that is available for assessing the conformational space of a peptide. In this regard, the GALA method relies on the geometrical analysis of ^{2}H quadrupolar splittings from deuterated alanines to find τ and ρ values that agree with the experimental data.[1,45] Typically, multiple alanine side chains are needed to specify a unique peptide topology. This approach requires that the peptide contain alanine residues and it is not possible to investigate the entirety of the peptide as needed to obtain an atomic-level structure and identify any distortion from ideality.

The complementarity of MD and OS ssNMR was illustrated in this chapter with the investigation of membrane-active piscidin. The orientation and fluctuations of piscidin in lipid bilayers that mimic bacterial cells have functional implications. For instance, the insertion of the peptide in the bilayer leads to disruptive effects such as membrane thinning and bending of the acyl chains. A detailed knowledge of peptide orientation and location in the bilayer is crucial to understanding its ability to perturb bilayers, form pores or micellar aggregates and ultimately lyse cells under native conditions. It has been shown for piscidin that varying the composition of the bilayer results in changes in peptide orientation and depth of insertion in the bilayer, and these effects are sequence dependent. In particular, increased anionic content of the bilayer results in less insertion in the bilayer and this may explain the lower activity of the peptide in bilayers that are more anionic.[26]

MD simulations provide time-resolved descriptions of a system that benefit NMR experiments by corroborating results, reducing ambiguities, and identifying difficult-to-measure properties. MD simulations are becoming more accessible through webservers such as CHARMM-GUI[46] and CHARMMing,[47] and programs such as VMD.[48] In addition, computer hardware and faster algorithms are allowing MD simulations to reach the scales of microsecond on current supercomputers or even milliseconds on specialized hardware, such as Anton.[49] This increased accessibility allows experimentalists to fit their data to a mathematical model, which can be visualized and manipulated to improve intuition about the system, reduce incorrect assumptions, and result in more precise hypotheses for future testing. A limitation is that force field parameters may not exist for a molecule of interest. These parameters are based on gas phase quantum mechanical calculations, which derive bond lengths, angles, and point charges for the atoms, followed by liquid state simulations of model compounds. Attempts have been made to automate the process through webservers or programs. One example is ParamChem, which produces topology and parameter files for the CHARMM General Force Field.[50] In addition to parameterizing a molecule, it also selects atom and residue identifiers that do not conflict with preexisting parameters.

In conclusion, specific examples have been presented here to illustrate the advantages of combining OS ssNMR and MD to investigate membrane-bound peptides. The agreement of simulation and experiment for DC and CS provides a validation of the simulation, while agreement for orientation arguably confirms inferences from experiment. Furthermore, results from the two methods that are cross-validated support the use of MD to determine physicochemical properties of the peptide and lipids that are not directly or easily accessible experimentally. The accuracy of the atomic-level orientation achievable by this approach is attractive for study of a plethora of peptides and proteins that are active at lipid membranes.

References

1. S. Ozdirekcan, C. Etchebest, J. A. Killian and P. F. J. Fuchs, *J. Am. Chem. Soc.*, 2007, **129**, 15174.
2. S. K. Straus, W. R. P. Scott and A. Watts, *J. Biomol. NMR*, 2003, **26**, 283.
3. I. Chandrasekhar, W. F. van Gunsteren, G. Zandomeneghi, P. T. F. Williamson and B. H. Meier, *J. Am. Chem. Soc.*, 2006, **128**, 159.
4. S. K. Kandasamy, D.-K. Lee, R. P. R. Nanga, J. Xu, J. S. Santos, R. G. Larson and A. Ramamoorthy, *Biochim. Biophys. Acta*, 2009, **1788**, 686.
5. L. Shi, A. Cembran, J. Gao and G. Veglia, *Biophys. J.*, 2009, **96**, 3648.
6. Y. H. Lam, A. Hung, R. S. Norton, F. Separovic and A. Watts, *Proteins*, 2010, **78**, 858.
7. A. Vogel, G. Reuther, M. B. Roark, K.-T. Tan, H. Waldmann, S. E. Feller and D. Huster, *Biochim. Biophys. Acta*, 2010, **1798**, 275.
8. A. Holt, L. Rougier, V. Réat, F. Jolibois, O. Saurel, J. Czaplicki, J. A. Killian and A. Milon, *Biophys. J.*, 2010, **98**, 1864.
9. A. Ramamoorthy, S. K. Kandasamy, D.-K. Lee, S. Kidambi and R. G. Larson, *Biochemistry*, 2007, **46**, 965.
10. S. Esteban-Martín, E. Strandberg, G. Fuertes, A. S. Ulrich and J. Salgado, *Biophys. J.*, 2009, **96**, 3233.
11. M. Weingarth, C. Ader, A. J. S. Melquiond, D. Nand, O. Pongs, S. Becker, A. Bonvin and M. Baldus, *Biophys. J.*, 2012, **103**, 29.
12. S. Jo and W. Im, *Biophys. J.*, 2011, **100**, 2913.
13. C. H. Wu, A. Ramamoorthy and S. J. Opella, *J. Magn. Reson., Ser. A*, 1994, **109**, 270.
14. R. Fu and T. A. Cross, *Annu. Rev. Biophys. Biomol. Struct.*, 1999, **28**, 235.
15. S. J. Opella and F. M. Marassi, *Chem. Rev.*, 2004, **104**, 3587.
16. F. M. Marassi, B. B. Das, G. J. Lu, H. J. Nothnagel, S. H. Park, W. S. Son, Y. Tian and S. J. Opella, *Methods*, 2011, **55**, 363.
17. F. M. Marassi and S. J. Opella, *J. Magn. Reson.*, 2000, **144**, 150.
18. F. M. Marassi and S. J. Opella, *J. Biomol. NMR*, 2002, **23**, 239.
19. J. Wang, J. Denny, C. Tian, S. Kim, Y. Mo, F. Kovacs, Z. Song, K. Nishimura, Z. Gan, R. Fu, J. R. Quine and T. A. Cross, *J. Magn. Reson.*, 2000, **144**, 162.

20. J. K. Denny, J. Wang, T. A. Cross and J. R. Quine, *J. Magn. Reson.*, 2001, **152**, 217.
21. A. A. Nevzorov, M. F. Mesleh and S. J. Opella, *Magn. Reson. Chem.*, 2004, **42**, 162.
22. M. F. Mesleh and S. J. Opella, *J. Magn. Reson.*, 2003, **163**, 288.
23. M. F. Mesleh, S. Lee, G. Veglia, D. S. Thiriot, F. M. Marassi and S. J. Opella, *J. Am. Chem. Soc.*, 2003, **125**, 8928.
24. S. Kim and T. A. Cross, *J. Magn. Reson.*, 2004, **168**, 187.
25. R. C. Page, S. Kim and T. A. Cross, *Structure*, 2008, **16**, 787.
26. B. S. Perrin Jr., Y. Tian, R. Fu, C. V. Grant, E. Y. Chekmenev, W. E. Wieczorek, A. E. Dao, R. M. Hayden, C. M. Burzynski, R. M. Venable, M. Sharma, S. J. Opella, R. W. Pastor, M. L. Cotten, DOI: 10.1021/ja411119m.
27. C. D. Schwieters, J. J. Kuszewski, N. Tjandra and G. M. Clore, *J. Magn. Reson.*, 2003, **160**, 65.
28. W. Mai, W. Hu, C. Wang and T. A. Cross, *Protein Sci.*, 1993, **2**, 532.
29. Q. Teng and T. A. Cross, *J. Magn. Reson.*, 1989, **85**, 439.
30. B. R. Brooks, C. L. Brooks III, A. D. MacKerell Jr, L. Nilsson, R. J. Petrella, B. Roux, Y. Won, G. Archontis, C. Bartels, S. Boresch, A. Caflisch, L. Caves, Q. Cui, A. R. Dinner, M. Feig, S. Fischer, J. Gao, M. Hodoscek, W. Im, K. Kuczera, T. Lazaridis, J. Ma, V. Ovchinnikov, E. Paci, R. W. Pastor, C. B. Post, J. Z. Pu, M. Schaefer, B. Tidor, R. M. Venable, H. L. Woodcock, X. Wu, W. Yang, D. M. York and M. Karplus, *J. Comput. Chem.*, 2009, **30**, 1545.
31. R. Bertram, T. Asbury, F. Fabiola, J. R. Quine, T. A. Cross and M. S. Chapman, *J. Magn. Reson.*, 2003, **163**, 300.
32. S. H. Park, A. A. Mrse, A. A. Nevzorov, A. A. De Angelis and S. J. Opella, *J. Magn. Reson.*, 2006, **178**, 162.
33. R. R. Ketchem, W. Hu and T. A. Cross, *Science*, 1993, **261**, 1457.
34. E. Chekmenev, B. Vollmar and M. Cotten, *Biochim. Biophys. Acta*, 2010, **1798**, 228.
35. K. Levenberg, *Q. Appl. Math.*, 1944, **2**, 164.
36. D. Marquardt, *SIAM J. Appl. Math.*, 1963, **11**, 431.
37. R. Ishima and D. A. Torchia, *Nat. Struct. Biol.*, 2000, **7**, 740.
38. G. Bouvignies, P. R. L. Markwick and M. Blackledge, *ChemPhysChem*, 2007, **8**, 1901.
39. R. W. Pastor, R. M. Venable, M. Karplus and A. Szabo, *J. Chem. Phys.*, 1988, **89**, 1128.
40. G. Lipari and A. Szabo, *J. Am.Chem. Soc.*, 1982, **104**, 4546.
41. W. Im, S. Jo and T. Kim, *Biochim. Biophys. Acta*, 2012, **1818**, 252.
42. J. B. Klauda, R. M. Venable, A. D. MacKerell Jr. and R. W. Pastor, *Curr. Topics Membr.*, 2008, **60**, 1.
43. O. Guvench and A. D. MacKerell, *Methods Mol. Biol.*, 2008, **443**, 63.
44. R. W. Pastor and A. D. MacKerell, *J. Phys. Chem. Lett.*, 2011, **2**, 1526.
45. B. Bechinger and E. S. Salnikov, *Chem. Phys. Lipids*, 2012, **165**, 282.
46. S. Jo, T. Kim, V. G. Iyer and W. Im, *J. Comput. Chem.*, 2008, **29**, 1859.

47. B. T. Miller, R. P. Singh, J. B. Klauda, M. Hodoščeк, B. R. Brooks and H. L. Woodcock, *J. Chem. Inf. Model.*, 2008, **48**, 1920.
48. W. Humphrey, A. Dalke and K. Schulten, *J. Mol. Graphics*, 1996, **14**, 33–38.
49. E. Shaw, M. M. Deneroff, R. O. Dror, J. S. Kuskin, R. H. Larson, J. K. Salmon, C. Young, B. Batson, K. J. Bowers, J. C. Chao, M. P. Eastwood, J. Gagliardo, J. P. Grossman, C. R. Ho, D. J. Ierardi, I. Kolossvary, J. L. Klepeis, T. Layman, C. McLeavey, M. A. Moraes, R. Mueller, E. C. Priest, Y. Shan, J. Spengler, M. Theobald, B. Towles and S. C. Wang, in *Proceedings of the 34th Annual International Symposium on Computer Architecture*, ACM, San Diego, CA, 2007, p. 1.
50. K. Vanommeslaeghe, E. Hatcher, C. Acharya, S. Kundu, S. Zhong, J. Shim, E. Darian, O. Guvench, P. Lopes, I. Vorobyov and A. D. MacKerell Jr., *J. Comput. Chem.*, 2010, **31**, 671.

CHAPTER 3

^{13}C–^{13}C Distance Measurements by Polarization Transfer Matrix Analysis of ^{13}C Spin Diffusion in a Uniformly ^{13}C-Labeled Molecular Complex under Magic Angle Spinning

AYAKO EGAWA, HIDEO AKUTSU AND TOSHIMICHI FUJIWARA*

Institute for Protein Research, Osaka University, 3-2 Yamadaoka, Suita 565-0871, Japan
*Email: tfjwr@protein.osaka-u.ac.jp

3.1 Introduction

^{13}C Spin diffusion experiments for uniformly ^{13}C-labeled compounds can provide a number of internuclear distances over a broad range because the diffusion is due to dipolar couplings that depend on distances. This NMR technique has been applied to the structural analysis of ^{13}C-labeled biomolecules in the solid state.[1–9] ^{13}C–^{13}C Polarization transfer under magic angle spinning (MAS) can be caused by dipolar couplings for ^{13}C and ^{1}H spins under several conditions, such as ^{1}H-driven, RF-driven, and rotor-driven polarization transfer.[10,11] The ^{1}H-driven spin diffusion (PDSD) has

New Developments in NMR No. 3
Advances in Biological Solid-State NMR: Proteins and Membrane-Active Peptides
Edited by Frances Separovic and Akira Naito

Published by the Royal Society of Chemistry, www.rsc.org

advantages over other diffusion schemes. PDSD does not need RFfield irradiation, which may heat samples and dissipate the total ^{13}C polarization.[12] It does not need rotational resonance conditions that may restrict the effective bandwidth for the transfer. Here we deal with the distance dependence of PDSD. The precision and range of the obtained distances govern the accuracy of the determined structure. Therefore, it is important to improve the measurement quality of the distances.

The following factors affect the distance information obtained from ^{13}C–^{13}C polarization transfer of uniformly ^{13}C-labeled samples. Firstly, strong short-range couplings between covalently bonded carbons suppress the long-range weak dipolar couplings. This phenomenon is known as dipolar truncation and impedes measurement of the long-range distances.[2,13–22] Secondly, it is difficult to distinguish between relayed and direct polarization transfer.[23] Since the relayed polarization-transfer rates depend on the transfer pathways, the rates do not give accurate distances without analysis of the pathways; direct polarization transfer gives distances more accurately. Thirdly, the approximation used for converting rates to distances also affects the accuracy of the distances. ^{1}H-Driven ^{13}C spin diffusion is enabled by line broadening, due to ^{13}C–^{1}H dipolar couplings, chemical shift anisotropies (CSA) and ^{1}H–^{1}H dipolar couplings, which leads to overlap of a pair of ^{13}C spins having different chemical shifts. Thus, ^{13}C spin diffusion depends on the orientation of molecules and the anisotropy of dipolar and CSA tensors under MAS.[24,25] Internuclear distances can be calculated using time-dependent perturbation theory of ^{13}C spin diffusion under MAS. The polarization transfer rate is related to zero-quantum (ZQ) lineshape functions, which can be estimated from the experimental single-quantum (SQ) lineshapes.[26] Fourthly, assignment of distance correlations to specific spin pairs is not straightforward, especially in molecular complexes, *e.g.* distinction between intra- and intermolecular distances. This can be addressed by the simultaneous assignment of distance correlations and structure optimization.[1]

We deal with the effects of the first three factors on the accuracy and range of the ^{13}C distances. For this purpose we analyze the experimental ^{1}H-driven spin-diffusion data used for the structure determination of uniformly ^{13}C-labeled bacteriochlorophyll *c* (BChl *c*) assembly (Figure 3.1).[1] The quantitative distance analysis was conducted on the basis of the polarization-transfer matrix analysis in combination with perturbation theory using ZQ dipolar spectral lineshapes.

3.2 Theory

We have employed the following rate equation for ^{13}C polarization transfer to derive the rate matrix **R** from experimental 2D PDSD spectra:[23,24,27,28]

$$\mathbf{M}_{\text{cal}}(\tau_{\text{mix}}) = [\exp(-\mathbf{R}\tau_{\text{mix}})]\mathbf{M}_{\text{exp}}(0) \tag{3.1}$$

Here, matrices $\mathbf{M}_{\text{cal}}(\tau_{\text{mix}})$ and $\mathbf{M}_{\text{exp}}(\tau_{\text{mix}})$ having 30×30 elements, respectively, give the normalized signal intensities of the calculated and

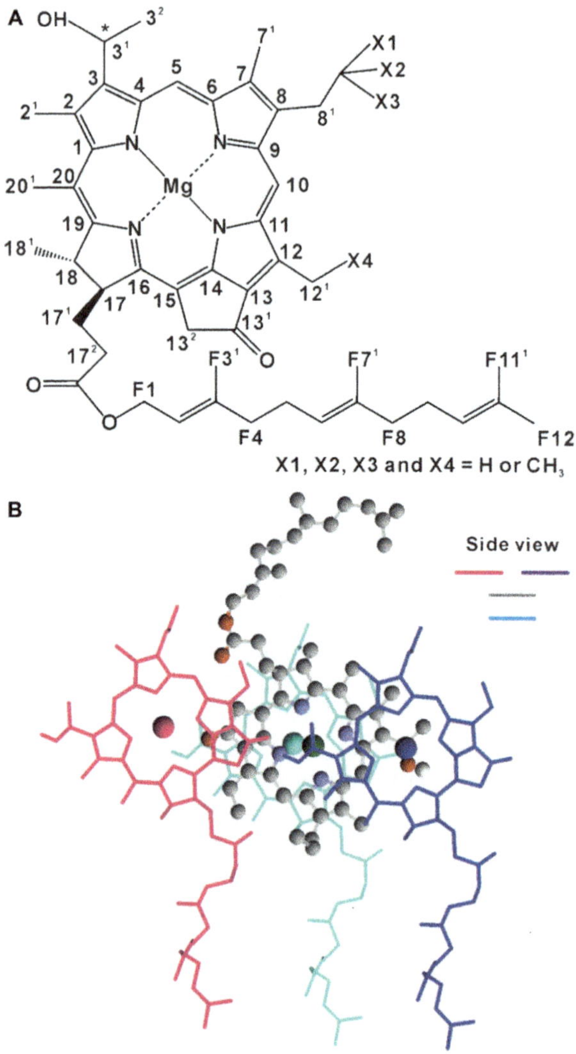

Figure 3.1 Structures of (A) BChl *c* and (B) BChl *c* assembly. (A) BChl *c* has two isomers with chirality at $C3^1$ and homologues with different side chains bonding to C8 and C12. (B) A top view of the unit structure in the assembly. Coloured molecules are stacked as illustrated by the side view. The central molecule in the unit is shown with *gray* balls and sticks. The *gray* and *dark blue* molecules form a piggyback dimer.

experimental peaks as a function of mixing time τ_{mix} for C1–C20, $C2^1$, $C3^1$, $C3^2$, $C7^1$, $C8^1$, $C12^1$, $C13^1$, $C13^2$, $C18^1$ and $C20^1$ in BChl *c* (Figure 3.1A). The matrix **R** was determined by minimizing the RMSD:

$$F = \left(\sum_{\tau_{mix},i,j} w_{\tau_{mix},i,j} \left(M_{cal,i,j}(\tau_{mix}) - M_{exp,i,j}(\tau_{mix}) \right)^2 \right)^{1/2} \tag{3.2}$$

by the nonlinear least-squares method using the Levenberg–Marquardt algorithm. Here, $w_{\tau_{mix},i,j}$ is a weighting for a correlation peak between spins i and j at τ_{mix}. The initial **R** was estimated linearly from experimental signal intensities at a short mixing time as $\mathbf{R} = -\ln(\mathbf{M}_{\text{exp}}(\tau_{\text{mix}})/\mathbf{M}(0))/\tau_{\text{mix}}$.[23] Experimental intensities were obtained from 2D PDSD spectra recorded under MAS at a frequency of 12.5 kHz with a contact time of 2 ms for ramped cross polarization from ^{1}H to ^{13}C. Parts of the spectra are shown in Figure 3.2 with the spectral slice at C3[1]. Overlapping signals in the slice spectra were deconvoluted to measure the intensity of each spin (Figure 3.2B). Signal intensity matrices were calculated from the volume integrals of the cross-peaks by using the data reduction software Felix95. The assignments of all the carbons in BChl *c* are given in Egawa *et al.*[1]

Distances were calculated from $R_{j,k}$ by using perturbation theory of ^{13}C spin diffusion under MAS at spinning frequency ω_R.[24] The rate $R_{j,k}$ for the direct polarization transfer can be expressed as:

$$R_{j,k} = \frac{\gamma^4 \hbar^2}{15\, {r^{\text{eff}}_{j,k}}^{6}} L^{(j,k)}_{\text{ZQ}} \tag{3.3}$$

where

$$L^{(j,k)}_{\text{ZQ}} = K^{(j,k)}_{\text{ZQ}}(\omega_R) + K^{(j,k)}_{\text{ZQ}}(-\omega_R) + \frac{1}{2}K^{(j,k)}_{\text{ZQ}}(2\omega_R) + \frac{1}{2}K^{(j,k)}_{\text{ZQ}}(\ 2\omega_R)$$

and

$$\left(\frac{1}{r^{\text{eff}}_{j,k}}\right)^6 = \left(\sum_{m=1}^{m_{\max}} \sum_{l=1}^{n} \frac{1}{n\, r^6_{j,k,l,m}}\right)$$

where $K^{(j,k)}_{\text{ZQ}}$ and $r_{j,k,l,m}$ are the ZQ lineshape function and the distance between spin j in molecule l and spin k in molecule m, respectively. Integer n is the number of nonequivalent molecules in the unit structure of the complex consisting of $m_{\max}$ (= 18) molecules. A schematic structure of the molecular complex can be found in Egawa *et al.*[1]

Experimental SQ spectra to calculate the ZQ lineshapes were obtained from 2D chemical-shift-resolved spectra with and without CH dipolar couplings. The pulse sequence for obtaining these spectra is the same as the PDSD sequence at $\tau_{\text{mix}} = 0$ ms, except for modification of ^{1}H decoupling in the t_1 period. The atomic coordinates determined by solid-state NMR[1] were used in the analysis of the obtained experimental distances.

3.3 Results

3.3.1 Build-up Curve Analysis by the Rate Matrix R

Both direct and relayed polarization-transfer processes give cross-peak intensities. Since the polarization-transfer matrix provides rates for the direct

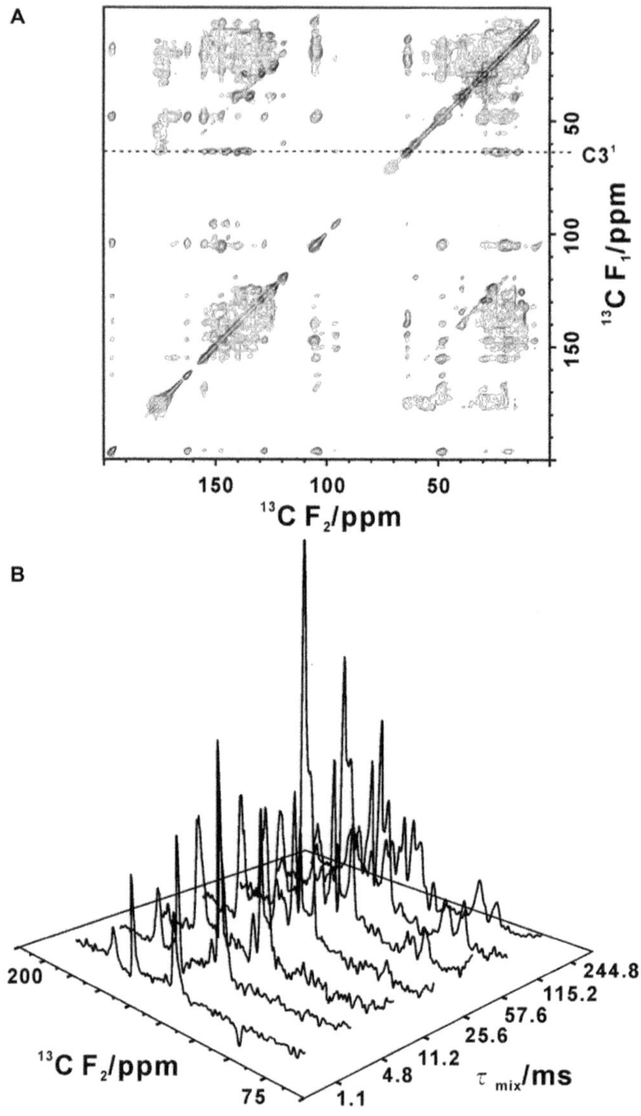

Figure 3.2 Proton-driven spin diffusion ^{13}C–^{13}C correlation NMR spectra of the uniformly ^{13}C-labeled BChl *c* assembly. (A) 2D spectrum at $\tau_{mix} = 115.2$ ms. (B) mixing-time-dependent spectra for the correlations of $C3^1$ at $F_1 = 64$ ppm.

transfer,[23] distances due to direct ^{13}C–^{13}C dipolar couplings can be obtained from **R**.

We have fit the calculated to the experimental signal intensities in the polarization-transfer matrix analysis. Examples of the build-up curve fits are presented for correlations with $C3^1$ in Figure 3.3. The build-up curves for ^{13}C spins separated by one, two, and three covalent bonds from the $C3^1$ atom are shown in Figure 3.3B–D, respectively. In Figure 3.3B, the polarization-transfer rate for the C3^1–C3 correlation is 5.4 s^{-1}. Exponential increases with large amplitude at a short mixing time of about 10 ms result from direct magnetization transfer at a distance of about 1.5 Å. The rate for the C3^1–C1 correlation at a distance of about 3.5 Å is 0.14 s^{-1} (Figure 3.3D). Longer distances are obtained from the slower build-up curves with small amplitudes (Figure 3.3E). The increase after a lag-time indicates indirect relayed transfer (Figure 3.3F) with a direct polarization-transfer rate $R_{i,j}$ of 0 s^{-1}.

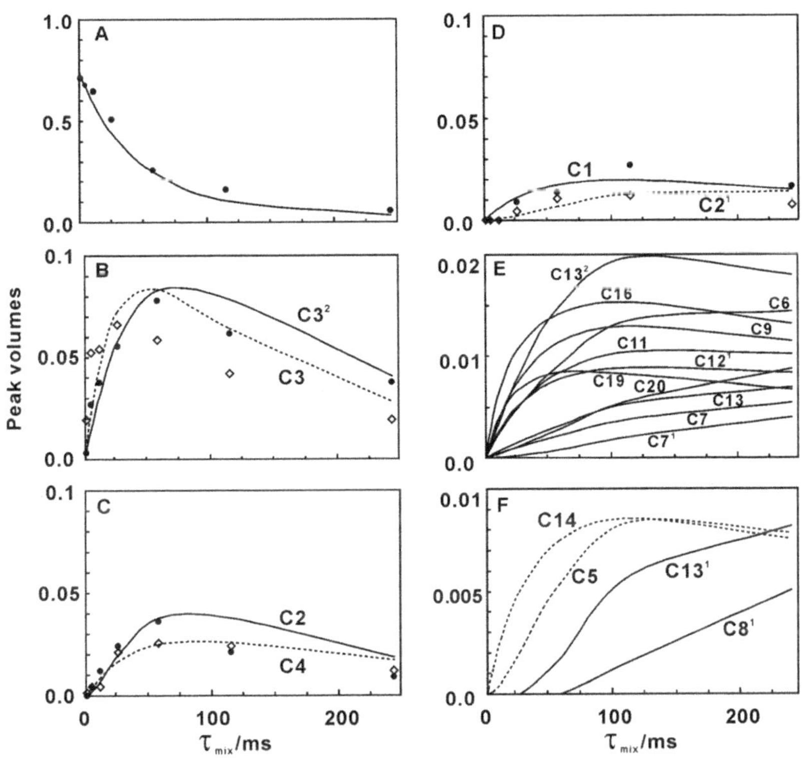

Figure 3.3 Experimental (*symbols*) and simulated (*lines*) build-up curves for the correlations of labeled carbons with C3^1 from ^{13}C spin-diffusion. *Solid* and *dashed* lines were respectively obtained from *closed circles* and *open diamonds* in A, B, C and D. (A) Decay curve for the C3^1 diagonal peak. (B) Build-up curves for pairs of carbons covalently bonded with lengths of about 1.5 Å. (C) Curves for carbons connected by two covalent bonds with distances of ~2.5 Å. (D) Curves for carbons connected by three covalent bonds with distances of ~3.5 Å. (E) Curves for intermolecular correlations for C3^1 with distances of ~4.5 Å. (F) Curves for relayed build-up curves with polarization-transfer rates of 0 s^{-1} (*solid lines*) and intermolecular polarization transfer with distances >4.5 Å (*dashed lines*). Noise level is about 0.002.

The experimental transfer curves agree with calculated ones with standard deviation (SD) of 5%, 19%, and 20% for diagonal peaks, cross-peaks with rate 3.0 s^{-1} and cross-peaks with rate 0.3 s^{-1} on average, respectively. Here, each SD is normalized by the maximum intensity of the corresponding time-dependent peak. Thus, the PDSD signals under MAS are approximately expressed by the rate equation (Eqn 3.1). These deviations affect the rates as described in Section 3.4.1 below.

3.3.2 Polarization Transfer Rates $R_{i,j}$

We show the statistics for the rate matrix **R**. The cross peaks observed at $\tau_{mix} = 244.5$ ms account for 47% of the off-diagonal elements of intensity matrix **M**. However, ~9% of the elements are explained by the polarization transfer rate 0 s^{-1}, which indicate that the cross peaks are due only to indirect polarization transfer. Thus about 38% of the total off-diagonal elements of the rate matrix provide the distance information.

The frequency distribution of the polarization-transfer rates $R_{i,j}$ obtained from the analysis of the spectra is presented in Figure 3.4. The obtained $R_{i,j}$ range from 30 s^{-1} to 0.003 s^{-1}. The correlations for intramolecular distances of about 1.5, 2.5, and 3.5 Å give polarization-transfer rates distributed over 2–20, 0.7–3, and 0.3–1.0 s^{-1}, respectively, as shown in Figure 3.4. Intermolecular correlations give longer distances of 3.1–5.9 Å with polarization-transfer rates of 0.003–2.7 s^{-1}. The rates for the BChl *c* assembly are compatible with the rates of about 0.7 s^{-1} for durene at ^{13}C distances of 2.5–3.5 Å under MAS with a spinning frequency of 4.5 kHz and resonance frequency difference of about 7.5 kHz.[29]

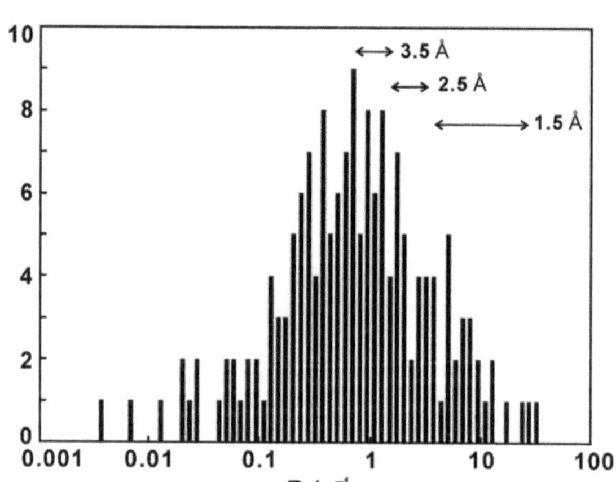

Figure 3.4 Frequency distribution of polarization-transfer rates $R_{i,j}$ obtained by rate matrix analysis. The *arrowed lines* indicate regions of $R_{i,j}$ which correspond to distances of about 3.5, 2.5, and 1.5 Å.

3.3.3 Calculation of Distances from Polarization Transfer Rates

3.3.3.1 Zero-Quantum Spectral Density

As shown in Eqn (3.3), we need ZQ spectral intensity $K_{ZQ}^{i,j}(\omega)$ to calculate distances $r_{i,j}$ from the polarization transfer rates $R_{i,j}$. The ZQ lineshape function was computed as:

$$K_{ZQ}^{(j,k)}(\Delta\omega) = \frac{1}{2\pi}\int_{-\infty}^{\infty} F_j(\Delta\omega - \omega)F_k(\omega)d\omega \tag{3.4}$$

where $F_j(\omega)$ is the normalized SQ dipolar lineshape function of spin j under C–H and H–H dipolar couplings.[26] The ^{13}C SQ lineshapes were computed from chemical shift spectra under CH dipolar fields and those under a ^1H decoupling RF field. As shown in Figure 3.5B and D, the CH dipolar couplings have a stronger effect on the C3^1 resonance than on the C3 one, because the C3^1 atom covalently bonds to ^1H but the C3 atom does not. The experimental SQ lineshapes without ^1H decoupling fields are primarily determined by CH dipolar couplings, but influenced also by inhomogeneous broadening. Since the CH couplings under ^1H–^1H couplings are the major driving force for the spin diffusion,[24] $F_j(\omega)$ was obtained by subtracting the inhomogeneous broadening from the experimental CH dipolar lineshapes. This inhomogeneous broadening due to chemical shifts was estimated from the ^1H-decoupled SQ lineshapes. Since the dipolar signal intensities decreased slowly with off-resonance frequency $\Delta\omega_z$ as $\sim 1/\Delta\omega_z{}^2$, the SQ dipolar lineshapes were better fit to Lorentzian than Gaussian functions. Thus, we expressed $F_j(\omega)$ as a Lorentzian function with the linewidth in the range between $\Delta\nu_{couple}$ and $\Delta\nu_{couple} - \Delta\nu_{decouple}$, where $\Delta\nu_{decouple}$ and $\Delta\nu_{couple}$ are linewidths observed with and without ^1H decoupling, respectively. These analyses gave the upper and lower limits of the distance, respectively. The differences between the limits corresponded to less than 30% of the distance. Here, we employed not the experimental signals but the fitted Lorentzian lineshapes to avoid inaccuracies due to noise and baseline distortion.

The experimental ZQ lineshape factor L_{ZQ} as a function of resonance frequency differences is presented in Figure 3.6A. Generally, C–C pairs without an attached proton gave a factor less than about 0.1×10^{-6}. The strong CH dipolar couplings with attached protons increased L_{ZQ}. The factor has large values, especially at frequency differences of 12.5 ± 0.3 kHz near the side-bands owing to the rotational resonance effect, which leads to 8–15 Å longer distances than the correct values. This inaccuracy originates from the limitations of perturbation theory.[30] The number of C–C correlations in this range for rotational resonance was only two out of 166 correlations.

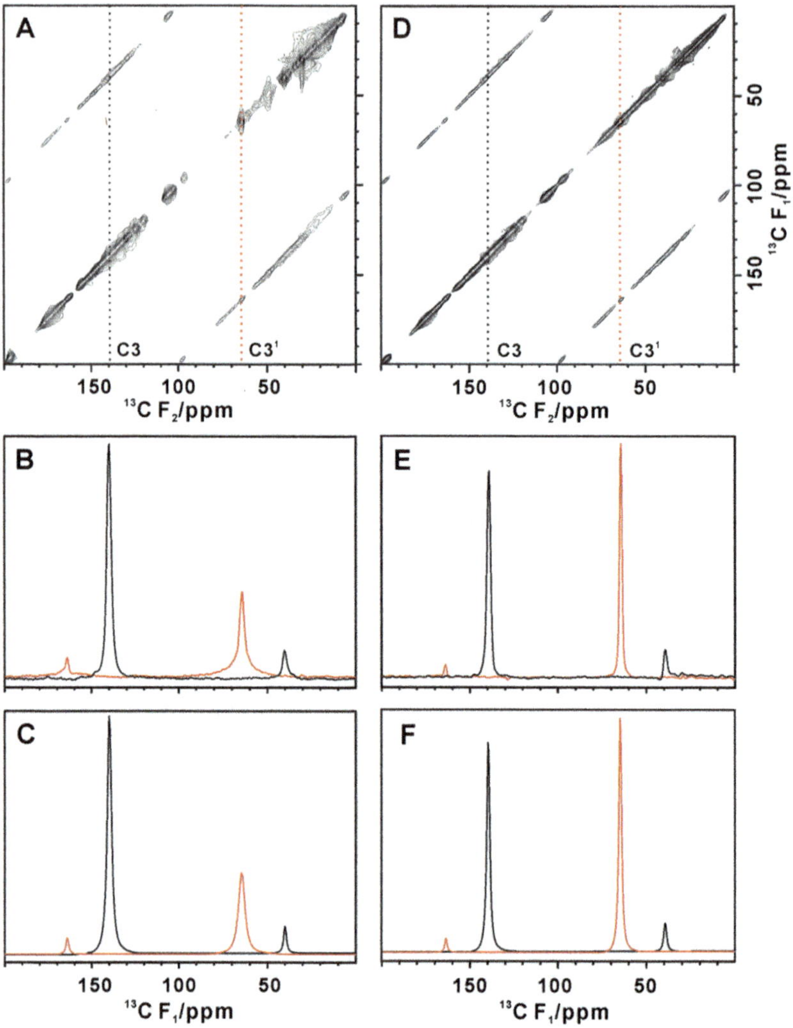

Figure 3.5 SQ lineshapes for calculation of ZQ lineshape functions. (A) Chemical-shift resolved CH dipolar spectrum observed without ^{1}H decoupling. (B) Cross-sections along F_1 at C3 and C3^1 resonances in F_2 of spectrum in A. (C) Calculated Lorentzian lineshapes fitted to cross sections in B. (D) Chemical-shift resolved spectrum observed under ^{1}H decoupling. (E) Cross-sections along F_1 of spectrum D. (F) Calculated lineshapes fitted to sections in E.

3.3.3.2 Distances Obtained by Perturbation Theory

Figure 3.7 shows the correlation between the distances obtained and those in the experimentally determined structure.[1] The distances corrected by the ZQ lineshapes are also presented. The SDs for the two distances give a measure of the accuracy for the distance measurements. The distances agree

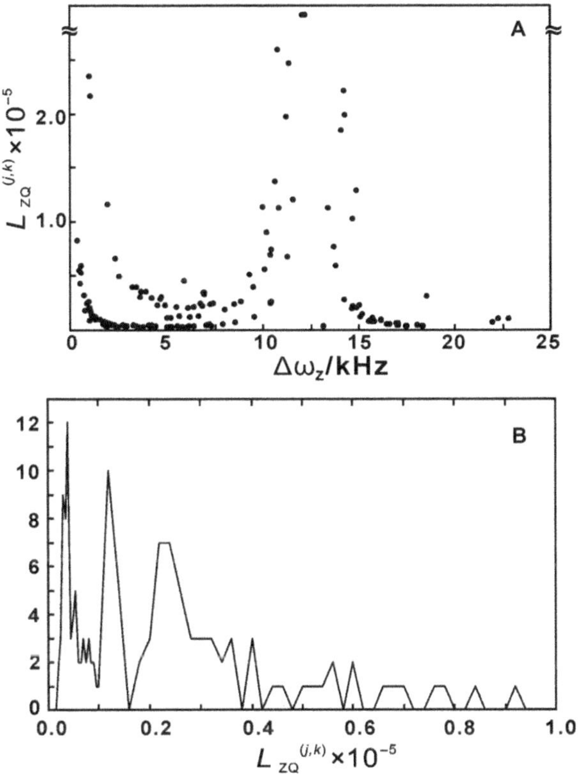

Figure 3.6 The ZQ lineshape function L_{ZQ}. (A) Function values of 228 spin pairs dependent on the resonance frequency difference $\Delta\omega_z$ at a spinning frequency of 12.5 kHz. (B) Frequency distribution of $L_{ZQ}^{(j,k)}$.

with the known distances with a SD of 25% for 68 intramolecular distances of C–C pairs connected by one to three covalent bonds. These distances are restricted by chemical bonds and do not depend on the structure optimization. The distances were obtained with a SD of 29% when the ZQ lineshape functions in Eqn (3.3) were replaced with a constant value of 0.5×10^{-6} that minimized the SD. The ZQ functions calculated with Eqn (3.4), therefore, improved the accuracy of the distances, as shown by the reduction in SD from 29% to 25%. These improvements confirm the validity of the approximation used for calculating ZQ lineshapes. Note that we obtained the distances from the rates without using reference distances.

Figure 3.7B shows the correlation between the intermolecular distances experimentally obtained and those in the structure optimized using internuclear distances and potential energies. Thus, those structural distances are biased toward the corrected distances used for the optimization. The SDs of 53% for corrected distances are less than those for the uncorrected distances, 69%. The uncorrected distances are generally shorter than the

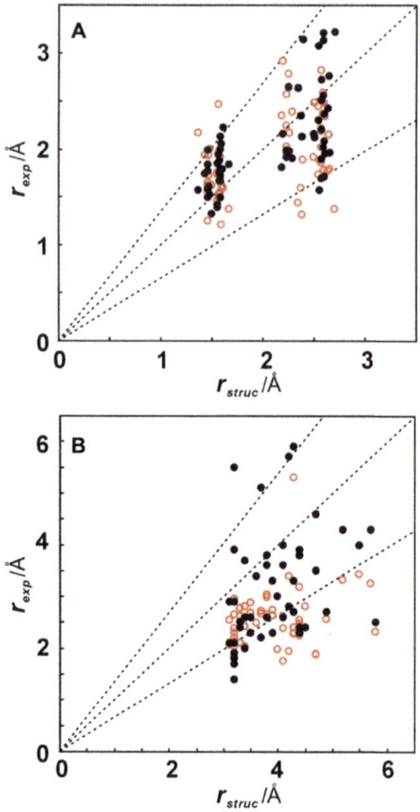

Figure 3.7 Correlation between experimental ^{13}C–^{13}C distances r_{exp} and distances in optimized model structures r_{struc}. (A) Correlation for intramolecular distances. (B) Correlation for intermolecular distances. Correlations for distances (●) corrected with ZQ lineshape functions and those for uncorrected distances (○) are shown.

distances in the energy optimized structure and would provide a structure under strong repulsive van der Waals forces.

3.4 Discussion

3.4.1 Factors Affecting Distance Precision

The precision of distances is affected by several factors: accuracy in quantifying spectral intensities, fitting of experimental to calculated intensities based on the rate equation, and estimation of the ZQ lineshape functions. We estimated errors in distance due to quantification of experimental spectral intensities and quantified all the 2D signals above the noise level. Overlapping signals were deconvoluted by optimizing the numerical signals defined by the chemical shifts, linewidths, and heights. These errors were about 40 Hz in linewidth and about 15% in signal intensity, and affect the accuracy of the distances by about 3% for signals with an S/N ratio larger than 5.

The polarization-transfer matrix was obtained from the signal intensities calculated with the rate equation (Eqn 3.1). This equation is derived by averaging the equation of motion over the powder orientations. Since the PDSD signals under MAS were expressed by the rate equation with a SD of $\sim$15% in the present study, this formalism is approximately valid for ^{13}C spin diffusion.[24] Note that this approximation does not hold for the initial build-up regime of ^{1}H spin diffusion in ^{13}C-labeled organic compounds under MAS owing to the coherency of the transfer.[31]

Precision of fitting of the rate equation affects the distances. As long as we used an initial rate **R** calculated linearly from short τ_{mix} in the least-squares optimization, we obtained the rates with a precision of 20%, corresponding to 4% precision in distance. The rate matrix analysis also can provide accurate rates for slow build-up curves. In contrast, calculation of the rates only from a single spin-diffusion spectrum can give erroneous values. For example, the polarization-transfer rate for the C3^{1}–C3 correlation was 5.4 s^{-1}. The slopes at $\tau_{mix} = 1.1$ and 57.6 ms in Figure 3.3B gave the rates 17.7 and 1.0 s^{-1}, respectively. These differences in the rates lead to more than about 50% deviations in distance. This problem occurs in the analysis of the transfer rate in the initial build-up regime, especially at longer τ_{mix}. The diffusion experiments at longer τ_{mix} are, however, necessary to detect long-distance correlations having slow build-up rates.

Some procedures gave better fits in the minimization of RMSD in Eqn (3.2). Ambiguous intensities and outliers were discounted by reducing their weights $w_{\tau_{mix},i,j}$. In the calculation for the C3^{1}–C8 correlation, the weight of the signal intensities for the minimization was 0.1 at 50 ms $\leq \tau_{mix}$ and 1.0 at $\tau_{mix} <$ 50 ms. This is because the cross-peaks for C3^{1}–C3 and C3^{1}–C8 have the same chemical shifts. These experimental peak intensities at the chemical shifts can chiefly be assigned to the intramolecular C3^{1}–C3 spin pair at a distance of 1.5 Å for $\tau_{mix} <$ 50 ms. However, the C3^{1}–C8 spin pair would contribute to the intensity for 50 ms $\leq \tau_{mix}$.

There are errors in the estimation of the ZQ lineshapes from the experimental CH dipolar spectra. These errors were mainly caused by the precision of the linewidth measurements, about 50 Hz in SD. This variation in linewidth alters the ZQ spectral density by a factor of about 1.3, which corresponds to $\sim$4% deviation in distances. The ZQ lineshape factor L_{ZQ} is distributed at the level of about $0.25 \times 10^{-6} \pm 0.15 \times 10^{-6}$ (Figure 3.6B). Thus, L_{ZQ} can modify the distance by about $\pm$10%.

The correlations between the two distances in Figure 3.7 give a measure of the accuracy in distance measurement and structural determination. The deviation for intermolecular distances, 53% SD, was larger than that for intramolecular distances, 25%. This could be due to the read error of small signal intensities for intermolecular distances that are generally longer than intramolecular distances, and also due to the accuracy and uniqueness of the determined structure. These factors have a contribution of ~46% [$=(53^2-25^2)^{1/2}$] if the squares of the SDs are additive as in stochastic processes.[32] The BChl *c* assembly, however, should assume a unique structure, as indicated by the RMSD of 0.2 Å of the determined structure and the inhomogeneous ^{13}C linewidths of less than about 1 ppm.[1]

We analyzed the SD for intramolecular distances, 25%, which is not affected by the structure optimized under distance constraints. The precisions in the signal quantification and transfer rates affected the distances by ~5%. The precision of $\boldsymbol{L}_{ZQ}$ modified the distances by ~10%, although correction of the distance with $\boldsymbol{L}_{ZQ}$ improved the distances. Thus SD of about 20% for intramolecular distances is probably due to the other factors. Thermal motion may deviate NMR distances from X-ray distances. This effect is conspicuous for shorter distances of 1.5 Å. The NMR distances obtained by solid-state NMR are known to be longer than those for X-ray by about 3%.[33] Signal overlap also reduces the accuracy of the rates owing to the difficulty in quantifying the intensities. Stronger dipolar couplings for distances less than 2.5 Å dominate spin diffusion for uniformly ^{13}C-labeled organic compounds and reduce the effect of smaller couplings. This problem can be alleviated by partial labeling[2] and frequency-selective homonuclear dipolar recoupling.[16–19,30] The present analysis of spin diffusion using the rate equation with $\boldsymbol{L}_{ZQ}$ is based on approximations. For example, the relative orientations of anisotropic interactions are not considered explicitly in $\boldsymbol{L}_{ZQ}$ for powder averaging to obtain the rate equation. This simplification can lead to additional deviations in distance calculation. Experimental imperfections may also affect the signal intensities by ~10%, *e.g.* neglect of small sideband intensity and incomplete excitation by cross polarization that are known to give asymmetric cross-peak intensities.[34]

3.4.2 Spin Diffusion in the Molecular Complex

2D Spin diffusion experiments reveal the polarization transfer between two ^{13}C spins through direct and indirect pathways. We focus on the spin diffusion from protonated carbon C3^1 and non-protonated carbon C7. The peak volume intensities are plotted for spin pairs including C3^1 and C7 as a function of the C–C distance (Figure 3.8). At $\tau_{mix}=0$ ms, only diagonal peaks were observed. The peak intensities for pairs of spins separated by one or two covalent bonds appeared by $\tau_{mix}=26$ ms (Figure 3.8A–E). After $\tau_{mix}=50$ ms, the intermolecular correlations were observed at distances 3.0–5.5 Å (Figure 3.8F). In $\tau_{mix}>115$ ms, the differences between diagonal and cross-peak intensities reduced. At $\tau_{mix}=245$ ms, the diagonal peaks had volume intensities comparable to those of the cross-peaks (Figure 3.8H). The diagonal peaks decreased to a stationary level of the cross-peaks at intensities of about 5% of the initial polarization, indicating the system reached a quasi-equilibrium state in a range of 6 Å. The source spins lost most of their polarization to transfer. Therefore, additional intermolecular correlations were not observed at $\tau_{mix} > 245$ ms.

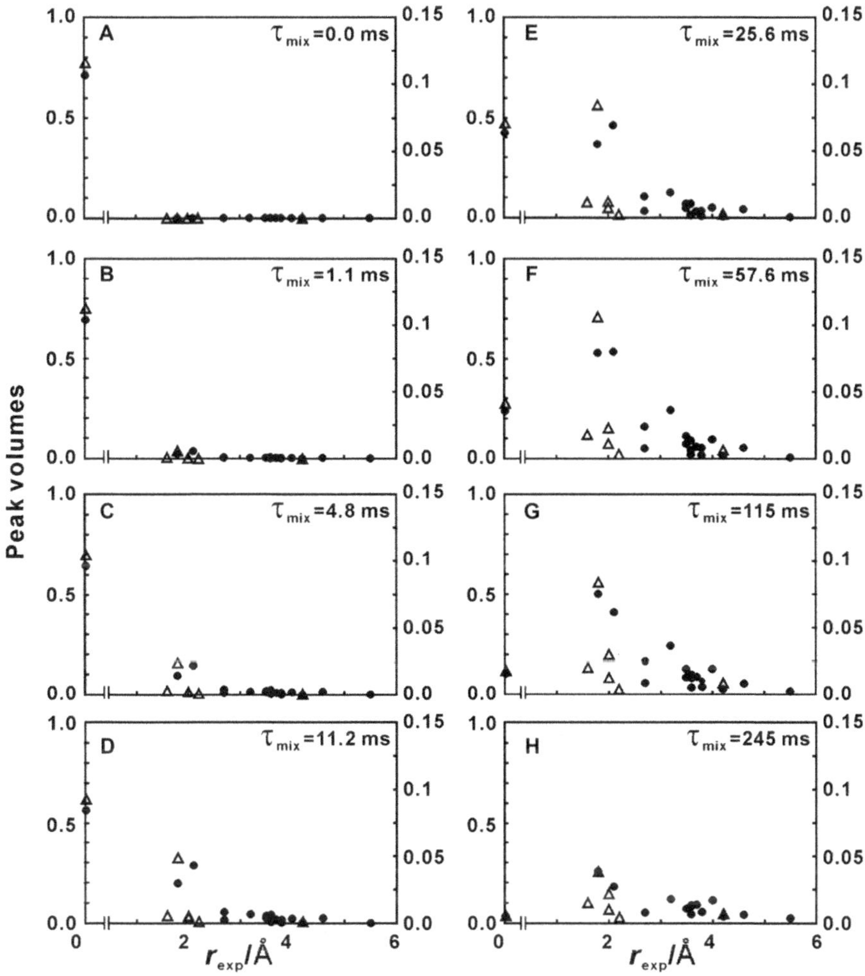

Figure 3.8 Cross-peak intensities as a function of experimental distances, showing spin diffusion depending on τ_{mix}. *Circles* and *triangles* indicate correlations for C3[1] and C7, respectively. Diagonal peak intensities shown at $r_{struc} = 0$ Å are given by the scale on the *left*. Cross-peak intensities shown in $r_{struc} > 0$ Å are given by the scale on the *right*.

The polarization transfer in the uniformly ^{13}C-labeled molecular complex should be expressed by the diffusion equation:

$$M_z(r, t) = C(Dt)^{-\frac{n}{2}} \exp\left(-\frac{r^2}{4Dt}\right) \tag{3.5}$$

that gives polarization at the distance r from the source spin with diffusion coefficient D, dimension n, and constant C. Fitting of this equation to the experimental transfer in Figure 3.8 gave $D \approx 0.15 \times 10^{-3}$ nm^2 ms^{-1} and $n \approx 1.2$. Abragam theoretically estimated $D = 0.2$ nm^2 ms^{-1} for a static system with a ^{1}H internuclear distance of 0.2 nm, which approximately agrees with experimental coefficients.[35,36] Substitution of the gyromagnetic ratio of ^{1}H with that of ^{13}C in Abragam's formula gives $D = 13 \times 10^{-3}$ nm^2 ms^{-1} for a ^{13}C static solid. In our experiments, MAS and resonance frequency differences larger than ^{13}C–^{13}C dipolar couplings should reduce the polarization transfer rates under C–H and H–H dipolar couplings. For example, homonuclear dipolar coupling strengths are effectively decreased by a factor of about 0.1 under strong C–H dipolar couplings and MAS.[31] Thus the experimental coefficient D is much smaller than the theoretical one for a static solid. The dimension $n \approx 1.2$ would indicate that ^{13}C spin diffusion is not isotropic as $n = 3$ on the timescale of 0.2 s, corresponding to ~4 Å in distance. On such a distance scale at atomic resolution, the ^{13}C spin diffusion would have a preferable direction in which the polarization is transferred rapidly by the dipolar couplings for spin pairs with C–C bond lengths.

3.4.3 Efficiency of Polarization Transfer

The PDSD experiments should be compared with the ^{13}C–^{1}H dipolar-assisted rotational resonance (DARR) experiments in the efficiency of polarization transfer.[4,5,25] Enhancement of the polarization-transfer rates for DARR is observed under recoupled dipolar interactions with the protons. However, in our PDSD experiments it is unnecessary to enhance the polarization-transfer rates. The polarization is transferred between spins before longitudinal relaxation of the spins under $\tau_{mix} \ll T_1$. Thus, the slower transfer rates in PDSD can be offset by increasing τ_{mix}. In our experiments the transferred polarization was almost saturated at $\tau_{mix} = 245$ ms, much shorter than $T_1 = 3$ s in our target sample. Proton-assisted recoupling (PAR) may give distance information less susceptible to relayed transfer from the combination of C–H dipolar couplings, although the theoretical calculation of the transfer rates is not tractable.[9,21]

PDSD experiments of site-directed ^{13}C-labeled molecules yield long-range distances up to about 7 Å.[2] PDSD experiments with $\tau_{mix} \approx 200$ ms provide many long-range C–C distances up to ~6 Å for uniformly ^{13}C-labeled proteins and biomolecular complexes.[1,8] Also, dipolar truncation, that suppresses weak couplings for long-range correlations, was not conspicuous in our spin-diffusion experiments.[22] Important factors in obtaining long distances are higher resolution and sensitivity of the spectra. We have obtained PDSD spectra with S/N rations of about 500 (Figure 3.3). Thus we could detect signals with a dynamic range greater than 10^2. In our study, signals with intensities less than ~5% of the initial diagonal intensities give long distances at longer τ_{mix}. These signals were observed generally at lower precision owing to the small intensity.

In conclusion, C–C distances up to ~6 Å were obtained through ^{13}C polarization-transfer rates in PDSD experiments for a uniformly ^{13}C-labeled molecular complex. The rate matrix analysis allowed distinction between direct and relayed polarization-transfer and ameliorated the accuracy of the transfer rates. The conversion of rates to distances using ZQ lineshape functions also improved the distance accuracy, as verified experimentally. These quantitative distances, followed by simultaneous assignment of distance correlations and structure optimization, contribute to determination of the long-range order of molecular complexes.

Acknowledgments

We thank Professor Y. Koyama, Dr T. Mizoguchi, and Dr Y. Kakitani for providing the BChl *c* assembly in chlorosomes. This work was partly supported by Grants-in-Aid from the Ministry of Education, Culture, Sports, Science and Technology of Japan and by SENTAN and Targeted Protein Projects from the Japan Science and Technology Agency.

References

1. A. Egawa, T. Fujiwara, T. Mizoguchi, K. Kakitani, Y. Koyama and H. Akutsu, *Proc. Natl. Acad. Sci. U.S.A.*, 2007, **104**, 790.
2. F. Castellani, B. van Rossum, A. Diehl, M. Schubert, K. Rehbein and H. Oschkinat, *Nature*, 2002, **420**, 98.
3. E. Crocker, A. B. Patel, M. Eilers, S. Jayaraman, E. Getmanova, P. J. Reeves, M. Ziliox, H. G. Khorana, M. Sheves and S. O. Smith, *J. Biomol. NMR*, 2004, **29**, 11.
4. S. G. Zech, A. J. Wand and A. E. McDermott, *J. Am. Chem. Soc.*, 2005, **127**, 8618.
5. D. Marulanda, M. L. Tasayco, M. Cataldi, V. Arriaran and T. Polenova, *J. Phys. Chem. B*, 2005, **109**, 18135.
6. K. Iwata, T. Fujiwara, Y. Matsuki, H. Akutsu, S. Takahashi, H. Naiki and Y. Goto, *Proc. Natl. Acad. Sci. U.S.A.*, 2006, **103**, 18119.
7. M. Baldus, *Curr. Opin. Struct. Biol.*, 2006, **16**, 618.
8. T. Manolikas, T. Hermann and B. H. Meier, *J. Am. Chem. Soc.*, 2008, **130**, 3959.
9. I. Bertini, A. Bhaumik, G. D. Paëpe, R. G. Griffin, M. Lelli, J. R. Lewandowski and C. Luchinat, *J. Am. Chem. Soc.*, 2010, **132**, 1032.
10. B. H. Meier, *Advances in Magnetic and Optical Resonance*, ed. W. S. Warren, Academic Press, San Diego, 1994, vol. 18, p. 1.
11. N. M. Szeverenyi, M. J. Sullivan and G. E. Maciel, *J. Magn. Reson.*, 1982, **47**, 462.
12. M. Hohwy, C. M. Rienstra, C. P. Jaroniec and R. G. Griffin, *J. Chem. Phys.*, 1999, **110**, 7983.
13. R. G. Griffin, *Nat. Struct. Biol.*, 1998, **5**, 508.
14. P. Hodgkinson and L. Emsley, *J. Magn. Reson.*, 1999, **139**, 46.
15. S. Kühne, M. A. Metha, J. A. Stringer, D. M. Gregory, J. C. Shiels and G. Drobny, *J. Phys. Chem. A*, 1998, **102**, 2274.
16. M. Hohwy, C. M. Rienstra and R. G. Griffin, *J. Chem. Phys.*, 2002, **117**, 4973.

17. Y. Matsuki, H. Akutsu and T. Fujiwara, *J. Magn. Reson.*, 2003, **162**, 54.
18. A. K. Paravastu and R. Tycko, *J. Chem. Phys.*, 2006, **124**, 194303.
19. I. Marin-Montesinos, G. Mollica, M. Carravetta, A. Gansmüller, G. Pileio, M. Bechmann, A. Sebald and M. H. Levitt, *Chem. Phys. Lett.*, 2006, **432**, 572.
20. N. Khaneja and N. C. Nielsen, *J. Chem. Phys.*, 2008, **128**, 015103.
21. G. De Paëpe, J. R. Lewandowski, A. Loquet, A. Böckmann and R. G. Griffin, *J. Chem. Phys.*, 2008, **129**, 245101.
22. A. Grommek, B. H. Meier and M. Ernst, *Chem. Phys. Lett.*, 2006, **427**, 404.
23. T. Kato, S. Endo, T. Fujiwara and K. Nagayama, *J. Biomol. NMR*, 1993, **3**, 653.
24. A. Kubo and C. A. McDowell, *J. Chem. Soc., Faraday Trans. 1*, 1988, **84**, 3713.
25. K. Takegoshi, S. Nakamura and T. Terao, *J. Chem. Phys.*, 2003, **118**, 2325.
26. P. M. Henrichs, M. Linder and J. M. Hewitt, *J. Chem. Phys.*, 1986, **85**, 7077.
27. B. Elena and L. Emsley, *J. Am. Chem. Soc.*, 2005, **127**, 9140.
28. B. Elena, G. Pintacuda, N. Mifsud and L. Emsley, *J. Am. Chem. Soc.*, 2006, **128**, 9555.
29. D. Reichert, G. Hempel, R. Poupko, Z. Luz, Z. Olejniczak and P. Tekely, *Solid State Nucl. Magn. Reson.*, 1998, **13**, 137.
30. M. H. Levitt, D. P. Raleigh, F. Creuzet and R. G. Griffin, *J. Chem. Phys.*, 1990, **92**, 6347.
31. H. Takahashi, H. Akutsu and T. Fujiwara, *J. Chem. Phys.*, 2008, **129**, 154504.
32. P. G. Hoel, *Introduction to Mathematical Statistics*, Wiley, New York, 5th edn, 1984, pp. 121–161.
33. Y. Ishii, T. Terao and S. Hayashi, *J. Chem. Phys.*, 1997, **107**, 2760.
34. S. Caldarelli and L. Emsley, *J. Magn. Reson.*, 1998, **130**, 233.
35. A. Abragam, *Principles of Nuclear Magnetism*, Oxford University Press, New York, 1961, pp. 136–144.
36. K. Schmidt-Rohr and H. W. Spiess, *Multidimensional Solid-State NMR and Polymers*, Academic Press, New York, 1994, pp. 402–439.

CHAPTER 4

Demonstration of the Equivalence of Solid-State NMR Orientational Constraints from Magnetic and Rotational Alignment of the Coat Protein in a Filamentous Bacteriophage

STANLEY J. OPELLA* AND BIBHUTI B. DAS

University of California, San Diego, 9500 Gilman Drive, La Jolla, CA 92093-0307, USA
*Email: sopella@ucsd.edu

4.1 Introduction

The appearance of the NMR spectra of proteins is largely defined by the timescales, directions, and amplitudes of their inter- and intramolecular motions. Global and local dynamics have a dominant effect on the line-shapes, linewidths, intensities, and, in some cases, the frequencies of the resonances observed in the spectra of proteins. Soluble, globular proteins that undergo isotropic reorientation faster than $\sim 10^8$ Hz give narrow single-line resonances in solution NMR spectra because the motion effectively

New Developments in NMR No. 3
Advances in Biological Solid-State NMR: Proteins and Membrane-Active Peptides
Edited by Frances Separovic and Akira Naito

Published by the Royal Society of Chemistry, www.rsc.org

averages the chemical shift and dipole–dipole interactions to their trace values for ^{1}H, ^{13}C, and ^{15}N, whose powder patterns typically have frequency spans of $\sim 10^4$ Hz. The situation is more complex for proteins that are part of supramolecular assemblies, such as virus particles or membranes. These proteins may not yield signals at all in solution NMR spectra because their overall reorientation is too slow, with the possible exception of uniaxial rotational diffusion, and that is insufficient to narrow resonances for detection with solution NMR experiments and instrumentation.

Solid-state NMR techniques are essential for the study of proteins in biological supramolecular assemblies. Some highly anisotropic motions, such as rotational diffusion, dramatically influence the lineshapes of the resonances in unoriented samples that are revealed by decoupling of the heteronuclear dipolar interactions that are the sources of the most severe line broadenings. The motional averaging of powder pattern lineshapes by rotational diffusion about a defined single axis results in "rotational alignment". This affects the observed lineshapes of powder patterns from dipole–dipole and chemical shift interactions in unoriented "powder" samples in an angular-dependent fashion that yields orientational constraints equivalent to those obtained from the frequencies measured from uniaxially aligned samples. The ability to turn the motional averaging "on" and "off" by altering the temperature, concentration, or some other experimental parameter provides the potential for accurate analysis of the orientation-dependent motional averaging for both dipole–dipole and chemical shift interactions. This results in the integration of oriented sample (OS) NMR and magic angle spinning (MAS) solid-state NMR approaches to protein structure determination in rotationally aligned (RA) solid-state NMR.[1–3]

Sample alignment is an essential component of nearly all contemporary NMR studies of proteins, whether weak alignment for the measurement of residual dipolar couplings in solution[4,5] or complete alignment that can be obtained through mechanical or magnetic interactions with assemblies of proteins. For example, we have exploited the near-perfect alignment of filamentous bacteriophages such as fd[6] and Pf1[7] in the magnetic field to determine the three-dimensional structures of their coat proteins by OS solid-state NMR. However, in some situations it may not be desirable or even possible to prepare well-aligned samples of protein complexes. Using the well-studied fd bacteriophage as an example, here we examine how rapid rotational diffusion of virus particles in unoriented samples provides the same structural information that is obtained from stationary uniaxially aligned samples of the same virus particles. The only requirements are that the axis of rotation and alignment are the same, and that the rotational diffusion occurs more frequently than the $\sim 10^4$ Hz breadth of the relevant heteronuclear dipole–dipole and chemical shift interaction powder patterns. High-concentration (>50 mg mL^{-1}) solutions of filamentous bacteriophages do not align in the magnetic field; thus by comparing results obtained from similar but slightly less concentrated samples that are aligned by the field, we can experimentally demonstrate the equivalence of OS solid-state NMR and RA solid-state NMR on a DNA–protein complex with a mass >10^6 Daltons.

It is well established in solid-state NMR that when rotational diffusion about the long axis of particles is slower than $\sim 10^2$ Hz, then the resonances appear as rigid lattice non-axially symmetric powder patterns. With fast rotational diffusion the powder patterns are axially symmetric because of the effects of the motional averaging; and the direction of diffusion relative to the principal axes of the chemical shift and dipolar-coupling tensors can be readily determined by simple angular dependence. Significantly, the parallel edge of the axially symmetric powder pattern in the rotationally aligned sample has the same frequency as the single-line resonance in the magnetically aligned sample. The study of unoriented samples also opens up possibilities for incorporating MAS methods into the experiments.

4.2 Results

Figure 4.1 compares simulated chemical shift anisotropy (CSA) powder lineshapes for rigid and rotationally averaged ^{13}C′ (Figure 4.1A and B) and ^{15}N amide (Figure 4.1C and D) sites in peptide bonds. The simulated spectra are for amino acid residues in helices undergoing rotational diffusion about an angle near the direction of alignment. Simulated spinning side-band spectra from MAS NMR spectra are aligned below their respective powder patterns. The spinning frequency is 7 kHz for ^{13}C′ sites in Figure 4.1A and B and 3 kHz for the ^{15}N amide sites in Figure 4.1C and D. Fast (on the relevant NMR timescale) rotational diffusion about a single axis always produces axially symmetric powder patterns, regardless of whether the rigid-lattice

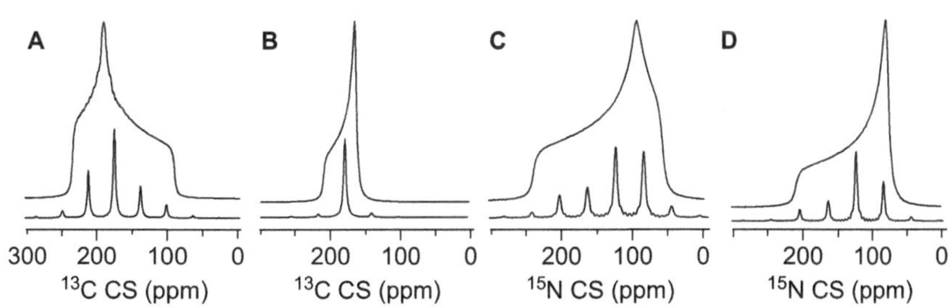

Figure 4.1 Simulated chemical shift powder patterns of carbonyl carbon and amide nitrogen nuclei with and without uniaxial rotational diffusion. Below the static powder patterns are the spinning side-band patterns under 7 kHz and 3 kHz spinning speeds for ^{13}C and ^{15}N nuclei, respectively. (A) ^{13}C chemical shift spectrum of a rigid lattice powder pattern reconstructed with $\delta_{11} = 246$ ppm, $\delta_{22} = 190$ ppm, $\delta_{33} = 88$ ppm, $\delta_{iso} = 175$ ppm, and $\eta = 0.65$. (B) Motionally averaged ^{13}C powder lineshape for $\delta_{\parallel} = 205$ ppm, $\delta_{\perp} = 160$ ppm, $\delta_{iso} = 175$ ppm, and $\eta = 0$. (C) ^{15}N rigid powder pattern with $\delta_{11} = 58$ ppm, $\delta_{22} = 93$ ppm, $\delta_{33} = 233$ ppm, $\delta_{iso} = 128$ ppm, and $\eta = 0.33$. (D) Motionally averaged ^{15}N powder pattern with $\delta_{\parallel} = 205$ ppm, $\delta_{\perp} = 89.5$ ppm, $\delta_{iso} = 128$ ppm, and $\eta = 0$.

spectrum is axially symmetric or asymmetric.[8,9] The properties of such a motionally averaged powder pattern, in terms of width and sign of the anisotropy, can be determined using the three principal values of the rigid-lattice chemical shift tensors (δ_{11}, δ_{22}, δ_{33}) and their orientation with respect to the diffusion axis. The anisotropic spread of the motionally averaged powder spectrum, $\Delta\delta^{R} = \delta_{\parallel} - \delta_{\perp}$, is calculated using the equation:

$$\Delta\delta^{R} = \frac{1}{2}(3\cos^2\beta - 1)\left[\delta_{33} - \frac{1}{2}(\delta_{11} + \delta_{22})\right] + \frac{3}{4}(\delta_{11} - \delta_{22})\sin^2\beta\cos 2\alpha \quad (4.1)$$

where α and β are the Euler angles describing the orientation of the principal axis system of the spin interaction tensors with respect to the rotational diffusion axis.[9]

4.2.1 ^{13}C NMR Experiments

Both stationary and MAS experiments were carried out on a concentrated sample of ~100 mg mL^{-1} 25% randomly ^{13}C and 98% uniformly ^{15}N labeled Y21M fd bacteriophage at 65 °C; at this temperature the bacteriophage particles undergo rapid rotational diffusion around their long axis. When this rotational diffusion is faster than the ~10^4 Hz timescale of the frequency breadths of the chemical shift and dipole–dipole interactions, characteristic motional averaged powder lineshapes result, such as those simulated in Figure 4.1. The motion of the phage particles is so slow or limited in amplitude at low temperatures (~ − 8 °C) that the ^{13}C CSA and ^{1}H–^{15}N dipolar coupling powder patterns essentially match the rigid lattice limit shown in Figure 4.1. Experimental measurements on ^{13}C′ (carbonyl carbon) chemical shift lineshapes shown in Figure 4.2 demonstrate that they are quite different (Figure 4.2A and D) at low and high temperatures. Notably, the MAS NMR spectra obtained with 7 kHz spinning rate (Figure 4.2B and C) show significant differences in sideband intensities. The 25% random labeling of the protein's carbon sites results in negligible line broadening due to homonuclear dipolar couplings for both static and spinning experiments. The simulated data were generated based on both the static powder pattern lineshape and the slow spinning sideband intensities.[10] The CSA powder pattern for a rigid carbonyl carbon was measured to be 155 ± 5 ppm (~29 kHz), and the rotationally averaged CSA powder pattern appears to have a ~30 ppm linewidth at half-maximum intensity. Notably, the isotropic chemical shift resonance at high temperature (Figure 4.2C) has a linewidth 2–3 times broader than the comparable spectrum recorded at low temperature. The spectrum of the 25% randomly ^{13}C labeled fd coat protein has an isotropic signal that results from overlapping intensities from >40 carbonyl groups in the protein. The three-dimensional spectrum of a stationary sample at low temperature (Figure 4.2A) has overlapping intensities from the signals from the many carbonyl, carboxylic, aromatic, and side-chain carbons in the protein. This spectrum is useful in qualitatively

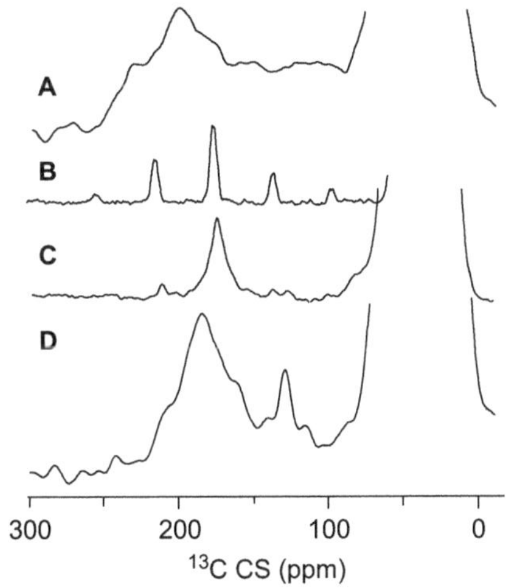

Figure 4.2 Experimental ^{1}H decoupled one-dimensional $^{13}C'$ (carbonyl carbon) chemical shift powder patterns of unoriented fd bacteriophage particles. The coat proteins were labeled with 25% random labeling for ^{13}C and 100% for ^{15}N. Phage particles were concentrated to ~100 mg mL^{-1}. (A) Static solid-state NMR spectrum at −8 °C. (B) 7 kHz MAS solid-state NMR spectrum at −8 °C. (C) 7 kHz MAS solid-state NMR spectrum at 65 °C. (D) Static solid-state NMR spectrum at 65 °C.

demonstrating an absence of substantial motional narrowing; however, the overlapping intensities make it difficult to discern the precise frequencies of the edges of the powder patterns. Assuming all the carbonyl carbons have the same or very similar chemical shift frequencies at low temperature, a two-dimensional ROCSA[11] experiment was carried out under MAS to recouple the CSA in an indirect dimension, correlating the chemical shift powder patterns with isotropic chemical shifts in the direct dimension. A well-defined anisotropic chemical shift powder pattern extracted at the average isotropic chemical shift frequency of 175 ppm from the two-dimensional ROCSA experiment provides the principal values of tensor elements at frequencies $\delta_{11} = 246$ ppm, $\delta_{22} = 190$ ppm, and $\delta_{33} = 88$ ppm.

Isotopic labeling of individual sites simplifies the spectrum for quantitative analysis of the motional averaging of the chemical shift powder lineshape. Figure 4.3 shows experimental results obtained on a sample of fd bacteriophage with $^{13}C'$ labeled at only the four valine residues in the coat protein. The one-dimensional powder pattern spectrum of concentrated, unoriented fd at 65 °C shows a chemical shift span between 160 ppm and 210 ppm (Figure 4.3A). In contrast, a low-temperature powder pattern spectrum (Figure 4.3B) shows a much larger frequency span of ~147 ppm. Either no side bands or a family of side bands (Figure 4.3C) are observed under 7 kHz MAS, demonstrating indirectly the presence of small and large frequency spans as a function of high and low temperatures, respectively.

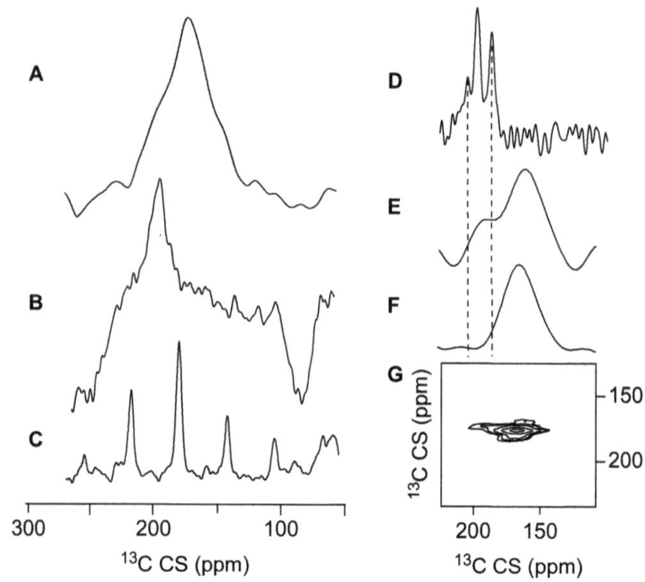

Figure 4.3 $^{13}C'$ chemical shift spectra recorded for [^{13}C]Val-labeled fd. (A) One-dimensional static powder spectrum of unoriented fd recorded at 65 °C. (B) Stationary spectrum recorded at −8 °C. (C) 7 kHz MAS spectrum at −8 °C. (D) Static spectrum of magnetically oriented fd recorded at 65 °C. (E, F) Slices extracted from a ROCSA two-dimensional NMR spectrum at isotropic chemical shifts of 175 and 181 ppm, respectively. (G) Two-dimensional ROCSA spectrum at 65 °C.

Solid-state NMR experiments were performed on a stationary sample of ~ 35 mg mL^{-1} fd, where the bacteriophage particles are uniaxially aligned in the magnetic field. Three out of the four labeled Val sites are distinguishable with chemical shifts at 184, 197, and 210 ppm in the spectrum in Figure 4.3D. The resolution results from difference in the angles between the principal axes of the $^{13}C'$ chemical shift tensors and the magnetic field. Motionally averaged powder lineshapes were obtained from an unoriented sample with relatively high concentration of fd bacteriophage particles. A two-dimensional ROCSA spectrum correlating isotropic chemical shifts with anisotropic powder lineshapes is shown in Figure 4.3G. Spectral slices obtained at the isotropic chemical shifts of 175 ppm and 181 ppm are shown in Figure 4.3E and F. The parallel edges of the recoupled CSA powder lineshapes provide the equivalent shift frequencies of 184 ppm and 210 ppm with respect to the oriented spectrum shown in Figure 4.3D.

Similar studies were carried out on a sample of selectively [^{13}C,^{15}N]Leu-labeled fd. The coat protein has two leucine residues, at position 41 in the hydrophobic helix and at position 14 in the amphipathic helix. The one-dimensional static spectrum at high temperature in Figure 4.4A shows a chemical shift span from 150 to 225 ppm. A significant increase in the span of the powder pattern to >145 ppm (100 to 245 ppm) is observed for the

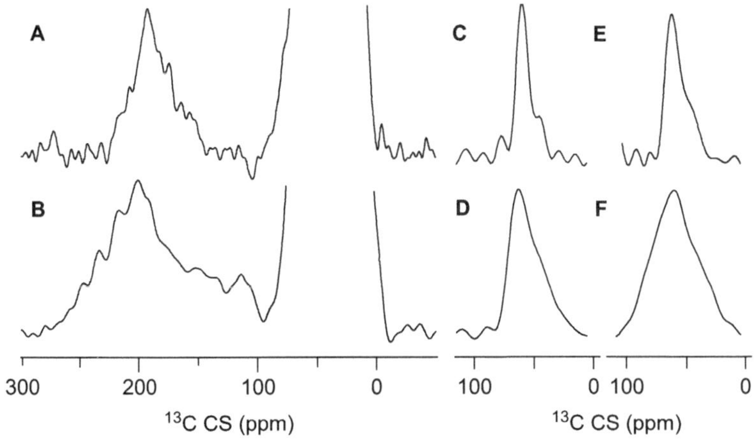

Figure 4.4 ^{13}C′ and ^{13}Cα chemical shift spectra recorded for uniformly [^{13}C,^{15}N]Leu-labeled fd coat protein. (A, B) One-dimensional ^{13}C′ static spectra recorded at 65 °C and −8 °C, respectively. (C–F) One-dimensional ^{13}Cα chemical shift slices obtained from a two-dimensional ROCSA experiment performed at 65 °C. (C, E) 55 and 55.7 ppm for L14 and L41 Cα isotropic shifts, respectively, recorded at 65 °C. (D, F) As for (C) and (E) except recorded at −8 °C.

static spectrum recorded at low temperature (Figure 4.4B). Individual CSA lineshapes for the α-carbons were also recorded using two-dimensional ROCSA experiments (Figure 4.4C–F). The isotropic chemical shifts at 55 ppm and 55.7 ppm were assigned to Leu14 and Leu41, respectively. One-dimensional spectral slices extracted from two-dimensional ROCSA data are shown for Leu14 (Figure 4.4C and D) and Leu41 (Figure 4.4E and F). The low-temperature data show a chemical shift span of ~50 ppm (Figure 4.4D and F). For comparison, the equivalent experiment at high temperature shows ~20 ppm span with an axially symmetric powder pattern lineshape (Figure 4.4C and E).

4.2.2 ^{15}N NMR Experiments

One-dimensional ^{15}N chemical shift MAS spectra were obtained with both ^{1}H and ^{13}C decoupling for aligned and unoriented powder samples of selectively [^{15}N]Leu-labeled fd coat protein. The spectra are shown in Figure 4.5. The unoriented fd sample labeled at two Leu residues shows two well-resolved isotropic chemical shift frequencies of 120 ppm and 125 ppm. The static powder pattern spectrum shown in Figure 4.5A has a chemical shift span between 217 ppm and 71 ppm. The one-dimensional static oriented spectrum is shown in Figure 4.5B. The two distinguishable frequencies are 217 ppm and 184 ppm. The recoupled chemical shift spectra at isotropic shifts of 120 ppm (Figure 4.5C) and 125 ppm (Figure 4.5D) were acquired using ROCSA. The measured frequencies at the perpendicular and

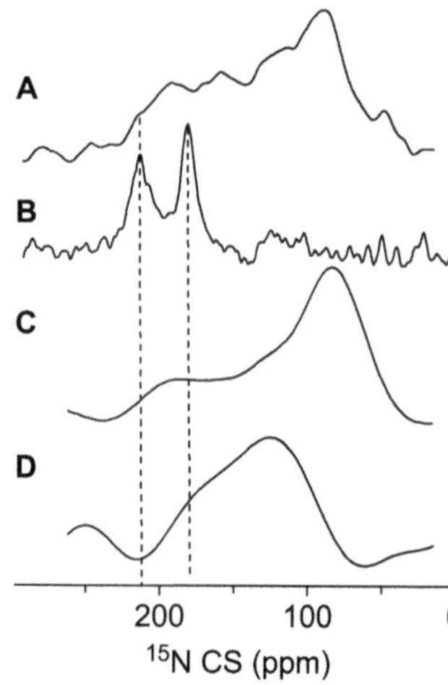

Figure 4.5 ^{15}N chemical shift spectra of [^{15}N]Leu-labeled fd coat protein recorded at a sample temperature of 65 °C. (A) Static powder spectrum. (B) Static spectrum of magnetically oriented fd. (C, D) Slices extracted at isotropic chemical shifts of 120 and 125 ppm from a two-dimensional ROCSA spectrum.

parallel edges in Figure 4.5C are: $\delta_{\parallel} = 217 \pm 5$ ppm and $\delta_{\perp} = 78 \pm 8$ ppm, and in Figure 4.5D are $\delta_{\parallel} = 185 \pm 5$ ppm and $\delta_{\perp} = 100 \pm 10$ ppm.

Spectra obtained from a [^{15}N]Trp-labeled fd sample are shown in Figure 4.6. A one-dimensional, 3 kHz MAS spectrum obtained at 65 °C is shown in Figure 4.6C. Two isotropic chemical shifts at 120.3 ppm and 127 ppm with families of side bands were observed. A two-dimensional experiment with recoupling of CSA under moderately fast spinning (11.111 kHz) is shown in Figure 4.6E. The two-dimensional spectrum represents correlation of the anisotropic powder pattern with isotropic shifts. Two distinguishable peaks are observed at 77 ppm and 122.5 ppm, which correspond to the perpendicular edges of the axially symmetric powder patterns. The one-dimensional powder spectrum extracted at 127 ppm is shown in Figure 4.6D. Using the experimental data in Figure 4.6, CSA powder lineshapes were simulated for the amide nitrogen with $\delta_{\parallel} = 208$ ppm, $\delta_{\perp} = 77$ ppm, $\delta_{iso} = 120.3$ ppm, and $\eta = 0$ (Figure 4.6A) and for the side-chain indole nitrogen with $\delta_{\parallel} = 136$ ppm, $\delta_{\perp} = 122.5$ ppm, $\delta_{iso} = 127$ ppm, and $\eta = 0$ (Figure 4.6B). The results shown in the right column (Figure 4.6H, I) are the same as the left column (Figure 4.6C–E) except that the experiments were carried out at a sample temperature of −8 °C.

^{1}H–^{15}N Dipolar couplings were measured under static and MAS conditions. In Figures 4.7–4.9 the two-dimensional SLF NMR spectra obtained for samples with Leu, Trp, and uniformly ^{15}N-labeled fd bacteriophages are shown. The results for a static oriented fd sample labeled at two Leu residues are shown in Figure 4.7C. The measured dipolar couplings are 5.1 kHz and 6.9 kHz for chemical shift frequencies at 184 ppm and 217 ppm, which

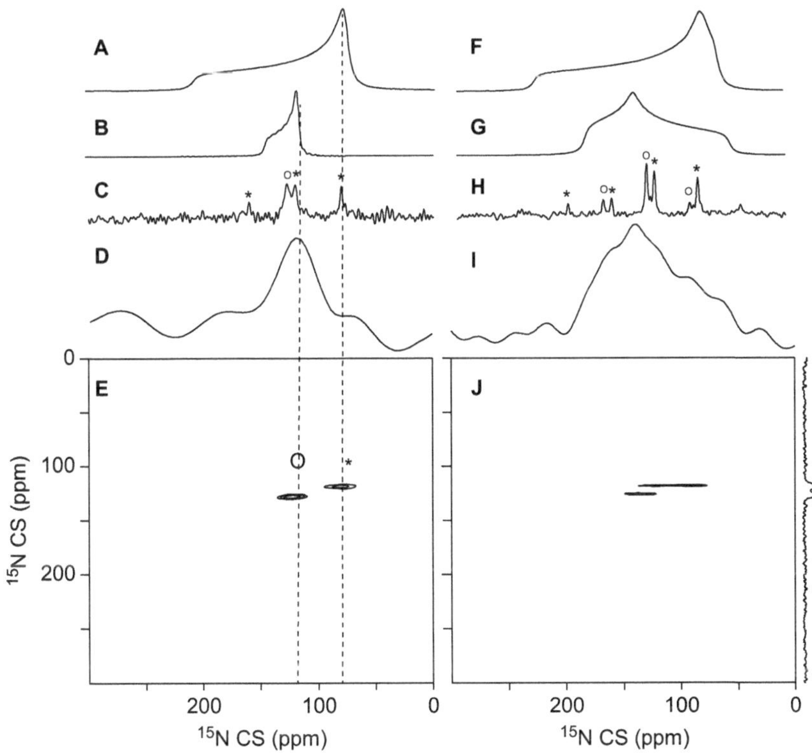

Figure 4.6 ^{15}N NMR spectra of unoriented [^{15}N]Trp-labeled fd. (A, B) Reconstructed powder lineshapes for motionally averaged chemical shift anisotropy of backbone (*) and indole ^{15}N nuclei (o), respectively. (C) With 3 kHz MAS at 65 °C. (D) Spectral slice at 128 ppm obtained from the two-dimensional ROCSA data in (E) recorded at 65 °C. (F, G) Unaveraged rigid powder lineshapes reconstructed for ^{15}N nuclei in the backbone and side chain. (H) 3 kHz MAS at −8 °C. (I) Spectral slice at 128 ppm obtained from the two-dimensional ROCSA data in (J) recorded at -8 °C. The parameters used in the simulations are as follows: (A) $\delta_{\parallel} = 208$ ppm, $\delta_{\perp} = 77$ ppm, $\delta_{iso} = 120.3$ ppm, and $\eta = 0$. (B) $\delta_{\parallel} = 139$ ppm, $\delta_{\perp} = 122.5$ ppm, $\delta_{iso} = 128$ ppm, and $\eta = 0$. (F) $\delta_{11} = 58$ ppm, $\delta_{22} = 72$ ppm, $\delta_{33} = 230$ ppm, and $\eta = 0.15$. (G) $\delta_{11} = 182$ ppm, $\delta_{22} = 131$ ppm, $\delta_{33} = 69$ ppm, $\delta_{iso} = 127$ ppm, and $\eta = 0.88$.

correspond to the residues at Leu14 and Leu41.[6] The 11 kHz PITANSEMA-MAS SLF two-dimensional spectrum for an unoriented fd sample at 65 °C is shown in Figure 4.7B. The two isotropic chemical shifts at 120 ppm and 125 ppm show dipolar couplings of 6.9 kHz and 5.1 kHz, respectively. In contrast, 10 kHz dipolar frequencies were observed for both leucine sites (Figure 4.7A) at low temperature. The maximum dipolar frequencies were attributed to rigid ^{1}H–^{15}N dipole–dipole frequencies, which are similar to those obtained for a polycrystalline *N*-acetylleucine sample.

Results from similar experiments performed on an unoriented [^{15}N]Trp-labeled fd sample are shown in Figure 4.8. The Trp residue has two ^{1}H–^{15}N

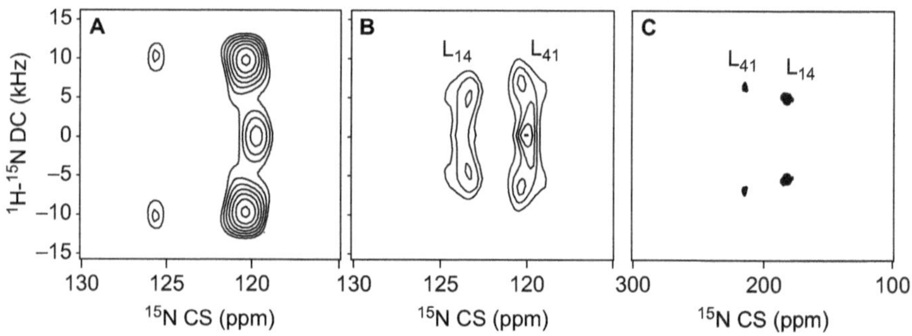

Figure 4.7 ^{1}H–^{15}N SLF spectra of [^{15}N]Leu-labeled fd. (A) Two-dimensional 11.111 kHz MAS SLF (PITANSEMA) spectrum obtained at −8 °C. (B) As (A), but recorded at 65 °C. (C) Two-dimensional SLF spectrum of magnetically aligned fd at 65 °C.

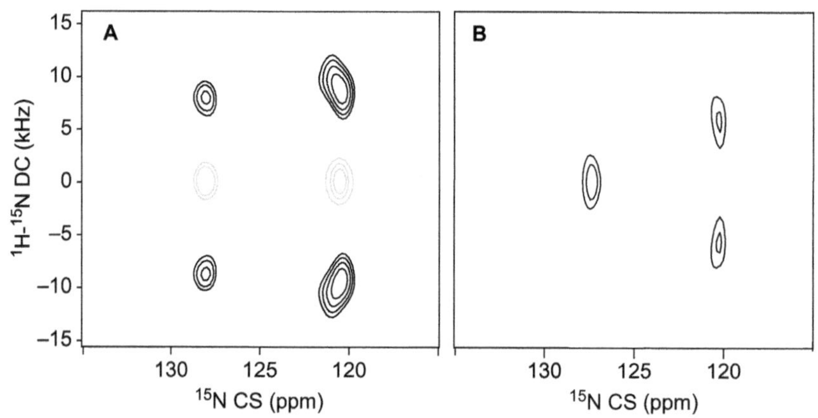

Figure 4.8 ^{1}H–^{15}N SLF spectra of [^{15}N]Trp-labeled fd. (A) Two-dimensional 11.111 kHz MAS SLF (PITANSEMA) spectrum obtained at −8 °C. (B) As (A) but recorded at 65 °C.

bonds, one from the backbone amide and one from the side-chain indole sites. The high-temperature SLF spectrum yields a dipolar coupling of 5.4 kHz at 120 ppm and 2 kHz linewidth, with almost no resolution at 127 ppm (Figure 4.8B). SLF data at low temperature showed dipolar frequencies of 10 kHz and 9.6 kHz for the backbone and side-chain sites, respectively.

Heteronuclear dipolar recoupling on a uniformly ^{15}N-labeled sample is shown in Figure 4.9. A large spread of dipolar frequencies ranging from 0 to 10.5 kHz (Figure 4.9D) is observed. There are >40 chemical shifts overlapping within a range of 110–130 ppm. Residue numbers are marked following the shifts observed in selectively labeled fd phage. Several dipolar slices at various chemical shifts extracted from the two-dimensional spectrum are shown in Figure 4.9A–C. Dipolar/shift data were measured with frequency splitting (peak to peak) of (A) 15.9 kHz/112 ppm, (B) 16, 10.2 kHz/125 ppm, and (C) 2.0 kHz/127 ppm. The SLF experiment was carried out using $R18^7_1$

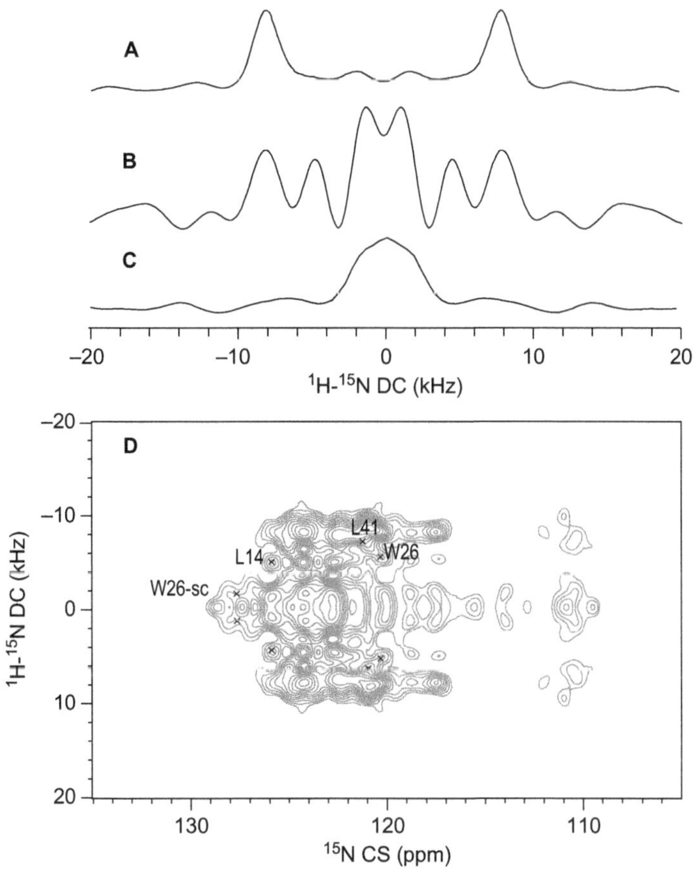

Figure 4.9 MAS ^1H–^{15}N SLF spectra of uniformly ^{15}N-labeled fd. (A–C) Dipolar frequency slices obtained from (D) at 112, 126, and 128 ppm, respectively. (D) Two-dimensional MAS SLF spectrum recorded at 65 °C and 11.111 kHz spinning speed.

symmetry pulse scheme. Data were zero filled to a matrix of 8192×512 prior to Fourier transformation in both dimensions. The dipolar frequency axis was rescaled to full dipolar spread using scaling factor of 0.17 determined experimentally for a ^{15}N-labeled *N*-acetylleucine polycrystalline sample. The maximum dipolar frequency splitting for a one-bond ^1H–^{15}N dipole–dipole interaction was set to 10 kHz.

4.3 Experimental Methods

4.3.1 Sample Preparation

Uniformly ^{15}N-labeled Y21M fd bacteriophage particles were obtained following the previously published protocol.[6] Selectively labeled samples were obtained by using M9 media supplemented with one ^{15}N-labeled amino acid and the other amino acids in natural isotopic abundance. The 25% randomly labeled ^{13}C, 98% ^{15}N-labeled sample was grown in specially prepared Bioexpress media. The samples for the stationary solid-state NMR experiments were prepared with a concentration of ~40 mg mL^{-1}. The pH of the sample was adjusted to 8.0 using 5 mM sodium borate buffer containing 0.1 mM sodium azide. A 5-mm thin wall glass tube containing ~150 μL of sample by volume was used for all stationary sample experiments. Under these conditions the bacteriophage particles align with their long axis parallel to the direction of the applied magnetic field. For MAS solid-state NMR experiments, bacteriophage particles from the same batch were concentrated to ~100 mg mL^{-1} using ultracentrifugation at 50 000 rpm for 6–8 hours. About 30 μL of fully hydrated phage particles at pH 8.0 were transferred to a 3.2 mm zirconium rotor. At this high concentration the bacteriophage particles no longer align in the magnetic field but do undergo rotational diffusion along their long axis.

4.3.2 NMR Experiments

The MAS solid-state NMR experiments were performed on a Bruker Avance spectrometer with a ^{1}H resonance frequency of 749.87 MHz. A triple resonance 3.2 mm Bruker E-free MAS probe was used, and ^{13}C and ^{15}N detection experiments were carried out at resonance frequencies of 188.459 MHz and 75.945 MHz, respectively. ^{15}N-Detected one- and two- dimensional experiments on magnetically aligned samples were carried out on a Bruker Avance spectrometer with a ^{1}H resonance frequency of 700.1 MHz equipped with a homebuilt double-resonance 5 mm MAGC probe,[12,13] except for the ^{13}C-detected experiment, which was performed on a 500 MHz spectrometer equipped with a homebuilt triple-resonance 5 mm MAGC probe.[14,15]

For one-dimensional experiments on stationary samples, CP-MOIST[16] was employed for cross-polarization, typically with 50 kHz radiofrequency (RF) applied to the ^{1}H and ^{13}C or ^{15}N channels. SPINAL-16 modulation was used for ^{1}H decoupling[17] and continuous wave irradiation was used for either ^{13}C or ^{15}N decoupling. Two-dimensional high-resolution separated local field spectra were obtained with the SAMMY pulse sequence[18] that correlates the ^{15}N or ^{13}C chemical shift with the associated ^{1}H–^{15}N or ^{1}H–^{13}C dipolar coupling, respectively. 43 kHz RF was used during magic sandwich pulses and cross polarization. For Leu-labeled fd, the two-dimensional data were collected with 500 transients co-added for each of 24 t_1 points with a dwell time of 70 μs and a 4 s recycle delay. The aligned bacteriophage sample temperature was maintained at 65 °C for optimal resolution.[19]

The MAS solid-state NMR experiments were performed with spinning rates of 11.111 kHz or 7 kHz, unless otherwise specified. The unoriented samples were prepared with ~100 mg mL^{-1} concentration of bacteriophage particles and were packed in a 3.2 mm rotor. The sample temperature was calibrated externally using ethylene glycol and water signals in the phage samples. The sample temperature was maintained at 65 °C for high-temperature experiments and −8 °C for low-temperature experiments, unless otherwise stated. Proton decoupling during acquisition was performed with SPINAL-16 under 78 kHz RF irradiation. Cross-polarization from ^{1}H to either ^{13}C or ^{15}N nuclei was carried out using CP with 50% amplitude modulated ^{1}H RF. Typically, 300–500 μs contact times were used for $^{13}C\alpha$ signal acquisition and 1–2 ms contact times for $^{13}C'$ signal acquisition. ^{1}H–^{15}N Dipolar recoupling experiments were carried out with polarization inversion time-averaged nutation spin-exchange at magic angle (PITANSEMA) at 11.111 kHz sample spinning.[20] A nutation frequency of 62.5 kHz at a resonance-offset frequency of 44.2 kHz was used during the frequency switched Lee–Goldburg (LG) cycles. An effective nutation frequency of 76.5 kHz was calculated for a LG dwell time of 13.07 μs. A time modulation of 15.4 μs and 7.7 μs for +LG and −LG periods was used to average the nutation to one-third. The Hartmann–Hahn match to ^{15}N was accomplished with an RF field match to the first-order sidebands. Amplitude modulation on the ^{15}N channel was used at nutation frequencies of 36.5 kHz and 14.3 kHz during the phase alternated LG irradiations. The experimentally determined scaling factor of 0.37 was taken into account in data processing. For selectively ^{15}N-labeled samples, the two-dimensional spectra resulted from signal averaging of ~1000 transients for each of 16 t_1 real points with a 46.2 μs dwell time and 4 s recycle delay for high-temperature experiments. The low-temperature experiments with better sensitivity gain were carried out with 100 scans for each of 32 t_1 data points. For uniformly ^{15}N-labeled samples, a R-SLF two-dimensional experiment was used to record the spectrum at a spinning speed of 11.111 kHz and 65 °C sample temperature. Two consecutive π pulses of 5 μs duration each and phase shifted by 70 and −70 were used in the $R18^{7}{}_{1}$ pulse scheme.[21] A dwell time of one rotor period, 4 s recycle delay, and 128 transients were co-added for a total dipolar evolution period of 3.42 ms.

Recoupling of the CSA was accomplished in a two-dimensional manner that correlates the isotropic chemical shift in the direct dimension with the anisotropic powder spectrum in the indirect dimension. Two different pulse schemes were used to recouple the CSA (ROCSA).[11,22] The two-dimensional SUPER spectrum was recorded with 4 kHz sample spinning at the magic angle. Typically, 3000 scans for signal averaging and 14 complex points were acquired in each two-dimensional spectrum. Two-dimensional ROCSA experiments were carried out under 11.111 kHz sample spinning with a 47 kHz RF field irradiation on $^{13}C/^{15}N$ during CSA evolution under proton decoupling. SPINAL-16 pulses with 105 kHz RF irradiation were used for proton decoupling during t_1 evolution. Theoretical scaling factors of 0.27 for ROCSA and 0.13 for SUPER were experimentally verified with standard samples and utilized in all data processing. Typically, 1000–2000 transients were co-added for each of 16–24 t_1 complex points, and 2–4 s recycle delays were used for carbon and nitrogen detection. Experimental data were processed either in Bruker Topspin or using NMR-Pipe[23] software. Simulations were carried out with the SIMPSON simulation package.[24]

4.4 Discussion

Bacteriophages uniaxially align in the magnetic field at a concentration <50 mg mL^{-1}. Above this concentration they do not align, but retain their fast rotational diffusion along their long axis. The motion can be switched off and on simply by varying the temperature. This effect is shown experimentally in Figure 4.2. The motion is suppressed at low temperatures and a rigid lattice powder spectrum can be observed. For simplicity, simulated powder lineshapes are shown in Figure 4.1. At the rigid limit, the full anisotropic powder spectrum for $^{13}C'$ (Figure 4.1A) and ^{15}N (Figure 4.1C) are shown. Axially symmetric powder pattern spectra for $^{13}C'$ (Figure 4.1B) and ^{15}N (Figure 4.1D) are observed in the presence of rotational motion. Slow spinning provides a family of side bands for low-temperature data compared to high temperatures where only low intense sidebands are observed. In static solid-state NMR of a $^{13}C/^{15}N$-labeled phage particle, we observed a span of 30 ppm at half maximum for measurements carried out at a high temperature. The reduction in CSA by fivefold compared to low-temperature data ($\sim$150 ppm spread) suggests that the partial averaging of carbonyl CSA depends upon the tilt of the helix relative to the long axis of rotation. A quantitative measurement at slow spinning speed is shown in Figure 4.2B and C. The reduction in $^{13}C'$ CSA is a clear indication that the phage particles are undergoing rapid motion. The chemical shift dispersion in carbonyl carbon isotropic shifts was qualitatively analyzed to discern the effects on line broadening due to motion and RF irradiation. The full width at half maximum linewidth of 12 ppm (2.2 kHz) was measured at high temperature compared to the 5 ppm linewidth (1 kHz) for isotropic shifts at low temperature for $^{13}C'$ in uniformly $^{13}C,^{15}N$-labeled fd. This effect is also observed in selectively isotopically labeled proteins, for example where only two methionine residues at positions 21 and 28 are labeled. The isotropic chemical shifts are indistinguishable and give a linewidth of $\sim$3.8 ppm at high temperature compared to 1.6 ppm at low temperature. The line broadening in isotropic chemical shifts at high temperature is apparently due to the interference of motion with the MAS and RF irradiation close to the amplitude of motion. This is a well-known case for shortening of T_2 due to the motion at a rate comparable to the shift anisotropy and RF irradiation.[25,26] The introduction of coherent averaging (spatial or spin-space) applied to the spin system undergoing motion near the frequency of applied modulation becomes less efficient and MAS will no longer narrow the line. Following the equation given by Suewlack *et al.*[27] and as described by Lewis *et al.*,[25] based on the linewidth/T_2, qualitative measurement of the correlation time for the uniaxial rotational diffusion was found to be near 10^{-5} s. Both the excess linewidth and the correlation time inferred suggest that the rotational diffusion of the phage particle does not quite satisfy the fast limit condition $\tau_C \ll 13$ μs. Motion with a correlation time τ_C of 10^{-5} s will reduce the $T_{1\rho}$ of ^1H spins and may result in low efficiency of cross-polarization dynamics, which has been reported earlier.[25,28] We also observed this low efficiency in cross-polarization of magnetization from ^{1}H to ^{13}C or ^{15}N nuclei.

The averaging effect was then quantified using two separate protein samples labeled selectively at valine and leucine residues (Figures 4.3 and 4.4, respectively). Solid-state NMR data for proteins labeled at carbonyl sites of Val-only residues are shown in Figure 4.3. The chemical shift frequencies observed for magnetically aligned phage particles in static solid-state NMR are overlaid with respect to the powder lineshapes obtained from two-dimensional ROCSA experiment in MAS solid-state NMR of unoriented phage particles. The oriented chemical shift frequencies and the parallel edges of the recoupled CSA powder pattern shown in Figure 4.3 are complementary and provide motional averaging of anisotropic spread ranging from 16 ppm to 60 ppm. Similar is the case observed for the Leu-labeled sample, where the spread is about 75 ppm for the carbonyl carbon (Figure 4.4A, B). The two-dimensional ROCSA spectrum shows a dramatic reduction in linewidth at high temperature compared to the low-temperature measurement (Figure 4.4C–F). The difference in the spans of the powder patterns can be attributed to different tilt angles at different residues in the helix. This is in good agreement with earlier studies.[6] The structure of the protein has been solved using ^{15}N solid-state NMR of an oriented sample. It was reported that the protein at residues 8–39 has a tilt angle of 23° and a 17° tilt for residues 40–50. Therefore, motionally averaged powder spectra of carbonyl carbons can be used to determine the secondary structure and topology of a protein with reasonable accuracy.

As the structure of the fd coat protein has been extensively studied with ^{15}N NMR in an oriented system, the current results with unoriented samples under MAS show a one-to-one correlation between the two. Starting with the ^{15}N-Leu labeled sample in Figure 4.5, the well-resolved chemical shifts in oriented fd and the parallel edges of the recoupled CSA spectra of unoriented fd at two isotropic shifts show equivalent frequencies. Equivalent frequencies in ^{15}N CSA of the two leucine sites in fd implies that angular constraints of a molecule in their native topology can be derived from an unoriented system provided that the molecule undergoes uniaxial rotational diffusion around a common axis, which in this case is the long axis of the bacteriophage. The current result with [^{15}N]tryptophan-labeled fd shows two distinguishable isotropic chemical shifts, one in the backbone and the other in the indole functional group at residue 26. The two isotropic frequencies in the same residue show a dramatic difference in their CSA. This can be observed either by spinning the sample at a relatively slower rate and the modulations in their sideband intensities or by recoupling the anisotropic chemical shifts at moderate spinning rate in a two-dimensional fashion. The one-dimensional chemical shift spectra recorded at 3 kHz spinning are shown in Figure 4.6C and H for 65 °C and −8 °C, respectively. High-temperature data show noticeable intensities of the first pair of sidebands for backbone ^{15}N shifts, whereas the indole functional group ^{15}N shows no sideband intensities even for a slower spinning rate at 2 kHz (data not shown). To obtain reliable CSA powder lineshapes, two-dimensional CSA recoupling experiments were performed. The perpendicular edges of the recoupled axially symmetric powder patterns are shown in Figure 4.6E. The spectral slice obtained for the indole group shows a much narrower lineshape compared to the backbone amide nitrogen. Based on the slow spinning sideband intensities and ROCSA data, reliable powder lineshapes were reconstructed and are shown in Figure 4.6A and B for the backbone and indole group ^{15}N nuclei. On the other hand, the one-dimensional spectrum at lower temperature shows variation in intensities even up to second and third sidebands for both ^{15}N sites. The two-dimensional ROCSA and one-dimensional chemical shift slice for the indole functional group are shown in Figure 4.6I and J. The reconstructed powder lineshapes are shown in Figure 4.6F and G for the backbone and indole group, respectively. Notably, the sense of direction observed in the high-temperature spectrum is reversed compared to the low-temperature one, as shown in Figure 4.6. This is a well-known fact that in the presence of rotational motion the anisotropic chemical shift powder pattern changes its lineshape from its asymmetry to axially symmetric.[9] The averaging of CSA and sense of direction implies a complementary angular restraint to individual residues in a secondary structure and topology.

To determine the orientation of the helix with high angular precision, the most sensitive nuclear spin interactions are the ^{1}H–^{15}N dipolar coupling and ^{15}N CSA, as both unique axes of the tensors are roughly parallel to the helix axis. In an oriented sample of bacteriophage particles, the long axis lies almost parallel to the external magnetic field, which is the rotational diffusion axis. In this case, the dipolar couplings in the tilted helix with respect to the long axis must show equivalent coupling frequencies in both oriented and unoriented phage samples. The dipolar coupling measurements in oriented and unoriented phage particles, with Leu-labeled fd proteins, are shown in Figure 4.7. At high temperatures, the ^{1}H–^{15}N dipolar frequencies obtained from magnetically oriented phage (Figure 4.7C) and the recoupled dipolar frequencies at isotropic chemical shifts (Figure 4.7B) show equivalent coupling frequencies for both cases. This is consistent with the previous report on magnetically oriented fd phage by Zeri *et al.*,[6] with leucine sites at residues 14 and 41 showing a dipolar splitting of 10.2 kHz and 13.8 kHz, respectively. At low temperature, the motion is expected to be completely frozen and the particles to behave like a powder sample with all possible orientations and yield the rigid limit dipolar frequencies. As expected, the dipolar frequencies in both sites show nearly equal coupling strength (assuming similar bond lengths) compared to the inequivalent and averaged dipolar frequencies measured in the presence of uniaxial motion.

The unoriented tryptophan-labeled fd phage sample shows two distinct ^{1}H–^{15}N dipolar frequencies measured under MAS at high temperature (Figure 4.8B). The dipolar frequency at 128 ppm, corresponding to the side-chain ^{15}N shift, shows almost no resolution with 2 kHz linewidth at half maximum. However, the backbone ^{15}N shift at 120 ppm shows a dipolar splitting of 10.8 kHz. Similar results in oriented fd have been reported previously.[6,29] On the other hand, the equivalent experiment performed at low temperature provides the rigid limit dipolar frequencies (Figure 4.8A). The small difference in dipolar frequencies for the side chain compared to the backbone is probably due to the difference in ^{1}H and ^{15}N bond length. The differences in bond length up to 4 pm for side chains have been reported for polycrystalline amino acids.[21,30] Separated local field spectroscopy of the uniformly labeled unoriented sample shows a spread of dipolar coupling frequencies ranging from 0 to 10.5 kHz for residue-specific ^{15}N shifts. Nearly 40% of the nitrogen sites are well resolved for a uniformly ^{15}N-labeled bacteriophage fd coat protein. The one-to-one match in dipolar and anisotropic shift frequencies in multiply labeled proteins suggests that rotational motion along the long axis of the phage particle retains the angular equivalence in both oriented and unoriented systems.

The presence of fast rotational diffusion in fd filamentous bacteriophage particles is concentration dependent. At high concentrations the bacteriophage particles are immobile on the relevant NMR timescales, as determined by the frequency span of the powder pattern. A dramatic reduction in ^{13}C′ CSA, ^{15}N amide CSA, and ^{1}H–^{15}N dipolar couplings is observed due to the motional averaging.

All of the effects demonstrated here using the filamentous bacteriophage fd have also been observed in other systems. These include the filamentous bacteriophage Pf1 and a range of membrane proteins in phospholipid bilayers.[31] Taken together, these examples place solid-state NMR spectroscopy alongside solution NMR spectroscopy and X-ray crystallography as a method for determining the structures of proteins at atomic resolution.

Acknowledgments

The research described in this chapter was supported by NIH grants R01GM099986, RO1EB005161, and P01AI074805. It was performed at the Biotechnology Resource Center for Molecular Imaging of Proteins supported by P41EB002031.

References

1. S. H. Park, B. B. Das, A. A. De Angelis, M. Scrima and S. J. Opella, *J. Phys. Chem. B*, 2010, **114**, 13995–14003.
2. B. B. Das, H. J. Nothnagel, G. J. Lu, W. S. Son, Y. Tian, F. M. Marassi and S. J. Opella, *J. Am. Chem. Soc.*, 2012, **134**, 2047–2056.
3. F. M. Marassi, B. B. Das, G. J. Lu, H. J. Nothnagel, S. H. Park, W. S. Son, Y. Tian and S. J. Opella, *Methods*, 2011, **55**, 363–369.
4. N. Tjandra and A. Bax, *Science*, 1997, **278**, 1111–1114.
5. J. H. Prestegard and A. I. Kishore, *Curr. Opin. Chem. Biol.*, 2001, **5**, 584–590.

6. A. C. Zeri, M. F. Mesleh, A. A. Nevzorov and S. J. Opella, *Proc. Natl. Acad. Sci. U. S. A.*, 2003, **100**, 6458–6463.
7. D. S. Thiriot, A. A. Nevzorov, L. Zagyanskiy, C. H. Wu and S. J. Opella, *J. Mol. Biol.*, 2004, **341**, 869–879.
8. H. S. Gutowsky and G. E. Pake, *J. Chem. Phys.*, 1950, **18**, 162–170.
9. M. Mehring, R. Griffin and J. Waugh, *J. Chem. Phys.*, 1971, **55**, 746.
10. J. Herzfeld and A. E. Berger, *J. Chem. Phys.*, 1980, **73**, 6021–6030.
11. J. C. C. Chan and R. Tycko, *J. Chem. Phys.*, 2003, **118**, 8378.
12. C. V. Grant, C. H. Wu and S. J. Opella, *J. Magn. Reson.*, 2010, **204**, 180–188.
13. C. V. Grant, Y. Yang, M. Glibowicka, C. H. Wu, S. H. Park, C. M. Deber and S. J. Opella, *J. Magn. Reson.*, 2009, **201**, 87–92.
14. E. C. Lin and S. J. Opella, *J. Magn. Reson.*, 2011, **211**, 37–44.
15. E. C. Lin, C. H. Wu, Y. Yang, C. V. Grant and S. J. Opella, *J. Magn. Reson.*, 2010, **206**, 105–111.
16. M. Levitt, D. Suter and R. Ernst, *J. Chem. Phys.*, 1986, **84**, 4243.
17. B. Fung, A. Khitrin and K. Ermolaev, *J. Magn. Reson.*, 2000, **142**, 97–101.
18. A. A. Nevzorov and S. J. Opella, *J. Magn. Reson.*, 2003, **164**, 182–186.
19. W. M. Tan, R. Jelinek, S. J. Opella, P. Malik, T. D. Terry and R. N. Perham, *J. Mol. Biol.*, 1999, **286**, 787–796.
20. K. Yamamoto, V. Ermakov, D. Lee and A. Ramamoorthy, *Chem. Phys. Lett.*, 2005, **408**, 118–122.
21. X. Zhao, M. Edén and M. H. Levitt, *Chem. Phys. Lett.*, 2001, **342**, 353–361.
22. S. Liu, J. Mao and K. Schmidt-Rohr, *J. Magn. Reson.*, 2002, **155**, 15–28.
23. F. Delaglio, S. Grzesiek, G. W. Vuister, G. Zhu, J. Pfeifer and A. Bax, *J. Biomol. NMR*, 1995, **6**, 277–293.
24. M. Bak, J. T. Rasmussen and N. C. Nielsen, *J. Magn. Reson.*, 2000, **147**, 296–330.
25. B. Lewis, G. Harbison, J. Herzfeld and R. G. Griffin, *Biochemistry*, 1985, **24**, 4671–4679.
26. W. Rothwell and J. Waugh, *J. Chem. Phys.*, 1981, **74**, 2721.
27. D. Suwelack, W. Rothwell and J. Waugh, *J. Chem. Phys.*, 1980, **73**, 2559.
28. M. D. Sefcik, J. Schaefer, E. O. Stejskal, R. A. McKay, J. F. Ellena, S. W. Dodd and M. F. Brown, *Biochem. Biophys. Res. Commun.*, 1983, **114**, 1048–1055.
29. T. Cross and S. Opella, *J. Am. Chem. Soc.*, 1983, **105**, 306–308.
30. A. Ramamoorthy, C. Wu and S. Opella, *J. Am. Chem. Soc.*, 1997, **119**, 10479–10486.
31. S. J. Opella and F. M. Marassi, *Chem. Rev.*, 2004, **104**, 3587–3606.

CHAPTER 5

Membrane Protein Interactions

JAMES A. JARVIS AND PHILIP T. F. WILLIAMSON*

Centre for Biological Sciences/Institute for Life Sciences, University of Southampton, Highfield Campus, Southampton, SO17 1BJ, UK
*Email: P.T.Williamson@soton.ac.uk

5.1 Introduction

Biological membranes are ubiquitous structures in biology, serving as a barrier between the cell and its environment and providing discrete compartments within the cell to isolate particular cellular processes. Whilst providing an impermeable barrier, biological membranes also have to provide the cell with the ability to transfer information and material both across the membrane and between different cellular compartments, a task largely accomplished through the presence of integral membrane proteins located within the membrane. A key step to understanding how the cell, and in particular the cell membrane, functions is not only our understanding of the structure of integral membrane proteins but also how they interact with each other, the surrounding lipid bilayer and a multitude of other soluble proteins and small molecules. It is these properties that play a vital role in regulating the processes integral membrane proteins support.

The major challenge associated with structural and functional analysis of biological membranes stems from their heterogeneous and dynamic nature as they contain a multitude of components, including lipids and proteins that are held together by a plethora of interactions to form the dynamic assembly that is the biological membrane. Typically, structural biologists have taken a reductionist approach to the study of biological membranes, isolating the constituent membrane proteins before their analysis by

New Developments in NMR No. 3
Advances in Biological Solid-State NMR: Proteins and Membrane-Active Peptides
Edited by Frances Separovic and Akira Naito

Published by the Royal Society of Chemistry, www.rsc.org

conventional structural biology techniques, including crystallography and liquid-state NMR. In doing so, however, significant information is lost relating to how the individual components of the biological membrane interact with one another to form such a macromolecular assembly and regulate the processes that the biological membrane supports. To address these issues a number of alternative techniques have been utilized that can provide valuable insights into molecular interactions between lipids and integral membrane proteins, with techniques such as fluorescence spectroscopy[1,2] and electron spin resonance[3,4] proving particularly useful. Recently, however, solid-state NMR has emerged as a powerful technique for studying biological membranes, permitting analysis at the molecular level of membrane proteins, either peripheral or integral, whilst they are located on or within the lipid bilayer. The possibility of studying such complex macromolecular assemblies provides a valuable route to the identification of interactions between the individual components of the biological membrane and their subsequent characterization at the molecular level. This chapter seeks to describe how solid-state NMR studies have been employed to probe interactions within the lipid bilayer. We highlight studies where these techniques have been applied to integral membrane proteins to study their interactions with soluble ligands, as well as protein/lipid and protein/protein interactions, demonstrating how the data obtained have contributed to our understanding of a number of cellular and disease states.

5.2 Solid-State NMR Methods Used for the Study of Membrane Protein Interactions

An analysis of the components that constitute biological membranes and the interactions between them has proven to be a significant challenge to structural biology, with difficulties at the level of expression, purification and reconstitution of the proteins of interest through to the intractability of membrane protein complexes to modern structural biology techniques. Indeed, despite recent developments in the crystallography of integral membrane proteins and many notable successes,[5] the elucidation of structures is far from routine. Similarly, improvement in liquid-state NMR methodology has led to significant increases in the size of systems which can be studied;[6–9] however, integral membrane proteins embedded in detergent micelles still remain a challenging proposition. Despite the introduction of a range of novel membrane mimetics,[10,11] the analysis of large multimeric complexes and the interactions between their component parts by liquid-state NMR methods still remains challenging.

5.2.1 Solid-State NMR

Many of the interactions experienced by the nuclear spins that give rise to NMR spectra are anisotropic in nature. In the liquid-state, where molecules undergo rapid molecular motions, these interactions are motionally averaged, resulting in the classical "high-resolution"-like spectra dominated by the chemical shift and *J*-couplings between individual nuclei, with anisotropic contributions giving rise to fluctuating local fields which manifest themselves as relaxation effects in the spectra. By contrast, in the solid-state, rapid molecular tumbling is absent and the anisotropic interactions give rise to a distribution of resonances reflecting the distribution of molecular orientations within the sample (Figure 5.1A). Analysis of these distributions can provide valuable insights into the anisotropic interactions experienced by the nuclear spin, with the chemical shift anisotropy and (for spins where $I>1/2$) the quadrupolar interaction giving insights into the local electron distribution around a nucleus, whilst the dipolar interactions between individual spins can impart information about their relative proximity. Furthermore, analysis of the dynamic averaging of these interactions can provide a wealth of information pertaining to the motions experienced by particular sites within the molecules. It is possible to exploit the structural and dynamic information encoded in the interactions experienced by the

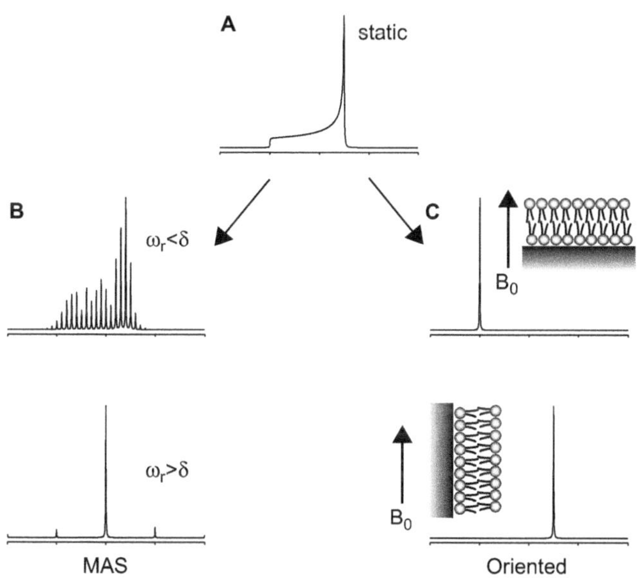

Figure 5.1 (A) The NMR spectrum of a static disordered solid gives rise to a broad distribution of resonances, with each molecular orientation giving a resonance. (B) The application of MAS results in the mechanical averaging of the anisotropic interactions, giving rise to an isotropic resonance when $\omega_r>\delta$ or an isotropic resonance and a family of spinning sidebands when $\omega_r<\delta$, where δ is the size of the anisotropic interaction. (C) Macroscopic alignment of the molecules within the sample can also result in improved resolution, with each resonance characterizing the orientation of an anisotropic interaction with respect to the magnetic field.

nuclear spin to determine how the structure and dynamics present within membrane proteins relates to their function, and how these molecular species interact within the membrane.

The information encoded into these anisotropic interactions does, however, come at the expense of increased spectral complexity, with the broad distribution of resonances making challenging the resolution and assignment of sites within a given molecule. Many of the developments in solid-state NMR methodology over the preceding years have focused on how to simplify the spectra such that site-specific resolution is achieved whilst simultaneously allowing the underlying anisotropic interactions to be characterized and the structural and dynamic information they contain to be extracted. Typically this has been achieved through the preparation of macroscopically aligned systems and/or the application of magic-angle spinning.

5.2.2 Magic-Angle Spinning NMR

Magic-angle spinning (MAS) NMR seeks to mechanically average the anisotropic interactions to obtain "high-resolution"-like spectra where resonances are resolved on the basis of their isotropic chemical shifts and, in favourable cases, their *J*-couplings.[12] To achieve this the sample is rotated rapidly (kHz) in the static magnetic field at an angle of 54.7°. As the anisotropic interactions that give rise to the broadening observed in the NMR spectra of solid materials posses a $\frac{1}{2}(3\cos^2\theta - 1)$ dependence, a scaling of the anisotropic interaction is observed (see Figure 5.1B). When the sample rotation frequency exceeds the size of the anisotropic interaction, this results in complete averaging of the anisotropic interaction and a resonance is observed at its isotropic position. When the sample rotation frequency is lower or a similar size to the anisotropic interaction, incomplete averaging occurs which is manifest in the solid-state NMR spectrum as a family of resonances separated from the isotropic resonance by multiples of the rotation frequency (Figure 5.1B). The significant improvements in resolution afforded by MAS come at the expense of the information that is contained within the anisotropic interactions. A number of methods have since been developed that permit the selective reintroduction of these anisotropic interactions to permit their analysis. These methods and their application to biological systems have been widely reviewed elsewhere.[13–16] A number of these methods will be discussed below, highlighting how they have been applied to the study of interactions in biological membranes.

5.2.3 Oriented Sample NMR

A simplification of the spectra of biological membranes can also be achieved through the macroscopic alignment of the anisotropic interactions. The macroscopic alignment of the interactions results in a significant reduction in the spectral complexity that in favourable circumstances results in the presence of a single narrow resonance whose frequency provides information on the orientation of the particular anisotropic interaction with respect to the magnetic field (Figure 5.1C). Frequently the orientation of a given anisotropic interaction is known with respect to a chemical moiety within a system, allowing the orientation of the chemical moiety to be determined with respect to the magnetic field. A number of techniques can be employed to align biological membranes, including magnetic alignment[11,17] in a range of membrane mimetics and mechanical alignment on glass/ Melanex sheets.[18]

5.3 Studying Molecular Interactions by Solid-State NMR

As highlighted above, integral membrane proteins experience interactions with a range of other molecular species. An important class of interactions are those between integral membrane proteins and soluble compounds, including signalling molecules, metabolites or drugs. These interactions play a vital role in cell signalling and the transport of material across the cell membrane. However, it has also become clear that there are a number of important interactions that occur within the lipid bilayer. Interactions between lipids and proteins at the lipid/protein interface have been shown to play an important role in the regulation of integral membrane proteins;[19] more recently, these interactions have been shown to play a major role in the regulation of protein structure and oligomeric state with implications in a range of signalling and trafficking events.[20–22] Finally, interactions between integral membrane proteins also play a key role in regulating a number of cellular processes, including the regulation of integral membrane protein activity, signal transduction events and the co-localization of membrane proteins in oligomeric complexes to facilitate enzymatic activity.[23–25] In each of these areas, solid-state NMR has proven to be a valuable tool to study these interactions and provide valuable molecular insights into how these systems function. Below, we will describe some selected examples of solid-state NMR studies that have contributed to this understanding whilst highlighting the types of techniques and technologies employed.

5.3.1 Interactions with Soluble Ligands

Integral membrane proteins are responsible for a number of cellular processes that rely on their interactions with soluble molecules, such as metabolites requiring transport across the cellular membrane and small regulatory molecules involved in regulating cell signalling events. Solid-state NMR has been exploited to study these interactions as it can provide high-resolution structural information about both the conformation of the bound molecule and how it interacts with the surrounding protein.[13,26–29] Such information has the potential to play a key role in the development of novel therapeutics that target this class of protein.[13,26–29]

The method of analysis employed to study the interaction of soluble ligands bound to integral membrane proteins is determined largely by their affinity. Ligands with weaker affinities for their target protein typically possess exchange rates that render them accessible to liquid-state NMR methods such as transferred-NOE/RDC, water LOGSY or saturation transfer difference spectroscopy, where binding information can be obtained from an analysis of the free population.[30–32] In contrast, those molecules that have a higher affinity and/or a long residency time to an integral membrane protein have typically proven to be ideal systems to study using solid-state NMR based methods. Oriented sample NMR has proved a valuable technique for the analysis of the orientation and dynamics of ligands covalently attached to integral membrane proteins, with extensive studies performed on bacteriorhodopsin, rhodopsin, Band III and the acetylcholine receptor.[33–38] MAS NMR has also been used to study covalently bound ligands;[39–50] however, the fully hydrated nature of the sample under study also permits the study of high-affinity ligands that are in slow exchange between their bound and free state (residency time >10's ms). Such MAS NMR studies, in addition to providing insights into the nature of ligand binding, can also inform as to a ligand's affinity/residency time on the integral membrane protein.[51–53]

One receptor whose high-affinity ligand interactions have been extensively studied by solid-state NMR methods is the nicotinic acetylcholine receptor, a member of the pentameric ligand-gated ion channel family of proteins.[26,54,55] The nicotinic acetylcholine receptor is a 280 kDa integral membrane protein that plays an important role in synaptic transmission and, in addition to its important role in muscle contraction, is also thought to play a major role in diseases such as schizophrenia, Alzheimer's disease and Parkinson's disease. The nicotinic acetylcholine receptor is found in abundance in the electroplax tissue of the electric ray and has been the subject of extensive biochemical, biophysical and pharmacological studies, becoming an important paradigm for other members of this important family of ligand-gated channels, which includes $GABA_A$, 5-HT_3 and ionotropic glutamate receptors.

A number of studies have focused on characterizing the interaction of the agonist, acetylcholine, with its respective binding site on the nicotinic acetylcholine receptor in order to better understand the pharmacology of the receptor. Early studies by Williamson *et al.*[56] focused on determining the dynamics and orientation of bromoacetylcholine within the agonist-binding site. Bromoacetylcholine reacts selectively with cysteine residues that are located at the agonist-binding site. The introduction of deuterium into the quaternary ammonium group of the bromoacetylcholine allows the acquisition of deuterium spectra that are dominated by the large quadrupolar interaction. Analysis of the deuterium spectra of nicotinic acetylcholine receptor membranes labelled with deuterated bromoacetylcholine as a function of temperature provides insights into the degree of motion experienced by the quaternary ammonium group within the agonist-binding site. The rather broad nature of the deuterium spectra compared to those obtained from crystalline deuterated acetylcholine and acetylcholine-analogue salts at equivalent temperatures indicates that the quaternary ammonium group is experiencing restricted dynamics within the binding site, with rotation of the methyl and quaternary ammonium groups significantly hindered. This suggests that the quaternary ammonium group is tightly constrained, with a high degree of complementarity between the ligand and the receptor binding site.[56]

To determine the orientation of the quaternary ammonium group of the bromoacetylcholine within the agonist-binding site, native nicotinic acetylcholine receptor membranes were macroscopically aligned on Melanex sheets using an isopotential spin-drying technique.[18] The resulting membrane is aligned with the membrane normal aligned perpendicular to the sheet with a distribution of orientations around the membrane directors of 20–25°.[18,56] Deuterium NMR spectra of the aligned membranes acquired with the membrane normal aligned both parallel and perpendicular to the magnetic field were subsequently analysed to obtain the orientation of the quaternary ammonium group with respect to the static magnetic field and the membrane normal (Figure 5.2). This analysis revealed that the

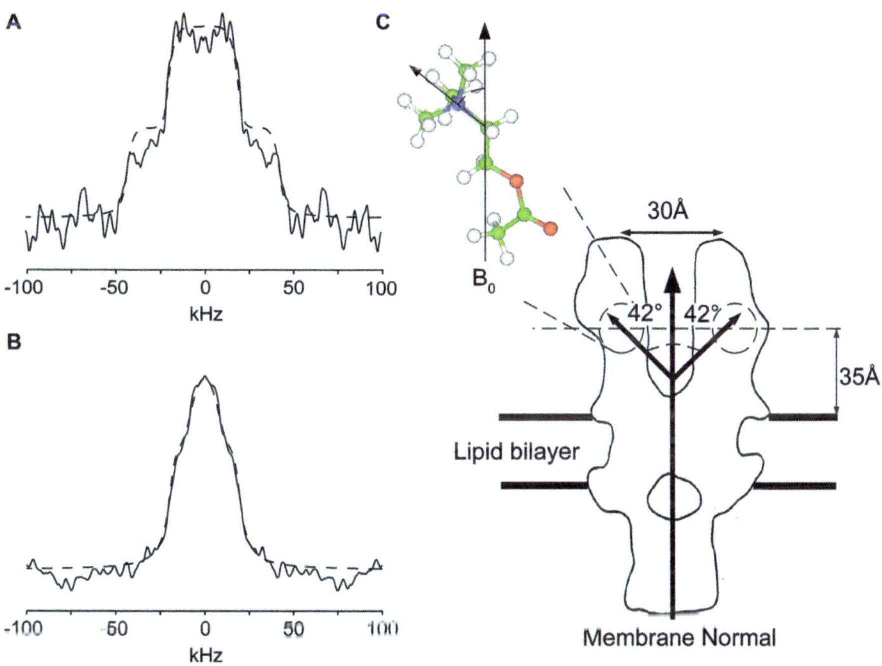

Figure 5.2 Deuterium NMR spectra of oriented AChR membranes labelled with d_9-bromoacetylcholine with the membrane normal aligned at (A) 0° and (B) 90° with respect to the magnetic field. The corresponding numerical simulations (*dashed line*) are consistent with the alignment of the quaternary ammonium group at an angle of 42°, as demonstrated in the schematic of the receptor (C).
(Adapted from Williamson *et al.*,[56] with permission from the National Academy of Sciences.)

quaternary ammonium group of the bromoacetylcholine is oriented at 42° with respect to the magnetic field,[56] consistent with the orientation of the binding pocket identified in electron diffraction studies.[57,58]

Oriented sample studies have been complemented by a series of MAS-NMR studies that sought to resolve the agonist, acetylcholine, within the agonist-binding site and study its affinity for the receptor, pharmacophores involved in binding and the conformation of the bound ligand. Initial studies focused on characterizing pharmacologically relevant binding of the ligand within its binding site.[59] These studies exploited cross-polarization as a motional filter to selectively observe ^{13}C-labelled acetylcholine within the agonist-binding site. Cross-polarization (CP) is a technique typically used as a mechanism to enhance the signal of low-γ nuclei, transferring the magnetization from the abundant protons with their high gyromagnetic ratio and favourable relaxation rates to the low-γ nucleus. The efficiency of this transfer is highly dependent on the strength of the dipolar couplings between the protons and the low-γ nuclei and is, therefore, sensitive to the degree of motion experienced by a molecule. Accordingly, when CP is applied to ^{13}C-labelled ligands bound to integral membrane proteins the experiment selectively observes only those ligands that are immobilized through their binding to the receptor and thus have stronger ^{1}H–^{13}C dipolar couplings than those in free solution.[51,59–63] Using CP as a motional filter, resonances attributable to ^{13}C labels incorporated into the quaternary ammonium group were resolved within the CP-MAS spectra, the pharmacological relevance of which was demonstrated by inhibition studies using α-bungarotoxin, a selective inhibitor of the agonist binding site of the nicotinic acetylcholine receptor.[59]

The differing motional properties of ligands in free solution and those bound to the membrane-bound receptor have also been exploited to study the rate of ligand exchange.[52,64] Typically, ligands in solution have a longer transverse relaxation rate (T_2) than those bound to the receptor and a simple spin echo can thus be employed to create a population where only free ligands possess magnetization. Experiments have been developed exploiting this differential relaxation behaviour where these ligands are free to exchange with those bound to the receptor, a process that can subsequently be probed by their detection using CP. Such experiments allow the rate of ligand exchange to be determined (or a limit placed upon it) through the observation of the rate at which the signals from the bound ligand reappear, which will be dependent on either (a) the rate of exchange between the bound and free ligand or (b) T_1 relaxation. When applied to acetylcholine, the rate of recovery of signal attributable to the bound acetylcholine recovered at a similar rate to other membrane components, indicating that signal recovery was limited not by ligand exchange but T_1 relaxation, suggesting that no exchange between the bound and free pool of agonist was on the T_1 timescale (100's ms).[59]

Resonances within the CP-MAS spectra that are attributable to the pharmacologically relevant ligands bound to their integral membrane proteins can also provide a wealth of information relating to the interaction between them. This information arises primarily through the change in electrostatic environment that occurs upon ligand binding, which can cause significant perturbations in chemical shift. Such studies have been greatly facilitated through the use of broadband recoupling sequences. Although CP-MAS studies selectively observe components whose motion is restricted, natural abundance ^{13}C signals arising from the proteins and lipids within the sample still make significant contributions to the spectra, frequently obscuring signals from the ligand under study. Broadband recoupling experiments such as proton-driven spin diffusion (PDSD)[65] and DARR[66,67] and their double quantum counterparts (POST-C7, SPC5, DREAM)[68–70] seek to reintroduce the dipolar couplings between ^{13}C labelled sites, which are averaged under MAS, to permit the acquisition of correlation spectra where cross-peaks appear from ^{13}C sites in close proximity. When applied to uniformly labelled ligands bound to integral membrane protein, correlations arise primarily from the labelled molecules where the probability of adjacent ^{13}C labels is higher than that in natural abundance material. Furthermore, using dipolar recoupling sequences which generate double quantum coherence between the labelled sites, such as POST-C7, SPC5 and DREAM,[68–71] permits the selection of only those resonances arising from the labelled molecules and the suppression of strong natural abundance resonances that are typically present on the diagonal of broadband recoupling experiments such as PDSD/DARR.[72,73]

Proton-driven spin-diffusion experiments have been applied to uniformly ^{13}C labelled acetylcholine bound to the nicotinic acetylcholine receptor to facilitate the identification of signals from the bound ligand from the intense signals present from the natural abundance ^{13}C within the sample (Figure 5.3B). Upon binding, significant perturbations in chemical shift were observed from ^{13}C resonances close to the ester linkage and the quaternary ammonium. The ^{13}C chemical shifts are sensitive to a number of properties, including conformation, local electrostatic environment and ring current effects.[59,74] Through a careful consideration of the factors which may affect the resonance frequency at each site it was concluded that the perturbations were consistent with the binding of the quaternary ammonium group close to extended aromatic systems through a number of cation–π interactions and the formation of a strong H-bond between the carbonyl oxygen of the ester linkage in the acetylcholine and an H-bond donor in the agonist.[59,74] Interestingly, these observations are in agreement with a large body of pharmacological studies which highlight the importance of these groups in the high-affinity binding of acetylcholine to this particular subclass of the nicotinic acetylcholine receptor.[59,74,75]

The identification of the pharmacophores responsible for the recognition and binding of ligands to membrane receptors is valuable information in the development of novel pharmaceuticals that target them. The spatial positioning of these groups further facilitates the design process and is information that can be attained through structural studies of the ligand within its binding site.[27,76,77] As highlighted above, dipolar recoupling techniques

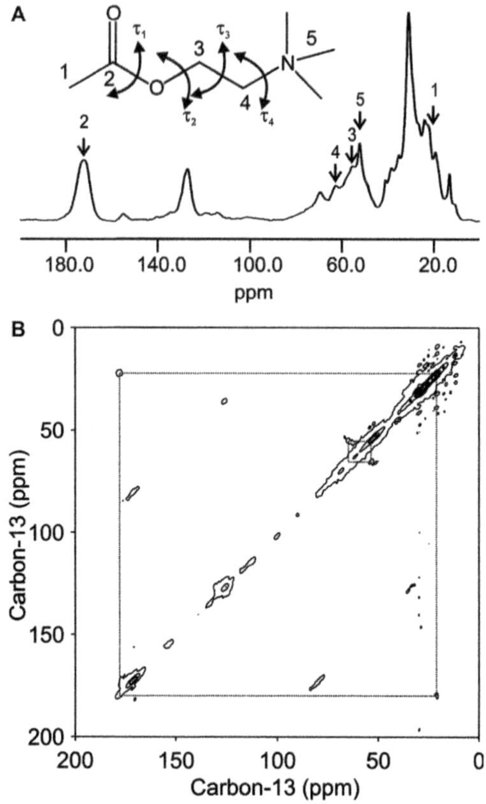

Figure 5.3 (A) CP-MAS spectra of 40 nmol of [U-^{13}C]acetylcholine bound to the nicotinic acetylcholine receptor. The *inset* shows the structure of acetylcholine together with the assignment of the resonances. (B) Corresponding 2D proton-driven spin-diffusion spectra of the labelled ligand bound to the receptor, with off-diagonal cross peaks highlighted between the adjacent carbon atoms aiding in the assignment of the bound molecule against the strong natural abundance ^{13}C signals. (Adapted from Williamson *et al.*,[74] with permission from the National Academy of Sciences.)

that seek to reintroduce dipolar couplings when studies are conducted under MAS can provide information on the distance between labelled sites within the ligand. Potentially, broadband recoupling schemes, such as proton-driven spin diffusion[65] or DARR,[66,67] can provide distance constraints between multiple sites within the ligand.[65] Typically, however, the transfer of magnetization between the labelled sites in broadband recoupled spectra is complex and the extraction of accurate and quantitative information is challenging, particularly with the signal-to-noise that is afforded in studies of membrane systems. For larger systems, however, multiple low-accuracy distance constraints obtained using broadband recoupling schemes used in conjunction with chemical shift data can be sufficient to facilitate the determination of the global fold of a molecule. For smaller molecules, where only a small number of distances can be measured and where accuracy is essential to restrain the conformation of the molecule, the low-resolution structural constraints from broadband recoupling experiments are of limited use. Selective dipolar recoupling schemes, where a specific dipolar coupling is reintroduced between a particular spin-pair, significantly simplifies the analysis of the transfer of magnetization between the two sites and makes possible a quantitative and accurate measurement.[78–80] When used in conjunction with specific isotope labelling analysis of the magnetization exchange dynamics, this is sufficiently robust to accurately determine internuclear distances and has been widely employed to study the conformation of ligands bound to integral membrane proteins.[60,62] However, using selective recoupling techniques, such as rotational-resonance or frequency selective REDOR, the magnetization can be driven between an actively recoupled spin-pair and the resulting magnetization exchange dynamics analysed to obtain quantitative and accurate distance information even in uniformly labelled molecules.[81–83]

Frequency selective studies proved particularly effective for studies of the nicotinic acetylcholine receptor, where rotational-resonance measurements have been employed to determine the geometry of the agonist, acetylcholine, within its binding site. Rotational-resonance, a frequency selective homonuclear recoupling scheme, reintroduces the dipolar coupling between two labelled sites when the spinning speed matches the chemical shift separation between the two peaks.[80,86] A series of measurements were conducted by selectively recoupling each spin pair in turn and measuring the rate of magnetization exchange; distances between particular spin pairs were determined with an accuracy of less than 0.5 Å for as little as 20 nmol of [U-^{13}C]acetylcholine bound to the nicotinic acetylcholine receptor[74] (Figure 5.4A). Using these distances as constraints, a complete search of the conformational space was conducted which identified two symmetrical families of structures.[74] In agreement with previous structure/activity studies, the structures revealed a distance of 5.1 Å between the quaternary ammonium group and the oxygen of the carbonyl ester, the two groups essential for high-affinity binding to this sub-family of the nicotinic acetylcholine receptor.[74] The resulting structures were used as the basis for docking studies to an acetylcholine binding protein with homology to the ligand-binding domain of the nicotinic acetylcholine receptor. Whilst applying NMR-derived restraints to the acetylcholine, docking studies were performed with an unrestrained protein to identify how the receptor responds to the binding of acetylcholine. This study revealed that upon docking the acetylcholine resided within the agonist binding site, with the quaternary ammonium group positioned tightly within an aromatic "cage", the carbonyl group aligned well with a potential H-bond donor within the binding site and the quaternary ammonium group aligned at 42° with what would be the membrane normal. These data are largely in agreement with independent analysis of the chemical shift data and orientational deuterium NMR studies.

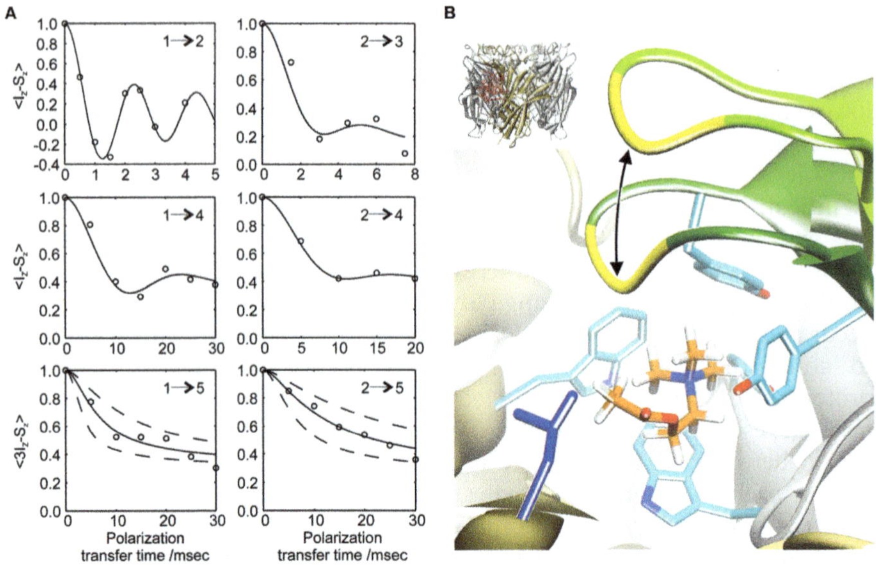

Figure 5.4 (A) Rotational-resonance magnetization exchange curves measured for [U-^{13}C]acetylcholine bound to the nicotinic acetylcholine receptor; numbering refers to the assignment given in Figure 5.3. Each curve represents a measurement between two of the sites within the labelled acetylcholine. Distances are extracted by fitting the experimental data (*circles*) to magnetization exchange curves calculated for the two actively recoupled spins under the rotational-resonance condition whilst accounting for the presence of other perturbing spins (*solid lines*). Where exchange occurs between a single site and the three magnetically equivalent carbon atoms on the quaternary ammonium group, fits of the experimental data are bracketed by the exchange curves expected for the calculated distance ± 0.5 Å (*dashed line*). (B) The distance constraints were used to refine the structure of the bound acetylcholine and the resulting structure docked to the crystal structure of an acetylcholine binding protein. The docking analysis reveals the identity of the pharmacophores involved in ligand binding, which are in excellent agreement with those proposed from an analysis of chemical shift data.
(Adapted from Williamson *et al.*,[74] with permission from the National Academy of Sciences.)

Interestingly, upon docking of the acetylcholine to the acetylcholine binding protein, the C-loop undergoes a significant motion enveloping the ligand within the binding pocket (Figure 5.4B). A number of crystallographic studies have subsequently highlighted the importance of this structural transition, not only in ligand binding but the subsequent activation of the receptor.[84]

These studies serve to highlight how molecular descriptions of the ligand within its binding site can not only provide the structural insights necessary for the rational development of drugs that target these receptors, but can make a significant contribution to our understanding of ligand recognition and subsequent receptor activation.

5.3.2 Interactions at the Lipid/Protein Interface

Since early reconstitution experiments, it has become clear that the activity of integral membrane proteins is dependent on the lipid composition of the bilayer in which they are embedded.[2,19,85] These studies clearly demonstrated that the activity of integral membranes is sensitive to the types of fatty acyl chains present, the types of lipid headgroups and whether the bilayer contains cholesterol.[2,19,85] Subsequently, a number of elegant electron spin resonance (ESR) and fluorescence studies identified that integral membrane proteins possess a layer of lipids, whose motion is restricted within the bilayer through their interaction with the integral membrane protein, and are now referred to as "annular" lipids.[4,86] The composition of the annular lipids is highly dependent upon the particular membrane protein and is closely correlated to the activity of the protein. Traditionally, NMR methods have not been widely used to study annular lipids, for although the exchange between the annular lipids and the bulk lipids is slow on the ESR timescale, allowing the identification of two motionally distinct components, the exchange between these two pools is too fast on the NMR timescale for the bound population to be resolved.[87,88]

Recently, however, a number of elegant saturation transfer difference MAS (STD-MAS) experiments have been implemented that facilitate the identification of preferential interactions between particular lipids present within the bilayer and integral membrane proteins, without necessitating the introduction of potentially perturbing groups such as spin labels or fluorophores.[89] To characterize the interaction of lipids with integral membrane proteins using STD-MAS experiments, protein resonances are selectively saturated by a train of selective rf pulses (Figure 5.5A). The magnetization is subsequently redistributed throughout the protein *via* spin diffusion. Transfer to the surrounding lipids is mediated by the presence of strong ^{1}H–^{1}H dipolar couplings between the protons within the protein and those of proximal lipids. During the extended period of the saturation pulses, lipids freely exchange between the annular and bulk lipid. If the rate of exchange of lipid is comparable to the length of the train of saturation pulses, then a significant population of the lipids will become saturated, resulting in an attenuation of the lipid signal intensity in the spectrum. The reduction of the signal intensity arising from a particular lipid is, therefore, a reflection of the proximity of a particular lipid to the protein and its residency time at the lipid/protein interface.[89,90] Since signals from saturated and unsaturated acyl chains of fatty acids can be resolved, this method can determine a protein's preference for interaction with either type. To identify the headgroup specificity, STD-MAS can be combined with CP which transfers magnetization from the protons to the phosphate in the lipid headgroup (Figure 5.5E). As the source of proton magnetization transferred to the phosphate by CP is predominantly from protons within the lipid, any saturation of the lipid protons by the train of selective rf pulses will result in a corresponding decrease in the ^{31}P signal in the MAS spectrum. As ^{31}P MAS

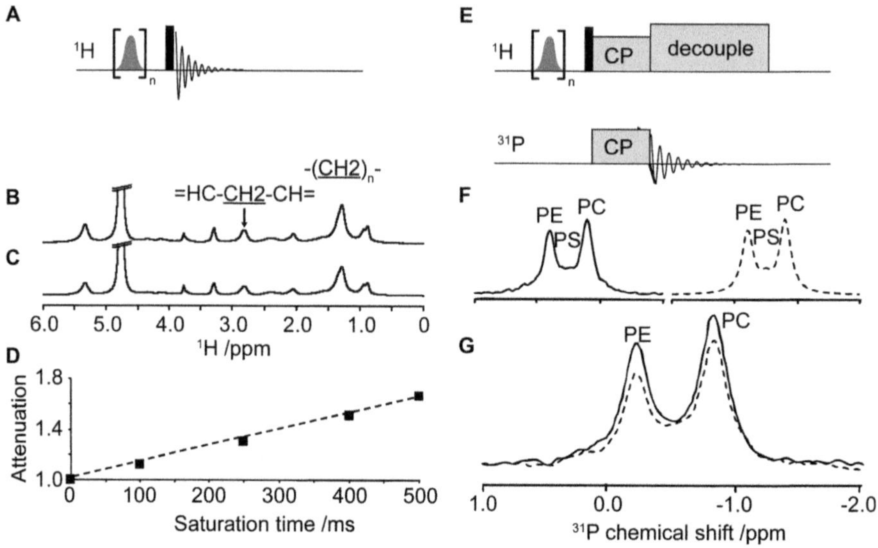

Figure 5.5 (A) Pulse sequence for ^{1}H saturation MAS NMR. Protein resonances saturated by a train of Gaussian pulses centred on the amide region of the ^{1}H spectrum. The proton spectrum of rod outer segments (B) without and (C) with magnetization saturation of the rhodopsin resonances with the attenuation of the signal from the CH_2 present in the unsaturated chains compared to those in the saturated chains, highlighting the preferred interaction between rhodopsin and polyunsaturated chains. (D) The degree of signal attenuation is dependent on the duration of saturation pulses. Headgroup specificity can be resolved through the transfer of the proton magnetization to the phosphate in the lipid headgroup following the saturation pulse sequence in (E). (F) MAS spectra of rod outer segment membranes resolve the three major lipid species PC, PS and PE, with corresponding simulation in *dashed line.* (G) Presaturation of the rhodopsin protons leads to a preferential attenuation of the PE signal, highlighting the preferred interaction between PE and rhodopsin.
(Adapted from Soubias *et al.*,[90] with permission from the American Society for Biochemistry and Molecular Biology.)

can resolve many phospholipids present in natural membranes, a STD-MAS-CP experiment can identify the preferential binding of particular classes of phospholipids.[90]

STD-MAS has been successfully used to study interactions between lipids and the G-protein coupled receptor, rhodopsin.[89,90] Reconstitution experiments had previously demonstrated that the structural transitions associated with the progression of rhodopsin through its photocycle and subsequent G-protein activation are critically dependent on the lipid composition of the lipid bilayer. In particular, the presence of the polyunsaturated fatty acid docosahexanoic acid (DHA), as well as phosphatidylethanolamine and phosphatidylserine headgroups, significantly enhance the lifetime of the Meta-II state, the photointermediate responsible for the activation of the associated G-protein transducin.[91–95]

STD-MAS experiments were applied to rod outer-segment discs, permitting an analysis of rhodopsin/lipid interactions in the native membrane. The ^{1}HSTD-MAS revealed that, in each case, transfer of magnetization from the rhodopsin was preferentially to the DHA, indicating that rhodopsin has a strong preference for this class of acyl chain (Figure 5.5B–D). Similarly, ^{31}P STD-MAS-CP experiments revealed a strong preference for phosphatidylethanolamine (Figure 5.5F, G). Interestingly, studies conducted as the rhodopsin proceeded through its photocycle showed that particular intermediates showed different affinities for the lipids, with the Meta-III state showing the greatest attenuation of the DHA signal.[90] STD-MAS studies were also performed on rhodopsin reconstituted into synthetic bilayers composed of binary mixtures of phosphatidylcholine with differing ratios of phosphatidylethanolamine or phosphatidylserine. Analysis of the signal attenuation in a range of lipid mixtures indicates that each of the lipids exhibits specific and distinct binding sites on the surface of the protein, contrasting with the perception that binding of lipids to the protein is rather non-specific in nature.[90,96]

Alternative strategies have also been proposed that permit the identification of the amino acid residues implicated in the binding of annular lipids to the surface of the protein. In a series of experiments analogous to those used to study ligand exchange rates, Kijac *et al.* exploited the relatively long T_2 of the lipids when compared to the integral membrane protein to generate a sample where magnetization is only present on the lipids.[97] Following a period where magnetization can "diffuse" from the lipids to protons on the surface of the protein, CP is subsequently used to observe those amino acids to which magnetization has been transferred. As the transfer between lipid and protein is dependent on the through-space dipolar couplings, the transfer of magnetization is a direct reporter on the proximity of lipid to protein. Using a combination of ^{1}H–^{13}C and ^{1}H–^{13}C–^{13}C correlation spectra, such studies have provided valuable insights into how lipids interact with the membrane scaffolding protein.[97]

In addition to the well-established presence of free bulk lipids and annular lipids, it has recently become apparent that a third population of lipid exists. These lipids are typically referred to as non-annular lipids.[2,85,98–100] Non-annular lipids were first described on the basis of their appearance in the crystal structures of several membrane proteins, which highlighted the relatively tight nature of the interaction between the lipid and protein surface. Subsequently, fluorescence-quenching studies have provided a valuable tool in characterizing both the affinity and the specificity of lipids for the non-annular binding sites on the surface of integral membrane proteins.[99,100]

The high affinity of non-annular lipids for their binding sites on integral membrane proteins indicates that, in contrast to annular lipids, the rate of exchange should be sufficiently slow to permit their detection by solid-state

NMR, assuming sufficient spectral resolution is achieved. Indeed, in a series of ^{31}P MAS studies performed on the potassium channel KcsA from *Streptomyces lividians*, whose activity is known to depend on the binding of anionic lipids to the non-annular lipid binding sites, such non-annular lipids have been resolved.[100] These studies revealed that, upon reconstitution into lipid vesicles composed of phosphatidylcholine and phosphatidylglycerol, a third resonance appeared in the ^{31}P spectrum whose intensity would be consistent with a population of phosphatidylglycerol molecules bound to the non-annular site and whose chemical shift had been perturbed through the change in electrostatic environment upon binding to the protein (Figure 5.6A).[87] On the basis of molecular dynamics studies, positively

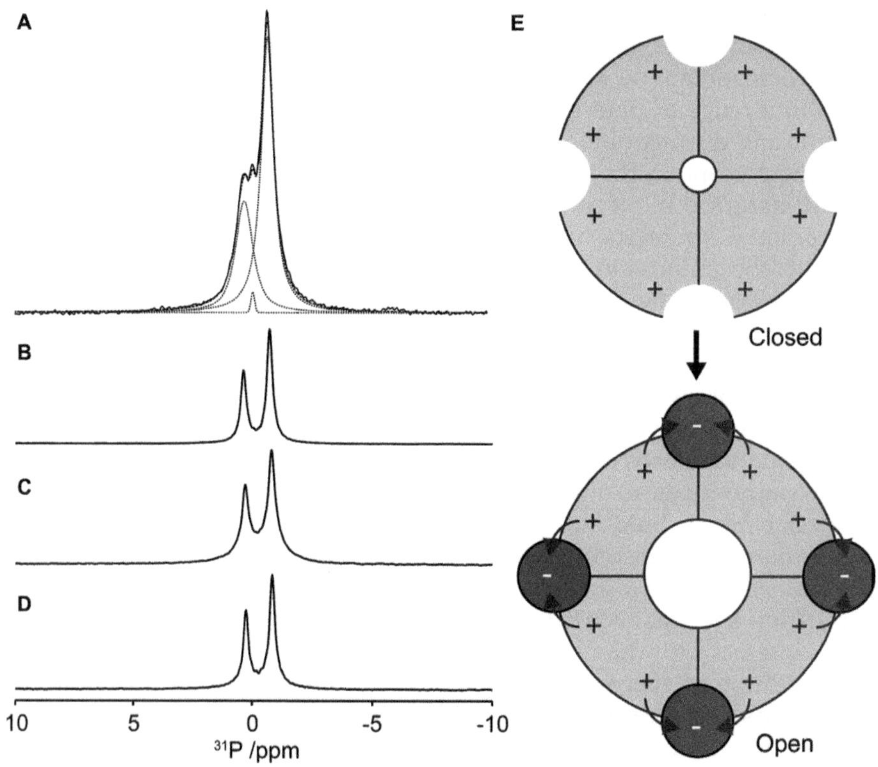

Figure 5.6 MAS ^{31}P NMR studies of the potassium channel KcsA reconstituted into vesicles composed of POPC and POPG (2 : 1). (A) The spectrum of the wild-type protein contains an additional resonance located between the intense signals of the bulk lipids (*dashed line* shows spectral deconvolution). Mutation of the arginines (B) R64, (C) R89 and (D) R64/R89 to leucines in the anionic lipid binding site results in the abolition of this resonance. In conjunction with electrophysiology data, these studies suggest that electrostatic interactions between the arginines and the negatively charged lipid headgroup help stabilize the open conformation (E).
(Adapted from Marius *et al.*,[87] with permission from Elsevier.)

charged arginine residues in the non-annular binding site were mutated that resulted in the abolition of this resonance, suggesting that the binding of the anionic lipids to the non-annular binding site is, at least in part, mediated through electrostatic interactions between the non-annular binding site and the negatively charged lipid headgroup (Figure 5.6B–D).[87] Interestingly, functional electrophysiology recordings further highlighted the importance of these residues, with the mutation of the arginines to leucines and the abolition of the electrostatic interactions resulting in the absence of a discrete conducting state.[87] These studies suggest that the formation of electrostatic interactions between R64–phosphatidylglycerol–R89 serve to stabilize the open form of the channel (Figure 5.6E).[87]

The importance of such interactions was further highlighted in an elegant study by Weingarth *et al.*[101] Molecular dynamics studies of the wild-type KcsA and a KcsA–Kv1.3 chimera, in which 11 residues between TM helices 1 and 2 of KcsA are replaced with corresponding residues from Kv1.3, revealed differences in the affinity of negatively charged lipids for the non-annular binding sites. Significantly, KcsA shows a greater preference for negatively charged lipids that reside in the non-annular binding site for significantly longer periods of time. Interestingly, in the chimera KcsA–Kv1.3, where R64 is replaced by a D64, a tight inter-subunit salt bridge is formed, an interaction confirmed by rotational-resonance experiments[78,79] performed on uniformly labelled KcsA–Kv1.3.[101] The formation of this salt bridge appears to narrow the non-annular binding site whilst the negative charge of D64 appears to repel the anionic lipids, both of which serve to limit binding of lipids to this site.

Interestingly, however, broadband ^{13}C–^{13}C correlation spectra[102,103] acquired on wild-type KcsA revealed a number of cross peaks consistent with contacts between the protein and lipid tightly bound in the non-annular binding site.[101] Their presence in the spectra indicates that their residence time in the non-annular binding site is in the order of μs or greater.[101] Together, these studies consistently highlight the importance of electrostatic interactions in the binding of anionic lipids to the non-annular binding site and their role in stabilizing the open form of the channel.

These studies serve to emphasize that, despite the rather transient nature of the interactions between lipid and protein and the lipid/protein interface, a range of methods are now available which can provide molecular insights into the nature of these interactions, data which complement that which is available through ESR and fluorescence studies.

5.3.3 Protein/Protein Interactions

Interactions between integral membrane proteins have been implicated in a number of cellular processes, including receptor signalling, regulation of enzymatic activity and protein processing and trafficking. In many cases, interactions between integral membrane proteins is mediated through interactions between their transmembrane domains.[23–25] Owing to their importance, these interactions have been the subject of intense investigations to identify interaction sites within the transmembrane domain.[104] Recently, it has become clear that the interactions between the transmembrane domains of integral membrane proteins are highly dependent on the surrounding lipid environment and in many cases they are readily disrupted upon extraction from the lipid bilayer.[105] The potential of solid-state NMR to study these interactions in complex bilayer mixtures and provide insights into how the bilayer modulates the structural and dynamic properties of the membrane protein has made it a valuable tool, providing insights into helix packing[106–113] and its role in signal transduction.[112]

One system where solid-state NMR studies of protein/protein interactions within the bilayer have made a significant contribution towards our understanding of the underlying processes within the cell is the processing of amyloid precursor protein (APP), the precursor protein that contributes to the formation of amyloid-β peptide (Aβ), a species implicated in the onset of Alzheimer's disease. The 770 residue APP protein is a Type-I integral membrane protein involved in a number of processes, including cell adhesion, cell signalling and development (Figure 5.7A). Within the cell the ectodomain is shed *via* the cleavage of the APP by either the α- or β-secretases, resulting in the formation of C83 and C99, both C-terminal membrane embedded peptides. Both of these peptides are substrates for the γ-secretase.[114] Cleavage of C99 results in the formation of the Aβ(1–40/42) peptide implicated in Alzheimer's disease, whilst cleavage of C83 results in the formation of the truncated species p3.[114] Interest in the role of the lipid bilayer in the processing of APP stems in part from the location of the cleavage sites of the α- and γ-secretases, which are located close to and within the lipid bilayer, respectively. Furthermore, a number of studies have revealed that patients suffering from Alzheimer's disease exhibit significant variations in their brain phospholipid compositions compared to a "normal" patient.[115] This, together with the elevated levels of oxidized lipids present in Alzheimer disease patients, highlighted that these changes may alter either the activity of the individual secretases or their accessibility to their substrate and cleavage sites.[115]

Early solid-state NMR studies by Marenchino *et al.* focused on identifying how the bilayer thickness altered the oligomeric state of the APP within the bilayer and the accessibility of the α-cleavage site located close to the bilayer surface.[116] A peptide corresponding to APP(684–726) was synthesized with the site-specific isotope labels at d_3-Ala713 and ^{15}N-Ala692. Following the reconstitution of APP(684–726) into lipid bilayers of differing lipid composition, MAS deuterium NMR studies were conducted to study the degree of motional averaging of the quadrupolar interaction. Below the liquid-crystalline to gel phase transition temperature of the lipids the sideband family was consistent with an immobile peptide within a lipid bilayer. However, above the phase transition temperature a significant reduction in the quadrupolar interaction was observed in bilayers composed of 1,2-dimyristoylphosphatidylcholine, 1,2-distearoylphosphatidylcholine and sphingomyelin, consistent with a helical peptide undergoing rapid rotation

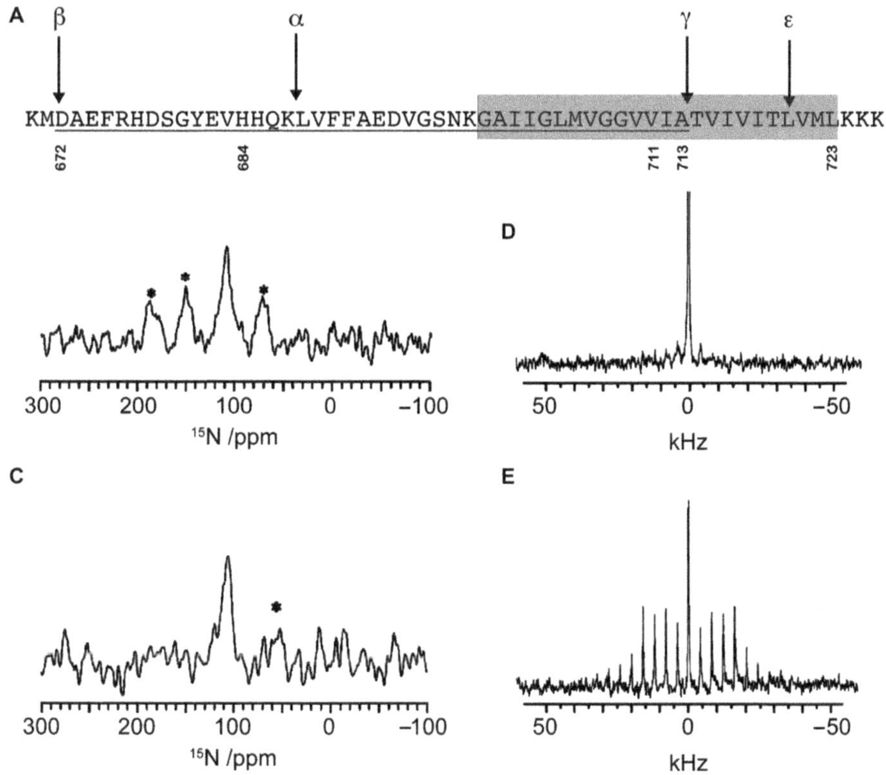

Figure 5.7 (A) Sequence of amyloid precursor protein corresponding to the transmembrane domain (*shaded area*) and flanking regions whose cleavage results in the formation of amyloid-β peptide (*underlined sequence*). Cleavage at the α-, β-, γ- and ε-cleavage sites are highlighted by the corresponding α, β, γ and ε. (B, C) MAS ^{15}N CP spectra of APP labelled at Ala692 close to the α-cleavage site. The increase in bilayer thickness brought about by the transition from the liquid-crystalline (C) to gel phase (B) results in a reduction in mobility, as evidenced by the increased sideband intensity (marked with a *), suggesting the cleavage site can be occluded for thicker bilayers. (D) Deuterium MAS spectra of APP labelled at Ala713 show high rotational mobility in saturated bilayers of DMPC, which is significantly reduced in unsaturated bilayers composed of egg PC (E), suggesting that fatty acid composition can alter the oligomeric state and thus cleavage site accessibility of protease cleavage sites.
(Adapted from Marenchino *et al.*,[116] with permission from Cell Press.)

within the bilayer. In contrast, the spectra of APP(684–726) reconstituted into unsaturated lipid bilayers composed of egg phosphatidylcholine and 1-palmitoyl-2-oleoylphosphatidylcholine still showed significant quadrupolar interaction (Figure 5.7D, E). The reduced rotational correlation time observed is indicative of a larger complex forming within the bilayer,[116,117] potentially mediated by the multiple GXXXG/A dimerization motifs present in APP, suggesting that the degree of unsaturation in the fatty acyl chains within the bilayer may play a significant role in determining the oligomeric state of the proteins.

To assess the effect of bilayer composition upon the accessibility of the α-cleavage site, the dynamics of Ala692 were determined through an analysis of the motional averaging of the chemical shielding anisotropy.[116] Slow-spinning ^{15}NCP-MAS revealed that only in the case of 1,2-distearoylphosphatidylcholine in the gel phase were significant sidebands observed, indicating in the other cases the α-cleavage site was sufficiently remote from the bilayer surface to exhibit enough motion to dynamically average the chemical shielding anisotropy (Figure 5.7B, C). In conjunction with enzymatic assays, these findings suggest that only when the bilayer exceeds a thickness of 3.0 nm does the α-cleavage site become masked from the α-secretase.[116]

Subsequent studies by Sato *et al.*[118] focused on the γ- and ε-cleavage sites that are located in the transmembrane domain. Using a family of synthetic peptides corresponding to APP(618–660) (APP695 numbering), ^{13}C labels were introduced at glycine residues located in the GxxxG putative dimerization motifs within the membrane-spanning region and in regions corresponding to the γ- and ε-cleavage sites. Analysis of chemical shifts, a property readily related to protein secondary structure, obtained from peptides labelled at amino acids at the γ-cleavage site and the glycines within the GxxxG motifs, indicated the presence of a helical structure similar to that proposed by Marenchino *et al.*[116] and complementary liquid-state NMR studies by Beel *et al.*[119,120] of APP in detergent micelles. To identify dimer interfaces, AAP(618–660) peptide was synthesized with glycines labelled at positions 625, 629 and 633. In each case, two peptides were synthesized, one labelled at the Cα and the other at the CO. The resulting peptides were then reconstituted into the lipid bilayer at a 1:1 ratio. Two-dimensional ^{13}C–^{13}C correlation spectra were recorded using DARR[67,121] to recouple homonuclear dipolar couplings under MAS. The presence of cross-peaks in these 2D spectra is a clear indication of the close proximity of the two glycines and the formation of a stable homodimer. These studies revealed significant cross-peaks in the case of Gly625, Gly629 and Gly633, indicating that each of these three residues is present at the dimer interface, as predicted on the basis of their inclusion in the GxxxG motif.[118] Sato *et al.* also investigated the secondary structure at the ε-cleavage site at the membrane interface at the C-terminal end of the peptide, again through an analysis of the chemical shifts observed.[118] In contrast to the γ-cleavage site, these residues exhibited chemical shifts characteristic of an unfolded state, suggesting that this site may well be more accessible to cleavage than a more structured region of the peptide. When this region was replaced by a tract of three leucines, a stable helical structure was observed; interestingly, this led to a reduction in the amount of Aβ produced and increased the amount of C83 (α-CTF), a non-Aβ forming fragment, produced.

Studies have also been performed by Lu *et al.*[122] to investigate the structure of APP both at the membrane boundaries and close to the γ-cleavage site in more complex membrane environments. Synthetic peptides containing uniformly $^{13}C/^{15}N$-labelled amino acids at Ala701 and Val721 (the membrane interface) and Gly709, Val711, Ile712, Ala713 and Thr714 (the γ-cleavage site) were incorporated into lipid vesicles composed of a range of lipid environments, including 1-palmitoyl-2-oleoylphosphatidylcholine, 1,2-dioleoylphosphatidylglycerol, brain lipid extract and a neuronal lipid mixture. Two-dimensional ^{13}C–^{13}C correlation spectra were recorded using DARR[67,121] to recouple homonuclear dipolar couplings under MAS. The resulting 2D spectra permitted the assignment of the labelled sites within APP and the measurement of their chemical shift, which provided insight into the proteins' secondary structure.[123,124]

Analysis of these chemical shifts revealed that, at lower temperatures (−68 °C), APP forms a stable helix in both simple single-component bilayers and in complex mixtures that are derived from or mimic neuronal membranes. However, at more physiological temperatures (24 °C) it appears that all labelled sites exhibited structural heterogeneity, with evidence of both helical and non-helical conformations at residues close to the γ-cleavage site and the C-terminal membrane interface in all membranes except 1,2-dioleoylphosphatidylglycerol.[122] The presence of two distinct sets of resonances in the spectra suggests that any exchange between these two populations must be slow (>ms).[122]

From this last study, under conditions that most closely mimic those found in neuronal membranes, APP clearly exists in multiple conformations. The presence of non-helical conformations clearly has a bearing on the processing of the APP. A number mutagenesis studies have shown that the oligomeric state can play an important role in determining the fate of the APP as it is processed, with the potential to either increase or decrease the Aβ(1–40/42) load. Similarly, the absence of secondary structure is thought to be a prerequisite to efficient cleavage by any of the secretases; thus any change in lipid conformation which serves to disrupt any secondary structure at the cleavage site again may well favour cleavage by particular families of proteases.

What these studies do clearly show, however, is the necessity to study integral membrane proteins in an environment as close to that found *in vivo* as is possible. As these examples clearly highlight, the lipid bilayer is not a simple homogenous structure composed of a limited number of lipids. Rather, it contains a multitude of different lipid species that possess a complex phase behaviour that potentially leads to the formation of domains rich in particular lipid types, which potentially affects both the lateral segregation of proteins within bilayers and their intrinsic structures, properties that can significantly affect their function and processing.

5.4 Conclusion

We hope the above examples have provided an overview of the types of information that solid-state NMR can contribute to our understanding of molecular interactions within cell membranes. Recent advances in solid-state NMR methodology are now providing us with unrivalled opportunities to study integral membrane proteins, the interactions they experience and how they modulate their function, processing and localization within the cell. Key to these successes is the ability of solid-state NMR to study integral membrane proteins under near-native conditions within the lipid bilayer. This enables us to study not only the interactions between the integral membrane protein and the surrounding lipids, but also to ascertain how the lipid bilayer modulates the protein structure and dynamics and the effect this has on the structure and function of the individual component. It is clear from the studies highlighted above that if we are to fully understand how integral membrane proteins function *in vivo* it is essential that we begin to understand how the complex environment within the cell impacts on membrane proteins. Solid-state NMR methodologies possess inherent drawbacks, with sensitivity being a major consideration. Recently, however, there have been a number of advances that may serve to ameliorate these issues, including the development of technologies such as dynamic nuclear polarization with the potential to bring orders of magnitude enhancements in signal-to-noise ratios. Already such approaches are proving enabling, facilitating the analysis of the structure of more complex molecules bound to integral membrane protein complexes, where previously sensitivity considerations would have precluded a study[125] or significantly compromised the quality of the data.[126] Although the application of these technologies is recent and some of the methods are still undergoing development, together with advances in our preparation of the systems and the development of novel pulse sequences for the characterization of the structure and dynamics of these assemblies, it would appear that solid-state NMR can make a valuable contribution to our understanding of the complex membrane assemblies.

Acknowledgements

The authors would like to acknowledge financial support from the Wellcome Trust (Equipment Grant 090658/Z/09/Z and a Career Development Fellowship) and the University of Southampton (PTFW, JJ) during the preparation of this manuscript.

References

1. J. Carney, J. M. East, S. Mall, P. Marius, A. M. Powl, J. N. Wright and A. G. Lee, *Curr. Protoc. Protein Sci.*, 2006, ch. 19, unit 19.12.
2. A. G. Lee, *Biochem. Soc. Trans.*, 2011, **39**, 761.
3. D. Marsh, *Eur. Biophys. J.*, 2010, **39**, 513.
4. D. Marsh, *Methods*, 2008, **46**, 83.
5. T. Gonen and G. Waksman, *Curr. Opin. Struct. Biol.*, 2012, **22**, 467.
6. A. Gautier and D. Nietlispach, *Methods Mol. Biol.*, 2012, **914**, 25.
7. D. Nietlispach and A. Gautier, *Curr. Opin. Struct. Biol.*, 2011, **21**, 497.

8. A. Gautier, H. R. Mott, M. J. Bostock, J. P. Kirkpatrick and D. Nietlispach, *Nat. Struct. Mol. Biol.*, 2010, **17**, 768.
9. A. Gautier, J. P. Kirkpatrick and D. Nietlispach, *Angew. Chem. Int. Ed.*, 2008, **47**, 7297.
10. J. L. Popot, *Annu. Rev. Biochem.*, 2010, **79**, 737.
11. I. Marcotte and M. Auger, *Concepts Magn. Reson., Part A*, 2005, **24A**, 17.
12. E. R. Andrew, A. Bradbury and R. G. Eades, *Nature*, 1958, **182**, 1659.
13. A. Watts, I. J. Burnett, C. Glaubitz, G. Grobner, D. A. Middleton, P. J. R. Spooner and P. T. F. Williamson, *Eur. Biophys. J. Biophys. Lett.*, 1998, **28**, 84.
14. R. G. Griffin, *Nat. Struct. Biol.*, 1998, **5**, 508.
15. M. Baldus, *Angew. Chem. Int. Ed.*, 2006, **45**, 1186.
16. M. Baldus, D. G. Geurts and B. H. Meier, *Solid State Nucl. Magn. Reson.*, 1998, **11**, 157.
17. D. E. Warschawski, A. A. Arnold, M. Beaugrand, A. Gravel, E. Chartrand and I. Marcotte, *Biochim. Biophys. Acta*, 2011, **1808**, 1957.
18. G. Grobner, A. Taylor, P. T. F. Williamson, G. Choi, C. Glaubitz, J. A. Watts, W. J. deGrip and A. Watts, *Anal. Biochem.*, 1997, **254**, 132.
19. A. G. Lee, *Biochim. Biophys. Acta*, 2003, **1612**, 1.
20. P. Marius, Y. M. Leung, T. J. Piggot, S. Khalid and P. T. Williamson, *Eur. Biophys. J.*, 2012, **41**, 199.
21. P. Marius, J. N. Wright, I. S. Findlow and P. T. Williamson, *Protein Expression Purif.*, 2010, **72**, 107.
22. D. A. Holdbrook, Y. M. Leung, T. J. Piggot, P. Marius, P. T. Williamson and S. Khalid, *Biochemistry*, 2010, **49**, 10796.
23. E. E. Matthews, M. Zoonens and D. M. Engelman, *Cell*, 2006, **127**, 447.
24. D. Schneider and D. M. Engelman, *J. Mol. Biol.*, 2004, **343**, 799.
25. K. Talbert-Slagle, S. Marlatt, F. N. Barrera, E. Khurana, J. Oates, M. Gerstein, D. M. Engelman, A. M. Dixon and D. Dimaio, *J. Virol.*, 2009, **83**, 9773.
26. P. T. F. Williamson, B. H. Meier and A. Watts, *Eur. Biophys. J. Biophys. Lett.*, 2004, **33**, 247.
27. A. Watts, *Nat. Rev. Drug Discovery*, 2005, **4**, 555.
28. P. J. Judge and A. Watts, *Curr. Opin. Chem. Biol.*, 2011, **15**, 690.
29. S. Tapaneeyakorn, A. D. Goddard, J. Oates, C. L. Willis and A. Watts, *Biochim. Biophys. Acta*, 2011, **1808**, 1462.
30. G. C. K. Roberts, *Drug Discovery Today*, 2000, **5**, 230.
31. J. W. Peng, J. Moore and N. Abdul-Manan, *Prog. Nucl. Magn. Reson. Spectrosc.*, 2004, **44**, 225.
32. C. B. Post, *Curr. Opin. Struct. Biol.*, 2003, **13**, 581.
33. A. S. Ulrich, I. Wallat, M. P. Heyn and A. Watts, *Nat. Struct. Biol.*, 1995, **2**, 190.
34. P. T. Williamson, J. A. Watts, G. H. Addona, K. W. Miller and A. Watts, *Proc. Natl. Acad. Sci. U. S. A.*, 2001, **98**, 2346.
35. G. Grobner, I. J. Burnett, C. Glaubitz, G. Choi, A. J. Mason and A. Watts, *Nature*, 2000, **405**, 810.

36. A. M. Taylor, G. Grobner, P. T. Williamson and A. Watts, *Biochemistry*, 1999, **38**, 11172.
37. G. Grobner, G. Choi, I. J. Burnett, C. Glaubitz, P. J. Verdegem, J. Lugtenburg and A. Watts, *FEBS Lett.*, 1998, **422**, 201.
38. A. S. Ulrich, A. Watts, I. Wallat and M. P. Heyn, *Biochemistry*, 1994, **33**, 5370.
39. W. C. Lai, N. McLean, A. Gansmuller, M. A. Verhoeven, G. C. Antonioli, M. Carravetta, L. Duma, P. H. M. Bovee-Geurts, O. G. Johannessen, H. J. M. de Groot, J. Lugtenburg, L. Emsley, S. P. Brown, R. C. D. Brown, W. J. DeGrip and M. H. Levitt, *J. Am. Chem. Soc.*, 2006, **128**, 3878.
40. P. J. R. Spooner, J. M. Sharples, S. C. Goodall, P. H. M. Bovee-Geurts, M. A. Verhoeven, J. Lugtenburg, A. M. A. Pistorius, W. J. DeGrip and A. Watts, *J. Mol. Biol.*, 2005, **345**, 1295.
41. A. F. L. Creemers, P. H. M. Bovee-Geurts, W. J. DeGrip, J. Lugtenburg and H. J. M. de Groot, *Biochemistry*, 2004, **43**, 16011.
42. A. F. L. Creemers, S. Kiihne, P. H. M. Bovee-Geurts, W. J. DeGrip, J. Lugtenburg and H. J. M. de Groot, *Proc. Natl. Acad. Sci. U. S. A.*, 2002, **99**, 9101.
43. X. Feng, P. J. E. Verdegem, Y. K. Lee, D. Sandstrom, M. Eden, P. BoveeGeurts, W. J. deGrip, J. Lugtenburg, H. J. M. deGroot and M. H. Levitt, *J. Am. Chem. Soc.*, 1997, **119**, 6853.
44. J. C. Lansing, M. Hohwy, C. P. Jaroniec, A. F. Creemers, J. Lugtenburg, J. Herzfeld and R. G. Griffin, *Biochemistry*, 2002, **41**, 431.
45. A. E. McDermott, F. Creuzet, R. Gebhard, K. van der Hoef, M. H. Levitt, J. Herzfeld, J. Lugtenburg and R. G. Griffin, *Biochemistry*, 1994, **33**, 6129.
46. M. Han, M. Groesbeek, T. P. Sakmar and S. O. Smith, *Proc. Natl. Acad. Sci. U. S. A.*, 1997, **94**, 13442.
47. M. Han and S. O. Smith, *Biochemistry*, 1995, **34**, 1425.
48. S. O. Smith, H. de Groot, R. Gebhard and J. Lugtenburg, *Photochem. Photobiol.*, 1992, **56**, 1035.
49. S. O. Smith, J. Courtin, H. de Groot, R. Gebhard and J. Lugtenburg, *Biochemistry*, 1991, **30**, 7409.
50. S. O. Smith, I. Palings, M. E. Miley, J. Courtin, H. de Groot, J. Lugtenburg, R. A. Mathies and R. G. Griffin, *Biochemistry*, 1990, **29**, 8158.
51. P. J. Spooner, N. G. Rutherford, A. Watts and P. J. Henderson, *Proc. Natl. Acad. Sci. U. S. A.*, 1994, **91**, 3877.
52. P. T. F. Williamson, G. Grobner, P. J. R. Spooner, K. W. Miller and A. Watts, *Biochemistry*, 1998, **37**, 10854.
53. S. G. Patching, A. R. Brough, R. B. Herbert, J. A. Rajakarier, P. J. Henderson and D. A. Middleton, *J. Am. Chem. Soc.*, 2004, **126**, 3072.
54. L. Krabben, B. J. van Rossum, S. Jehle, E. Bocharov, E. N. Lyukmanova, A. A. Schulga, A. Arseniev, F. Hucho and H. Oschkinat, *J. Mol. Biol.*, 2009, **390**, 662.

55. L. Krabben, B. J. van Rossum, F. Castellani, E. Bocharov, A. A. Schulga, A. S. Arseniev, C. Weise, F. Hucho and H. Oschkinat, *FEBS Lett.*, 2004, **564**, 319.
56. P. T. F. Williamson, J. A. Watts, G. H. Addona, K. W. Miller and A. Watts, *Proc. Natl. Acad. Sci. U. S. A.*, 2001, **98**, 2346.
57. N. Unwin, *J. Mol. Biol.*, 2005, **346**, 967.
58. N. Unwin, *FEBS Lett.*, 2003, **555**, 91.
59. P. T. Williamson, G. Grobner, P. J. Spooner, K. W. Miller and A. Watts, *Biochemistry*, 1998, **37**, 10854.
60. D. A. Middleton, R. Robins, X. Feng, M. H. Levitt, I. D. Spiers, C. H. Schwalbe, D. G. Reid and A. Watts, *FEBS Lett.*, 1997, **410**, 269.
61. C. Glaubitz, A. Groger, K. Gottschalk, P. Spooner, A. Watts, S. Schuldiner and H. Kessler, *FEBS Lett.*, 2000, **480**, 127.
62. D. A. Middleton, S. Rankin, M. Esmann and A. Watts, *Proc. Natl. Acad. Sci. U. S. A.*, 2000, **97**, 13602.
63. J. A. Watts, A. Watts and D. A. Middleton, *J. Biol. Chem.*, 2001, **276**, 43197.
64. P. J. R. Spooner, W. J. O'Reilly, S. W. Homans, N. G. Rutherford, P. J. F. Henderson and A. Watts, *Biophys. J.*, 1998, **75**, 2794.
65. A. Grommek, B. H. Meier and M. Ernst, *Chem. Phys. Lett.*, 2006, **427**, 404.
66. K. Takegoshi, S. Nakamura and T. Terao, *J. Chem. Phys.*, 2003, **118**, 2325.
67. K. Takegoshi, S. Nakamura and T. Terao, *Chem. Phys. Lett.*, 2001, **344**, 631.
68. M. Hohwy, C. M. Rienstra, C. P. Jaroniec and R. G. Griffin, *J. Chem. Phys.*, 1999, **110**, 7983.
69. M. Hohwy, H. J. Jakobsen, M. Eden, M. H. Levitt and N. C. Nielsen, *J. Chem. Phys.*, 1998, **108**, 2686.
70. R. Verel, M. Baldus, M. Ernst and B. H. Meier, *Chem. Phys. Lett.*, 1998, **287**, 421.
71. T. Westfeld, R. Verel, M. Ernst, A. Bockmann and B. H. Meier, *J. Biomol. NMR*, 2012, **53**, 103.
72. S. Luca, J. F. White, A. K. Sohal, D. V. Filippov, J. H. van Boom, R. Grisshammer and M. Baldus, *Proc. Natl. Acad. Sci. U. S. A.*, 2003, **100**, 10706.
73. J. J. Lopez, A. K. Shukla, C. Reinhart, H. Schwalbe, H. Michel and C. Glaubitz, *Angew. Chem. Int. Ed.*, 2008, **47**, 1668.
74. P. T. F. Williamson, A. Verhoeven, K. W. Miller, B. H. Meier and A. Watts, *Proc. Natl. Acad. Sci. U. S. A.*, 2007, **104**, 18031.
75. A. L. Cashin, E. J. Petersson, H. A. Lester and D. A. Dougherty, *J. Am. Chem. Soc.*, 2005, **127**, 350.
76. D. A. Middleton, *Biochem. Soc. Trans.*, 2007, **35**, 985.
77. A. Watts, *Curr. Opin. Biotechnol.*, 1999, **10**, 48.
78. M. H. Levitt, D. P. Raleigh, F. Creuzet and R. G. Griffin, *J. Chem. Phys.*, 1990, **92**, 6347.

79. D. P. Raleigh, M. H. Levitt and R. G. Griffin, *Chem. Phys. Lett.*, 1988, **146**, 71.
80. T. Gullion and J. Schaefer, *J. Magn. Reson.*, 1989, **81**, 196.
81. P. T. F. Williamson, A. Verhoeven, M. Ernst and B. H. Meier, *J. Am. Chem. Soc.*, 2003, **125**, 2718.
82. A. Verhoeven, P. T. F. Williamson, H. Zimmermann, M. Ernst and B. H. Meier, *J. Magn. Reson.*, 2004, **168**, 314.
83. C. P. Jaroniec, B. A. Tounge, J. Herzfeld and R. G. Griffin, *J. Am. Chem. Soc.*, 2001, **123**, 3507.
84. S. B. Hansen, G. Sulzenbacher, T. Huxford, P. Marchot, Y. Bourne and P. Taylor, *J. Mol. Neurosci.*, 2006, **30**, 101.
85. A. G. Lee, *J. Biol.*, 2009, **8**, 86.
86. D. Marsh, D. Kurad and V. A. Livshits, *Chem. Phys. Lipids*, 2002, **116**, 93.
87. P. Marius, M. R. de Planque and P. T. Williamson, *Biochim. Biophys. Acta*, 2012, **1818**, 90.
88. P. L. Yeagle and D. Kelsey, *Biochemistry*, 1989, **28**, 2210.
89. O. Soubias and K. Gawrisch, *J. Am. Chem. Soc.*, 2005, **127**, 13110.
90. O. Soubias, W. E. Teague and K. Gawrisch, *J. Biol. Chem.*, 2006, **281**, 33233.
91. T. S. Wiedmann, R. D. Pates, J. M. Beach, A. Salmon and M. F. Brown, *Biochemistry*, 1988, **27**, 6469.
92. Y. Wang, A. V. Botelho, G. V. Martinez and M. F. Brown, *J. Am. Chem. Soc.*, 2002, **124**, 7690.
93. N. J. Gibson and M. F. Brown, *Biochemistry*, 1993, **32**, 2438.
94. A. V. Botelho, N. J. Gibson, R. L. Thurmond, Y. Wang and M. F. Brown, *Biochemistry*, 2002, **41**, 6354.
95. I. D. Alves, G. F. Salgado, Z. Salamon, M. F. Brown, G. Tollin and V. J. Hruby, *Biophys. J.*, 2005, **88**, 198.
96. K. Gawrisch, O. Soubias and M. Mihailescu, *Prostaglandins, Leukotrienes Essent. Fatty Acids*, 2008, **79**, 131.
97. A. Kijac, A. Y. Shih, A. J. Nieuwkoop, K. Schulten, S. G. Sligar and C. M. Rienstra, *Biochemistry*, 2010, **49**, 9190.
98. A. G. Lee, *Trends Biochem. Sci.*, 2011, **36**, 493.
99. A. M. Powl, J. M. East and A. G. Lee, *Biochemistry*, 2008, **47**, 4317.
100. P. Marius, M. Zagnoni, M. E. Sandison, J. M. East, H. Morgan and A. G. Lee, *Biophys. J.*, 2008, **94**, 1689.
101. M. Weingarth, A. Prokofyev, E. A. van der Cruijsen, D. Nand, A. M. Bonvin, O. Pongs and M. Baldus, *J. Am. Chem. Soc.*, 2013, **135**, 3983.
102. M. Weingarth, P. Tekely, R. Bruschweiler and G. Bodenhausen, *Chem. Commun.*, 2010, **46**, 952.
103. M. Weingarth, G. Bodenhausen and P. Tekely, *J. Am. Chem. Soc.*, 2009, **131**, 13937.
104. S. O. Smith and B. J. Bormann, *Proc. Natl. Acad. Sci. U. S. A.*, 1995, **92**, 488.

105. P. Marius, Y. M. Leung, T. J. Piggot, S. Khalid and P. T. F. Williamson, *Eur. Biophys. J. Biophys. Lett.*, 2012, **41**, 199.
106. L. K. Thompson, *Curr. Opin. Struct. Biol.*, 2002, **12**, 661.
107. B. Isaac, G. J. Gallagher, Y. S. Balazs and L. K. Thompson, *Biochemistry*, 2002, **41**, 3025.
108. O. J. Murphy, 3rd, F. A. Kovacs, E. L. Sicard and L. K. Thompson, *Biochemistry*, 2001, **40**, 1358.
109. W. Liu, J. Z. Fei, T. Kawakami and S. O. Smith, *Biochim. Biophys. Acta*, 2007, **1768**, 2971.
110. W. Liu, E. Crocker, S. N. Constantinescu and S. O. Smith, *Biophys. J.*, 2005, **89**, 1194.
111. K. P. Howard, W. Liu, E. Crocker, V. Nanda, J. Lear, W. F. Degrado and S. O. Smith, *Protein Sci.*, 2005, **14**, 1019.
112. S. O. Smith, C. Smith, S. Shekar, O. Peersen, M. Ziliox and S. Aimoto, *Biochemistry*, 2002, **41**, 9321.
113. S. O. Smith, T. Kawakami, W. Liu, M. Ziliox and S. Aimoto, *J. Mol. Biol.*, 2001, **313**, 1139.
114. H. Li, M. S. Wolfe and D. J. Selkoe, *Structure*, 2009, **17**, 326.
115. J. W. Pettegrew, K. Panchalingam, R. L. Hamilton and R. J. McClure, *Neurochem. Res.*, 2001, **26**, 771.
116. M. Marenchino, P. T. Williamson, S. Murri, G. Zandomeneghi, H. Wunderli-Allenspach, B. H. Meier and S. D. Kramer, *Biophys. J.*, 2008, **95**, 1460.
117. P. T. Williamson, G. Zandomeneghi, F. J. Barrantes, A. Watts and B. H. Meier, *Mol. Membr. Biol.*, 2005, **22**, 485.
118. T. Sato, T. C. Tang, G. Reubins, J. Z. Fei, T. Fujimoto, P. Kienlen-Campard, S. N. Constantinescu, J. N. Octave, S. Aimoto and S. O. Smith, *Proc. Natl. Acad. Sci. U. S. A.*, 2009, **106**, 1421.
119. A. J. Beel, C. K. Mobley, H. J. Kim, F. Tian, A. Hadziselimovic, B. Jap, J. H. Prestegard and C. R. Sanders, *Biochemistry*, 2008, **47**, 9428.
120. A. J. Beel and C. R. Sanders, *Cell. Mol. Life Sci.*, 2008, **65**, 1311.
121. C. R. Morcombe, V. Gaponenko, R. A. Byrd and K. W. Zilm, *J. Am. Chem. Soc.*, 2004, **126**, 7196.
122. J. X. Lu, W. M. Yau and R. Tycko, *Biophys. J.*, 2011, **100**, 711.
123. D. S. Wishart, B. D. Sykes and F. M. Richards, *Biochemistry*, 1992, **31**, 1647.
124. D. S. Wishart and B. D. Sykes, *J. Biomol. NMR*, 1994, **4**, 171.
125. L. Reggie, J. J. Lopez, I. Collinson, C. Glaubitz and M. Lorch, *J. Am. Chem. Soc.*, 2011, **133**, 19084.
126. W. T. Franks, A. H. Linden, B. Kunert, B. J. van Rossum and H. Oschkinat, *Eur. J. Cell Biol.*, 2012, **91**, 340.

CHAPTER 6

Magnetic Liposomes and Bicelles: New Tools for Membrane-Peptide Structural Studies

ERICK J. DUFOURC,*[a] NICOLE HARMOUCHE,[a] CÉCILE LOUDET-COURRÈGES,[a] REIKO ODA,[a] ANNA DILLER,[c] BENOIT ODAERT,[a] AXELLE GRÉLARD[a] AND SÉBASTIEN BUCHOUX[b]

[a] Institute of Chemistry & Biology of Membranes & Nanoobjects, UMR 5248, Université Bordeaux, CNRS, Institut Polytechnique Bordeaux, Allée Geoffroy Saint Hilaire, 33600 Pessac, France; [b] Enzyme & Cell Engineering (GEC), UMR6022, Université Picardie Jules Verne, CNRS, Amiens, France; [c] Institute of Experimental Genetics/Institute of Structural Biology, Helmholtz Zentrum, München, Germany
*Email: e.dufourc@cbmn.u-bordeaux.fr

6.1 Introduction

The structure at atomic resolution and the topology (orientation) of peptides or proteins in membranes are essential for understanding their function. Although X-rays give the atomic resolution structure in the crystalline state, so far NMR is the only technique that may lead to the

New Developments in NMR No. 3
Advances in Biological Solid-State NMR: Proteins and Membrane-Active Peptides
Edited by Frances Separovic and Akira Naito

Published by the Royal Society of Chemistry, www.rsc.org

atomic structure within the natural medium, the lipid membrane. The easiest approach, in general, resides in dissolving the water-insoluble membrane protein or peptide with surfactants into micelles and using the classical tools of solution NMR.[1,2] However, the use of stringent surfactants, such as SDS, and the highly curved micelle surface may prevent obtaining the "active" structure and topology. It is preferable to embed the peptide/protein within lipids that better represent an adequate natural anisotropic environment (the bilayer membrane) in which the structure and topology may be obtained from solid-state NMR.[3] Indeed, the mechanical orientation between glass plates of lipid membranes containing peptides and use of wide-line solid-state NMR are capable of yielding the topology/orientation within the membrane.[4–6] Hydration of such systems is, however, low and may not represent the natural cell hydration; pH and salinity are also difficult to monitor in between glass plates.

Recently, a new membrane model system for which the hydration is controlled at will has become very popular, representing an intermediate between the lipid vesicles and classical micelles. These model systems are based on the pioneering work on "bilayered micelles", now called bicelles.[7] Owing to their remarkable features such as alignment in the magnetic field, B_0, bicelles have become a valuable tool for NMR applications.[8–10] Bicelles are a mixture of aliphatic long-chain lipids (between 12 to 18 carbons) and short-chain lipids (6 to 8 carbons). Their morphology is fairly versatile, depending on composition, temperature and hydration. The most recognized organization is a nano-disc with the long-chain lipids present mainly in the disc plane and the short-chain lipids distributed mainly in the torus of the disc (Figure 6.1b).

Conditions to build such nano-discs are now well defined and temperature–composition–hydration phase diagrams are available.[10–13] A large number of molecules, including peptides and proteins, can be embedded within these nano-discs. However, some molecules may perturb the phase diagram and destroy the bicelle, so the model is not universal. Alignment with respect to the magnetic field, B_0, is in principle not obtained with liposomes (simple water lipid dispersions, Figure 6.1a), which prevents defining the topology of peptides with these easily made membranes. There are, however, examples for which magnetic deformation of liposomes has been reported,[14,15] opening an avenue for studies in systems that can be deformed in such a way that topology information can be obtained.

In this chapter we will focus on phospholipid systems that are well hydrated, relatively easy to prepare, and that can be used for determination of the structure and topology of peptides and proteins embedded in such membranes. Two systems will be presented: (i) magnetically alignable bicelles, and (ii) magnetically deformable liposomes. The properties of such lipid assemblies are presented and examples of topology/structure determination provided.

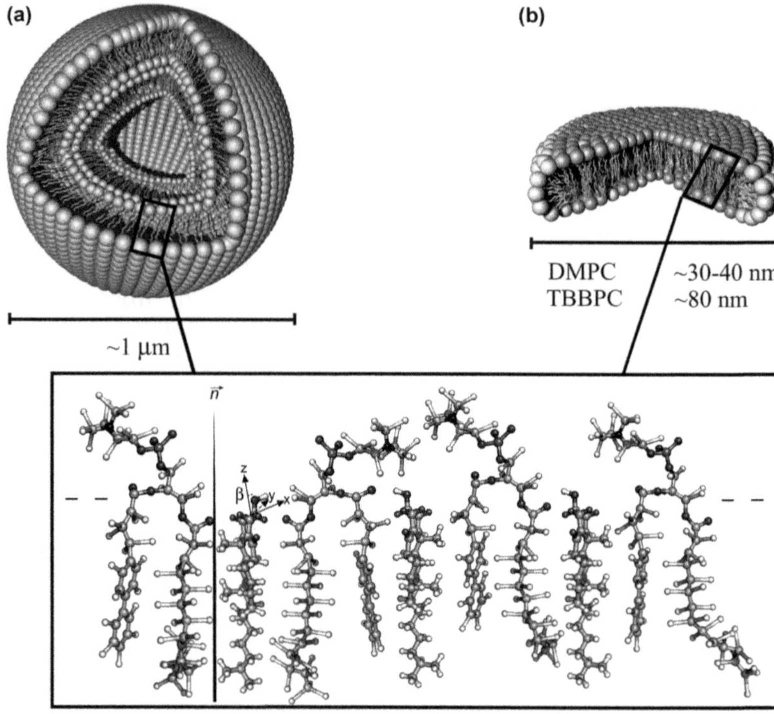

Figure 6.1 Artistic representation of (a) micrometre-size multilamellar dispersions of lipids (MLVs) made of saturated, unsaturated or biphenyl-containing phospholipids and cholesterol, and (b) nanometer-size phospholipid bicelles made with similar lipids. Unsonicated MLVs, also called liposomes, are prepared by lipid hydration in excess water (>80% w/w) and performing several freeze–thaw–shaking cycles. Bicelles are made of long-chain lipids (C_{14}–C_{18} acyl chain length) that are mainly located in the disc plane and short-chain lipids (C_6–C_8 acyl chain length) that are mostly present in the disc torus. Bicelles are prepared in the same way as liposomes, with the correct long-chain lipid to short-chain lipid ratio ensuring the disc formation at specific temperatures.

6.2 Magnetic Properties of Lipids and Lipid Assemblies; Magnetic Orientation

6.2.1 Magnetic Properties of Molecules

Magnetic fields of a few Tesla have little influence on isolated biological molecules in terms of magnetically induced macroscopic orientation. In general, small diamagnetic molecules undergo isotropic tumbling in solution, leading to an orientation independent spectrum: the "isotropic" liquid state spectrum. There are, however, examples where some large proteins in solution can be slightly oriented in very large magnetic fields of 20 T or more.[16] This is due to the anisotropic magnetic susceptibility of iron atoms (paramagnetic) that are bound to the heme-containing protein. Very small dipolar or quadrupolar splittings (<1 Hz) can then be measured. NMR of partially aligned liquids has also been reported for a collection of aromatic molecules dissolved in ethanol.[17] Macroscopic orientations in these cases are also very weak and represent a residual of 10^{-6} with respect to the total interaction that is sampled (such as quadrupolar or dipolar) and is only detected for magnetic fields greater than 10 T. There are also many examples of field-driven macroscopic orientation of proteins in solution as promoted by aligning media: purple membranes, phages, solvent mixtures, *etc.*[18,19] In what follows we will describe magnetic orientation effects on ensembles of lipid molecules forming liposomes or bicelles in a well-hydrated medium.

6.2.2 Orientation of Lipid Assemblies in Magnetic Fields: Liposomes

Macroscopic orientation of lipids in organic solution has not been reported so far. However, lipids in the form of liposomes can be oriented in high magnetic fields,[14,15] although the magnetic susceptibility of phospholipids, or the anisotropy, is very weak. In the pioneering work of Helfrich,[20] the volume magnetic susceptibility anisotropy, $\Delta\chi_V$, of unsaturated phospholipids (egg lecithin) was found to be -2.8×10^{-9} (c.g.s.-e.m.u. units), which converts into a magnetic susceptibility anisotropy per molecule $[\Delta\chi_{molecule} = \Delta\chi_V \times M_W/\rho N_A]$ of -3.8×10^{-35} m^3 (SI units). For comparison, the magnetic susceptibility of water is -2.7×10^{-28} m^3 and that of biphenyl in ethanol was found to be -2.2×10^{-33} m^3.[17] In high magnetic fields, vesicles made of saturated lipids such as DMPC can be deformed to prolate spheroids with the polar axis greater than the diameter ($c/a > 1$, Figure 6.2c, left). This reflects a negative magnetic susceptibility anisotropy ($\Delta\chi = \chi_{//} - \chi_{\perp} < 0$) and may easily be detected by wide-line solid-state ^{31}P NMR, with the shoulder near 20–30 ppm having a very low intensity. In other words, the lipid long axis tends to align perpendicularly to the magnetic field in order to minimize the potential magnetic energy.

Dispersions of phospholipids having a biphenyl moiety in one of their chains present a totally different ^{31}P NMR spectrum with the 20–30 ppm resonance very intense, dominating the entire spectrum. This reflects an oblate deformation with the polar axis smaller than the diameter ($c/a < 1$, Figure 6.2c, right) and is indicative of a positive magnetic susceptibility anisotropy ($\Delta\chi = \chi_{//} - \chi_{\perp} > 0$). Here, the lipid long axis tends to align parallel to the magnetic field. Such NMR lineshapes can be theoretically modeled by calculating the probability distribution of NMR resonances on the surface of a revolution ellipsoid.[14,21] It is noteworthy that very high fields provide a liposome deformation that may be used to look at molecular topologies within the membrane using either DMPC or TBBPC (*vide infra*).

6.2.3 Orientation of Lipid Assemblies in Magnetic Fields: Bicelles

On the other hand, it has also been reported that lipid mixtures that form 50–100 nm diameter nano-discs, or so-called bicelles, show very interesting orienting properties in magnetic fields. Bicelles are obtained by addition of

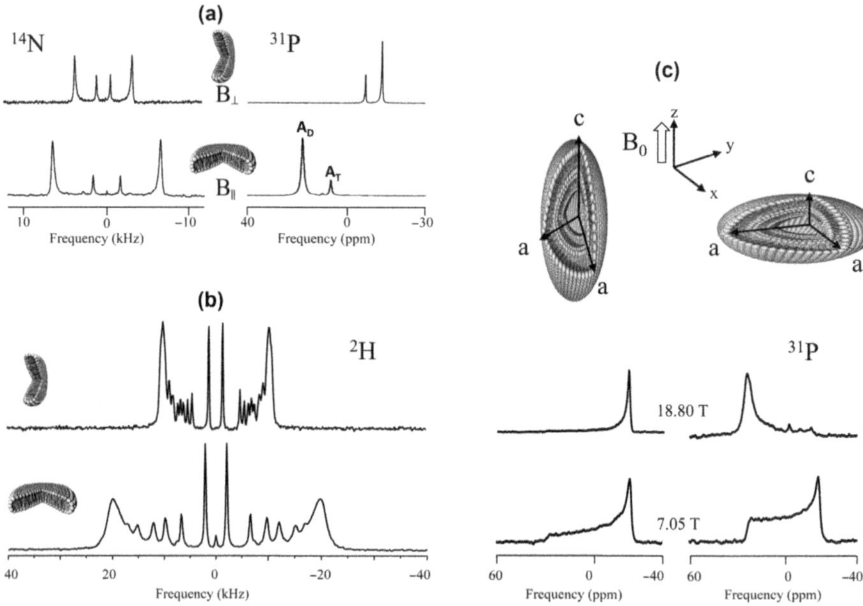

Figure 6.2 Solid-state NMR spectra of (a, b) bicelles and (c) liposomes. (a) DMPC/DCPC bicelles (*top*) as recorded using the ^{14}N quadrupolar interaction and the ^{31}P CSA of phosphocholine lipids; *bottom* spectra were recorded for the TBBPC/DMPC bicelle. A 11.8 T magnetic field-induced change in disc orientation is readily detected, as sketched alongside spectra. A_D and A_T represent the area of lipids in discs and in the torus, respectively. (b) [$^2H_{27}$]DMPC/DCPC (*top*) and [$^2H_{27}$]TBBPC/DCPC (*bottom*) bicelle spectra as recorded using the ^2H quadrupolar interaction and showing the 90° change in disc orientation produced by the 11.8 T magnetic field (quadrupolar splitting doubled for each labelled position). (c) ^{31}P Chemical shielding interaction to record DMPC (*left*) and TBBPC (*right*) liposome deformation as induced by an 18.8 T magnetic field, together with the corresponding sketches of prolate (*left*) and oblate (*right*) MLV. Note the small lineshape modification indicating little deformation, if any, in a field of 7.1 T.
(Adapted from Loudet *et al.*,[10,14] with permission.)

the required amount of short-chain lipids, DCPC (C_6 chain length), that stabilize the disc torus, to long-chain lipids, DMPC (C_{14} chain length), in a 70–95% hydration medium (see Figure 6.1).[10–14] These nano-objects have been imaged by electron microscopy and show very interesting remnant orientation effects; they may stay oriented for days outside magnetic fields once having been oriented for a few minutes in the field.[14,22]

X-Ray studies can be performed with these materials, as shown in Figure 6.3 where bicelles made of TBBPC and DCPC show a very long remnant orientation outside the field. From the experiments shown in Figure 6.3 (oriented in magnetic field and further transferred to a SAXS machine for X-rays observation), it is clear that bicelles made of saturated

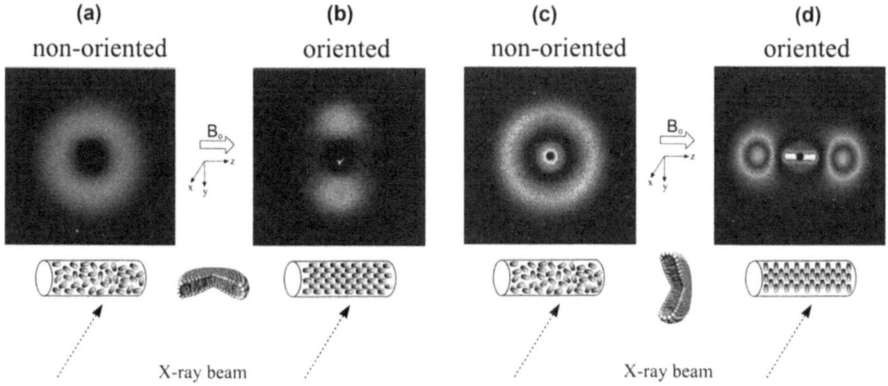

Figure 6.3 SAXS of bicelles before and after 11.8 T magnetic field orientation at 35°C. DMPC/DCPC bicelles (a) prior and (b) 15 min after field orientation; the annulus of the diffracted signal (*left*) depicts the random orientation of bicelles and the two spots (*right*) depict the bicelle normal orientation at 90° with respect to the field (sketches are to help clarify the position of the capillary tube in the X-ray setup and in the magnetic field). TBBPC/DCPC bicelles (c) prior and (d) after magnetic field alignment: again the annulus (*left*) shows the random orientation and the two spots (*right*) the bicelle normal orientation at 0° with respect to the field.
(Adapted from Loudet *et al.*[10,14] and Loudet-Courreges *et al.*,[22] with permission.)

lipids (DMPC/DCPC) orient with their plane parallel to the magnetic field, whereas biphenyl bicelles (TBBPC/DCPC) orient with their plane perpendicular to the magnetic field. The latter appears to provide oriented materials for days. As shown in Figure 6.2a and 6.2b, saturated bicelles or biphenyl bicelles lead to quasi perfectly magnetically oriented samples, as can be detected by wide-line solid-state NMR. Disc orientations perpendicular and parallel to the magnetic field are obtained by using DMPC or TBBPC, respectively. The orientation is readily detected by ^{31}P, ^{14}N and ^{2}HNMR as magnetic interactions scale as $P2(\cos\theta)$, where θ is the angle between the disc plane and the magnetic field. This is reflected by a collection of sharp lines showing a scaling factor of 2 between DMPC ($P2 = -1/2$) and TBBPC ($P2 = 1$) bicelles (Figure 6.2a and b) and a change in sign of overall spectral shapes for the spectrum dominated by the chemical shielding anisotropy (CSA). This behaviour reflects the change in sign of magnetic susceptibility anisotropy for the long-chain lipid. Two lines are observed using ^{31}P NMR, one for each lipid location in the disc plane and in the torus, respectively, the short-chain lipid DCPC being mostly located in the torus. Line areas (A_D and A_T) allow determination of disc size in very good agreement with electron microscopy measurements.[11] ^{14}N NMR leads to four lines (quadrupolar interaction), again a doublet corresponding to each location in the nano-disc, plane and torus. ^{2}H NMR of DMPC perdeuterated in the *sn*-2 acyl chain shows well-resolved doublets corresponding to the motional non-equivalent C–^{2}H bonds. This allows an easy determination of the gradient of order parameters along the acyl chain and, hence, across the lipid bilayer. The same is observed with deuterium-labelled TBBPC, the myristoyl chain in position *sn*-1 being perdeuterated. Here, the spectrum is two times wider, allowing a greater accuracy in quadrupolar splitting measurement.

6.3 Structures, Topologies in Magnetically Aligned Membranes

6.3.1 Membrane Proteins

Determination of membrane protein structure and topology in a natural environment, the lipid bilayer, becomes feasible using NMR but still represents a challenge. Most NMR experiments of insoluble membrane proteins have been performed by dissolving the membrane in a more or less mild detergent system that maintains membrane proteins in an isotropic environment and, hence, allows the use of liquid-state NMR. Advantages and drawbacks of micelles, isotropic bicelles, amphipols, fluorinated surfactants and organic solvent mixtures have been well discussed in recent reviews.[23–25] It is highlighted that the main problem resides in protein misfolding and the ability to control biological function in such media. Membrane protein topology in lipid membranes has been addressed by the pioneering work of the Opella group,[26–28] who developed sample alignment experiments and has shown importance both for solution and solid-state NMR studies of proteins. Such experiments indeed offer access to orientational constraints for the calculation of atomic-resolution protein structures and topologies with respect to the membrane plane. Figure 6.4 summarizes some of these approaches by comparing 1D and 2D solid-state NMR experiments of uniformly ^{15}N-labelled Pf1 coat protein embedded in lipid bilayers either sandwiched in-between glass plates (Figure 6.4a), in biphenyl bicelles (Figure 6.4b) or in "flipped" saturated DMPC bicelles (Figure 6.4c).[29]

The latter, which usually orient with their disc planes parallel to the magnetic field, are flipped by 90° after addition of lanthanides such as ytterbium(III) chloride, $YbCl_3$. Such lanthanides have a positive magnetic susceptibility anisotropy that prevails over the negative anisotropy of saturated bicelles. All three experiments are designed such that the bilayer normal is parallel to the magnetic field, which is much desirable to determine the orientation of the helical parts of Pf1. The two bicelle systems are oriented by the magnetic field action and the glass plate sample is mechanically oriented. The 1D ^{15}N spectrum of the Pf1 coat protein embedded in membranes sandwiched in-between glass plates shows much lesser resolution than the other two samples. This is mainly due to imperfect orientation (glass plate defects, glass plate miss-orientation in the field, *etc.*) and low hydration that provides greater membrane rigidity. The two bicelle

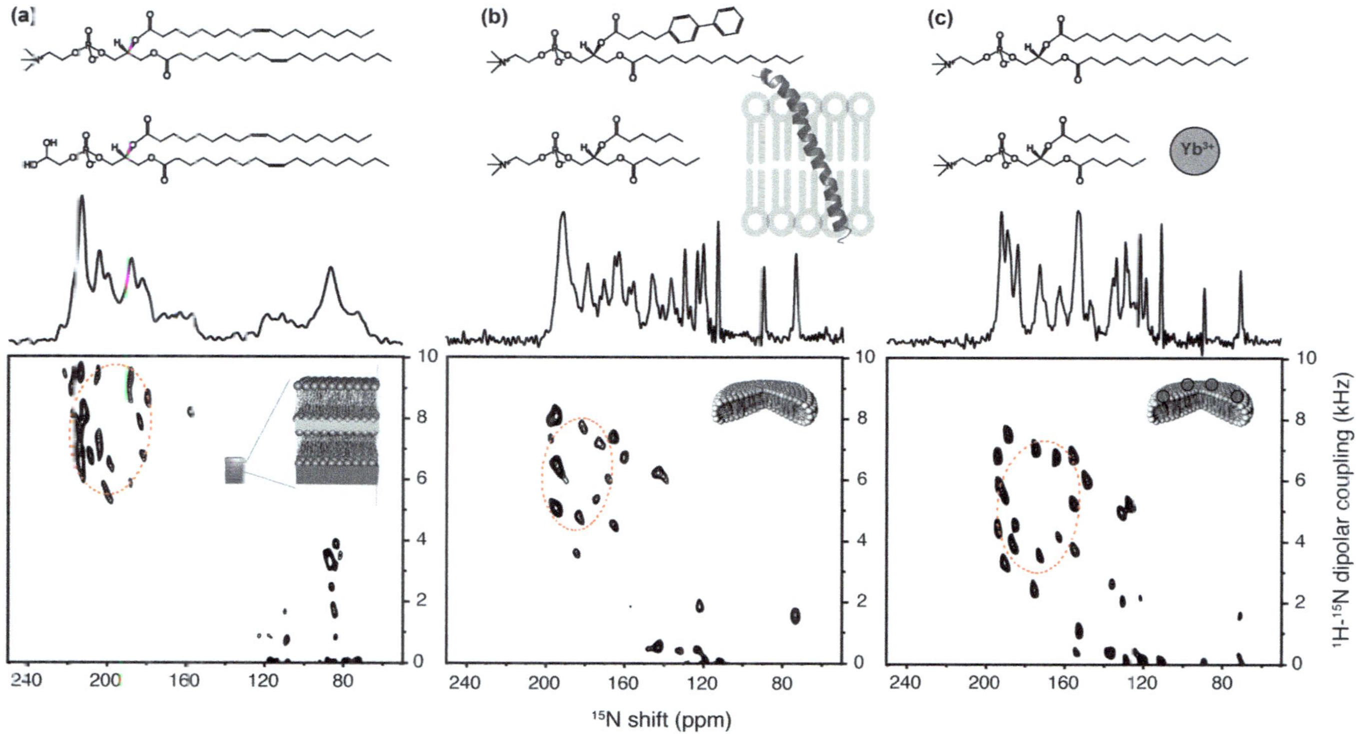

Figure 6.4 One-dimensional ^{15}N chemical shift and two-dimensional ^{15}N chemical shift/$^{1}H-^{15}N$ dipolar coupling separated local field (SLF) solid-state NMR spectra. The Pf1 coat protein is uniformly ^{15}N labelled and embedded in membranes aligned with their bilayer normal parallel with respect to the applied magnetic field (17.6 T); various lipid bilayer environments are represented on *top*. (a) DOPC/DOPG bilayers aligned mechanically on glass plates at 23 °C; (b) TBBPC/DCPC bicelles at 40 °C; and (c) DMPC/DCPC flipped with 3 mM $YbCl_3$ at 40 °C. The molar ratios of lipid bilayers to protein concentration (*L*/*P*) are (a) 88, (b) 50 and (c) 54. Superimposed on the experimental data are the ideal PISA wheels for an α-helix with uniform dihedral angles ($\phi = -61°$ and $\psi = -45°$) tilted at (a) 20° with $\delta = 1.0$, (b) 21° with $\delta = 0.83$, and (c) 27° with $\delta = 0.8$, where δ is the order parameter of the transmembrane peptide helix.
(Adapted from Park *et al.*,[29] with permission.)

samples lead to very high spectral resolution, with many resonances within the range associated with tilted transmembranous helices (160–220 ppm). Frequency distribution of resonances for the two bicelle samples is, however, slightly different, suggesting that the well-known shift reagent effect of Yb^{3+} ions may perturb line positions. In the 2D separated local field (SLF) experiment shown in Figure 6.4 (bottom), almost complete spectral resolution is achieved: each amide site leads to a correlation resonance characterized by a unique spot at the intersection of the ^{1}H–^{15}N dipolar coupling and the ^{15}N chemical shift frequencies. The characteristic PISA wheel patterns (dotted ellipses) corresponding to helical wheel projections are formed by the resonances from residues in the transmembrane helix part of the protein.[30–32] From the breadth of the ellipses a *ca.* 20–30° tilt can be inferred.

6.3.2 Peptides

Membrane interacting peptides play essential roles in cellular biology. They may kill cells or help them in adapting to external constraints, and increasing resistance of bacteria to conventional antibiotics has pressed for the development of new modes of treatment. Antimicrobial peptides (AMPs) have been presented as a potential solution to this problem.[33] Whereas classical antibiotics are specifically targeting biosynthetic pathways, AMPs can directly destabilize the lipid membrane and constitute a promising alternative strategy for fighting microorganisms. Antimicrobial peptides are rather small and were first identified in the hemolymph of insects and in the secretion of immune cells. Insects produce such peptides as part of a systemic response induced by microorganisms.[34,35] Membrane interacting peptides have been recently reviewed in terms of structure and mode of action on lipid membranes.[36–38] In this chapter we will highlight how solid-state NMR has helped in deciphering the mode of action of two peptides in membranes: surfactin (SF) and methionine enkephalin (MENK). Surfactin belongs to the lipopeptide family by displaying a hydrocarbon chain in complement to the peptidic small cyclic backbone, whereas MENK is a short linear peptide involved in neurotransmission found in the human central nervous system, especially in regions of the brain and spine associated with diffuse pain pathways.[36] Surfactin is secreted by *Bacillus subtilis* and is composed of a ring-shaped peptidic backbone of seven amino acids (Glu-Leu-Leu-Val-Asp-Leu-Leu) in a chiral sequence of L/D enantiomers.[39] The ring is closed by a β-hydroxy fatty acid with different aliphatic chains of length ranging from 12 to 14 carbon atoms. The presence of both lipophilic and hydrophilic/negatively charged sections confers powerful surfactant capabilities on SF, such as emulsification and foaming. It has a very large spectrum of biological activities and, surprisingly, this negatively charged amphipathic peptide is very active against negatively charged membranes at physiological pH. Addition of SF to negatively charged μm-sized multilamellar liposomes (MLVs) converts them into very small unilamellar vesicles (SUVs) of 30–50 nm.[40]

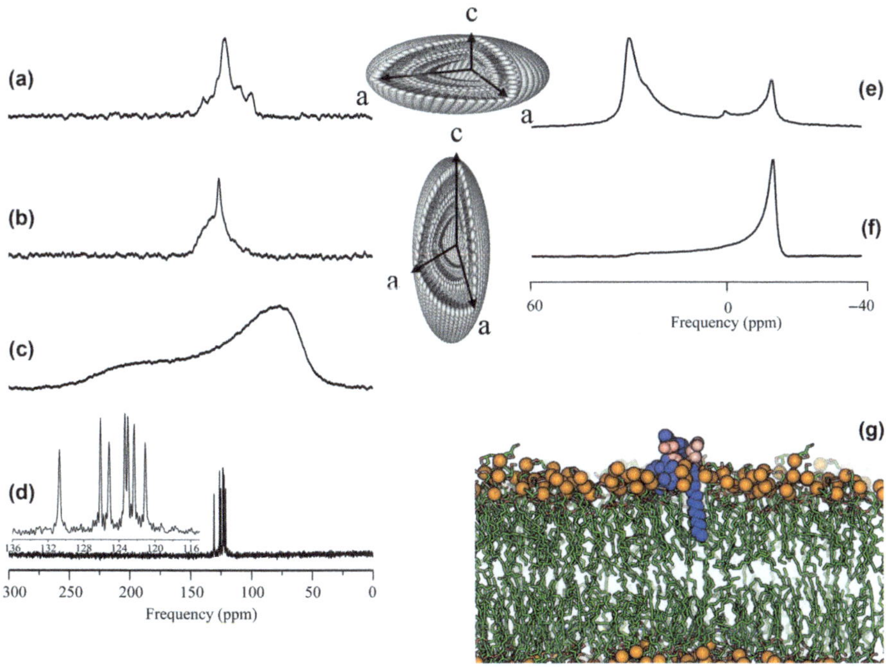

Figure 6.5 Surfactin topology using magnetically deformed liposomes. Solid-state ^{1}H cross-polarized ^{15}N NMR spectra of uniformly labelled ^{15}N-surfactin recorded at 18.8 T and 35 °C in (a) TBBPC liposomes, (b) DMPC liposomes, (c) dry powder and (d) organic solution using liquid-state NMR (*insert* is a horizontal expansion of amine resonances). Solid-state ^{1}H cross-polarized ^{31}P NMR spectra of ^{15}N-surfactin recorded at 18.8 T and 35°C in (e) TBBPC liposomes and (f) DMPC liposomes. *Sketches* represent an artist's view of liposome deformation as induced by a high magnetic field. (g) Molecular dynamics simulation of surfactin in a well-hydrated phosphatidylcholine bilayer (water molecules are not shown for clarity). Lipid chains are in *green* and phosphates shown as *orange-gold* van der Waals spheres. Hydrophobic surfactin amino acids and the C-14 acyl chain are shown in *blue*. Negatively charged amino acids are in *light pink* and point out towards the water layer, whereas the C-14 acyl chain is well embedded in the bilayer hydrophobic core.
(Adapted from Dufourc *et al.*,[36] with permission.)

Magnetic liposomes that can be aligned/deformed by high magnetic fields have been used to probe SF topology on membranes. As shown in Figure 6.5, uniformly ^{15}N-labelled SF was applied to DMPC and TBBPC liposomes that were extremely deformed as prolates or oblates in a magnetic field of 18.8 T. ^{31}P Wide-line solid-state NMR was used to show that SF did not change much the strong liposome deformation produced by the magnetic field: intense peaks at *ca.* −15 ppm and 30 ppm are detected as respective indicators of prolate or oblate deformation (Figure 6.5e and f). Spectra of ^{15}N-surfactin in solution (Figure 6.5d) show seven sharp peaks in the

121–131 ppm range, corresponding to the amine protons of the seven amino acid residues present in the cyclic peptide. The solid-state spectrum of SF dry powder displays a characteristic pattern of unoriented material, the so-called powder pattern, showing the quasi-axial symmetry of the chemical shielding tensor whose principal axis is almost collinear with the ^{15}N–H bond. An overall CSA of 180 ± 10 ppm is in good agreement with individual literature values[41,42] obtained for the amino acids present in SF. Surfactin embedded in DMPC and TBBPC magnetically deformed liposomes (Figure 6.5b and a) are clearly different. For TBBPC, individual lines are almost resolved and span a 95–140 ppm range. For DMPC, less resolution is observed and the lines span the 110–145 ppm range. These results clearly indicate that SF is well embedded in the membrane whose orientation changes with the deformation imposed on liposomes by the magnetic field. Of course, resolution is still relatively low due to imperfect alignment (liposomes are not infinitively deformed), but spectra are in close agreement with the calculated orientation of SF in membranes.[36] Further calculations and measurements are clearly needed to validate this novel and simple way to explore peptide topology on membranes.

The five-residue neuropeptide MENK (Tyr-Gly-Gly-Phe-Met) was added to DMPC/DCPC bicelles and the pseudo "high-resolution" ^{1}H NMR spectrum recorded as shown in Figure 6.6a. As expected, a featureless spectrum is

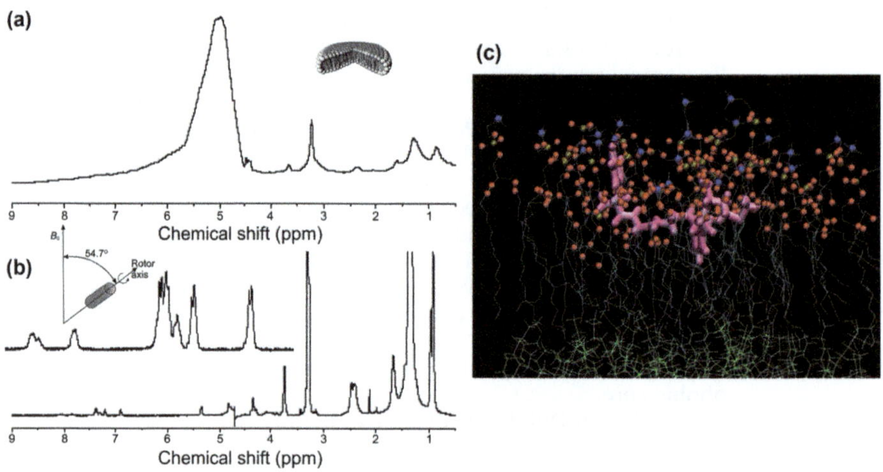

Figure 6.6 ^{1}H NMR spectra of DMPC/DCPC bicelles containing the neuropeptide MENK (lipid-to-peptide molar ratio of 25) at 25°C: (a) static and (b) with a MAS speed of 10 kHz. A sketch of the MAS set-up together with an expansion of the amide resonances are also shown. (c) Position of MENK at the bilayer interface as reconstructed from 2D-NMR constraints using molecular dynamics. Colour code: peptide in *pink*, lipids with oxygen atoms of fatty acid chains in *red*, phosphorus in *orange-gold*, nitrogen in *blue* and carbon skeleton in *light blue*; water molecules are not represented for clarity.
(Adapted from Diller *et al.*,[9] with permission.)

seen, displaying a strong and broad water signal and a few lipid resonances at low ppm values. Broadening originates mainly from residual dipolar ^1H–^1H interactions that are not averaged to zero in the anisotropic and field-oriented bicelle sample. Of interest, no isotropic signal is detected, indicating that the peptide is mainly associated/exchanging with the membrane. The bicelle sample was spun at the magic angle at a frequency of 10 kHz and the resulting spectrum obtained is shown in Figure 6.6b. A high-resolution spectrum is observed where lipid resonances dominate the spectrum (water resonance has been suppressed with a selective pulse). However, although the peptide to lipid molar ratio is 1:25, amide resonances corresponding to the five amino acids are readily detected (see insert with increased vertical scale).

A similar MAS experiment could have been performed in liposomes, but extra resolution is brought by the bicelle medium which is slightly more disordered than MLVs.[43,44] With such resolution, the use of classical multidimensional liquid-state NMR sequences (TOCSY, NOESY), together with molecular modelling, allows determination of both the 3D structure of MENK and its localization at the membrane interface (Figure 6.6c).[9] This structure confirms earlier works that reported the structure of a parent enkephalin, Leu-enkephalin, using transferred NOE experiments, which are more difficult to set up.[45] In solution, the peptide does not show any structure, whereas MENK displays a bent fold in the bilayer membrane. The peptide sits at the membrane interface with, however, its hydrophobic amino acids pointing towards the membrane core. Such a topology is in agreement with the mechanism called "membrane catalysis", where the polar headgroups of the lipids at the cell membrane surface interact with such small peptides, which enter into the membrane *via* hydrophobic interactions and undergo a conformational change. According to this hypothesis, enkephalins would migrate to the receptor site with a suitable structure for binding. Results obtained from MAS ^1H NMR of bicelles confirm this assumption.

6.4 Conclusion

Phospholipid bicelles are particularly suitable tools for hosting membrane proteins and peptides in a relevant membrane environment. Membrane peptide/protein topology and atomic structure may thus be obtained relatively easily using the entire collection of NMR techniques, from wide-line NMR to MAS of bicelles. Modest magnetic fields are required for disc alignment. Varying the lipid magnetic susceptibility anisotropy leads to disc orientation parallel or perpendicular to the magnetic fields *z*-direction, thus conferring to the bicelle a "quality" as a "molecular goniometer".[9] However, bicelle disc stability and, hence, alignment may be perturbed by the addition of peptide or proteins, so not all membrane components can be studied using these systems. As magnetic fields increase in intensity (20–25 T), "classical" liposomes may represent an interesting and simpler alternative. The lipid magnetic susceptibility anisotropy indeed leads to liposome deformation in such a way that considerable, though imperfect, membrane alignment with the field may be obtained. Here again, playing with the lipid magnetic properties, which directly depend on molecular structures, allows mastering liposome deformation as either prolate or oblate. Although such experiments need to be improved because they lead to a lesser spectral resolution (until now), they signify great hope due to the simplicity in sample preparation. Both bicelle and liposomes need to be further explored as representative membrane substrates for peptide and protein structure determination.

Abbreviations

ρ	density in g cm^{-3}
χ_V	volume magnetic susceptibility (dimensionless in SI units)
χ_{mol}	molar magnetic susceptibility in m^{3}mol^{-1}
$\chi_{molecule}$	molecule magnetic susceptibility = χ_{mol}/N_A in m^3
$\Delta\chi = \chi_{//} - \chi_{\perp}$	magnetic susceptibility anisotropy
AMP	antimicrobial peptide
CSA	chemical shielding anisotropy
DCPC	1,2-dihexanoyl-*sn*-glycero-3-phosphocholine (6:0/6:0)
DMPC	1,2-dimyristoyl-*sn*-glycero-3-phosphocholine (14:0/14:0)
DOPC/DOPG	1,2-dioleoyl-*sn*-glycero-3-phosphocholine/phosphoglycerol (18:1/18:1)
MAS	magic angle sample spinning
MENK	Tyr-Gly-Gly-Phe-Met (methionin enkephalin)
MLVs	multilamellar vesicles
NOESY/TOCSY	NOE/total correlation spectroscopy
POPC	1-palmitoyl-2-oleoyl-*sn*-glycero-3-phosphocholine (16:0/18:1)
SAXS	small-angle X-ray scattering
S_{CD}	carbon deuterium bond order parameter
SDS	sodium dodecyl sulfate
SF	Glu-Leu-Leu-Val-Asp-Leu-Leu-myr (surfactin myristoylated)
SUVs	small unilamellar vesicles
TBBPC	1-tetradecanoyl-2-[4-(4-biphenyl)butanoyl]-*sn*-glycero-3-phosphocholine (14:0/BB).

Acknowledgements

Financial support from the TGIR-RMN-THC FR3050 CNRS for conducting the research is gratefully acknowledged. The University Bordeaux and the Aquitaine government are thanked for funding the NMR platform.

References

1. A. Arora, F. Abildgaard, J. H. Bushweller and L. K. Tamm, *Nat. Struct. Biol.*, 2001, **8**, 334–338.
2. C. Fernández, K. Adeishvili and K. Wüthrich, *Proc. Natl. Acad. Sci. U. S. A.*, 2001, **98**, 2358–2363.
3. E. J. Dufourc, in *Chemical Biology*, ed. B. Larijani, R. Woscholski and C. A. Rosser, Wiley, Chichester, 2006, pp. 113–131.
4. F. M. Marassi and S. J. Opella, *Protein Sci.*, 2003, **12**, 403–411.
5. F. M. Marassi, A. Ramamoorthy and S. J. Opella, *Proc. Natl. Acad. Sci. U. S. A.*, 1997, **94**, 8551–8556.
6. B. Bechinger and C. Sizun, *Concepts Magn. Reson.*, 2003, **18A**, 130–145.
7. N. E. Gabriel and M. F. Roberts, *Biochemistry*, 1984, **23**, 4011–4015.
8. C. R. Sanders and R. S. Prosser, *Struct. Fold. Des.*, 1998, **6**, 1227–1234.
9. A. Diller, C. Loudet, F. Aussenac, G. Raffard, S. Fournier, M. Laguerre, A. Grelard, S. J. Opella, F. M. Marassi and E. J. Dufourc, *Biochimie*, 2009, **91**, 744–751.
10. C. Loudet, S. Manet, S. Gineste, R. Oda, M. F. Achard and E. J. Dufourc, *Biophys. J.*, 2007, **92**, 3949–3959.
11. A. Arnold, T. Labrot, R. Oda and E. J. Dufourc, *Biophys. J.*, 2002, **83**, 2667–2680.
12. G. Raffard, S. Steinbruckner, A. Arnold, J. H. Davis and E. J. Dufourc, *Langmuir*, 2000, **16**, 7655–7662.
13. C. Loudet, L. Khemtemourian, F. Aussenac, S. Gineste, M. F. Achard and E. J. Dufourc, *Biochim. Biophys. Acta, Gen. Subj.*, 2005, **1724**, 315–323.
14. C. Loudet, A. Diller, A. Grelard, R. Oda and E. J. Dufourc, *Prog. Lipid Res.*, 2010, **49**, 289–297.
15. J. Seelig, F. Borle and T. A. Cross, *Biochim. Biophys. Acta*, 1985, **814**, 195–198.
16. L. Banci, I. Bertini, J. G. Huber, C. Luchinat and A. Rosato, *J. Am. Chem. Soc.*, 1998, **120**, 12903–12909.
17. P. C. M. Van Zijl, B. H. Ruessink, J. Bulthuis and C. MacLean, *Acc. Chem. Res.*, 1984, **17**, 172–180.
18. M. R. Hansen, L. Mueller and A. Pardi, *Nat. Struct. Biol.*, 1998, **5**, 1065–1074.
19. J. Sass, F. Cordier, A. Hoffmann, A. Cousin, J. G. Omichinski, H. Lowen and S. Grzesiek, *J. Am. Chem. Soc.*, 1999, **121**, 2047–2055.
20. E. Boroske and W. Helfrich, *Biophys. J.*, 1978, **24**, 863–868.
21. T. Pott and E. J. Dufourc, *Biophys. J.*, 1995, **68**, 965–977.
22. C. Loudet-Courreges, F. Nallet, E. J. Dufourc and R. Oda, *Langmuir*, 2011, **27**, 9122–9130.
23. A. A. De Angelis, D. H. Jones, C. V. Grant, S. H. Park, M. F. Mesleh and S. J. Opella, in *Nuclear Magnetic Resonance of Biological Macromolecules, Part C*, ed. T. L. James, Elsevier Academic, San Diego, 2005, vol. 394, pp. 350–382.

24. R. S. Prosser, F. Evanics, J. L. Kitevski and M. S. Al-Abdul-Wahid, *Biochemistry*, 2006, **45**, 8453–8465.
25. C. R. Sanders and K. Oxenoid, *Biochim. Biophys. Acta, Biomembr.*, 2000, **1508**, 129–145.
26. T. A. Cross and S. J. Opella, *Curr. Opin. Struct. Biol.*, 1994, **4**, 574–581.
27. S. J. Opella, *Nat. Struct. Biol.*, 1997, **4**, 845–848.
28. S. J. Opella and F. M. Marassi, *Chem. Rev.*, 2004, **104**, 3587–3606.
29. S. H. Park, C. Loudet, F. M. Marassi, E. J. Dufourc and S. J. Opella, *J. Magn. Reson.*, 2008, **193**, 133–138.
30. F. M. Marassi and S. J. Opella, *J. Magn. Reson.*, 2000, **144**, 150–155.
31. J. Wang, J. Denny, C. Tian, S. Kim, Y. Mo, F. Kovacs, Z. Song, K. Nishimura, Z. Gan, R. Fu, J. R. Quine and T. A. Cross, *J. Magn. Reson.*, 2000, **144**, 162–167.
32. A. Naito, *Solid-State Nucl. Magn. Reson.*, 2009, **36**, 67–76.
33. R. M. Epand and H. J. Vogel, *Biochim. Biophys. Acta, Biomembr.*, 1999, **1462**, 11–28.
34. J. L. Dimarcq, P. Bulet, C. Hetru and J. Hoffmann, *Biopolymers*, 1998, **47**, 465–477.
35. M. G. Giovannini, L. Poulter, B. W. Gibson and D. H. Williams, *Biochem. J.*, 1987, **243**, 113–120.
36. E. J. Dufourc, S. Buchoux, J. Toupe, M. A. Sani, F. Jean-Francois, L. Khemtemourian, A. Grelard, C. Loudet-Courreges, M. Laguerre, J. Elezgaray, B. Desbat and B. Odaert, *Curr. Protein Pept. Sci.*, 2012, **13**, 620–631.
37. B. Bechinger, *J. Membr. Biol.*, 1997, **156**, 197–211.
38. B. Bechinger, *Biochim. Biophys. Acta, Biomembr.*, 1999, **1462**, 157–183.
39. K. Arima, A. Kakinuma and G. Tamura, *Biochem. Biophys. Res. Commun.*, 1968, **31**, 488–494.
40. S. Buchoux, J. Lai-Kee-Him, M. Garnier, P. Tsan, F. Besson, A. Brisson and E. J. Dufourc, *Biophys. J.*, 2008, **95**, 3840–3849.
41. J. R. Brender, D. M. Taylor and A. Ramamoorthy, *J. Am. Chem. Soc.*, 2001, **123**, 914–922.
42. D. K. Lee, J. S. Santos and A. Ramamoorthy, *Chem. Phys. Lett.*, 1999, **309**, 209–214.
43. F. Aussenac, M. Laguerre, J. M. Schmitter and E. J. Dufourc, *Langmuir*, 2003, **19**, 10468–10479.
44. C. Carlotti, F. Aussenac and E. J. Dufourc, *Biochim. Biophys. Acta, Biomembr.*, 2002, **1564**, 156–164.
45. A. Milon, T. Miyazawa and T. Higashijima, *Biochemistry*, 1990, **29**, 65–75.

CHAPTER 7

Membranes and Their Lipids: A Molecular Insight into Their Organization and Function

MARTIN LIDMAN,[a] MARCUS WALLGREN[b] AND GERHARD GRÖBNER*[a]

[a] Department of Chemistry, Umeå University, 90187 Umeå, Sweden;
[b] Department of Medical Biochemistry and Biophysics, Umeå University, 90187 Umeå, Sweden
*Email: gerhard.grobner@chem.umu.se

7.1 Introduction: Lipid Membranes in Life

Virtually every process in living organisms involves biological membranes and their constituents. However, understanding the function of biological membranes relies on an atomic-level insight into their three-dimensional structural and dynamic organization, ranging from the lipid matrix to their membrane proteins and lipid/protein assemblies, one of the most challenging tasks in structural biology.[1–4] Fortunately, non-perturbing solid-state NMR methods have emerged in recent decades to provide a molecular insight into the fundamental processes associated with these unique lipid-based two-dimensional assemblies,[5–15] which exert numerous tasks in the cell and are involved in various disorders. In general, biological membranes are composed of lipid molecules forming the typical bilayer leaflet, as well as of integral and peripheral proteins responsible for

New Developments in NMR No. 3
Advances in Biological Solid-State NMR: Proteins and Membrane-Active Peptides
Edited by Frances Separovic and Akira Naito

Published by the Royal Society of Chemistry, www.rsc.org

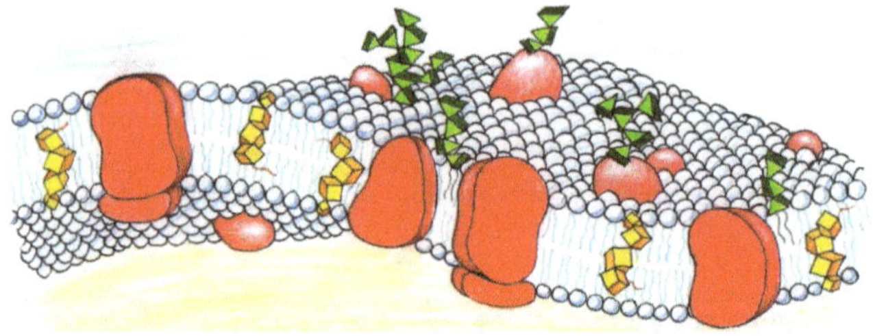

Figure 7.1 Schematic picture of a biological membrane composed of lipids, sterols (*yellow*), glycosylated molecules (*green*) and membrane proteins (*red*). (Adapted from Liljas *et al.*,[16] with permission)

numerous functions.[1–3] A schematic picture reflecting this basic organization is shown in Figure 7.1.

Owing to their unique structural and functional organization, membranes act not only as cellular compartmental barriers but also tightly control, as in the case of the plasma membrane, the communication of a cell with its near surroundings. During evolution, eukaryotic cells optimized their internal working organization by distributing tasks to numerous intracellular organelles.[1,4,16] These organelles need to be separated from each other to be able to execute their specialized functions simultaneously; this is achieved by wrapping them in a lipid-based membrane envelope.[4,17] To fulfill the diverse functions of a cell and its organelles, membranes are highly optimized. By varying the lipid composition and associated proteins, biological membranes can fine-tune their physiological properties so as to modulate all these functions tightly. While parts of their physiological features are regulated by the constituent proteins, other properties such as fluidity and permeability are mainly determined by the lipid composition, as well as the specific structural and dynamic organization of the relevant membrane layer.[16–21] Most eukaryotic membranes use phospholipids as their major structural constituents.[1,22–24] These amphiphilic molecules have geometric and physicochemical features which provide them with the unique ability to arrange spontaneously into bilayers in an aqueous environment (as visualized in Figure 7.1), where the lipid molecules are held together solely by non-covalent physical and thermodynamic driving forces.[2,19,25,26] Due to the amphiphilic features of the individual lipid molecules, the generated lipid bilayer leaflet possesses a hydrophilic surface and a hydrophobic membrane interior. Therefore, lipid membranes can interact *via* their hydrophilic interface with their aqueous environment and soluble proteins, ions and other molecular species.[14,27–31] Simultaneously, *via* their hydrophobic interior they control membrane fluidity and also largely the permeability for ions and specific molecules. To optimize their function with respect to the membrane interface and interior properties, lipid compositions can be varied by altering the nature of the lipid headgroups and the type of fatty acids.[4,17,19,22] Because of this huge variability, an average eukaryotic cell contains 1200–1500 different lipid species for its specific purposes, while in humans the total number of lipids may exceed tens of thousands.[4,22,32] For a long time this huge lipid diversity was mysterious, since lipids were assumed to be merely simple amphiphilic building blocks for the basic bilayer structure, with proteins carrying out all the physiological tasks. Therefore, a new area in lipid membrane research has emerged, "lipidomics", with a focus on providing a basic understanding of this lipid variation in a cellular context.[17,19] With an increasing interest in lipids, these molecules have revealed important functional and physiological roles. Their tasks cover areas such as "raft" formation, involvement in signaling, which also includes membrane protein activation, and even regulation of ion channels. Therefore, lipids also have potential as novel biomarkers for diseases, especially in the area of cancer diagnosis, prognosis and potential therapeutic indications.[17,21,33,34]

7.2 Solid-State NMR: Insight into the Organization of Lipid Bilayers

7.2.1 General

To provide atomic-level insight into the structural and dynamic organization of membranes and changes due to physiological activities, NMR spectroscopy has emerged as a powerful tool. Compared to other biophysical techniques that often rely on bulky and disruptive reporter molecules, NMR uses non-perturbing reporter nuclei inside the lipid molecule of interest and can, therefore, provide atomic resolution for different molecular segments.[11,13,35–40] By observing the protons within the lipid species by high-resolution solid-state MAS methods (^{1}H HR MAS NMR), the molecular organization can be monitored across the membrane, ranging from the hydrophilic headgroup region and the lipid backbone to the hydrophobic membrane core formed by the lipid acyl chains.[11,12,35,37] To obtain a more detailed view of the membrane interface and its response to physiological events like protein association, fusion or conversion into non-lamellar phases, the naturally abundant phosphorus-31 spin, which is located in the lipid headgroups of phospholipids, is exploited, and also the inherent nitrogen-14 in the choline region of phosphatidylcholine (PC) lipids.[5,6,8,13–15,41] In combination with MAS NMR methods, a molecular view can be provided for each lipid component individually in mixed lipid systems and even in natural membranes, which opens the potential for lipid profiling work of intact biological systems by natural abundance ^{31}P MAS NMR and ^{1}H HR MAS NMR.[42,43]

However, prior to the routine use of MAS NMR for spectral separation of individual lipid components, wide-line solid-state NMR was used to study mainly the basic phase behavior and dynamical features of PC membranes and changes induced by external membrane effectors such as proteins, ions, pH, temperature, *etc.*[5,6,27,44] Especially in these areas, wide-line static NMR spectra are still used nowadays.[45–47] However, as seen in Figure 7.2a and

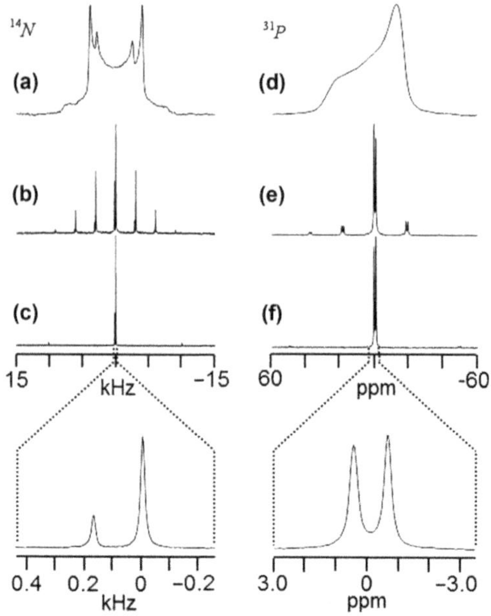

Figure 7.2 Concept of natural abundance solid-state ^{14}N and ^{31}P MAS NMR of multilamellar lipid vesicles (MLVs). *Left*: ^{14}N NMR spectra of dimyristoylphosphatidylcholine (DMPC)/didodecyldimethylammonium bromide (DDAB) vesicles at a 80/20 molar ratio: (a) static spectrum; (b) MAS spectrum at 3 kHz sample spinning frequency; and (c) MAS at 10 kHz. *Right*: ^{31}P NMR spectra of lipid vesicles composed of a mixture of DMPC/dimyristoylphosphatidylglycerol (DMPG) lipids at a 50/50 molar ratio: (d) static spectrum; (e) MAS spectrum at 750 Hz spinning speed; and (f) MAS spectrum at 2 kHz spinning speed. Experiments were performed at 308 K using Chemagnetics (USA) NMR spectrometers at 400 MHz proton frequency for ^{14}N NMR studies and at a 100 MHz proton frequency for ^{31}P NMR experiments.
(Reproduced from Lindström *et al.*,[64] with permission from the PCCP Owner Societies)

7.2d, respectively, the typical NMR lineshapes represent the general organization of a mixed lipid bilayer in a liquid-crystalline L_α phase, but lack sufficient spectral resolution to identify the different lipid species and their individual sub-spectra. To relieve this problem, deuterium nuclei were introduced by synthetic means in the headgroup or fatty acid region of the lipid component of interest.[7,48–52] This way, membrane interactions could be addressed more specifically by monitoring headgroup changes in response to variations in the membrane surface potential, *e.g.* upon association of peripheral proteins by electrostatic interactions.[51,53,54] Furthermore, by using deuterated acyl chains, the disordering of the hydrophobic core due to insertion of peptides and proteins, free fatty acids or other lipid components could be investigated.[48,55–60] In addition, changes in lipid dynamics and conformation occurring in the headgroup and fatty acid region can be probed. However, this technique is restricted to specifically deuterated lipids and only allows monitoring of this compound, with no direct information on the behavior of other lipid components in mixed bilayers. Additionally, natural membranes cannot be studied this way on an individual lipid level.

7.2.2 Exploiting Natural Abundance Nuclei in Lipids by MAS NMR

A major breakthrough in NMR of lipid bilayers occurred with the development of high-resolution MAS techniques. By combining MAS with the exploitation of the natural abundance ^{31}P in phospholipids, observations of the individual lipid components were possible since they were resolved in the NMR spectra according to their differing isotropic chemical shifts.[13,61,62] In the same way, the natural abundant ^{14}N nucleus in PC lipids could be used as an NMR probe.[13] By observing both nuclei in PC-containing membranes, the electrostatic surface potential of mixed membranes can be monitored in a kind of "natural voltmeter" approach[13,15] (see Figure 7.2 for more details). Figure 7.2 shows typical ^{14}N and ^{31}P NMR spectra from lipid bilayers in their physiologically relevant liquid-crystalline phase, where the individual membrane constituents possess considerable internal motional freedom and undergo fast axial rotation around their long axis.[44,48,63] The static ^{14}N NMR spectrum in Figure 7.2a reflects this situation well, since it reveals as expected a highly symmetric lineshape for a spin-1 system. The static ^{31}P NMR spectrum also reflects this situation with a typical spin-1/2 powder lineshape for vesicles of lipids undergoing axially symmetric fast motional processes.

In both cases, the lineshapes comprise a superposition of two sub-spectra, whose specific NMR properties can be deconvoluted by applying MAS conditions. In the case of ^{14}N, the sub-spectra of DMPC and DDAB are both dominated by the size and orientation dependence of the anisotropic quadrupolar interactions of the nitrogen nuclei with the applied magnetic field.[8,13,41,64] Already, applying a moderate spinning rate of 3 kHz reveals (Figure 7.2b) the chemical shift differences between the two quaternary ammonium groups of DMPC and DDAB. The size of their interactions, as seen in their different quadrupolar splitting values, reflects the fast rotation around the long axis of DDAB and DMPC, with the latter also including the tilt of the whole choline group relative to the averaging axis and the fast pre-averaging of the quadrupole moment around the C–N bond of the whole C_β–$N(CH_3)_3$ segment.[13] The magnitude of the quadrupole splitting $\Delta\nu_Q$ for the DMPC compound in the NMR spectra, therefore, can be directly related to the tilt of the P–N dipole in the DMPC headgroup with respect to the molecular long axis. This reflects the changes in tilt angle due to variation of the membrane surface potential, as systematically studied by us, with follow-up studies by colleagues.[13,15,65] Using the same approach for the chemical shift anisotropy of the phosphate group of DMPC (either static ^{31}P or analysis of slow-spinning MAS NMR spectra as in Figure 7.2e), the change in phosphate orientation can be monitored in response to the membrane surface potential. Applying MAS conditions at higher spinning frequencies (Figure 7.2c and 7.2f), the corresponding isotropic ^{14}N and ^{31}P chemical values reflect the size of the electric field at the choline respective phosphate region.[13,51]

This non-invasive NMR approach of exploiting the naturally occurring ^{14}N and ^{31}P nuclei as reporters, in combination with spectral separation of the individual membrane components under MAS conditions, provides a unique ability to obtain new insight at a molecular level into numerous biological processes where lipids and membranes play crucial roles. Applications in this area of monitoring the membrane surface potential by ^{31}P/^{14}N MAS NMR range from electrostatically driven association of proteins and peptides to their target membranes (including antimicrobial proteins), to the effect of drugs and ions on membrane interface properties, and finally the impact of specific lipid compounds on the membrane organization. In all of these cases the negatively charged phosphate regions and the positively charged quaternary N-choline groups at the interface of PC-containing membrane systems can be probed directly.

Especially, study of the phosphate region in lipids by ^{31}P MAS NMR has been widely utilized in investigating the role of specific lipid components and the membrane surface charge in interactions with peripheral proteins and peptides involved in numerous cellular processes. In general, these interactions are controlled by electrostatic interactions between negatively charged membrane surfaces and positively charged residues of the proteins.[61,66–68] In most cases, lipid-specific interactions are not required but only the presence of negatively charged lipids. Therefore, in many studies researchers replace, *e.g.* the negatively charged phosphatidylserine (PS) with phosphatidylglycerol (PG) lipids, without changing the affinity of the proteins of interest to their negatively charged target membranes. This has been shown in great detail for amyloidogenic proteins whose electrostatically driven membrane adsorption could induce accelerated aggregation due to an increase in local surface concentration.[69] By varying the amount of negatively charged lipids in the model membranes, the electrostatic nature of amyloid-β (Aβ) association is clearly seen, where an increase in membrane potential increased the propensity towards aggregation into β-sheet structured assemblies and fibrils with toxic properties.[68,69] By combining ^{31}P MAS NMR with circular dichroism (CD) measurements of protein structural arrangements, the folding behavior of amyloid proteins at membrane surfaces was shown to fundamentally differ from the situation in solution. This way an additional role of cellular membranes in the process of pathological conversions of amyloidogenic proteins, namely as two-dimensional misfolding/aggregation templates, could be established.[69] This additional function of (pathologically altered?) membranes, which may also play an important role in disease onset, has been observed in increasing numbers of amyloidogenic disorders including Parkinson's disease, diabetes mellitus and others.[70] Even in amyotrophic lateral sclerosis (ALS) the main culprit, the cytosolic soluble Cu-Zn superoxide dismutase 1 (SOD1) protein, was found to interact with charged membrane interfaces in its metal-free aggregation prone apo-state.[69]

While in the vast majority of cases the electrostatic association process depends only on the presence of membrane surface charge and not on the specific type of lipid, there are also cases where lipid specificity is important. Studying the detail of the membrane-mediated function of the antibiotic nisin, Bonev and co-workers could show, by exploiting ^{31}P MAS in combination with cross polarization (CP) NMR techniques (CP conditions enhance signal of specifically bound lipid component), that nisin exerts its antibacterial action by targeting specific pyrophosphate groups of lipid II, a cell wall precursor found in the outer leaflet of the bacterial plasma membrane.[62] In contrast, other lipid components of the mixed model membranes used in their experiments were not affected. They concluded that the highly conserved pyrophosphate moiety was not susceptible to bacterial adaptations that might introduce antimicrobial drug resistance. Since the lipid information is not encoded in the genome, unlike protein drug targets, lipid targeting antimicrobial peptides have been working successfully in our innate system for many thousand years without any signs of losing their effectiveness against bacterial strains.[71] By using a similar ^{31}P MAS NMR approach the group of Williamson identified the negatively charged PG as being present at high-affinity non-annular lipid binding sites of the potassium KcsA channel.[72] Changes in the chemical shift values for this lipid in the presence of the protein, upon incorporation into mixed lipid membranes, suggested a purely electrostatic interaction at these binding sites, which seems to be essential for the tight regulation of channel activity by its lipid environment. Removal of the PG lipid compound or site-directed removal of basic residues in lipid binding sites of the channel aggravates its function.

7.2.3 Tracking Lipids *ex vivo/in vivo* in Complex Systems by ^{31}P and ^{1}H MAS NMR

Obtaining comprehensive lipid profiles of intact biological systems and monitoring their individual behavior during physiological activities *in situ* in a non-invasive and non-destructive way at atomic resolution is one of the most challenging tasks in the area of lipidomics. Fortunately, solid-state (HR) MAS NMR spectroscopy has emerged as a powerful technique to study intact tissues, living cells, and mitochondria to elucidate specific biochemical processes by generating chemical profiles of the unique sets of macromolecules and metabolites.[34,42,43,73] By retaining system integrity, this NMR approach provides direct comparison with consecutive histopathological investigation of the same specimen. In addition, NMR can also evaluate biopsies prior to and after treatment. While ^{1}H HR MAS NMR methods are more widely used in the area of NMR metabolomics of intact specimens, direct focus on the ^{31}P nucleus enables a direct follow-up of the behavior of membrane phospholipids in a huge variety of intact biological systems, ranging from organelles to material originating from plants, tumors, *etc.*

By using fully functional, contaminant-free mitochondria as readily accessible systems in combination with a non-invasive, solid-state ^{31}P MAS NMR approach, changes in membrane morphology and the individual phospholipids in response to apoptotic stress can be monitored simultaneously.[42] This made it possible to follow the physiological track of the mitochondrion-specific phospholipid cardiolipin (CL) during the response period to apoptotic stimuli of living mitochondria. CL is known for exerting specific interactions with proteins in eukaryotic cells. This unique lipid with two negatively charged phosphate groups is intimately linked to cytochrome *c* function and apoptosis.[74] Prior to sensing of stress signals, the CL resonance in the corresponding NMR spectrum of intact mitochondria was invisible due to its strong affinity to the paramagnetic cytochrome *c* at the mitochondrial inner membrane. However, after induction of stress, this complex dissociates and the CL signal appeared in the NMR spectrum. This dissociaton event, which is followed by cytochrome *c* release from the mitochondria, is seen as the "point of no return" for a cell, ultimately leading to its death.

However, CL is also intimately involved in the regulation of opposing members of the Bcl-2 protein family, that meet at the outer mitochondrial membrane system where they tightly regulate the fate of the organelle and finally the cell.[75] Biophysical studies were carried out where CL was specificially incorporated into mitochondrial contact sites (CS), mimicking lipid vesicles.[29] By using ^{1}H and ^{31}P MAS NMR, pronounced perturbations in the membrane interface region by the membrane targeting Bax-α1 peptide – a functional important region of the pro-apopotic Bax protein – were only observed in the presence of CL. This interaction also promoted the peptide to adopt a predominantly α-helical structure, indicating a subsequent insertion into the membrane as part of the irreversible membrane perforation by activated Bax during its pro-apoptotic action.[76]

However, obtaining lipid profiles of intact biological material is not restricted to lipid headgroup signatures by ^{31}P MAS NMR. While ^{1}H HR MAS NMR can provide a general overview about all metabolites, including lipids and their composing fatty acids,[34,43] the ^{13}C nucleus can be exploited as an additional natural-abundance NMR reporter to provide very specific information about the profile of the fatty acid region of membranes. By using ^{1}H HR MAS NMR, Stenman *et al.* have provided metabolic profiles of intact prostate tissues.[73] Additional 2D ^{1}H–^{1}H and ^{1}H–^{13}C HR MAS NMR experiments yielded unambiguous identification of *n*–6 polyunsaturated fatty acids (PUFAs) in prostate cancer tissues. These *n*–6 PUFAs are known to promote prostate cancer.

By retaining system integrity, this non-invasive HR MAS NMR approach allows direct comparison with consecutive histopathological investigation of the same specimen.[43] Therefore, in the area of chemical fingerprinting of intact systems, ^{1}H HR MAS NMR is most widely used and, for example, is ideal to follow treatment success of cancer therapies.[77] However, by using ^{31}P and ^{13}C as additional reporters, more detailed information can be provided about specific groups of metabolites and macromolecules and, in the case of membranes, crucial information about their various phospholipids and the core fatty acid species. Nevertheless, this approach is not only restricted to the kind of static profiling of intact specimens. In the context of a "molecular system biology" approach, ^{31}P in combination with ^{1}H HR MAS NMR methods are also ideal to study physiological processes – from the lipid point of view – at a more complex level compared to using defined lipid mixtures. These solid-state NMR methods allow lipid profiling not only of intact systems, such as mitochondria or intact tissues, but also can monitor changes in the lipid patterns due to physiological activities or upon treatment, *e.g.* of cancer tissues. While ^{1}H HR MAS NMR is used routinely for this type of metabolic profiling, ^{31}P MAS NMR can provide additional diagnostic power. By combining with specific filtering methods, lipid-specific metabolic fingerprints can be provided together with changes in membrane morphology, such as formation of non-lamellar phases or membraneous fragments.[42] In particular, by exploiting the information inherent in these lipid profiles and their changes by multivariate analysis methods (which might include also complete proton derived metabolic profiles), not only the regulation of lipid metabolic pathways may be elucidated but lipids as novel potential biomarkers might be identified which could be useful in diagnostics and prognosis of tumors and tailored treatment strategies in a clinical environment.[78]

7.3 The Role of Oxidized Phospholipids in Membrane Function: Regulation of Mitochondria-Mediated Apoptosis

Under conditions of oxidative stress in the cell, various biological components are altered and damaged, including the membrane systems of the organelles.[33,79] Because of having oxidizable double bonds, (poly)-unsaturated diacyl- and alk(en)ylacyl-glycerophospholipid membranes are highly prone to react with reactive oxygen species (ROS) to form oxidized phospholipids (OxPls). These lipid species carry polar functional groups, *i.e.* aldehyde or carboxyl moieties, at the truncated *sn-2* acyl chain. Since PC is the most commonly found headgroup in mammalian cells, the majority of OxPls belongs to this class of lipids. When situated in biological membranes, the polar moiety at the end of the truncated *sn-2* fatty acid chain triggers a reorientation of this acyl chain towards the aqueous phase (see Figure 7.3) due to the energy penalty of embedding a polar group in the hydrophobic membrane core.[80,81] This dramatic change in the organization of individual

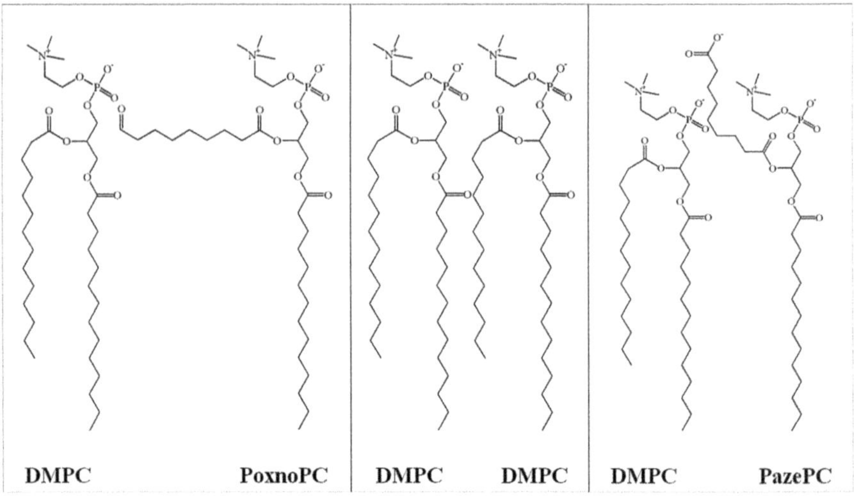

Figure 7.3 DMPC bilayers: schematic illustration of the different reorientation position of the *sn*-2 acyl chain of OxPls containing aldehyde [1-palmitoyl-2-(9′-oxononanoyl)-*sn*-glycero-3-phosphocholine (PoxnoPC)] or carboxyl [1-palmitoyl-2-azelaoyl-*sn*-glycero-3-phosphocholine (PazePC)] moieties at their ω-position. (Reproduced from ref. 84)

lipid species in the bilayer can exert dramatic effects on the structure and dynamics of OxPl-containing membranes, affecting both lipid–lipid and lipid–protein interactions, thereby altering the functions of membrane proteins.[81,82] It is, therefore, not surprising that OxPls have been shown to be involved in various biological processes, ranging from aging and anti-inflammatory processes to disorders such as atherosclerosis, Alzheimer's disease and diabetes.[83] Owing to these findings, interest in studying the impact of OxPls on the biophysical properties of membranes and their involvement in biochemical pathways has been growing in recent years.

The reversal of the truncated *sn-2* acyl chain can give rise to an increase in the average area per lipid and hence a reduction in the thickness of the bilayer.[85] Also, cholesterol has been shown to be sequestered from OxPl-rich regions, leading to formation of liquid ordered-phase-enriched domains.[86] Furthermore, the presence of OxPls enhances the bilayer penetration of water molecules,[85,87] thereby increasing the polarity of the interior of the bilayer. As a consequence, lipid transfer between both leaflets (flip-flop) can be greatly facilitated, independent of the action of scramblases.[87] Since oxidative stress is one of the hallmarks of apoptosis,[88,89] OxPls could play an important role in the onset and progression of this evolutionary conserved, programmed cell death process. Apoptosis is crucial for sustaining the health of multicellular organisms and is, for example, involved in the regulation of tissue homeostasis and the removal of harmful or damaged cells. Cells that undergo apoptosis are eliminated by the action of caspases, triggered either *via* the extrinsic (death receptor-regulated) or intrinsic (mitochondria-mediated) pathway, which cleave and degrade various intracellular key proteins. Indeed, NADPH oxidase-induced apoptosis is associated with the oxidation of PC, phosphatidylserine (PS) and phosphatidylethanolamine (PE).[90] Furthermore, oxidatively modified PS was suggested to be involved during the later stages of apoptosis, *i.e.* in the recognition of apoptotic cells during phagocytosis.[91] OxPls also appear to be able to activate sphingomyelinases, presumably leading to the formation of ceramides that subsequently can induce programmed cell death.[92,93] Mitochondria, the powerhouses of cells, are believed to be the major source of intracellular oxidative stress. This accumulation of ROS in the mitochondrial milieu can give rise to locally high concentrations of OxPls, which potentially can trigger the onset of the intrinsic apoptotic pathway. In support of this notion, OxPls has been shown to rapidly associate with mitochondria, giving rise to the release of caspase activating cytochrome *c* and execution of the intrinsic pathway.[94] Overexpression of the anti-apoptotic Bcl-2 protein Bcl-X_L (a major regulatory protein of the intrinsic pathway) was able to counter the detrimental effect of OxPls on both cells and isolated mitochondria.[95]

Nevertheless, to understand how OxPls are able to induce apoptosis and other cellular processes, including pathologies, molecular insights into how they modify the structure and dynamics of membranes are essential. To this end, differential scanning calorimetry (DSC) and solid-state NMR spectroscopy can be used as invaluable tools for gaining crucial knowledge of the biophysics of membranes.[84] While DSC reveals general macroscopic changes of lipid bilayers, NMR provides an additional atomic-level insight into the organization of these bilayers and the occurrence of changes during phase transition events. By using NMR, information is gained on the dynamics and conformational disorder of the different regions of the membrane on an individual molecular level, by simply probing the signals of defined nuclei inherent in the lipid compounds.

DSC profiles reveal the occurrence of thermotropic events of a membrane system as a function of temperature, in turn providing information of lipid organization in the bilayer. Since different lipid compositions display specific heat profiles, the impact of individual lipid species on the different membrane phases can be deduced. This is demonstrated in Figure 7.4, showing the different thermotropic phase behavior of phospholipid vesicles depending on the composition and the presence of OxPls. For this purpose, MLVs composed of either pure DMPC (Figure 7.4A, used as reference system) or mixtures of 1-palmitoyl-2-oleoyl-*sn*-glycero-phosphocholine (POPC), 1-palmitoyl-2-oleoyl-*sn*-glycero-3-phosphoethanolamine (POPE) and 1,1,2,2-tetramyristoylcardiolipin (TMCL) prior to and upon incorporation of the oxidized phospholipid PazePC were used. By comparing the thermograms obtained for DMPC, POPC : POPE : TMCL (43 : 36 : 21 molar ratio) and POPC : POPE : TMCL : PazePC (33 : 36 : 21 : 10 molar ratio) liposomes, the complexity of the melting events for vesicles mimicking mitochondrial CS prior to (Figure 7.4B) and upon onset of oxidative stress as simulated by the presence of PazePC (Figure 7.4C) is clearly visible.

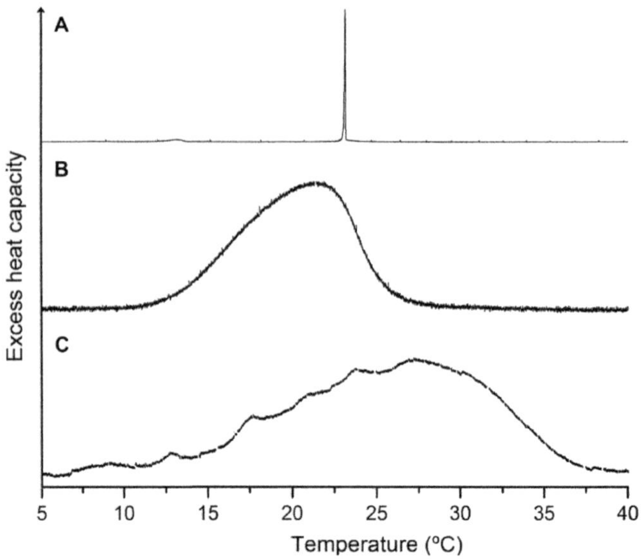

Figure 7.4 DSC thermograms of 3 mM multilamellar vesicles suspended in 25 mM sodium phosphate (pH 7.4), 100 mM KCl, 1 mM EDTA. MLV composition: (A) pure DMPC; (B) a mixture of POPC : POPE : TMCL at a 43 : 36 : 21 molar ratio; and (C) a mixture of POPC : POPE : TMCL : PazePC at a 33 : 36 : 21 : 10 molar ratio. The scanning range was from 5 to 45 °C at a scan rate of 2 °C h^{-1} and the thermograms were all scaled relative to each other with factor 75 for (B) and 200 for (C) with respect to the DMPC trace. These MLV samples were also used for the NMR experiments presented in Figures 7.5 and 7.6.

Figure 7.4A shows how phospholipid bilayers composed of only one component (here DMPC) possess a thermotropic phase behavior where two distinct phase transition events are observed. The pre-transition at lower temperatures corresponds to an increase in rotational freedom in the lipid headgroup regions, which reflects a transition from a gel phase state $L_{\beta'}$ to the so-called ripple-phase or $P_{\beta'}$ state.[2] Upon further increase in temperature, a melting of the fatty acid chains occurs, with the bilayer converting into the biologically significant, membrane liquid crystalline L_α phase, where the individual molecules also rotate around the long axis on a ns timescale and the bilayer has a high degree of segmental and chain disorder as well as dynamic flexibility.[2,44,63] For the CS mimicking lipid mixture (Figure 7.4B), these sharp phase transition events are no longer visible but broader peaks are seen in the thermogram, reflecting the different melting behavior of the lipid components in the mixture and also a potential domain formation where the individual lipid compositions can alter. Since mitochondria produce large amounts of ROS,[33,96] oxidative damage in the mitochondrial CS mimicking vesicles was simulated by replacing a fraction of the neutral POPC with the same molar amount of PazePC as an example of an oxidative lipid product. As seen in Figure 7.4C, the presence of this lipid

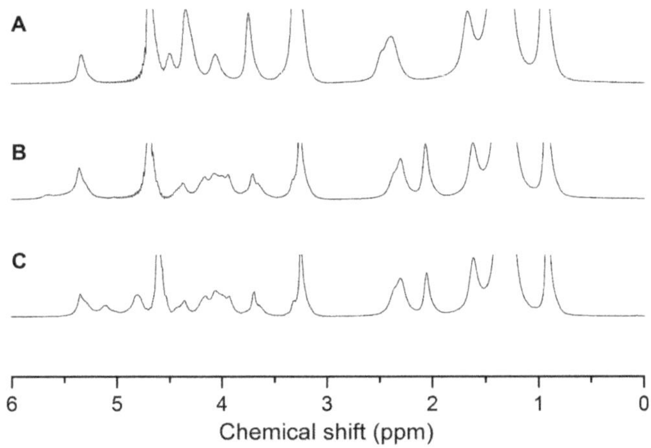

Figure 7.5 Expansion of a ^1H MAS NMR spectrum obtained at 5 kHz spinning rate on MLVs as prepared for the DSC experiments (Figure 7.4) and hydrated at a 60 : 1 D_2O-to-lipid molar ratio: (A) pure DMPC; (B) a mixture of POPC : POPE : TMCL at a 43 : 36 : 21 molar ratio; and (C) a mixture of POPC : POPE : TMCL : PazePC at a 33 : 36 : 21 : 10 molar ratio. The most intense peaks have been "cut" to allow comparison of the small resonances in the different spectra. All NMR spectra were collected at 308 K where all lipid bilayers are in the liquid-crystalline phase.

gives rise to a thermogram that has adopted a more asymmetric multicomponent shape, which reflects the large perturbation potential of OxPls due to their geometry and polar moieties at their chain ends.

In Figure 7.5 the ^1H MAS NMR spectra of these lipid bilayers provide a general overview of the whole membrane by producing resonances for protons in all parts of the individual lipid components. Since these resonances reflect the isotropic chemical shifts for specific groups, the signals can be readily assigned.[11,12,97] For example, signals from the acyl chains can be found in the spectral region ranging from 0 to 2.5 ppm. The signal at 2 ppm in Figure 7.5B and C corresponds to the hydrogen in the α-position relative to the double bond in the oleoyl acyl chain of POPC and POPE. NMR signals appearing above 3 ppm can be assigned to the lipid headgroup region mainly (one of the exceptions is the water resonance around 4.6–4.7 ppm). Signals from the choline group of the DMPC headgroup in Figure 7.5A appear at 3.3, 3.7 and 4.3 ppm and are clearly more intense than the glycerol signals, located at 4.1, 4.5 and 5.3 ppm, in the same spectrum. In the two other spectra the corresponding choline peaks are smaller with respect to their corresponding glycerol peaks, although a strong signal around 3.25–3.30 ppm is readily identified as belonging to the N-terminal methyl protons in the choline group. This reduction in intensity is not surprising, since the other two lipid components (TMCL and POPE) used in the mixtures for Figure 7.5B and C possess no choline group. In Figure 7.5B, the NMR signal at 5.4 ppm is an unresolved multicomponent peak originating from the glycerol-2 proton as well as the protons at the double bond between carbon-9 and carbon-10 in the oleoyl acyl chains, as have been observed for POPE/POPE mixtures. This peak exhibits an asymmetric lineshape with a left-hand shoulder. At most temperatures (not shown) this shoulder is characteristic for both mixtures of POPC : POPE : TMCL with and without PazePC, and presumably reflects that some of the above-mentioned overlapping signals are resolved in these lipid mixtures. However, as seen in Figure 7.5C, two new signals at 4.8 ppm and 5.1 ppm have appeared. Most likely these correspond to a shift of the signals due to the unsaturation.[97] Note that PazePC is analogous to POPC, and that the carboxyl group is located where the unsaturation in POPC would have been. Another possible explanation could be a difference in hydration pattern of the headgroups, and that a population of bound water is seen at 5.1 ppm, which is distinct from the water signal at 4.6 ppm. When analyzing ^{1}H NMR spectra at different temperatures (not shown), a pronounced nonlinearity in the temperature dependence of the water peak is observed for the mixture with PazePC at 308 K.

As seen in Figure 7.6, the ^{13}C CP MAS NMR spectra obtained for vesicles composed of mitochondrial model membranes, and doped with PazePC,

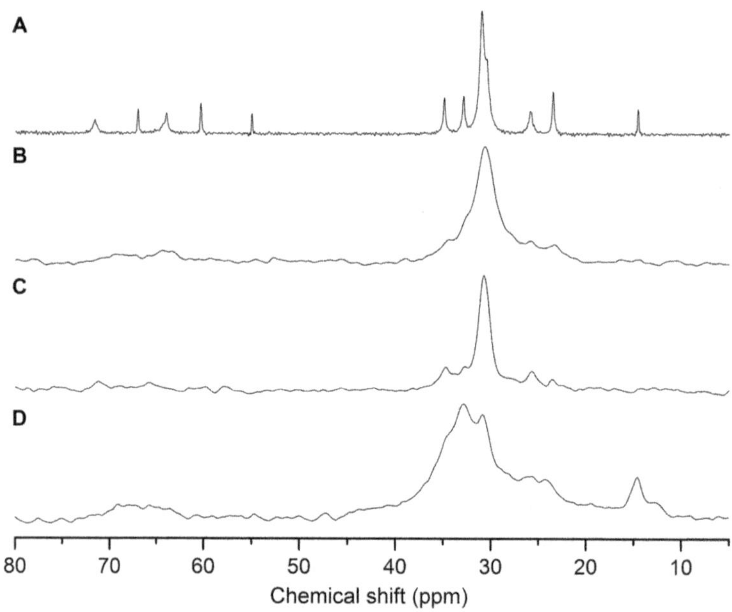

Figure 7.6 ^{13}C CP MAS NMR spectra obtained at 5 kHz spinning speed of MLVs as prepared for DSC experiments (see Figure 7.4) and hydrated at a 60 : 1 D_2O-to-lipid molar ratio: (A) pure DMPC bilayers in their liquid-crystalline phase at 308 K, used as reference spectrum. Spectra of a mixture of POPC : POPE : TMCL : PazePC at a 33 : 36 : 21 : 10 molar ratio were obtained at (B) 308 K, (C) 298 K, and (D) 288 K, respectively. The NMR spectrum in (D) was multiplied by a factor of two to compensate for the effect on the splitting of the main fatty acid CH_2 peaks at 30–34 ppm caused by two different phases.

display pronounced temperature dependence. Changes in these spectra can be correlated with the corresponding DSC trace in Figure 7.4C and provide the necessary molecular information about alterations in these membranes and their physicochemical properties. A comparison of the NMR spectra at 308 K (Figure 7.6B), 298 K (Figure 7.6C) and 288 K (Figure 7.6D) reveals immediately that some resonances broaden and disappear. This broadening reflects a change in the timescale of various motional processes that occur in the membrane.[98] Reducing the temperature can slow down the faster intra- and intermolecular motions, which then can contribute significantly to spin–spin relaxation and, therefore, to line broadening.[99] At the different temperature points studied, the different parts of the lipid molecules will be affected by different amounts in overall dynamics. The glycerol and head-group regions give rise to NMR signals between 54 and 72 ppm,[99,100] as seen in spectra Figure 7.6B and D but not in spectrum Figure 7.6C. The methyl signal at 14.6 ppm is only apparent in the NMR spectrum at 288 K. The signals from carbons ω–1 (23.5 ppm) and C3 (25.8 ppm) on the acyl chains lose intensity as the temperature increases from 288 to 308 K. The signal from ω–2 carbons at 32.8 ppm is reduced at 298 K compared to the higher temperature in the spectrum of Figure 7.6C. For most lipid bilayers in their liquid-crystalline state, the majority of acyl chain carbons, C4 to ω–3, display a peak at 30 ppm.[100] However, at the lowest temperature a population of these alkylic carbons has a signal which is shifted to around 32 ppm, corresponding to a membrane in the gel phase.[99] Another indication of the gel phase is that the methylene peaks at 14.6 ppm at 288 K are broadened, reflecting an insufficient averaging of their chemical shift anisotropy interactions due to reduced dynamics.

The results presented here demonstrate the strong influence of OxPls on the membrane organization. Recently, we showed that the incorporation of the same OxPl as used in this study (PazePC) triggered the soluble pro-apoptotic Bcl-2 family protein Bax to become persistently bound to the membrane.[101] Bax is one of the major executors of the intrinsic pathway and is believed to be mainly cytosolic in healthy cells, while during conditions that trigger cell death it changes conformation to incorporate into the mitochondrial outer membrane (MOM) where it forms pores through which cytochrome *c* can escape.[102,103] The presence of PazePC lipids clearly shifted the equilibrium of Bax molecules from being predominantly cytosolic to become largely membrane bound. Membrane-bound anti-apoptotic Bcl-2 family proteins, *e.g.* Bcl-2 or Bcl-X_L, can prevent the pore-forming activity of Bax until a certain amount of activated Bax is inserted into the MOM. The presence of OxPls could, therefore, be a key step in oxidative stress-induced apoptosis by increasing the population of membrane-associated Bax, a key factor in our novel concept of apoptotic regulation at the mitochondrion during apoptotic stress, as presented in Figure 7.7. When a certain critical level is reached, probably facilitated by activating BH3-only proteins (also belonging to the Bcl-2 protein family), activated Bax can overcome the inhibitory anti-apoptotic Bcl-2 proteins. The remaining membrane-bound

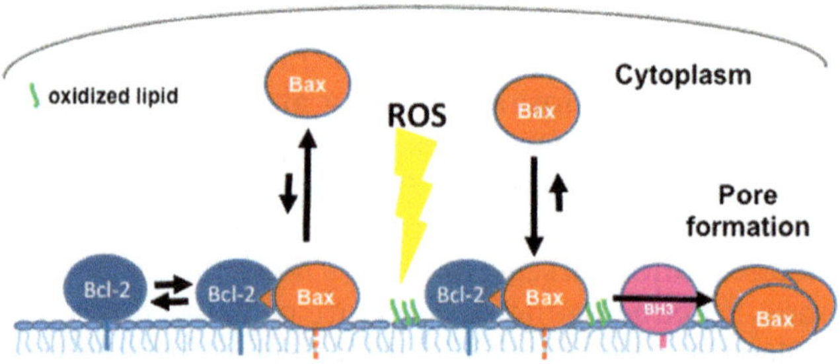

Figure 7.7 Our concept of the MOM as an active partner in apoptotic regulation. Membrane-mediated recognition of Bcl-2 by Bax and its apoptotic consequences is based on the "embedded together model".[105] *Left*: in a normal cell, Bcl-2 inhibits the few primed Bax molecules upon membrane binding; *middle*: ROS-induced production of OxPls (*green*) increases the population of primed Bax,[101,106] which can overcome the inhibitory Bcl-2 effect by direct binding[107] and induces, *via* membrane-bound BH3-only proteins, activation of mitochondrial outer membrane permeabilization (MOMP) and finally cell death (*right*).

fraction of Bax would then, upon activation by membrane-located BH3-only proteins such as tBid (truncated Bid), oligomerize into cytochrome *c* releasing pores, a mechanism suggested in 2008,[104] with further support recently based on a Bax crystal structure which indicates the mode of its oligomerization into the MOM, a key step in inducing apoptosis.[76]

Acknowledgements

We thank Martin Hof, Paavo Kinnunen, Quoc Dat Pham, Konrad Cyprych, Lenka Beranovka, Jan Procek and Khanh Linh for all their contributions in the area of oxidized lipids. We thank all colleagues and the KBC (Chemical Biological Centre) for their support. This work was financially supported by the Swedish Research Council, the Kempe Foundation, the Knut and Alice Wallenberg foundation ("NMR for Life" Programme), the EuroMEMBRANE programme (ESF) and Umeå Insamlingsstiftelse.

References

1. M. Luckey, *Membrane Structural Biology*, Cambridge University Press, New York, 2008.
2. *The Structure of Biological Membranes*, ed. P. L. Yeagle, CRC Press, Boca Raton, 3rd edn., 2011.
3. D. M. Engelman, *Nature*, 2005, **438**, 578.
4. G. van Meer, D. R. Voelker and G. W. Feigenson, *Nat. Rev. Mol. Cell Biol.*, 2008, **9**, 112.
5. B. de Kruijff, P. R. Cullis and G. K. Radda, *Biochim. Biophys. Acta*, 1975, **406**, 6.
6. P. L. Yeagle, *Acc. Chem. Res.*, 1978, **11**, 321.

7. H. U. Gally, W. Niederberger and J. Seelig, *Biochemistry*, 1975, **14**, 3647.
8. T. M. Rothgeb and E. Oldfield, *J. Biol. Chem.*, 1981, **256**, 6004.
9. F. Adebodun, J. Chung, B. Montez, E. Oldfield and X. Shan, *Biochemistry*, 1992, **31**, 4502.
10. J. Villalain, *Eur. J. Biochem.*, 1996, **241**, 586.
11. J. Forbes, C. Husted and E. Oldfield, *J. Am. Chem. Soc.*, 1988, **110**, 1059.
12. L. L. Holte and K. Gawrisch, *Biochemistry*, 1997, **36**, 4669.
13. F. Lindström, P. T. Williamson and G. Gröbner, *J. Am. Chem. Soc.*, 2005, **127**, 6610.
14. A. Ramamoorthy, D.-K. Lee, J. S. Santos and K. A. Henzler-Wildman, *J. Am. Chem. Soc.*, 2008, **130**, 11023.
15. J. P. Doux, B. A. Hall and J. A. Killian, *Biophys. J.*, 2012, **103**, 1245.
16. A. Liljas, L. Liljas, J. Piskur, G. Lindblom, P. Nissen and M. Kjeldgaard, *Textbook of Structural Biology*, World Scientific, New Jersey, 2009.
17. M. Oresic, V. A. Hanninen and A. Vidal-Puig, *Trends Biotechnol.*, 2008, **26**, 647.
18. A. Rietveld and K. Simons, *Biochim. Biophys. Acta, Rev. Biomembr.*, 1998, **1376**, 467.
19. O. G. Mouritsen, *Life – As a Matter of Fat: The Emerging Science of Lipidomics*, Springer, Berlin, 2005.
20. M. Bohn, E. Heinz and S. Luthje, *Arch. Biochem. Biophys.*, 2001, **387**, 35.
21. P. V. Escriba, J. M. Gonzalez-Ros, F. M. Goni, P. K. J. Kinnunen, L. Vigh, L. Sanchez-Magraner, A. M. Fernandez, X. Busquets, I. Horvath and G. Barcelo-Coblijn, *J. Cell. Mol. Med.*, 2008, **12**, 829.
22. D. Marsh, *Handbook of Lipid Bilayers*, CRC Press, Boca Raton, 2nd edn., 2013.
23. K. W. J. Wahle, *Proc. Nutr. Soc.*, 1983, **42**, 273.
24. D. Ardail, J. P. Privat, M. Egret-Charlier, C. Levrat, F. Lerme and P. Louisot, *J. Biol. Chem.*, 1990, **265**, 18797.
25. E. V. Brink-van der Laan, J. A. Killian and B. de Kruijff, *Biochim. Biophys. Acta, Biomembr.*, 2004, **1666**, 275.
26. P. A. Janmey and P. K. J. Kinnunen, *Trends Cell Biol.*, 2006, **16**, 538.
27. M. F. Brown and J. Seelig, *Nature*, 1977, **269**, 721.
28. A. Watts, *J. Bioenerg. Biomembr.*, 1987, **19**, 625.
29. M. A. Sani, E. J. Dufourc and G. Gröbner, *Biochim. Biophys. Acta, Biomembr.*, 2009, **1788**, 623.
30. I. Marcotte, K. L. Wegener, Y. H. Lam, B. C. S. Chia, M. R. R. de Planque, J. H. Bowie, M. Auger and F. Separovic, *Chem. Phys. Lipids*, 2003, **122**, 107.
31. N. Weizenmann, D. Huster and H. A. Scheidt, *Biochim. Biophys. Acta, Biomembr.*, 2012, **1818**, 3010.
32. O. G. Mouritsen, *Phys. Chem. Chem. Phys.*, 2011, **13**, 19195.
33. G. O. Fruhwirth, A. Loidl and A. Hermetter, *Biochim. Biophys. Acta, Mol. Basis Dis.*, 2007, **1772**, 718.

34. T. E. Sjobakk, R. Vettukattil, M. Gulati, S. Gulati, S. Lundgren, I. S. Gribbestad, S. H. Torp and T. F. Bathen, *Int. J. Mol. Sci.*, 2013, **14**, 2104.
35. W. Guo and J. A. Hamilton, *Biochemistry*, 1995, **34**, 14174.
36. M. Auger, *Biophys. Chem.*, 1997, **68**, 233.
37. S. Everts and J. H. Davis, *Biophys. J.*, 2000, **79**, 885.
38. K. Gawrisch, N. V. Eldho and I. V. Polozov, *Chem. Phys. Lipids*, 2002, **116**, 135.
39. D. E. Warschawski, M. Traikia, P. F. Devaux and G. Bodenhausen, *Biochimie*, 1998, **80**, 437.
40. V. Zorin, F. Ciesielski, D. C. Griffin, M. Rittig and B. B. Bonev, *Magn. Reson. Chem.*, 2010, **48**, 925.
41. D. J. Siminovitch, M. F. Brown and K. R. Jeffrey, *Biochemistry*, 1984, **23**, 2412.
42. M. A. Sani, O. Keech, P. Gardeström, E. J. Dufourc and G. Gröbner, *FASEB J.*, 2009, **23**, 2872.
43. O. Beckonert, M. Coen, H. C. Keun, Y. Wang, T. M. Ebbels, E. Holmes, J. C. Lindon and J. K. Nicholson, *Nat. Protoc.*, 2010, **5**, 1019.
44. E. J. Dufourc, C. Mayer, J. Stohrer, G. Althoff and G. Kothe, *Biophys. J.*, 1992, **61**, 42.
45. S. Morein, J. A. Killian and M. M. Sperotto, *Biophys. J.*, 2002, **82**, 1405.
46. D. I. Fernandez, A. P. Le Brun, T. C. Whitwell, M. A. Sani, M. James and F. Separovic, *Phys. Chem. Chem. Phys.*, 2012, **14**, 15739.
47. V. Zhendre, A. Grelard, M. Garnier-Lhomme, S. Buchoux, B. Larijani and E. J. Dufourc, *PloS One*, 2011, **6**(9), e23859.
48. J. H. Davis, *Biochim. Biophys. Acta*, 1983, **737**, 117.
49. N. J. Ryba, C. E. Dempsey and A. Watts, *Biochemistry*, 1986, **25**, 4818.
50. J. Seelig, P. M. Macdonald and P. G. Scherer, *Biol. Chem.*, 1987, **368**, 1272.
51. M. Roux, J. M. Neumann, R. S. Hodges, P. F. Devaux and M. Bloom, *Biochemistry*, 1989, **28**, 2313.
52. P. M. Macdonald, J. Leisen and F. M. Marassi, *Biochemistry*, 1991, **30**, 3558.
53. M. R. Morrow, S. Taneva, G. A. Simatos, L. A. Allwood and K. M. W. Keough, *Biochemistry*, 1993, **32**, 11338.
54. J. A. G. Areas, G. Gröbner, C. Glaubitz and A. Watts, *Biochemistry*, 1998, **37**, 5582.
55. K. Weisz, G. Gröbner, C. Mayer, J. Stohrer and G. Kothe, *Biochemistry*, 1992, **31**, 1100.
56. L. L. Holte, S. A. Peter, T. M. Sinnwell and K. Gawrisch, *Biophys. J.*, 1995, **68**, 2396.
57. B. B. Bonev and M. R. Morrow, *Phys. Rev. E*, 1997, **55**, 5825.
58. K. A. Henzler-Wildman, G. V. Martinez, M. F. Brown and A. Ramamoorthy, *Biochemistry*, 2004, **43**, 8459.
59. F. Lindström, S. Thurnhofer, W. Vetter and G. Gröbner, *Phys. Chem. Chem. Phys.*, 2006, **8**, 4792.

60. A. Vogel, G. Reuther, K. Weise, G. Triola, J. Nikolaus, K. T. Tan, C. Nowak, A. Herrmann, H. Waldmann, R. Winter and D. Huster, *Angew. Chem. Int. Ed.*, 2009, **48**, 8784.
61. M. A. Carbone and P. M. Macdonald, *Biochemistry*, 1996, **35**, 3368.
62. B. B. Bonev, E. Breukink, E. Swiezewska, B. De Kruijff and A. Watts, *FASEBJ.*, 2004, **18**, 1862.
63. J. Seelig, *Q. Rev. Biophys.*, 1977, **10**, 353.
64. F. Lindström, M. Bokvist, T. Sparrman and G. Gröbner, *Phys. Chem. Chem. Phys.*, 2002, **4**, 5524.
65. J. S. Santos, D. K. Lee and A. Ramamoorthy, *Magn. Reson. Chem.*, 2004, **42**, 105.
66. T. J. T. Pinheiro and A. Watts, *Biochemistry*, 1994, **33**, 2459.
67. B. B. Bonev, W. C. Chan, B. W. Bycroft, G. C. K. Roberts and A. Watts, *Biochemistry*, 2000, **39**, 11425.
68. M. Bokvist, F. Lindström, A. Watts and G. Gröbner, *J. Mol. Biol.*, 2004, **335**, 1039.
69. C. Aisenbrey, T. Borowik, R. Byström, M. Bokvist, F. Lindström, H. Misiak, M. A. Sani and G. Gröbner, *Eur. Biophys. J.*, 2008, **37**, 247.
70. S. M. Butterfield and H. A. Lashuel, *Angew. Chem. Int. Ed.*, 2010, **49**, 5628.
71. Y. Shai, *Biopolymers*, 2002, **66**, 236.
72. P. Marius, M. R. de Planque and P. T. Williamson, *Biochim. Biophys. Acta, Biomembr.*, 2012, **1818**, 90.
73. K. Stenman, J. B. Hauksson, G. Gröbner, P. Stattin, A. Bergh and K. Riklund, *Magn. Reson. Mat. Phys. Biol. Med.*, 2009, **22**, 327.
74. F. Gonzalvez and E. Gottlieb, *Apoptosis*, 2007, **12**, 877.
75. S. Cory and J. M. Adams, *Nat. Rev.*, 2002, **2**, 647.
76. P. E. Czabotar, D. Westphal, G. Dewson, S. Ma, C. Hockings, W. D. Fairlie, E. F. Lee, S. G. Yao, A. Y. Robin, B. J. Smith, D. C. S. Huang, R. M. Kluck, J. M. Adams and P. M. Colman, *Cell*, 2013, **152**, 519.
77. M. D. Cao, B. Sitter, T. F. Bathen, A. Bofin, P. E. Lonning, S. Lundgren and I. S. Gribbestad, *NMR Biomed.*, 2012, **25**, 369.
78. J. K. Nicholson, E. Holmes, J. M. Kinross, A. W. Darzi, Z. Takats and J. C. Lindon, *Nature*, 2012, **491**, 384.
79. V. A. Tyurin, Y. Y. Tyurina, P. M. Kochanek, R. Hamilton, S. T. DeKosky, J. S. Greenberger, H. Bayir and V. E. Kagan, *Methods Enzymol.*, 2008, **442**, 375.
80. L. Beranova, L. Cwiklik, P. Jurkiewicz, M. Hof and P. Jungwirth, *Langmuir*, 2010, **26**, 6140.
81. H. Khandelia and O. G. Mouritsen, *Biophys. J.*, 2009, **96**, 2734.
82. F. H. Greig, S. Kennedy and C. M. Spickett, *Free Radical Biol. Med.*, 2012, **52**, 266.
83. N. Leitinger, *Subcell Biochem.*, 2008, **49**, 325.
84. M. Wallgren, L. Beranova, Q. D. Pham, K. Linh, M. Lidman, J. Procek, K. Cyprych, P. K. J. Kinnunen, M. Hof and G. Gröbner, *Faraday Discuss.*, 2013, **161**, 499.

85. J. Wong-Ekkabut, Z. Xu, W. Triampo, I. M. Tang, D. P. Tieleman and L. Monticelli, *Biophys. J.*, 2007, **93**, 4225.
86. R. F. Jacob and R. P. Mason, *J. Biol. Chem.*, 2005, **280**, 39380.
87. R. Volinsky, L. Cwiklik, P. Jurkiewicz, M. Hof, P. Jungwirth and P. K. Kinnunen, *Biophys. J.*, 2011, **101**, 1376.
88. D. D. Newmeyer and S. Ferguson-Miller, *Cell*, 2003, **112**, 481.
89. H. U. Simon, A. Haj-Yehia and F. Levi-Schaffer, *Apoptosis*, 2000, **5**, 415.
90. A. Arroyo, M. Modriansky, F. B. Serinkan, R. I. Bello, T. Matsura, J. Jiang, V. A. Tyurin, Y. Y. Tyurina, B. Fadeel and V. E. Kagan, *J. Biol. Chem.*, 2002, **277**, 49965.
91. H. R. Horvitz, *Cancer Res.*, 1999, **59**, 1701s.
92. V. Ganesan, M. N. Perera, D. Colombini, D. Datskovskiy, K. Chadha and M. Colombini, *Apoptosis*, 2010, **15**, 553.
93. A. Loidl, E. Sevcsik, G. Riesenhuber, H. P. Deigner and A. Hermetter, *J. Biol. Chem.*, 2003, **278**, 32921.
94. R. Chen, L. Yang and T. M. McIntyre, *J. Biol. Chem.*, 2007, **282**, 24842.
95. R. Chen, A. E. Feldstein and T. M. McIntyre, *J. Biol. Chem.*, 2009, **284**, 26297.
96. J. Montero, M. Mari, A. Colell, A. Morales, G. Basanez, C. Garcia-Ruiz and J. C. Fernandez-Checa, *Biochim. Biophys. Acta, Bioenerg.*, 2010, **1797**, 1217.
97. W. Willker and D. Leibfritz, *Magn. Reson. Chem.*, 1998, **36**, S79.
98. E. Oldfield, F. Adebodun, J. Chung, B. Montez, K. D. Park, H. B. Le and B. Phillips, *Biochemistry*, 1991, **30**, 11025.
99. R. E. Jacobs and E. Oldfield, *Prog. Nucl. Magn. Reson. Spectrosc.*, 1980, **14**, 113.
100. G. Gröbner, C. Glaubitz and A. Watts, *J. Magn. Reson.*, 1999, **141**, 335.
101. M. Wallgren, M. Lidman, Q. D. Pham, K. Cyprych and G. Gröbner, *Biochim. Biophys. Acta, Biomembr.*, 2012, **1818**, 2718.
102. K. G. Wolter, Y. T. Hsu, C. L. Smith, A. Nechushtan, X. G. Xi and R. J. Youle, *J. Cell Biol.*, 1997, **139**, 1281.
103. R. Eskes, B. Antonsson, A. Osen-Sand, S. Montessuit, C. Richter, R. Sadoul, G. Mazzei, A. Nichols and J. C. Martinou, *J. Cell Biol.*, 1998, **143**, 217.
104. J. F. Lovell, L. P. Billen, S. Bindner, A. Shamas-Din, C. Fradin, B. Leber and D. W. Andrews, *Cell*, 2008, **135**, 1074.
105. B. Leber, J. Lin and D. W. Andrews, *Apoptosis*, 2007, **12**, 897.
106. B. Schellenberg, P. B. Wang, J. A. Keeble, R. Rodriguez-Enriquez, S. Walker, T. W. Owens, F. Foster, J. Tanianis-Hughes, K. Brennan, C. H. Streuli and A. P. Gilmore, *Mol. Cell*, 2013, **49**, 959.
107. M. Wallgren, M. Lidman, A. Pedersen, K. Brännström, B. G. Karlsson and G. Gröbner, *PloS One*, 2013, **8**, e61452.

CHAPTER 8

Structural Studies of Small Bioactive Compounds Interacting with Membranes and Proteins

SHIGERU MATSUOKA AND MICHIO MURATA*

JST · ERATO Lipid Active Structure Project, Project Research Centre for Fundamental Science, Osaka University, and Department of Chemistry, Graduate School of Science, Osaka University, 1-1 Machikaneyama, Toyonaka, Osaka 560-0043, Japan
*Email: murata@chem.sci.osaka-u.ac.jp

8.1 Introduction

Small molecules comprise the great majority of commercial drugs, as well as potential therapeutics in preclinical and clinical development, because of advantages in manufacturing and delivery.[1,2] To improve drug efficacy, it is vital to obtain the 3D structure of a lead compound when taking a biologically active form in complex with the target biomolecule. For this purpose, X-ray crystallography and solution NMR have been widely utilized to elucidate the structure of a drug–protein complex in co-crystals and solutions, respectively. However, the majority of the drug targets are non crystalline solids, such as cell membranes, cell walls and protein aggregates, where X-ray crystallography and solution NMR are rather

New Developments in NMR No. 3
Advances in Biological Solid-State NMR: Proteins and Membrane-Active Peptides
Edited by Frances Separovic and Akira Naito

Published by the Royal Society of Chemistry, www.rsc.org

powerless. To complement these two methods, solid-state NMR is gaining importance as a high-resolution spectroscopy for non-crystalline solid samples.

Biological systems that can be studied by solid-state NMR include lipid bilayers, protein aggregates, protein complexes/polymers, cell walls and even whole cells.[3–8] Among those, lipid bilayers and protein aggregates that are formed by using conventional methodologies are particularly problematic in structural studies. For instance, the structure of a drug–lipid complex in the lipid bilayer is usually dependent on bulk membrane lipids because the complex structure is significantly affected by various factors, including lipid composition and protein/lipid ratio. To obtain biologically relevant structures, experiments ideally should be carried out at a low drug concentration in the presence of an excess amount of membrane lipids, *e.g.* the binding sites for amyloid-specific probes in protein aggregates give rise to a similar problem as the binding sites cannot be isolated or concentrated. These conditions are unfavourable for X-ray crystallography. Solid-state NMR is the only methodology to access complex structures in lipid bilayers and protein aggregates at atomic resolution because the complex with an arbitrary ratio can be subject to NMR measurements.[9–11]

In this chapter, we introduce the basic methodologies and applications of solid-state NMR in structural studies of small bioactive compounds in membrane-bound and protein-aggregate forms.

8.2 Detection of Molecular Interactions of Small Compounds in Biological Solids

Interactions between small molecules and biological solids can be detected either by observing the perturbation of intramolecular magnetic interactions or by the direct measurement of intermolecular dipolar interactions. The former includes isotropic chemical shift, chemical shift anisotropy (CSA) and quadrupolar coupling. In the case of multiply labelled compounds, isotropic chemical shifts are essential for many NMR experiments. In addition, isotropic shifts can be utilized to identify a binding site on a solid target on the basis of chemical shift changes caused by small-molecule binding.[12,13] However, anisotropic magnetic interactions are strongly correlated with molecular motion and orientation, and often used to examine the mobility change of small molecules and/or target biological solids before and after complexation.[14–22] Particularly, anisotropic coupling is a powerful tool in investigating lipid bilayers: unlike free ligands in solution, membrane-bound small molecules have restricted molecular motion and orientation with respect to the membrane normal, which eventually produces anisotropic interactions with the external magnetic field. Moreover, the main constituents of bilayer membranes are small molecule lipids and, thus, intramolecular anisotropic couplings are also useful to examine drug binding to a membrane. On the other hand, intermolecular dipolar couplings provide direct evidence of intermolecular interactions between a ligand and lipid.[23–29] Analysis of the magnitude of an observed dipolar coupling may further provide an interatomic distance, which can be used as a restraint for modelling a complex structure. The accumulation of intermolecular dipolar couplings would thus be one of the most important experiments of solid-state NMR. In this chapter we briefly introduce basic solid-state NMR methodologies to obtain the NMR parameters required to elucidate interactions between small molecules and non-crystalline biological solids.

8.2.1 Isotropic Chemical Shift of Membrane Lipids and Small Compounds

To obtain isotropic chemical shifts, high-resolution solid-state NMR methodology,[30] which can be achieved by combining high-power proton decoupling[31,32] and magic angle sample spinning (MAS),[33,34] is essential. Without high-resolution techniques, solid samples give rise to extremely broad NMR lines, which are difficult to assign and, in most cases, even to observe. On the other hand, NMR signals from amorphous solids, including membranes and aggregates, contain more structural information than those from solution. Figure 8.1 shows magnetic interactions for spin-$\frac{1}{2}$ nuclei, comprising the Zeeman (with external magnetic field), chemical shift (shielding effect by electrons), *J*-coupling (nuclear–nuclear indirect interaction through bond electrons) and dipolar interactions (nuclear–nuclear direct interaction through space). For nuclear spins with $I>\frac{1}{2}$, such as ^{2}H and ^{17}O, the quadrupolar interaction (nuclear–electron magnetic field) also causes line broadening. The Zeeman interaction, the isotropic part of the chemical shift and the *J*-coupling are independent of the molecular orientation relative to the external magnetic field. The anisotropic part of chemical shift (chemical shift anisotropy, CSA), dipolar interaction and quadrupolar interaction depend on the molecular orientation relative to the

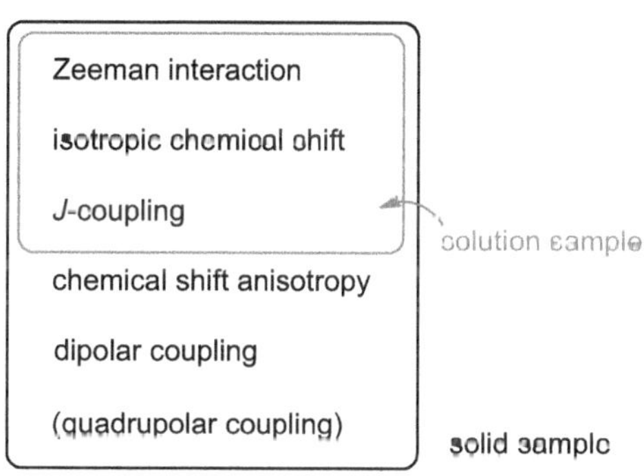

Figure 8.1 Observable NMR interactions in solution and solid samples.

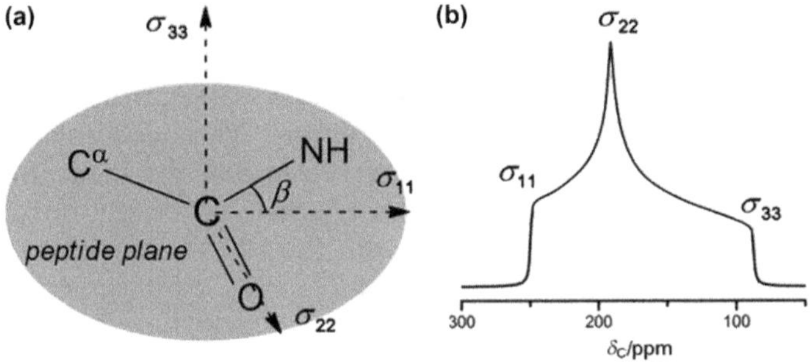

Figure 8.2 (a) ^{13}C chemical shift anisotropy tensor orientation in the molecular frame of the amide carbonyl carbon (δ_{11} = 249 ppm, δ_{22} = 189 ppm, δ_{33} = 88 ppm). (b) Ideal powder-type ^{13}C spectrum simulated with the CSA tensor of the amide carbonyl carbon.

external magnetic field, as well as on the orientation of a relevant interatomic vector to the molecular rotation axis within a membrane system. These effects are called anisotropic interactions. Amongst the above six interactions, three are observable in solution NMR where dissolved molecules of interest undergo fast motion, resulting in the cancellation of anisotropic magnetic interactions. However, in solid systems, the additional three interactions are effective due to the anisotropic nature of solids, thus contributing to further line broadening. For example, the effect of CSA on the ^{13}C carbonyl lineshape is shown in Figure 8.2.[35] In addition to chemical shifts, dipolar and quadrupolar interactions are also anisotropic and, thus, the sum of these interactions gives rise to extremely broad signals.

For the purpose of recording high-resolution solid-state NMR, anisotropic interactions need to be cancelled. To simplify the case, we examine ^{13}C, which is the most important spin-$\frac{1}{2}$ nucleus in organic and biological samples. Problematic anisotropic interactions in ^{13}C NMR are CSA and dipolar interactions. High resolution is attained in a solution state by intrinsic fast molecular reorientation, which cancels anisotropic interactions. To compensate for the lack of molecular motion, MAS and high-power proton decoupling are adopted in solid-state NMR experiments. Figure 8.3 presents the ^{13}C NMR spectra of an α-form crystal of [U-$^{13}C_2$,^{15}N]glycine measured under various conditions. Figure 8.3a shows the spectrum without high-resolution techniques, namely proton non-decoupled static ^{13}C NMR. The signals are extremely broadened by CSA and dipolar interactions and become unobservable. The static ^{13}C NMR of the same sample, but with high-power proton decoupling, is shown in Figure 8.3b. The proton-decoupling field was set to 100 kHz, which is effective to remove ^{1}H–^{13}C dipolar couplings (approximately 20 kHz). The resulting broad-line spectrum represents the anisotropic tensors of CSA and dipolar interactions between non-decoupled low-γ nuclei such as ^{13}C–^{13}C and ^{13}C–^{15}N. In addition to high-power ^{1}H decoupling, MAS was applied to remove the CSA and dipolar interactions between low-γ nuclei (Figure 8.3c,d).

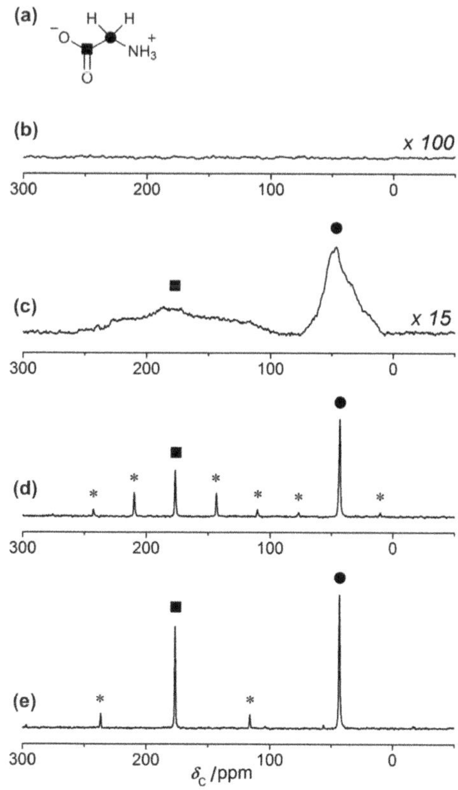

Figure 8.3 Solid-state ^{13}C NMR spectra at 150 MHz. (a) 5% [1,2-$^{13}C_2$,^{15}N]Gly was crystallized from aqueous ethanol and ground into a powder sample; (b) ^{13}C NMR recorded without ^{1}H decoupling or MAS; (c) ^{13}C NMR recorded with ^{1}H decoupling at 130 W; (d) ^{13}C NMR recorded with ^{1}H decoupling at 130 W and MAS at 5000 Hz; and (e) ^{13}C NMR recorded with ^{1}H decoupling at 130 W and MAS at 9091 Hz. Cross-polarization was used in all spectra to enhance sensitivity.
(Reproduced from *Yakugaku Zasshi*, 2012, **132**, 9, with permission from The Pharmaceutical Society of Japan.)

Another improvement in sensitivity can be implemented by the introduction of cross-polarization (CP).[36,37] ^{13}C NMR of solid systems bearing a long T_1 relaxation time requires a long time interval between two transients until thermal equilibrium is reached. Compared to ^{13}C, a ^{1}H nucleus has a larger polarization and a significantly shorter T_1 relaxation time. The $^{13}C\{^{1}H\}$ CP shortens the time interval and increases the efficiency of ^{13}C NMR measurements by transferring magnetization from ^{1}H to ^{13}C through the ^{13}C–^{1}H dipolar interaction. All spectra in Figure 8.3 were measured under $^{13}C\{^{1}H\}$ CP conditions.

High-resolution solid-state NMR provides isotropic chemical shifts regardless of the sample status. Particularly for analysis of multiply labelled samples, high resolution is required for resonance assignments.

8.2.2 Intramolecular Anisotropic Interactions of Membrane Lipids and Small Compounds

8.2.2.1 Static NMR of Non-oriented Membrane Systems

As described in the previous section, a static solid-state NMR spectrum embodies intact anisotropic interactions, which contain information about molecular mobility and orientation (Figure 8.4). However, the static spectrum of a non-labelled sample is usually too complicated to be analysed due to extremely broadened signals and severe signal overlap owing to multiple anisotropic interactions. To gain information about molecular mobility and orientation the spectrum needs to be simplified, usually by selective labelling with a stable isotope such as ^{13}C or ^{15}N.

In static solid-state NMR, deuterium is a spin-1 nucleus showing a quadrupolar interaction and regarded as a useful tool for investigating molecular motion and orientation. Because the natural abundance of ^{2}H (0.015%) gives virtually no background signals, a ^{2}H-labelled site exclusively

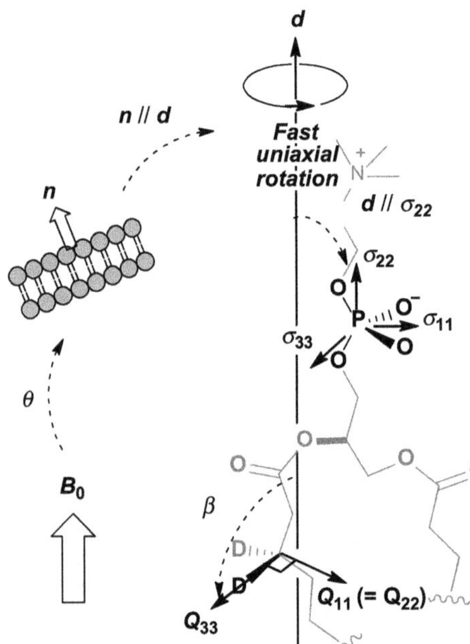

Figure 8.4 An example for the scaling of intramolecular anisotropic interactions by the fast uniaxial rotation of a phospholipid molecule in a lipid bilayer. θ is the orientation angle between the bilayer normal (n) and the external magnetic field (B_0). Vector d is the direction of uniaxial rotation of a phospholipid, which is parallel to the bilayer normal, n. In lipid bilayers, d is collinear to the intrinsic σ_{22} component of the ^{31}P CSA of the phosphodiester moiety. β is an orientation angle between d and a C–D quadrupolar tensor component Q_{33}. The order parameters are defined as $S_{BN} = (3\cos^2\theta - 1)/2$ and $S_{CD} = (3\cos^2\beta - 1)/2$.

produces a signal. ^{2}H NMR is particularly useful for studying soft materials and, therefore, is extensively applied to lipid bilayer membranes.[38] In most cases, either a membrane lipid or a small molecule is often selectively labelled with ^{2}H and incorporated into lipid bilayers. The lineshape of the signals is referred to as a Pake doublet (Figure 8.5a),[39] the general shape obtained from powder (non-oriented) samples, which usually becomes narrower by molecular mobility in a membrane. The size of quadrupolar coupling is experimentally obtained as the magnitude of the signal splitting $\Delta\nu_Q$ (Figure 8.5). The local dynamics are evaluated as an order parameter S_{CD}, which is defined by the following equations:[40,41]

$$\Delta\nu_Q = (3/2)A_Q \cdot S_{BN} \cdot S_{CD} \quad (8.1)$$

$$S_{BN} = \langle(3\cos^2\theta - 1)/2\rangle \quad (8.2)$$

$$S_{CD} = \langle(3\cos^2\beta - 1)/2\rangle \quad (8.3)$$

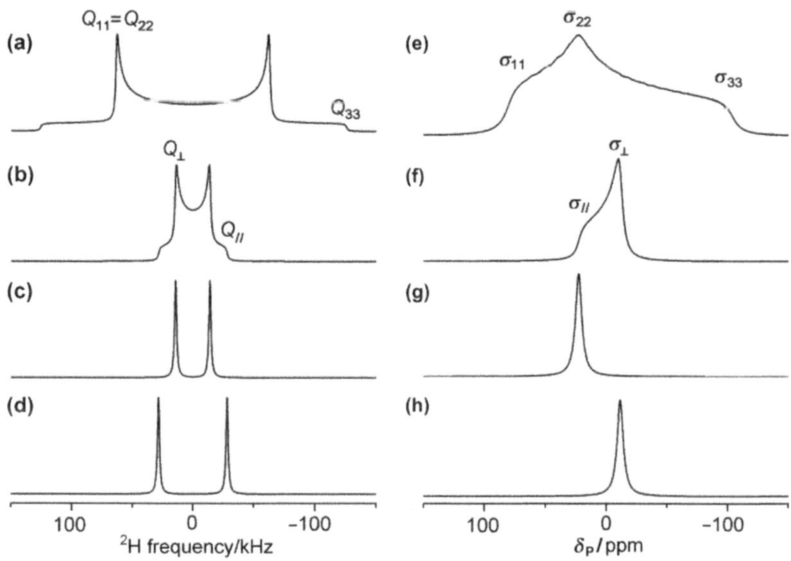

Figure 8.5 Patterns of static solid-state NMR spectrum for (a–d) ^{2}H quadrupolar coupling and (e–h) ^{31}P CSA. The sample system is shown in Figure 8.4. Spectra (a) and (e) are powder-pattern spectra representing tensor components of ^{2}H and ^{31}P, respectively, which are measured on dried powders. In the liposome sample, the ^{2}H quadrupolar tensor and ^{31}P CSA tensor are motionally averaged into narrower uniaxial symmetric tensors with components parallel ($Q_\parallel$ and $\sigma_\parallel$) and perpendicular ($Q_\perp$ and $\sigma_\perp$) to the bilayer normal (b and f). When the bilayer normal, n, is vertical to the applied magnetic field B_0, sharp signals are obtained at resonance frequencies corresponding to perpendicular components (c and g). When n is collinear with B_0, resonances are observed at the frequencies of parallel components (d and h).

where θ and β are orientation angles defined in Figure 8.4, and A_Q is the static quadrupolar coupling constant that is equal to 168 kHz for a C–^{2}H bond in lipid acyl chains.[42] The order parameters S_{NB} and S_{CD} reflect the average orientation of the bilayer normal with respect to the external magnetic field and the average orientation of C–^{2}H vectors relative to the bilayer normal, respectively. Membrane lipids in a fluid phase can be regarded to rapidly rotate about an axis parallel to the bilayer normal, in which the angle between the membrane normal and magnetic field is θ to give S_{BN}. S_{CD} basically depends on the angle between C–^{2}H and the molecular rotational axis and also on the wobbling of the axis. The apparent size of quadrupolar coupling, therefore, provides information about the mobility of ^{2}H-labelled small molecules and/or dynamics in the hydrophobic region of a bilayer formed by ^{2}H-labelled lipids.[14,18,19,21]

^{31}P is another informative nucleus for static solid-state NMR of lipid bilayers. Phospholipids, major constituents of the lipid bilayers, have a phosphodiester group at the polar head region. ^{31}P is a spin-$\frac{1}{2}$ isotope of 100% natural abundance and, therefore, has high NMR sensitivity. Furthermore, ^{31}P in a phosphate group has a large CSA that can be utilized for detecting the motion and orientation of polar headgroups. The approximate orientation of the ^{31}P CSA tensor with respect to the phosphodiester moiety of the headgroup is shown in Figure 8.5.[43,44] The principal values of CSA are $\sigma_{11} \approx -81$ ppm, $\sigma_{22} \approx -23$ ppm and $\sigma_{33} \approx 104$ ppm.[45,46] In hydrated lipid bilayers the ^{31}P CSA tensor is motionally averaged to display an axially symmetric pattern, which is defined by two parameters, parallel ($\sigma_{\parallel} \approx \sigma_{22}$) and perpendicular ($\sigma_{\perp} \approx (\sigma_{11} + \sigma_{33})/2$) to the bilayer normal (Figure 8.5b). When a drug binds to the membrane surface and modulates the orientation of the phosphodiester moiety, the interaction can cause alteration of the motionally averaged ^{31}P tensor. Therefore, the lineshape of static ^{31}P NMR has often been used to investigate the effects of drug binding to the hydrophilic surface region of the lipid bilayers.[14,15]

8.2.2.2 Static NMR of Oriented Membrane Systems

Static solid-state NMR combined with oriented membrane preparation is also an effective method to investigate anisotropic interactions.[47–52] When the labelled molecule, as is the case for membrane lipids, has a particular orientation relative to the external magnetic field, the anisotropic interaction converges to a particular single value and is consequently observed as a narrower peak with stronger intensity (Figure 8.5c,d,g,h). These experimental conditions can be established by fixing the bilayer orientation with respect to the external magnetic field. Oriented lipid bilayers are prepared mainly by a mechanical method or molecular alignment in a magnetic field. One of the advantages of an oriented membrane is the enhancement in spectrum resolution; for nuclear spins possessing a similar isotropic chemical shift, NMR signals are separated depending on the local orientation. This signal separation permits multiple resonance assignments based on the order parameters. By taking advantages of these techniques, the chain order of membrane lipids and orientation of membrane-bound helical peptides have been extensively studied using oriented membranes.[53,54]

8.2.3 Intermolecular Dipolar Coupling of Small Compounds

The intermolecular dipolar coupling between a small compound and its target provides direct evidence of a molecular contact, which often leads to a molecular image of their complex structure. The size of the dipolar coupling is inversely proportional to the third power of the distance between a pair of nuclear spins, thus making dipolar coupling-based methodology sensitive to molecular contact. Intermolecular dipolar interactions are usually far weaker than intramolecular anisotropic interactions and, consequently, selective observations by a recoupling technique under MAS conditions are required.Various types of dipolar recoupling methods have been developed and applied to biological solids.[55,56] We will briefly explain two recoupling methods, rotational-echo double-resonance (REDOR)[57,58] and dipolar-assisted rotational resonance (DARR),[59] which are frequently applied to structural studies of small molecules binding to lipid membranes and protein aggregates.

8.2.3.1 REDOR

REDOR, one of the most widely used solid-state NMR methods of distance measurements in biological systems,[60–64] is based on dipole-recoupling effects; a similar method, TEDOR, has also been developed and utilized in biological solids.[65] The REDOR method selectively reintroduces heteronuclear dipolar couplings under MAS conditions. As described previously, heteronuclear dipolar interactions can be averaged to 0 under MAS. The modulation of the size of dipolar couplings by MAS and the resulting phase evolution of a magnetization vector in the rotating frame are depicted in Figure 8.6. REDOR experiments are executed by rotor-synchronous irradiation of 180° rf pulses that flip the sign of irradiated nuclear spins and eventually that of heteronuclear dipolar couplings. This rotor-synchronized inversion of spins disrupts the MAS-averaging effect to reintroduce heteronuclear dipolar couplings (Figure 8.6a). As a result, the magnetization vector of the observed nucleus undergoes a phase evolution in accordance with the size of the heteronuclear dipolar coupling (Figure 8.6b,c). The size of the dipolar coupling depends on the interatomic distance and the orientation of a dipolar vector with respect to the external magnetic field. In a non-crystalline solid with a random molecular orientation, the magnitude of the dipolar coupling is variable and, thus, the phase evolution of a magnetization vector is dispersed after MAS rotation under REDOR conditions. The recoupling of heteronuclear dipolar interactions for the selected pair of observed/irradiated nuclei is detected as a decrease in the signal intensity in REDOR spectra. The size of the signal decay in the spectra depends on the

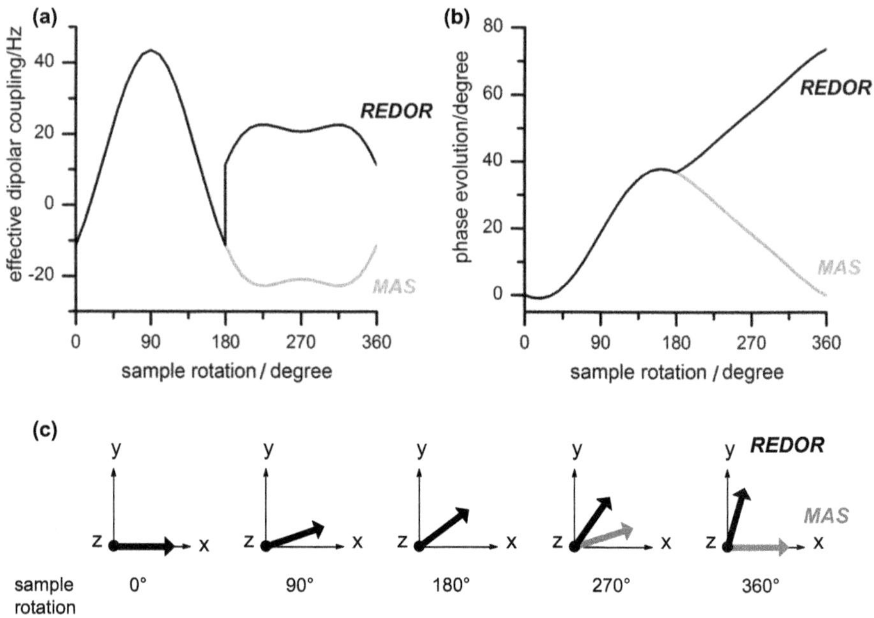

Figure 8.6 (a) Modulation of effective dipolar coupling under MAS conditions. The calculation was performed with a spin pair which has a dipolar coupling of 100 Hz with a polar angle $\beta = 45°$. When the sample rotates from 0° to 360° along the MAS axis, the sign of the effective dipolar coupling is modulated from positive to negative under MAS conditions. Under REDOR conditions, the sign of the effective dipolar coupling is flipped at the middle of the rotor cycle by application of a 180° pulse on the dephaser channel. (b) The phase evolution of magnetization of the observed nucleus by dipolar coupling. After 360° sample rotation, the dipolar phase evolution is averaged to 0 under MAS conditions. However, the phase evolution is kept positive under REDOR conditions by the sign-flip of the effective dipolar coupling at the middle of the rotor period. (c) Dipolar phase evolution of the observed nucleus in vector representation. The magnetization vector stays along the *x*-axis after a rotor cycle under MAS conditions, while the magnetization vector evolves under REDOR conditions.
(Reproduced from *Yakugaku Zasshi*, 2012, **132**, 9, with permission from The Pharmaceutical Society of Japan.)

magnitude of the heteronuclear dipolar coupling that is inversely proportional to the third power of the interatomic distance and, consequently, a larger signal decay is detected for observed nuclei closer to the irradiated nucleus (Figure 8.7a).

REDOR decay makes it possible to assign an intermolecular contact site. In addition, under the appropriate conditions, quantitative analysis of REDOR decay can provide an accurate interatomic distance, which can be directly used as a distance constraint for molecular modelling.[66,67] Quantitative analysis is performed by the theoretical fitting of a REDOR dephasing

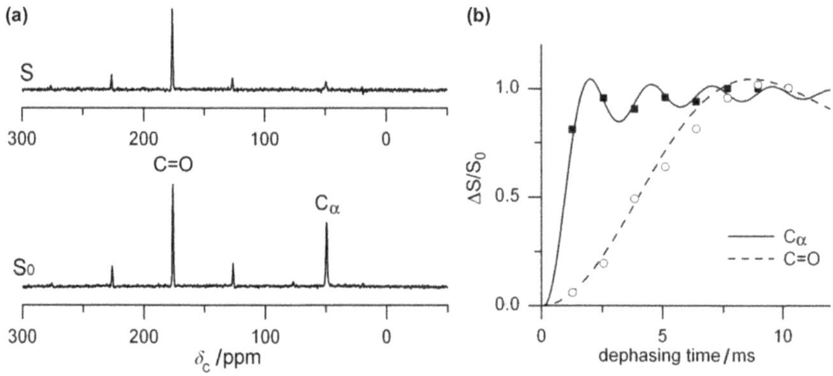

Figure 8.7 (a) Reference full-echo (S_0, *bottom*) and REDOR (S, *top*) spectra of [1,2-$^{13}C_2$,^{15}N]Ala. The spectra were obtained with a dephasing time of 2.56 ms. (b) Build-up of REDOR difference ($\Delta S/S_0$). *Dots* represent the experimental data (*filled squares*: Cα, *open circles*: C=O) and *lines* show the simulated values (*solid line*: Cα, *dotted line*: C=O). The data were measured using the DQF-REDOR sequence.
(Reproduced from S. Matsuoka and J. Schaefer, *J. Magn. Reson.*, 2006, **183**, 252–258, with permission from Elsevier.)

curve, where the normalized size of REDOR dephasing is plotted against the duration of dephasing time (Figure 8.7b).[68]

8.2.3.2 DARR

The solid-state NMR method, DARR, is designed to recouple *homonuclear* dipolar couplings under MAS, in contrast to REDOR that was designed to detect *heteronuclear* interactions. DARR should be a suitable method to efficiently detect the interaction between low-γ nuclei with the aid of ^{1}H–X dipolar coupling.[59] Currently, DARR is one of the most important methods for assigning ^{13}C signals and obtaining distance restraints for multiply labelled biomolecules in solid systems.[69,70]

Polarization transfer by DARR recoupling is based on rotational resonance (R^2).[71–78] R^2 is a selective recoupling method for a homonuclear dipolar interaction under MAS. This recoupling becomes most efficient when an integral multiple of the MAS frequency matches the difference of isotropic chemical shifts between a pair of homonuclear-coupled nuclei, where one of the spinning sidebands of the spin overlaps with the isotropic chemical shift of the other spin. The R^2 method has been applied widely for accurately measuring the distance between selectively labelled homonuclear pairs in biological systems.[79] However, the labelling strategy is restricted by isotropic chemical shifts of a pair-labelled site, which should have the appropriate frequency difference to fulfil R^2 conditions under particular MAS spinning frequency.

DARR recoupling can compensate for these drawbacks and, in fact, has been developed as a broadband homonuclear recoupling method for

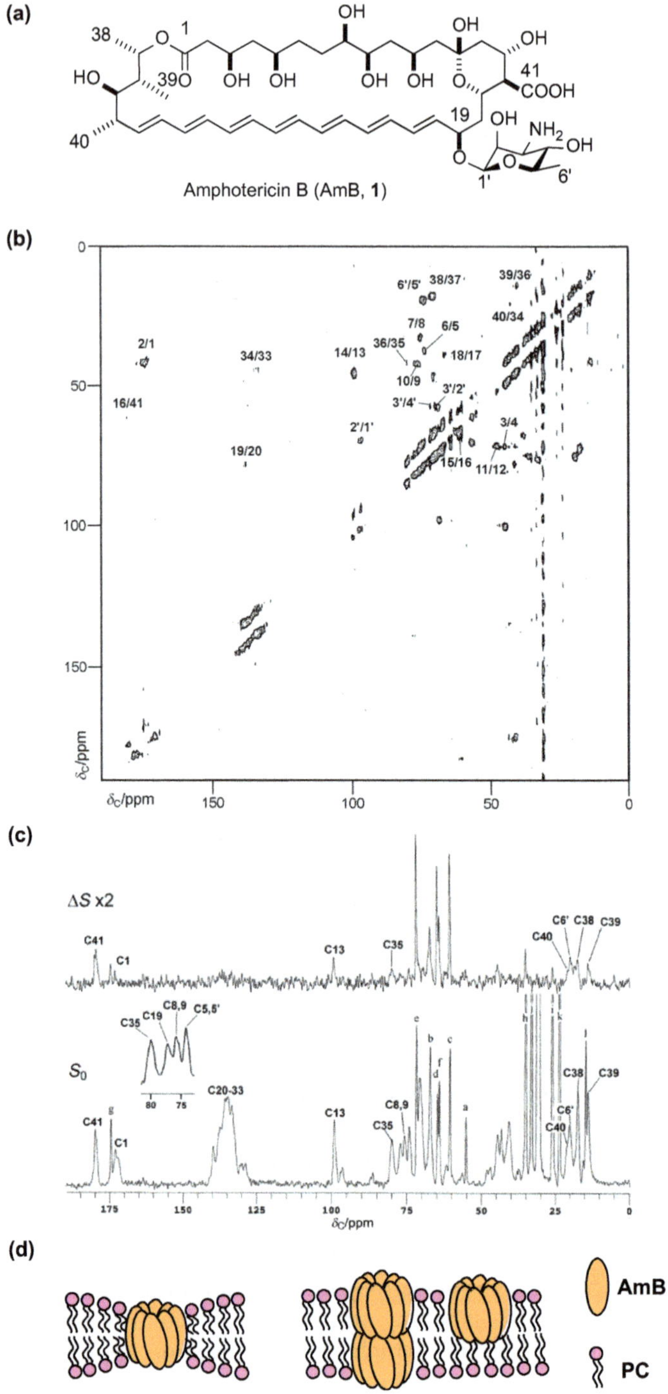
(a)
Amphotericin B (AmB, 1)
(b)
δC/ppm
(c)
ΔS x2
S0
δC/ppm
(d)
AmB
PC

multiply labelled samples. In DARR recoupling, an rf pulse is applied to a ^{1}H resonance when a low-γ nucleus X is not observed to recouple the X–^{1}H dipolar interactions, which causes a broadening of X resonance lines to achieve signal overlap necessary for R^2-type polarization transfer (Figure 8.8).[59] Fewer limitations in experimental conditions make DARR a valuable tool for structural studies of biomolecules by providing multiple distance restraints.

Long-range intermolecular dipolar interactions are far weaker than short-range intramolecular dipolar interactions. This situation causes dipolar truncation, where the desired polarization transfers through intermolecular dipolar interactions are cut off in the presence of undesired stronger intramolecular dipolar couplings.[80] To determine intermolecular distances using uniformly labelled samples, recoupling methods should have a suitable mechanism to observe weak dipolar couplings in the presence of undesired strong dipolar couplings. DARR is based on a second-order mechanism, which can avoid dipolar truncations and can be used as a long-range distance measurement methodology.[81] DARR has been applied to detect the interaction sites of small molecules and amyloid fibrils by using multiply ^{13}C-labelled samples.[25]

8.3 Examples of Applications to Lipid Membranes

The cell membrane is an attractive target in the development of drugs that aim to combat infections caused by antibiotic resistant strains and slow-growing bacteria, because the modification of membrane components such as lipids is less likely than the mutation of proteins and damage to the cell membrane is effective even for slow-growing cells.[82–85] Furthermore, lipid "rafts" or microdomains formed in cell membranes have been revealed to play an important role in viral proliferation and, thus, are expected to be potential therapeutic targets in the treatment of virus infections, including HIV.[86,87] Understanding the structure of molecular complexes formed in the

Figure 8.8 (a) Chemical structure of amphotericin B (AmB). (b) ^{13}C–^{13}C correlation spectrum. The DLPC membrane contained ^{13}C-enriched AmB at the AmB/DLPC molar ratio of 1:8 and 50 wt% 10 mM HEPES/D_2O buffer (pH 7.0). The spectra were obtained with MAS at 7 kHz at 30°C. Assignments of cross peaks are provided in the spectrum with *m*/*n*; *m* corresponds to an F1 (vertical) signal, while *n* to an F2 (horizontal). (c) ^{13}C–^{31}P RDX spectrum of uniformly ^{13}C-labelled AmB in a hydrated DLPC membrane with MAS at 7 kHz at 30°C. The number of scans accumulated for the full echo spectrum, S_0, was 55,552. The top trace is the REDOR difference spectrum, ΔS. The *insets* show the expanded spectra for 72–82 ppm. (d) Schematic models for membrane interaction of AmB with DLPC bilayer membranes as single-length (*left*) and double-length (*right*) assemblies.
(Reproduced in part from Matsuoka *et al.*,[90] with permission from Elsevier.)

cell membrane is essential for the rational design of new membrane-targeted drugs.

The main problems in the structural study on drug–membrane complexes arise from sample heterogeneity and the low concentration of drug molecules in an active structure. Unlike the ligand binding sites of proteins, which have definite structures and stoichiometry, a drug-binding site on lipid bilayers is structured upon drug binding and alters the stoichiometry depending on the drug concentration. Moreover, the concentration of drugs taking an active conformation is insufficient for solid-state NMR measurements because a considerable number of lipid molecules are required for mimicking cell membrane environments, where excess addition of a drug increases the portion of the free drug with an inactive structure and may cause undesirable membrane disruption. Consequently, selective observation of a low-concentration of drugs in the active structure must be achieved by various contrivances, such as labelling of small molecules, background suppression during sample preparation, and pulse techniques for signal selection. We now introduce some solid-state NMR experiments for determination of the membrane-associated structure of small compounds.

8.3.1 Amphotericin B

^{13}C Chemical shift assignments of small compounds, including natural products, were implemented by 2D chemical shift correlation spectra under high-resolution solid-state NMR conditions, *e.g.* the ^{13}C–^{13}C chemical shift correlation spectrum of uniformly ^{13}C-labelled erythromycin A was recorded using a crystal solid sample,[88] in which the ^{13}C–^{13}C magnetization transfer was performed by a dipolar-based double quantum mixing sequence, CMR7. The same methodology was applied to membrane-bound amphotericin B (AmB, Figure 8.8a), a polyene macrolide antibiotic that forms an ion-permeable channel across a lipid bilayer.[89] For the structural study of the AmB–membrane lipid complex, the ^{13}C NMR chemical shifts of the antibiotics in a membrane-bound form are required. AmB was uniformly ^{13}C-enriched at 15% by feeding [U-$^{13}C_6$]glucose to *Streptomyces nodosus.*[90] At this ^{13}C-labelling yield, ^{13}C–^{13}C cross peaks were observed only for single-bond correlations with an incorporation pattern of biosynthetic precursors, acetates and propionates. All ^{13}C NMR signals of AmB in hydrated 1,2-dilauroyl-*sn*-glycero-3-phosphocholine (DLPC) membranes were assigned in the 2D plane of the ^{13}C–^{13}C CMR7 spectrum, as shown in Figure 8.8b.

With the chemical shift assignment in hand, the interaction between amphotericin B and the phosphate group of DLPC was evaluated on the basis of the ^{13}C–^{31}P dipolar coupling.[28,29,90] Conventional ^{13}C{^{31}P} REDOR pulse sequences[57,91] were not directly applicable to uniform [^{13}C]AmB because of homonuclear coupling interactions arising from ^{13}C–^{13}C spin networks. While chemical shifts and heteronuclear *J*-coupling are refocused by a Hahn echo, homonuclear *J*-coupling is still active, which causes phase modulations in the REDOR spectra. Among other modified methods, REDOR for X clusters (RDX) turned out to be the most suitable for this purpose.[92] By using RDX, it was possible to compare the relative distances from phosphorus among labelled ^{13}C nuclei. In the spectrum shown in Figure 8.8c, significant REDOR dephasing effects were observed for the ^{13}C NMR signals near both termini of the AmB molecule in addition to signals due to the DLPC head group. In contrast, the broad peak due to 14 double-bond carbons (C20–C33) showed only a marginal dephase. These data imply that AmB spans DLPC membranes as depicted in Figure 8.8d. The lack of dephasing at C8, C9 and most of the heptaene carbon atoms clearly indicates that the middle of the AmB molecule is embedded in the membrane interior. Therefore, the ion-permeable channel assembled by AmB is assumed to be one molecule long rather than of double length (Figure 8.8d) in the DLPC membrane.

The antibiotic efficacy of AmB is thought to be manifested by a membrane-disrupting activity that is selective for fungal membranes over human tissues, which is thought to be due to more efficient formation of an ion-channel assembly in the presence of fungal ergosterol over human cholesterol.[93] Thus, the elucidation of the mode of molecular interactions between AmB and ergosterol in bilayers is particularly important. $^{13}C\{^{19}F\}$ REDOR experiments were performed with C14-^{19}F-labelled AmB and ^{13}C-labelled ergosterol at the dimethyl terminus (C26/C27).[24] The spectra unambiguously demonstrated the presence of a "head-to-tail" orientation for AmB–ergosterol pairing (Figure 8.9a,b). Furthermore, $^{13}C\{^{31}P\}$ REDOR, using the labelled ergosterol and the phosphorus atom of the POPC headgroup, gave rise to significant REDOR dephasing for the C26/C27 signal of ergosterol in the presence of AmB, but not in the absence of AmB, clearly indicating that the side-chain terminus of ergosterol in the AmB complex comes close to the bilayer surface (Figure 8.9a).[24]

8.3.2 Catechin

Polyphenols are bioactive natural products that seem to act against a wide range of pathologies. Catechins exhibit an antibacterial effect, which is thought to be caused by their binding to the cell membrane.[94] To understand the molecular mobility of catechins in cells, the interactions of catechins with lipid membranes were examined by solid-state NMR. Epigallocatechin gallate (EGCg), the ester of epigallocatechin and gallic acid, is one of the major catechins found in green tea polyphenols. The molecular mobility of EGCg in a DMPC bilayer was monitored by static ^{2}H NMR of [4-^{2}H]EGCg (Figure 8.10a).[18,19] The spectra of [4-^{2}H]EGCg in a multi-lamellar vesicle (MLV) suspension demonstrated the coexistence of free and membrane-bound EGCgs. From the size of the quadrupolar coupling of the membrane-bound components (approximately 19.5 kHz), the tilt angle of the C–^{2}H bond of [4-^{2}H]EGCg to the molecular rotation axis (β) was evaluated to be 48° (Figure 8.10b). Further investigation of the orientation of the

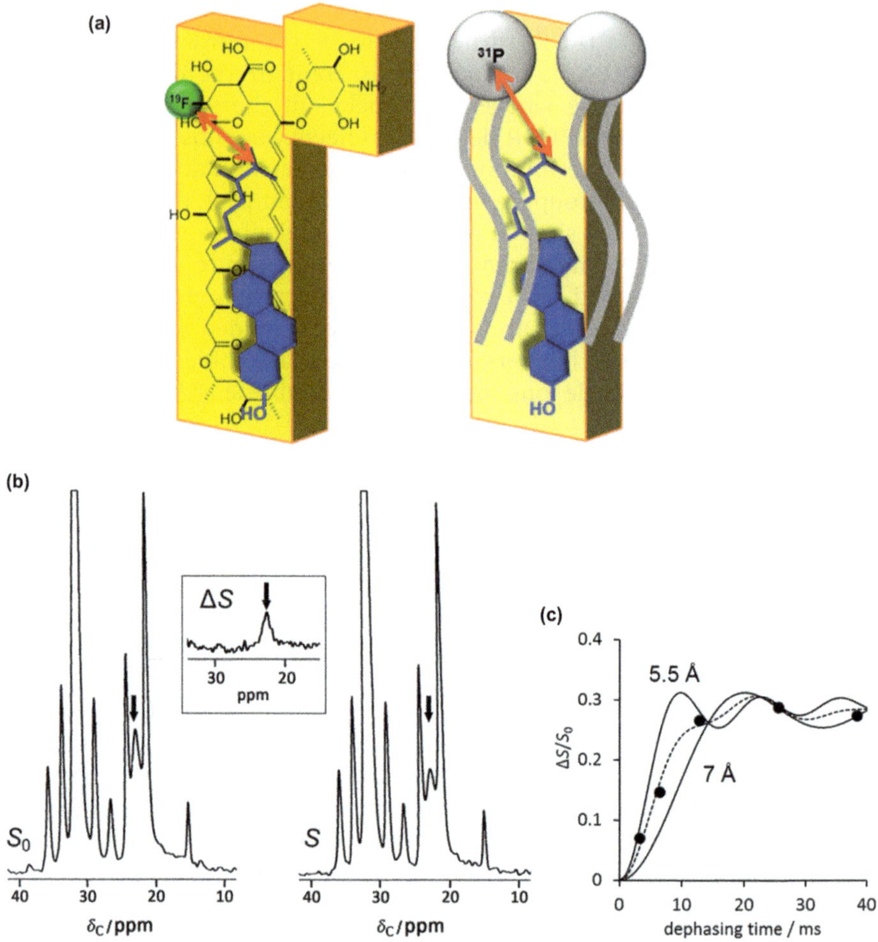

Figure 8.9 (a) AmB (*yellow boxes*) bound to bilayers interacts with ergosterol (*blue*) in a "head-to-tail" orientation (*left*). The terminus of an ergosterol side-chain comes close to the head groups of POPC (*silver*) in the presence of AmB (*right*). *Red arrows* point to the pair of labelled atoms that gave rise to REDOR dephasing. (b) $^{13}C\{^{19}F\}$ REDOR spectra of non-irradiated (full-echo) (S_0), ^{19}F-irradiated experiments (S) and difference spectrum (ΔS) for 14-F AmB shown in (a) and [26,27-$^{13}C_2$]ergosterol at the ratio of 14-F AmB/[26,27-$^{13}C_2$]ergosterol/POPC (1:1:9); hydrated POPC bilayers were prepared in 10 mM HEPES/D_2O (50 wt%). The data were obtained at 38°C with 29 488 scans. *Arrows* depict the C26/C27 signals. (c) Experimental $^{13}C\{^{19}F\}$ REDOR dephasing values ($\Delta S/S_0$) for the C26/C27 signal at 23.8 ppm (●) and simulation curves (*solid lines*) for the ^{13}C–^{19}F distances of 5.5 and 7.0 Å, assuming that 30% of [26,27-$^{13}C_2$]ergosterol is involved in a "head-to-tail" pairing. The simulation curve (*dashed line*) derived from two different distances, 5.5 Å and 7.0 Å (1:1), nicely fits the experimental data.
(Adapted from Umegawa *et al.*,[24] with permission from the American Chemical Society.)

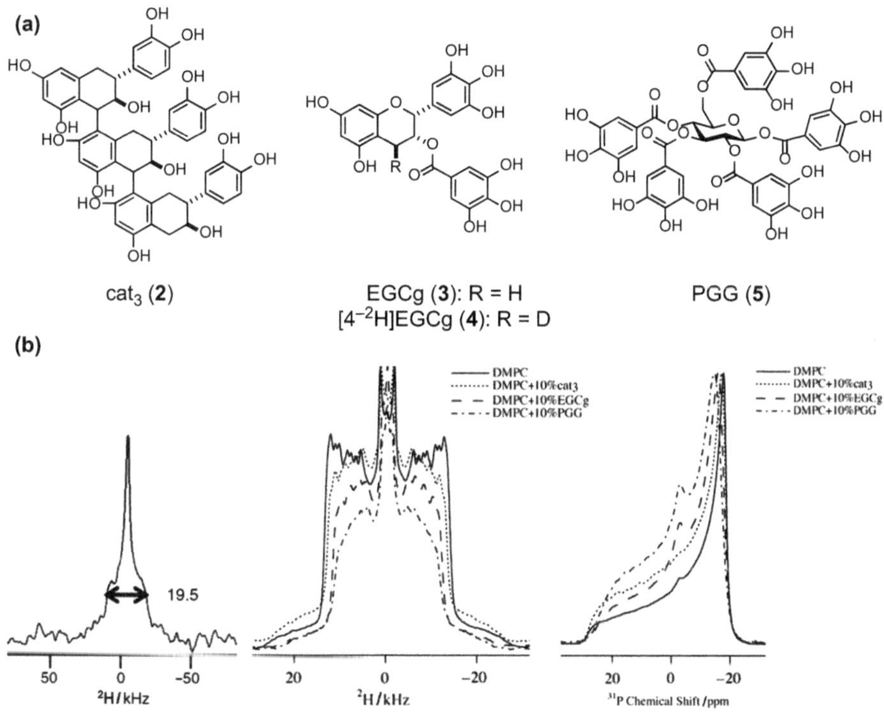

Figure 8.10 Chemical structures of polyphenols and solid-state NMR spectra of membrane-bound samples. (a) Chemical structures of cat$_3$ (**2**), EGCg (**3**), [4-^{2}H]EGCg (**4**) and PGG (**5**). (b) The static ^{2}H NMR spectra for [4-^{2}H]EGCg in DMPC MLVs at 40°C. Splitting of the perpendicular component is approximately 19.5 kHz. (c) ^{2}H NMR static spectra of DMPC MLVs containing 10 mol% of three polyphenols (cat3, EGCg and PGG, respectively) at 35°C. (d) ^{1}H-Decoupled ^{31}P NMR static spectra of DMPC MLVs containing 10 mol% of three polyphenols (cat3, EGCg and PGG, respectively) at 35°C.
(Reproduced in part from Kajiya *et al.*[18] and Yu *et al.*,[15] with permissions from Wiley and the American Chemical Society, respectively.)

molecular rotary axis relative to the bilayer normal was carried out using a magnetically oriented bilayer system. [4-^{2}H]EGCg bound to DMPC/DHPC bicelles showed a quadrupolar splitting of *ca.* 15.5 kHz, which was nearly the same as the width of the perpendicular component in the MLV. Hence, the [4-^{2}H]EGCg molecules rotate about the bilayer normal with a constant tilt angle of the axis.

In regard to lipids, perturbations of the membrane dynamics caused by polyphenols were examined by static ^{2}H and ^{31}P NMR of polyphenol-bound MLVs.[95] Three structurally diverse polyphenols, (–)-EGCg, a proanthocyanidin trimer comprising catechin-(4–8)-catechin-(4–8)-catechin (cat$_3$), and a hydrolysable tannin, 1,2,3,4,6-penta-*O*-galloyl-β-D-glucopyranose (PGG), were mixed into ^{2}H-labelled DMPC MLVs (Figure 8.10a). ^{31}P NMR was performed to determine the dynamics of the headgroup region and of the lipid bilayer upon addition of polyphenols. The CSA width of the ^{31}P NMR spectra decreased upon addition of polyphenols (Figure 8.10c). In particular, PGG induced a dramatic reduction in the CSA width, whereas addition of cat_3 barely reduced the CSA width. ^{2}H Solid-state NMR experiments were carried out to obtain the dynamics of the hydrophobic acyl chains. The ^{2}H quadrupolar splitting of the lipids also decreased upon addition of polyphenols (Figure 8.10d). At the same concentration, PGG substantially decreased the quadrupolar splitting, whereas cat_3 barely did. From a calculation of the order parameters of the acyl chain region of the lipid bilayer, the hydrophobic part of the lipid bilayer was perturbed by PGG, whereas cat_3 did not cause large perturbations. The data suggest that the polarity of the polyphenols affects their affinity to membranes.

8.3.3 Curcumin

Curcumin is the active ingredient of turmeric powder and is known to possess various types of bioactivity, such as antioxidant, anti-carcinogenic, anti-mutagenic and anti-inflammatory properties (Figure 8.11a). Studies have shown that curcumin influences structurally unrelated membrane proteins across several signalling pathways.[96] The actions of membrane proteins can be modulated indirectly by changing the physical properties of the membrane rather than by the direct binding of curcumin to the protein.[97]

Solid-state NMR was employed to study the interaction of curcumin with membranes by observing intramolecular anisotropic interactions of phospholipids in magnetically oriented bilayers.[98] In this study, the dynamics of non-isotopic labelled DMPC was investigated by means of ^{14}N quadrupolar coupling and ^{13}C–^{1}H dipolar coupling. Compared to most cases, the ^{14}N electric field gradient tensor of a choline lipid headgroup is considerably reduced by the near tetrahedral symmetry and is expedient for NMR observation. The ^{14}N of the choline group in DMPC can serve as an effective probe for changes in the electrostatic environment of the headgroup region of the membrane.[99,100] The addition of curcumin to DMPC modestly increases the ^{14}N quadrupolar coupling to $\sim$8.5 kHz from 7.8 kHz. This small increase in ^{14}N quadrupolar coupling is explained by the distortion of the near tetrahedral symmetry of the choline group upon addition of anionic amphiphiles (Figure 8.11b).[101]

^{13}C–^{1}H Dipolar coupling was measured by proton-detected local field (PDLF) spectra[102] to provide information about the hydrophobic region. Within the hydrophobic core, the effect of curcumin on the membrane was uneven, with the carbons close to the glycerol region affected more than those toward the middle of the membrane (Figure 8.11c). Interestingly, this ordering effect is absent at the terminus of the acyl chain. The ordering profile indicates curcumin inserts into the bilayer deeply, most likely penetrating nearly to the end of the DMPC molecule. The solid-state NMR

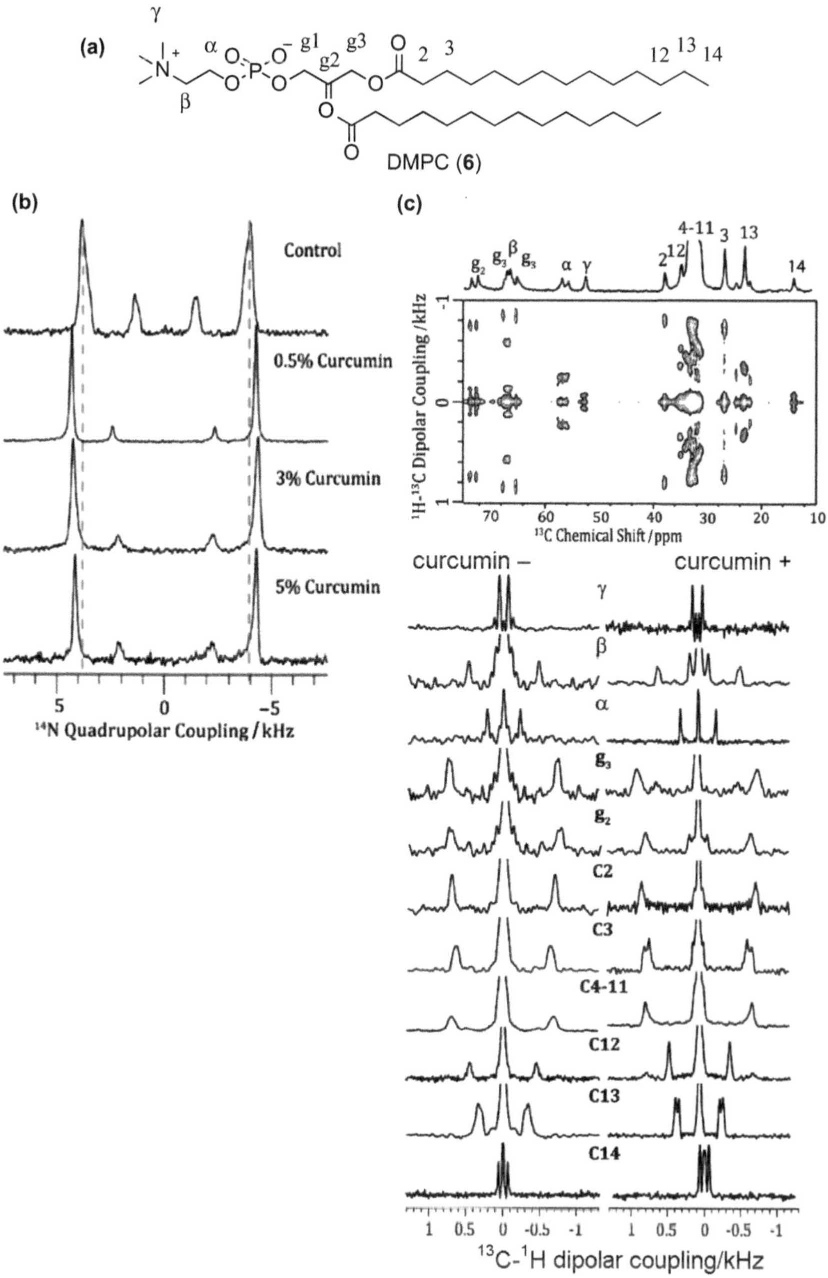

Figure 8.11 (a) Chemical structure of DMPC. (b) ^{14}N Spectrum of magnetically aligned 9:2 DMPC/DHPC bicelles at various curcumin concentrations. (c) 2D PDLF spectrum of magnetically aligned bicelles in the absence of curcumin (*top*) and C–H dipolar coupling spectral slices extracted from 2D PDLF spectra without (*bottom left*) and with 5 mol% curcumin (*bottom right*).
(Adapted from Barry *et al.*,[98] with permission from the American Chemical Society.)

methods shown in this study do not require isotope labelling and may have a wider application in studies on the molecular interactions between membrane lipids and small compounds.

8.4 Examples of Applications to Protein Aggregates and Polymers

Protein aggregates, including amyloids, are another important research area of solid-state NMR. Unlike folded proteins, most aggregated proteins rarely form crystals. However, despite the apparent random structures of protein aggregates, some small fluorescent molecules, so-called amyloid-specific dyes like thioflavin T and Congo Red, recognize amyloid structures. These small-molecule probes have been used in the diagnosis of amyloids for many years without knowing the clear mechanism of staining. Structural information of this type of small-molecule complex is in high demand, not only to understand the molecular mechanism of staining, but also to develop an amyloid-specific molecular probe as a diagnostic and therapeutic tool.

Proteins often polymerize and depolymerize spontaneously. Actin, myosin and tubulin are well-known examples. The structures of these polymerized proteins have often been elucidated by X-ray reflection methods. However, as is the case with aggregated protein, co-crystallography of the complex between these polymeric proteins and drugs is extremely difficult. In this section we describe three examples of structural studies of the binding complex of protein aggregates[103] and polymeric proteins with small compounds.

8.4.1 Curcumin

In vitro studies have shown that curcumin not only inhibits Aβ aggregation but also disaggregates Aβ fibrils.[104,105] *In vivo* studies using Alzheimer disease transgenic mice have shown that chronic dietary curcumin lowered Aβ deposition in the brain,[106] and intravenously injected curcumin crossed the blood/brain barrier and bound to the amyloid plaques.[105,107] Interestingly, the systemic treatment of mice with curcumin reduced new amyloid accumulation as well as previously deposited amyloid plaques. This is because curcumin not only inhibits the aggregation of Aβ42 but also binds to its aggregates (fibrils), resulting in disaggregation. However, the mechanism of interaction between curcumin and the Aβ42 fibrils remains unclear.

The binding mode of curcumin to Aβ42 fibrils was analysed by solid-state NMR using DARR, from which intermolecular ^{13}C–^{13}C dipolar couplings were observed (Figure 8.12).[25] The quality of the 2D DARR data was improved by processing with the covariance NMR method, which enabled the detection of weak cross-peaks between carbons of curcumin and those of the Aβ42 fibrils. The observed ^{13}C–^{13}C cross-peaks indicated that curcumin interacts with the 12th and 17th–21st residues included in the β-sheet structure of the Aβ42 fibrils. Interestingly, aromatic carbons adjacent to the

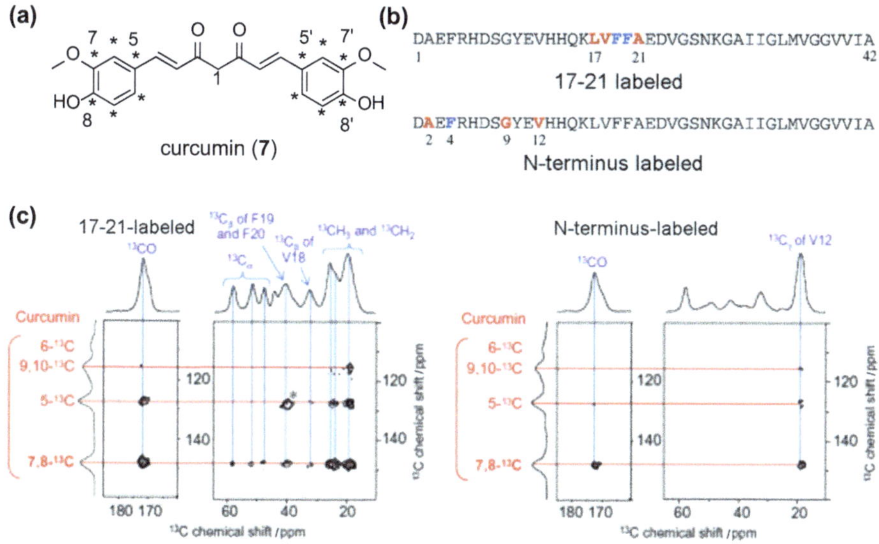

Figure 8.12 (a) Chemical structure of curcumin (7). The aromatic carbons are ^{13}C-labelled and marked with *asterisks*. (b) Two selective labelling schemes for Aβ42. Amino acid residues shown in *red* are uniformly labelled with ^{13}C and ^{15}N, and those shown in *blue* are labelled with ^{13}C at the Cβ position. (c) 2D DARR spectra at a mixing time of 500 ms. Aβ42-curcumin (7) intermolecular cross-peaks are enlarged. The experiments were performed for Aβ42 labelled at 17–21 residues (*left*) and Aβ42 fibrils labelled at the N-terminus (*right*). The *asterisked* cross-peak might be due to intramolecular dipolar coupling between ^{13}Cβ and naturally abundant aromatic ^{13}C in F19 and/or F20 of the Aβ42 fibrils.
(Reproduced from Masuda *et al.*,[25] with permission from Elsevier.)

methoxy and/or hydroxy groups of curcumin showed clear cross-peaks with the Aβ42 fibrils. This suggested that these functional groups of curcumin play an important role in its interaction with the Aβ42 fibrils.

8.4.2 Thioflavin T

Thioflavin T (ThT) is one of the most important amyloid-specific dyes and has been used for more than 50 years (Figure 8.13a).[108] It has been a significant challenge to study dye–peptide complex structures due to the insolubility and heterogeneity of the peptide aggregates. Furthermore, potential multiplicity of the binding modes and low concentration of the binding sites in fibrils make the situation even more problematic. To gain insight into the local structure of the ThT binding site in a peptide aggregate, intermolecular ^{2}H–^{13}C dipolar coupling was measured between [3-Me-d_3]ThT and octaglutamine peptide Q_8 model peptides by means of ^{13}C{^{2}H} REDOR.[23] Q_8 Peptides were chemically synthesized and stained positive to ThT. Two types of ThT-binding sites, a high-affinity site (k_{d1} = 0.1–0.17 μM)

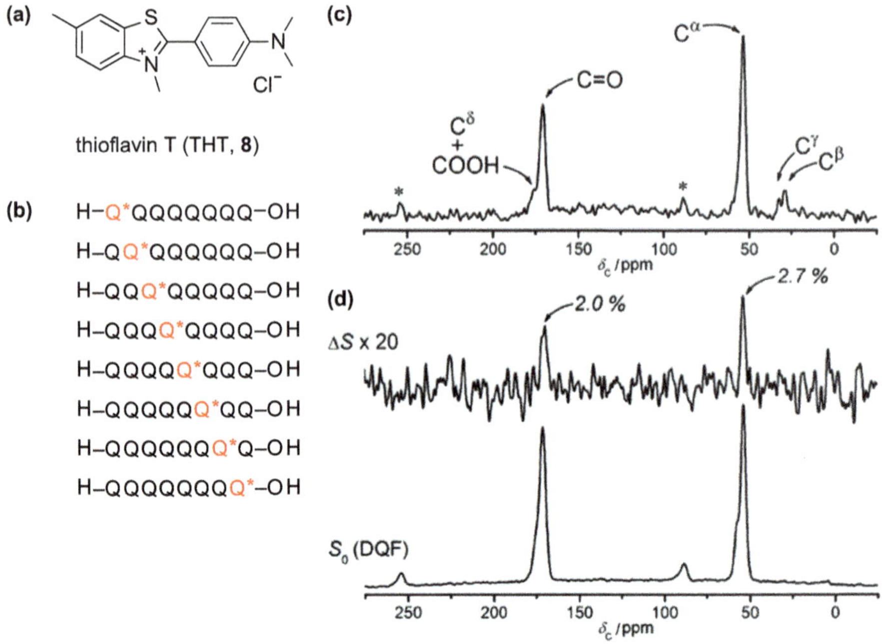

Figure 8.13 (a) Chemical structure of thioflavin T (ThT, **8**). (b) [1,2-$^{13}C_2$]Gln-incorporated H-Q_8-OH peptides. Labelled residues are marked with *asterisks* and shown in *red*. (c) ^{13}C CP-MAS spectrum of [3-Me-d_3]ThT (**8**)/H-Q*Q_7-OH (1/50 mol/mol) complex. (d) $^{13}C\{^2H\}$ DQF-REDOR spectra of the [3-Me-d_3]ThT (**8**)/H-Q*Q_7-OH complex. DQF full-echo ^{13}C NMR spectrum (S_0) is shown at the *bottom*, and the REDOR difference (ΔS) is shown on the *top* with 20× magnification. Spectra were recorded with a dephasing time of 9.6 ms (80 rotor cycles) at a MAS speed of 8333 Hz. The number of scans was 100 000. The spinning side bands are marked with an *asterisk*.

and a low-affinity site ($k_{d2} = 5.7$–7.4 μM), were observed in short polyQs ($n = 6$–9). $^{13}C\{^2H\}$ REDOR difference spectra of the [3-Me-d_3]ThT/natural abundance Q_8 complex indicated that all of the five carbon atoms of the glutamine residue participated in the formation of ThT binding sites. $^{13}C\{^2H\}$ DQF-REDOR experiments of [3-Me-d_3]ThT residue-specific [1,2-$^{13}C_2$]-labelled Q_8 complexes (1:50 mol/mol) demonstrated that the N-terminal glutamine residue had direct contact with the ThT molecule at the high-affinity ThT binding sites (Figure 8.13b). Site-selective observation of the $^{13}C_2$-labelled site and intermolecular distance measurement by ^{13}C–^2H dipolar coupling enabled the extraction of the local structure of the ThT binding site despite severe background signal overlap.

8.4.3 Paclitaxel Binding to Microtubules

Paclitaxel (Figure 8.14b, 9), a terpenoid isolated from the bark of the Pacific yew tree, *Taxus brevifolia*, is one of the most widely used

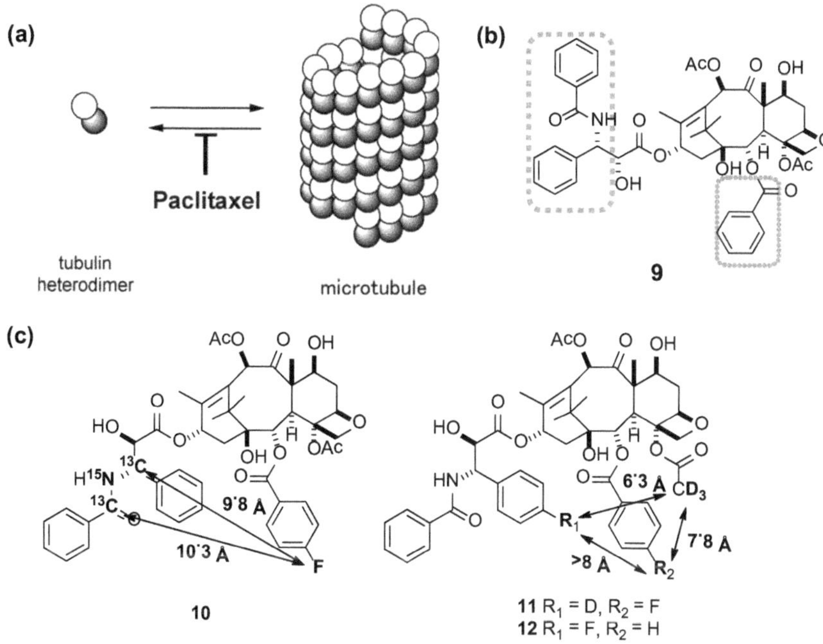

Figure 8.14 (a) Paclitaxel inhibits tubule depolymerization. (b) Structure of paclitaxel with two flexible aromatic side-chains depicted by *boxes*. (c) Isotope-labelled paclitaxel and interatomic distances determined by REDOR.

anticancertherapeutic drugs.[109,110] The drug binds to the β-subunit of tubulin in a polymeric form, preventing mitosis through stabilization of the polymeric state, called microtubules (Figure 8.14a).[111,112] This microtubule binding activity is believed to be directly related to clinical efficacy, where the side-chain atoms of paclitaxel are known to play an essential role in binding to the target. The first structural study of the paclitaxel–microtubule binding complex was performed by cryo-electron microscopy at a resolution of 3.7 Å,[113] and later refined to 3.5 Å.[114] However, the resolution was not sufficiently high to gain an image of the conformation of the bound drug. Determination of the conformation was hampered by the flexibility of the molecule. Three types of taxol-binding conformations were possible, according to the 3D orientation between the two aromatic rings, as shown in Figure 8.14b.[115] REDOR distance measurements successfully provided an answer to this question. The drug was labelled with ^{19}F, ^{13}C and ^{15}N according to the β-lactam synthon method.[116] β-Lactam labelled with $^{13}C_2$ and ^{15}N was coupled with a fluorinated baccatin III derivative to afford a [$^{13}C_2$,^{15}N,^{19}F]paclitaxel derivative (**10**). Bovine tubulin was polymerized by the isotope-labelled paclitaxel **10** to form microtubules. Distance measurement was performed using $^{13}C\{^{19}F\}$ REDOR with $^{13}C\{^{15}N\}$ REDOR used as a background filter. The $^{13}C\{^{15}N\}$ REDOR difference was used for signal assignment and estimation of the full-echo intensity of the ^{13}C–^{15}N–^{13}C spin cluster by filtering out the large ^{13}C background signals from the protein and other components. Two interatomic distances (Figure 8.14c) were obtained from $^{13}C\{^{19}F\}$ REDOR with data collection over three months. The structural confirmation was made by $^{2}H\{^{19}F\}$ REDOR of two [^{2}H,^{19}F]-labelled compounds, **11** and **12**, in which the additional three ^{2}H–^{19}F distance restraints (Figure 8.14c) allowed a single binding conformation. The REDOR distances provide direct evidence that the drug adopts the conformation (Figure 8.14c) in its tubulin-binding state.[117]

8.5 Conclusion

This review described basic concepts of solid-state NMR and its applications to small compounds in biological solids. Solid-state NMR is regarded as the sole method by which atomic-resolution structural information for non-crystalline solids can be obtained. However, the number of restraints obtainable from solid-state NMR measurements is very limited compared to those from high-resolution solution NMR. In addition, preparation of a relatively large amount of isotope-labelled materials has hampered the wider utilization of solid-state NMR to biological systems, limiting its applications to cases where X-ray or solution NMR does not work. NMR observation in solids is very informative when observing the local structure of a biological matrix surrounding a ligand or drug. By using a natural product labelled with NMR-active nuclei, intermolecular REDOR experiments are able to highlight the structure of the binding site in a multiply labelled biological matrix. REDOR is a powerful and reliable method for distance measurements in any types of solids, including protein aggregates, cell membranes, cell walls and even whole cells. In particular, small molecules in membranes usually undergo fast lateral movement, but have a limited mobility with respect to the membrane normal. Solid-state NMR is increasingly utilized for these highly mobile anisotropic systems, which will lead to better understanding of the molecular basis of structure–function relationships of lipids and drugs in and around biomembranes. Hardware improvements, such as high magnetic fields with high-speed MAS and dynamic nuclear polarization (DNP), can boost the sensitivity and resolution of solid-state NMR. With such improvements, biological solid-state NMR will open new ways for *in situ* structural studies of small compounds in their biologically active state.

References

1. A. D. G. Lawson, *Nat. Rev. Drug Discovery*, 2012, **11**, 519.
2. J. A. Wells and C. L. McClendon, *Nature*, 2007, **450**, 1001.
3. P. J. Judge and A. Watts, *Curr. Opin. Chem. Biol.*, 2011, **15**, 690.
4. A. Loquet, C. Gardiennet and A. Böckmann, *C. R. Chim.*, 2010, **13**, 423.
5. A. Naito, *Solid State Nucl. Magn. Reson.*, 2009, **36**, 67.

6. M. Baldus, *J. Biomol. NMR*, 2007, **39**, 73.
7. M. Hong, *J. Phys. Chem. B*, 2007, **111**, 10340.
8. S. J. Opella and F. M. Marassi, *Chem. Rev.*, 2004, **104**, 3587.
9. R. Tycko, *Annu.Rev. Phys. Chem.*, 2011, **62**, 279.
10. A. McDermott, *Annu. Rev. Biophys*, 2009, **38**, 385.
11. B. Bechinger, R. Kinder, M. Helmle, T. C. B. Vogt, U. Harzer and S. Schinzel, *Biopolymers*, 1999, **51**, 174.
12. J. M. Lopez del Amo, U. Fink, M. Dasari, G. Grelle, E. E. Wanker, J. Bieschke and B. Reif, *J. Mol. Biol.*, 2012, **421**, 517.
13. T. Kimura, K. Cheng, K. C. Rice and K. Gawrisch, *Biophys. J.*, 2009, **96**, 4916.
14. C. Fotakis, G. Megariotis, D. Christodouleas, E. Kritsi, P. Zoumpoulakis, D. Ntountaniotis, M. Zervou, C. Potamitis, A. Hodzic, G. Pabst, M. Rappolt, G. Mali, J. Baldus, C. Glaubitz, M. G. Papadopoulos, A. Afantitis, G. Melagraki and T. Mavromoustakos, *Biochim. Biophys. Acta*, 2012, **1818**, 3107.
15. X. Yu, S. Chu, A. E. Hagerman and G. A. Lorigan, *J. Agric. Food Chem.*, 2011, **59**, 6783.
16. J. Barry, M. Fritz, J. R. Brender, P. E. S. Smith, D.-K. Lee and A. Ramamoorthy, *J. Am. Chem. Soc.*, 2009, **131**, 4490.
17. N. Matsumori, K. Tahara, H. Yamamoto, A. Morooka, M. Doi, T. Oishi and M. Murata, *J. Am. Chem. Soc.*, 2009, **131**, 11855.
18. K. Kajiya, S. Kumazawa, A. Naito and T. Nakayama, *Magn. Reson. Chem.*, 2008, **46**, 174.
19. S. Kumazawa, K. Kajiya, A. Naito, H. Saito, S. Tuzi, M. Tanio, M. Suzuki, F. Nanjo, E. Suzuki and T. Nakayama, *Biosci. Biotechnol. Biochem.*, 2004, **68**, 1743.
20. M. J. Paquet, I. Fournier, J. Barwicz, P. Tancrède and M. Auger, *Chem. Phys. Lipids*, 2002, **119**, 1.
21. A. W. Hing, J. Schaefer and G. S. Kobayashi, *Biochim. Biophys. Acta*, 2000, **1463**, 323.
22. E. J. Dufourc, I. C. Smith and H. C. Jarrell, *Biochim. Biophys. Acta*, 1984, **776**, 317.
23. S. Matsuoka, M. Murai, T. Yamazaki and M. Inoue, *Org. Biomol. Chem.*, 2012, **10**, 5787.
24. Y. Umegawa, Y. Nakagawa, K. Tahara, H. Tsuchikawa, N. Matsumori, T. Oishi and M. Murata, *Biochemistry*, 2012, **51**, 83.
25. Y. Masuda, M. Fukuchi, T. Yatagawa, M. Tada, K. Takeda, K. Irie, Y. Monobe, T. Imazawa and K. Takegoshi, *Bioorg. Med. Chem.*, 2011, **19**, 5967.
26. Y. Umegawa, N. Matsumori, T. Oishi and M. Murata, *Biochemistry*, 2008, **47**, 13463.
27. Y. Kasai, N. Matsumori, Y. Umegawa, S. Matsuoka, H. Ueno, H. Ikeuchi, T. Oishi and M. Murata, *Chem.–Eur. J*, 2008, **14**, 1178.
28. S. Matsuoka, H. Ikeuchi, N. Matsumori and M. Murata, *Biochemistry*, 2005, **44**, 704.

29. L. Cegelski, C. V. Rice, R. D. O'Connor, A. L. Caruano, G. P. Tochtrop, Z. Y. Cai, D. F. Covey and J. Schaefer, *Drug Dev. Res*, 2005, **66**, 93.
30. J. Schaefer and E. O. Stejskal, *J. Am. Chem. Soc.*, 1976, **98**, 1031.
31. B. M. Fung, A. K. Khitrin and K. Ermolaev, *J. Magn. Reson.*, 2000, **142**, 97.
32. A. E. Bennett, C. M. Rienstra, M. Auger, K. V. Lakshmi and R. G. Griffin, *J. Chem. Phys.*, 1995, **103**, 6951.
33. E. R. Andrew, A. Bradbury and R. G. Eades, *Nature*, 1958, **182**, 1659.
34. I. J. Lowe, *Phys. Rev. Lett.*, 1959, **2**, 285.
35. F. Cisnetti, K. Loth, P. Pelupessy and G. Bodenhausen, *ChemPhysChem*, 2004, **5**, 807.
36. S. R. Hartmann and E. L. Hahn, *Phys. Rev.*, 1962, **128**, 2042.
37. A. Pines, M. G. Gibby and J. S. Waugh, *J. Chem. Phys.*, 1973, **59**, 569.
38. J. H. Davis, *Biochim. Biophys. Acta*, 1983, **737**, 117.
39. G. E. Line and J. Pake, *Chem. Phys.*, 1948, **16**, 327.
40. L. S. Vermeer, B. L. de Groot, V. Réat, A. Milon and J. Czaplicki, *Eur. Biophys. J.*, 2007, **36**, 919.
41. J.-P. Douliez, A. Leonard and E. J. Dufourc, *Biophys. J.*, 1995, **68**, 1727.
42. L. J. Burnett and B. H. Muller, *J. Chem. Phys.*, 1971, **55**, 5829.
43. M. J. Potrzebowski, S. Kaźmierski, H. Kassassir and B. Miksa, *Annu. Rep. NMR Spectrosc*, 2010, **70**, 35.
44. C. Kim, J. Spano, E.-K. Park and S. Wi, *Biochim. Biophys. Acta*, 2009, **1788**, 1482.
45. R. G. Griffin, L. Powers and P. S. Pershan, *Biochemistry*, 1978, **17**, 2718.
46. J. Herzfeld, R. G. Griffin and R. A. Haberkorn, *Biochemistry*, 1978, **17**, 2711.
47. G. A. Lorigan, P. C. Dave, E. K. Tiburu, K. Damodaran, S. Abu-Baker, E. S. Karp, W. J. Gibbons and R. E. Minto, *J. Am. Chem. Soc.*, 2004, **126**, 9504.
48. K. A. H. Wildman, D. K. Lee and A. Ramamoorthy, *Biochemistry*, 2003, **42**, 6545.
49. S. J. Opella, *Nat. Struct. Biol.*, 1997, **4**, 845.
50. J. Seelig, *J. Am. Chem. Soc.*, 1970, **92**, 3881.
51. R. S. Prosser, F. Evanics, J. L. Kitevski and M. S. Al-Abdul-Wahid, *Biochemistry*, 2006, **45**, 8453.
52. C. R. Sanders II, B. J. Hare, K. P. Howard and J. H. Prestegard, *Prog. Nucl. Magn. Reson. Spectrosc.*, 1994, **26**, 421.
53. B. Bechinger, *Biochim. Biophys. Acta*, 1999, **1462**, 157.
54. T. A. Cross and S. J. Opella, *Curr. Opin. Struct. Biol.*, 1994, **4**, 574.
55. D. A. Middleton, E. Hughes and J. Madine, *J. Am. Chem. Soc.*, 2004, **126**, 9478.
56. I. Schnell, *Prog. Nucl. Magn. Reson. Spectrosc.*, 2004, **45**, 145.
57. T. Gullion and J. Schaefer, *J. Magn. Reson.*, 1989, **81**, 196.
58. T. Gullion and J. Schaefer, *Adv. Magn. Reson.*, 1989, **13**, 58.
59. K. Takegoshi, S. Nakamura and T. Terao, *Chem. Phys. Lett.*, 2001, **344**, 631.

60. O. Toke and L. Cegelski, in *Solid-State NMR Studies of Biopolymers*, ed. A. E. McDermott and T. Polenova, Wiley, New York, 2010, p. 473.
61. S. Matsuoka and M. Inoue, *Chem. Commun.*, 2009, 5664.
62. T. Gullion, *Annu. Rep. NMR Spectrosc.*, 2009, **65**, 111.
63. S. L. Grage and A. Watts, *Annu. Rep. NMR Spectrosc.*, 2006, **60**, 191.
64. J. Schaefer, in *Recent Trends in Molecular Recognition*, ed. F. Diederich and H. Kunzer, Springer, Berlin, 1998, p. 26.
65. A. W. Hing, S. Vega and J. Schaefer, *J. Magn. Reson.*, 1992, **96**, 205.
66. A. Naito and H. Saito, in *Encyclopedia of Nuclear Magnetic Resonance*, ed. D. M. Grant and R. K. Harris, Wiley, Chichester, 2002, vol. 9, p. 283.
67. A. Naito, K. Nishimura, S. Kimura, S. Tuzi, M. Aida, N. Yasuoka and H. Saito, *J. Phys. Chem.*, 1996, **100**, 14995.
68. K. T. Mueller, T. P. Jarvie, D. J. Aurentz and B. W. Roberts, *Chem. Phys. Lett.*, 1995, **242**, 535.
69. I. Bertini, A. Bhaumik, G. De Paepe, R. G. Griffin, M. Lelli, J. R. Lewandowski and C. Luchinat, *J. Am. Chem. Soc.*, 2010, **132**, 1032.
70. T. I. Igumenova, A. E. McDermott, K. W. Zilm, R. W. Martin, E. K. Paulson and A. J. Wand, *J. Am. Chem. Soc.*, 2004, **126**, 6720.
71. M. H. Levitt, D. P. Raleigh, F. Creuzet and R. G. Griffin, *J. Chem. Phys.*, 1990, **92**, 6347.
72. D. P. Raleigh, F. Creuzet, S. K. Das Gupta, M. H. Levitt and R. G. Griffin, *J. Am. Chem. Soc.*, 1989, **111**, 4502.
73. D. P. Raleigh, M. H. Levitt and R. G. Griffin, *Chem. Phys. Lett.*, 1988, **146**, 71.
74. M. G. Colombo, B. H. Meier and R. R. Ernst, *Chem. Phys. Lett.*, 1988, **146**, 189.
75. B. H. Meier and W. L. Earl, *J. Am. Chem. Soc.*, 1987, **109**, 7937.
76. D. P. Raleigh, S. G. Harbison, J. E. Neiss, J. E. Roberts and R. G. Griffin, *Chem. Phys. Lett.*, 1987, **138**, 285.
77. E. R. Andrew, S. Clough, L. F. Farnell, T. A. Gledhill and I. Roberts, *Phys. Lett.*, 1966, **21**, 505.
78. E. R. Andrew, A. Bradbury, R. G. Eades and V. T. Wynn., *Phys. Lett.*, 1963, **4**, 99.
79. S. O. Smith, eMagRes, 2007; DOI: 10.1002/9780470034590.emrstm0477.
80. M. J. Bayro, M. Huber, R. Ramachandran, T. C. Davenport, B. H. Meier, M. Ernst and R. G. Griffin, *J. Chem. Phys.*, 2009, **130**, 114506.
81. K. Takegoshi, S. Nakamura and T. Terao, *J. Chem. Phys.*, 2003, **118**, 2325.
82. J. G. Hurdle, A. J. O'Neill, I. Chopra and R. E. Lee, *Nat. Rev. Microbiol.*, 2011, **9**, 62.
83. Van Bambeke, M. P. Mingeot-Leclercq, M. J. Struelens and P. M. Tulkens, *Trends Pharmacol. Sci.*, 2008, **29**, 124.
84. C. Sundhahar, R. M. Haney, Y. Xue and R. V. Stahelin, *Curr. Drug Targets*, 2008, **9**, 603.
85. A. B. Hendrich and K. Michalak, *Curr. Drug Targets*, 2003, **4**, 23.

86. S. P. Verma, *Curr. Drug Targets*, 2009, **10**, 51.
87. D. H. Nguyen and J. E. K. Hildreth, *J. Virol.*, 2000, **74**, 3264.
88. C. M. Rienstra, M. E. Hatcher, L. J. Mueller, B. Sun, S. W. Fesik and R. G. Griffin, *J. Am. Chem. Soc.*, 1998, **120**, 10602.
89. B. De Kfuijff, W. J. Gerritsen, A. Oerlemans, R. A. Demel and L. L. M. van Deenen, *Biochim. Biophys. Acta*, 1974, **339**, 30.
90. S. Matsuoka, H. Ikeuchi, Y. Umegawa, N. Matsumori and M. Murata, *Bioorg. Med. Chem.*, 2006, **14**, 6608.
91. D. J. Hirsh, N. Lazaro, L. R. Wright, J. M. Boggs, T. J. McIntosh, J. Schaefer and J. Blazyk, *Biophys. J.*, 1998, **75**, 1858.
92. A. K. Mehta, L. Cegelski, R. D. O'Connor and J. Schaefer, *J. Magn. Reson.*, 2003, **163**, 182.
93. J. D. Readio and R. Bittman, *Biochim. Biophys. Acta*, 1982, **685**, 219.
94. K. Kajiya, H. Hojo, M. Suzuki, F. Nanjo, S. Kumazawa and T. Nakayama, *J. Agric. Food Chem.*, 2004, **52**, 1514.
95. X. Yu, S. Chu, A. E. Hagerman and G. A. Lorigan, *J. Agric. Food Chem.*, 2011, **59**, 6783.
96. J. G. Bilmen, S. Z. Khan, M. U. H. Javed and F. Michelangeli, *Eur. J. Biochem.*, 2001, **268**, 6318.
97. H. I. Ingolfsson, R. E. Koeppe and O. S. Andersen, *Biochemistry*, 2007, **46**, 10384.
98. J. Barry, M. Fritz, J. R. Brender, P. E. S. Smith, D.-K. Lee and A. Ramamoorthy, *J. Am. Chem. Soc.*, 2009, **131**, 4490.
99. A. Ganesan, *Curr. Opin. Chem. Biol.*, 2008, **12**, 306.
100. P. D. Leeson and B. Springthorpe, *Nat. Rev. Drug Discovery*, 2007, **6**, 881.
101. P. G. Scherer and J. Seelig, *Biochemistry*, 1989, **28**, 7720.
102. S. Dvinskikh, U. Dürr, K. Yamamoto and A. Ramamoorthy, *J. Am. Chem. Soc.*, 2006, **102**, 6326.
103. J. M. L. del Amo, U. Fink, M. Dasari, G. Grelle, E. E. Wanker, J. Bieschke and B. Reif, *J. Mol. Biol.*, 2012, **421**, 517.
104. F. Yang, G. P. Lim, A. N. Begum, O. J. Ubeda, M. R. Simmons, S. S. Ambegaokar, P. P. Chen, R. Kayed, C. G. Glabe, S. A. Frautschy and G. M. Cole, *J. Biol. Chem.*, 2005, **280**, 5892.
105. K. Ono, K. Hasegawa, H. Naikia and M. Yamada, *J. Neurosci. Res.*, 2004, **75**, 742.
106. G. P. Lim, T. Chu, F. Yang, W. Beech, S. A. Frautschy and G. M. Cole, *J. Neurosci.*, 2001, **21**, 8370.
107. M. Garcia-Alloza, L. A. Borrelli, A. Rozkalne, B. T. Hyman and B. J. Bacskai, *J. Neurochem.*, 2007, **102**, 1095.
108. H. LeVine III, *Methods Enzymol.*, 1999, **309**, 274.
109. E. K. Ronwinsky, *Annu. Rev. Med.*, 1997, **48**, 353.
110. M. C. Wani, H. L. Taylor, M. E. Wall, P. Coggon and A. T. MacPhail, *J. Am. Chem. Soc.*, 1971, **93**, 2325.
111. P. B. Schiff, J. Fant and S. B. Horwitz, *Nature*, 1979, **277**, 665.
112. P. B. Schiff and S. B. Horwitz, *Proc. Natl. Acad. Sci. U. S. A.*, 1980, 77, 1561.

113. E. Nogales, S. G. Wolf and K. H. Downing, *Nature*, 1998, **391**, 199.
114. J. Lowe, H. Li, K. H. Downing and E. Nogales, *J. Mol. Biol.*, 2001, **313**, 1045.
115. J. P. Snyder, J. H. Nettles, B. Cornett, K. H. Downing and E. Nogales, *Proc. Natl. Acad. Sci. U. S. A.*, 2001, **98**, 5312.
116. I. Ojima, S. D. Kuduk, J. C. Slater, R. H. Gimi and C. M. Sun, *Tetrahedron*, 1996, **52**, 209.
117. Y. Paik, C. Yang, B. Metaferia, S. Tang, S. Bane, R. Ravindra, N. Shanker, A. A. Alcaraz, S. A. Johnson, J. Schaefer, R. D. O'Connor, L. Cegelski, J. P. Snyder and D. G. I. Kingston, *J. Am. Chem. Soc.*, 2007, **129**, 361.

CHAPTER 9

Lipopolysaccharide Induces Raft Domain Expansion in a Cholesterol-Containing Membrane

KAORU NOMURA* AND SHOICHI KUSUMOTO

Suntory Institute for Bioorganic Research, 1-1-1 Wakayamadai, Shimamoto-Cho, Mishima-Gun, Osaka 618–8503, Japan
*Email: nomura@sunbor.or.jp

9.1 Introduction

Innate immunity is the first-line defense system of higher animals which operates upon infection by pathogens before the system of adaptive immunity goes into action. Such immediate protective mechanisms are initiated by molecules which are components of pathogens, like viruses and bacteria. Lipopolysaccharide (LPS) is a major constituent of the outer leaflet of the outer membrane of Gram-negative bacteria (Figure 9.1a) and is known as a potent inducer of the first defense system of the innate immunity.[1,2] When Gram-negative bacteria invade higher animals, immunocompetent cells such as macrophages and dendritic cells in host animals are exposed to LPS once it is separated from the bacteria. Then, LPS activates the innate immune receptor Toll-like receptor-4 (TLR-4),[3] and finally proinflammatory cytokines, such as tumor necrosis factor alpha (TNF-α) and those of the interleukin family, are released. If bacterial infections are extensive, LPS

New Developments in NMR No. 3
Advances in Biological Solid-State NMR: Proteins and Membrane-Active Peptides
Edited by Frances Separovic and Akira Naito

Published by the Royal Society of Chemistry, www.rsc.org

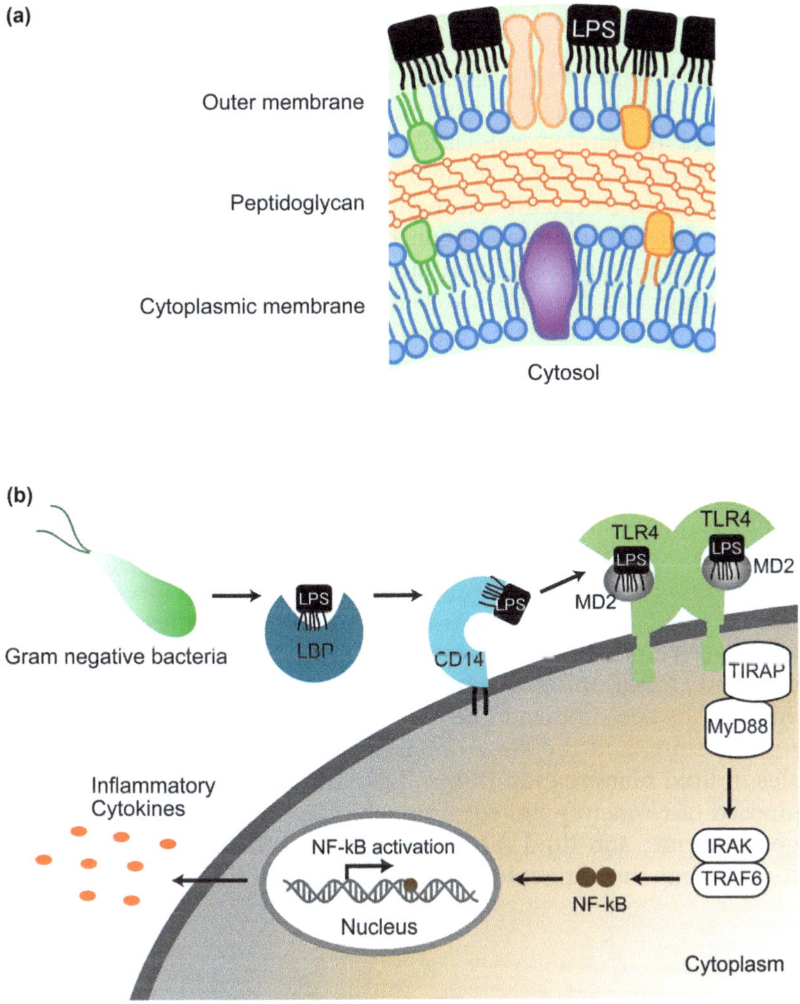

Figure 9.1 (a) Schematic diagram of the outer surface of a Gram-negative bacterium. (b) Mechanism of the innate immune response induced by LPS.

excessively stimulates immunocompetent cells, leading to multiple organ damage, high fever and septic shock. To develop promising therapeutic strategies against LPS-induced excessive immune responses, the mechanisms of the innate immune response must be understood in more detail. The current understanding of the cellular activation mechanisms by LPS is as follows: LPS initially binds to the LPS binding protein (LBP)[4] and then glycosylphosphatidylinositol (GPI)-anchored protein CD14 detects LPS/LBP on the surface of immunocompetent cells.[5] Then, LPS is transported to the MD2/TLR 4 complex,[6] where TLR4 recognizes the LPS/MD2 complex at the extracellular domain. Following that, the receptor complex activates the signal transduction cascade consisting of MyD88, IRAK, TRAF6, and

NF-κB inside the cells. Finally, the cells synthesize and release cytokines (Figure 9.1b).

In the early stage of this cascade, the target cell membrane plays a substantial role. Recent studies, in particular, have shown the importance of the sphingolipid and cholesterol-based membrane microdomain formation for cellular activation by LPS.[7] These microdomains, called the liquid-ordered phase (L_o), are segregated from the liquid-disordered phase (L_d) due to immiscibility between membrane constituents.[8–10] These "raft" domains are emerging as the sites of cellular signaling, protein and lipid sorting, and virus budding, and they are involved in the onset of diseases such as Alzheimer and prion.[11] Certain receptor proteins involved in the innate immune response, such as CD14 and heat shock proteins (hsp) 70 and 90, are constitutively found in raft domains.[7] By contrast, the LPS-cellular activation proteins, TLR4 and chemokine receptor 4 (CXCR4), are not. Rather, they are gathered there by LPS stimulation.[7] Thus, LPS stimulation induces the integration of receptor proteins into lipid rafts upon the onset of the innate immune response and can facilitate interactions between receptor proteins and the subsequent signal transduction system. In this way, lipid rafts play a significant role in the LPS-activated innate immune response; however, little is known about the effect of LPS insertion on membranes and its relationship between lipid raft formations. We examined these in pure phospholipids and raft-forming membranes by using solid-state NMR. This technique has been shown to be useful for probing membrane morphology, the phase state, and the structure and dynamics of membrane-active molecules in lipid bilayers even in insoluble and noncrystaline samples. LPS is composed of covalently linked three main parts, *i.e.* O-polysaccharide, core oligosaccharide, and lipid A (Figure 9.2a).[2] Although the O-polysaccharide part represents the antigenic property of the bacteria, it plays no significant

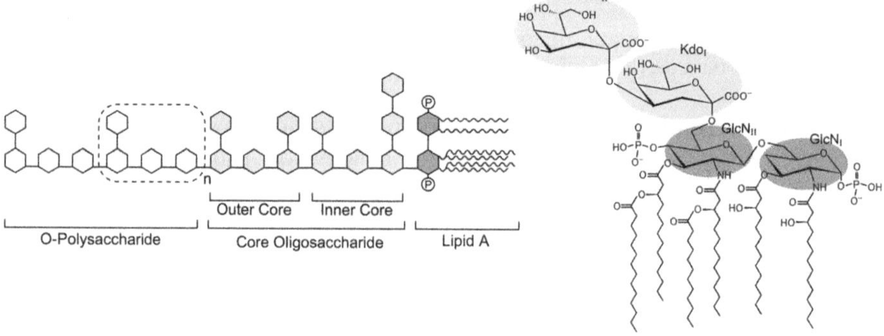

Figure 9.2 (a) Schematic structures of LPS from Gram-negative bacteria. The O-polysaccharide region consists of up to 50 repeating units. (b) Chemical structure of ReLPS. Abbreviations: GlcN, D-glucosamine; Kdo, 3-deoxy-D-*manno*-octulosonic acid; P, phosphate.

role for the endotoxic activity of LPS. In addition, the O-polysaccharide part has a lot of variation in chain length, which complicates NMR spectra. Therefore, ^{13}C-labeled, homogeneous LPS molecules are required for NMR studies.

Lipid A is the active site of LPS and Re-type LPS (ReLPS) is the smallest natural LPS, lacking most of the polysaccharide part, *i.e.* it is composed of lipid A and 3-deoxy-D-*manno*-2-octulosonic acid (Kdo) disaccharide, the latter being linked to the 6′-position of the former (Figure 9.2b). ReLPS retains full endotoxic activities, including potent antitumor and cytokine-inducing activity.[12] Furthermore, ReLPS is the sole endotoxic LPS derivative which can be obtained in a homogeneous form from natural sources. For these reasons, we used ReLPS in this study. Non-labeled ReLPS is now commercially available, but the ^{13}C-labeled specimen had to be prepared by biosynthesis by a deep rough *Escherichia coli* K-12 mutant.[13] The procedure for the purification of ^{13}C-labeled ReLPS is described in the experimental section.

The effects of ReLPS on the membrane morphology of various lipid membranes were first examined by ^{31}P NMR.[14] Next, we observed the ^{13}C magic angle spinning (MAS) NMR spectra of the 1,2-dielaidoyl-*sn*-glycero-3-phosphoethanolamine (DEPE)/cholesterol (Chol) lipid bilayers to determine the membrane phase state from the DEPE C=O signal structure. Then, using this method, we determined the phase state of DEPE and DEPE/sphingomyelin(SM)/Chol membranes in the presence and absence of ReLPS.[15] Finally, the molecular motions of the ReLPS and DEPE phospholipids in these membranes were evaluated from the ^{1}H spin–lattice relaxation time in the rotating frame ($T_{1\rho}{}^{H}$) and the temperature dependence of the spectral line widths.[15]

9.2 Experimental

9.2.1 Materials

Egg-PC, 1-palmitoyl-2-oleoyl-*sn*-glycero-3-phosphoglycerol (POPG), DEPE, *E. coli* total lipids extract, *E. coli*-PE, brain-PS, bovine-SM, egg-SM, plant-derived cholesterol, and deep rough mutant ReLPS from *E. coli* strain WBB06 were purchased from Avanti Polar Lipids (Alabaster, USA). Uniformly ^{13}C-labeled Bioexpress cell growth media (U-^{13}C, 98%) was obtained from Cambridge Isotope Laboratory (Andover, USA). Unlabeled ReLPS was used after conversion to the triethylammonium (TEN) salt form, as described in the following section. Other chemicals were used without further purification. Octyl β-D-glucoside (OG) was purchased from Nacalai Tesque (Kyoto, Japan). Dowex 50W-X8 cation-exchange resin (H^{+} form) was obtained from Muromachi Technos (Tokyo, Japan). Dowex MAC-3 weak acid cation-exchange resin (H^{+} form) was purchased from Sigma-Aldrich. Sephadex LH-20 was obtained from GE (Uppsala, Sweden). Tris buffer (20 mMTris-HCl, 100 mMNaCl, pH 7.6) was used for all experiments.

9.2.2 Sample Preparation

Uniformly ^{13}C-labeled ReLPS was prepared as follows.[15] After cultivation of the heptose-deficient *E. coli* strain WBB06[13] in a uniformly ^{13}C-labeled cell growth media, LPS was extracted from the dried cells by the phenol/chloroform/petroleum ether (PCP) method.[16] Since an ReLPS derivative with an additional phosphoethanolamine (pEtN) substitution at position 7 of the outer Kdo residue was mixed as a minor component,[17] to remove it, purification was performed using cation-exchange column chromatography (Dowex 50W-X8, H^+ form) with a two-phase solvent system (chloroform/methanol/ethanol/H_2O, 8:2:1:2; v/v/v/v). Next, exchange of the phosphoryl residues to their TEN-salt form was accomplished by cation-exchange column chromatography (MAC-3 cation-exchange resin, TEN-salt form). Finally, purification by liquid–liquid partition column chromatography using a two-phase solvent system (butan-1-ol/tetrahydrofuran/methanol/H_2O, 12:4.4:1:40) was performed on Sephadex LH20 gel to remove the lipid A formed during the purification procedure. After removal of the solvent, pure ReLPS was obtained as an amorphous solid.

Multilamellar vesicles (MLVs) were prepared with Tris buffer, as previously described.[14,15] For ^{31}P NMR measurements under static conditions, the water content was adjusted to about 80% (w/w). Samples were then transferred to 5-mm NMR glass tubes which were cut to an appropriate size (about 4 cm). The tubes were capped to prevent dehydration. For ^{13}C MAS experiments, the dry powder MLVs were rehydrated to 35% water by weight and transferred to a glass tube insert. The tube was sealed with Araldite standard epoxy resin (Huntsman Advanced Materials, USA) to prevent dehydration and inserted into an NMR rotor.

9.2.3 Solid-State NMR Spectroscopy

All solid-state NMR experiments were performed on a CMX 300 Infinity NMR spectrometer (Chemagnetics, Varian, Palo Alto, USA). All static ^{31}P spectra were obtained using a 90° excitation pulse width of 5 μs, 30 kHz proton decoupling, 50 μs dwell time, a recycle delay of 5 s, and 1024–60 000 data points. Spectra were processed with 30 Hz line broadening and externally referenced relative to 85% H_3PO_4 (0 ppm). All ^{13}C direct polarization MAS NMR spectra were acquired using a 90° excitation pulse width of 4 μs, 75 kHz proton decoupling, 60 μs dwell time, a recycle delay of 5 s, 1000–12 000 data points, and under 5 kHz MAS speed. Spectra were processed using 10 Hz line broadening for Figures 9.5 and 9.8, and 3 Hz for Figure 9.6. The actual temperature in the sample under 5 kHz MAS and 75 kHz ^{1}H CW decoupling was calibrated by using chemical shift differences between water and the methyl peak in the DEPE 35% hydrated sample.[18] All ^{13}C spectra were externally referenced relative to the methine carbon of adamantane (29.5 ppm).

9.3 Results and Discussion

9.3.1 ReLPS Membrane Interactions

To examine the effects of incorporated ReLPS on the morphology of lipid membranes, the ^{31}P NMR spectra of various lipid membranes in the absence and presence of ReLPS were measured (Figure 9.3) with a ReLPS/lipid molar ratio of 1:10 under static conditions above T_m of the respective lipids. Figure 9.3a shows ^{31}P NMR spectra of ReLPS/egg-PC (1:10) membranes ($T_m \approx -5$ °C)[19] at 0 and 30 °C. In the presence of ReLPS, an isotropic component superimposed on a bilayer component was observed at 0 °C. This isotropic component may be attributed to fast isotropic motion of the phospholipids, which may arise from the formation of micelles, small unilamellar vesicles, small discoidal bilayers, or a cubic phase. At 30 °C, the proportion of isotropic component became larger, suggesting that micellization of ReLPS owing to the high mobility of the acyl chains of egg-PC was enhanced at the higher temperature. This isotropic component increased for

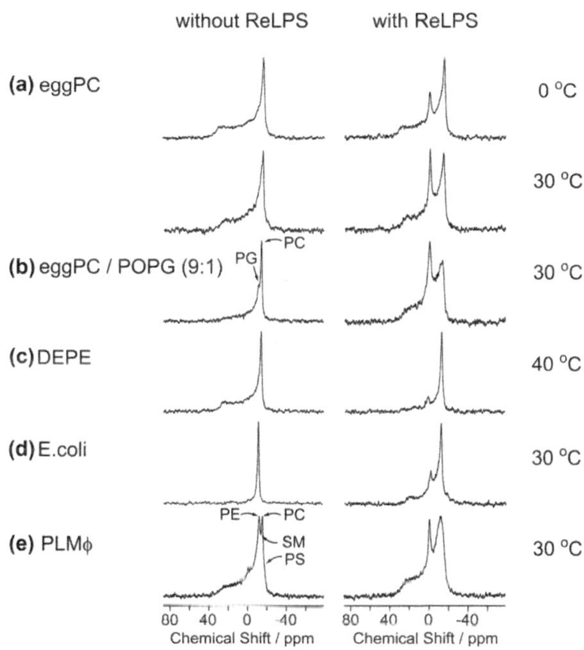

Figure 9.3 Effect of ReLPS on the ^{31}P NMR spectra of phospholipids in various membranes composed of (a) egg-PC, (b) egg-PC/POPG (9 : 1), (c) DEPE, (d) *E. coli* polar lipid extract, and (e) $PL_{M\phi}$. The spectra on the *right* have a molar ratio of ReLPS/lipids = 1:10. For spectra without ReLPS, (b) and (e), the resonances were assigned by comparison with egg-PC, egg PC, *E. coli*-PE, bovine-SM, or brain-PS.

ReLPS/egg-PC/POPG (1:9:1) membranes (Figure 3b). The ReLPS partially incorporated into the membrane and became a form of aggregate on the surface.[14] This is probably due to electrostatic repulsion between POPG and ReLPS which increases the energy required for mixing. On the other hand, in the case of ReLPS/DEPE membranes at 40 °C, the isotropic signal became much smaller in the presence of ReLPS (Figure 9.3c) than for ReLPS/egg-PC (Figure 9.3a) and ReLPS/egg-PC/POPG (9:1) (Figure 9.3b) membranes. These differences are probably related to their molecular shape differences. Whereas PC is cylindrical due to its bulky choline head group, PE has a conical molecular shape. Schromm *et al.* revealed that ReLPS has a conical shape like PE.[20] Owing to the similarity of the molecular shapes, ReLPS may be more readily integrated into PE-rich rather than PC-rich membranes, producing smaller perturbations to the bilayer structure of the former.

Since LPS is a typical component in the outer membrane of Gram-negative bacterial cells, the effect of ReLPS on a membrane reconstituted from components of bacterial cells is of interest. Figure 9.3d shows the ^{31}P NMR spectra of *E. coli* model membranes composed of a phospholipid mixture extracted from *E. coli.* Similar to the case of ReLPS/DEPE membranes (Figure 9.3c), the observed isotropic signal superimposed on a bilayer component is very small, even in the presence of ReLPS. This suggests that almost all ReLPS molecules intercalated into the *E. coli* model membrane, as for the DEPE membrane. The phospholipid mixture extracted from *E. coli* contains approximately 57.5% PE, 15.1% PG, 9.8% cardiolipin, and 17.6% other lipids (w/w).[21] Thus, these results imply that, in PE-rich membranes, ReLPS hardly disturbs the membrane morphology and forms phospholipid bilayers. The high content of PE in Gram-negative bacterial membranes may help prevent the bacterial cell from micellization.

Finally, we examined how ReLPS affects membranes mimicking that of the host macrophage cell composed of a molar phospholipid mixture ($PL_{M\phi}$) of [egg-PC]:[*E. coli*-PE]:[bovine-SM]:[brain-PS] = 1 : 0.7 : 0.5 : 0.4.[22] ^{31}P NMR spectra obtained from $PL_{M\phi}$ membranes at 30 °C confirmed a lamellar phase, and signals from PC, PE, SM, and PS appeared at different chemical shift values (Figure 9.3e, left). By contrast, an isotropic peak appeared superimposed on a bilayer component in the presence of ReLPS (Figure 9.3e, right). Owing to the low proportion of PE in the $PL_{M\phi}$ membrane, the intensity of the isotropic peak relative to the bilayer components was higher than for ReLPS/DEPE (Figure 9.3c) or ReLPS/*E. coli* model membranes (Figure 9.3d). However, the value was lower than those of ReLPS/egg-PC (Figure 9.3a) and ReLPS/egg-PC/POPG (Figure 9.3b) membranes. This suggests that when Gram-negative bacteria invade higher animals, ReLPS partially micellizes the host macrophage cell. In addition, the phospholipid peaks observed in the absence of ReLPS (Figure 3d, left) were merged in the presence of ReLPS, indicating that ReLPS induced mixing of phospholipids in the $PL_{M\phi}$ membrane.[23]

9.3.2 Concentration Dependence of ReLPS–DEPE Lipid Bilayer Interaction

Since micellization of membranes by ReLPS was suppressed in PE-rich membranes, we used DEPE-based membranes in subsequent experiments. In order to examine at what ratio ReLPS/DEPE does not generate partial micellization of the lipid bilayer, we acquired ^{31}P NMR spectra of lipid bilayers composed of various ratios of DEPE and ReLPS at 40°C under static condition (Figure 9.4a–d). Without ReLPS, an axially symmetric powder pattern was observed, representing a bilayer membrane (Figure 9.4a). At a molar ratio of ReLPS/DEPE = 1:20, the spectrum also showed the axially symmetric powder pattern (Figure 9.4b). This clearly indicated that the

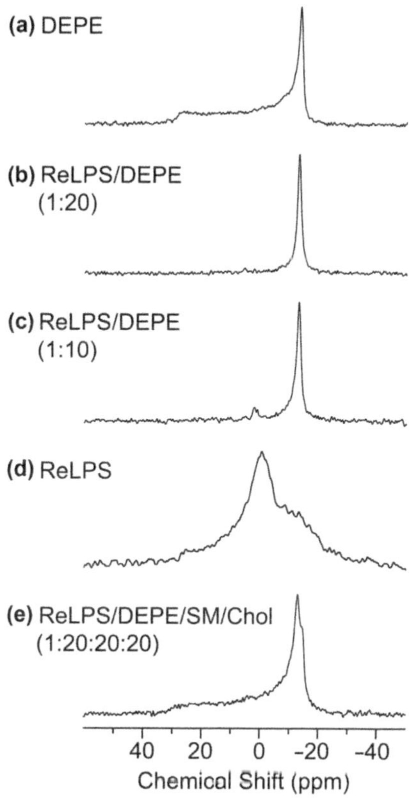

Figure 9.4 (a–d) Effect of ReLPS concentration on ^{31}P NMR spectra of DEPE membranes at 40 °C. The gel–lamellar phase transition temperature T_c of DEPE is ~35 °C; the lamellar-inverted hexagonal phase (H_{II}) phase transition temperature T_h of DEPE is ~60 °C;[45] and the gel–lamellar phase transition temperature T_m of ReLPS (ethanolamine-salt form) reported by Seydel *et al.* is ~33 °C.[24] The ReLPS/DEPE molar ratios are (a) 0:100, (b) 1:20, (c) 1:10, and (d) 100:0. (e) ^{31}P NMR spectrum of ReLPS/DEPE/SM/Chol (1:20:20:20).

ReLPS was incorporated into the DEPE membranes and did not disturb the membrane morphology. By contrast, for ReLPS/DEPE = 1:10, a small isotropic signal appeared, superimposed on the bilayer component, indicating the partial micellization of the membrane (Figure 9.4c). Thus a critical ratio at which all ReLPS is uniformly incorporated into DEPE membranes is ReLPS/DEPE = 1:20, and this ratio was used in all subsequent experiments. For membranes made only of ReLPS, the spectrum showed a complicated pattern (Figure 9.4d), suggesting that ReLPS forms a randomly associated mixture of isotropic and lamellar phases.[24]

For the model of a raft-forming membrane, we chose the DEPE/SM/Chol (1:1:1) ternary system. Owing to the highly saturated tightly packed acyl chains of SM and the ability of Chol to facilitate packing interactions between such saturated lipids, both Chol and SM form the L_o phase in the membrane. Figure 9.4e shows the ^{31}P NMR spectrum of ReLPS/DEPE/SM/Chol (1:20:20:20) membranes as a superposition of axially symmetric powder patterns and no isotropic signal, which suggests that all the ReLPS is completely incorporated into the lipid bilayer.

9.3.3 ^{13}C NMR Spectra of Uniformly ^{13}C-Labeled ReLPS in Membranes

Figures 9.5a and 9.5b show the ^{13}C direct polarization NMR spectra of ReLPS/DEPE (1:20) membranes, where ReLPS is (a) uniformly ^{13}C-labeled and (b) unlabeled. By comparing Figure 9.5a with 9.5b, the sharp signals in Figure 9.5a can readily be distinguished from ReLPS signals and assigned as DEPE signals.[25] Since both ReLPS and DEPE have acyl chains, many peaks in Figure 9.5a, especially those in the aliphatic region, are overlapping, except for $GlcN_{1,2}$ C_2 (54 ppm), Kdo_2 $C_{4,5}$ (69 ppm), $GlcN_1$ C_1 (94 ppm), and $Kdo_{1,2}$ C_2 (100 ppm). By referring to solution NMR chemical shifts, these four resonances of ReLPS were able to be assigned.[26,27] Figures 9.5c and 9.5d show the ^{13}C spectra of ReLPS/DEPE/SM/Chol (1:20:20:20), where ReLPS is (c) uniformly ^{13}C-labeled and (d) unlabeled. The DEPE, SM, and Chol signals are also sharp in these spectra. Although the ReLPS $GlcN_{1,2}$ C_2 (54 ppm) signal is overlapped by the SM and Chol signals, the other three peaks can barely be distinguished from the DEPE, SM, and Chol peaks. In the ReLPS/DEPE (1:20) membrane, the chemical shift value of the ReLPS $(CH_2)_n$ peak was 31.0 ppm (Figure 9.5a), whereas it shifted to 32.3 ppm in the ReLPS/DEPE/SM/Chol membrane (Figure 9.5c). This low-field shift shows that the acyl chain adopts a *trans-gauche* conformation in ReLPS/DEPE membranes and an all-*trans* conformation in ReLPS/DEPE/SM/Chol membranes.[28] These results suggest that almost all the ReLPS exists in the L_d phase in ReLPS/DEPE membranes and in the L_o phase in ReLPS/DEPE/SM/Chol membranes.

Figure 9.6a shows the carbonyl region of the ^{13}C MAS NMR spectra of both DEPE and DEPE/Chol/SM membranes in the presence and absence of

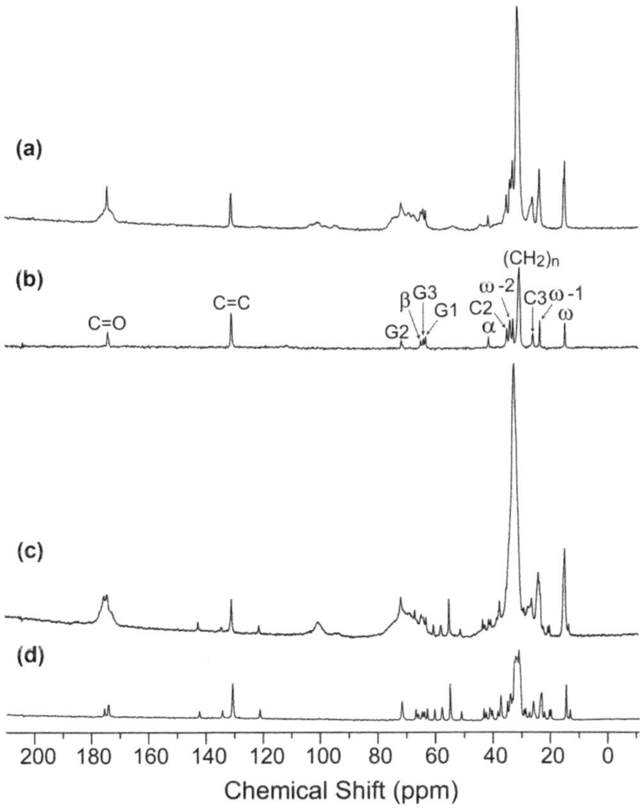

Figure 9.5 ^{13}C NMR spectra of DEPE (a, b) and DEPE/SM/Chol (c, d) membranes in the presence (a, c) and absence (b, d) of ReLPS under MAS with a spinning speed of 5 kHz measured at 37 °C. The DEPE ^{13}C assignment is indicated.

ReLPS. The insertion of ReLPS into the DEPE lipid bilayer changes the DEPE C=O chemical shift value from 173.99 to 174.08 ppm. Hydrogen bonding has been known to alter the chemical shifts of C=O to lower fields. Therefore, the low-field shift in the presence of ReLPS indicates that either hydroxy or amide groups of ReLPS form a hydrogen bond with the DEPE C=O or conformational changes of DEPE molecules occur in the presence of ReLPS.[29] Similarly, a chemical shift change for C=O was due to ReLPS insertion in the DEPE/SM/Chol membrane; the SM C=O chemical shift value changed from 175.45 to 175.49 ppm. This change is also likely due to intermolecular hydrogen bonding between the hydroxy or amide group of ReLPS and the SM C=O or due to a conformational change of SM molecules. Whatever the reason, these results suggest that ReLPS penetrates both DEPE and DEPE/SM/Chol membranes and aligns with the phospholipid, where the hydroxy and/or amide groups of ReLPS reside near the C=O groups of DEPE and SM.

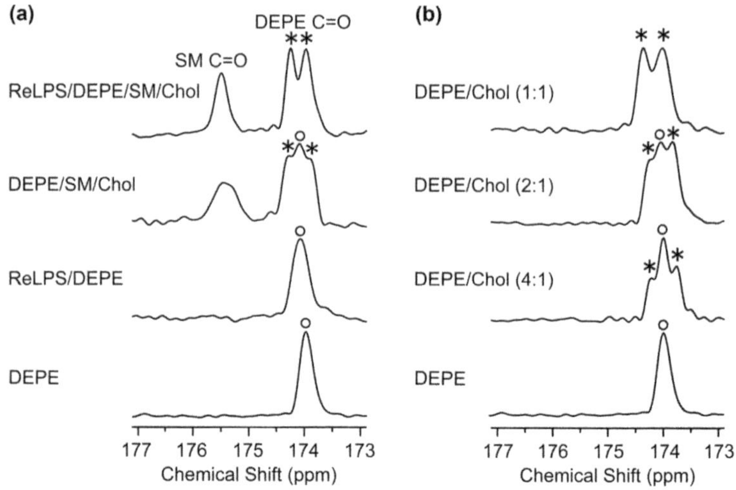

Figure 9.6 (a) Carbonyl carbon region of ^{13}C NMR spectra of various membranes. (b) Carbonyl carbon signal of the DEPE/Chol membrane with Chol contents as indicated. The central and both side peaks are marked with circles (○) and asterisks (✶), respectively. All spectra were obtained under MAS with a spinning speed of 5 kHz at 37 °C and processed with 3 Hz line broadening.

9.3.4 Characterization of DEPE/SM/Chol Membranes

In DEPE/SM/Chol membranes, the insertion of ReLPS also shifts the DEPE C=O signal to the lower field; however, since the C=O peak is a triplet in the absence of ReLPS and a doublet in its presence, direct estimation of the shift value was difficult. These complex line structures probably originate from the interaction between DEPE, ReLPS, and Chol. To examine the effect of Chol concentration on the DEPE C=O signal, the ^{13}C MAS NMR spectra of DEPE/Chol membranes with different Chol contents were observed (Figure 9.6b). Without Chol, the C=O signal was a singlet. At 20 mol% Chol, the C=O peak was split into a triplet. The intensity of the central peak (shown with circle, ○) gradually decreased with increasing Chol content. At 50 mol% Chol the central peak disappeared and the two side peaks (shown with asterisk, ✶) remained. The peak intensity dependence on the Chol content implies that the central peak represents the L_d phase and the left and right peaks represent the L_o phase. This is likely because, in the L_o phase, non-identical *sn*-1 and *sn*-2 C=O peaks in phospholipids can be resolved,[30] whereas in the L_d phase these two carbonyl carbons become indistinguishable. If we apply this approach to the spectra shown in Figure.9.6a, then only the L_d phase exists in DEPE and ReLPS/DEPE membranes. In DEPE/SM/Chol, both L_o and L_d phases coexist, whereas the ReLPS/DEPE/SM/Chol membrane is only in the L_o phase. These results show remarkably that incorporation of ReLPS into the raft-forming membrane

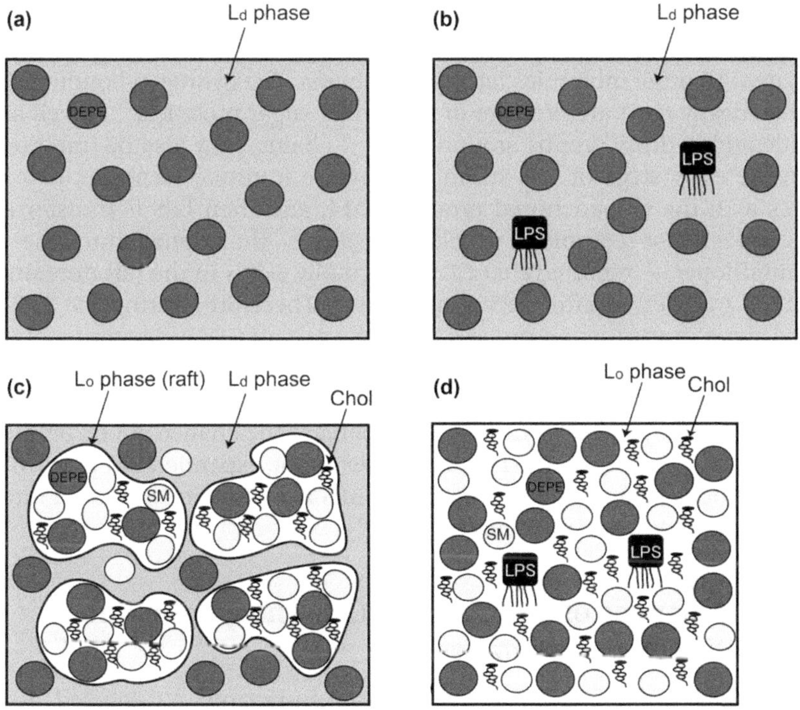

Figure 9.7 Schematic models of phase states of the (a) DEPE, (b) ReLPS/DEPE (1:20), (c) DEPE/SM/Chol (20:20:20), and (d) ReLPS/DEPE/SM/Chol (1:20:20:20) membranes. The L_d and L_o phase domains are *gray* and *white*, respectively.

induces the expansion of the raft area. Figure 9.7 illustrates the phase states of DEPE and DEPE/SM/Chol membranes with and without ReLPS.

Suzuki *et al.* have proposed that the clustering of a GPI-anchored protein induces rafts that trigger signal transduction.[31,32] They reported that the binding of CD59, a GPI-anchored receptor, with its ligand, C8, induces the clustering of about six CD59 molecules. The CD59 clusters induce Chol and glycolipid accumulation around them and form a raft. Then, intracellular signaling molecules are recruited to the raft and form a complex, leading to the inositol 1,4,5-triphosphate (IP_3)-Ca^{2+} signal. Our present study shows that ReLPS induces expansion of the raft area in the raft-forming membrane. This ability of LPS appears to be similar to that of CD59 clustering induced raft formation. The GPI anchor is a glycolipid composed of a phosphatidylinositolglycan core and phosphoethanolamine. The GPI anchor is linked to the C-terminal amino acid of a protein through the phosphoethanolamine bridge to form a so-called GPI-anchored protein. The acyl chains in the phosphatidylinositol moiety are all saturated, which is important for raft formation.[33,34] ReLPS also has saturated acyl groups linked to carbohydrate residues. This is assumed to be the most essential structural similarity of CD59 and ReLPS, underlying their raft-forming ability. The six saturated acyl chains in ReLPS are likely to have a stronger tendency than the GPI-anchored protein to associate with other molecules having acyl chains. The hydrogen-bonding ability and sterically rigid architecture of the ReLPS sugar moieties,[35] as well as the amide and hydroxy groups around the acyl chains, may also be important.

In the early stage of LPS-stimulated innate immune response, LPS associates with the GPI-anchored protein CD14, and then LPS is transferred to the MD2/TLR-4 complex, which transduces the signal into the cell. Triantafilou*et al.* reported that CD14 originally exists in the raft domain, but TLR4 is transferred after LPS stimulation.[7] Therefore, during the LPS-activated signal transduction process, as in the case of CD59, clustering of CD14 probably induces raft formation and then LPS, when incorporated into the cell membrane, most likely expands the raft area. Such consecutive events could bring about synergistic effects to increase the chance of LPS binding to the signaling receptor MD2/TLR-4 complex in the expanded raft area. In fact, Gutsmann*et al.* previously reported that LPS is readily incorporated into the cell membrane in the presence of LBP.[36]

9.3.5 Dynamics of ReLPS and DEPE Phospholipid in Different Membranes

Figure 9.8a shows the spectra for the uniformly ^{13}C-labeled ReLPS/DEPE(1 : 20) membrane at four different temperatures. Both the ReLPS and

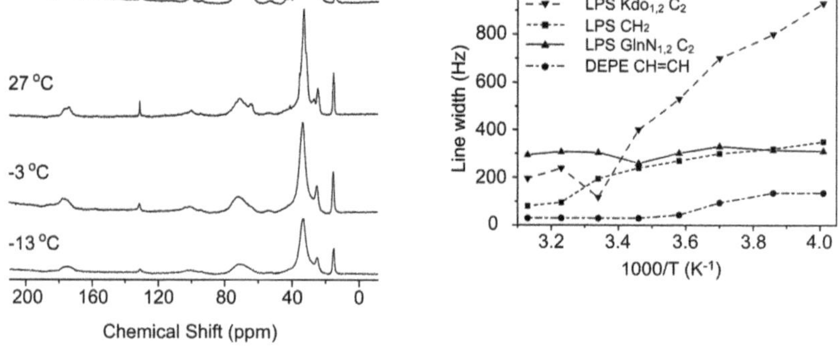

Figure 9.8 (a) ^{13}C NMR spectra of a ^{13}C-labeled ReLPS/DEPE (1:20) sample at 37, 27, −3, and −13 °C under MAS with a spinning speed of 5 kHz. (b) Temperature dependence of the linewidth of the $(CH_2)_n$ and Kdo C_4, C_5, GlcN C_2 in ReLPS and DEPE CH=CH peaks.

DEPE lines became broad with decreasing temperature. Figure 9.8b shows the relationship between temperature and the linewidth of distinct peaks, *i.e.* $(CH_2)_n$ (31 ppm), $Kdo_{1,2}$ C_2 (100 ppm), and $GlcN_{1,2}$ C_2 (54 ppm) of ReLPS and CH=CH (130 ppm) of DEPE. As the temperature decreased, the line width of the ReLPS $(CH_2)_n$, Kdo C_2, peaks, and the DEPE CH=CH peak increased. Thus, in this temperature range, the molecular motion became faster as the relaxation times increased. Besides, their motional rates are all faster than the timescale of the inverse of the proton decoupling field strength (75 kHz, 13 μs).[37] By contrast, the ReLPS GlcN C_2 peak (54 ppm) showed two maxima, suggesting the presence of a distribution of motional correlation times.[38]

To compare the motion of ReLPS and phospholipid in DEPE and DEPE/SM/Chol membranes, we evaluated the ^{1}H spin–lattice relaxation time in the rotating frame $T_{1\rho}^H$. The $T_{1\rho}^H$ values are sensitive to molecular motion, having the time scale of the inverse of the spin-lock field strength (60 kHz, 16 μs). The $T_{1\rho}^H$ values for ReLPS peaks in DEPE and DEPE/SM/Chol membranes are shown in Table 9.1. In both membranes, $T_{1\rho}^H$ values of these peaks are approximately 1–4 ms, suggesting that the motion of ReLPS is not much affected by the membrane environment. We also measured the $T_{1\rho}^H$ for DEPE peaks in the presence and absence of ReLPS in the DEPE membrane (Table 9.2). Without ReLPS, $T_{1\rho}^H$ values for DEPE peaks were

Table 9.1 ^{13}C $T_{1\rho}^H$relaxation times (ms) of ReLPS carbons in DEPE and DEPE/SM/Cholmembranes.[a]

	DEPE/ReLPS	*DEPE/SM/Chol/ReLPS*
$GlcN_{1,2}$ C_2	1.4 ± 0.1	–
Kdo_2 $C_{4,5}$	3.4 ± 0.1	3.1 ± 0.5
$GlcN_1$ C_1	1.7 ± 0.2	1.2 ± 0.1
$Kdo_{1,2}$ C_2	1.0 ± 0.1	–

[a] $T_{1\rho}^H$ values are for 35% hydrated samples under MAS ($\omega_r = 5$ kHz).

Table 9.2 ^{13}C $T_{1\rho}^H$ relaxation times (ms) of DEPE carbons in DEPE or DEPE/SM/Chol membranes in the presence and absence of ReLPS.[a]

	DEPE	*ReLPS/DEPE*	*DEPE/SM/Chol*	*ReLPS/DEPE/SM/Chol*
α	67.0 ± 4.2	104.9 ± 6.9	101 ± 11	135.1 ± 1.0
β	84.6 ± 3.2	24.1 ± 2.0		–
G3	91.9 ± 4.9	34.0 ± 2.4	–	–
G2	91.7 ± 2.2	32.8 ± 4.9	108 ± 11	109.5 ± 7.8
G1	197.3 ± 5.0	64.8 ± 4.1	–	–
C1	99.0 ± 2.9	67.2 ± 9.1	–	–
C2	118.0 ± 5.4	39.6 ± 1.4	74 ± 19	246.2 ± 18.4
C3	157.1 ± 22.4	47.3 ± 2.9	125 ± 2	169.0 ± 13.2
C=C	462.4 ± 62.8	160.3 ± 34.0	692 ± 137	no relaxation
CH_2	146.0 ± 13.3	125.5 ± 21.2	127 ± 4	731.9 ± 111.9
ω–2	354.7 ± 30.9	133.7 ± 24.6	163.7 ± 14	257.8 ± 25.7

[a] $T_{1\rho}^H$ values are for 35% hydrated samples under MAS ($\omega_r = 5$ kHz).

approximately 60–500 ms, whereas in the presence of ReLPS the $T_{1\rho}{}^{H}$ values became shorter (20–200 ms). From this, we can infer that the motion of DEPE in the DEPE membranes is much faster than that of ReLPS and becomes slower when ReLPS is in the membrane. Since both ReLPS and phospholipids have acyl chains, we can compare their dynamics from relaxation times. The differences in the rate of motion between ReLPS and DEPE in both membranes can be attributed to the difference in their molecular sizes. The molecular weight of ReLPS is about three times larger than that of DEPE. Since the diffusion rate in lipid membranes is slower with increasing molecular sizes,[39,40] then ReLPS moves slower than DEPE. The smaller $T_{1\rho}{}^{H}$ values of DEPE carbons in the presence of ReLPS suggest that the slow movement of ReLPS disturbs the DEPE lateral diffusion. This observation is consistent with our previous results by microscopic observations regarding the lipid lateral diffusion coefficient measurement at various molar ratios of egg-PC/ReLPS.[14]

$T_{1\rho}{}^{H}$ values for DEPE carbons in the DEPE/SM/Chol system are also summarized in Table 9.2. Without ReLPS, $T_{1\rho}{}^{H}$ values for the acyl chain region were smaller than those in DEPE membranes, whereas the values for the head group became larger. Thus, in the DEPE/SM/Chol membrane, the DEPE acyl chain region moves slower and the head group moves faster than in the membrane composed of DEPE alone. A partitioning of Chol between the acyl chain moiety of the DEPE molecules in the bilayer is probably the reason for the restriction of the DEPE acyl chain motion.[41] On the other hand, the DEPE head group can move relatively freely around the molecular axis, because the cholesterol molecule does not interact with the DEPE head group. The effect of ReLPS on the motion of DEPE in the DEPE/SM/Chol membrane is opposite to that in the DEPE membrane. In the presence of ReLPS, the $T_{1\rho}{}^{H}$ values of both the acyl chain and the head group of DEPE became larger, suggesting that the presence of ReLPS reduces the packing effect of Chol and frees the DEPE acyl chain from Chol. ReLPS also moves more slowly than DEPE in the DEPE/SM/Chol membrane. Thus, the rapid motion of raft-forming membranes is expected to further increase the chance of LPS binding to its receptor, the MD2/TLR4 complex, in the raft region.

9.4 Conclusions

We investigated the mutual interactions of LPS, a ligand molecule of innate immune response of higher animals, and phospholipids in various membranes. Three main findings in this study are: (1) ReLPS has higher miscibility with PE than PC membranes. Therefore, if membrane bilayers contain much PE, the membranes tolerate the micellization due to ReLPS. (2) In the raft-forming membrane, ReLPS induced the expansion of the raft area. (3) The membrane motion in the raft is accelerated by the ReLPS itself. Result (1) explains why PE occupies more than half of the lipid constituting the Gram-negative bacterial membrane, in which large amounts of LPS are originally included. Results (2) and (3) imply that, when higher animals are infected by Gram-negative bacteria, LPS acts autocatalytically in immunocompetent cells and facilitates the innate immune recognition in addition to its role as a ligand. In addition to LPS, other pathogen-associated molecules (PAMPs), such as ssRNA, dsRNA, fragelin, and lipoprotein, are also ligands for the innate immune response of higher animals. Although the PAMPs have broad structural and physicochemical characteristics, their respective receptors in the Toll-like family[42,43] specifically recognize these PAMPs to induce the host response. Among these PAMPs, only the bacterial lipoprotein and its N-terminal active entity, lipopeptide, have structural similarities with LPS, because they contain saturated acyl chains linked to hydrophilic head groups. Besides, intercalation of lipopeptides, such as Pam2CSK4, Pam3CSK4, and lipolan, into liposomes has been observed by FRET measurements.[44] Thus only lipoproteins (lipopeptides) have the possibility to expand the raft area and accelerate the motion of phospholipids in rafts as does LPS. Result (2) also provides a methodology to distinguish the membrane phase state from the C=O line shape of the ^{13}C NMR spectrum of the lipid bilayer. Since the central peak in the three C=O peaks originates from the carbonyl carbon existing in the L_d phase and both side peaks originate from that in the L_o phase, the percentage of phospholipid in the L_o phase (raft area) in the membrane could be estimated. This method may be used to determine the phase state of phospholipid-containing membranes.

Acknowledgements

This work was supported by a Grant-in-Aid for Scientific Research (#19550174) to KN from the Ministry of Education, Culture, Sports, Science and Technology of Japan.

References

1. H. Brade, S. M. Opal, S. N. Vogel and D. C. Morrison, *Endotoxin in Health and Disease*, Dekker, New York, 1999.
2. S. Kusumoto and K. Fukase, *Chem. Rec.*, 2006, **6**, 333.
3. A. Poltorak, X. He, I. Smirnova, M.-Y. Liu, C. Van Huffel, X. Du, D. Birdwell, E. Alejos, M. Silva, C. Galanos, M. Freudenberg, P. Ricciardi-Castagnoli, B. Layton and B. Beutler, *Science*, 1998, **282**, 2085.
4. P. S. Tobias, K. Soldau and R. J. Ulevitch, *J. Exp. Med.*, 1986, **164**, 777.
5. R. R. Schumann, *Res. Immunol.*, 1992, **143**, 11.
6. R. Shimazu, S. Akashi, H. Ogata, Y. Nagai, K. Fukudome, K. Miyake and M. Kimoto, *J. Exp. Med.*, 1999, **189**, 1777.
7. M. Triantafilou, K. Miyake, D. T. Golenbock and K. Triantafilou, *J. Cell Sci.*, 2002, **115**, 2603.
8. D. A. Brown and E. London, *J. Biol. Chem.*, 2000, **275**, 17221.
9. K. Simons and E. Ikonen, *Nature*, 1997, **387**, 569.
10. S. J. Singer and G. L. Nicolson, *Science*, 1972, **175**, 720.

11. J. Fantini, N. Garmy, R. Mahfoud and N. Yahi, *Expert Rev. Mol. Med.*, 2002, **4**, 1.
12. C. R. Raetz, T. A. Garrett, C. M. Reynolds, W. A. Shaw, J. D. Moore, D. C. Smith, Jr., A. A. Ribeiro, R. C. Murphy, R. J. Ulevitch, C. Fearns, D. Reichart, C. K. Glass, C. Benner, S. Subramaniam, R. Harkewicz, R. C. Bowers-Gentry, M. W. Buczynski, J. A. Cooper, R. A. Deems and E. A. Dennis, *J. Lipid Res.*, 2006, **47**, 1097.
13. W. Brabetz, S. Müller-Loennies, O. Holst and H. Brade, *Eur. J.Biochem.*, 1997, **247**, 716.
14. K. Nomura, T. Inaba, K. Morigaki, K. Brandenburg, U. Seydel and S. Kusumoto, *Biophys.J.*, 2008, **95**, 1226.
15. K. Nomura, M. Maeda, K. Sugase and S. Kusumoto, *Innate Immun.*, 2011, **17**, 256.
16. C. Galanos, O. Lüderitz and O. Westphal, *Eur. J. Biochem.*, 1969, **9**, 245.
17. M. I. Kanipes, S. Lin, R. J. Cotter and C. R. Raetz, *J. Biol. Chem.*, 2001, **276**, 1156.
18. S. V. Dvinskikh, V. Castro and D. Sandström, *Magn. Reson. Chem.*, 2004, **42**, 875.
19. J. Seelig and N. Waespe-Sarcevic, *Biochemistry*, 1978, **17**, 3310.
20. A. B. Schromm, K. Brandenburg, H. Loppnow, U. Zähringer, E. T. Rietschel, S. F. Carroll, M. H. Koch, S. Kusumoto and U. Seydel, *J. Immunol.*, 1998, **161**, 5464.
21. M.J. Newman and T.H. Wilson, *J. Biol. Chem.*, 1980, **255**, 10583.
22. E. E. Kröner, B. A. Peskar, H. Fischer and E. Ferber, *J. Biol. Chem.*, 1981, **256**, 3690.
23. T. L. Pukala, M. P. Boland, J. D. Gehman, L. Kuhn-Nentwig, F. Separovic and J. H. Bowie, *Biochemistry*, 2007, **46**, 3576.
24. U. Seydel, H. Labischinski, M. Kastowsky and K. Brandenburg, *Immunobiology*, 1993, **187**, 191.
25. R. D. Yang, K. M. Patel, H. J. Pownall, R. D. Knapp, L. A. Sklar, R. B. Crawford and J. D. Morrisett, *J. Biol. Chem.*, 1979, **254**, 8256.
26. S. M. Strain, S. W. Fesik and I. M. Armitage, *J. Biol. Chem.*, 1983, **258**, 13466.
27. W. Wang, H. J. Sass, U. Zähringer and S. Grzesiek, *Angew. Chem. Int. Ed.*, 2008, **47**, 9870.
28. M. Matsui, Y. Yamane, H. Kimura, S. Kuroki, I. Ando, K. Fu and J. Watanabe, *J. Mol. Struct.*, 2003, **650**, 175.
29. H. Saito, *Magn. Reson. Chem.*, 1986, **24**, 835.
30. W. Guo and J.A. Hamilton, *Biochemistry*, 1995, **34**, 14174.
31. K. G. Suzuki, T. K. Fujiwara, M. Edidin and A. Kusumi, *J. Cell Biol.*, 2007, **177**, 731.
32. K. G. Suzuki, T. K. Fujiwara, F. Sanematsu, R. Iino, M. Edidin and A. Kusumi, *J. Cell Biol.*, 2007, **177**, 717.
33. M. Fujita, M. Umemura, T. Yoko-o and Y. Jigami, *Mol. Biol. Cell*, 2006, **17**, 5253.

34. Y. Maeda, Y. Tashima, T. Houjou, M. Fujita, T. Yoko-o, Y. Jigami, R. Taguchi and T. Kinoshita, *Mol. Biol. Cell*, 2007, **18**, 1497.
35. J. M. Boggs, *Biochim. Biophys. Acta*, 1987, **906**, 353.
36. T. Gutsmann, A. B. Schromm, M. H. Koch, S. Kusumoto, K. Fukase, M. Oikawa, U. Seydel and K. Brandenburg, *Phys. Chem. Chem. Phys.*, 2000, **2**, 4521.
37. W. P. Rothwell and J. S. Waugh, *J. Chem. Phys.*, 1981, **74**, 2721.
38. T. Doherty, A. J. Waring and M. Hong, *Biochemistry*, 2008, **47**, 1105.
39. C. C. Lee, M. Revington, S. D. Dunn and N. O. Petersen, *Biophys. J.*, 2003, **84**, 1756.
40. C. Liu, A. Paprica and N. O. Petersen, *Biophys. J.*, 1997, **73**, 2580.
41. H. M. McConnell and M. Vrljic, *Annu. Rev. Biophys. Biomol. Struct.*, 2003, **32**, 469.
42. A. Aderem and R. J. Ulevitch, *Nature*, 2000, **406**, 782.
43. S. Akira, S. Uematsu and O. Takeuchi, *Cell*, 2006, **124**, 783.
44. A. B. Schromm, J. Howe, A. J. Ulmer, K. H. Wiesmüller, T. Seyberth, G. Jung, M. Rössle, M. H. Koch, T. Gutsmann and K. Brandenburg, *J. Biol. Chem.*, 2007, **282**, 11030.
45. R. M. Epand, *Biochemistry*, 1985, **24**, 7092.

CHAPTER 10

Deuterium NMR of Mixed Lipid Membranes

SHERRY S. W. LEUNG[a] AND JENIFER THEWALT*[a,b,c]

[a] Department of Physics, Simon Fraser University, British Columbia, Canada; [b] Department of Molecular Biology and Biochemistry, Simon Fraser University, British Columbia, Canada; [c] Department of Dermatology and Skin Science, University of British Columbia, British Columbia, Canada

*Email: jenifer_thewalt@sfu.ca

10.1 Introduction

Membranes having defined compositions, *i.e.* "model" membranes, have been instrumental in allowing researchers to develop a sophisticated level of understanding of the physical, chemical and thermodynamic forces governing lipid organization. Many aspects of membrane behaviour were first discovered in model membranes, including thermotropic and lyotropic phase changes, the formation of the liquid ordered phase and domain organization. Model lipid membranes can be as simple as a single type of phospholipid in water or as complicated as desired, for example model skin barrier membranes can contain a myriad of different lipids mimicking those found in the upper layer of the epidermis. Model membranes intrinsically differ from biological membranes in that they are almost always studied at equilibrium and, for example, cannot regulate their lipid composition in response to stress. Biological membranes, of course, have the capacity to self-regulate and perform many other life-sustaining functions. Even so, they

New Developments in NMR No. 3
Advances in Biological Solid-State NMR: Proteins and Membrane-Active Peptides
Edited by Frances Separovic and Akira Naito

Published by the Royal Society of Chemistry, www.rsc.org

evolved in the physical world and are subject to the same forces as model membranes: it is advantageous for a living system to exploit its environment and save energy by taking advantage of spontaneous tendencies whenever possible. Model membranes allow us to fully explore fundamental lipid–lipid interactions in a controlled system.

A wealth of different techniques has been used to measure the properties of lipid membranes. An incomplete list of these includes calorimetry, scattering/diffraction of X-rays or neutrons, infrared and Raman spectroscopy, electron microscopy, atomic force microscopy, fluorescence microscopy and spectroscopy, electron paramagnetic resonance and NMR. Many of these experimental approaches are complementary, yielding information about membrane structure and dynamics in different spatial and temporal regimes. ^{2}H NMR is unusual in that the signal is very directly interpretable: the acquired spectrum immediately tells the spectroscopist about the membrane's phase state. Examples of the ^{2}H NMR spectra of different membrane phases are shown in Figure 10.1. When a membrane is not in a pure phase, the spectrum will be a superposition of the signals from each phase present. The signal is directly proportional to the amount of labelled lipid in the particular phase, allowing a quantitative calculation of the phases present and their compositions.

This chapter does not describe ^{2}H NMR in detail. The interested reader is referred to reviews by pioneers in this area, Joachim and Anna Seelig.[1,2] We are proud to note that Canadians have had a disproportionately high impact on the study of lipid membranes using ^{2}H NMR. Some of the pioneering Canadian researchers in this area are: Ian Smith and Myer Bloom, the intellectual fathers of big families of researchers applying NMR to biological systems;[3,4] Jim Davis, who realized that the spin-1 of the deuteron meant that quickly-decaying NMR signals from deuterated lipids in membranes could be refocused with the solid echo pulse sequence;[5,6] and Michael Morrow, who extended the technique towards biology by using ^{2}H NMR to study a variety of model lung surfactant membranes.[7]

In this chapter we highlight some of the recent advances in our understanding of lipid membrane organization that have been made possible by ^{2}H NMR. First, we discuss acyl chain order in a number of *binary* lipid membranes in the liquid crystalline lamellar phase; this section is quite accessible for readers unfamiliar with ^{2}H NMR. We tabulate and compare the chain order of each component of several different binary lipid membranes. Second, we review the use of ^{2}H NMR to map phase diagrams of *ternary* lipid membranes and compare the results with those obtained *via* fluorescence microscopy. Finally, we provide an overview of our work on model skin barrier membranes, highlighting results that have implications for the practical problem of transdermal drug delivery. This work involves membranes containing four lipids. In our opinion, the advantages of studying model membranes diminish significantly when the number of lipid components exceeds four.

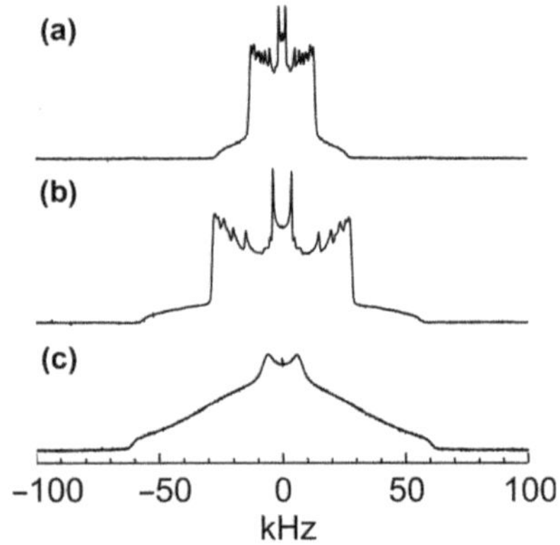

Figure 10.1 Three examples of ^2H NMR spectra obtained from lamellar membranes containing acyl chain-perdeuterated lipids. Each spectrum represents a particular phase discussed in this chapter. (a) The liquid crystalline lamellar L_α phase, also referred to as the liquid disordered, **ld**, phase. In this phase, the spectrum is a superposition of Pake doublets arising from deuterons all along the rapidly rotating acyl chain. The spectrum is symmetric about "0" kHz (the frequency axis is referenced to the Larmor frequency). The quadrupolar splitting, $\Delta\nu_Q$, is the frequency separation of the sharp peaks found in each Pake doublet, and is proportional to the carbon–deuteron order parameter. The smallest quadrupolar splitting in this spectrum is due to the methyl deuterons; this pair of peaks, near the centre of the spectrum, is separated by approximately 4 kHz. Methylene deuterons give rise to Pake doublets with increasingly large $\Delta\nu_Q$ values as we move up the chain from the methyl group towards the lipid/water interface. The maximum $\Delta\nu_Q$ is approximately 30 kHz and corresponds to several methylene deuterons attached to carbon atoms near the acyl chain ester linkage; these experience similar motional freedom. These deuterons form the "plateau" in order parameter profiles. Liquid disordered phases are observed in membranes at temperatures above the main transition temperature T_m. (b) The liquid ordered, **lo**, phase. The **lo** phase spectrum is qualitatively similar to that of the **ld** phase, but the contributing Pake doublets have $\Delta\nu_Q$ values approximately twice those in the **ld** phase, reflecting the suppression of *trans-gauche* isomerization of acyl chain methylenes. Liquid ordered phase spectra are nearly insensitive to temperature (over a range of roughly 50 °C) and are observed in membranes that contain a large amount, *e.g.* >30 mol%, of cholesterol or similar sterol. (c) The gel or solid ordered, **so**, phase. In this phase, the spectrum has a rather featureless bell-shape extending to ±63 kHz. This shape is very distinct from both the **ld** and **lo** spectra. The **so** phase labelled lipids do not rotate rapidly on the NMR timescale (~10 μs), so quadrupolar splittings are not easily measurable from these spectra.

10.2 Simple Mixed Lipid Membranes: Binary Systems

Cell membranes are made up of many different types of proteins and lipids, and are thus difficult to study. Ternary mixtures of lipids, typically comprising one high-melting or saturated lipid, one low-melting (*i.e.*, unsaturated) lipid, and cholesterol (or other sterol), are often used as model plasma membranes with the aim of understanding how membrane "rafts" may spontaneously form. We can further simplify the study of molecular interactions within membranes by first looking at pairwise interactions between lipids. In this section we set out to compare the physical environment of the two lipids in a binary membrane. When studying multi-lipid membrane systems, deuteration can be carried out on any component. By chain-deuterating one of the lipids at a time, the chemical environment of each lipid can be monitored separately and the phase behaviour of each can be observed independently. By alternately deuterating the two lipid components in a binary membrane, we can compare the organization of their acyl chains and, in principle, find out if the two lipids experience similar or different environments.

The orientation of phospholipids in multilamellar vesicles is randomly distributed with respect to the external magnetic field, leading to the characteristic Pake doublet spectrum. If chain-deuterated lipids are rotating rapidly about the normal to the membrane surface, the quadrupolar splittings in ^{2}H NMR spectra are related to the orientational order parameters of the individual C–D bonds, S_{CD}, on the lipid chain:

$$S_{CD} = \frac{4}{3A_Q}\Delta\nu_Q \tag{10.1}$$

where $\Delta\nu_Q$ is the quadrupolar splitting, and A_Q, the static quadrupolar coupling constant, is equal to 167 kHz.[8] The static quadrupolar coupling constant can be easily determined from the Pake doublet spectrum of a dry lipid powder. The order parameter describes the amplitudes of the angular excursions of the C–D bond and is related to the average structure of the lipid. It can be measured for lipids with fully deuterated lipid chains or selectively deuterated lipids, but it is much easier to synthesize perdeuterated chains than to label specific positions along the lipid chain. The time-averaged order parameter, $\langle S_{CD}\rangle$, is equal to $-\frac{1}{2}$ for an all-*trans* rotating hydrocarbon chain, but conformational fluctuations reduce this and such fluctuations increase toward the middle of the bilayer. Order parameter profiles describe the variation in the absolute value of the segmental order as a function of acyl chain position. For perdeuterated chains, smoothed order parameter profiles can be obtained by assuming a monotonic decrease in chain order as we move away from the lipid head group. These profiles do not take into account the detailed behaviour of the chain near the head group, but can still be used to characterize the chain as a whole.[9]

For a liquid crystalline membrane in a single phase, one might expect the two lipids to have similar degrees of chain ordering, but this has not always been observed.[10–14] Even when the phase diagram shows that the lipids are ideally mixed, differences in chain ordering occur. Thus, one cannot conclude that the lipids are in compositionally distinct domains when the two lipids' labelled chains display different order parameter profiles.[12]

In this section, we compile several examples of binary lipid systems from the literature and compare the respective ordering of the components. Our primary question: which physical characteristic – backbone or main transition temperature T_m – causes the components in a well-mixed binary membrane to have different chain order? To uncover the fundamental physical forces at play, binary mixtures are categorized into one of three groups: phosphoglycerolipid–phosphoglycerolipid, sphingolipid–sphingolipid and phosphoglycerolipid–sphingolipid mixtures. We have restricted our comparison to phosphoglycerolipids with a phosphorylcholine head group and sphingolipids with either a phosphorylcholine head group (*i.e.*, sphingomyelin, Sm) or a hydroxyl head group (*i.e.*, ceramide, Cer). We are not considering mixing of lipids with a net charge since the chain order of charged lipids is heavily influenced by the characteristics of the aqueous medium, such as salt concentration and pH. It would be interesting to extend this comparison to include mixtures combining phosphoglycerolipids with different head group sizes, such as phosphatidylethanolamine.

One of the variables we are investigating, the gel to liquid crystalline transition temperature, is tangled with variations in chain length and degree of unsaturation. Tables 10.1 and 10.2 summarize the chain lengths and T_m for the phosphoglycerolipids and sphingolipids, respectively, in the binary

Table 10.1 Phosphoglycerolipid chain lengths and main transition temperatures.

	Chain length		*Main transition temperature (°C)*		*Ref.*	
Lipid	*sn-1*	*sn-2*	*Protonated*	*Deuterated*	*Protonated*	*Deuterated*
POPC	16	18:1c9	−2	−6[a]	15	16
DMPC	14	14	24	19[b]	17–19	10
DPPC	16	16	41	37[b]	17,18,20	21,22
DSPC	18	18	55	51[b]	17–19	10

[a]POPC is perdeuterated only on the palmitoyl (16:0) chain.
[b]Other phosphoglycerolipids are perdeuterated on both chains.

Table 10.2 Sphingolipid chain lengths and main transition temperatures.

	Chain length	*Main transition temperature (°C)*		*Ref.*	
Lipid	*N-Linked chain*	*Protonated*	*Deuterated*	*Protonated*	*Deuterated*
ESm	mainly 16[a]	39	–	23,24	–
PSm	16	41	39[b]	25	14,27
PCer	16	90	>85[b,c]	26	–[d]

[a]Sm from natural sources are heterogeneous in composition. ESm fatty acid composition is primarily 16:0 but contains a significant amount of longer chains as well.[16]
[b]Sphingolipids are deuterated on the N-linked chain.
[c]Data for PCer-D31 is available up to 85 °C, at which temperature the PCer-D31 was only starting to melt.
[d]J. Thewalt, unpublished results.

mixtures being compared. DMPC, DPPC and DSPC are saturated phosphoglycerolipids with phosphorylcholine head groups, glycerol backbones and symmetric acyl chains of length 14, 16 and 18, respectively; the melting temperature increases with increasing chain length. POPC has one saturated 16-carbon chain and one mono-*cis*-unsaturated 18-carbon chain, whereas DOPC has two mono-*cis*-unsaturated 18-carbon chains. In general, saturated lipids pack well and have much higher T_m than unsaturated lipids since the fixed kink of the *cis* double bond prevents close packing. Sphingolipids have sphingosine backbones that include a polar region at the lipid/water interface as well as a long hydrophobic moiety that performs the role of an acyl chain. Sphingolipids from natural sources, such as egg Sm (ESm), have a distribution of N-linked acyl chain lengths and also some variation in the length of the sphingosine backbone. The synthetic sphingolipids PSm and PCer both have 16-carbon N-linked chains attached to the defined 18-carbon sphingosine backbone. Cer has a very small head group and it is in some sense misleading to include Cer in Table 10.2 as Cer is a non-swelling lipid: on its own, in the presence of excess water, it melts from a solid to a very disordered phase that most likely takes the form of oily droplets. However, in mixed membranes, Cer can participate in lamellar phases and is known to encourage gel phase formation and order membranes.

10.2.1 Order Parameters of Lipids in Two-Component Membranes

The plateau order parameters for the components of a number of binary mixtures are presented in Table 10.3 and the pairwise comparisons between the two components of a given mixture are in Figure 10.2. Motions of phospholipid chains near the membrane/water interface are quite restricted; the peaks from the deuterons in this region overlap in a ^{2}H NMR spectrum. This manifests as a plateau in an order parameter profile. The plateau order parameters presented in Table 10.3 are based on averaging the order parameters from C4 to C6. C2 and C3 are excluded because their added proximity to the phospholipid backbone can impose further restrictions on their motion, resulting in order parameters that are uncorrelated with the rest of the lipid chain.[7] For example, the lipid chain becomes considerably more disordered during heating but C2 and C3 order parameters remain relatively constant. Typically, two samples, each with a single deuterium-labelled lipid, are compared. The percent difference between their plateau order parameters are shown in Table 10.3 and in Figure 10.2. These differences in plateau order parameters are representative of the differences seen throughout the entire order parameter profile.

Phosphoglycerolipid sphingolipid mixtures display a much larger difference in plateau order parameter than either phosphoglycerolipid–phosphoglycerolipid or sphingolipid–sphingolipid mixtures (Figure 10.2).

Table 10.3 Plateau order parameter, determined by averaging the C4 to C6 order parameters, for various liquid crystalline binary membranes containing phosphatidylcholine and/or sphingomyelin and/or ceramide.

Concentration			*Order parameters*			
[DMPC]	*[DSPC]*	*Temp (°C)*	*DMPC*[b]	*DSPC*	*% Difference*	*Ref.*
25	75	60	–	0.192	–	10
30	70	60	0.184	–	4.3%[a]	10
50	50	60	0.180	0.188	4.3%	10
75	25	60	0.173	0.186	7.2%	10
[PSm]	*[PCer]*		*PSm*[c]	*PCer*		
60	40	75	0.270	0.264	2.4%	14
70	30	75	0.249	0.253	1.7%	14
80	20	75	0.247	0.240	2.7%	14
90	10	75	0.227	0.227	0.22%	14
[POPC]	*[PSm]*		*POPC*	*PSm*		
33	67	48	0.206	0.238	14%	12
50	50	40	0.197	0.247	23%	27
50	50	50	0.203	0.242	18%[d]	13
67	33	48	0.195	0.229	16%	12
[POPC]	*[DPPC]*		*POPC*	*DPPC*		
50	50	35	0.206	0.223	8.1%	–[f]
50	50	40	0.198	0.211	6.7%	–[f]
67	33	48	0.185	0.198	6.8%[e]	12
80	20	35	0.200	0.214	7.0%	–[f]
[POPC]	*[PCer]*		*POPC*	*PCer*		
85	15	57	0.189	0.211	11%	28
90	10	57	0.181	0.204	12%	28

[a]Comparison between 30:70 DMPC-D54/DSPC and 25:75 DMPC/DSPC-D70.
[b]Owing to the short, 14-carbon, chain length of DMPC, the order parameter of C6 is quite a bit smaller than C4 and C5. It is not clear if C6 can be considered a part of the plateau. C6 has been included in the plateau order parameters of DMPC in this table for consistency. It has been verified that comparisons of average order parameter from C4 to C5 of DMPC with average order parameter from C4 to C6 of DSPC would yield even smaller differences than those shown here (data not shown).
[c]Distinct C2 peaks were identified in all PSm/PCer membranes, but distinct C3 peaks were only identified in 90:10, 80:20 and 70:30 PSm-D31/PCer.
[d]Comparison between 50:50 PSm-D31/POPC and 50:50 ESm/POPC-D31.
[e]Order parameters are for the *sn*-2 chain of DPPC.
[f]J. Thewalt, unpublished results.

Thus, the heightened conformational rigidity that is observed in sphingolipid chains compared with analogous phosphoglycerolipid chains, *e.g.* PSm *vs.* DPPC, is maintained in mixtures.[12] The two phosphoglycerolipid–sphingolipid pairs tabulated are POPC/Cer and POPC/PSm. ^{2}H NMR revealed that POPC and Cer complete their melting at the same temperature (30 ± 2°C) in POPC/Cer membranes, consistent with a well-defined liquidus in their phase diagram.[11] Despite being in a homogenous liquid crystalline phase at 57 °C, however, POPC-D31 and PCer-D31 have quite different order

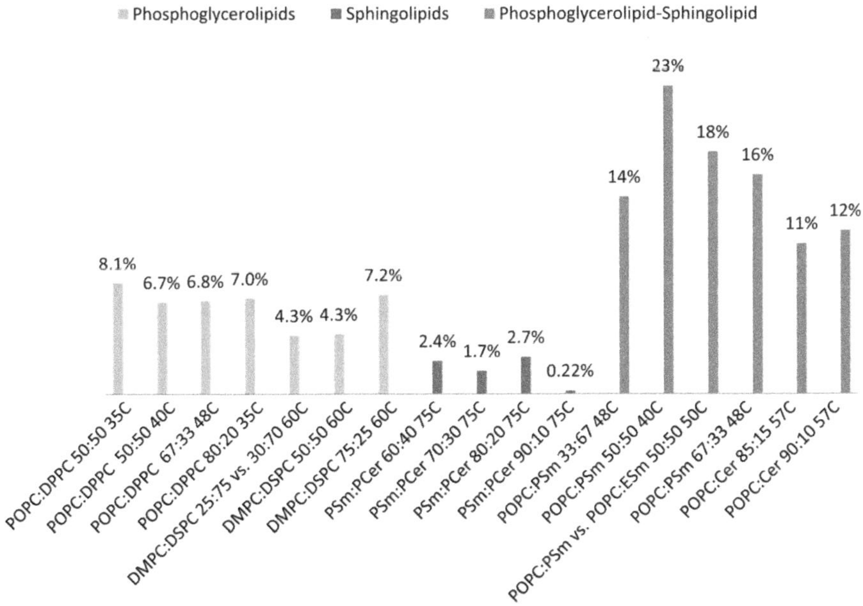

Figure 10.2 Percent difference in the plateau order parameter between the components of binary mixtures. To measure this, two samples are prepared, each containing one chain-labelled lipid. Bars are colour coded for mixtures which contain only phosphoglycerolipids (*light grey*), only sphingolipids (*dark grey*) and both phosphoglycerolipids and sphingolipids (*grey*). Membranes are identified by their composition and temperature at which the plateau order parameter was measured (below each bar). Plateau order parameters are measured from C4 to C6. The percent difference is stated above each bar. Within a given lipid pairing, bars are organized from left to right in increasing concentration of the first lipid, where the first lipid is the lipid with the lower melting temperature, followed by increasing temperature order. Vertical lines segregate data with different lipid pairs. Some comparisons are inexact due to the limited availability of published data: DMPC-D54/DSPC (30:70) is compared with DMPC/DSPC-D70 (25:75) at 60°C and POPC/PSm-D31 (50:50) is compared with POPC-D31/ESm (50:50).

parameter profiles (figure 8 in Hsueh *et al.*[11]). It is interesting to note that, even though T_m of POPC and PSm are much closer than the T_m of POPC and the melting temperature of PCer, the plateau chain order differences are less for POPC/PCer than for POPC/PSm. The high intrinsic chain order of PSm is evident even when it is mixed with a low-melting PC.

Two lipid pairs are phosphoglycerolipid–phosphoglycerolipid combinations: POPC/DPPC and DMPC/DSPC. Since all four of these lipids have phosphorylcholine head groups, the differences between their chain order parameters arise strictly from their T_m and chain structures. Differences in the plateau order parameters illustrated in Figure 10.2 are consistent with higher T_m lipids displaying higher chain order even in mixtures. However, compared with the POPC/sphingolipid mixture, these differences are comparatively small. Figure 10.2 reveals that changes in the plateau order parameter as a function of concentration are gradual in DMPC/DSPC membranes.[10] In fact, a compositional difference of 5% is insignificant for this system. Thus, when data for equivalent compositions is unavailable, we are justified in comparing two slightly compositionally different membranes, *i.e.* 30:70 DMPC-D54/DSPC and 25:75 DMPC/DSPC-D70. Note that the difference in chain order between these two compositionally distinct DMPC/DSPC samples is much less than the difference in chain order observed for phosphoglycerolipid–sphingolipid mixtures.

To our knowledge, the only sphingolipid–sphingolipid combination available for comparison is PSm/PCer. The PSm/PCer phase diagram determined using ^{2}H NMR shows that for up to at least 50 mol % PCer, PSm and PCer mix ideally in the liquid crystalline phase.[14] Even so, the two sphingolipids do not have identical chain order parameter profiles in the liquid crystalline phase (figure 3 in Leung *et al.*[14]). The fact that these differences are minor is reflected in Figure 10.2. Indeed, of all lipid pairs examined, PSm and PCer exhibit the smallest differences. Note that the differences in plateau order in POPC/PCer are much greater than those in PSm/PCer.

Differences in order parameter between the components of a binary mixture can indicate local inhomogeneities, such as nanodomains, or can simply be an intrinsic property of the binary mixture.[27] To distinguish between these scenarios with confidence, a partial phase diagram should be determined. In the absence of such a phase diagram, it is clear that differences in chain order up to approximately 20% should not be taken as indicative of a membrane containing coexisting phases or domains, especially if the two lipids have different backbones. The data in Figure 10.2 show conclusively that the lipid backbone is a significant contributor to the differences observed in the local environment of the deuterated lipid chain of each component in a binary mixture.

10.2.1.1 Are These Comparisons Valid Given That Lipid Chain Deuteration Affects T_m*?*

Perdeuteration of lipid chains is known to decrease the main transition temperatures of pure lipids. Thus, a binary mixture composed of 75% deuterated lipid A and 25% lipid B is intrinsically different from the mixture composed of 75% lipid A and 25% deuterated lipid B. Ideally, then, to keep the overall level of deuteration constant, comparisons should only be made in mixtures where both lipids are deuterated or in 50:50 mixtures.[10] The former option would obviously make peak identification difficult if the deuterated chains of the two lipids experience similar environments. To relate "alternately labelled" comparisons and "dually labelled" comparisons, Lu *et al.* compared singly deuterated and dually deuterated 50 : 50 DMPC/DSPC at 60°C.[10] A weighted sum of the two singly deuterated spectra yielded a spectrum similar to that of the dually deuterated membrane. Lu *et al.* suggested that, for a wide range of compositions, comparing pairs of spectra *at a given temperature* is perhaps equivalent to comparing them at some calibrated separation from some effective transition temperature.

Exploring the available data to test this suggestion, we find two 50 : 50 pairings: DMPC/DSPC at 60°C and DPPC/POPC at 35 and 40°C. The difference in plateau order parameter in both of these samples is comparable to those at different lipid ratios. In fact, for DMPC/DSPC, DPPC/POPC and POPC/PSm the differences in order parameter are, for a particular lipid pair, relatively independent of the ratio of the two lipids. For example, the plateau order parameters of POPC and PSm are more than 14% different in both 33 : 67 and 67 : 33 POPC/PSm membranes. This reassures us that comparisons made at a fixed temperature are valid, in agreement with Lu *et al.*'s proposal. The trends in differences between the plateau order parameter between partnering lipids are as reported in Table 10.3 and Figure 10.2, despite the differences in main transition temperature of the individual lipids due to deuteration.

We have described in some detail the surprisingly non-trivial characterization of acyl chain order in liquid crystalline model membrane composed of just two lipids in excess water. While such membranes can have interesting biological analogues, for example Sm is converted to Cer during apoptosis, this is admittedly rather rare. One perennial problem, under investigation for decades but still not resolved, is the phase behaviour of mixtures of lipids with cholesterol. Since mammalian cell plasma membranes are generally replete with cholesterol, understanding its effect on a variety of phosphoglycerolipids and sphingolipids is of great interest. There is a wealth of published research on binary mixtures of a single lipid plus cholesterol but the field has become somewhat Balkanized, with several variably entrenched views regarding data interpretation and membrane organization. Rather than attempt to summarize this subject in what would inevitably be an unsatisfactory and perhaps unconsciously biased manner, we have instead decided to focus on systems that are intrinsically slightly more complex.

10.3 Membranes with Three Lipid Components

Model membranes composed of a low T_m lipid, a high-T_m lipid and cholesterol make attractive simplified versions of the lipids found in the plasma membrane outer leaflet. Besides cholesterol, this leaflet contains high-T_m sphingolipids (chiefly Sm) and low-T_m zwitterionic phosphoglycerolipids (chiefly PCs). Transient compositional heterogeneities in plasma membrane lipids are postulated to allow the cell to organize active membrane components such as receptors in an energy-efficient fashion. These compositional heterogeneities may be fluctuations or nanodomains having distinct physical characteristics, similar to the liquid ordered phase.[29,30] Many phase diagrams of ternary mixtures of lipids have been mapped in an attempt to fully understand the interactions governing the physical organization of such model plasma membranes.

In accordance with the Gibbs phase rule, ternary membranes can have up to three phases in coexistence. For a good introduction to this topic, see Veatch and Keller.[31] Briefly, a common representation of a ternary membrane phase diagram is a Gibbs triangle: a triangular slice mapping the membrane's organization at a fixed temperature, with each point within the triangle corresponding to a particular ternary lipid composition. The vertices represent pure single-component systems, while each side corresponds to a binary mixture of the two components named at the vertices the side connects. Intermediate concentrations are linearly interpolated. Within the triangle, single-phase regions as well as two- and three-phase coexistence regions are separated by phase boundaries. In a two-phase coexistence region, tie-lines connect "end-points" (located on phase boundaries) that correspond to the compositions of the individual phases. At each point along the tie-line the membrane spectrum is a weighted sum of the spectra of the end-point phases. Three-phase coexistence regions are triangular, with each vertex representing the composition of a single phase. Points within the three-phase triangle correspond to a membrane containing some proportion of each of these three phases. Model plasma membrane ternary mixtures studied at room temperature often have a region at low cholesterol concentrations where liquid ordered (**lo**), liquid disordered (**ld**) and solid ordered (**so**) phases coexist.

^{2}H NMR has been used to characterize ternary membranes such as POPC/PSm/Chol and Sm/POPC/Chol and, more recently, in the construction of DOPC/DPPC-D62/Chol phase diagrams.[13,27,32,33] Mapping ternary phase diagrams with ^{2}H NMR is a Herculean task, but the results are arguably superior to other methods since no extrinsic probes of membrane phase are required. The spectrum from a ternary membrane in a region of micron-scale phase coexistence reflects the amount of labelled lipid in each phase; the signal is directly proportional to the composition. Additional confidence in the accuracy of a ternary phase diagram is achieved by making samples that match, for example, the composition of tie-line end-points. ^{2}H NMR is an excellent technique for studying **lo** + **ld** + **so** three-phase coexistence because the signal from labelled lipid in each phase has features that are distinguishable even when the constituent spectra are superimposed. In the **lo** and **ld** phases, prominent methyl peaks with small quadrupolar splittings can always be discerned, and the broad shoulders of the **so** (gel) phase spectra extend beyond the liquid spectral signals to $\pm$63 kHz.

Figure 10.3 shows an overlay of two recently published partial phase diagrams for DOPC/DPPC-D62/Chol, produced completely independently.[32,33] As noted by Davis *et al.*, the two phase diagrams are in broad agreement.[33] This is a remarkable achievement, given the complexity of the phase diagram and attests to the robustness of ^{2}H NMR. In particular, it is a difficult process to locate the three-phase region so the discrepancies between the red triangles in Figure 10.2 are in fact rather minor. In the **lo** + **ld** two-phase

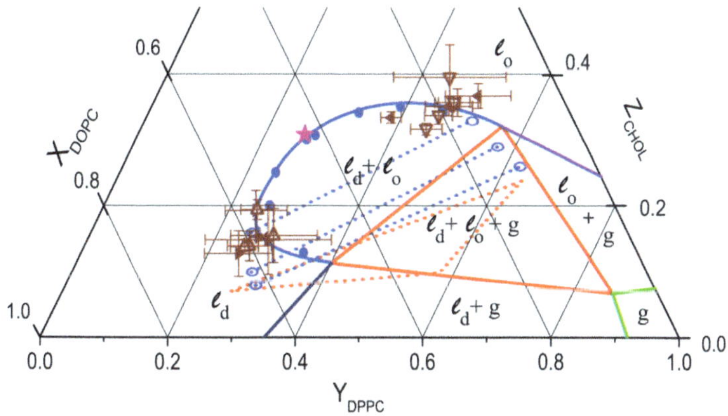

Figure 10.3 Overlay of two DOPC/DPPC-D62/Chol phase diagrams mapped out using ^{2}H NMR by two independent research groups. The three-phase triangle at 20°C (*red dotted lines*), **lo + ld** tie-lies and end-points at 22.5°C (*blue dotted lines* and *blue open circles*) are described in tables 1 and 2 in the supplementary information of Veatch *et al.*[32] These features are superimposed on the 22°C phase diagram of Davis *et al.* (figure 9b).[33] All one-, two- and three-phase region labels, solid lines and filled-in symbols are from Davis's phase diagram. The triangular symbols are Davis *et al.*'s **lo + ld** end-points determined from spectral subtraction and the *magenta star* is Davis *et al.*'s estimated critical composition. The Davis *et al.* tie-lines (not shown) are almost parallel to the left edge of the three phase triangle.[34]
(Reprinted from Davis *et al.*,[33] with permission from Elsevier.)

region the tie-lines and critical points determined by the two groups are rather similar at this temperature.[34] Given that the phase diagram was constructed using information from just one of the three lipid components, the similarities are even more striking. Recently, techniques for measuring the order parameters of non-deuterated lipid chains have been successfully applied to mixtures of POPC/Chol and such an approach could potentially be used to supplement ^{2}H NMR data when determining ternary membrane phase diagrams.[35]

A wide range of spectroscopic and imaging techniques has been applied to the study of membrane phase behaviour, providing complementary information on different length and time scales. Although ^{2}H NMR gives unambiguous information regarding the amount and nature of the phases present in a given model membrane, mapping phase diagrams is an extremely time-consuming and expensive endeavour. As well, ^{2}H NMR cannot readily be used to study membrane organization in living cells. Many of these limitations are avoided by using fluorescence, a popular family of techniques used to study phase coexistence in lipid membranes. It is necessary, though, to carefully compare the results obtained *via* fluorescence with those obtained *via* ^{2}H NMR spectroscopy.

The advantages of fluorescence techniques are many. Fluorescence microscopy yields valuable information about the morphology of coexisting phases in a model membrane, uses little material and is quick; fluorescent probes are also widely available. Fluorescence resonance energy transfer (FRET), between an appropriate energy acceptor–donor probe pair, is capable of detecting phase boundaries of coexisting domains when their sizes are much smaller than the wavelength of visible light.[36,37] Steady-state fluorescence anisotropy can distinguish membrane phase states since lipid and probe rotational dynamics are more restricted in the **lo** phase.[38] Fluorescence lifetime imaging microscopy (FLIM) can also be used in the presence of a probe whose fluorescence lifetime is different between the **ld** and **lo** phases.[36] Relying on phase-dependent differential partitioning of particular probes, fluorescence microscopy can be used to visualize coexisting phases in giant unilamellar vesicles (GUV), and has been used, alone or in conjunction with other techniques, to map out ternary phase diagrams.[39,40] A number of groups have begun to develop algorithms for analyzing 3D microscopy data to obtain area fractions of coexisting domains for ternary lipid membranes so as to make correlations with thermodynamic quantities possible: Fidorra *et al.* confirmed that the domains observed in microscopy obey the lever rule; Husen *et al.* showed that tie-lines can be measured from microscopy images alone; and Juhasz *et al.* showed that tie-lines obtained from fluorescence microscopy are in quantitative agreement with ^{2}H NMR.[34,41–43]

Fluorescence microscopy is done with GUV as they serve as good models for cell membranes due to their size and shape. Their diameters are roughly between 5 and 200 μm,[44] which conveniently allows micron sized phase separation to be observed visually. Since membrane curvature is known to affect phase behaviour, this similarity in size is also an important requirement for the use of GUV as models of cell membranes. One concern regarding comparison of results from fluorescence microscopy of GUV and ^{2}H NMR of multilamellar vesicles (MLV) is whether couplings between bilayers in MLV may stabilize distinct thermodynamic phases; GUV, in contrast, are free standing uncoupled bilayers.[41,45,46] In the presence of fluorescent probes, ^{2}H NMR of MLV and fluorescence microscopy experiments on GUV yield the same **lo** + **ld** miscibility transition temperature (T_{misc}) for a 35:35:30 DOPC/DPPC/Chol membrane.[47] Whether such agreement between GUV and MLV can be generalized for all lipid mixtures remains to be explored, but this observation at least validates the experimental comparisons described below.

10.3.1 Comparisons between ^{2}H NMR and Fluorescence Microscopy Investigations of Ternary Membranes

Veatch *et al.* performed detailed studies on DOPC/DPPC/Chol using ^{2}H NMR and fluorescence microscopy.[32,40] The same phases and general phase diagram appearance as that of Figure 10.3 were found using the two different techniques, but the positions of the phase boundaries do not agree.

There are two major differences between the phase diagrams. First, the **lo** + **ld** coexistence region extends to much higher cholesterol and DPPC concentrations in the fluorescence phase diagram than in the ^{2}H NMR phase diagram. Possible reasons for this will be discussed below. Second, the position of the **lo** + **ld** + **so** three-phase triangle is different in the two phase diagrams and, in the fluorescence phase diagram, is in fact predicted, not measured. The probe used in this study "preferred" the **ld** phase and did not discriminate between the **lo** and the **so** phases. This meant that the **lo** and **so** phases were both dark and impossible to distinguish unless domain shapes could be observed (solid or gel phases typically have irregular shapes). The authors put forward the following additional possible explanations for the second discrepancy: (1) the optical resolution of microscopy (approximately 300 nm) limited the detection of microscopic phase separation;[48] and (2) viscous phases that have elongated into thin lines are particularly hard to observe because it takes a long time for the domains to become large enough for detection.[40] Macroscopically phase-separated domains are needed for reproducible measurements of relative area fractions of the domains. Thus, fluorescence microscopy is ill-suited for determination of the location of the three-phase region.

Fluorescence operates on the assumption that the amount of probe added is small enough such that its presence does not affect the system under study. In order to investigate this assumption and to attempt to resolve some of the discrepancy between the high T_{misc} found by Veatch *et al.* using fluorescence microscopy and ^{2}H NMR for 35:35:30 DOPC/DPPC/Chol, we performed a ^{2}H NMR study on the effects of trace amounts of DiIC12 on ternary membrane phase behaviour.[49] We found that just 0.5 mol% of this fluorescent probe is enough to alter the miscibility transition temperature of 35:35:30 DOPC/DPPC-D62/Chol membranes by as much as 6°C.[47] Even as little as 0.05 mol% has an effect. We can now explain the expanded **lo** + **ld** coexistence region in the DOPC/DPPC/Chol phase diagram measured using fluorescence microscopy as being a consequence of the probe's tendency to stabilize phase coexistence. Further insight into the effects of the fluorescent probe DiIC12 on the DOPC/DPPC-D62/Chol membrane was obtained by studying how the probe influenced chain order. In the ^{2}H NMR spectra, three distinct methyl splittings are observed when there is **lo** + **ld** coexistence: two of these are from inequivalent methyl group deuterons on the two deuterated acyl chains of the **lo** phase DPPC-D62, and the third splitting is from the **ld** phase DPPC-D62. The methyl group splittings are very different for membranes with **lo** + **ld** phase coexistence in the presence and absence of DiIC12; DiIC12 caused the **ld** methyl splitting to narrow and the **lo** splittings to widen. Thus, DiIC12 disorders the **ld** phase and orders the **lo** phase. In the same study it was also found that a different probe, DiOC18, had a smaller effect on methyl splittings and T_{misc}. DiOC18 partitions more evenly between the **ld** and the **lo** phases than does DiIC12, leading us to hypothesize that probe partitioning behaviour causes probe-induced changes to membrane phase behaviour. Probe effects may be particularly problematic near **lo** + **ld** phase boundaries, since a small proportion of **ld** phase means that the number of **ld**-preferring probe molecules in these membrane domains can reach a very significant level.[40] More recent research has proven that probe effects on ternary membrane phase behaviour are not rare. Juhasz *et al.* performed a systematic confocal fluorescence microscopy study on a variety of probes in 35:35:30 and 32:48:20 DOPC/DPPC/Chol membranes and also found that various probes increased T_{misc} with a variation of 4 °C amongst the probes.[50]

Discrepancies between fluorescence microscopy of GUV labelled with TR-DPPE and spectroscopy experiments using FRET in PSm/POPC/cholesterol LUV with NBD-DPPE and Rhodamine-DOPE have also been documented.[41] Concerns have been raised in the literature regarding the use of fluorescence microscopy in membrane phase studies. When domains are reconstructed in 3D microscopy for the purpose of domain area determination, the point spread function is asymmetric in the *z*-axis, since images are acquired as *xy*-slices, which leads to artefacts in the data.[41] Microscopy relies on differential probe partitioning. Dual labelling has been used in some cases to verify the identity of a domain, but most fluorescent probes partition away from ordered – both **lo** and **so** – domains. Moreover, recent systematic studies on probe behaviour found that probe behaviour can be altered by the presence of other probes.[50–52] Fluorescent probes are sensitive to photo-oxidation and photo-oxidation has been shown to induce phase separation in microscopy experiments.[53,54] Attaching fluorophores to raft-associated lipid can cause the modified species to partition away from rafts.[55] Some fluorescent lipid analogues display different partitioning preferences, depending on their chain lengths.[51] One has to be careful when making fluorescence measurements because some probes exhibit probe concentration-dependent self-quenching.[56] Examples of membrane composition dependent probe partitioning behaviour have also been recorded.[51] There is evidence that electrostatic interactions between charged fluorophores and lipid molecules are particularly problematic. Atomistic MD simulations found that DPPC bound electrostatically to TR-DPPE, which has a net charge of −1 and is twice the mass of DPPC, at 4.7 mol% TR-DPPE. This results in reduced diffusion of DPPC and, more importantly, a predicted increase in the T_{misc} of a generic ternary lipid system of saturated lipids, unsaturated lipids and cholesterol.[57] This predicted increase in T_{misc} is in agreement with experimental results and this observation, combined with reports that 0.2–2% TR-DPPE had no effect on T_{misc} of DOPC/DPPC/Chol, means that 0.2% TR-DPPE is already sufficient to modulate the phase coexistence behaviour of this membrane.[40,58]

In summary, fluorescence is fast, sensitive and certainly capable of defining the general features of a lipid membrane phase diagram or of exploring heterogeneities in cell membranes.[59] However, fluorescent probes, which are, after all, comparable in size to lipid molecules, are often not inert reporters of the lipid state but instead significantly affect phase behaviour. It is likely that probe-related distortions to membrane phase behaviour will be most significant for lipid compositions near phase boundaries or critical points. While there is currently no practical alternative to fluorescence techniques, especially for studies of phase behaviour in biological systems, it is important to exercise caution in the interpretation of results.

10.4 ^{2}H NMR Studies of Model Skin Barrier Membranes

The skin barrier is the top layer of the epidermis and consists of terminally differentiated keratinocytes surrounded by layers of essentially protein-free lipids. The skin barrier lipids are chiefly ceramides, saturated fatty acids and cholesterol. The former two classes of lipid are products of concerted lipid metabolism in the viable epidermis; the end result of this metabolic activity is a mixture of lipids unlike those found in any other mammalian tissue. These lipids create a tight barrier through which water penetrates very slowly and which completely blocks most molecules. Early work on a simplified model barrier membrane consisting of equimolar bovine brain ceramide, perdeuterated palmitic acid and cholesterol, BBCer/PA-D31/Chol, yielded ^{2}H NMR spectra that were dominated by a Pake doublet of maximum width (*i.e.*, the sharp peaks were separated by $\frac{3}{4}A_Q = 126$ kHz), indicating solid, nearly motionless PA-D31 that was confirmed by X-ray diffraction to be packing as an orthorhombic crystal.[60–62] This solid signal was reduced upon increasing the pH from 5.2 to 7.4, consistent with the observations that: (1) barrier lipids in skin have an effective pH environment of approximately 5, at which level the fatty acids are protonated, and (2) hydration of the skin by occlusion both raises its pH, implying that negatively charged fatty acids will be present, and weakens the barrier.[63] Heating the model barrier membrane resulted in conversion of the solid signal to an **lo** phase spectrum, consistent with the large amount of cholesterol present. If the PA-D31 is replaced with a longer fatty acid, this melting transition occurs at higher temperatures, indicating that the "skin temperature" model barrier membrane is stabilized when the fatty acid is lengthened.[64] Since fatty acid elongation is another important lipid modification that occurs in the skin, we are tempted to conclude that epidermal metabolism is in some sense governed by the same rules that determine model membrane structure.

There is considerable interest in understanding the molecular organization of the skin barrier lipids. One practical motivation for this interest is that in order to facilitate the transdermal delivery of medications one must reversibly weaken the skin barrier. Empirically, a number of compounds have been found to enhance the permeability of drugs through skin. These "penetration enhancers" can be lipid-like, such as oleic acid (C18:1 *cis*Δ9), or solvents such as ethanol. When model skin barrier membranes are prepared in the presence of these penetration enhancers, the membrane phase behaviour is significantly affected.[65,66] A prominent narrow peak appears in the centre of the ^{2}H NMR spectrum, implying that the labelled lipid is experiencing rapid isotropic motion. The most likely explanation is that the penetration enhancers solubilize some of the model skin barrier. If this happens *in vivo*, it is clear that penetration enhancers will disrupt barrier membranes, allowing drugs to pass through resulting defects in their structure.

Skin barrier membranes *in vivo*, though, are likely to be arranged in compositionally and structurally asymmetric layers, possibly reflecting the compositional asymmetry in the leaflet of their precursor membranes. Model membranes are generally unable to capture this aspect of lipid organization.

10.5 Conclusion

^{2}H NMR studies of mixed lipid membranes have had a big influence on the membrane biophysics field. The usually straightforward nature of ^{2}H NMR data interpretation means that published findings are confidently received by the research community. Improving commercial availability of deuterated lipids ensures that the design of future studies will not be hampered by lack of material, although lack of research funds can still be a problem. Model membrane composition will increasingly be able to be tailored to the biological membrane system of interest. Model membrane samples used for ^{2}H NMR can be directly used in infrared spectroscopy due to the lower frequencies of C–D bond vibrations compared to those of C–H bonds. Similarly, neutron scattering employs deuteration to exploit the large scattering cross-section of the deuteron. Thus, collaborations between researchers with complementary expertise can employ the same samples, reducing cost and eliminating uncertainties introduced with different sample preparation methods. Model membranes are not restricted to lipids alone; many studies of plasma membrane organization employ proteins or protein fragments. ^{2}H NMR is a powerful method to study these more complicated systems and we anticipate a continued stream of quantitative, unambiguous ^{2}H NMR model membrane data in the years to come.

Acknowledgements

We are grateful for support from the Canadian Natural Sciences and Engineering Research Council.

References

1. J. Seelig, *Q. Rev. Biophys.*, 1977, **10**, 353–418.
2. J. Seelig and A. Seelig, *Q. Rev. Biophys.*, 1980, **13**, 19–61.
3. G. W. Stockton and I. C. P. Smith, *Chem. Phys. Lipids*, 1976, **17**, 251–263.
4. M. Bloom, E. Evans and O. G. Mouritsen, *Q. Rev. Biophys.*, 1991, **24**, 293–397.

5. J. H. Davis, K. R. Jeffrey, M. Bloom, M. I. Valic and T. P. Higgs, *Chem. Phys. Lett.*, 1976, **42**, 390–394.
6. J. H. Davis, *Biochim. Biophys. Acta*, 1983, **737**, 117–171.
7. A. S. Dico, J. Hancock, M. R. Morrow, J. Stewart, S. Harris and K. M. W. Keough, *Biochemistry*, 1997, **36**, 4172–4177.
8. J. H. Davis and K. R. Jeffrey, *Chem. Phys. Lipids*, 1977, **20**, 87–104.
9. M. Lafleur, B. Fine, E. Sternin, P. R. Cullis and M. Bloom, *Biophys. J.*, 1989, **56**, 1037–1041.
10. D. Lu, I. Vavasour and M. R. Morrow, *Biophys. J.*, 1995, **68**, 574–583.
11. Y.-W. Hsueh, R. Giles, N. Kitson and J. Thewalt, *Biophys. J.*, 2002, **82**, 3089–3095.
12. T. Mehnert, K. Jacob, R. Bittman and K. Beyer, *Biophys. J.*, 2006, **90**, 939–946.
13. T. Bartels, R. S. Lankalapalli, R. Bittman, K. Beyer and M. F. Brown, *J. Am. Chem. Soc.*, 2008, **130**, 14521–14532.
14. S. S. W. Leung, J. V Busto, A. Keyvanloo, F. M. Goñi and J. Thewalt, *Biophys. J.*, 2012, **103**, 2465–2474.
15. M. Fidorra, L. Duelund, C. Leidy, A. C. Simonsen and L. A. Bagatolli, *Biophys. J.*, 2006, **90**, 4437–4451.
16. J. Thewalt, C. Hanert, F. Linseisen, A. Farrall and M. Bloom, *Acta Pharm.*, 1992, **42**, 9–23.
17. A. Blume, *Biochemistry*, 1983, **22**, 5436–5442.
18. T. P. McMullen and R. N. McElhaney, *Biochim. Biophys. Acta*, 1995, **1234**, 90–98.
19. H. M. Seeger, M. Fidorra and T. Heimburg, *Macromol. Symp.*, 2005, **219**, 85–96.
20. D. P. Kharakoz, A. Colotto, K. Loher and P. Laggneri, *J. Phys. Chem.*, 1993, 9844–9851.
21. M. R. Vist and J. H. Davis, *Biochemistry*, 1990, **29**, 451–464.
22. B. B. Bonev and M. R. Morrow, *Can. J. Chem.*, 1998, **76**, 1512–1519.
23. J. Sot, L. A. Bagatolli, F. M. Goñi and A. Alonso, *Biophys. J.*, 2006, **90**, 903–914.
24. A. Filippov, B. Munavirov, G. Gröbner and M. Rudakova, *Magn. Reson. Imaging*, 2012, **30**, 413–421.
25. P. R. Maulik and G. G. Shipley, *Biochemistry*, 1996, **2960**, 8025–8034.
26. J. Shah, J. M. Atienza, A. V. Rawlings and G. G. Shipley, *J. Lipid Res.*, 1995, **36**, 1945–1955.
27. A. Bunge, P. Müller, M. Stöckl, A. Herrmann and D. Huster, *Biophys. J.*, 2008, **94**, 2680–2690.
28. Y. W. Hsueh, R. Giles, N. Kitson and J. Thewalt, *Biophys. J.*, 2002, **82**, 3089–3095.
29. S. L. Veatch, P. Cicuta, P. Sengupta, A. Honerkamp-Smith, D. Holowka and B. Baird, *ACS Chem. Biol.*, 2008, **3**, 287–293.
30. D. Lingwood and K. Simons, *Science*, 2010, **327**, 46–50.
31. S. L. Veatch and S. L. Keller, *Biochim. Biophys. Acta*, 2005, **1746**, 172–185.

32. S. L. Veatch, O. Soubias, S. L. Keller and K. Gawrisch, *Proc. Natl. Acad. Sci. U. S. A.*, 2007, **104**, 17650–17655.
33. J. H. Davis, J. J. Clair and J. Juhasz, *Biophys. J.*, 2009, **96**, 521–539.
34. J. Juhasz, F. J. Sharom and J. H. Davis, *Biochim. Biophys. Acta*, 2009, **1788**, 2541–2552.
35. T. M. Ferreira, F. Coreta-Gomes, O. H. S. Ollila, M. J. Moreno, W. L. C. Vaz and D. Topgaard, *Phys. Chem. Chem. Phys.*, 2013, **15**, 1976–1989.
36. R. F. M. de Almeida, L. M. S. Loura and M. Prieto, *Chem. Phys. Lipids*, 2009, **157**, 61–77.
37. M. L. Frazier, J. R. Wright, A. Pokorny and P. F. F. Almeida, *Biophys. J.*, 2007, **92**, 2422–2433.
38. A. Gidwani, D. Holowka and B. Baird, *Biochemistry*, 2001, **40**, 12422–12429.
39. G. W. Feigenson and J. T. Buboltz, *Biophys. J.*, 2001, **80**, 2775–2788.
40. S. L. Veatch and S. L. Keller, *Biophys. J.*, 2003, **85**, 3074–3083.
41. M. Fidorra, A. Garcia, J. H. Ipsen, S. Härtel and L. A. Bagatolli, *Biochim. Biophys. Acta*, 2009, **1788**, 2142–2149.
42. P. Husen, L. R. Arriaga, F. Monroy, J. H. Ipsen and L. A. Bagatolli, *Biophys. J.*, 2012, **103**, 2304–2310.
43. N. Bezlyepkina, R. S. Gracià, P. Shchelokovskyy, R. Lipowsky and R. Dimova, *Biophys. J.*, 2013, **104**, 1456–1464.
44. F. M. Menger and J. S. Keiper, *Curr. Opin. Chem. Biol.*, 1998, **2**, 726–732.
45. V. Kiessling, C. Wan and L. K. Tamm, *Biochim. Biophys. Acta, Biomembr.*, 2009, **1788**, 64–71.
46. C. M. Funkhouser, M. Mayer, F. J. Solis and K. Thornton, *J. Chem. Phys.*, 2013, **138**, 024909.
47. S. L. Veatch, S. S. W. Leung, R. E. W. Hancock and J. L. Thewalt, *J. Phys. Chem. B*, 2007, **111**, 502–504.
48. S. D. Connell, G. Heath, P. D. Olmsted and A. Kisil, *Faraday Discuss.*, 2013, **161**, 91–111.
49. S. L. Veatch, I. V Polozov, K. Gawrisch and S. L. Keller, *Biophys. J.*, 2004, **86**, 2910–2922.
50. J. Juhasz, J. H. Davis and F. J. Sharom, *Biochem. J.*, 2010, **430**, 415–423.
51. T. Baumgart, G. Hunt, E. R. Farkas, W. W. Webb and G. W. Feigenson, *Biochim. Biophys. Acta*, 2007, **1768**, 2182–2194.
52. T. Baumgart, S. T. Hess and W. W. Webb, *Nature*, 2003, **425**, 821–824.
53. A. G. Ayuyan and F. S. Cohen, *Biophys. J.*, 2006, **91**, 2172–2183.
54. H. Bouvrais, T. Pott, L. a. Bagatolli, J. H. Ipsen and P. Méléard, *Biochim. Biophys. Acta*, 2010, **1798**, 1333–1337.
55. J. E. Shaw, R. F. Epand, R. M. Epand, Z. Li, R. Bittman and C. M. Yip, *Biophys. J.*, 2006, **90**, 2170–2178.
56. V. Coste, N. Puff, D. Lockau, P. J. Quinn and M. I. Angelova, *Biochim. Biophys. Acta*, 2006, **1758**, 460–467.
57. M. J. Skaug, M. L. Longo and R. Faller, *J. Phys. Chem. B*, 2011, **115**, 8500–8505.
58. S. L. Veatch, personal communication.

59. S. J. Sahl, M. Leutenegger, M. Hilbert, S. W. Hell and C. Eggeling, *Proc. Natl. Acad. Sci. U. S. A.*, 2010, **107**, 6829–6834.
60. J. Thewalt, N. Kitson, C. Araujo, A. MacKay and M. Bloom, *Biochem. Biophys. Res. Commun.*, 1992, **188**, 1247–1252.
61. N. Kitson, J. Thewalt, M. Lafleur and M. Bloom, *Biochemistry*, 1994, **33**, 6707–6715.
62. J. A. Bouwstra, J. Thewalt, G. S. Gooris and N. Kitson, *Biochemistry*, 1997, **36**, 7717–7725.
63. S. Bjorklund, J. Engblom, K. Thuresson and E. Sparr, *J. Control. Release*, 2010, **143**, 191–200.
64. X. Chen, S. Kwak, M. Lafleur, M. Bloom, N. Kitson and J. Thewalt, *Langmuir*, 2007, **23**, 5548–5556.
65. A. C. Rowat, N. Kitson and J. L. Thewalt, *Int. J. Pharm.*, 2006, **307**, 225–231.
66. S. Kwak, E. Brief, D. Langlais, N. Kitson, M. Lafleur and J. Thewalt, *Biochim. Biophys. Acta, Biomembr.*, 2012, **1818**, 1410–1419.

CHAPTER 11

Membrane Interactions of Amphiphilic Peptides with Antimicrobial Potential: A Solid-State NMR Study

MATTHIEU FILLION,[a] NORMAND VOYER[b] AND MICHÈLE AUGER*[a]

[a] Department of Chemistry, Regroupement québécois de recherche sur la fonction, la structure et l'ingénierie des protéines (PROTEO), Centre de recherche sur les matériaux avancés (CERMA), Centre québécois sur les matériaux fonctionnels (CQMF), Université Laval, Québec, QC, Canada, G1V 0A6; [b] Department of Chemistry, PROTEO, Université Laval, Québec, QC, Canada, G1V 0A6

*Email: michele.auger@chm.ulaval.ca

11.1 Introduction

11.1.1 Bacterial Resistance to Antibiotics

The abuse and inappropriate use of antibiotics has caused bacteria to acquire resistance mechanisms against antibiotics. This has resulted in a decrease in the efficacy of antibiotics to fight infections. This is now a very important clinical problem with the appearance of multiresistant bacteria in hospital environments and the increase in the number of infections due to resistant bacteria.[1] Several pathogenic bacteria, such as methicillin-resistant

New Developments in NMR No. 3
Advances in Biological Solid-State NMR: Proteins and Membrane-Active Peptides
Edited by Frances Separovic and Akira Naito

Published by the Royal Society of Chemistry, www.rsc.org

Staphylococcus aureus (MRSA) and vancomycin-resistant *Enterococcus*, are resistant to antibiotics. In addition, the number of novel antimicrobial agents approved by the US Food and Drug Administration (FDA) is continuously decreasing, with the introduction of 16 new antimicrobial agents between 1983 and 1987 but only five between 2003 and 2007.[2] It is also important to note that new antimicrobial agents are often analogues of known antibiotics and their mechanisms of action are often similar. Therefore, very few novel classes of antibiotics have been developed in the last 30 years.[3] It is, therefore, of great importance to develop antimicrobial agents with novel modes of action and antimicrobial peptides are an interesting alternative.

11.1.2 Natural Antimicrobial Peptides

Natural antimicrobial peptides are a class of molecules that are part of the immune systems of several organisms such as mammals, plants, insects, amphibians and bacteria.[4] Despite the great diversity of these peptides, they share several common characteristics.[5] In particular, they are composed of 12–45 amino acids, possess a positive charge (varying from +2 to +9) and have a marked amphiphilic character. In addition, they present several advantages compared to conventionally used antibiotics. More specifically, they cause a rapid disruption of the pathogenic agent, have a large spectrum of activity (both against Gram negative and Gram positive bacteria), induce little bacterial resistance and present a synergic action with other antibiotics. In addition to their antibacterial activity, some peptides will also lead to the destruction of fungi, parasites, viruses and cancer cells. However, some peptides are toxic to eukaryotic cells and are said to be haemolytic. Natural antimicrobial peptides are involved at two levels in the immune response.[6] They can be modulators of the innate immunity or act as agents that destroy pathogenic agents by inducing membrane deformations or by attacking intracellular targets.

11.1.3 Membrane Selectivity

It is generally admitted in the literature that the bacterial cellular membrane is the main target for cationic antimicrobial peptides. Therefore, the membrane selectivity is very important.[7] Several elements are involved in the preferential interactions of antimicrobial peptides with the cellular membranes of prokaryotes as opposed to eukaryotic membranes. More specifically, both electrostatic and hydrophobic interactions are involved in the interaction of cationic antimicrobial peptides with bacterial cell membranes, since these membranes are rich in anionic phospholipids. When a cationic antimicrobial peptide is placed in contact with eukaryotic cell membranes, hydrophobic interactions may dominate since these membranes are rich in zwitterionic phospholipids. In addition, the presence of cholesterol will cause an increased rigidity of eukaryotic membranes and, therefore, limit membrane interactions. Hence, it is the combination of the cationic character of antimicrobial peptides and the large proportion of anionic phospholipids in bacterial cell membranes that will cause the peptide selectivity.

11.1.4 Mechanisms of Action

The mechanisms by which antimicrobial peptides lead to the destruction of pathogenic agents are still not completely understood. However, several models have been proposed to explain this mechanism of action, namely the carpet-like, the barrel-stave, the toroidal pore and the sinking raft models.[8] All these models of membrane perturbation have as the initial step the adsorption of the peptide at the membrane surface. In the carpet-like mechanism, the peptide action is similar to that of a detergent and leads to the formation of micelles. For the barrel-stave and toroidal pore mechanisms, the peptides adopt a transmembrane orientation. In the barrel-stave mechanism the peptides form a structure similar to that of an ion channel and this mechanism of action is favoured for long and hydrophobic peptides. For shorter peptides having a larger positive charge, the toroidal pore mechanism will be favoured since each peptide is surrounded by phospholipids, which limits electrostatic repulsions between charged peptides. Finally, the sinking raft model is similar to the toroidal pore model with the difference that the peptide sits at the surface of the membrane. In addition, the peptides induce a positive curvature to the membrane in these last two models.

11.2 Synthetic Peptides

11.2.1 Introduction

The understanding of the mechanisms of action of antimicrobial peptides is essential for the design and synthesis of novel antimicrobial agents. Several approaches can be taken to determine the effects of different molecular determinants on the peptide antimicrobial activity, haemolytic character and conformation. In particular, one of the approaches entails developing model synthetic peptides with characteristics similar to those of natural antimicrobial peptides and that perturb the integrity of cell membranes. Then, different modifications can be made to the peptide primary sequence to generate several analogues with diverse characteristics. Several molecular determinants can, therefore, be investigated, such as the peptide charge, amphiphilic character, conformation, hydrophobicity and polar angle on the membrane interaction.[9]

11.2.2 Base 14-mer Peptide

In order to better understand the hydrophobic interactions involved in the haemolytic activity of membrane-active peptides, we have first investigated a

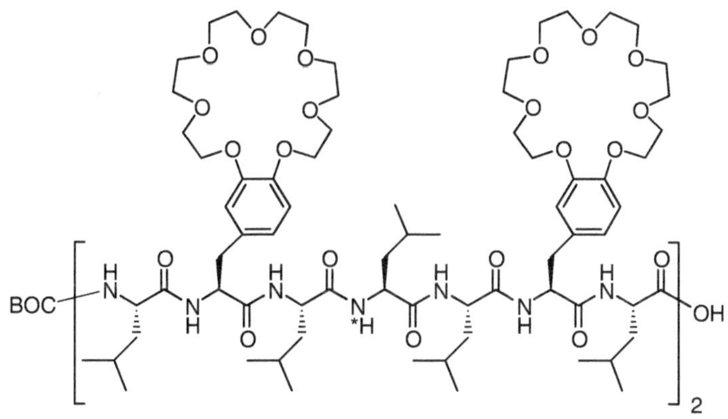

Figure 11.1 Monoprotected base 14-mer peptide. The ^{15}N-labelled leucine residue marked by an *asterisk* is located in the first heptamer at position 4. (Reproduced from Ouellet *et al.*,[16] with permission.)

model synthetic peptide, the so-called base 14-mer peptide. More specifically, we have designed a synthetic helical amphipathic peptide that is an oligomer of a repeating unit of five leucine residues and two synthetic 21-crown-7-phenylalanines appropriately positioned so that the hydrophilic crown ethers align on one side of the hydrophobic helical axis, to form a neutral amphipathic 14-mer peptide (Figure 11.1).[10] Neutral peptides are rarely studied because antimicrobial activity is known to increase with cationic charge.[11] However, from a mechanistic perspective, an uncharged peptide with membrane disruption activity allows studies mainly of the effect of hydrophobic forces, which seem to dominate the haemolytic activity of antibacterial peptides, on the interaction with lipids without the influence of strong electrostatic forces. The 14-mer peptide constitutes a suitable model to study the interaction between lipids and membrane-active peptides since it is short and can be easily chemically modified.

11.2.3 Charged Analogues

In order to confer selectivity to the base 14-mer peptide, we have systematically replaced one or two leucine residues by positively charged lysine residues.[12] Results obtained by fluorescence spectroscopy indicated that some of these cationic peptides lost their ability to significantly permeabilize zwitterionic dimyristoylphosphatidylcholine (DMPC) vesicles and, therefore, are selective towards negatively charged dimyristoylphosphatidylglycerol (DMPG) vesicles, while other cationic 14-mer peptides remain active against DMPC vesicles[12] and are, therefore, nonselective. The position(s) of leucine to lysine substitutions dictate this selectivity. We have also demonstrated that selective peptides do not adopt an amphipathic α helical structure, like the base 14-mer peptide,[10] but are structured as non-amphipathic intermolecular β-sheets both in solution and in the presence of DMPC or DMPG vesicles.[12] These findings question the importance of helicity and amphipathicity in the selectivity of antimicrobial peptides.

11.2.4 Goals of the Study

The goal of the present study was, therefore, to determine by solid-state NMR spectroscopy the membrane interactions and topology of both the uncharged base 14-mer peptide and its positively charged analogues. A combination of ^{31}P and ^{2}H solid-state NMR spectroscopies on both multilamellar vesicle samples and lipid bilayers oriented between glass plates were used to determine the effect of the peptides on the lipid dynamics and orientation of the lipids, while the topology of the base 14-mer peptide was determined by ^{15}N solid-state NMR spectroscopy.

11.3 Interactions of the Base 14-mer Peptide with Membranes

11.3.1 Effect on Lipid Orientation

Since the phospholipid headgroup contains a phosphorus-31 nucleus with a 100% natural isotopic abundance, ^{31}P NMR is a powerful technique to monitor changes occurring in the polar region of the bilayer.[13–15] We have used ^{31}P NMR spectroscopy to investigate changes occurring in the headgroup region of mechanically aligned bilayers of DMPC upon the addition of the base 14-mer peptide.[16] More specifically, static ^{31}P solid-state NMR spectra provide information on the conformation of the polar headgroup and on the quality of the alignment of phospholipid bilayers obtained from changes in the chemical shift and in the chemical shift anisotropy (CSA) throughout the entire spectrum.

Figure 11.2 shows the ^{31}P NMR spectra of pure DMPC oriented bilayers and in the presence of the 14-mer peptide. The spectrum of the pure lipid

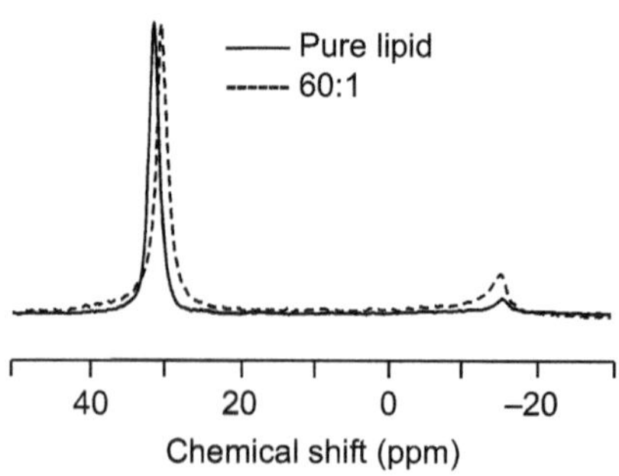

Figure 11.2 ^{31}P NMR spectrum of mechanically oriented lipid bilayers of DMPC at 37 °C in the absence and presence of the base 14-mer peptide at a lipid/peptide molar ratio of 60:1.
(Adapted from Ouellet *et al.*,[16] and reproduced with permission.)

system is characteristic of well-aligned bilayers in the fluid phase with the bilayer normal parallel to the external magnetic field. More specifically, a single sharp resonance is observed at 31.1 ppm. A small contribution of lipids aligned at 90° is also observed, with the acyl chains perpendicular to the magnetic field direction. From the ratio of the spectral area of the single resonance over the entire spectra, the proportion of aligned *versus* non-aligned lipids was evaluated at 81%.[16]

In the presence of the base 14-mer peptide, the CSA is decreased, as reflected by a shift of the high-frequency resonance to 30.1 ppm. In addition, a new component characteristic of 90° oriented bilayers appears at −15.3 ppm and the spectral intensity between 30.1 and −15.3 ppm is slightly increased upon peptide addition, resulting in a degree of 0° alignment of 64% compared to 81% for pure DMPC bilayers.[16]

The decrease in the CSA observed in DMPC bilayers upon the addition of the 14-mer peptide can be explained by an interaction of the peptide at the polar headgroup interface. The positioning of the crown ethers at sites 2, 6, 9 and 13 results in all the crown ethers well sequestered on one side of the α-helix, conferring amphiphilicity to the 14-mer peptide, with the crown ethers facing the aqueous phase and the hydrophobic leucine residues facing the hydrophobic region of the bilayer. An interfacial location would then affect the conformation of the bilayers. It is interesting to note that a similar decrease in the lipid ^{31}P CSA has also been observed in DMPC vesicles.[17]

11.3.2 Effect on Lipid Acyl Chain Order

We have also investigated the effect of the 14-mer peptide on the lipid order in bilayers stacked between glass plates by measuring the quadrupolar splittings upon peptide addition.[16] The increase (decrease) in the quadrupolar splitting ($\Delta\nu_Q$) for a C–D bond is associated with order (disorder) in the deuterated chains and is related to an order parameter S_{CD}:

$$\Delta\nu_Q = \frac{3}{4}\frac{e^2qQ}{h}\left(3\cos^2\theta - 1\right)S_{CD} \qquad (11.1)$$

where (e^2qQ/h) is the quadrupole coupling constant for C–D bonds (~167 kHz) and θ is the angle between the bilayer normal and B_0.[13,18] The quadrupolar splitting is reported here for the plateau region of the acyl chains ($\Delta\nu_P$) and the terminal methyl group ($\Delta\nu_M$).

The ^2H NMR spectra of DMPC bilayers mechanically oriented between glass plates are shown in Figure 11.3 in the absence and presence of the 14-mer peptide. The spectrum of pure DMPC in the parallel orientation shows a quadrupolar splitting of 52.5 kHz at the plateau region and 6.3 kHz at the terminal methyl group. Upon the addition of the 14-mer peptide, $\Delta\nu_P$ increased by 3.0 kHz while the methyl group was not significantly affected. In addition, a powder pattern at ± 10 kHz was observed upon the addition of the peptide, which has been determined to originate from non-oriented lipids,[16] as also observed in the ^{31}P NMR spectra of the same system. The

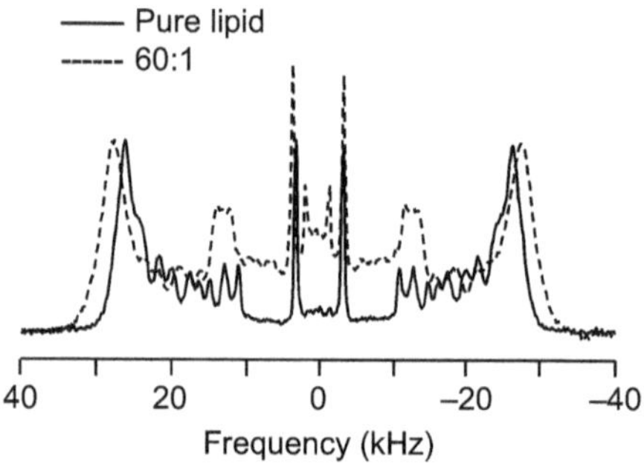

Figure 11.3 ^{2}H NMR spectra of mechanically oriented lipid bilayers of DMPC at 37 °C in the absence and presence of the base 14-mer peptide at a lipid/peptide molar ratio of 60:1.
(Adapted from Ouellet *et al.*,[16] and reproduced with permission.)

ordering effect in the plateau region of mechanically oriented DMPC bilayers upon the 14-mer addition supports the possible surface location of the peptide. It is interesting to note that a similar increase in the lipid chain order has also been observed in DMPC vesicles.[17]

11.3.3 Orientation of the Peptide in Membranes

We have used solid-state ^{15}N NMR spectroscopy to investigate the membrane orientation of the 14-mer peptide in DMPC bilayers mechanically oriented between glass plates. This can be done using a ^{15}N-labelled peptide incorporated into oriented bilayers. The 14-mer peptide was synthesized with ^{15}N incorporated into the leucine residue at position 4 in the amino acid sequence (Figure 11.1). Owing to the helical conformation of the 14-mer peptide, and the size and orientation of the σ_{33} element of the ^{15}N chemical shift tensor, it is possible to correlate the ^{15}N chemical shift with the helix orientation relative to the magnetic field direction from oriented ^{15}N NMR spectra.[19,20] Chemical shifts below 100 ppm or higher than 200 ppm are expected for helical peptides oriented in an in-plane or transmembrane fashion, respectively.

A chemical shift of 70 ppm was measured in the ^{15}N NMR spectrum of the peptide incorporated in DMPC bilayers mechanically oriented between glass plates,[16] which corresponds to a surface orientation of the 14-mer peptide. It is also interesting to note that similar results are seen for bilayers with lesser and greater thicknesses,[16] indicating that the orientation of the peptide is independent of bilayer thickness. The surface location of the 14-mer peptide determined by ^{15}N NMR spectroscopy has also been confirmed by measuring the distance between the [^{15}N]leucine residue in the 14-mer peptide and the ^{31}P nucleus in the phospholipid headgroup. A mean distance of 7.6 ± 0.6 Å was measured, which supports a model in which the peptide is located near the phospholipid headgroups.[16,21]

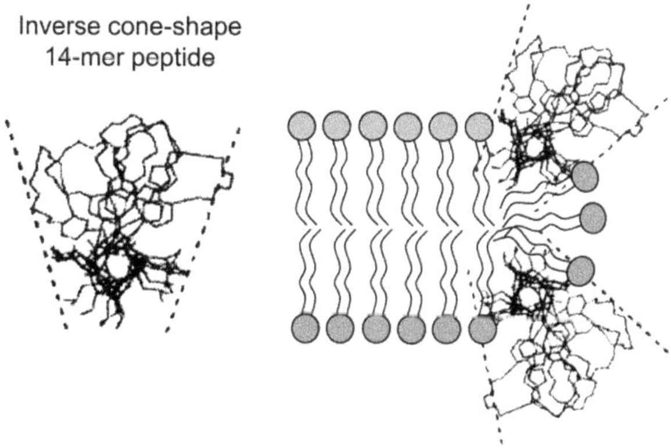

Figure 11.4 Positive curvature strain and membrane destabilization imposed by the base 14-mer peptide laid at the bilayer surface.
(Adapted from Ouellet *et al.*,[16] and reproduced with permission.)

11.3.4 Mechanism of Membrane Perturbation

The results obtained with the base 14-mer peptide in lipid bilayers stacked between glass plates suggest that the peptide adopts an in-plane orientation and perturbs the membrane *via* the induction of a positive curvature strain.[16] When the bilayer undergoes a positive curvature strain, the lipid molecules that surround the peptide are constrained to tilt relative to the bilayer normal. These lipids would then contribute to the powder pattern signal observed in both ^{31}P and ^{2}H NMR spectra. This model of interaction, illustrated in Figure 11.4, is similar to the sinking raft model.[8]

11.4 Interactions of Cationic 14-mer Peptides with Membranes

11.4.1 Interactions with Lipid Vesicles

We have also investigated the interactions of cationic 14-peptides in which one or two leucines are substituted by positively charged lysine residues at each position,[12,22] as shown schematically in Figure 11.5 As mentioned above, we demonstrated that the position of substitution dictates the peptide selectivity, with nonselective peptides adopting an α-helical structure while an intermolecular β-sheet structure is observed for selective peptides.[12] More specifically, we used a combination of ^{31}P and ^{2}H solid-state NMR spectroscopy to determine how the peptide interacts with the lipid headgroups and acyl chains, respectively.[22] This study was performed using both zwitterionic DMPC bilayers, mimicking eukaryotic membranes, and negatively charged DMPG bilayers, mimicking bacterial membranes.

The ^{31}P NMR spectra in the absence and presence of both nonselective and selective peptides (Figure 11.6) reveal lineshapes that are typical of fluid systems with axial symmetry.[15] In addition, the shape of the DMPC ^{31}P NMR

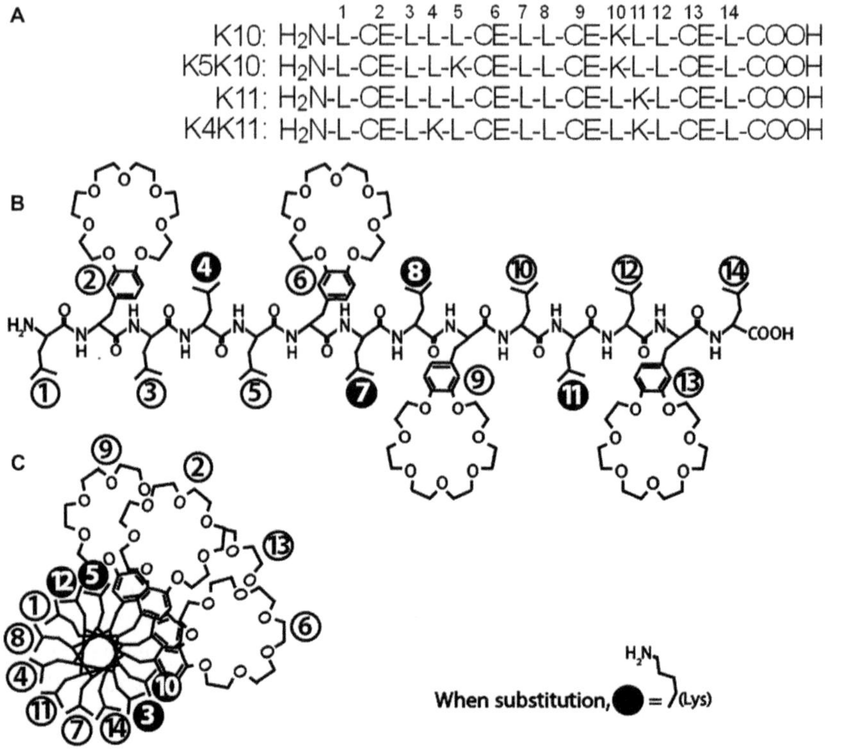

Figure 11.5 (A) Examples of positively charged analogues used in this study. The sequences of other positively charged analogues are designed on the same concept. (B) β-Strand diagram and (C) helical wheel projection showing the distribution of amino acid side chains of the base 14-mer peptide. *Black circles* indicate the leucine to lysine substitution(s). (Reproduced from Lorin *et al.*,[22] with permission.)

spectra indicates that the DMPC vesicles are ellipsoidal while the DMPG vesicles are more spherical. The absence of significant isotropic peaks in the spectra obtained in the presence of peptides indicates that the peptides do not perturb the membranes *via* a detergent-like mechanism.[23,24] This was also confirmed by dynamic light scattering experiments.[12] However, the results presented in Figure 11.6 indicate that the presence of selective and nonselective peptides induces different effects on the ^{31}P chemical shift anisotropies of DMPG vesicles, with a CSA decrease in the presence of nonselective peptides and a CSA increase in the presence of selective peptides. For DMPC vesicles, a CSA decrease is obtained in the presence of both selective and nonselective peptides. These results indicate that the peptides interact with both the DMPC and DMPG headgroups but that the interaction of selective peptides with negatively charged membranes induces ordering. This behavior has also been observed with other antimicrobial peptides adopting β-sheet structures.[25,26]

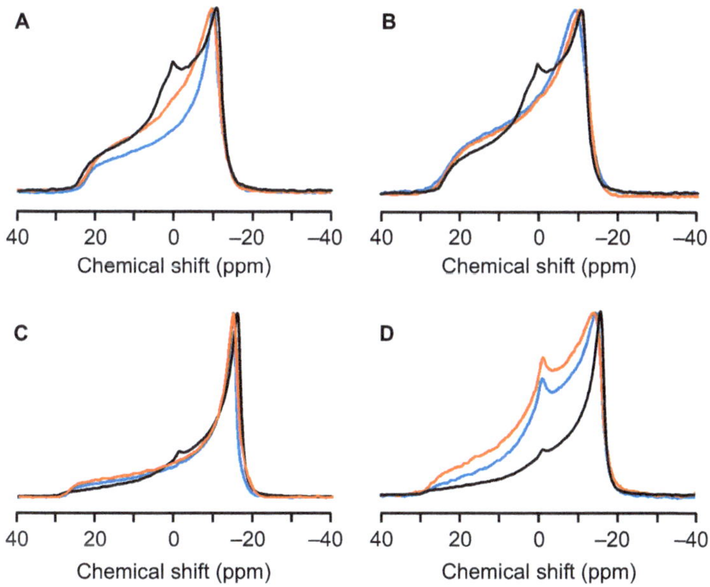

Figure 11.6 ^{31}P NMR spectra of multilamellar vesicles of: (A) DMPG in the absence (*black*) and presence of two nonselective peptides, K10 (*red*) and K5K10 (*blue*); (B) DMPG in the absence (*black*) and presence of two selective peptides, K11 (*red*) and K4K11 (*blue*); (C) DMPC in the absence (*black*) and presence of two nonselective peptides, K10 (*red*) and K5K10 (*blue*); and (D) DMPC in the absence (*black*) and presence of two selective peptides, K11 (*red*) and K4K11 (*blue*). The measurements were performed at 37 °C and at a lipid/peptide molar ratio of 60:1. (Adapted from Lorin *et al.*,[22] and reproduced with permission.)

We have also performed static ^{2}H NMR experiments to determine the effects of both nonselective and selective peptides on DMPC and DMPG bilayers.[22] The results indicate that both nonselective and selective peptides induce an increase of the quadrupolar splittings at the plateau ($\Delta\nu_P$) and the terminal methyl ($\Delta\nu_M$) regions of DMPG bilayers, while nonselective peptides induce the same effect on the acyl chains of DMPC bilayers. However, the quadrupolar splitting of DMPC bilayers do not increase in the presence of selective peptides. Therefore, peptides inducing leakage of DMPC vesicles (*i.e.*, nonselective peptides) increase the order of the DMPC acyl chains while peptides that do not induce leakage have no significant effect. The ^{2}H NMR results thus indicate that leakage of vesicles induced by 14-mer peptides is accompanied by an increase of the order of the lipid acyl chains.[22]

11.4.2 Interactions with Mechanically Oriented Bilayers

We have also investigated the effects of two nonselective peptides (K3 and K5K10) and of two selective peptides (K4 and K4K11) on the orientation of DMPC bilayers mechanically oriented between glass plates. A lipid-to-peptide

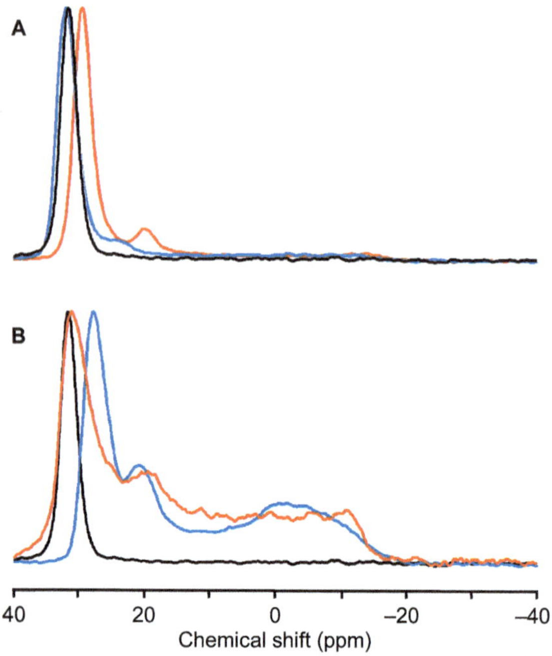

Figure 11.7 ^{31}P NMR spectra of DMPC bilayers macroscopically oriented between glass plates of: (A) in the absence (*black*) and presence of two nonselective peptides, K3 (*red*) and K5K10 (*blue*); and (B) in the absence (*black*) and presence of two selective peptides, K4 (*red*) and K4K11 (*blue*). The measurements were performed at 37 °C and at a lipid/peptide molar ratio of 20:1.

molar ratio of 20:1 was used in this case to increase the intensity of the signal resulting from peptide perturbation. The results presented in Figure 11.7 first indicate an ideal alignment of the lipid bilayers in the absence of peptides. The addition of both nonselective and selective peptides induces a loss of orientation, with a greater effect observed with the selective peptides.

The effect of nonselective peptides on the lipid orientation is very similar to that observed with the base 14-mer peptide. More specifically, the peptides induce the appearance of a powder pattern between about 30 ppm and –15 ppm that is attributed to the presence of a small proportion of non-oriented lipids, possibly through the sinking raft mechanism as illustrated in Figure 11.4. A small peak is also present at around 20–25 ppm in the spectra in the presence of peptides, which can be attributed to membrane thinning.[27] The results obtained with the nonselective peptides indicate that peptides structured as α-helices disturb DMPC bilayers in a similar manner, whether they contain a positive charge or not. Therefore, the secondary structure of the peptide appears to be the driving force in the membrane interactions of the peptide in the absence of significant electrostatic interactions.

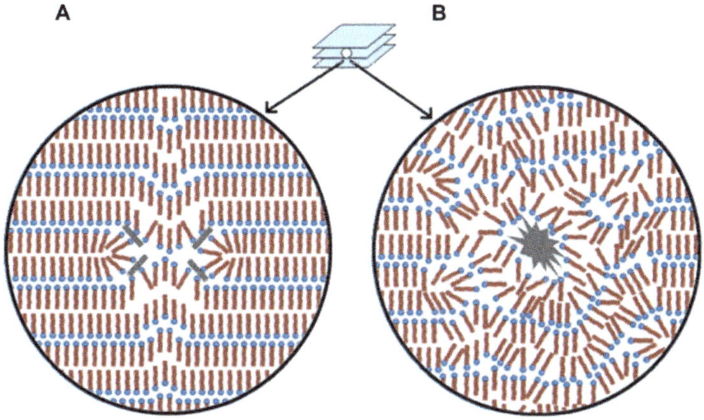

Figure 11.8 Schematic representation or disorder induced by: (A) membrane pores and (B) peptide aggregates on macroscopically oriented lipid bilayers.

The greater effect observed with the selective peptides is surprising since these peptides do not induce calcein release in DMPC vesicles. This result could be explained if the selective peptides did not induce pore formation but rather aggregated at the surface of the membrane, affecting the lipid organization in a greater manner as illustrated schematically in Figure 11.8. Indeed, Jean-Francois *et al.*[25] suggest that multimerization of cateslytin peptides at some regions of the bilayer surface induces an ordering of lipid acyl chains but without pore formation. Leakage is then caused bymembrane defects induced by the thickness difference between free and peptide-associated membrane regions. This hypothesis also was proposed to explain leakage induced by two alpha/beta antimicrobial peptides.[28] Hence, another hypothesis is that the selective cationic 14-mer peptides could result in the formation of domains of higher order in which permeabilization would be due to defects at the domain borders.

11.5 Conclusions

We have investigated the membrane interactions and topology of a base 14-mer amphipathic peptide and of several positively charged analogues in model membranes using solid-state NMR. The results indicate that the base 14 mer peptide adopts an in-plane orientation and perturbs the membrane *via* the induction of a positive curvature strain. In addition, both non-selective and selective cationic peptides interact with the headgroups of both DMPC and DMPG bilayers. However, an ordering of the hydrophobic core of lipid bilayers is observed only for peptides inducing calcein leakage in these bilayers, suggesting that the ability of the cationic 14 mer peptides to permeabilize lipid vesicles is related to their ability to increase the order of the bilayer hydrophobic core. In addition, the interaction of selective peptides with DMPC bilayers induces significant perturbation in the lipid orientation but without pore formation. These results bring significant insights into understanding the mechanisms of action of the new amphipathic 14-mer peptides investigated in the present study and could lead to design of peptides with greater selectivity towards bacterial membranes.

Acknowledgements

The authors would like to thank Marise Ouellet, Aurélien Lorin, Mathieu Noël and Marie-Ève Provencher for their significant input in the work presented in this manuscript and Pierre Audet for his technical assistance in the solid-state NMR measurements. This work was supported by the Natural Sciences and Engineering Research Council of Canada (NSERC), the Fonds de recherche du Québec – Nature et Technologies (FRQ-NT), the Regroupement québécois de recherche sur la structure, la fonction et l'ingénierie des protéines (PROTEO), the Centre de recherche sur les matériaux avancés (CERMA) and the Centre québécois sur les matériaux fonctionnels (CQMF).

References

1. G. Taubes, *Science*, 2008, **321**, 356.
2. B. Spellberg, R. Guidos, D. Gilbert, J. Bradley, H. W. Boucher, W. M. Scheld, J. G. Bartlett and J. Edwards, *Infectious Diseases Society of America Clin. Infect. Dis.*, 2008, **46**, 155.
3. B. Hamad, *Nat. Rev. Drug Discovery*, 2010, **9**, 675.
4. R. E. W. Hancock and D. S. Chapple, *Antimicrob. Agents Chemother.*, 1999, **43**, 1317.
5. R. E. W. Hancock and R. Lehrer, *Trends Biotechnol.*, 1998, **16**, 82.
6. R. E. W. Hancock and H.-G. Sahl, *Nat. Biotechnol.*, 2006, **24**, 1551.
7. K. Matsuzaki, *Biochim. Biophys. Acta, Biomembr.*, 1999, **1462**, 1.
8. D. I. Chan, E. J. Prenner and H. J. Vogel, *Biochim. Biophys. Acta, Biomembr.*, 2006, **1758**, 1184.
9. M. R. Yeaman and N. Y. Yount, *Pharmacol. Rev.*, 2003, **55**, 27.
10. Y. R. Vandenburg, B. D. Smith, E. Biron and N. Voyer, *Chem. Commun.*, 2002, 1694.
11. M. Dathe, H. Nikolenko, J. Meyer, M. Beyermann and M. Bienert, *FEBS Lett.*, 2001, **501**, 146.
12. A. Lorin, M. Noël, M.-E. Provencher, V. Turcotte, C. Masson, S. Cardinal, P. Lagüe, N. Voyer and M. Auger, *Biochemistry*, 2011, **50**, 9409.
13. J. Seelig and A. Seelig, *Q. Rev. Biophys.*, 1980, **13**, 19.
14. J. Seelig, *Biochim. Biophys. Acta*, 1978, **515**, 105.
15. I. C. P. Smith and I. H. Ekiel, in *Phosphorus-31 NMR: Principles and Applications*, ed. D. G. Gorenstein, Academic Press, London, 1984, p. 447.
16. M. Ouellet, J. D. Doucet, N. Voyer and M. Auger, *Biochemistry*, 2007, **46**, 6597.

17. M. Ouellet, G. Bernard, N. Voyer and M. Auger, *Biophys. J.*, 2006, **90**, 4071.
18. J. Seelig, *Q. Rev. Biophys.*, 1977, **10**, 353.
19. B. Bechinger and C. Sizun, *Concepts Magn. Reson.*, 2003, **18A**, 130.
20. B. Bechinger, *Mol. Membr. Biol.*, 2000, **17**, 135.
21. D. J. Hirsh, J. Hammer, W. L. Maloy, J. Blazyk and J. Schaefer, *Biochemistry*, 1996, **35**, 12733–12741.
22. A. Lorin, M. Noël, M.-E. Provencher, V. Turcotte, S. Cardinal, P. Lagüe, N. Voyer and M. Auger, *Biophys. J.*, 2012, **103**, 1470.
23. G. Anderluh, M. Dalla Serra, G. Viero, G. Guella, P. Macek and G. Menestrina, *J. Biol. Chem.*, 2003, **278**, 45216.
24. T. Pott, M. Paternostre and E. J. Dufourc, *Eur. Biophys. J.*, 1998, **27**, 237.
25. F. Jean-Francois, S. Castano, B. Desbat, B. Odaert, M. Roux, M. H. Metz-Boutigue and E. J. Dufourc, *Biochemistry*, 2008, **47**, 6394.
26. A. Ramamoorthy, S. Thennarasu, A. Tan, K. Gottipati, S. Sreekumar, D. L. Heyl, F. Y. An and C. E. Shelburne, *Biochemistry*, 2006, **45**, 6529.
27. S. Wi and C. Kim, *J. Phys. Chem. B*, 2008, **112**, 11402.
28. R. F. Epand, M. A. Schmitt, S. H. Gellman and R. M. Epand, *Biochim. Biophys. Acta*, 2006, **1758**, 1343.

CHAPTER 12

Investigations of the Structure, Topology and Dynamics of Membrane-Associated Polypeptides by Solid-State NMR Spectroscopy

EVGENIY S. SALNIKOV, CHRISTOPHER AISENBREY, JESUS RAYA AND BURKHARD BECHINGER*

Université de Strasbourg/CNRS, UMR7177, Institut de Chimie, 1 rue Blaise Pascal, 67070 Strasbourg, France
*Email: bechinger@unistra.fr

12.1 Introduction

Much progress has been made during recent years in the structural analysis of membrane proteins, and novel structures are regularly added to the protein structural database. Nevertheless, many questions about their mechanisms of action and their dynamic features remain obscure and this is where NMR spectroscopy can make important contributions. A number of membrane-mimetic systems have been developed for solution NMR investigations, such as micelles,[1–3] nanodiscs[4–12] or small bicelles.[13,14] Furthermore, structural studies of polypeptides in lipid bilayers or in larger bicelles[14–18] can be performed using solid-state NMR spectroscopy.[19–22] The latter approach is of particular interest for peptides and proteins when their

New Developments in NMR No. 3
Advances in Biological Solid-State NMR: Proteins and Membrane-Active Peptides
Edited by Frances Separovic and Akira Naito

Published by the Royal Society of Chemistry, www.rsc.org

lipid interactions and/or their topological analysis are important to understand their functional mechanisms.[23–26]

The solid-state NMR investigations of polypeptides inserted into extended lipid bilayers follow two different approaches: either spinning of the sample around the magic angle (MAS),[27–30] or orienting the bilayers relative to the magnetic field direction,[22,31–34] and each has distinct advantages. On the one hand, MAS provides sharp lines, higher signal-to-noise and isotropic chemical shift information which has been correlated to the protein secondary structure.[35] However, for more detailed structural investigations it is often required that distance-related cross-peaks are obtained, which make the spectra considerably more crowded and thereby complicating the assignment. On the other hand, oriented samples are usually characterized by broader lines which are, however, spread over the much larger spectral width of the chemical shift anisotropy. As both the linewidth and the chemical shift dispersion are increased by about an order of magnitude compared to isotropic MAS solid-state NMR spectra, the signal-to-noise ratio is correspondingly lower. Furthermore, assignment of oriented solid-state NMR spectra from uniformly labeled samples can be quite difficult, as the chemical shifts correlate much less with the amino acid type, and the lines are usually broader and tend to overlap. However, the chemical shift position and other anisotropic interactions that can be extracted from these spectra provide direct structural information about the alignment of the polypeptide bonds at the corresponding sites and also about their conformational and topological heterogeneity and dynamics.[22,36,37] Therefore, if selectively or specifically labeled samples are used and/or the assignment of the resonances can be performed, such oriented solid-state NMR spectra are highly informative.

12.2 Oriented Solid-State NMR

Here we focus on the second approach, where oriented bilayer samples are investigated by solid-state NMR spectroscopy. A common strategy involves the preparation of polypeptides labeled with stable isotopes, such as ^{15}N, ^{13}C and/or ^{2}H, their reconstitution into phospholipid bilayers and the recording of the solid-state NMR spectra of these nuclei. Whereas peptides can be conveniently prepared by solid-phase peptide synthesis and thereby labeled at one or a few selected sites, an alternative approach consists in the bacterial over expression and biochemical purification of proteins or peptides. In the latter case, various labeling schemes exist where specific isotopes are incorporated into all sites (uniformly) or, depending on the composition of the growth medium, only a selected set of atoms or amino acid residues.[17,38–40] In the meantime, a number of different labeling schemes exist which, by feeding cocktails of labeled/unlabeled amino acids or partially labeled metabolic precursors, result in simplified NMR spectra with improved resolution and that are easier to assign. The preparation method and concomitantly the labeling scheme have profound consequences on which of the solid-state NMR experiments are best suited for the investigation, and usually the design of the labeling scheme and the choice of NMR experiments form a tightly linked entity. For example, the alignment of samples that have been uniformly labeled has to be optimized for the best resolution, and 2D separated local-field solid-state NMR spectra have been obtained for transmembrane helical peptides in magnetically oriented bicelles as well as in mechanically oriented membranes.[14,15,17,41] For uniformly ^{15}N labeled peptides the spectra monitor correlations between the ^{15}N chemical shift of the backbone amides and the corresponding dipolar couplings.[42] The corresponding patterns can be directly analyzed in terms of helical tilt angles, irregularities within this secondary structure and, provided that the peaks can be assigned by additional experiments and assumptions, also the helical pitch angle.[41,43,44] The approach has been successfully applied to transmembrane helical sequences, including phospholamban and its interaction partners,[45,46] alamethicin,[41] the influenza M2 sequence of hemagglutinin,[31] a twin-arginine translocase,[17] antimicrobial peptides[47] and larger proteins.[48–50]

Many of the above-cited studies have been performed on transmembrane helices that have been reconstituted into mixtures of long-chain phospholipids and detergent-like molecules (*e.g.* short-chain phospholipids) as these bicellar mixtures align within the magnetic field of the NMR spectrometer. Although bicelles can provide well-resolved NMR spectra, they are less well suited for helices that are not integrated into the hydrophobic core of the lipid bilayer. This is due to the reduced membrane-partitioning constants of amphipathic domains concomitant with fast exchange between the water phase and the membrane surface. The resulting averaging of the anisotropy of NMR interactions may make this combination of water-soluble peptide sequences and bilayers unsuitable for oriented solid-state NMR investigations. Also the choice of lipid composition is somewhat restricted when magnetically aligned membranes (bicelles) are investigated.

As an alternative, membranes can be oriented along mechanical supports and hydrated through the gas phase at defined humidity, *e.g.* 95% RH.[51,52] This condition ensures that the lipids are well hydrated, while at the same time avoiding the formation of a bulk water phase. Therefore, the peptides accumulate in the proximity of the membrane surface and the high local concentration favours the membrane-associated state similar to the accumulation of cationic peptides at negatively charged surfaces.[53,54] This set-up has been used to investigate the structure, topology and dynamics of a great number of polypeptides, including cationic antimicrobial peptides,[34,37,55–59] peptaibols,[60–63] signal sequences,[64,65] surfactant peptide B,[66] the Huntingtin amphipathic amino-terminal domain,[67] β-sheet peptides[68,69] and model peptides,[70–74] as well as larger proteins.[50,75–77]

12.3 Orientational Restraints from Solid-State NMR Parameters

In the following we present how solid-state NMR spectra of oriented membranes provide angular constraints using concepts that have been published previously.[22] Thereafter we extend previous reviews on the subject and discuss the accuracy that can be obtained by such angular constraints and the experimental and systematic errors that may be associated with such measurements. Deriving structural information from oriented solid-state NMR is based on the dependence of most NMR interactions on the alignment of bonds and molecules relative to the magnetic field direction. This anisotropy is averaged in solution by Brownian tumbling or in MAS solid-state NMR by mechanically spinning the samples. Anisotropy becomes obvious in the absence of such motions or when they are reduced.

In order to calculate the angular restraints from solid-state NMR spectra it is necessary to know with accuracy the tensor describing the interactions of nuclei with the magnetic field, *i.e.* the strength and the directionality of these interactions. This includes the size of the main tensor elements and also the alignment of this tensor within the molecular coordinate system. The interaction tensors of a number of biologically interesting nuclei have been characterized, including the ^{13}C[78–80] or the ^{15}N chemical shift interaction of peptide bonds (reviewed[81]). Notably, when compared to the ^{15}N chemical shift dispersion in solution, what becomes obvious when a HSQC spectrum of a uniformly labelled protein is analyzed, is that the static ^{15}N chemical shift anisotropy of each site is typically an order of magnitude higher (*e.g.* figure 1 in Bechinger *et al.*[82]). In the static case, the measured chemical shift, σ_{zz}, is calculated according to:

$$\sigma_{zz} = \sigma_{11}\,\sin^2\Theta\,\cos^2\Phi + \sigma_{22}\,\sin^2\Theta\,\sin^2\Phi + \sigma_{33}\,\cos^2\Theta,$$

where σ_{11}, σ_{22} and σ_{33} are the main tensor elements, and Θ and Φ the Euler angles which position the chemical shift tensor relative to B_0. The main tensor elements of the ^{15}N amide bond have been characterized and found to be on average 55, 80 and 225 ppm ($\pm$5 ppm) and the alignment of these vectors relative to the molecular coordinate system is described in detail in the literature (reviewed[81]).

The dipolar coupling is another interaction that can provide valuable structural information. It is averaged in isotropic solution but can dominate solid-state NMR spectra. The dipolar coupling is calculated according to:

$$D_{zz} = \frac{\mu_0}{4\pi}\frac{\gamma_1\gamma_2 h}{r^3}\left(3\cos^2\Theta - 1\right) \tag{12.1}$$

(secular part), where r is the distance between the two interacting nuclei and Θ the angle between the distance vector and B_0. Within this expression the dipolar coupling constant is expressed by

$$\frac{\mu_0}{4\pi}\frac{\gamma_1\gamma_2 h}{r^3}$$

and depends on the gyromagnetic ratios, γ_1 and γ_2, of the two interacting nuclear dipoles, and where μ_0 represents the free magnetic inductivity. When changing the orientation of the dipolar interaction vector Θ relative to

B_0, the dipolar interaction of the amide bond ^{1}H–^{15}N nuclei covers values between −20 kHz and 10 kHz (detailed elsewhere[82]).

The ^{2}H quadrupolar splitting of the deuterated alanine methyl group provides another parameter that can be analyzed to obtain orientational constraints and used to align polypeptides relative to the magnetic field direction/membrane normal. Fast rotation around the Cα–Cβ bond ensures that at ambient temperatures the three deuterons are equivalent[83] and a single intensity is recorded from the three sites. Thereby the deuterated alanine methyl group provides a direct indicator of the Cα–Cβ orientation relative to B_0 and therefore, also of the polypeptide backbone, of which the Cα-atoms are an integral part. Hence, measuring the alignment the Cα–Cβ axis is a direct indicator of the topology of the corresponding polypeptide domain. In helical conformations the methyl group exhibits a well-defined alignment relative to the helix long axis (59.4°; *cf.* Vostrikov *et al.*[84]). Measuring deuterium quadrupolar splittings alone has been used to analyze the topology of helical peptides labeled with [C-$^{2}H_3$]alanines successively at selected positions (GALA method[85]). Furthermore, the alanine quadrupolar splitting provides highly complementary information to the ^{15}N labeled backbone sites to restrict the tilt/rotational angular pairs and thereby a much-reduced number of labeled sites is required. An additional advantage of using a combination of different isotopes is that a single peptide can be labeled at two selected sites without causing assignment problems. Furthermore, whereas the GALA method relies on alanines being present in the sequence at different positions, the ^{15}N nucleus can be incorporated into a much wider range of amino acids. Another quadrupolar nucleus, ^{14}N, which occurs at high natural abundance, has also been explored for structural investigations, but fewer studies exist at the present time.[86]

Owing to fast rotation around the Cα–Cβ bond, the static quadrupolar tensor for the alanine methyl group is scaled by a factor of −1/3 (*cf.* below) and the resulting averaged tensor aligned parallel to this rotation axis. The quadrupolar splitting is calculated according to:

$$\Delta\nu_Q = \frac{3}{2}\left(\frac{e^2qQ}{3h}\right)\left(\frac{3\cos^2\Theta - 1}{2}\right) \tag{12.2}$$

with Θ being the angle between the Cα–Cβ bond and the magnetic field direction B_0. The static quadrupolar coupling constant, e^2qQ/h, is about 167 kHz for aliphatic C–^{2}H bonds.[83,87] However, when room temperature deuterium solid-state NMR spectra are recorded of dry peptide powders of [C–$^{2}H_3$]alanine or [C–$^{2}H_3$]-labeled Plexiglas (poly(methyl methacrylate)), maximum splittings of ∼74 kHz are observed (instead of expected value of ∼84 kHz), which suggests additional averaging by librations of the deuterated sites.

Such motional contributions dominate the ^{2}H solid-state NMR spectra of deuterated fatty acyl chain segments within phospholipid bilayers. Therefore, the structure of these chains are described by segment-specific C–D order parameters S^i_{CD} which are directly linked to the C–$^2H_2$ deuterium quadrupole splittings $\Delta\nu^i$ according[88] to:

$$\Delta\nu^i = \frac{3}{2}\left(\frac{e^2qQ}{h}\right)\left(\frac{3\cos^2\gamma - 1}{2}\right)S^i_{CD} \tag{12.3}$$

This measurement is equivalent to comparing the measured deuterium quadrupole splitting to the static maximal splitting. When the quadrupolar splittings are measured between the intensity maxima in a non-oriented sample, this corresponds to half of the averaged splitting constant since the $\gamma = 90°$ alignment from the lipids being aligned perpendicular to the magnetic field direction, dominates the so-called powder pattern. The corresponding quadrupolar coupling constants (e^2qQ/h) are 167 kHz for the aliphatic chain segments and 175 kHz in the context of a double bond,[88,89] meaning that a static C–^{2}H segment would exhibit a splitting between the two main peaks in non-oriented samples (*i.e.*, 90° orientation) of up to ~125 kHz, to which the measured values are compared. The deuterium order parameter is a measure of the motional anisotropy of the selected C–^{2}H bond and defined as:

$$S^i_{CD} = \frac{1}{2}\overline{\left(3\cos^2\beta_i - 1\right)} \tag{12.4}$$

where β_i is the instantaneous angle between the C–^{2}H vector and the bilayer normal and the bar indicates the time average.[90,91]

The concept of order parameters is quite useful as various motional regimes with significantly different time frames can be incorporated one after the other by calculating each of the resulting order parameters separately:

$$\Delta\nu^i = \frac{3}{2}\left(\frac{e^2qQ}{h}\right)S_{bilayer}S_{mol}S^i_{loc}\left(\frac{3\cos^2\beta_i - 1}{2}\right) \tag{12.5}$$

In this equation, $S_{bilayer}$ represents the degree of order and alignment of the bilayer normal (including wobbling motions or the global angle at which the bilayers are aligned), S_{mol} takes into account motions and alignment of the molecule or a segment thereof, S^i_{loc} the local order (*e.g.* librational motions and the alignment of a particular bond), and β_i the angle between the C–^{2}H bond and the axis of motion (usually the membrane normal).[88,89] Thus at $\gamma_{bilayer} = 90°$, rotation around the bilayer normal causes a reduction of the quadrupolar splitting of the fatty acyl chain C–^{2}H by a factor of −1/2, or with a tetrahedral angle of 109.4° the free rotation of a methyl group reduces the quadrupolar splitting by $S_{loc} = -1/3$.[83,88] It should be noted that the concept of order and scaling parameters has been used by a number of groups, albeit sometimes with somewhat differing reference systems and definitions.[92–95]

For some peptide sequences carrying methyl-deuterated alanine residues, relatively small quadrupolar splittings have been measured and the above concept of analyzing the deuterium NMR spectra by order parameters seems useful. A spectrum with a small quadrupolar splitting can be indicative of motional averaging or of a static angle of the Cα–Cβ vector that is close to the magic angle, or situations where both contribute. For example, a perfectly transmembrane helix oriented parallel to the magnetic field direction would result in a quadrupolar splitting of only around 10 kHz (depending on the detailed helix structure), even in the absence of wobbling motions. Alanines that are part of a non-structured region are expected to result in even smaller quadrupolar splittings. In the case of isotropic motions, the splitting disappears completely and, if the polypeptide motions are restricted by H-bonding or similar-sized interactions and by its localization in the narrow inter-membrane space of stacked bilayers,[96] small values of only a few kHz are expected to be observed. In these cases, motional and angular information intermix and giving an order parameter for the Cα–Cβ bond:

$$S^i_{CD} = \frac{1}{2}\overline{\left(3\cos^2\alpha_i - 1\right)} \tag{12.6}$$

and providing S_{CD} seems a suitable alternative for a structural description (*cf.* also[97,98]).

12.4 Supported Lipid Bilayers

Each of the above-listed NMR interactions can provide information that allows one to restrict the possible alignment of the backbone atoms and thereby of the whole molecule relative to the magnetic field direction (B_0). Therefore, it is convenient to orient the samples with the membrane normal parallel to B_0, an arrangement where lateral and rotational diffusion around the bilayer normal does not change the relative alignment of the polypeptides with respect to the magnetic field, *i.e.* the rotational symmetry of the lipid bilayer system and of the B_0 field coincide.

Figure 12.1 shows proton-decoupled ^{15}N and ^{2}H solid-state NMR spectra of a hydrophobic model peptide, [^{15}N-Leu17, C-$^{2}H_3$-Ala16]-hΦ19W,[99] reconstituted into oriented DMPC/DMPG 3 : 1 mol/mole phospholipid bilayers that are oriented with the membrane normal parallel to the magnetic field direction. Owing to its specific properties, the amide ^{15}N chemical shift has been shown directly correlated to the approximate tilt angle of the peptide relative to the membrane normal, where the value of 196 ppm (Figure 12.1A) is indicative of a transmembrane alignment of the peptide.[82]

Additional and highly complementary constraints are obtained from methyl-deuterated alanines. The corresponding spectrum is shown in Figure 12.1B, where the inner signal with a linewidth of 1.5 kHz is from natural abundance deuterons of membrane-associated water and the outer splitting of 12 kHz arises from the alanine site.

In a next step, the data of Figures 12.1A and 12.1B were analyzed in more detail by testing which orientations of the helix fit the solid-state NMR measurements. A coordinate system was defined where the tilt constitutes the angle between the long axis of the helix and the membrane normal and the pitch the angle between the membrane normal and the line within an arbitrary plane of the peptide helical wheel projection. Whenever an angular

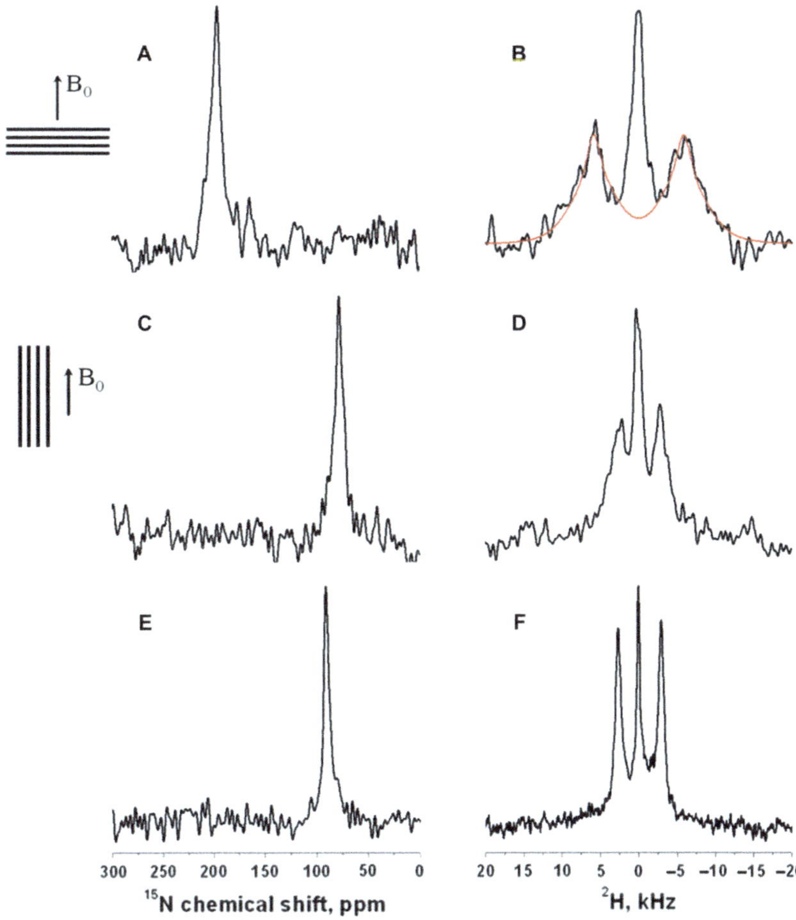

Figure 12.1 Proton-decoupled ^{15}N and ^{2}H solid-state NMR spectra of [^{15}N-Leu17,C-^{2}H$_3$-Ala16]-hΦ19W in DMPC/DMPG (3 : 1 mol/mol) membranes oriented on glass plates (A–D) and within DMPC/DMPG/di-C7 : 0-PC bicelles (2.4 : 0.8 : 1 molar ratio; $q = 3.2$; total lipid content 28% (w/v)) (E,F). The *thin red line* in (B) shows a simulated spectrum arising from a 3° Gaussian distribution for the peptide alignment and a spectral line width of 2.5 kHz. The orientation of the glass plates is shown in the sketch on the *left.*

pair fits the measured orientation-dependent NMR parameter, the corresponding alignment is labeled in black (^{15}N chemical shift) or red (^{2}H quadrupolar splitting). Figure 12.2A shows the corresponding traces by taking into account experimental errors of 3 ppm and 3 kHz, respectively. The ^{15}N and ^{2}H traces intersect at four regions, representing tilt angles between 10° and 35° and additional data are required to unambiguously determine which of these tilt/azimuthal angular pairs represent the polypeptide alignment.

Whereas here we have used an ideal α-helix ($\phi = -65°$, $\psi = -45°$) to calculate the alignment of hΦ19W (Figures 12.1 and 12.2), such solid-state NMR

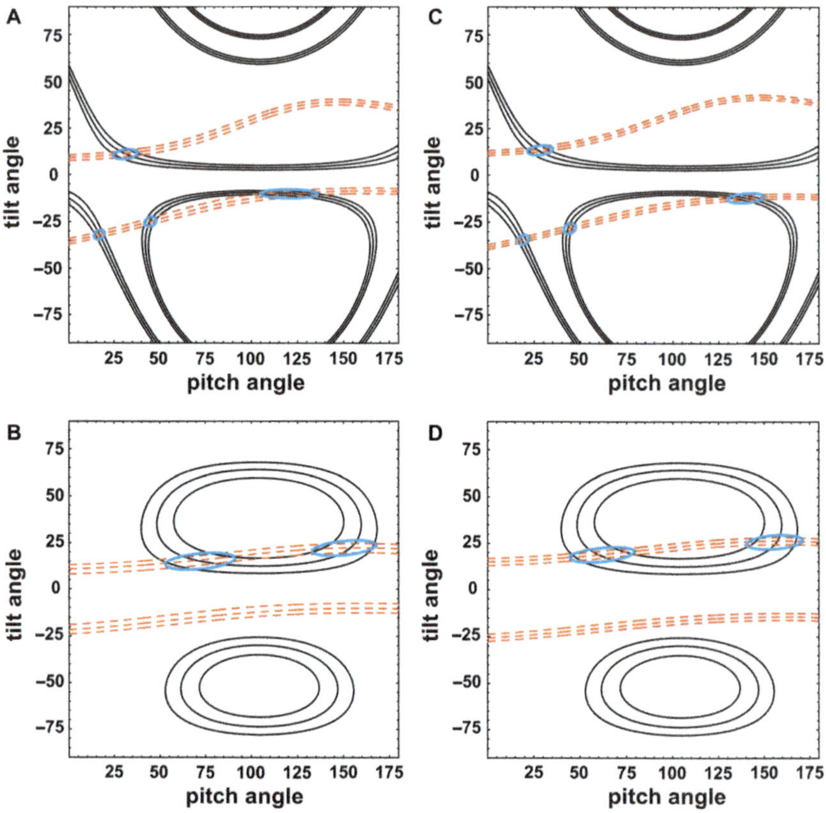

Figure 12.2 Angular restrictions obtained from the oriented solid-state NMR spectra shown in Figures 12.1A and 12.1B. Here an analysis of systematic errors is performed by variation of the main tensor elements as well as modifications of the motions that lead to spectral averaging. The possible alignments of an ideal α-helix ($\phi = -65°$, $\psi = -45°$) are represented by their helical tilt and the rotational pitch angles. The *solid black lines* represent angular pairs that agree with the experimental ^{2}H quadrupolar splitting obtained from [$^{2}H_3$]Ala16 (12 ± 2 kHz, Figure 12.1B) and the *hatched red lines* that from ^{15}N chemical shifts of [^{15}N]Leu17 (196 ± 3 ppm, Figure 12.1A). Restriction plots obtained when using the ^{15}N main tensor elements (54, 79, 221 ppm) are shown in A and B, and those using the tensor (51, 75, 229 ppm) in C and D. Panels A and C assume a static peptide, whereas B and D were calculated with wagging fluctuations of the peptide orientations within a 12° Gaussian distribution and azimuthal fluctuations of 80°. This motional regime was previously published for another transmembrane peptide, WALP23, in DMPC membranes.[132]

data can also be used to verify and align more detailed models, *e.g.* PDB structural data files that arise from X-ray structural analysis, NMR spectroscopy in micellar or membrane-mimetic environments, or other structural elements.[41,67,100,101] Initially, the oriented solid-state NMR approach was used to compare and align, within lipid bilayers, structures that have been obtained by solution-state NMR methods.[102,103]

In a first step, the NMR spectra of the labels have to qualitatively fit the structural model, *i.e.* NMR values in the random coil region should be at or close to the isotropic values, whereas anisotropic values are indicative of structured regions. This has to be done with caution, as apparently isotropic values can also be due to structured regions where the alignment comes close to the magic angle. Nevertheless, the qualitative comparison allows one to eliminate some of the models which do not match these criteria.

Thereafter the residues that form a domain (such as a helix) in the structural model, as well as with lipid bilayers, are tested in more detail. In our experience, having three NMR measurements provides a quite stringent test for a given structural model as many times the experimental parameters do not coincide in a single tilt/pitch angular pair. The situation becomes more complex as some of the data may come to fit when motions are taken into consideration, but not when an interaction devoid of wobbling and rocking motions is assumed. In previous analyses of transmembrane model peptides it has been found that six parameters (^{15}N, ^{13}C and ^{2}H from two differently labeled peptides) are sufficient to determine tilt and pitch angle and their Gaussian distribution due to motions.[104] Furthermore, the structure and orientational distribution of a single site from a transmembrane peptide, alamethicin, labeled with ^{15}N and/or ^{2}H, was determined from solid-state NMR parameters,[98,105] and this combination has also allowed the refining of the structure of an amphipathic membrane anchor of Huntingtin, taking into consideration rocking and wobbling motions around the helical domain.[67,106]

12.5 Magnetic Alignment of Bicelles

The spectra shown in Figures 12.1A and 12.1B were obtained from oriented membranes supported on a number of ultrathin glass plates and equilibrated at 95% RH.[51,52] Although this protocol results in stacks of close to fully hydrated lipid bilayers, there is no bulk water in between the membranes. On the one hand, this shifts the equilibria characterized by low to moderate membrane partitioning constants of amphipathic polypeptides towards the bilayer-associated state. On the other hand, bulk extra-membranous domains of larger proteins may be restricted in their motions and/or interfere with the alignment process. An alternative method exists in magnetically orienting phospholipid bilayers when these occur in so-called bicellar phases.[14,107–109] On a structural level, some ambiguity exists if the magnetic alignment phase represents small patches of bilayers that are stabilized by a rim of detergent or detergent-like molecules[110] or if these phases rather consist of perforated sheets of membranes.[111] Interestingly, bicellar phases can be obtained from phospholipid mixtures with detergents, short-chain lipids or with amphipathic peptides,[108,109,112,113] and they can be made from purely zwitterionic or from mixed phospholipid bilayers.[109,114] Furthermore, bicelles have been formed from membrane compositions that align with their normal parallel to the magnetic field direction,[16,115] rather than the original perpendicular alignment.[107–109]

Figures 12.1E and 12.1F show NMR spectra that were obtained from the hΦ19W peptide when inserted in a bicellar mixture made of the same bilayer lipids as used to prepare the samples shown in Figures 12.1A and 12.1B. In addition, the short-chain lipid and deuterium-depleted water were added in the right proportions to obtain a composition that allows magnetic alignment. The magnetic interactions cause the bilayer component to be oriented perpendicular to the magnetic field direction, *i.e.* at 90° relative to the sample alignment shown in Figures 12.1A and 12.1B. Therefore, the ^{15}N chemical shift resonances of the transmembrane peptide appear at 91 ppm in this tilted arrangement (Figure 12.1E). For comparison, the spectrum of the glass plate sample was also recorded at this tilted orientation, giving a ^{15}N chemical shift of 78 ppm (Figure 12.1C). The reduced anisotropy of the bicellar sample can be explained by changes in the alignment when going from one system to the other, due to differences in hydration and composition. However, more importantly, wobbling and rocking motions of the bicelles or undulations of the wormlike micelles in this more dilute system scale the chemical shift anisotropy and can be taken into account by an order parameter, S_{bilayer}. This order parameter has been found to be in the range of 0.5 to close to unity in previous publications, depending on the experimental conditions such as the composition of the bicellar mixture and temperature.[92–95,116] Furthermore, the strength of the magnetic field (B_0) can be relevant as the magnetic alignment is $\propto \Delta\chi B_0^2$, where $\Delta\chi$ is the anisotropy of the magnetic susceptibility of the assembly.[116]

From the same samples the ^2H solid-state NMR spectra were also recorded and reveal quadrupolar splittings of 6 and 5.5 kHz, respectively (Figures 12.1D and 12.1F). Relative to Figure 12.1B, the splittings are scaled by a factor of −1/2 that arises from changing the bilayer alignment γ from 0° to 90°. Notably, the degree of ordering appears considerably improved in the bicellar sample, which becomes most obvious when the ^2H spectra are compared to each other (Figures 12.1D and 12.1F). Knowing that the bilayer phase undergoes considerable motions that cause $S_{\text{bilayer}} < 1$, the sharper lines are probably due to fast motional averaging of orientational and/conformational inhomogeneities. In the more densely packed mechanically supported bilayer samples, misalignment of the membranes, peptides and the slower exchange of conformational heterogeneities cause inhomogenous line broadening effects.

12.6 Error Analysis and Sample Mosaicity

The experimental errors that come into play in solid-state NMR spectra from oriented samples are uncertainties in determining the chemical shifts and quadrupolar splittings. Despite the acquisition of several thousand FIDs (typically, spectral acquisitions take hours to overnight), the spectra are relatively broad and tend to be noisy. Therefore, measurements of the chemical shifts and the quadrupolar splittings are associated with experimental errors that cannot be neglected. These experimental uncertainties can be taken into account during the restriction analysis by including those angles that agree with a range of chemical shifts or quadrupolar splittings. Typically, the width at half-height, at 80% height or constant estimates of 3 ppm and 2 kHz, are taken into consideration for the ^{15}N chemical shift and ^{2}H quadrupolar splitting, respectively (Figure 12.2). Similarly, the experimental error of dipolar coupling measurements depends on the data set (typically in the range of 300–1000 Hz). By combining measurements from a number of restraints, either from different labels within the same residue[98] or from different sites along the sequence of the polypeptide,[67] or both,[104] the topological analysis becomes better defined when at the same time the experimental errors are averaged.[99,101] In this context it should be mentioned that some of the orientational constraints are more complementary than others. For example, on the one hand the unique main axis of the ^{15}N chemical shift and the ^{1}H–^{15}N dipolar interaction tensors are close to parallel and, therefore, provide related restriction analyses.[22,104] Also, the ^{2}H quadrupolar splitting of methyl-deuterated alanines and the amide bond ^{15}N chemical shift are highly complementary (*cf.* Figure 12.2).[22,99]

An additional source of error may arise from inaccuracies in orienting the sample relative to the magnetic field direction. This can be a problem for mechanically oriented samples where the glass plates do not stack in a perfect parallel fashion and/or the sample as a whole is slightly tilted with respect to B_0. In this case, repeated measurements as well as using different coils/NMR probe heads[117] (*e.g.* for ^{2}H and ^{15}N experiments) are beneficial for averaging such experimental errors.

Figure 12.1 also shows that when compared to the ^{15}N chemical shift, the ^{2}H quadrupolar splitting can be more sensitive to the orientational heterogeneity of the labeled site and the mosaicity of the sample. This has been analyzed systematically[99] and in Figures 12.3A and 12.3B, and is evident when Figures 12.1C and 12.1E as well as 12.1D and 12.1F are compared. Whereas in Figure 12.1 the ^{15}N line of the bicellar sample is about twice as sharp as for the glass plate sample, the difference is much bigger for the ^{2}H solid-state NMR spectra. Although this effect makes it more difficult to record oriented deuterium spectra, they can be analyzed to reveal the conformational and topological mosaicity more accurately (Figures 12.3C and 12.3D). Because the quadrupolar splitting changes fast with the angle and in a non-linear manner (Figures 12.3A and 12.3B), the ^{2}H NMR spectra should be simulated to obtain the mean for the quadrupolar splitting rather than merely reading the distance between the intensity maxima (*cf.* Figure 12.3C and Aisenbrey and Bechinger[99]).

The comparison of solid-state NMR measurements allows one to monitor relative changes with high accuracy. By contrast, the absolute structural details are much less precise due to systematic errors that are associated with the approach. For example, the tensor values of ^{15}N chemical shift, *i.e.*

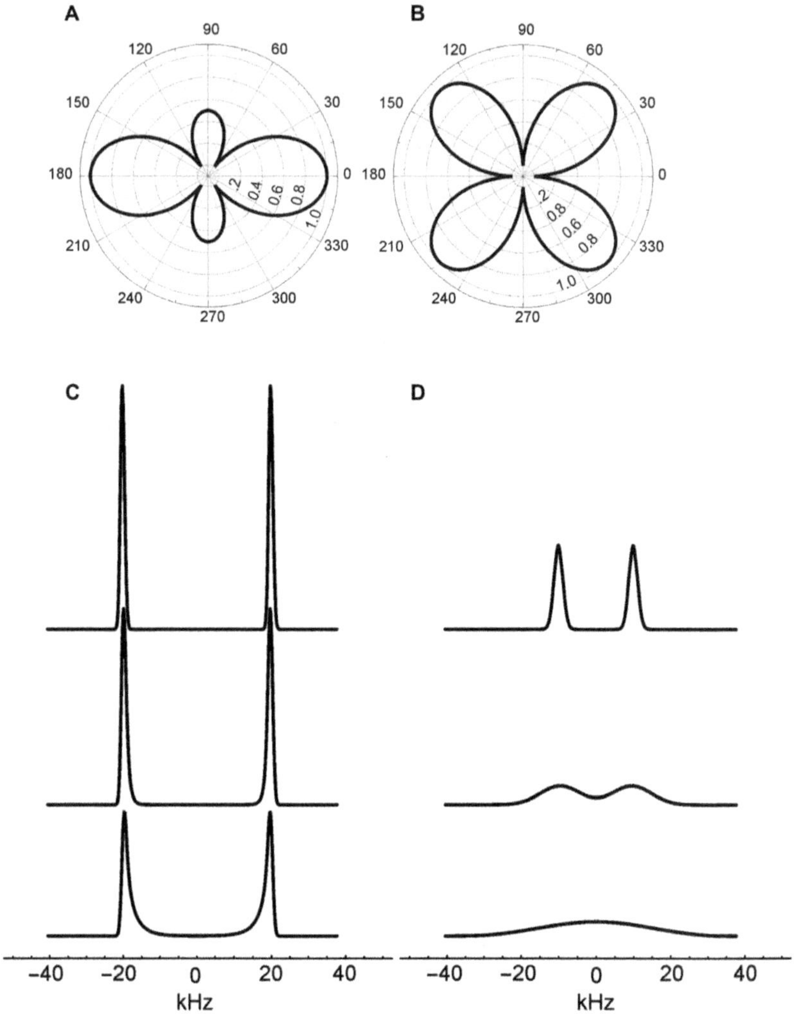

Figure 12.3 (A) Polar plot of the quadrupolar splitting as a function of the angle α between the Cα–Cβ bond and the magnetic field. The *circles* are indicative of the measured quadrupolar splitting (in units of the coupling constant), which is a function of the angle α given on the outside. (B) Polar plot of the first derivative of the quadrupolar splitting as a function of the Cα–Cβ orientation. Here the *circles* represent changes in the coupling constant per radian. (C, D) Simulation of the effect of a Gaussian mosaic spread for an orientation between the Cα–Cβ bond and the magnetic field of $\alpha = 45°$ (C) and $\alpha = 90°$ (D). From the top going down the mosaic the spread is 1°, 5° and 10°.

the description of the nuclear interactions with the magnetic field, typically show variations of 5–7 ppm (reviewed and tabulated in Salnikov *et al.*[81]). This translates to uncertainty in the tilt angle determination of a few degrees, depending on the chemical shift value itself (Figures 12.2A and 12.2C).

Ideally, the chemical shift tensor of each site is measured and then used for the calculations.[118,119] Whereas the main tensor elements are extracted from chemical powder patterns, determining their alignment within the molecular frame is considerably more difficult.[120–122] Measuring the isotropic value by tilting the sample provides at least insight into the isotropic values which, when very different, are an indicator of the necessity for large tensor readjustments for the labeled sites.[106] For example, an isotropic chemical shift of 118 ppm was obtained for hΦ19W, also investigated in Figures 12.1 and 12.2, from the proton-decoupled ^{15}N MAS spectrum of the peptide powder. The same value was obtained when the trace was calculated from the ^{15}N spectra obtained from two orthogonal alignments shown in Figures 12.1A and 12.1C (196 ± 3 ppm and 78 ± 2 ppm). Similar considerations may apply for the tensor of the deuterium quadrupole tensor of alanine,[83] although these may be less dependent on the local peptide environment than the ^{15}N chemical shift.[123]

When the dipolar couplings are converted into angles, the length of covalent bonds such as the N–H vector is usually taken as a fixed value, although this may need further investigation, in particular due its involvement in hydrogen bonding interactions. Furthermore, PISEMA spectra are scaled by a theoretical value of 0.82 (when the experiment performs in a perfect manner), but this number depends on the correct setting of the experiment. For example, we have observed that the powder spectra of a peptide bond can have variable scaling factor along the ^{15}N chemical shift axis (Figure 12.4), probably due to differential inhomogeneity within the B_1 field distribution in the sample.

Another important source of systematic errors can arise from motions that are not always easy to be taken into consideration. This can be undulations of the membranes or motions of the bicelles as a whole. In general, the order parameter of bicelles is taken into consideration during analysis of solid-state NMR spectra (compare Figures 12.1C–12.1F), but this calibration depends on how well S_{bilayer} has been determined. Previously the errors were estimated to be in the 0.05 range,[92] but this value also depends on the shimming of the magnet and the measurement/analysis of the corresponding ^{31}P NMR spectra.

Motions of the polypeptides within the membrane can be another source of errors. For samples oriented with the membrane normal parallel to B_0, rotational diffusion around the membrane normal has *a priori* no effect on the analysis. In contrast, rotation around the helix long axis, or wobbling motions, average the anisotropy and, therefore, need to be taken into consideration when analyzing structural constraints. Whereas the ^{15}N chemical shift is a quite robust measure of helical tilt angles even in the presence of such motions, the ^{2}H quadrupolar splitting of methyl-deuterated alanines can be strongly affected[124] and needs to be taken into account for the correct evaluation of tilt and pitch angles.[125,126]

The effect of motions on the topological analysis are illustrated in Figure 12.2 for the transmembrane model sequence hΦ19W, and has been

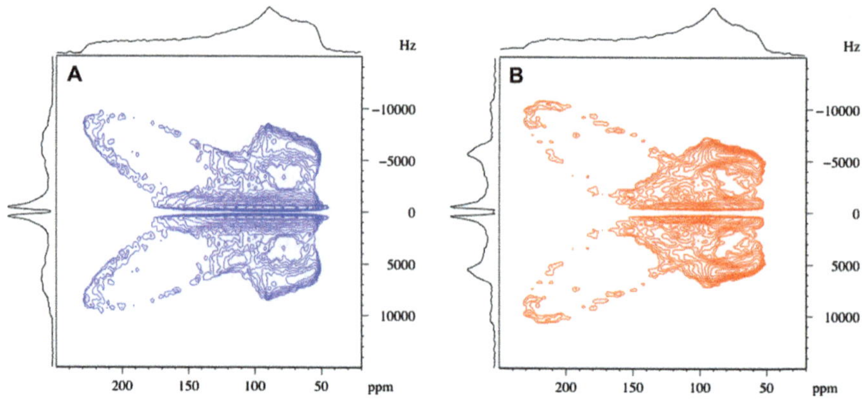

Figure 12.4 PISEMA separated local-field spectra[42] of a lyophilized powder of [^{15}N]acetylleucine[133] recorded at a ^{1}H frequency of 750 MHz using different commercial probe heads. Possibly the difference is in the (A) imbalanced and (B) balanced ^{1}H and ^{15}N fields. The correct butterfly shape where the dipolar projection corresponds to the Pake spectrum is obtained when the hardware and settings work correctly as in (B), whereas the scaling factors in the dipolar dimension vary with the ^{15}N chemical shift in (A). In the projections, only the correct settings provide an undistorted ^{15}N chemical shift powder pattern or the dipolar Pake doublet, respectively. In both acquisitions the effective B_1 fields were 50 kHz during the Lee–Goldberg spin lock and the repetition time was 10 s.

published elsewhere for an helical domain of Huntingtin that is oriented approximately parallel to the membrane surface.[106] With only a limited data set, a motional model has to be assumed that can be related to peptides of similar size and alignment, but, based on our present knowledge, it seems reasonable to assume that such motions considerably average the NMR interactions in fluid lipid membranes. Comparison of Figures 12.2A and 12.2C with 12.2B and 12.2D indicates that the topological analysis results in a different set of tilt/pitch angular pairs when motions are assumed. Whereas the tilt angle remains transmembrane (in the range of 15–25°), quite different pitch angles are obtained, not surprising as extensive rotational averaging around the helix long axis was assumed to occur when calculating the topology plots of Figures 12.2B and 12.2D.

Interestingly, based on four orientational constraints, the amphipathic helix of residues 5–17, which composes the Huntingtin membrane anchor domain, could be aligned quite accurately ($\pm 5°$, due to systematic errors) and, in particular, only one of the 20 lowest-energy configurations obtained from solution NMR in DPC micelles fits well the complete solid-state NMR data set, which can be used to further refine the solution NMR structure.[45,106]

When tilting the samples of peptides and smaller proteins, such that the alignment of the membrane normal is perpendicular to the magnetic field direction, averaging is very often sufficiently fast for a single line (or a single quadrupolar splitting) to be observed, although a circular distribution of alignments relative to B_0 is expected (Figures 12.1C and 12.1D).[100,101,106,127,128] Under such conditions of fast rotational averaging around the membrane normal, the shape of the resulting powder patterns depends much on the molecular alignment and, therefore, it is possible to obtain orientational information also from the non-oriented samples.[82,129–131]

12.7 Conclusions

Angular constraints obtained from oriented solid-state NMR spectroscopy are used to calculate the secondary structure, topology and orientational distribution of polypeptides in liquid disordered lipid bilayers. Furthermore, the same samples report on the dynamics of the labeled sites. The technique is thereby unique in providing such information at atomic resolution in an environment that closely models biological membranes and, to date, most of the deposits in the protein database obtained in liquid disordered membranes have used this approach. In order to improve the precision of such structures, the potential experimental and systematic errors need to be understood, carefully evaluated and, thereby, kept to a minimum.

References

1. C. Fernandez, C. Hilty, G. Wider, P. Guntert and K. Wuthrich, *J. Mol. Biol.*, 2004, **336**, 1211.
2. B. Liang and L. K. Tamm, *Proc. Natl. Acad. Sci. U. S. A.*, 2007, **104**, 16140.
3. S. Hiller, R. G. Garces, T. J. Malia, V. Y. Orekhov, M. Colombini and G. Wagner, *Science*, 2008, **321**, 1206.
4. M. A. Schuler, I. G. Denisov and S. G. Sligar, in *Lipid–Protein Interactions: Methods and Protocols*, ed. J. H. Kleinschmidt, Springer, New York, 2013, vol. 974, p. 415.
5. F. Hagn, M. Etzkorn, T. Raschle and G. Wagner, *J. Am. Chem. Soc.*, 2013, **135**, 1919.
6. C. Tzitzilonis, C. Eichmann, I. Maslennikov, S. Choe and R. Riek, *PLoS One*, 2013, **8**, e54378.
7. T. Chew, S. Abraham, S. Elvington and M. Maduke, *FASEB J.*, 2012, **26**, 604.1.
8. M. Wadsater, S. Maric, J. B. Simonsen, K. Mortensen and M. Cardenas, *Soft Matter*, 2013, **9**, 2329.
9. C. Roos, M. Zocher, D. Muller, D. Munch, T. Schneider, H. G. Sahl, F. Scholz, J. Wachtveitl, Y. Ma, D. Proverbio, E. Henrich, V. Dotsch and F. Bernhard, *Biochim. Biophys. Acta, Biomembr.*, 2012, **1818**, 3098.
10. J. M. Gluck, M. Wittlich, S. Feuerstein, S. Hoffmann, D. Willbold and B. W. Koenig, *J. Am. Chem. Soc.*, 2009, **131**, 12060.
11. J. Borch and T. Hamann, *Biol. Chem.*, 2009, **390**, 805.

12. A. Z. Kijac, Y. Li, S. G. Sligar and C. M. Rienstra, *Biochemistry*, 2007, **46**, 13696.
13. R. R. Vold, R. S. Prosser and A. J. Deese, *J. Biomol. NMR*, 1997, **9**, 329.
14. U. H. N. Durr, M. Gildenberg and A. Ramamoorthy, *Chem. Rev.*, 2012, **112**, 6054.
15. K. Bertelsen, B. Vad, E. H. Nielsen, S. K. Hansen, T. Skrydstrup, D. E. Otzen, T. Vosegaard and N. C. Nielsen, *J. Phys. Chem. B*, 2011, **115**, 1767.
16. C. Loudet, A. Diller, A. Grelard, R. Oda and E. J. Dufourc, *Prog. Lipid Res.*, 2010, **49**, 289.
17. T. H. Walther, S. L. Grage, N. Roth and A. S. Ulrich, *J. Am. Chem. Soc.*, 2010, **132**, 15945.
18. K. R. Mote, T. Gopinath, N. J. Traaseth, J. Kitchen, P. L. Gor'kov, W. W. Brey and G. Veglia, *J. Biomol. NMR*, 2011, **51**, 339.
19. K. Yamamoto, S. Vivekanandan and A. Ramamoorthy, *J. Phys. Chem. B*, 2011, **115**, 12448.
20. T. Jacso, W. T. Franks, H. Rose, U. Fink, J. Broecker, S. Keller, H. Oschkinat and B. Reif, *Angew. Chem. Int. Ed.*, 2012, **51**, 432.
21. D. E. Warschawski, A. A. Arnold, M. Beaugrand, A. Gravel, E. Chartrand and I. Marcotte, *Biochim. Biophys. Acta*, 2011, **1808**, 1957.
22. B. Bechinger, J. M. Resende and C. Aisenbrey, *Biophys. Chem.*, 2011, **153**, 115.
23. A. Grelard, C. Loudet, A. Diller and E. J. Dufourc, *Methods Mol. Biol.*, 2010, **654**, 341.
24. B. Bechinger and E. S. Salnikov, *Chem. Phys. Lipids*, 2012, **165**, 282.
25. J. R. Brender, S. Salamekh and A. Ramamoorthy, *Accounts Chem. Res.*, 2012, **45**, 454.
26. O. Soubias and K. Gawrisch, *Biochim. Biophys. Acta, Biomembr.*, 2012, **1818**, 234.
27. W. T. Franks, A. H. Linden, B. Kunert, B. J. van Rossum and H. Oschkinat, *Eur. J. Cell. Biol.*, 2012, **91**, 340.
28. X. Ding, X. Zhao and A. Watts, *Biochem. J.*, 2013, **450**, 443.
29. M. Renault, A. Cukkemane and M. Baldus, *Angew. Chem. Int. Ed.*, 2010, **49**, 8346.
30. M. Hong and Y. Su, *Protein Sci.*, 2011, **20**, 641.
31. T. V. Can, M. Sharma, I. Hung, P. L. Gor'kov, W. W. Brey and T. A. Cross, *J. Am. Chem. Soc.*, 2012, **134**, 9022.
32. R. Verardi, L. Shi, N. J. Traaseth, N. Walsh and G. Veglia, *Proc. Natl. Acad. Sci. U. S. A.*, 2011, **108**, 9101.
33. S. Bhattacharjya and A. Ramamoorthy, *FEBS J.*, 2009, **276**, 6465.
34. M. Cotten, W. E. Wieczorek, M. Sharma, M. Truong, B. S. Vollmar, E. D. Gordon, R. M. Venable, R. W. Pastor and R. Q. Fu, *Biophys. J.*, 2011, **100**, 496.
35. G. Cornilescu, F. Delaglio and A. Bax, *J. Biomol. NMR*, 1999, **13**, 289.
36. M. K. Pandey, S. Vivekanandan, S. Ahuja, K. Pichumani, S. C. Im, L. Waskell and A. Ramamoorthy, *J. Phys. Chem. B*, 2012, **116**, 7181.

37. A. Ramamoorthy, D. K. Lee, T. Narasimhaswamy and R. P. R. Nanga, *Biochim. Biophys. Acta*, 2010, **1798**, 223.
38. F. Castellani, B. van Rossum, A. Diehl, M. Schubert, K. Rehbein and H. Oschkinat, *Nature*, 2002, **420**, 98.
39. C. Aisenbrey, M. Cusan, S. Lambotte, P. Jasperse, J. Georgescu, U. Harzer and B. Bechinger, *ChemBioChem*, 2008, **9**, 944.
40. D. Nand, A. Cukkemane, S. Becker and M. Baldus, *J. Biomol. NMR*, 2012, **52**, 91.
41. E. S. Salnikov, H. Friedrich, X. Li, P. Bertani, S. Reissmann, C. Hertweck, J. D. O'Neil, J. Raap and B. Bechinger, *Biophys. J.*, 2009, **96**, 86.
42. A. Ramamoorthy, Y. Wei and D. Lee, *Ann. Rep. NMR Spectrosc.*, 2004, **52**, 1.
43. J. Wang, J. Denny, C. Tian, S. Kim, Y. Mo, F. Kovacs, Z. Song, K. Nishimura, Z. Gan, R. Fu, J. R. Quine and T. A. Cross, *J. Magn. Reson.*, 2000, **144**, 162.
44. M. Gustavsson, N. J. Traaseth, C. B. Karim, E. L. Lockamy, D. D. Thomas and G. Veglia, *J. Mol. Biol.*, 2011, **408**, 755.
45. L. Shi, N. J. Traaseth, R. Verardi, A. Cembran, J. Gao and G. Veglia, *J. Biomol. NMR*, 2009, **44**, 105.
46. N. J. Traaseth, L. Shi, R. Verardi, D. Muller, G. Barany and G. Veglia, *Proc. Natl. Acad. Sci. U. S. A.*, 2009, **106**, 10165.
47. R. Verardi, N. J. Traaseth, L. Shi, F. Porcelli, L. Monfregola, S. De Luca, P Amodeo, G. Veglia and A. Scaloni, *Biochim. Biophys. Acta, Biomembr.*, 2011, **1808**, 34.
48. C. Li, P. Gao, H. Qin, R. Chase, P. L. Gor'kov, W. W. Brey and T. A. Cross, *J. Am. Chem. Soc.*, 2007, **129**, 5304.
49. M. Gustavsson, N. J. Traaseth and G. Veglia, *Biochemistry*, 2011, **50**, 10367.
50. S. Lambotte, P. Jasperse and B. Bechinger, *Biochemistry*, 1998, **37**, 16.
51. C. Aisenbrey, P. Bertani and B. Bechinger, in *Antimicrobial Peptides*, ed. A. Guiliani and A. C. Rinaldi, Humana Press/Springer, New York, 2010, p. 209.
52. C. Aisenbrey, M. Michalek, E. S. Salnikov and B Bechinger, in *Lipid–Protein Interactions: Methods and Protocols*, ed. J. H. Kleinschmidt, Springer, New York, 2013, p. 357.
53. J. Seelig, *Biochim. Biophys. Acta*, 2004, **1666**, 40.
54. B. Bechinger, *Crit. Rev. Plant Sci.*, 2004, **23**, 271.
55. B. Bechinger, *J. Pept. Sci.*, 2011, **17**, 306.
56. A. J. Mason, P. Bertani, G. Moulay, A. Marquette, B. Perrone, A. F. Drake, A. Kichler and B. Bechinger, *Biochemistry*, 2007, **46**, 15175.
57. M. Sugawara, J. M. Resende, C. M. Moraes, M. H. Metz-Boutigue and B. Bechinger, *FASEB J.*, 2010, **24**, 1737.
58. R. Heinzmann, S. L. Grage, C. Schalck, J. Burck, Z. Banoczi, O. Toke and A. S. Ulrich, *Eur. Biophys. J.*, 2011, **40**, 463.
59. E. Salnikov and B. Bechinger, *Biophys. J.*, 2011, **100**, 1473.

60. C. L. North, M. Barranger-Mathys and D. S. Cafiso, *Biophys. J.*, 1995, **69**, 2392.
61. K. Bertelsen, J. M. Pedersen, B. S. Rasmussen, T. Skrydstrup, N. C. Nielsen and T. Vosegaard, *J. Am. Chem. Soc.*, 2007, **129**, 14717.
62. E. S. Salnikov, M. De Zotti, F. Formaggio, X. Li, C. Toniolo, J. D. O'Neil, J. Raap, S. A. Dzuba and B. Bechinger, *J. Phys. Chem. B*, 2009, **113**, 3034.
63. B. Bechinger, D. A. Skladnev, A. Ogrel, X. Li, N. V. Swischewa, T. V. Ovchinnikova, J. D. O'Neil and J. Raap, *Biochemistry*, 2001, **40**, 9428.
64. C. Aisenbrey, U. Harzer, G. Bauer-Manz, G. Bär, I. N. Husmal Chotimah, P. Bertani, C. Sizun, A. Kuhn and B. Bechinger, *FEBS J.*, 2006, **273**, 817.
65. A. Ramamoorthy, S. K. Kandasamy, D. K. Lee, S. Kidambi and R. G. Larson, *Biochemistry*, 2007, **46**, 965.
66. P. Bertani, V. Vidovic, T. C. Yang, J. Rendell, L. M. Gordon, A. J. Waring, B. Bechinger and V. Booth, *Biochim. Biophys. Acta, Biomembr.*, 2012, **1818**, 1165.
67. M. Michalek, E. S. Salnikov, S. Werten and B. Bechinger, *Biochemistry*, 2013, **52**, 847.
68. E. Salnikov, C. Aisenbrey, S. V. Balandin, M. N. Zhmak, A. Y. Ovchinnikova and B. Bechinger, *Biochemistry*, 2011, **50**, 3784.
69. M Tang, A. J. Waring, R. I. Lehrer and M. Hong, *Biophys. J.*, 2006, **90**, 3616.
70. N. J. Gleason, V. V. Vostrikov, D. V. Greathouse, C. V. Grant, S. J. Opella and R. E. Koeppe, *Biochemistry*, 2012, **51**, 2044.
71. U. Harzer and B. Bechinger, *Biochemistry*, 2000, **39**, 13106.
72. B. Bechinger, *J. Mol. Biol.*, 1996, **263**, 768.
73. B. Bechinger, *Curr. Opin. Chem. Biol.*, 2000, **4**, 639.
74. C. Aisenbrey, R. Kinder, E. Goormaghtigh, J. M. Ruysschaert and B. Bechinger, *J. Biol. Chem.*, 2006, **281**, 7708.
75. A. Chenal, L. Prongidi-Fix, A. Perier, G. Vernier, S. Lambotte, G. Fragneto, B. Bechinger, D. Gillet, V. Forge and M. Ferrand, *J. Mol. Biol.*, 2009, **391**, 872.
76. C. Aisenbrey, L. Prongidi-Fix, A. Chenal, D. Gillet and B. Bechinger, *J. Am. Chem. Soc.*, 2009, **131**, 6340.
77. G. Grobner, G. Choi, I. J. Burnett, C. Glaubitz, P. J. E. Verdegem, J. Lugtenburg and A. Watts, *FEBS Lett.*, 1998, **422**, 201.
78. F. Separovic, R. Smith, C. S. Yannoni and B. A. Cornell, *J. Am. Chem. Soc.*, 1990, **112**, 8324.
79. N. Takeda, S. Kuroki, H. Kurosu and I. Ando, *Biopolymers*, 1999, **50**, 61.
80. Y. Wei, D. K. Lee and A. Ramamoorthy, *J. Am. Chem. Soc.*, 2001, **123**, 6118.
81. E Salnikov, P. Bertani, J. Raap and B. Bechinger, *J. Biomol. NMR*, 2009, **45**, 373.
82. B. Bechinger and C. Sizun, *Concepts Magn. Reson.*, 2003, **18A**, 130.
83. L. S. Batchelder, H. Niu and D. A. Torchia, *J. Am. Chem. Soc.*, 1983, **105**, 2228.

84. V. V. Vostrikov, B. A. Hall, D. V. Greathouse, R. E. Koeppe 2nd and M. S. Sansom, *J. Am. Chem. Soc.*, 2010, **132**, 5803.
85. P. C. Van Der Wel, E. Strandberg, J. A. Killian and R. E. Koeppe, *Biophys. J.*, 2002, **83**, 1479.
86. A. Ramamoorthy, D. K. Lee, J. S. Santos and K. A. Henzler-Wildman, *J. Am. Chem. Soc.*, 2008, **130**, 11023.
87. J. H. Davis, *Biochim. Biophys. Acta*, 1983, **737**, 117.
88. J. Seelig, *Q. Rev. Biophys.*, 1977, **10**, 353.
89. M. P. Marsan, I. Muller, C. Ramos, F. Rodriguez, E. J. Dufourc, J. Czaplicki and A. Milon, *Biophys. J.*, 1999, **76**, 351.
90. J. Seelig and A. Seelig, *Q. Rev. Biophys.*, 1980, **13**, 19.
91. L. S. Vermeer, B. L. de Groot, V. Reat, A. Milon and J. Czaplicki, *Eur. Biophys. J.*, 2007, **36**, 919.
92. C. R. Sanders and J. P. Schwonek, *Biochemistry*, 1992, **31**, 8898.
93. F. Aussenac, M. Laguerre, J. M. Schmitter and E. J. Dufourc, *Langmuir*, 2003, **19**, 10468.
94. T. B. Cardon, P. C. Dave and G. A. Lorigan, *Langmuir*, 2005, **21**, 4291.
95. K. Yamamoto, R. Soong and A. Ramamoorthy, *Langmuir*, 2009, **25**, 7010.
96. B. Bechinger, in *Encyclopedia of Biophysics*, ed. G. Roberts, Springer, New York, 2013, vol. 1.
97. E Goormaghtigh, V. Raussens and J. M. Ruysschaert, *Biochim. Biophys. Acta, Rev. Biomembr.*, 1999, **1422**, 105.
98. K. Bertelsen, B. Paaske, L. Thogersen, E. Tajkhorshid, B. Schiott, T. Skrydstrup, N. C. Nielsen and T. Vosegaard, *J. Am. Chem. Soc.*, 2009, **131**, 18335.
99. C. Aisenbrey and B. Bechinger, *Biochemistry*, 2004, **43**, 10502.
100. C. Aisenbrey, U. S. Sudheendra, H. Ridley, P. Bertani, A. Marquette, S. Nedelkina, J. H. Lakey and B. Bechinger, *Eur. Biophys. J.*, 2007, **36**, 451.
101. J. M. Resende, C. Mendonca Moraes, V. H. D. O. Munhoz, C. Aisenbrey, R. M. Verly, P. Bertani, A. Cesar, D. Pilo-Veloso and B. Bechinger, *Proc. Natl. Acad. Sci. U. S. A.*, 2009, **106**, 16639.
102. V. Wray, R. Kinder, T. Federau, P. Henklein, B. Bechinger and U. Schubert, *Biochemistry*, 1999, **38**, 5272.
103. K Mattila, R. Kinder and B. Bechinger, *Biophys. J.*, 1999, 77, 2102.
104. A. Holt, L. Rougier, V. Reat, F. Jolibois, O. Saurel, J. Czaplicki, J. A. Killian and A. Milon, *Biophys. J.*, 2010, **98**, 1864.
105. E. Strandberg, S. Esteban-Martin, J. Salgado and A. S. Ulrich, *Biophys. J.*, 2009, **96**, 3223.
106. M. Michalek, E. Salnikov and B. Bechinger, 2013, *Biophys. J.*, **105**, 699.
107. P. Ram and J. H. Prestegard, *Biochim. Biophys. Acta*, 1988, **940**, 289.
108. C. R. Sanders II, B. J. Hare, K. P. Howard and J. H. Prestegard, *Prog. Nucl. Magn. Reson. Spectrosc.*, 1994, **26**, 421.
109. J. Struppe, J. A Whiles and R. R. Vold, *Biophys. J.*, 2000, **78**, 281.
110. R. R. Vold and R. S. Prosser, *J. Magn. Reson., Ser. B*, 1996, **113**, 267.

111. M. P. Nieh, V. A. Raghunathan, C. J. Glinka, T. A. Harroun, G. Pabst and J. Katsaras, *Langmuir*, 2004, **20**, 7893.
112. B. Bechinger, *Biochim. Biophys. Acta*, 2005, **1712**, 101.
113. T. K. Ritchie, Y. V. Grinkova, T. H. Bayburt, I. G. Denisov, J. K. Zolnerciks, W. M. Atkins and S. G. Sligar, *Methods Enzymol.*, 2009, **464**, 211.
114. L. MacEachern, A. Sylvester, A. Flynn, A. Rahmani and M. R. Morrow, *Langmuir*, 2013, **29**, 3688.
115. R. S. Prosser, S. A. Hunt, J. A. DiNatale and R. R. Vold, *J. Am. Chem. Soc.*, 1996, **118**, 269.
116. C. R. Sanders II and J. H. Prestegard, *Biophys. J.*, 1990, **58**, 447.
117. W. H. Goldmann, B. Bechinger and T. Lele, in *Planar Lipid Bilayers and Their Applications: Recent Progress*, ed. H. T. Tien and A. Ottova-Leitmannova, Elsevier, Amsterdam, 2008, vol. 6, p. 227.
118. N. D. Lazo, W. Hu and T. A. Cross, *J. Magn. Reson.*, 1995, **107**, 43.
119. H. Saito, I. Ando and A. Ramamoorthy, *Prog. Nucl. Magn. Reson. Spectrosc.*, 2010, **57**, 181.
120. T. G. Oas, C. J. Hartzell, F. W. Dahlquist and G. P. Drobny, *J. Am. Chem. Soc.*, 1987, **109**, 5962.
121. A. Shoji, T. Ozaki, T. Fujito, K. Deguchi and I. Ando, *Macromolecules*, 1987, **20**, 2441.
122. A. Poon, J. Birn and A. Ramamoorthy, *J. Phys. Chem. B*, 2004, **108**, 16577.
123. A. Shoji, T. Ozaki, T. Fujito, K. Deguchi, S. Ando and I. Ando, *J. Am. Chem. Soc.*, 1990, **112**, 4693.
124. E. Salnikov, C. Aisenbrey, V. Vidovic and B. Bechinger, *Biochim. Biophys. Acta*, 2010, **1798**, 258.
125. S. Ozdirekcan, C. Etchebest, J. A. Killian and P. F. Fuchs, *J. Am. Chem. Soc.*, 2007, **129**, 15174.
126. S. Esteban-Martin and J. Salgado, *Biophys. J.*, 2007, **93**, 4278.
127. C. Aisenbrey and B. Bechinger, *J. Am. Chem. Soc.*, 2004, **126**, 16676.
128. C. Aisenbrey, C. Sizun, J. Koch, M. Herget, U. Abele, B. Bechinger and R. Tampe, *J. Biol. Chem.*, 2006, **281**, 30365.
129. S. D. Cady, C. Goodman, C. D. Tatko, W. F. DeGrado and M. Hong, *J. Am. Chem. Soc.*, 2007, **129**, 5719.
130. L. Prongidi-Fix, P. Bertani and B. Bechinger, *J. Am. Chem. Soc.*, 2007, **129**, 8430.
131. J. Hirschinger, J. Raya, B. Perrone and B. Bechinger, *Chem. Phys. Lett.*, 2011, **508**, 155.
132. A. Holt, L. Rougier, V. Reat, F. Jolibois, O. Saurel, J. Czaplicki, J. A. Killian and A. Milon, *Biophys. J.*, 2010, **98**, 1864.
133. R. M. Herbst and D. Shemin, *Org. Synth., Coll.*, 1943, **Vol. II**, 11.

CHAPTER 13

NMR of Lipids and Lipid/ Peptide Mixtures

JAMES H. DAVIS,* MIRANDA L. SCHMIDT AND
IVANA KOMLJENOVIĆ

Department of Physics, University of Guelph, Guelph, ON, Canada N1G 2W1
*Email: jdavis@uoguelph.ca

13.1 Introduction to Lipids and Membranes

13.1.1 Membrane Lipids

Lipids are the fundamental building blocks of biological membranes. Phospholipids, sphingolipids and glycolipids are common classes of membrane lipids. Variety in headgroup compositions, and the length and level of unsaturation of the acyl chain(s), result in numerous lipid species. Despite the diversity of lipid species, all lipids are amphiphilic molecules with a polar headgroup and non-polar hydrocarbon chain(s). When mixed with water, lipids self-assemble into several types of structures, with the hydrophilic head groups closely interacting with the water and the hydrocarbon chains occupying regions where water is excluded. The characteristics of the structures formed depend on the structure of the lipids themselves.

13.1.2 The Structure of Membranes

Biological membranes are bilayer structures that can contain many different lipid species, often along with sterols and proteins. In 1972, Singer and

New Developments in NMR No. 3
Advances in Biological Solid-State NMR: Proteins and Membrane-Active Peptides
Edited by Frances Separovic and Akira Naito

Published by the Royal Society of Chemistry, www.rsc.org

Nicolson introduced the fluid mosaic model to describe the organization of cellular membranes. In this model the proteins are randomly distributed in a "sea" of lipids.[1] With some refinements as new information is obtained, this model still forms the basis of our understanding of membrane structure. Proteins can be incorporated into the membrane in different ways, either as integral or peripheral proteins. Integral proteins are permanently incorporated into the bilayer and can either span the bilayer (transmembrane proteins), or be partially inserted into the bilayer, *i.e.* interact with a single leaflet. Peripheral proteins are more loosely attached to the membrane.

13.1.3 Thermodynamic Phases Formed in the Presence of Water

Lipids tend to aggregate in water as a result of their amphiphilicity. These aggregates can take on many different forms, depending on the shape of the lipid molecules themselves, the amount of water present and the temperature. The preferred phase of a given lipid can often be inferred from the packing parameter $V/a_o l_c$, where V is the volume of the lipid, a_o is the optimal surface area of the headgroup, and l_c is the critical chain length.[2] Conical lipids, those with relatively large headgroups or a single acyl chain, have a packing parameter value $<1/3$ and form micellar phases. Cylindrical lipids such as phosphatidylcholines have a packing parameter of ~ 1 and form planar bilayers or lamellar phases. Inverted conical lipids, often those with a small headgroup and two kinked acyl chains, have a packing parameter >1 and form inverted micellar phases. Temperature also has an influence on lipid packing, with inverted phases being preferred at high temperatures when the acyl chains are more mobile, resulting in an expanded bilayer (greater area per lipid). Examples of some of the polymorphic phases that lipids can form are shown in Figure 13.1.[3]

Pure phospholipid bilayers will undergo a phase transition at a point known as the chain melting temperature, T_m, which depends on the acyl chain length and degree of unsaturation. Generally, T_m increases with chain length and decreases with the number of double bonds. When two or more lipid species are mixed, they can adopt other phases and multiple phase coexistence regions, depending on the properties of the individual lipids. Below T_m the phases are gel phases and above they are fluid liquid crystalline phases. There are several different gel phases which are distinguished by the bilayer shape and tilt of the lipids. These include the 3D lamellar crystal phase (L_c), the lamellar gel phase (L_β), the tilted lamellar gel phase ($L_{\beta'}$), and the rippled gel phase ($P_{\beta'}$).[4–6] Gel phases have limited acyl chain mobility, which leads to broadened NMR spectra. The most common fluid phase is the liquid crystalline (L_α), also referred to as the liquid disordered (l_d) phase. This phase is characterized by a high degree of flexibility in the acyl chains. When certain phospholipids are mixed with cholesterol, a

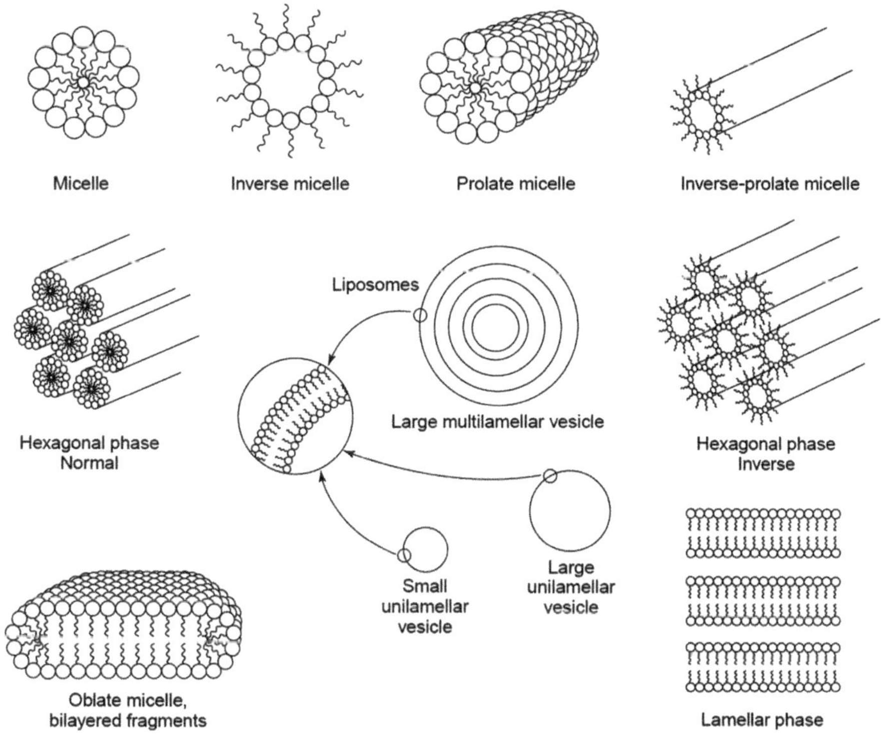

Figure 13.1 Examples of lipid aggregates which can form in water, depending on lipid shape, hydration level and temperature.
(Reproduced from Lasic,[3] with permission from Elsevier.)

second fluid phase is observed at some compositions and temperatures. This phase is called the liquid ordered (l_o) phase and, as the name suggests, it is more ordered than the l_d phase.[7–10]

13.1.4 The Dynamics of Lipids and Peptides in Membranes

The types of motions experienced by the molecular components of membranes and their time scales depend on the membrane composition, the thermodynamic phase or phases in which the membrane finds itself, and on the level of internal flexibility characteristic of the particular lipid or peptide. Lipids with long saturated chains tend to form a gel phase at physiological temperatures. The lipid motions in this phase include slow axial reorientation of the whole molecule about its long axis, slow lateral diffusion within the membrane, highly restricted *gauche/trans* isomerization along the hydrocarbon chains, and axial reorientation of the head group.[11] These same motions are present in the higher temperature fluid phase but are generally much more rapid and the population of *gauche* rotamers on the chains is significantly larger than in the gel phase. Undulations of the membrane bilayer can also be significant. The principal difference between the l_o and l_d phases seems to be the population of *gauche* rotamers.[10] The chains are much more highly ordered in the l_o phase than in the l_d phase. Rates of axial reorientation and lateral diffusion are comparable in the two phases.

For membranes with more than 20 mol% concentration of cholesterol there may be fluctuations in composition associated with proximity to a critical point, which can profoundly affect the NMR linewidths. This is especially true for ^{2}H NMR, since the ^{2}H quadrupolar splittings depend strongly on cholesterol concentration.[12,13]

The motions of peptides within the bilayer will depend on the nature of the interaction between the peptide and the bilayer itself. Transmembrane peptides typically show some degree of tilt with respect to the local bilayer normal and one expects this tilt angle to fluctuate within the orienting potential of the bilayer.[14] Unless one or both ends of a transmembrane peptide are "anchored" at the surface, one can expect the peptide to undergo axial reorientation about its long axis (or about its moment of inertia principal axis). The degree of internal flexibility of a peptide will depend on the strength of inter-residue hydrogen bonding and on steric hindrance. For example, peptides with a glycine residue may exhibit "hinging" at that point. Peptide side chains may also exhibit a considerable degree of flexibility, depending on the nature of the side chain. The type and degree of motion exhibited by the whole molecule or by some part of the molecule will be reflected in the splittings or anisotropic chemical shifts observed for the affected nuclei and measurements of these residual couplings provide information on molecular order and dynamics.

13.2 Introduction to NMR

13.2.1 The Visible Nuclei

Organic molecules are mainly comprised of a few elements: carbon, hydrogen, oxygen, nitrogen, and phosphorus. NMR-active nuclei have non-zero spin angular momentum. Because of their prevalence in biological molecules, the most commonly used nuclei for biological applications are ^{1}H, ^{2}H, ^{13}C, ^{14}N, ^{15}N, and ^{31}P; although it is possible to perform NMR on ^{17}O, it is very expensive. ^{1}H, ^{13}C, ^{15}N and ^{31}P are spin-$\frac{1}{2}$ nuclei, while ^{2}H and ^{14}N are quadrupolar nuclei both with spin-1. In addition, fluorine, ^{19}F, is a spin-$\frac{1}{2}$ nucleus which can be strategically substituted for hydrogen to study peptides.[15,16] There are several different interactions which can be investigated using NMR. These include the chemical shift, dipolar coupling, and quadrupolar coupling (for nuclei with spin $>\frac{1}{2}$). The relative magnitude of and our ability to obtain information about these different types of interactions depend on the type of nucleus and other sample factors. Details about some experiments that can probe the different interactions and the information that can be obtained regarding lipid and lipid/peptide systems are presented later in this chapter.

13.2.2 Oriented *vs.* Powder Samples in Solid-State NMR

There are two types of membrane samples used for solid-state NMR: powder and oriented. When discussing powder samples of model membrane systems, it must be noted that these samples are not dry powders, but rather hydrated mixtures in which all orientations of the lipids (and peptides) are equally likely. On the other hand, in most oriented samples (either mechanically or magnetically aligned) the majority of the sample has the bilayer normal oriented in a single direction. Bicelles and samples which coat the inside of nanopores (which are generally aligned with the pore axis parallel to the magnetic field) have the bilayer normal perpendicular to the field, though not aligned in a single direction. Bicelle samples can also be made to align with their normals parallel to the field. The lineshapes of NMR spectra from powder and oriented samples are profoundly different.

13.2.2.1 Lineshapes

NMR spectral shapes are governed by different nuclear interactions including dipolar couplings, chemical shielding and, in the case of nuclei with spin $>\frac{1}{2}$, quadrupolar interactions. Fully hydrated lipid and/or lipid/peptide powder samples result in NMR spectra that have contributions from all possible orientations, which generally leads to broad "powder pattern" spectra. Well oriented samples on the other hand, have a single direction for the bilayer normal of all the lipids. If the bilayer normal is an axis of symmetry for the motion, then this results in simplified NMR spectra which are due to a single orientation for a particular nucleus. Often, samples are not perfectly aligned but if most of the sample has the same director, it is possible to get useful oriented sample spectra.

The chemical shift is an anisotropic interaction which can lead to two different types of powder lineshapes, depending on whether or not the chemical shielding is axially symmetric. Figures 13.2(a) and 13.2(b) show the carbonyl region of ^{13}C spectra of an α-helical peptide with sequence $K_2GL_NK_2A$-amide (or peptide-*N*, with $N=16$), where the carbonyl carbon sites on all the leucine residues were ^{13}C-labelled. In (a), the chemical shift powder pattern is shown for a sample of dry peptide, while in (b), the peptide is incorporated into a 1-palmitoyl,2-oleoyl-*sn*-glycero-3-phosphocholine (POPC) lipid bilayer and undergoes rapid reorientation about its long axis, reducing the linewidth and resulting in an axially symmetric powder pattern. Phosphorus spectra are also dominated by chemical shift interactions. Figure 13.2(c) shows ^{31}P NMR spectra for a 1,2-dioleoyl-PC (DOPC)/1,2-dipalmitoyl-PC (DPPC)/cholesterol mixture (molar composition 37.5 : 37.5 : 25) aligned on glass plates, with the bilayer normal oriented either parallel (0°) or perpendicular (90°) to the external magnetic field, as indicated in the

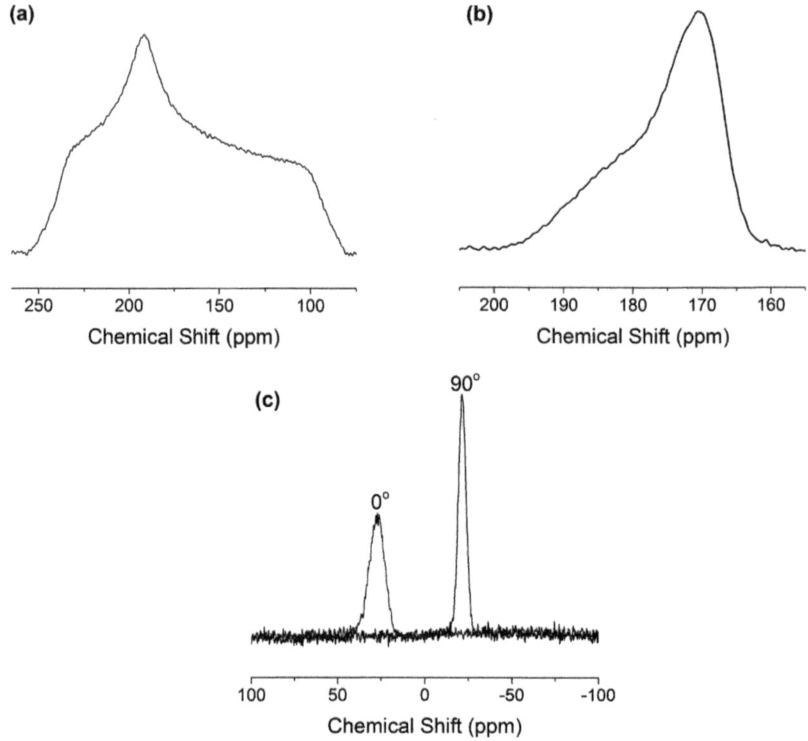

Figure 13.2 (a) An arbitrary chemical shift powder pattern from ^{13}C static NMR on a dry powder sample of peptide of composition $K_2GL_{16}K_2A$-amide; (b) an axially symmetric chemical shift powder pattern from ^{13}C static NMR on the same peptide in POPC multilamellar dispersions; and (c) ^{31}P static NMR spectra of a DOPC/DPPC/cholesterol mixture with molar composition 37.5 : 37.5 : 25 oriented on glass plates; the 0° orientation has the bilayer normal parallel to the external magnetic field, and the 90° orientation has the bilayer normal perpendicular to the external magnetic field. The experiments were performed on a 500 MHz spectrometer.

figure. In this example, the sample is very well aligned: the spectra correspond to a single orientation and there is no evidence of a powder pattern.

Figure 13.3 illustrates the difference between ^2H NMR spectra for powder and aligned samples. In this case, the sample is a mixture of DOPC, DPPC, and cholesterol at molar ratios of 30 : 45 : 25, and it is the DPPC that is deuterium labelled on both acyl chains. The spectra in Figures 13.3(a)–(c) are from powder samples, while those in 13.3(d)–(f) are from samples oriented on glass plates. In (a), (b), (d), and (e) the DPPC is specifically labelled with a single deuterated methylene group on each of the acyl chains at the C-13 position (for a total of four deuterons per lipid). In (c) and (f) the DPPC is chain perdeuterated, so there are 62 deuterons per lipid. The spectra in (b), (c), (e), and (f) are at comparable temperatures, given their degrees of

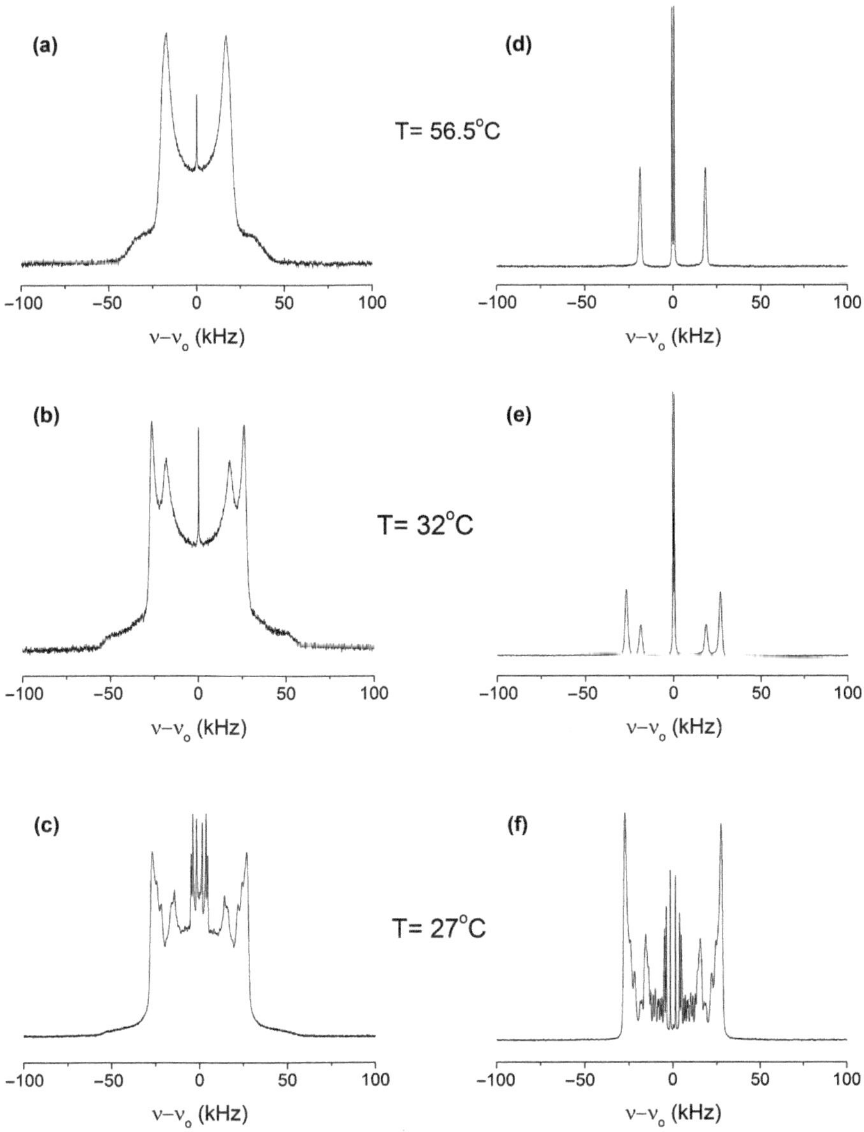

Figure 13.3 ^{2}H static NMR spectra at 76.77 MHz of DOPC/DPPC/cholesterol ternary mixtures with a molar ratio of 30:45:25. (a)–(c) Powder patterns: (a) and (b) have deuterons at the C-13 position methylene of each of the DPPC acyl chains, and (c) has chain-perdeuterated DPPC. (d)–(f) Glass plate oriented samples: (d) and (e) have deuterons at the C-13 position of each of the DPPC acyl chains, and (f) has chain-perdeuterated DPPC. The sharp central peak in (a) and (b), and the sharp central doublet in (d) and (e), arise from the residual deuterium in the buffer.

deuteration; note that the spectra are greatly simplified by orienting the samples, and by reducing the number of deuterons. Figure 13.3(a) depicts the typical l_d phase powder spectrum for lipid samples, and (b) and (c) show two fluid phase coexistence. An example of a gel phase lipid powder pattern can be seen in Figure 13.6(D).

The fluid l_d lipid bilayer phase is characterized by an intermediate degree of molecular orientational order. The flexibility of the lipid hydrocarbon chains results in C–D bond order parameters, $S_{CD} \approx 0.2$, or less, depending on the temperature, the type of lipid, the location of the C–D bond on the chain, and on the amount of cholesterol present in the system. This "order parameter" is defined in terms of the quadrupolar splitting, ν_{CD}, by:

$$\nu_{CD} = \frac{3e^2qQ}{8h}(3\cos^2\theta - 1)\left\langle\frac{1}{2(3\cos^2\beta - 1)}\right\rangle = \frac{3e^2qQ}{8h}(3\cos^2\theta - 1)S_{CD} \quad (13.1)$$

where the angle θ is the angle between the bilayer normal and the magnetic field, and β is the angle between the instantaneous orientation of the C–D bond and the local bilayer normal. The angular brackets denote an average over the rapid reorientation of the C–D bond vector relative to the local bilayer normal.

More rigid molecules, such as cholesterol or membrane spanning peptides or proteins, are likely to possess a higher degree of orientational order than the lipids themselves. The values of any residual couplings (anisotropic chemical shifts, dipolar couplings, or quadrupolar splittings) will depend on the amount of motion present at the site of interest as well as on the "orientation" of the coupling. For example, for the internuclear dipolar coupling in a ^{13}C–^{1}H bond, the orientation of the bond vector relative to the local bilayer normal (if the molecule experiences axially symmetric reorientation about the bilayer normal) also determines the magnitude of the coupling observed. For the chemical shift anisotropy (CSA), the orientation of the chemical shift tensor as well as any reorientation of that tensor caused by molecular motion determines the size of the CSA observed. For quadrupolar splittings it is the orientation of the electric field gradient which determines, together with the motion, what values of splittings will be observed.

The moments of ^{2}H NMR spectra can be measured very accurately. The nth moment of a ^{2}H NMR spectrum is defined as:

$$M_n = \frac{1}{A}\int_0^\infty \omega^n f(\omega)\mathrm{d}\omega \quad (13.2)$$

where $f(\omega)$ describes the lineshape as a function of frequency, ω, and A is the area of the spectrum which is centred about zero frequency, defined as the ^{2}H Larmor precession frequency.[17] The first moment is proportional to the C–D bond order parameter and as a result a large increase in the first moment is observed at the transition from the liquid crystalline to gel phase.

There are two approaches to applying solid-state NMR. The first involves the use of static (non-rotating) samples for the measurement of the motionally averaged residual second-rank tensor couplings such as the dipolar coupling between nuclei, the anisotropic chemical shift, or the quadrupolar interaction. The second is to use magic angle spinning (MAS) to eliminate these second-rank tensor couplings (all of which are zero at the magic angle of 54.7°). Although MAS offers the possibility of very high resolution (approaching that of solution NMR in some cases), the focus of this chapter is on static solid-state NMR and the use of the residual tensor couplings.

13.3 Phase Equilibria of Lipids and Lipid/Peptide Mixtures

The phase behaviour of numerous lipids and lipid mixtures has been investigated using several different physical and chemical techniques, including NMR. Here a selection of lipid phase diagrams determined by NMR and the methods used to obtain them are presented. For a complete survey of the phase transitions of single lipids and phase diagrams of binary and ternary mixtures determined prior to 2013, see Marsh's *Handbook of Lipid Bilayers.*[18]

13.3.1 Lipid/Water Mixtures

Single phospholipid/water mixtures are one of the simplest model membrane systems, since these mixtures spontaneously form aggregates whose properties mimic those of natural membranes. Determining the phases and transitions of these systems is important in order to understand how the lipids behave at different temperatures. Several phospholipid/water phase diagrams have been determined from NMR measurements.

Figure 13.4 shows a phase diagram of DPPC/D_2O obtained from interpretation of the measured quadrupolar splittings and asymmetry parameters from the ^{2}H NMR peaks of deuterated water in the mixture. The quadrupolar splittings of the ^{2}H NMR D_2O spectra are sensitive to the interactions between water and lipids and can be interpreted in terms of how the lipid phases affect the surrounding water. As a result, the D_2O spectra can be used to determine the phase diagram of the lipid aggregates. The amount of water dictates the types of phases and transition temperatures of the mixtures; not all phases are accessible for a given lipid/water ratio.[19] The interpretation of phase behaviour based on ^{2}H NMR of D_2O is quite challenging and most of the recent phase behaviour investigations are performed on systems with deuterated lipid acyl chains, thus studying the lipids directly.

The phase behaviour of 1,2-dilauroyl-PC (DLPC), 1,2-dimyristoyl-PC (DMPC), DPPC, and 1,2-distearoyl-PC (DSPC) bilayers were determined using quadrupolar splittings and the first moments of ^{2}H NMR spectra of chain deuterated lipids.[17,20–23]

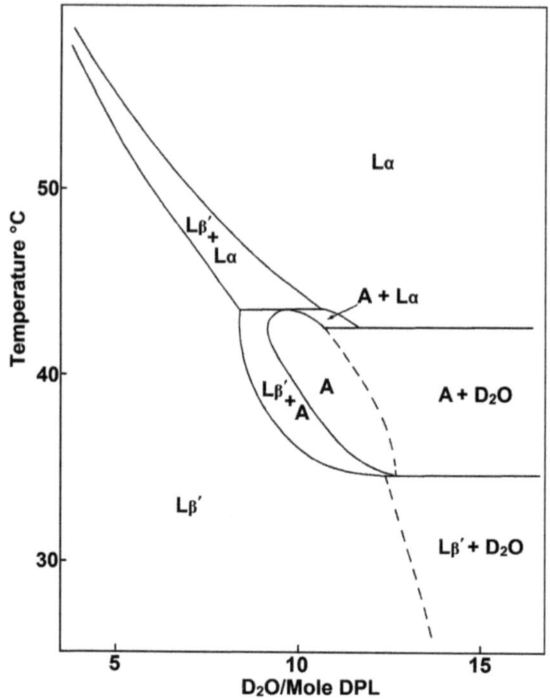

Figure 13.4 DPPC/D_2O phase diagram obtained from ^2H NMR at 15.4 MHz. (Reproduced from Ulmius *et al.*,[19] with permission from the American Chemical Society.)

13.3.2 Binary Lipid Mixtures

Since biological membranes contain many lipid species, it may be desirable to increase the complexity of model systems in order to mimic more accurately the characteristics of real membranes. This is done by mixing different types of lipids. The simplest mixtures contain two lipids. The phase behaviour of several different binary mixtures has been characterized using NMR and other techniques. Generally, model systems should be well hydrated in order for them to behave in a manner reminiscent of biological membranes which are in an environment with excess water. Figure 13.5 shows a phase diagram for a phospholipid with long saturated chains, DSPC, and DMPC, a lipid with medium-length, saturated chains.[24]

The boundaries of the two-phase region were determined by spectral subtraction. For binary lipid mixtures, two different compositions, which are both within a two-phase region at a given temperature, lie on a tie-line.

This means that their spectra can be decomposed into the same two end-point spectra *via* pair-wise subtractions.[25] By performing several sets of pair-wise subtractions, and using the lever rule, a phase diagram can be built up with estimates of the uncertainty of the compositions at the edges of the

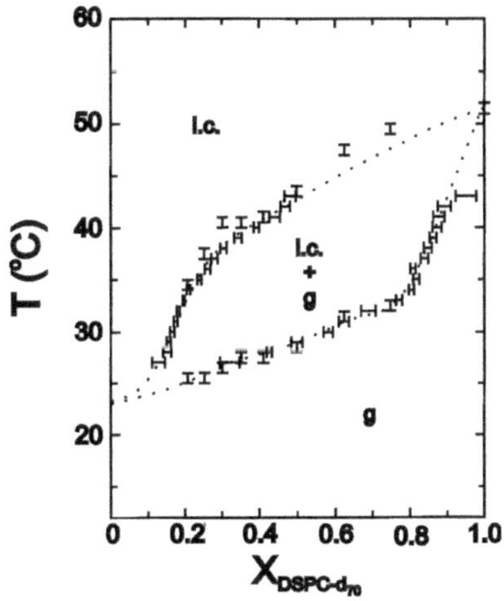

Figure 13.5 Phase diagram of a DSPC/DMPC binary mixture. *Vertical error bars* from phase transition observed directly from spectra as temperature was scanned; *horizontal error bars* from spectral subtractions. (Reproduced from Morrow *et al.*,[24] with permission from Elsevier.)

two-phase region(s). Two ^{2}H spectra from samples of DMPC/1,2-di-d_{35}-stearoyl-PC (DSPC-d_{70}) containing both fluid and gel phases, used for the determination of the phase diagram, and the corresponding fluid and gel phase end-point spectra obtained by spectral subtraction, are shown in Figure 13.6.[24]

Some other binary lipid phase diagrams that have been determined using NMR include DOPC and DPPC,[26] *N*-(lignoceroyl-d_{47})galactosylceramide (a glycolipid) and 1-stearoyl,2-oleoyl-PC (SOPC),[27] both using ^{2}H NMR and the spectral subtraction technique, and SOPC and 1-palmitoyl,2-oleoyl-*sn*-glycero-3-phosphoethanolamine (POPE) mixtures (using pulsed field gradient proton MAS NMR).[28]

13.3.3 The Effect of Sterols

Sterols are amphiphilic molecules that have a characteristic rigid ring structure, a polar headgroup, and a short hydrocarbon chain. Cholesterol is commonly found in animal cell membranes. Other members of the sterol family include stigmasterol, which is found in plant cells, and ergosterol, which is found in fungi. Sterols lower the melting temperature of lipids by inhibiting the tight packing of the phospholipid acyl chains required for the gel phase; thus they decrease the order of the gel phase lipid chains. On the other hand, sterols also increase the order of the liquid crystalline or l_d phase, by stiffening the chains of the phospholipids in this phase. This can induce the formation of the more ordered l_o fluid phase.[7–10,29] An example of a lipid/sterol phase diagram for DPPC and ergosterol, comprising data from ^{2}H NMR and differential scanning calorimetry, is shown in Figure 13.7.[30]

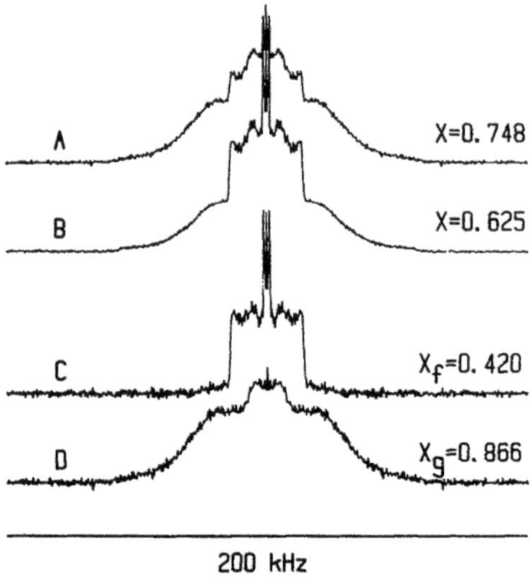

Figure 13.6 (A) and (B) are ^{2}H spectra of DMPC/DSPC-d_{70} at 40°C for two different molar ratios (X is the fraction of DSPC-d_{70}), (C) is the fluid phase end-point spectrum (X_f is the composition at the fluid phase boundary), and (D) is the gel phase end-point spectrum (X_g is the composition at the gel phase boundary). Both (C) and (D) were obtained from spectral subtractions of (A) and (B).
(Reproduced from Morrow *et al.*,[24] with permission from Elsevier.)

The coexistence of the l_o and l_d fluid phases can also be seen in phase diagrams of other binary phospholipid/cholesterol mixtures, *e.g.* DPPC/cholesterol[10] and DSPC/cholesterol.[29]

13.3.4 Ternary Lipid/Sterol Mixtures

The phase behaviour of ternary mixtures can be considerably more complicated than that of binary mixtures. While the spectral subtraction technique discussed above can still be used in regions of two-phase coexistence, finding two or more sample compositions which lie along the same tie-line may be difficult. Figure 13.8 shows ^{2}H spectra for three sample compositions of the ternary mixture DOPC/1,2-di-d_{31}-palmitoyl-PC (DPPC-d_{62})/cholesterol lying along a tie-line at 26°C.[13] In this particular case, spectral subtractions can be used to obtain the end-point compositions and, therefore, two points on the phase boundary separating the two-phase l_o/l_d coexistence region from the l_o phase on the right (high DPPC-d_{62} and high cholesterol concentration) and from l_d phase on the left (high DOPC and low cholesterol concentration). Unlike the situation in binary mixtures, the tie-lines in ternary systems do not need to be parallel at different temperatures.

Similarly, although the tie-lines at fixed temperature within a given two-phase region can never cross one another, they are not necessarily parallel and there is a gradual change in their slope as the composition is changed

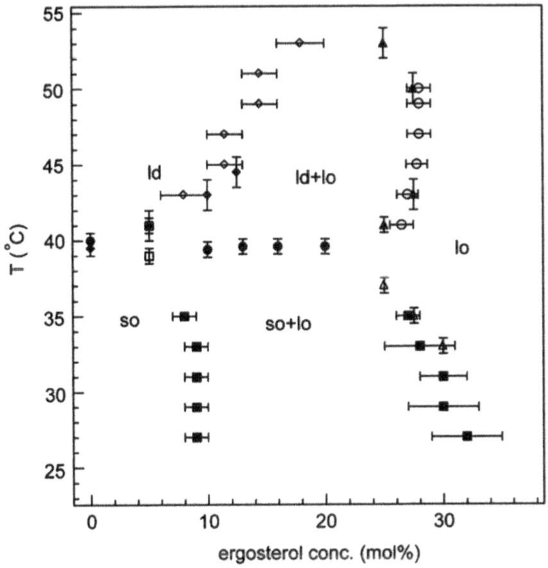

Figure 13.7 Phase diagram of a binary DPPC/ergosterol mixture. Phase transitions observed directly from dePaked spectra are indicated by *open diamonds* and *filled triangles.* Moment analysis was used to determine points marked with *filled* and *open circles*, *open squares*, and *open triangles.* Spectral subtractions were used to obtain points indicated by *filled squares.* DSC data points correspond to the *filled diamonds.* (Reproduced from Hsueh *et al.*,[30] with permission from the Biophysical Society.)

from being near the boundary with the three-phase coexistence region to being near the critical point (shown as a "star" in the isothermal plane in Figure 13.9[31]). Combining the results of spectral subtraction, analysis of the moments of the spectra, and identification of the temperatures where changes in the spectra are observed, one can construct isothermal planes such as that shown in Figure 13.9.

The circular points defining the periphery of the l_o/l_d coexistence region were determined through analysis of the spectra, whereas the triangular points and their error bars were obtained by spectral subtraction. The other solid lines complete the picture by comparing the observations made on the ternary system with those made on the binary systems of DOPC/DPPC-d_{62}[26] and DPPC-d_{62}/cholesterol.[10] Combining isothermal planes such as this, one can obtain the temperature/composition phase diagram.[13]

A number of ternary lipid systems have been studied both by NMR and other physical techniques. These include DOPC/DPPC-d_{62}/cholesterol;[12,13,31–34] DOPC/eSM/cholesterol (where eSM is egg sphingomyelin);[35] POPC/SM/cholesterol;[36–38] POPC/DSPC/cholesterol and SOPC/DSPC/cholesterol;[39] 1-d_{31}-palmitoyl,2-elaidoyl-PC (PEPC-d_{31})/SM/cholesterol and 1-d_{31}-palmitoyl,2-dodecahexaenoyl-PC (PDPC-d_{31})/SM/cholesterol.[40]

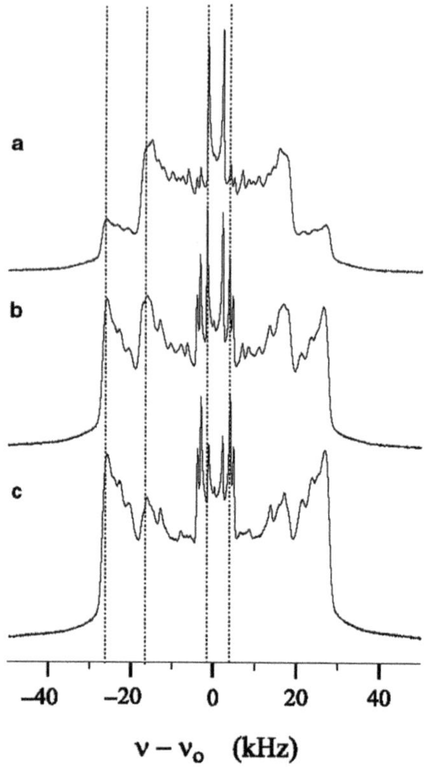

Figure 13.8 ^{2}H NMR (76.77 MHz) spectra of mixtures of DOPC/DPPC-d_{62}/cholesterol with molar proportions (a) 48:32:20, (b) 37.5 : 37.5 : 25, and (c) 28 : 42 : 30, all at 26°C. The *vertical lines* show that the splittings of major features occur at the same frequencies in all of these three spectra. These three samples all line on the same tie-line at this temperature.
(Reproduced from Davis *et al.*,[13] with permission from the Biophysical Society.)

13.3.5 Lipid/Peptide or Lipid/Protein Mixtures

The interactions between proteins and lipids affect both the structure and function of proteins and the membrane itself. The modification of membrane phase equilibria is a potentially important effect.[25] Integral membrane proteins typically have hydrophobic segments with lengths that approximately match the hydrophobic thickness of the lipid bilayer. Mismatch between the hydrophobic region of the protein and the lipid hydrophobic thickness could lead to exposure of the bilayer interior or the hydrophobic core of the protein to the aqueous medium. In order to minimize the free energy of the system, the protein and lipid bilayer readjust to a more favourable hydrophobic match. The changes in membrane phase equilibria reflect the changes in the total free energy of the system. The measurement

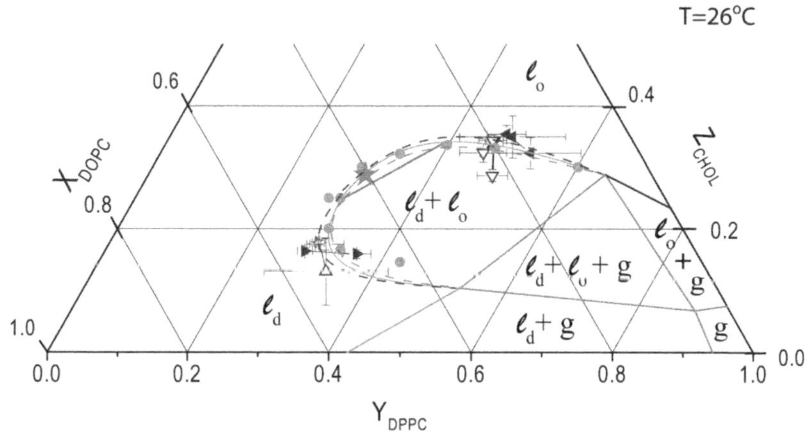

Figure 13.9 Isothermal plane of the ternary phase diagram of DOPC/DPPC-d_{62}/cholesterol at 26°C. Several tie-lines within the two-phase l_o/l_d coexistence region are shown.
(Adapted from Juhasz *et al.*,[31] with permission from Elsevier.)

of the phase boundaries provides information about the molecular interactions of the protein and lipids.[41,42]

Huschilt *et al.*[25] studied the phase boundaries of the fluid and gel phases of synthetic lipid/peptide mixtures. Figure 13.10 illustrates the phase boundaries of peptide-16 and peptide-24 mixed with DPPC-d_{62}, at various lipid/peptide molar concentrations. The phase diagrams of both lipid/peptide systems were similar in that both peptides were found to induce the formation of a two-phase region at temperatures below the melting temperature of DPPC. The two-phase curve separating the fluid from the gel/fluid two-phase region for peptide-16 lies slightly below that of peptide-24. The hydrophobic core of peptide-16 more closely matches the hydrophobic thickness of the lipid bilayer. The suppression of the chain melting transition by the addition of the peptides indicates that the more disordered fluid phase is preferred (by the peptides) as that phase is stabilized to lower temperature. The fact that peptide-16 lowers the transition temperature more than does peptide-24 shows that the longer hydrophobic core does not match the fluid phase as well as the shorter core of peptide-16. However, the two-phase curve separating the mixed phase from the gel phase did not differ significantly for the two peptides.

Morein *et al.* studied the thermotropic behaviour of mixtures of DPPC and KALP23 [Ace-GKK(LA)$_8$LKKA-amide], a synthetic peptide with eight repeating leucine-alanine residues in its core. A tentative phase diagram was outlined for the system, using differential scanning calorimetry and ^{31}P NMR.[43] It was observed that the phase behaviour was characterized by an extensive region below T_m, within which the system undergoes a gel–fluid separation. The ^{31}P NMR measurements indicated possible transition points between the gel phases.

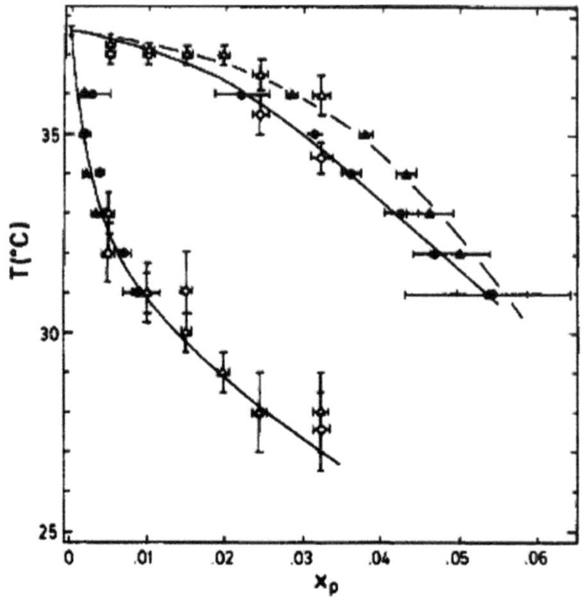

Figure 13.10 Phase diagram of two lipid/peptide systems. DPPC samples were mixed with various concentrations of peptide-24 (*triangles*) or peptide-16 (*circles*). These peptides differ in the length of their leucine-rich core. It is shown that the longer peptide increases the temperature of the two-phase curve separating the fluid and mixed phase regions due to the increased mismatch between the length of the hydrophobic core of the peptide and the lipid.
(Reproduced from Huschilt *et al.*,[25] with permission from the American Chemical Society.)

13.4 NMR of Oriented Samples

Integral membrane proteins play key roles in many biological processes essential for living cells and multicellular organisms. Membrane proteins are often found to be either non-functional or only partially functional following their purification into model membranes such as detergent micelles or lipid/detergent mixed micelles. When there is reduced functionality, this can be a reflection of the destabilization of a protein, its misfolding, or a perturbation of the membrane protein's structure in the model membrane relative to its native conditions.[44] Since it has been found that some model membrane systems such as micelles fail to fully support native membrane protein stability or structure, other model membrane systems such as bicelles and nanodiscs have been developed which more closely mimic the native membrane environment. Bicelle preparations, in particular, result in oriented samples which have the advantage of providing much sharper lines in NMR, as discussed in Section 13.2. Two other important methods for producing oriented lipid samples for NMR experiments are supported bilayers stacked on glass plates or lining the cylindrical nanopores of anodic aluminum oxide (AAO).

13.4.1 Orientation between Glass Plates

The orientation of lipid bilayers between glass plates is one of the most widely used methods for producing oriented lipid samples.[45,46] Briefly, alignment is achieved by depositing the sample in solution onto glass plates and letting it dry, then rehydrating the bilayers in a humid environment. This technique can result in very well oriented lipid and lipid–peptide samples; however, the sample composition and the presence of peptides can affect the ability to produce a perfectly oriented sample. For example, as the degree of unsaturation of the lipid acyl chains increases, the lipids become more fluid at ambient temperatures and this makes it much more difficult to keep these samples oriented as the lipids tend to flow off the plates when they are stacked (see Figure 13.15).[47]

The ^{31}P signal from phospholipid headgroups can be used to determine the degree to which a sample is oriented. Figure 13.2(c) shows the ^{31}P static spectrum of a 37.5 : 37.5 : 25 (molar proportions) mixture of DOPC/DPPC/cholesterol oriented on glass plates in both the 0° (bilayer normal parallel to the external field) and 90° (bilayer normal perpendicular to the external field) orientations. This is an example of an extremely well-aligned sample where there is no evidence of powder or non-oriented lipids. Examples of NMR experiments used to study lipids and peptides which are oriented on glass plates are discussed in further detail in Section 13.5.

13.4.2 Bicelles

Phospholipids typically assemble into bilayers and adopt multilamellar structures when mixed with water (as discussed in Section 13.1). These aggregates are too large for the use of solution-state NMR techniques for the study of membrane-associated proteins. In contrast, bicelle aggregate sizes can be tailored for solution- and solid-state NMR studies.

Bicelles typically comprise one or more long-chain phospholipids in combination with a short chain lipid. The short-chain lipids are found on the highly curved periphery of the bicelle while the long-chain lipids are predominantly found in the planar central region. Typical lipids used to prepare bicelles are 1,2-dicaproyl-PC (DCPC) (often referred to in the literature as DHPC for dihexanoyl-PC) and DMPC, although DMPC can be replaced with a variety of other saturated or unsaturated medium- or long-chain lipids.[48] Because of the variability in preparation, bicelles can mimic many different biomembranes.

The shape of the bicelle is dictated by the long- to short-chain phospholipid ratio, q, as well as the water concentration and the temperature. As a result, bicelles can be found as discs, vesicles, or perforated vesicles (as shown in Figure 13.11).[48]

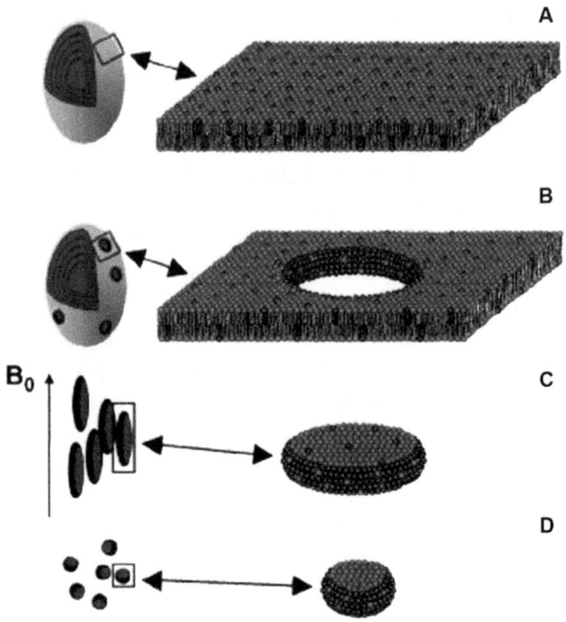

Figure 13.11 Various morphologies of bicelles: (A) large mixed multilamellar vesicles; (B) perforated multilamellar vesicles; (C) large magnetically aligned bicelles; and (D) small isotropic bicelles.
(Reproduced from Warschawski *et al.*,[48] with permission from Elsevier.)

At temperatures above the gel-to-fluid phase transition of DMPC, with lipid ratios $q>2.3$ and lipid concentrations of 3–60% w/v in aqueous solution, bicelles spontaneously align with their bilayer normal perpendicular to the external magnetic field [as shown in Figure 13.11(C)]. These aligned bicelles are frequently used to determine the orientation of transmembrane helices with respect to the bilayer normal, from small antimicrobial peptides to large transmembrane proteins.[49–52]

The perpendicular orientation of bicelles with respect to the magnetic field is not preferred for the structural studies of transmembrane peptides or proteins. The use of parallel oriented bicelles ensures that peptides or proteins incorporated into the bicelles which self-align with respect to the bilayer normal will also be aligned with respect to the magnetic field. This parallel configuration can be obtained by the addition of paramagnetic ions, which have the effect of changing the sign of the magnetic susceptibility anisotropy of the bicelles[44] or through the use of biphenyl lipids.[53–55]

Figure 13.12 illustrates the differences observed in 1D ^{15}N chemical shift spectra when a transmembrane helical peptide is introduced into bicelles. The transmembrane domain of Vpu (a small 81-residue membrane protein whose sequence is encoded in the genome of HIV-1) was added to bicelles comprising DCPC and either DMPC [Figures 13.12(A) and (C)] or DPPC

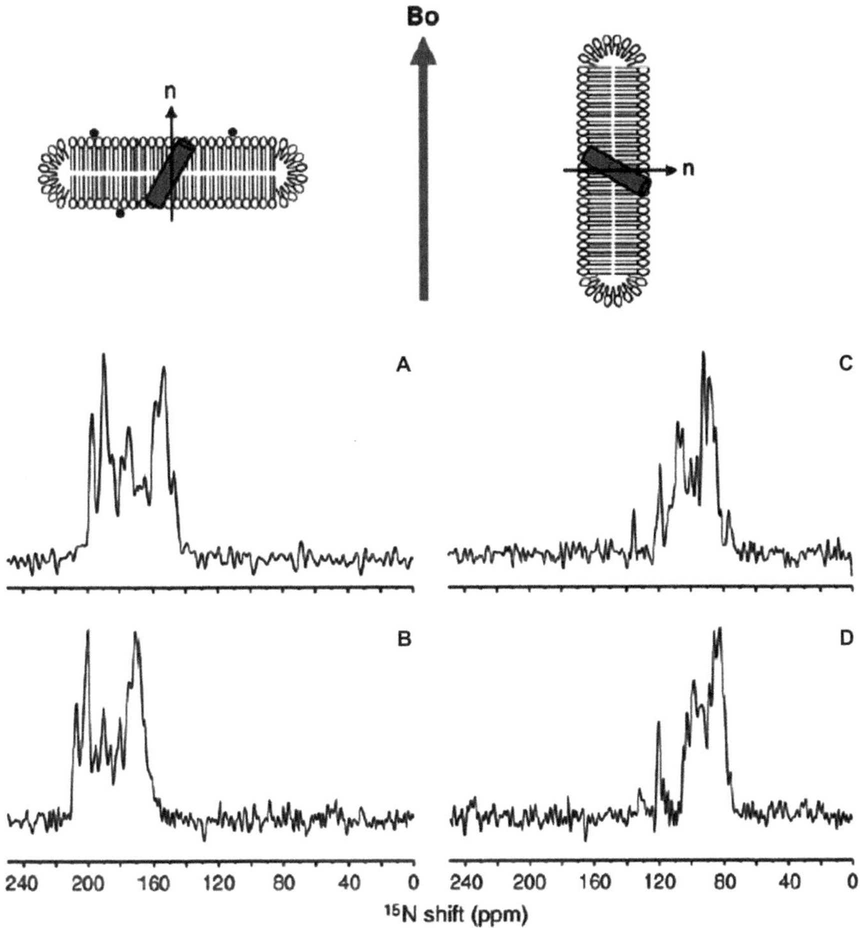

Figure 13.12 One-dimensional ^{15}N chemical shift NMR spectra of uniformly ^{15}N-labelled transmembrane domain of Vpu in aligned bicelles. The bicelles were mixtures of DMPC/DCPC (A and C) or DPPC/DCPC (B and D) and were aligned with the bilayer normal parallel (A and B) or perpendicular (C and D) to the external magnetic field. (Reproduced from Park *et al.*,[56] with permission from the Biophysical Society.)

[Figures 13.12(B) and (D)]. To study the effects of bicelle alignment, bicelles were either allowed to align spontaneously in the magnetic field (C and D) or were doped with Yb^{3+} to change the orientation to parallel (A and B). The only resonances visible are those of the transmembrane region residues. It was determined that the tilt angle for the transmembrane region of Vpu was 30° in the DMPC/DCPC bicelles, and 21° in the DPPC/DCPC bicelles. This shows that the peptide tilt is related to the hydrophobic mismatch between the length of the helix and the hydrophobic region of the phospholipids.[56]

13.4.3 Nanopores

As discussed above, lipid bilayer samples can be aligned either mechanically by assembling bilayers on planar surfaces such as glass plates, or magnetically with the use of bicelles. However, samples on glass plates have drawbacks in terms of limited signal-to-noise since most of the sample volume is occupied by the glass and the limited ability to control sample hydration, pH, ionic strength, and concentration of molecules in the aqueous phase. This leads to difficulty in exposing the sample to aqueous agents which can trigger changes in the membrane protein conformation or function. Preparations of nanopore-supported cylindrical lipid bilayers formed on the inside of AAO substrates have been introduced as a new way to align membrane protein samples for solid-state NMR experiments. The very large surface area accessible in such samples alleviates many of the drawbacks listed above.

In order to support conclusions about peptide and protein structure and function, it may be important that the peptide or protein be anchored by a single bilayer, and surrounded by excess aqueous phase. Gaede *et al.*[57] were able to deposit a single lipid bilayer in AAO nanopores to study the physical properties of the lipids. POPC and DMPC lipids were loaded by an extruder, and aliquots of water, buffer, or aqueous poly(ethylene glycol) (PEG) were flushed through the AAO nanopores to remove any remaining liposomes or additional bilayers that were loosely attached to the first bilayer at the AAO surface.

To determine the orientation and spread of the adsorbed lipids, Gaede *et al.* obtained ^{2}H NMR spectra of 1-d_{31}-palmitoyl,2-oleoyl-PC (POPC-d_{31}) adsorbed on AAO nanopores with the pore axis parallel and perpendicular to the external magnetic field (Figure 13.13). When the pores were oriented parallel to the external magnetic field, the bilayer normal was preferentially oriented perpendicular to the field, and when the pores were rotated to be perpendicular to the field, the bilayer normal exhibited a circular distribution and the spectrum adopted a more complex lineshape. ^{2}H NMR spectra of headgroup methylene labelled POPC-d_4 were acquired to investigate the interaction between the lipid headgroups and the AAO surface (Figure 13.14). In addition, Gaede *et al.* argued that the lipids adsorbed in a single bilayer, that the bilayer seal was extremely durable in the presence of shear forces induced by Pr^{3+}, and that the average thickness of the water layer trapped between the bilayer and AAO support was 3 nm.[57]

Studies done by Chekmenev *et al.* have shown that the AAO-supported bilayers are able to retain many biophysical properties of unsupported lipid vesicles.[58] 2D NMR of membrane proteins aligned with the help of nanoporous AAO substrates was demonstrated in this study. Some of the advantages of nanoporous AAO substrates include a higher thermal conductivity when compared to glass slides (which results in a suppression of thermal gradients across the bilayer sample), as well as accessibility of the bilayer surface to aqueous molecules.

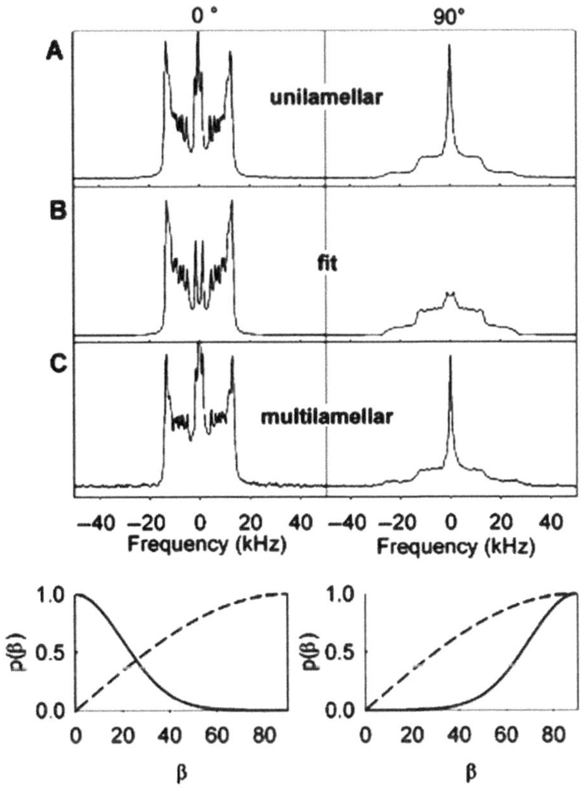

Figure 13.13 ^{2}H NMR spectra (76 MHz, 24.2°C) of POPC-d_{31} adsorbed on AAO, with the pores oriented parallel (0°) and perpendicular (90°) to the external magnetic field. (A) Unilamellar samples prepared by flushing the pores after lipid loading; (B) simulated spectra; and (C) multilamellar samples prepared with no flushing step after lipid loading. (Reproduced from Gaede *et al.*,[57] with permission from the American Chemical Society.)

13.5 Lipid/Peptide Orientational Order and Dynamics

Figure 13.15 shows the static ^{2}H NMR spectra of 5 mol% di-d_{27}-myristoyl-PC (DMPC-d_{54}) incorporated into oriented bilayers of a series of unsaturated phospholipids showing the effect of the increase in the level of unsaturation on lipid hydrocarbon chain order at a temperature of 298.4 K.[47] Not only is the average quadrupolar splitting (and, therefore, the average chain order parameter) reduced, but the variation of splittings with the chain position is dramatically modified. This behaviour indicates that there is significantly less inter-lipid interaction (hindrance to chain motion) in the polyunsaturated bilayer.

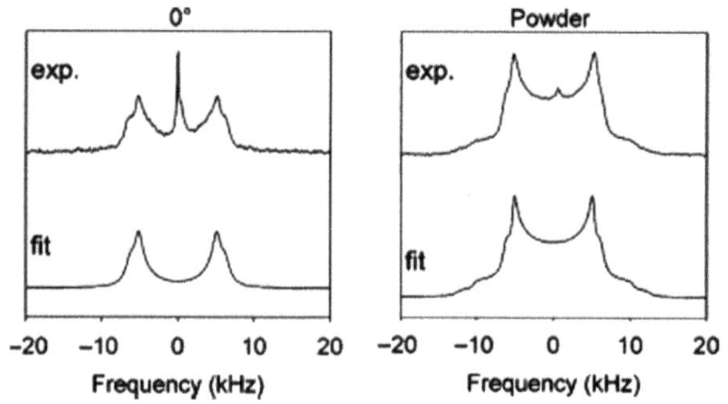

Figure 13.14 ^{2}H spectra (76 MHz, 24.2°C) of POPC-d_4 adsorbed on AAO pores oriented parallel to the external magnetic field (*left*) and POPC-d_4 multilamellar vesicles, together with simulated spectra. Both samples show quadrupolar splitting corresponding to the POPC choline α and β resonances.
(Reproduced from Gaede *et al.*,[57] with permission from the American Chemical Society.)

When a more rigid molecule, 2,2;3;4,4;6-d_6-cholesterol (cholesterol-d_6), is incorporated into bilayers having different levels of unsaturation, there is also a systematic decrease in the overall molecular order parameter of the rigid cholesterol ring structure as the level of unsaturation of the phospholipid is increased. Thus, the level of orientational order of the host lipid is transmitted even to more rigid molecules incorporated into the bilayers.[47] Perhaps, more importantly, the use of polyunsaturated lipids generally leads to more rapid molecular reorientation. This leads to sharper NMR lines and higher resolution. Finally, the more rapid motion generally results in longer relaxation times, allowing the NMR coherences to persist longer and making sophisticated NMR experiments more feasible.

Measures of molecular order can be obtained from unlabelled samples through the use of separated local field (SLF) spectroscopy experiments such as DROSS,[59] APM-CP,[60] and R-PDLF,[61] which use the high resolution in the ^{13}C dimension of natural abundance MAS NMR to resolve many different positions on the molecules without the necessity of isotopic labelling. Figure 13.16 shows an R-PDLF spectrum of fully hydrated POPC/cholesterol at a 40 : 60 molar ratio at 300 K.[61] The labelled peaks in the ^{13}C projection indicate the corresponding carbon positions on the cholesterol molecule. The ^{1}H–^{13}C dipolar coupling visible in the second dimension is used to determine the magnitude of the C–H bond order parameter, S_{CH}, which is completely analogous to the ^{2}H C–D bond order parameter. A scaling factor of 0.315 relates the "effective" dipolar splitting measured in this experiment to S_{CH}:

$$\nabla \nu_{CH} = 0.315 \times d_{CH} = 0.315 \times \frac{h}{2}\left(\frac{\mu_0}{4\pi}\right)\gamma_H \gamma_C \langle r^{-3} \rangle |S_{CH}| \qquad (13.3)$$

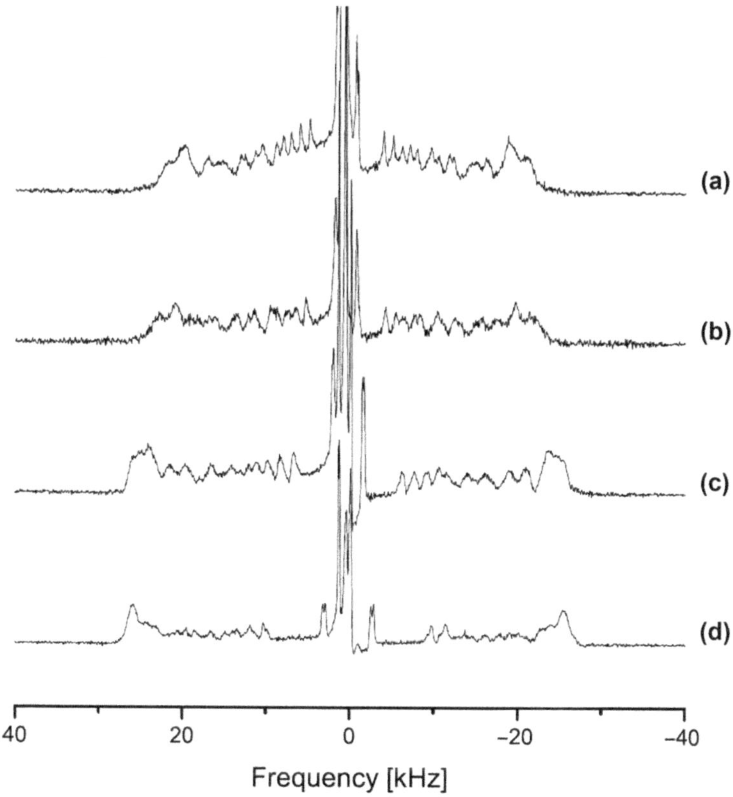

Figure 13.15 The decrease in lipid hydrocarbon chain order as the level of chain unsaturation increases. The spectra are from 95/5 mol% ratios of (a) di-22:6PC/DMPC-d_{54}, (b) di-20:4PC/DMPC-d_{54}, (c) di-18:2PC/DMPC-d_{54}, and (d) di-18:1PC/DMPC-d_{54} mixtures oriented between glass plates. Spectra were taken at a temperature of 298.4 K and at a ^{2}H frequency of 76.77 MHz.
(Reproduced from Braithwaite,[47] with permission from the University of Guelph.).

where h is Planck's constant, μ_0 is the permeability of free space, γ_H and γ_C are the hydrogen and carbon gyromagnetic ratios, and r is the C–H internuclear distance. It is important to note that in the case of "dipolar order parameters" the mean inverse cubed distance between nuclei is also an important factor, *i.e.* when the interaction is between nuclei which are not directly bonded to each other. The orientational average of the internuclear vector relative to the axis of symmetry of the motion or, in the static case, relative to the static magnetic field, is the other factor determining the value of the order parameter. Such spectra can provide a detailed quantitative description of molecular order even for systems where isotopic labelling is impractical, such as for natural membranes.

In the absence of motion, the quadrupolar or dipolar splittings, or the anisotropic chemical shift, can be used to obtain direct information on

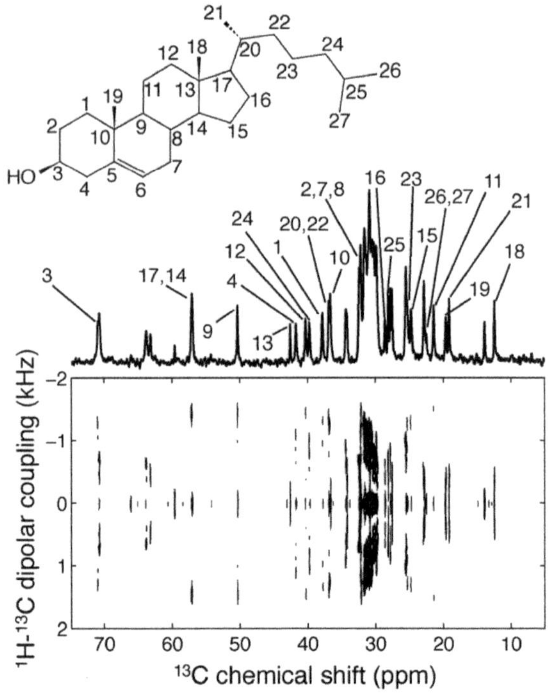

Figure 13.16 ^1H–^{13}C R-PDLF separated local field spectrum of a fully hydrated POPC/cholesterol multilamellar dispersion at 40:60 mole ratio at 300 K at a ^1H frequency of 500 MHz.
(Reproduced with permission from Ferreira *et al.*[61])

molecular structure or orientation. Figure 13.17 shows a trans-bilayer α-helical peptide juxtaposed to bilayer forming lipids in a disordered configuration, on the far left, and in a more highly ordered configuration at the centre of the figure.[62] The peptide on the right is tilted with respect to the bilayer normal (N) as a means of accommodating its hydrophobic length to the bilayer thickness. If the peptide undergoes no molecular reorientation, then the NMR spectra which we would observe for any given tilt angle τ and any polar angle ρ can be easily predicted. For example, if the α-carbons are labelled with ^2H, then from one amino acid to the next there is a change in the orientation of the C_α–^2H bond direction relative to the coordinate system fixed in the molecule and defined by the angles τ and ρ with respect to the bilayer normal fixed reference frame. Hence, the quadrupolar splitting observed for each of these labelled positions will follow a characteristic pattern.

Figure 13.18 illustrates this effect using the ^2H quadrupolar splittings obtained by substituting ^2H-labelled alanine at different positions in MSI-103 (KIAGKIAKIAGKIAKIAGKIA-NH_2), a synthetic antibiotic based on the sequence of PGLa from *Xenopus laevis* (GMASKAGAIAGKIAKVALKAL-NH_2). The methyl ^2H-labelled alanines were substituted at Ala-7, Ile-9, Ala-10, Gly-11, Ile-13, Ala-14, and Ala-17.[63] The quadrupolar splittings were

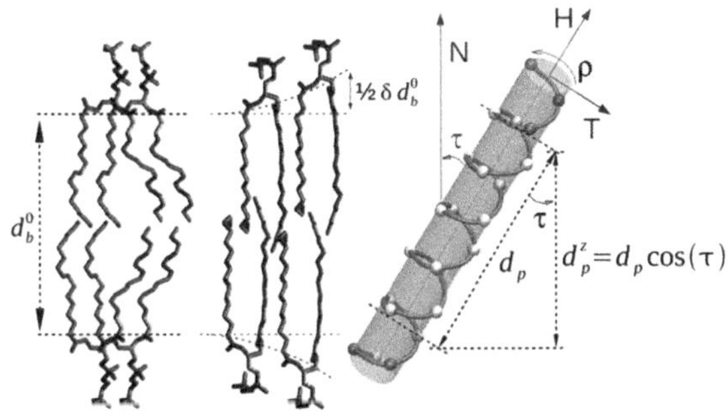

Figure 13.17 The orientation of a transmembrane α-helical peptide showing the tilt angle τ necessary to accommodate the hydrophobic mismatch between the length of the trans-bilayer peptide segment and the hydrophobic thickness of the lipid bilayer. Lipids in a more disordered phase (depicted on the *left*) have a thinner bilayer than those in a more ordered phase (*centre*). The peptide tilt angle adjusts to accommodate the local bilayer thickness. The polar angle ρ further defines the orientation of the peptide fixed coordinate system relative to the bilayer fixed system.
(Reproduced from Strandberg *et al.*,[62] with permission from Elsevier.)

measured for MSI-103 in three different lipid systems at a peptide/lipid mole ratio of 1 : 50. At this concentration the peptide can be either in a surface-bound state "S" or in a trans-bilayer state "T", depending on the lipid environment. Figure 13.18(A) shows the quadrupolar splittings for the labelled peptide in DOPC for the seven different label positions and compares them to a solid curve (splitting *vs.* polar angle) corresponding to angles $\tau = 94°$and $\rho = 133°$. Figure 13.18(B) shows the range of τ and ρ consistent with the uncertainty in the quadrupolar splittings observed. Thus, in DOPC the MSI-103 peptide appears to be on the bilayer surface, *i.e.* in the "S" state. In (C) and (D) are shown the data obtained for the same peptide in DLPC, where there is clear evidence that a much broader range of tilt angles τ and perhaps two values of ρ are consistent with the observed splittings. Here the solid curve is based on an average over the distribution of angles. Finally, Figures 13.18(E) and (F) show the results for the peptide in a 1 : 1 mixture of POPC and 1-lyso,2-oleoyl-PC (lyso-OPC), where again a broad range of tilt angles and what appear to be two polar angles are consistent with the data. In both of the environments the peptide seems to be predominantly in the trans-bilayer state (*i.e.*, the "T" state), with τ in the neighbourhood of 130°. It remains unclear, however, whether the distribution of angles consistent with the observed splittings is due to a dynamic equilibrium among different orientations or to a static heterogeneity in orientation. Nonetheless, these results illustrate how NMR can be used to study the orientation of peptides within a membrane as well as how the properties of the lipid bilayers can affect the orientation and the dynamics of peptides with which they interact. A similar example involves the comparison of the measured anisotropy of the chemical

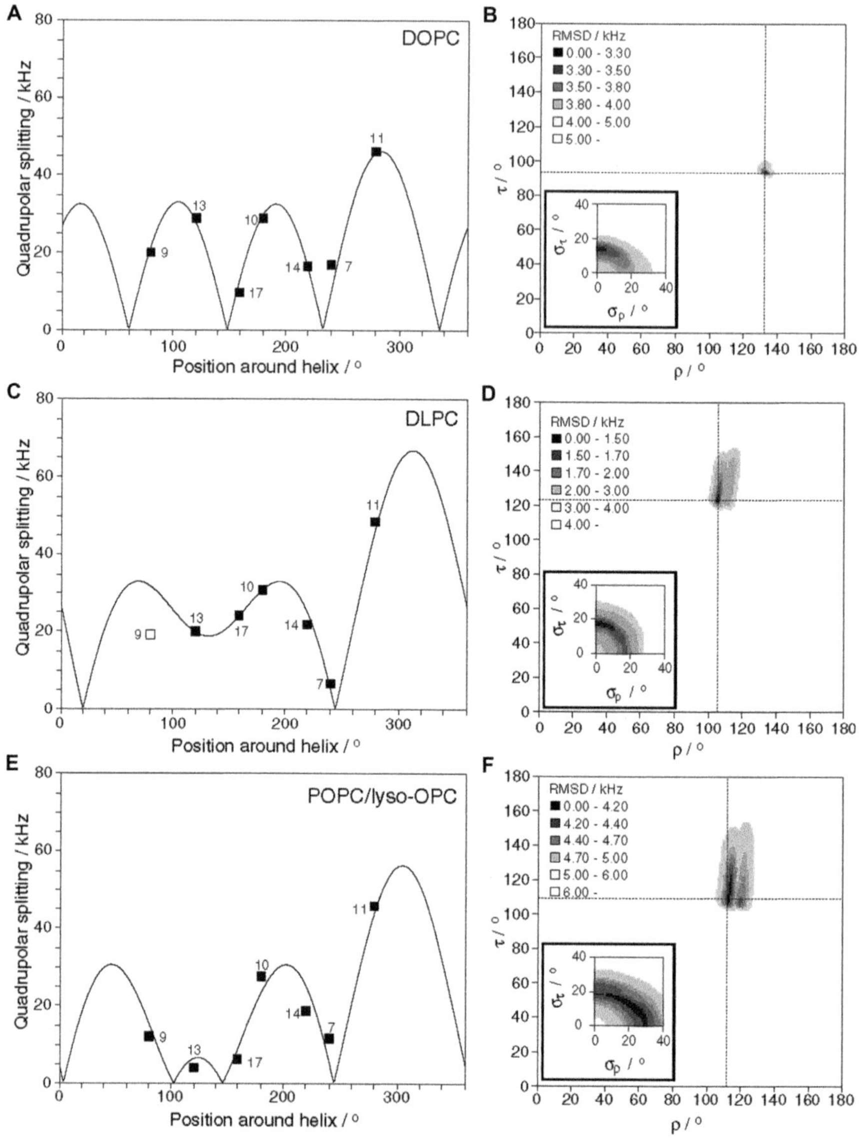

Figure 13.18 ^{2}H quadrupolar splittings of methyl-labelled alanines incorporated into MSI-103 in oriented bilayers of DOPC (A and B), DLPC (C and D), and POPC/lyso-OPC (E and F). The *solid curves* in A, C, and E are predicted dependence on the angles τ and ρ. The values of τ and ρ consistent with the uncertainties in the splittings are shown by the plots in B, D, and F. Data were taken using sample oriented with the bilayer normal parallel to the external magnetic field.
(Reproduced from Strandberg *et al.*,[63] with permission from Elsevier.)

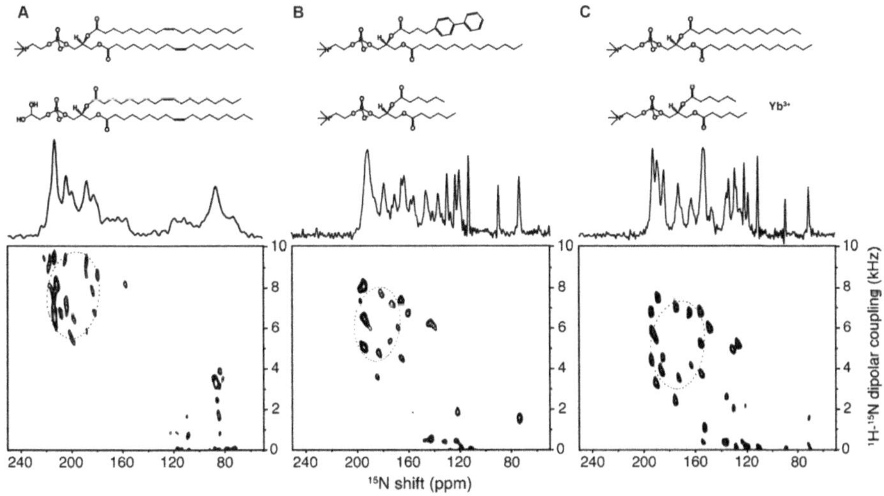

Figure 13.19 $^{15}N/^{15}N-^{1}H$ separated local field spectra of uniformly labelled Pf1 coat protein incorporated into lipid bilayers (A) and lipid bicelles (B) and (C) with the normal to the bilayer parallel to the magnetic field. The lipid bilayer in (A) consists of a 9:1 w/w ratio of DOPC and DOPG at 23°C. The bicelles in (B) are a TBBPC/DCPC mixture at a molar ratio of 8 : 1 at 40°C. The bicelles in (C) are a mixture of DMPC/DCPC at a molar ratio of 3.2 : 1 in water containing 3 mM $YbCl_3$ at 40°C. (Reproduced from Park *et al.*,[55] with permission from Elsevier.)

shift of ^{13}C labelled at different positions along backbone of different peptides, including lactoferrampin,[64] bonbolitin II,[65] and alamethicin.[66]

Separated local field spectroscopy (SLF) of uniformly ^{15}N labelled peptides in aligned lipid bilayers provides a potentially powerful tool for determining peptide orientation within the bilayer and for studying peptide dynamics. Figure 13.19 shows the ^{15}N chemical shift/$^{1}H-^{15}N$ dipolar coupling SLF spectra of ^{15}N-labelled Pf1 coat protein in samples formed with (A) DOPC/1,2-dioleoyl-*sn*-glycero-3-phosphoglycerol (DOPG) on glass plates, (B) 1-tetradecanoyl-2-[4-(4-biphenyl)butanoyl]-*sn*-glycero-3-PC (TBBPC)/DCPC, and (C) DMPC/DCPC bicelles containing 3 mM $YbCl_3$ (which results in alignment of the bicelle normal parallel to the magnetic field).[55] The use of bicelles results in significant resolution enhancement over glass plates. The dotted curves on these spectra shows the predicted locus of peaks for an α-helix with uniform dihedral angles ($\phi = -61°$ and $\psi = -45°$) with tilt angle $\tau$and overall system order parameter δ of (A) $\tau = 20°$, $\delta = 1.0$, (B) $\tau = 21°$, $\delta = 0.83$, and (C) $\tau = 27°$, $\delta = 0.8$. The peaks cluster around the predicted curve, but structural variations and internal motions can result in significant deviations from the precise peak positions expected for a perfect α-helix.

Figure 13.20 shows the changes in such SLF spectra predicted for an ideal α-helix tilted at (A) 0, (B) 10, (C) 20, and (D) 30° with respect to the magnetic field direction. The pattern expected for a 3_{10} helix tilted by 10° is shown in (E), while the predicted spectra of the three conformers of alamethicin based

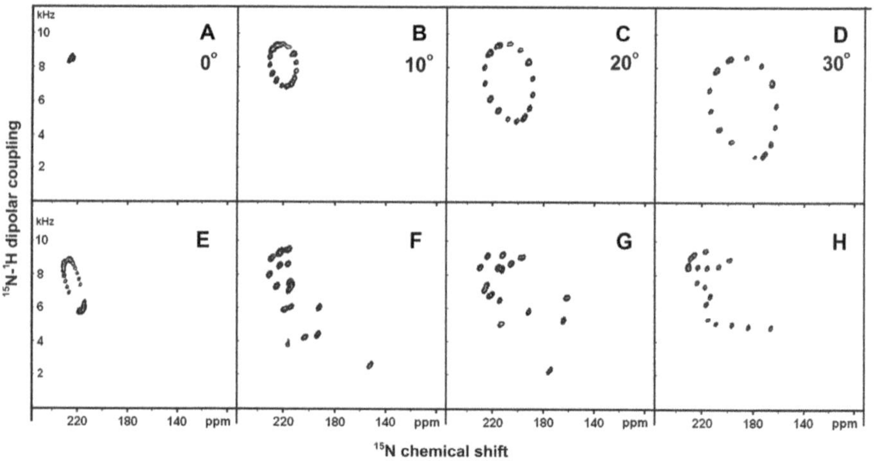

Figure 13.20 The changes observed in separated local field spectra of a uniform α-helix as tilt angle τ is changed: (A) $\tau = 0°$, (B) $\tau = 10°$, (C) $\tau = 20°$, and (D) $\tau = 30°$. Part (E) shows the spectrum of an ideal 3_{10} helix oriented at $\tau = 10°$, while the spectra predicted on the basis of three conformers obtained from the X-ray structure, tilted at $\tau = 10°$, are shown in (F), (G), and (H).
(Reproduced from Bechinger *et al.*,[68] with permission from Elsevier.)

on its crystal structure (PDB ID: 1AMT[67]) are shown for a 10° tilt in (F), (G), and (H). While there is a clustering of peaks in the region predicted, there is little correspondence with the simple ideal helices.[68] This example illustrates the effect of relatively small deviations from ideal structures, but does not illustrate the effect of molecular motion on the spectra. A more detailed discussion of the importance of motion is given by Farès and Davis.[69]

Rapid molecular motion generally reduces the residual couplings observed in NMR. For example, with a dry powder sample of ^{2}H-labelled phospholipid a static Pake doublet spectrum with a quadrupolar splitting of 126 kHz is observed (the difference in frequency between the two sharp 90° edges of the Pake doublet). In a hydrated lipid bilayer, where the only motion would be a rapid axial reorientation about the bilayer normal, the spectrum would be a Pake doublet with a 63 kHz splitting. If the chain methyl groups simultaneously reorient about the last C–C bond, their splitting would be 41 kHz. Each new type of motion reduces the splitting by some factor, depending on the nature of the motion. The same is true for dipolar couplings and chemical shift anisotropies.

If the molecular motions are very slow (*i.e.*, having a characteristic time longer than T_2), then there may be no observable effect on the spectrum, but it is possible to study the motion with relaxation methods. If the motions are on the intermediate time scale (of the same order as T_2), then the lineshape can change dramatically [see, for example, the gel phase spectrum shown in Figure 13.6(D)] and generally the observed resonances broaden. Sometimes

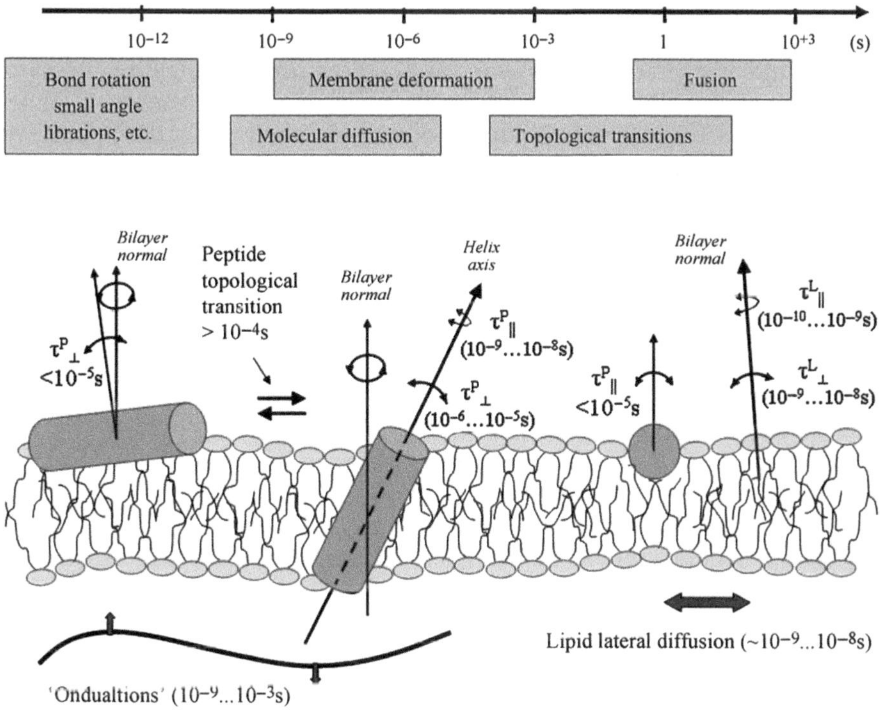

Figure 13.21 The types of molecular motions found in membranes and their approximate timescales. Lipid molecules form the bilayer and helical peptides are associated with the surface or are trans-bilayer. (Reproduced from Bechinger and Salnikov,[70] with permission from Elsevier.)

this broadening can be so severe that it becomes difficult to observe the spectrum at all. Figure 13.21 shows some of the kinds of motions characteristic of molecules in a membrane bilayer and some typical timescales for these motions.[70] These ranges of timescales are only approximate and in different specific cases there may be significant departure from the values shown in the figure.

NMR is well suited to study these molecular motions. Of course, that means that these motions also can profoundly affect the NMR spectra and usually with negative consequences. Since sensitivity and resolution are always key characteristics in NMR, intermediate timescale motions which broaden the lines and reduce their intensities need to be avoided as much as possible. The molecules in fluid-phase lipid bilayers undergo reorientation over a broad range of timescales, but different lipid bilayers can be quite different. The motion of peptides and proteins incorporated into lipid bilayers will reflect at least to some degree the motion of the lipid background. It should be possible to reduce to some extent the linewidths obtained for peptides by choosing appropriate lipids.

Acknowledgements

We would like to thank the Natural Sciences and Engineering Research Council of Canada, the Canadian Foundation for Innovation, and the Ontario Ministry of Research and Innovation for their financial support. We also want to thank the staff of the University of Guelph NMR Centre, Janos Juhasz, and Jesse Clair for their help with some of the results presented here.

References

1. S. J. Singer and G. L. Nicolson, *Science*, 1972, **175**, 720.
2. J. N. Israelachvili, *Intermolecular and Surface Forces*, Academic Press, Waltham, MA, 3rd edn, 2011, pp. 535–576.
3. D. D. Lasic, *Tibtech*, 1998, **16**, 307.
4. V. Luzzati, *Biological Membranes: Physical Fact and Function*, ed. D. Chapman, Academic Press, London, 1968, vol. 1, pp. 71–123.
5. A. Tardieu, V. Luzzati and F. C. Reman, *J. Mol. Biol.*, 1973, **75**, 711.
6. J. M. Seddon and R. H. Templer, *Handbook of Biological Physics*, ed. R. Lipowsky and E. Sackmann, North-Holland, Amsterdam, 1995, vol. 1A, pp. 97–160.
7. M. R. Vist, *Partial Phase Behaviour of Perdeuterated Dipalmitoylphosphatidylcholine-cholesterol Model Membranes*, MSc Thesis, University of Guelph, Canada, 1984.
8. J. H. Davis, *Physics of NMR Spectroscopy in Biology and Medicine*, Proceedings of the International School of Physics "Enrico Fermi", course 100, ed. B. Maraviglia, North-Holland, Amsterdam, 1988, pp. 302–312.
9. J. H. Ipsen, G. Karlstrom, O. G. Mouritsen, H. Wennerstrom and M. Zuckermann, *Biochim. Biophys. Acta*, 1987, **905**, 162.
10. M. R. Vist and J. H. Davis, *Biochemistry*, 1990, **29**, 451.
11. P. Meier, E. Ohmes and G. Kothe, *J. Chem. Phys.*, 1986, **85**, 3598.
12. S. L. Veatch, O. Soubias, S. L. Keller and K. Gawrisch, *Proc. Natl. Acad. Sci. USA*, 2007, **104**, 17650.
13. J. H. Davis, J. J. Clair and J. Juhasz, *Biophys. J.*, 2009, **96**, 521.
14. R. S. Prosser, J. H. Davis, C. Mayer, K. Weisz and G. Kothe, *Biochemistry*, 1992, **31**, 9355.
15. R. W. Glaser, C. Sachse, U. H. N. Dürr, P. Wadhwani, S. Afonin, E. Strandberg and A. S. Ulrich, *Biophys. J.*, 2005, **88**, 3329.
16. V. S. Kubyshkin, I. V. Komarov, S. Afonin, P. K. Mykhailiuk, S. L. Grage and A. S. Ulrich, *Fluorine in Pharmaceutical and Medicinal Chemistry: From Biophysical Aspects to Clinical Applications*, ed. V. Gouverneur and K. Müller, Imperial College Press, London, 2012, pp. 91–138.
17. J. H. Davis, *Biophys. J.*, 1979, **27**, 339.
18. D. Marsh, *Handbook of Lipid Bilayers*, CRC Press, Boca Raton, FL, 2nd edn, 2013.
19. J. Ulmius, H. Wennerström, G. Lindblom and G. Arvidson, *Biochemistry*, 1977, **16**, 5742.

20. B. Bonev and M. R. Morrow, *Biophys. J.*, 1996, **70**, 2727.
21. B. Bonev and M. R. Morrow, *Phys. Rev. E*, 1997, **55**, 5825.
22. M. R. Morrow, J. P. Whitehead and D. Lu, *Biophys. J.*, 1992, **63**, 18.
23. M. R. Morrow and J. H. Davis, *Biochim. Biophys. Acta*, 1987, **904**, 61.
24. M. R. Morrow, R. Srinivasan and N. Grandal, *Chem. Phys. Lipids*, 1991, **58**, 63.
25. J. C. Huschilt, R. S. Hodges and J. H. Davis, *Biochemistry*, 1985, **24**, 1377.
26. M. L. Schmidt, L. Ziani, M. Boudreau and J. H. Davis, *J. Chem. Phys.*, 2009, **131**, 175103.
27. M. R. Morrow, D. Singh, D. Lu and C. W. M. Grant, *Biochim. Biophys. Acta*, 1992, **1106**, 85.
28. I. V. Polozov and K. Gawrisch, *Biophys. J.*, 2004, **87**, 1741.
29. T.-H. Huang, C. W. B. Lee, S. K. Das Gupta, A. Blume and R. G. Griffin, *Biochemistry*, 1993, **32**, 13277.
30. Y.-W. Hsueh, K. Gilbert, C. Trandum, M. Zuckermann and J. Thewalt, *Biophys. J.*, 2005, **88**, 1799.
31. J. Juhasz, F. J. Sharom and J. H. Davis, *Biochim. Biophys. Acta*, 2009, **1788**, 2541.
32. J. A. Clarke, J. M. Seddon and R. V. Law, *Soft Matter*, 2009, **5**, 369.
33. J. Wolff, C. M. Marques and F. Thalmann, *Phys. Rev. Lett.*, 2011, **106**, 128104.
34. P. Uppamoochikkal, S. Tristram-Nagle and J. F. Nagle, *Langmuir*, 2010, **26**, 17363.
35. N. Bezlyepkina, R. S. Gracia, P. Shchelokovskyy, R. Lipowsky and R. Dimova, *Biophys. J.*, 2013, **104**, 1456.
36. I. V. Ionova, V. A. Livshits and D. Marsh, *Biophys. J.*, 2012, **102**, 1856.
37. T. Bartels, R. S. Lankalapalli, R. Bittman, K. Beyer and M. F. Brown, *J. Am. Chem. Soc.*, 2008, **130**, 14521.
38. R. F. M. de Almeida, A. Fedorov and M. Prieto, *Biophys. J.*, 2003, **88**, 2406.
39. F. A. Heberle, J. Wu, S. L. Goh, R. S. Petruzielo and G. W. Feigenson, *Biophys. J.*, 2010, **99**, 3309.
40. J. A. Williams, S. E. Batten, M. Harris, B. D. Rockett, S. R. Shaikh, W. Stillwell and S. R. Wassall, *Biophys. J.*, 2012, **103**, 228.
41. J. H. Davis, D. M. Clare, R. S. Hodges and M. Bloom, *Biochemistry*, 1983, **22**, 5298.
42. O. Mouritsen and M. Bloom, *Biophys. J.*, 1984, **46**, 141.
43. S. Morein, J. A. Killian and M. M. Speretto, *Biophys. J.*, 2002, **82**, 1405.
44. R. S. Prosser, F. Evanics, J. L. Kitevski and M. S. Al Abdul-Wahid, *Biochemistry*, 2006, **45**, 8453.
45. J. J. de Vries and H. J. C. Berendsen, *Nature*, 1969, **221**, 1139.
46. G. Lindblom, *Acta Chem. Scand.*, 1972, **26**, 1745.
47. I. M. Braithwaite, *Investigating the Molecular Order and Orientation of Cholesterol in Mixtures of Polyunsaturated Phospholipids*, MSc Thesis, University of Guelph, Canada, 2011.
48. D. E. Warschawski, A. A. Arnold, M. Beaugrand, A. Gravel, É. Chartrand and I. Marcotte, *Biochim. Biophys. Acta*, 2011, **1808**, 1957.

49. J. X. Lu, K. Damodaran and G. A. Lorigan, *J. Magn. Reson.*, 2006, **178**, 283.
50. S. Dvinskikh, U. Dur, K. Yamamoto and A. Ramamoorthy, *J. Am. Chem. Soc.*, 2006, **128**, 6326.
51. M. Triba, M. Zoonens, J. L. Popot, P. F. Deveaux and D. E. Warschawski, *Eur. Biophys. J.*, 2006, **35**, 268.
52. R. Mahalakshmi and F. M. Marassi, *Biochemistry*, 2008, **47**, 6531.
53. C. Tan, B. M. Fung and G. Cho, *J. Am. Chem. Soc.*, 2002, **124**, 11827.
54. C. Loudet, L. Khemtémourian, F. Aussenac, S. Gineste, M.-F Achard and E. J. Dufourc, *Biochim. Biophys. Acta*, 2005, **1724**, 315.
55. S. H. Park, C. Loudet, F. M. Marassi, E. J. Dufourc and S. J. Opella, *J. Magn. Reson.*, 2008, **193**, 133.
56. S. H. Park, A. A. De Angelis, A. A. Nevzorov, C. H. Wu and S. J. Opella, *Biophys. J.*, 2006, **91**, 3032.
57. H. C. Gaede, K. M. Luckett, I. V. Pozolov and K. Gawrisch, *Langmuir*, 2004, **20**, 7711.
58. E. Y. Chekmenev, J. Hu, P. L. Gor'kov, W. W. Brey, T. A. Cross, A. Ruuge and A. I. Smirnov, *J. Magn. Reson.*, 2005, **173**, 322.
59. J. D. Gross, D. E. Warschawski and R. G. Griffin, *J. Am. Chem. Soc.*, 1997, **119**, 796.
60. S. V. Dvinskikh, V. Castro and D. Sandstrom, *Phys. Chem. Chem. Phys.*, 2005, 7, 3255.
61. T. M. Ferreira, F. Coreta-Gomes, O. H. Samuli Ollila, M. J. Morena, W. L. C. Vaz and D. Topgaard, *Phys. Chem. Chem. Phys.*, 2013, **15**, 1976.
62. E. Strandberg, S. Esteban-Martín, A. S. Ulrich and J. Salgado, *Biochim. Biophys. Acta*, 2012, **1818**, 1242.
63. E. Strandberg, D. Tiltak, S. Ehni, P. Wadhwani and A. S. Ulrich, *Biochim. Biophys. Acta*, 2012, **1818**, 1764.
64. A. Tsutsumi, N. Javkhlantugs, A. Kira, M. Umeyama, I. Kawamura, K. Nishimura, K. Ueda and A. Naito, *Biophys. J.*, 2012, **103**, 1735.
65. S. Toraya, N Javkhlantugs, D. Mishima, K. Nishimura, K. Ueda and A. Naito, *Biophys. J.*, 2010, **99**, 3282.
66. A. Naito, *Solid State Nucl. Magn. Reson.*, 2009, **36**, 67.
67. R. O. Fox Jr. and F. M. Richards, *Nature*, 1982, **300**, 325.
68. B. Bechinger, J. M. Resende and C. Aisenbrey, *Biophys. Chem.*, 2011, **153**, 115.
69. C. Farès and J. H. Davis, *NMR of Ordered Liquids*, ed. E. E. Burnell and C. A. de Lange, Kluwer, Dordrecht, 2003, pp. 191–213.
70. B. Bechinger and E. S. Salnikov, *Chem. Phys. Lipids*, 2012, **165**, 282.

CHAPTER 14

NMR Investigations of the Structure and Dynamics of Antimicrobial Peptides: The Peptaibol Alamethicin

THOMAS VOSEGAARD*[a,b] AND NIELS CHR. NIELSEN[a]

[a] Center for Insoluble Protein Structures, Interdisciplinary Nanoscience Center and Department of Chemistry, Aarhus University, Denmark;
[b] School of Engineering and Department of Engineering, Aarhus University, Denmark
*Email: tv@chem.au.dk

14.1 Introduction

Antimicrobial peptides (AMPs) are part of the immune system of various organisms and for a long time have attracted considerable scientific interest for several reasons. Their high specificity for bacterial cells and little sensitivity towards bacterial mutations make them attractive alternatives to small-molecule-based antibiotics. Issues like protein degradation cause serious problems for the practical use of AMPs for treatment of bacterial infections, and have led researchers to search for AMPs that do not degrade in the presence of proteases. Peptaibols are a class of AMPs with a high content (~40%) of the rare amino acid Aib (α-aminoisobutyric acid), an amino alcohol in the C-terminal end and typically an N-terminal end protected by an acetyl group.[1] Peptaibols show antibacterial activity, and the high content of

New Developments in NMR No. 3
Advances in Biological Solid-State NMR: Proteins and Membrane-Active Peptides
Edited by Frances Separovic and Akira Naito

Published by the Royal Society of Chemistry, www.rsc.org

Aib makes the peptides less prone to protein degradation because of steric hindrance of the two methyl groups at the β-position for this amino acid.

Within the peptaibol family, alamethicin is probably the most studied molecule as representative of the archetypal pore-forming AMP.[2–8] Its crystal structure was determined more than 30 years ago by Fox and Richards (*cf.* Figure 14.1)[9] and, since then, numerous liquid-state[10–13] and solid-state

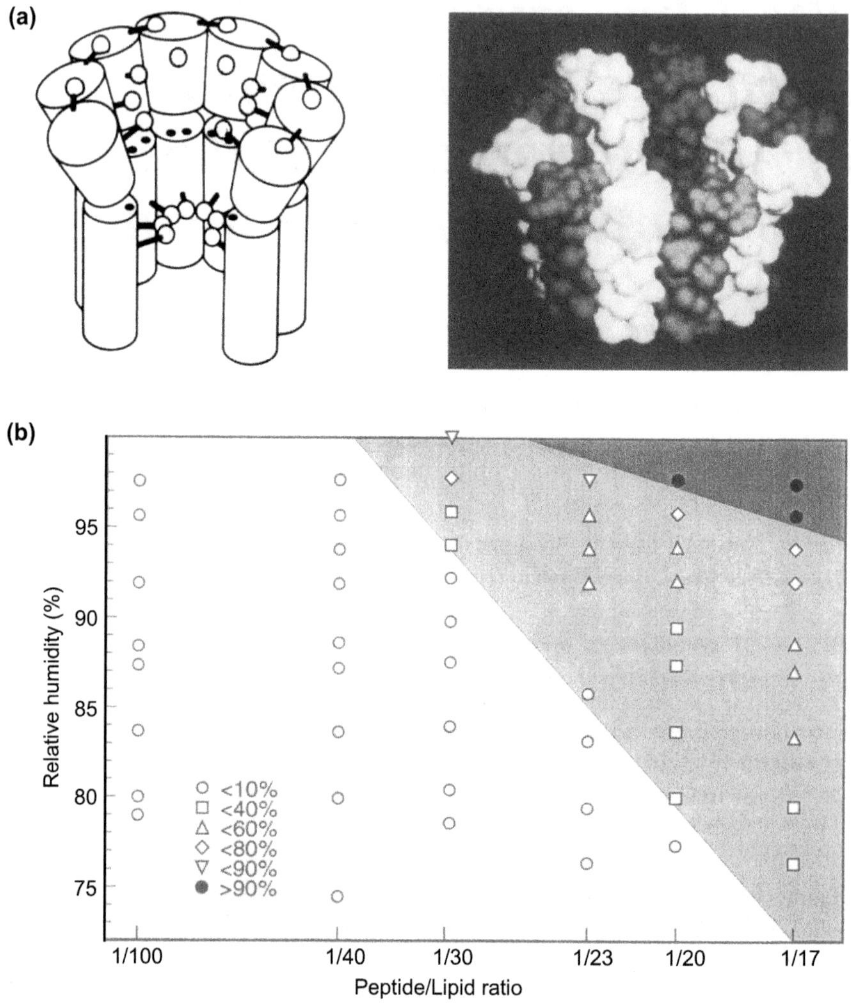

Figure 14.1 (a) Model of the alamethicin pore based on the X-ray diffraction structure. (b) Diagram showing the orientation of alamethicin in DPhPC bilayers as a function of the peptide/lipid ratio and the relative humidity, as determined by oriented sample circular dichroism measurements. Percentages indicate the portion of alamethicin aligned parallel to the bilayer normal.
(Part (a) reproduced from Fox and Richards,[9] with permission from Macmillan; part (b) adapted from He *et al.*,[5] with permission from Elsevier.)

NMR[14–24] studies have added knowledge on the structure. In addition to the numerous experimental studies on the alamethicin structure and channel-forming properties, a large number of molecular dynamics (MD) studies have also addressed these questions over the years.[25–32]

In this chapter, we review efforts to understand alamethicin function based on results from some of our previous studies of its structure and dynamics within lipid membranes using solid- and liquid-state NMR spectroscopy and MD simulations[13,17,19,20,22–24,32] and first attempts to modify its channel-forming properties by scaffolding alamethicin oligomers.[33]

14.2 Peptide Structure and Conformation

The structure of alamethicin was first determined in 1982 in an X-ray diffraction (XRD) study by Fox and Richards[9] (*cf.* Figure 14.1). Since then, numerous NMR studies have confirmed an overall α-helical structure with a kink around Pro14, consistent with the alamethicin amino acid sequence containing the well-known helix-kinking motif Gly-X-X-Pro. Following previous work on alamethicin in lipid bilayers,[14–16] in 2001 we initialized a series of studies of alamethicin in oriented lipid bilayers first involving orientational constraints in the form of orientation-dependent ^{15}N anisotropic chemical shifts and ^{1}H–^{15}N dipole–dipole couplings for three residues at different places in the sequence.[17] While these restraints were insufficient for a full structure determination, they allowed for modeling of the channel-forming state of alamethicin by comparing the orientational restraints with MD simulations of alamethicin. The essence of this study is shown in Figure 14.2, where the experimental oriented-sample ^{15}N solid-state NMR spectra are shown along with two different structural models, one assuming alamethicin to be a straight α-helix, the other assuming that the alamethicin monomers adopt the structure obtained by Fox and Richards.[9]

To obtain more orientational restraints for alamethicin in the membrane-bound state, we developed protocols for synthesis of ^{15}N, ^{2}H, or ^{19}F isotope-labeled Aib.[19] These efforts were based on previous work by Ogrel *et al.*[34] and a new automated procedure for solid-phase peptide synthesis of peptaibols to produce different isotope-labeled variants of alamethicin and other peptaibols.[35]

Having constructed Aib-d_6 (Aib with deuterated methyl groups) to be incorporated into alamethicin through solid-phase peptide synthesis, we could experimentally exploit the orientational restraints from the ^{2}H quadrupole coupling tensors of the two deuterated methyl groups to provide additional orientational information on the membrane-bound conformation of alamethicin.[19] Figure 14.3 outlines the procedure for obtaining such restraints, taking advantages of the fast rotational diffusion of peptides in lipid membranes,[15,36–38] in this case for alamethicin reconstituted in vesicles.[19] In particular, we noticed that the powder (non-oriented sample) ^{2}H magic-angle spinning (MAS) spectrum of alamethicin-Aib8 d_6 showed a reduction of the effective quadrupole coupling constant (C_Q) from 53 kHz in a lyophilized

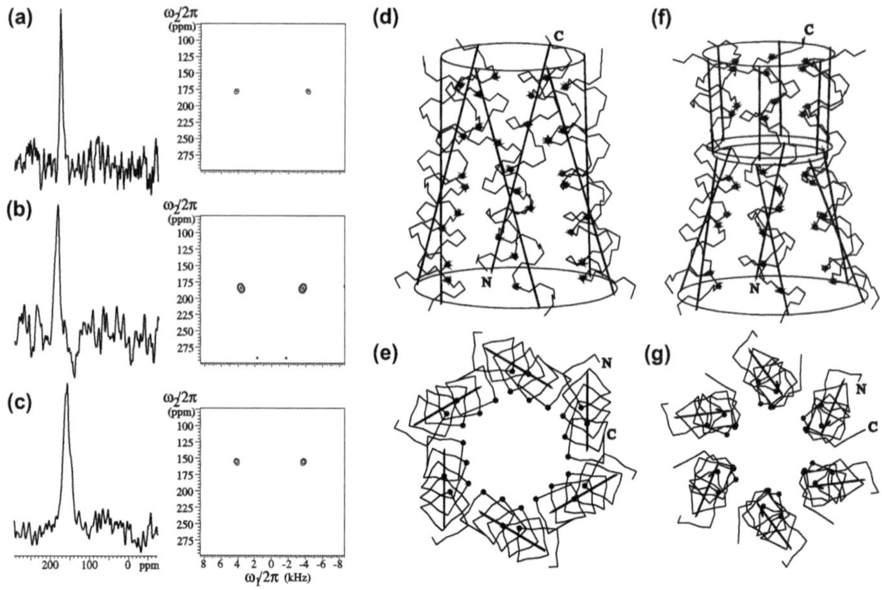

Figure 14.2 (a–c) ^{15}N cross polarization (CP) (*left column*) and 2D ^{1}H–^{15}N separated-local-field (SLF) (*right column*) solid-state NMR spectra of alamethicin labeled with ^{15}N isotopes at the (a) Ala6, (b) Val9, and (c) Val15 positions and oriented macroscopically in DMPC lipid bilayers. (d–g) Models of hexamericalamethicin ion channels generated from MD simulations of (d,e) tilted α-helices and (f,g) tilted and kinked α-helices compatible with the NMR data in (a–c). The putative ion channels are viewed from the side (d,f) and top (e,g). (Adapted from Bak *et al.*,[17] with permission from Elsevier.)

sample (Figure 14.3d) to 6–10 kHz in hydrated vesicles (Figure 14.3e). The reduced C_Q values can be transformed into geometric scaling factors reporting on the membrane-bound conformation of alamethicin, as shown by the restriction plot in Figure 14.3g. The restriction plot shows the possible membrane-bound conformations of alamethicin expressed as the helix tilt angle, τ, and the rotational pitch, ρ (see Bertelsen *et al.*[19] for details). For all τ,ρ values, the expected effective C_Q values are calculated through a number of coordinate transformations from the principal axis frame of the quadrupole coupling tensor *via* the bilayer system, where the axial averaging takes place, to the laboratory frame, where it is compared with the observed values of 6 and 10 kHz. The agreement between the calculated and experimental values is assessed by the restriction function, plotted in Figure 14.3g:

$$R(\tau,\rho) = \prod_{i=1}^{2} \exp\left\{ -4\ln 2 \left(\frac{C_{Q,i}^{calc}(\tau,\rho) - C_{Q,i}^{exp}}{\Delta C_Q)} \right)^2 \right\} \tag{14.1}$$

which ranges from 0, in the case of large disagreements between calculated and experimental quadrupole couplings, to 1 for complete agreement, and is

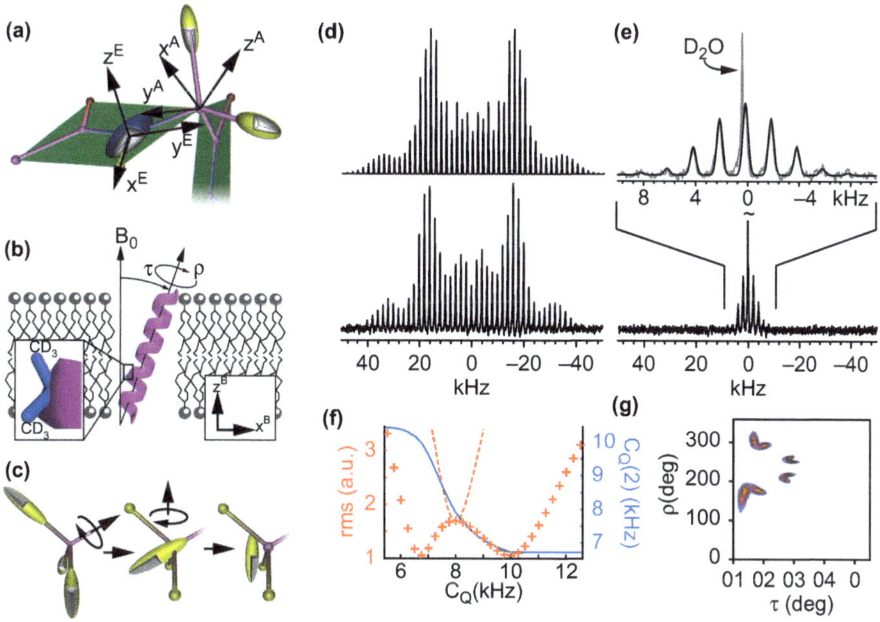

Figure 14.3 (a) Geometrical representation of a peptide plane and the side chain for an Aib residue highlighting the ^{15}N chemical shift tensor ellipsoid (*blue*), the ^{2}H quadrupole coupling tensor ellipsoids (*yellow*), and the two peptide planes connected at the Cα atom of the Aib residue (*green*). (b) Schematic representation of an α-helical peptide highlighting the two deuterated methyl groups of an Aib residue. (c) Illustration of the biaxial rotational averaging of the ^{2}H quadrupolar coupling interaction due to fast rotation around the Cβ–Cα internuclear axis (*left to middle figure*) and rotation around the bilayer normal (*middle to right figure*). (d,e) Experimental (*bottom*) and simulated (*top*) ^{2}H MAS NMR spectra of (e) lyophilized and (f) DMPC vesicle-reconstituted alamethicin with ^{2}H labeling of the Aib8 methyl groups. (f) Plot of the rms deviation between simulated and experimental spectra as a function of the ^{2}H quadrupole coupling constant (C_Q). (g) Restriction plot showing the conformations of alamethicin in agreement with the ^{2}H quadrupole couplings as a function of the peptide tilt angled τ and the rotational pitch ρ; see definitions in (b). (Adapted from Bertelsen *et al.*,[19] with permission from the American Chemical Society.)

a product of individual restrictions for the quadrupole couplings of the two deuterated methyl groups of Aib.

To provide a more complete structural model, we incorporated alamethicin into DMPC/DHPC bicelles and determined its structure by liquid-state ^{1}H NMR experiments.[13] To avoid complications due to dominating ^{1}H resonances from the lipids, commercially available deuterated versions of DMPC and DHPC were used. In these lipids, the hydrogens on the aliphatic tails are replaced by ^{2}H, but the headgroup hydrogens remain, leaving substantial residual ^{1}H signals of the lipids, which overlap with the peptide

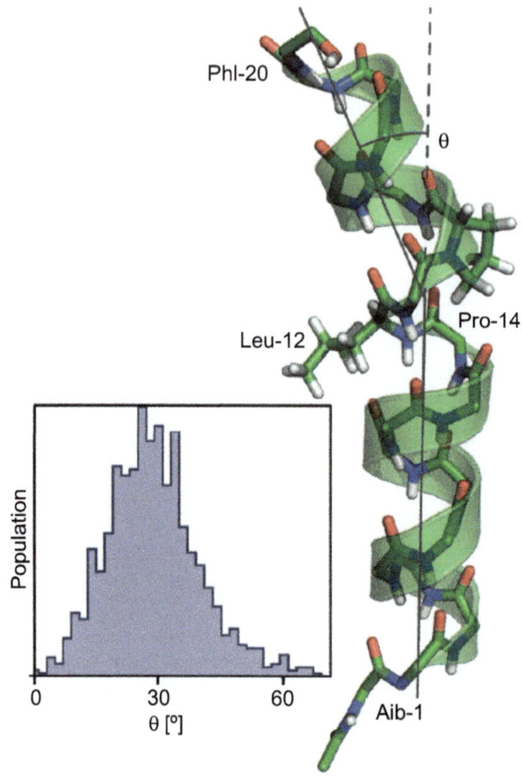

Figure 14.4 Structural model of alamethicin in bicelles. The model is obtained from liquid-state NMR and MD restraints using simulated annealing. The distribution of kink angles from the MD simulations is shown in the *insert*. The shown structural model exhibits a kink of 31°. (Adapted from Dittmer *et al.*,[13] with permission from the American Chemical Society.)

signals. In spite of this, analysis of the Hα chemical shifts as well as a number of medium-range NOE connectivities gave a good indication of a generally α-helical secondary structure. Based on the experimental data and data from MD, it was possible to obtain the structural model shown in Figure 14.4.[13] It was not possible to determine the kink angle from the experimental data alone, but the kink angles reported in Figure 14.4 are all consistent with the liquid-state ^{1}H NMR data.

14.3 Membrane Insertion and Peptide-Induced Membrane Mosaic Spread

Having established good structural models and characterization of the transmembrane nature of the membrane-bound conformation of alamethicin, we aimed at further understanding of the insertion of alamethicin in bilayers by combining NMR and MD simulation studies. Pursuing the

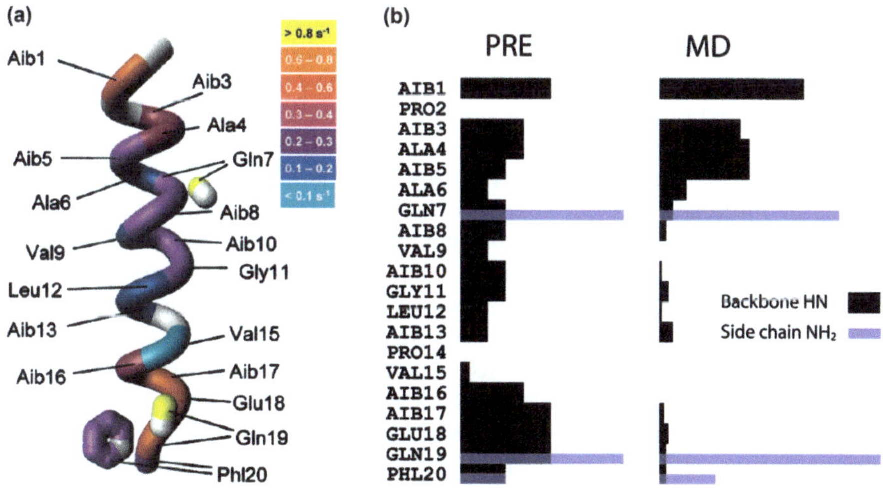

Figure 14.5 Peptide–water contacts measured by paramagnetic relaxation enhancement (PRE) ^{1}H liquid-state NMR experiments. (a) Alamethicin structure with the amino acid residues colored according to the PRE observed for the backbone H^N or side-chain H_2N signals. (b) Peptide–water contacts measured by PRE (*left column*) compared with number of occurrences that the protons in question are within a radius r < 15 Å of a water molecule, weighted by r^{-6} (*right column*). (Reproduced from Dittmer *et al.*,[13] with permission from the American Chemical Society.)

liquid-state NMR experiments, we measured paramagnetic relaxation enhancement (PRE) factors of the hydrogens in alamethicin induced by a soluble gadolinium complex present in the water phase surrounding the bicelles with alamethicin incorporated (Omniscan).[13] These experiments enabled us to map the solvent exposure of the different alamethicin residues. Figure 14.5 shows the PREs of the different amino acids in alamethicin and indicates that largely all residues on a statistical basis are exposed to the solvent to some extent. An immediate explanation could be the formation of water-containing channels. However, the most favored model with a hexameric water channel[25] would be too small to host the gadolinium complex. A comparison with MD simulations of alamethicin in DMPC/DHPC bicelles, however, could explain most of the PRE results (*cf.* Figure 14.5b) without abundant formation of water channels. The simulations revealed that the membrane anchoring of alamethicin is highly dynamic, even allowing most of the residues in the middle of the alamethicin sequence, those expected to be buried in the hydrophobic core of the bilayer, to make excursions to the bilayer surface.

Further exploring the potential of combining information from MD with the experimental data obtained from solid-state NMR experiments, we performed a series of MD simulations of alamethicin in a lipid bilayer using both all-atom (AA) and coarse-grained (CG) MD approaches.[32] CG simulations represent molecules by a few "beads", so the computational efforts are dramatically reduced when compared to the all-atom approach. For example, most amino acids in peptide chains may be represented by as little as a backbone bead and a side-chain bead. This allows one to obtain simulations for a longer timescale, thereby potentially revealing slower dynamics and the formation of long-term stable membrane-bound states. This is highly interesting in our search for more insight into the mechanism of cell lysis by AMPs such as alamethicin. As an example, the insertion of an alamethicin molecule into the lipid bilayer was observed (Figure 14.6a) and also the occurrence of water-leaking (see Figure 14.6c) and barrel-stave channels (Figure 14.6b) formed by alamethicin.[32]

Overall, the CG MD simulations corroborate the observations of the AA MD simulations and the PRE ^{1}H NMR experiments, namely that the alamethicin monomers are highly mobile and adopt many different conformations during extended time periods. Notably, Figure 14.6d shows examples of the diverse alamethicin structures involved in an active water channel. Adding to this picture, Figure 14.7c shows a snapshot from the MD simulation highlighting four alamethicin molecules in a lipid bilayer. Interestingly, this illustration also reveals that the lipid bilayer deviates significantly from a well-aligned bilayer but wobbles up and down, the so-called mosaic spread.

To further study the diversity of the alamethicin structures in lipid bilayers and identify potential membrane mosaic spread, we incorporated an [^{15}N]Aib8-labeled sample of alamethicin into oriented lipid bilayers.[20] For this sample we recorded ^{15}N spectra with ^{1}H heteronuclear decoupling (Figure 14.7a) and ^{1}H homonuclear FSLG decoupling (Figure 14.7b) during signal recording. In the former case a single peak is detected, while in the latter a doublet splitting of the ^{15}N signal is seen due to the ^{1}H–^{15}N dipole–dipole coupling. This splitting and, in particular, the differential line-broadening of the two peaks in the doublet report eventually on mosaic spread. In the presence of mosaic spread, a line narrowing of one of the doublet peaks is observed relative to what may be achieved using ^{1}H heteronuclear decoupling.[20] In the case of alamethicin, the experimental spectra were modeled convincingly (*cf.* Figure 14.7a and 14.7b) using a mosaic spread of 18°.

By inspecting the MD simulations, we calculated the helix alignments for the ^{15}N-labeled Aib8 residue in all frames of the MD simulation to produce a so-called conformation probability plot of the middle region of the alamethicin molecules.[20] This plot (Figure 14.7d) indicates that the alamethicin molecules cover a large conformational space, in agreement with the experimental observations, but, as we will address later, the distinction between dynamic and static disorder should be made.

To further detail the anchoring of alamethicin in membrane lipid bilayers, we followed the idea of Kim and Wi[39] and established a simple exchange model for the lipid dynamics, allowing us to probe different geometries for the peptide interaction with the lipid bilayers from ^{31}P spectra of oriented

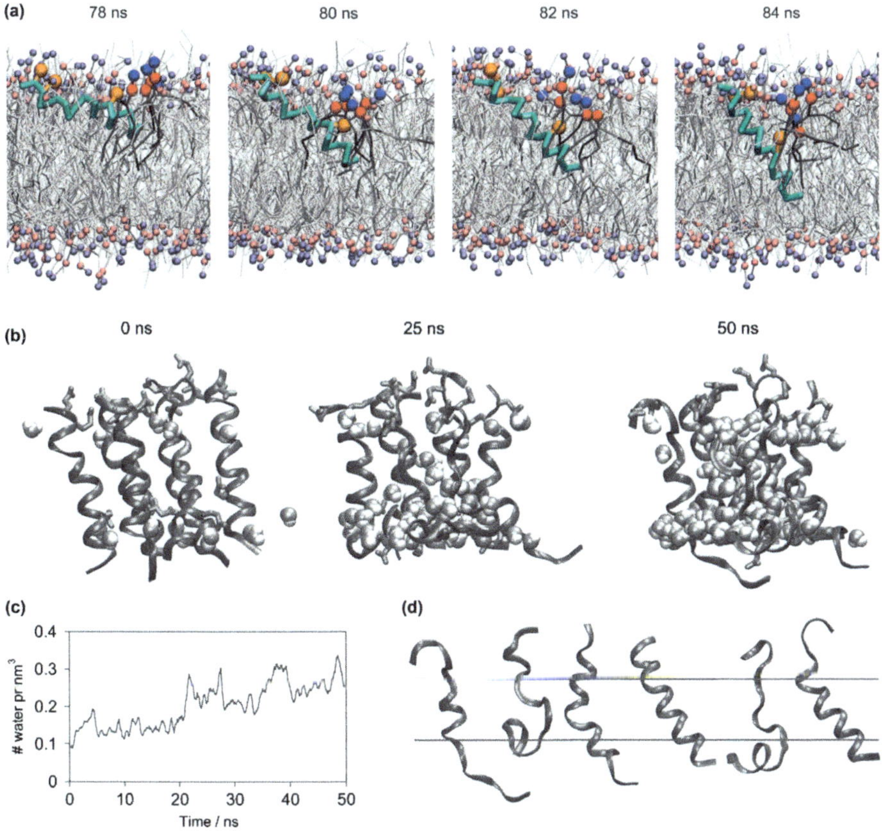

Figure 14.6 All-atom and coarse-grained MD simulations describing the mechanisms of alamethicin membrane-insertion and channel-induced water leakage. (a) Transition between the surface-bound and membrane-spanning states of alamethicin revealed by CG MD simulations. The three highlighted lipid molecules with their headgroups in *red/blue* facilitate the insertion of the peptide. (b) Formation of a leaky membrane by alamethicin peptides in an AA MD simulation. Water in the middle 20 Å of the bilayer and within 8 Å of the six peptides is shown. (c) Water content of the alamethicin channel. (d) Representative individual conformations of the peptides from the cluster shown in (b) after 50 ns simulation time.
(Reproduced from Thøgersen *et al.*,[32] with permission from Elsevier.)

samples.[24] For comparison, two peptides, alamethicin and novicidin, were investigated and assumed that they would perturb the lipid bilayer either by forming a barrel-stave ion channel or toroidal pore. These geometries and their expected perturbation of the lipids (thinning or formation of a curved toroidal pore) are shown in Figures 14.8a and 14.8b.

We recorded ^{31}P spectra for a number of mechanically oriented samples with varying peptide to lipid stoichiometry. These spectra were in agreement

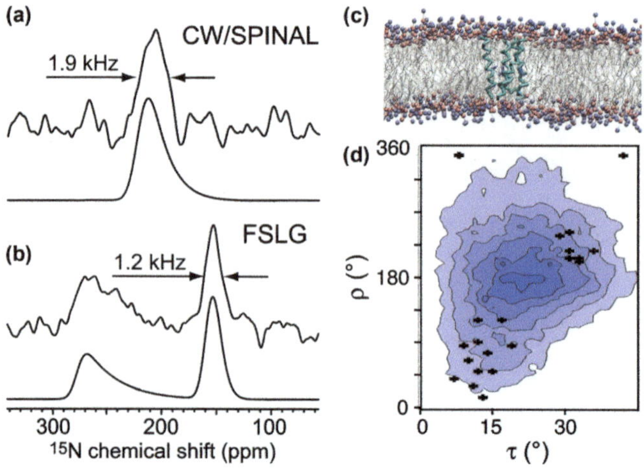

Figure 14.7 (a, b) Experimental (*top*) and simulated (*bottom*) ^{1}H–^{15}N spectra for a sample of [^{15}N]Aib8 alamethicin in oriented lipid bilayers. The spectra are obtained by sampling under conditions of: (a) heteronuclear SPINAL-64 ^{1}H decoupling, and (b) ^{1}H homonuclear FSLG decoupling. The simulations assume a single α-helix conformation and 18° mosaic spread. (c) Snapshot of an alamethicin channel-like aggregate with four alamethicin molecules resulting from a MD simulation. (d) Distribution of helix conformations in the MD simulation (definition of angles in Figure 14.3b). *Blue contours* represent conformation counts, while the *black points* represent average conformations back-calculated from averaging the nuclear spin interaction frequencies for each peptide conformation.
(Adapted from Vosegaard *et al.*,[20] with permission from the American Chemical Society.)

with a model of peptide insertion into the lipid bilayer in which alamethicin at low concentrations binds weakly to the surface, inducing a slight thinning of the bilayer, then at higher concentrations binds more tightly, and, at very high concentrations, inserts into the lipid bilayer creating a small fraction of slowly diffusing lipids around the alamethicin channels. This model is illustrated in Figure 14.8e and alamethicin at high concentrations adopts the bottom-left conformation. In Figure 14.8e, alamethicin is depicted as a monomer, since the present method cannot reveal its oligomerization state. However, other studies have well established that several alamethicin monomers team up in forming the ion channels.

This ^{31}P approach to study the peptide perturbation of the lipids is appealing due to the high abundance of lipids and the high sensitivity of ^{31}P NMR. The sensitivity is illustrated in Figures 14.8c and 14.8d that report the ^{15}N spectrum recorded in the course of hours and the ^{31}P spectrum recorded in a few minutes of an oriented sample containing alamethicin in a bilayer. This demonstrates that the present ^{31}P membrane-based "spy" method is highly sensitive and even allows the study of the membrane-bound conformation of peptides without any isotope labeling.

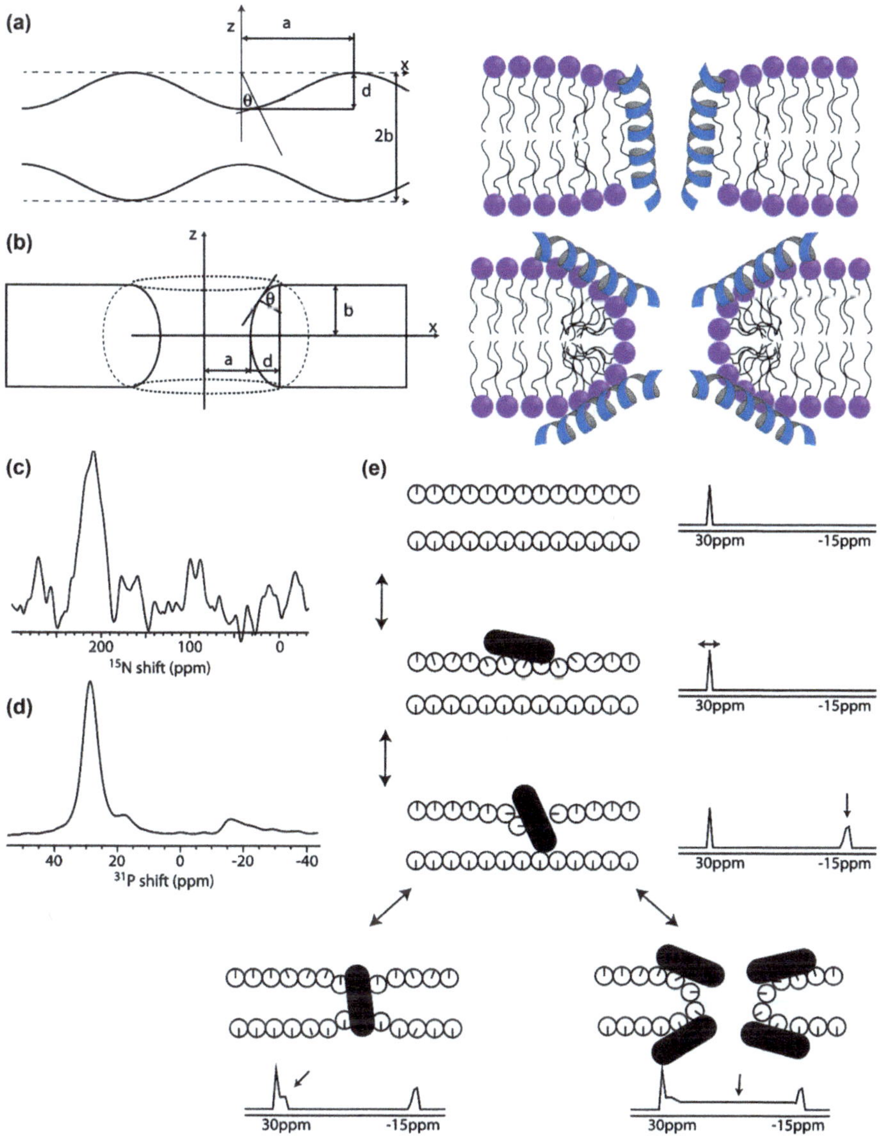

Figure 14.8 (a, b) Sketches of the geometries (*left column*) and examples of membrane/AMP systems described by the geometries (*right column*). The geometries are referred to as (a) a thinned bilayer and (b) a toroidal pore. (c, d) Oriented-sample solid-state (c) ^{15}N and (d) ^{31}P NMR spectra of [^{15}N]Aib8-labeled alamethicin in oriented DMPC bilayers. (e) Model of peptide interactions with the lipid membrane based on oriented-sample solid-state NMR ^{31}P spectra. The lipids are illustrated by *circles*, with the radial line illustrating the orientation of the ^{31}P headgroup. Alamethicin forms the barrel-stave channels (shown in the *bottom left panel*) without significant perturbation of the lipid. The peptide novicidin creates toroidal pores in the lipids (*bottom right panel*).
(Adapted from Bertelsen *et al.*,[24] with permission from the Public Library of Science.)

14.4 Membrane Systems and their Influence on Peptide Structure

There has been a long-lasting interest in finding cell-membrane models that are representative of naturally occurring membranes and simultaneously amenable for macroscopic alignment and sufficiently robust for long-term NMR studies.[41,42] With focus on oriented systems, mechanically oriented bilayers have been the prevailing method,[43] eventually in combination with MAS, so-called magic angle oriented sample spinning (MAOSS).[44,45]

Micelles consisting of small detergent molecules represent the most simple membrane model system, which generally works well with large membrane proteins that achieve their tertiary structure mainly through their own scaffold. Micelles provide a lipid-like environment, but they do not exhibit the bilayer structure.[46] To achieve this, while retaining the short correlation times essential for high-resolution liquid-state NMR, small bicelles have been used extensively to study membrane proteins.[47–52] The size of the bicelles may be controlled by the molar ratio of the short- and long-chained lipids, characterized by the q-value:

$$q = \frac{\text{Long}}{\text{Short}} \tag{14.2}$$

as illustrated in Figure 14.9. Interestingly, for large values of q (≥ 2), the bicelles will align spontaneously in the magnetic field,[48] making bicelles suitable for oriented-sample solid-state NMR experiments. Moreover, upon addition of lanthanide ions, the orientation of the large bicelles may be flipped, so their normal is parallel to the magnetic field.[53] The same effect may be achieved by using a special type of biphenyl lipids.[54] As a final method for providing oriented samples, so-called macro (or nano) discs show promising properties.[55]

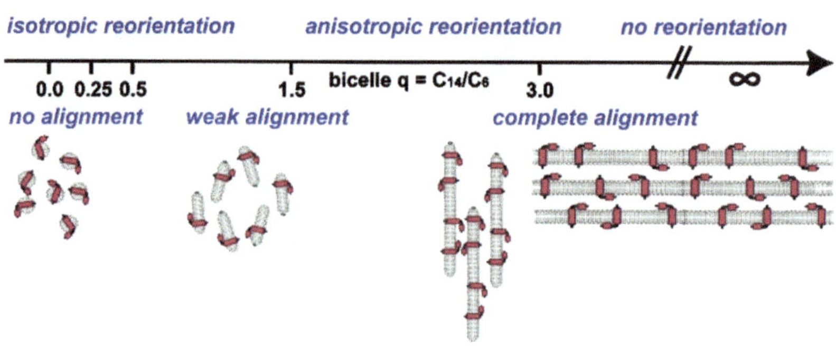

Figure 14.9 Overview of different oriented lipid systems and their degree of alignment. On the *left* are micelles and small isotropic bicelles; in the *middle* are larger bicelles with weak to complete alignment, characterized by their q value; and on the *right* are mechanically oriented bilayers.
(Reproduced from Opella and Marassi,[40] with permission from the American Chemical Society.)

Turning again our attention to alamethicin, we have characterized this molecule with solid-state NMR in both mechanically oriented bilayers and flipped and non-flipped bicelles. Interestingly, this molecule behaves structurally differently in different environments.[23] In oriented bilayers and flipped bicelles, the resonance frequency of ^{15}N for Aib8 is at ~200 ppm and the ^{1}H–^{15}N dipole–dipole coupling is *ca.* 7 kHz, corresponding to a transmembrane orientation. In non-flipped bicelles (*i.e.*, with the bilayer normal perpendicular to the magnetic field), the actual chemical shift was measured to ~110 ppm. Since it is difficult to compare the shifts for the media with the bilayer normal parallel and perpendicular to the magnetic field, Figure 14.10 shows a plot of the order-parameter-scaled spectrum, where the

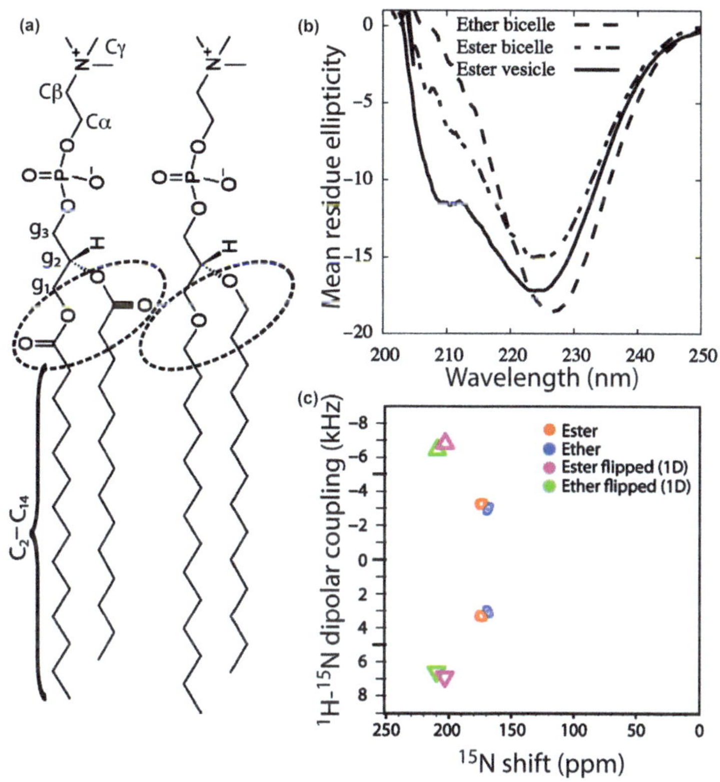

Figure 14.10 (a) Chemical structure of DMPC in its natural ester form with annotations (*left*) and an ether variant, 14-O-PC (*right*). The ester/ether groups are framed by *circles*. (b) CD spectra of alamethicin in ester vesicles, and ester and ether bicelles. (c) Resonance positions in order-parameter-scaled 2D ^{1}H–^{15}N SLF spectra of d_6-^{15}N-Aib8-labeled alamethicin in ester (*red* and *magenta symbols*) and ether (*blue* and *green symbols*) bicelles in standard non-flipped orientation with the bicelle normal perpendicular to the B_0 field (*red* and *blue symbols*) and in flipped bicelles (*magenta* and *green symbols*).
(Adapted from Bertelsen *et al.*,[23] with permission from the American Chemical Society.)

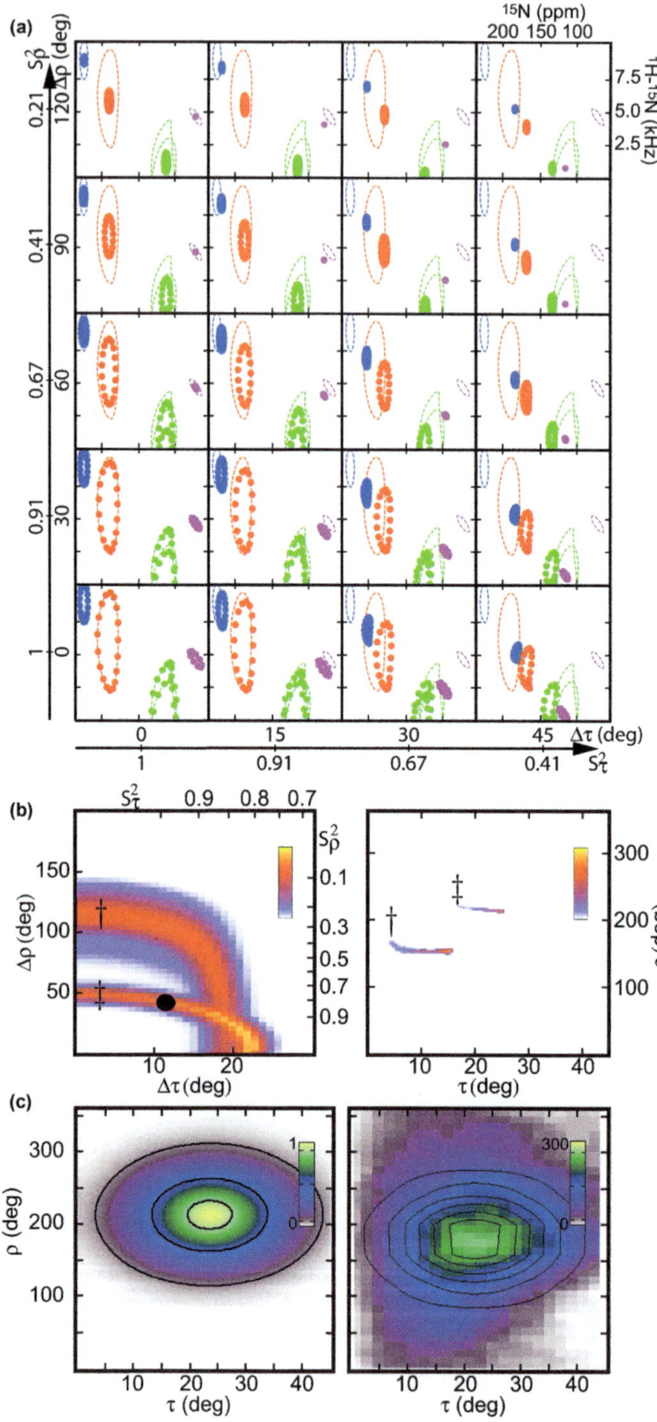
(a)
Sρ²
Δρ (deg)
0.21 120
0.41 90
0.67 60
0.91 30
1 0
¹⁵N (ppm)
200 150 100
¹H-¹⁵N (kHz)
7.5
5.0
2.5
0 15 30 45 Δτ (deg)
1 0.91 0.67 0.41 Sτ²
(b)
Sτ² 0.9 0.8 0.7
Sρ²
0.1 0.3 0.5 0.7 0.9
150 100 50
Δρ (deg)
10 20
Δτ (deg)
300 200 100
ρ (deg)
10 20 30 40
τ (deg)
(c)
300 200 100
ρ (deg)
1 0
10 20 30 40
τ (deg)
300 0
10 20 30 40
τ (deg)

axes are compensated according to the order parameter observed in a ^{31}P spectrum.[23] This means that the spectrum is inverted through the point located at the isotropic chemical shift and zero dipole–dipole coupling to compensate for the scaling of the chemical shift by the perpendicular orientation of the bicelle. This makes the different measurements represented in Figure 14.10c directly comparable.

We observed that alamethicin adopts the same membrane-bound conformation in bilayers and flipped bicelles of both normal ester lipids (DMPC : DHPC) and ether lipids (14-O-PC : 6-O-PC), whereas the conformation changes markedly in non-flipped bicelles (red and blue contours in Figure 14.10c). This suggests that the interaction between alamethicin and the lipids is weak and dynamic, thus implying that it may be necessary to include dynamics when investigating structural features of alamethicin.

14.5 Peptide Dynamics in Oriented Lipid Bilayers

Addressing the latter aspect, we may directly target the peptide dynamics within the lipid bilayer using solid-state NMR. Several studies have recently addressed the fact that small peptides are embedded in the lipid environment in a very dynamic way.[22,37,56–60] In a simple model, we assume that the peptide secondary structure is a rigid α-helix that wobbles in the membrane. The average helix conformation may be characterized by the helix tilt (τ) and rotational pitch (ρ), and fluctuations of the peptide around this average conformation are characterized by $\Delta\tau$ and $\Delta\rho$.

Figure 14.11a shows the influence of the dynamics on the polarity index slant angle (PISA) wheels[61,62] expected for a uniformly ^{15}N-labeled α-helical peptide at different tilt angles.[22] Increasing values of $\Delta\rho$ diminishes the diameter of the wheel without moving its center of gravity, while large values

Figure 14.11 (a) Simulated PISA wheels for ^{1}H–^{15}N SLF experiments of an α-helical peptide with different degrees of conformational dynamics as expressed by $\Delta\tau$ and $\Delta\rho$ or the corresponding order parameters S_τ^2 and S_ρ^2 (*filled circles*) and without dynamics (*dashed lines*). The wheels represent τ values of 10° (*blue*), 30° (*red*), 60° (*green*), and 90° (*magenta*). (b) Maximum score in the combined restriction plots (see Eqn 14.1) for the deuterium quadrupole coupling, the ^{15}N chemical shift, and ^{1}H–^{15}N dipole–dipole coupling as a function of $\Delta\tau$ and $\Delta\rho$ (*left*) and sum of the maximum score in all restriction plots (for different $\Delta\tau$, $\Delta\rho$ values) with scores larger than 0.5 (*right*). (c) Conformation probability plots. The *left panel* is obtained by convolution of the restriction plot with a Gaussian broadening function corresponding to $\Delta\tau$, $\Delta\rho = 15°$, 56°. The *right panel* shows a count of conformations of Aib8 in alamethicin in hydrated bilayers from a coarse-grained MD (same data as in Figure 14.7d).
(Adapted from Bertelsen *et al.*,[22] with permission from the American Chemical Society.)

of $\Delta\tau$ moves the resonances towards the ^{15}N isotropic chemical shift and zero dipole coupling. Overall, these parameters significantly change the resonance frequencies, suggesting the importance of including these effects in the analysis of oriented-sample solid-state NMR spectra.

To characterize any differential dynamics in the peptide, we developed a technique to provide residue-specific information about the dynamics of peptaibols, relying on the complementary structural information obtained through ^{15}N chemical shifts, ^{1}H–^{15}N dipole–dipole couplings, and ^{2}H quadrupole couplings for Aib residues.[22] A combined investigation of all these data demonstrated that alamethicin– or at least the fraction around Aib8 – exhibits significant dynamics in the lipid bilayer, as illustrated in Figure 14.11c. We note that this observation is in very good agreement with our previous MD studies demonstrating the large conformational space occupied by alamethicin.[32]

14.6 Altering Function: Peptide Nanochannels by Scaffolding

An ultimate goal of our detailed studies of alamethicin and other AMPs is to understand their function in detail and potentially alter their composition or function, so we can ultimately use them, *e.g.* as peptide alternatives to antibiotics. In order to control the ion-channel properties of alamethicin, we followed the line of several previous studies[25,63–66] using a scaffold to obtain more homogeneous ion channels.[33] Using click chemistry, we synthesized alamethicin multimers linked covalently to a cyclodextrin scaffold. In this way, we have been able to synthesize alamethicin hexamers and heptamers, as illustrated in Figure 14.12a.

Voltage-clamp experiments (Figure 14.12b) show the large diversity in channel properties obtained for different multimerization states of alamethicin in a lipid bilayer. The diversity is evident from the large number of open states of the channel. In contrast, a cyclodextrin-scaffolded hexameric alamethicin construct shows a single well-defined channel, which most interestingly shows a higher open probability and more than 20-fold longer average lifetime of the open state as evidenced by the voltage-clamp experiment in Figure 14.12c.

Further documentation for improved channel-forming properties of the scaffolded alamethicin nanochannel were obtained from dye release experiments in which calcein is entrapped in vesicles at a concentration high enough to cause self-quenching of its fluorescence. When alamethicin is added, the calcein may be released through the alamethicin channels, allowing the dye to escape into the solution (Figure 14.12d). From the kinetics of these experiments, a 10-fold increase of calcein release in C-terminally scaffolded alamethicin was observed and 100-fold increase for alamethicin scaffolded at its N-terminus. Also, highly notably, the peptide concentration needed to cause the calcein release decreases by several orders

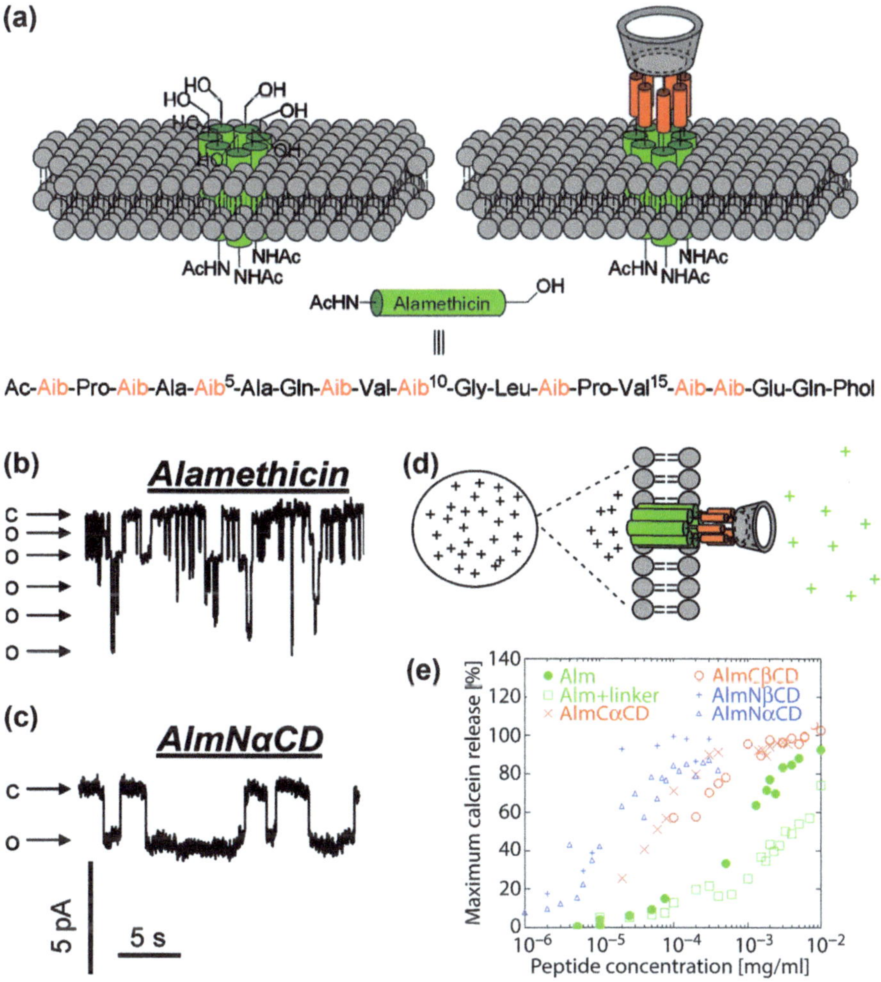

Figure 14.12 (a) Schematic representations of an alamethicin ion-channel formed by alamethicin monomers (*left*; peptide sequence in bottom) and alamethicin: cyclodextrin conjugate linked at the C-terminal (*right*). (b, c) Single channel current recordings of (b) alamethicin and (c) the alamethicin: CD conjugate AlmNαCD (*bottom*); C = closed, O = open. (d) Illustration of a calcein release experiment where a vesicle leaks calcein (+) through the alamethicin channel. (e) Calcein release experiments showing the maximum calcein release as a function of the peptide concentration.
(Adapted from Hjørringgaard *et al.*,[33] with permission from the American Chemical Society.)

of magnitude for scaffolded alamethicin (Figure 14.12e). These observations speak in favor of the barrel-stave mechanism for alamethicin insertion and channel formation, since the scaffolded peptides have a high local concentration (6–7 monomers very close) even at low overall concentration.

14.7 Conclusion

To conclude, we have described studies of the structure and dynamics of the AMP alamethicin using liquid- and solid-state NMR spectroscopy in combination with MD simulations. Our studies reveal the complexity and highly dynamic features of AMPs as being linked intimately to their membrane disruptive properties. The structural diversity, dynamics, and the strong dependence of the structure and membrane-bound conformation on the lipid composition renders such studies highly challenging. In addition to providing putative models for the alamethicin mode of action, we have demonstrated the possibility to improve function through scaffolding of peptides to cyclodextrin rings, thereby altering the peptide interactions and the channel formation propensity.

References

1. L. Whitmore and B. A. Wallace, *Nucleic Acids Res.*, 2004, **32**, D593–D594.
2. G. A. Woolley and B. A. Wallace, *Biochemistry*, 1993, **32**, 9819–9825.
3. M. S. Sansom, *Eur. Biophys. J.*, 1993, **22**, 105–124.
4. D. S. Cafiso, *Annu. Rev. Biophys. Biomol. Struct.*, 1994, **23**, 141–165.
5. K. He, S. J. Ludtke, W. T. Heller and H. W. Huang, *Biophys. J.*, 1996, **71**, 2669–2679.
6. H. Duclohier and H. Wroblewski, *J. Membr. Biol.*, 2001, **184**, 1–12.
7. B. Bechinger, C. Aisenbrey and P. Bertani, *Biochim. Biophys. Acta*, 2004, **1666**, 190–204.
8. B. Leitgeb, A. Szekeres, L. Manczinger, C. Vágvölgyi and L. Kredics, *Chem. Biodiversity*, 2007, **4**, 1027–1051.
9. R. O. Fox, Jr. and F. M. Richards, *Nature*, 1982, **300**, 325–330.
10. A. A. Yee and J. D. O'Neil, *Biochemistry*, 1992, **31**, 3135–3143.
11. J. C. Franklin, J. F. Ellena, S. Jayasinghe, L. P. Kelsh and D. S. Cafiso, *Biochemistry*, 1994, **33**, 4036–4045.
12. S. Jayasinghe, M. Barranger-Mathys, J. F. Ellena, C. Franklin and D. S. Cafiso, *Biophys. J.*, 1998, **74**, 3023–3030.
13. J. Dittmer, L. Thøgersen, J. Underhaug, K. Bertelsen, T. Vosegaard, J. M. Pedersen, B. Schiøtt, E. Tajkhorshid, T. Skrydstrup and N. C. Nielsen, *J. Phys. Chem. B*, 2009, **113**, 6928–6937.
14. S. O. Smith and O. B. Peersen, *Annu. Rev. Biophys. Biomol. Struct.*, 1992, **21**, 25–47.
15. C. L. North, M. Barranger-Mathys and D. S. Cafiso, *Biophys. J.*, 1995, **69**, 2392–2397.
16. B. Bechinger, *Biochim. Biophys. Acta*, 1999, **1462**, 157–183.
17. M. Bak, R. P. Bywater, M. Hohwy, J. K. Thomsen, K. Adelhorst, H. J. Jakobsen, O. W. Sørensen and N. C. Nielsen, *Biophys. J.*, 2001, **81**, 1684–1698.

18. B. Bechinger, D. A. Skladnev, A. Ogrel, X. Li, E. V. Rogozhkina, T. V. Ovchinnikova, J. D. O'Neil and J. Raap, *Biochemistry*, 2001, **40**, 9428–9437.
19. K. Bertelsen, J. M. Pedersen, B. S. Rasmussen, T. Skrydstrup, N. C. Nielsen and T. Vosegaard, *J. Am. Chem. Soc.*, 2007, **129**, 14717–14723.
20. T. Vosegaard, K. Bertelsen, J. M. Pedersen, L. Thøgersen, B. Schiøtt, E. Tajkhorshid, T. Skrydstrup and N. C. Nielsen, *J. Am. Chem. Soc.*, 2008, **130**, 5028–5029.
21. E. S. Salnikov, H. Friedrich, X. Li, P. Bertani, S. Reissmann, C. Hertweck, J. D. J. O'Neil, J. Raap and B. Bechinger, *Biophys. J.*, 2009, **96**, 86–100.
22. K. Bertelsen, B. Paaske, L. Thøgersen, E. Tajkhorshid, B. Schiøtt, T. Skrydstrup, N. C. Nielsen and T. Vosegaard, *J. Am. Chem. Soc.*, 2009, **131**, 18335–18342.
23. K. Bertelsen, B. Vad, E. H. Nielsen, S. K. Hansen, T. Skrydstrup, D. E. Otzen, T. Vosegaard and N. C. Nielsen, *J. Phys. Chem. B*, 2011, **115**, 1767–1774.
24. K. Bertelsen, J. Dorosz, S. K. Hansen, N. C. Nielsen and T. Vosegaard, *Plos One*, 2012, 7, e47745.
25. S. You, S. Peng, L. Lien, J. Breed, M. S. Sansom and G. A. Woolley, *Biochemistry*, 1996, **35**, 6225–6232.
26. J. Breed, P. C. Biggin, I. D. Kerr, O. S. Smart and M. S. Sansom, *Biochim. Biophys. Acta*, 1997, **1325**, 235–249.
27. N. Gibbs, R. B. Sessions, P. B. Williams and C. E. Dempsey, *Biophys. J.*, 1997, **72**, 2490–2495.
28. D. P. Tieleman, J. Breed, H. J. Berendsen and M. S. Sansom, *Faraday Discuss.*, 1998, 209–223.
29. D. P. Tieleman, H. J. Berendsen and M. S. Sansom, *Biophys. J.*, 2001, **80**, 331–346.
30. D. P. Tieleman, B. Hess and M. S. Sansom, *Biophys. J.*, 2002, **83**, 2393–2407.
31. Q. Zhu and M. W. Vaughn, *J. Phys. Chem. B*, 2005, **109**, 19474–19483.
32. L. Thøgersen, B. Schiøtt, T. Vosegaard, N. C. Nielsen and E. Tajkhorshid, *Biophys. J.*, 2008, **95**, 4337–4347.
33. C. U. Hjørringgaard, B. S. Vad, V. V. Matchkov, S. B. Nielsen, T. Vosegaard, N. C. Nielsen, D. E. Otzen and T. Skrydstrup, *J. Phys. Chem. B*, 2012, **116**, 7652–7659.
34. A. Ogrel, V. I. Shvets, B. Kaptein, Q. B. Broxterman and J. Raap, *Eur. J. Org. Chem.*, 2000, 857–859.
35. C. U. Hjørringgaard, J. M. Pedersen, T. Vosegaard, N. C. Nielsen and T. Skrydstrup, *J. Org. Chem.*, 2009, **74**, 1329–1332.
36. C. Aisenbrey and B. Bechinger, *J. Am. Chem. Soc.*, 2004, **126**, 16676–16683.
37. S. D. Cady, C. Goodman, C. D. Tatko, W. F. DeGrado and M. Hong, *J. Am. Chem. Soc.*, 2007, **129**, 5719–5729.

38. A. A. Nevzorov and S. J. Opella, *J. Magn. Reson.*, 2007, **185**, 59–70.
39. S. Wi and C. Kim, *J. Phys. Chem. B*, 2008, **112**, 11402–11414.
40. S. J. Opella and F. M. Marassi, *Chem. Rev.*, 2004, **104**, 3587–3606.
41. S. F. Poget and M. E. Girvin, *Biochim. Biophys. Acta*, 2007, **1768**, 3098–3106.
42. D. E. Warschawski, A. A. Arnold, M. Beaugrand, A. Gravel, É. Chartrand and I. Marcotte, *Biochim. Biophys. Acta*, 2011, **1808**, 1957–1974.
43. B. Bechinger and S. J. Opella, *J. Magn. Reson.*, 1991, **95**, 585–588.
44. C. Glaubitz and A. Watts, *J. Magn. Reson.*, 1998, **130**, 305–316.
45. C. Sizun and B. Bechinger, *J. Am. Chem. Soc.*, 2002, **124**, 1146–1147.
46. T. A. Cross and S. J. Opella, *Biochem. Biophys. Res. Commun.*, 1980, **92**, 478–484.
47. J. Bian and M. F. Roberts, *Biochemistry*, 1990, **29**, 7928–7935.
48. C. R. Sanders and J. P. Schwonek, *Biochemistry*, 1992, **31**, 8898–8905.
49. C. R. I. I. Sanders, B. J. Hare, K. P. Howard and J. H. Prestegard, *Progr. Nucl. Magn. Reson.*, 1994, **26**, 421–444.
50. R. R. Vold and R. S. Prosser, *J. Magn. Reson. B*, 1996, **113**, 267–271.
51. R. R. Vold, R. S. Prosser and A. J. Deese, *J. Biomol. NMR*, 1997, **9**, 329–335.
52. C. R. Sanders and R. S. Prosser, *Structure*, 1998, **6**, 1227–1234.
53. R. S. Prosser, S. A. Hunt, J. A. DiNatale and R. R. Vold, *J. Am. Chem. Soc.*, 1996, **118**, 269–270.
54. S. H. Park, C. Loudet, F. M. Marassi, E. J. Dufourc and S. J. Opella, *J. Magn. Reson.*, 2008, **193**, 133–138.
55. S. H. Park, S. Berkamp, G. A. Cook, M. K. Chan, H. Viadiu and S. J. Opella, *Biochemistry*, 2011, **50**, 8983–8985.
56. C. Aisenbrey, C. Sizun, J. Koch, M. Herget, R. Abele, B. Bechinger and R. Tampe, *J. Biol. Chem.*, 2006, **281**, 30365–30372.
57. S. H. Park, A. A. Mrse, A. A. Nevzorov, A. A. De Angelis and S. J. Opella, *J. Magn. Reson.*, 2006, **178**, 162–165.
58. R. C. Page, S. Kim and T. A. Cross, *Structure*, 2008, **16**, 787–797.
59. E. Strandberg, S. Esteban-Martìn, J. Salgado and A. S. Ulrich, *Biophys. J.*, 2009, **96**, 3223–3232.
60. A. Holt, L. Rougier, V. Réat, F. Jolibois, O. Saurel, J. Czaplicki, J. A. Killian and A. Milon, *Biophys. J.*, 2010, **98**, 1864–1872.
61. J. Wang, J. Denny, C. Tian, S. Kim, Y. Mo, F. Kovacs, Z. Song, K. Nishimura, Z. Gan, R. Fu, J. R. Quine and T. A. Cross, *J. Magn. Reson.*, 2000, **144**, 162–167.
62. F. M. Marassi and S. J. Opella, *J. Magn. Reson.*, 2000, **144**, 150–155.
63. H. Duclohier, *Biochim. Biophys. Acta*, 1999, **1420**, 14–22.
64. M. Sakoh, T. Okazaki, Y. Nagaoka and K. Asami, *Biochim. Biophys. Acta*, 2003, **1612**, 117–121.
65. H. Duclohier, G. Alder, K. Kociolek and M. T. Leplawy, *J. Pept. Sci.*, 2003, **9**, 776–783.
66. G. A. Woolley, *Chem. Biodiversity*, 2007, **4**, 1323–1337.

CHAPTER 15

Solid-State NMR Studies of Antimicrobial Peptide Interactions with Specific Lipid Environments

MARC-ANTOINE SANI AND FRANCES SEPAROVIC*

School of Chemistry and Bio21 Institute, University of Melbourne, VIC 3010, Australia
*Email: fs@unimelb.edu.au

15.1 Introduction

Antimicrobial peptides (AMPs) that target the lipid membrane of bacteria with higher affinity than eukaryotic cells are very promising antibiotic alternatives as they induce little resistance and are active against multi-drug resistant strains.[1–5] AMPs adopt their active conformation when they encounter a lipid surface and, therefore, study of their mode of action is performed in a specific model membrane or lipid environment.[2,6] Solid-state NMR has been widely applied to study model membranes and is a powerful technique to determine the interactions between these membrane-active peptides and membranes of specific lipid compositions.[7,8] Model membranes are often used as they are less prone to degradation during long experiments and because monitoring of specific interactions can be achieved. Phospholipids are the main components of lipid membranes, so diacylphosphatidyl lipids are commonly used to reproduce the bilayer

New Developments in NMR No. 3
Advances in Biological Solid-State NMR: Proteins and Membrane-Active Peptides
Edited by Frances Separovic and Akira Naito

Published by the Royal Society of Chemistry, www.rsc.org

structure of natural membranes. Phosphorus-31 NMR is frequently used to study the phospholipid headgroup region, while the hydrophobic region is investigated with deuterium (^{2}H) NMR by doping the membrane with acyl-chain deuterated lipids. NMR experiments can be performed to determine the perturbations induced by the AMPs on lipid organization and dynamics, based on the orientation dependence of the magnetic properties of the observed nuclei.[9–15]

15.2 NMR Interactions and Motional Timescales Relevant to Lipid Membranes

The chemical shift anisotropy, dipolar interactions and, for nuclei of spin $>\frac{1}{2}$, the quadrupolar interaction contribute to the shape and width of the NMR spectrum. In solution NMR, fast tumbling molecules display sharp resonances (full width at half maximum, FWHM $\ll$ 0.1 ppm) at the isotropic chemical shift, with scalar couplings splitting the resonances depending on the atomic chemical environment (for a good description, see Keeler[16] and Levitt[17]). In the solid-state, spectra are broadened and can span hundreds of ppm, so that the resonances from nuclei in different environments or from different chemical groups can overlap. Adequate resolution is often difficult to obtain without using special techniques to average the orientation-dependent interactions. An overview is here presented of the interaction between the nuclei and the magnetic field and high-resolution solid-state NMR methods used to study lipid membranes and AMPs.

15.2.1 Chemical Shift Anisotropy

Nuclei in a magnetic field experiencing different electronic environments resonate at different frequencies. The strength of the magnetic interaction is directly proportional to the spectrometer magnetic field, $\boldsymbol{B}_0$. The induced local fields on the nuclei are described by three principal axes and three principal values, δ_{11}, δ_{22} and δ_{33}, which define the chemical shift tensor. If two or more of the principal values are non-degenerate, the tensor is said to be anisotropic, resulting in chemical shift anisotropy (CSA). The isotropic chemical shift of a nucleus, $\delta_0 = (\delta_{11} + \delta_{22} + \delta_{33})/3$, whereas the chemical shift is angular dependent, according to:[9]

$$\delta(\theta, \phi) = I_z\, \delta_0 \left[3\cos^2\theta - 1 - \eta \sin^2\theta \cos(2\phi)\right] \tag{15.1}$$

where I_z is the nuclear spin, θ and ϕ are the Euler angles used to describe the tensor in the laboratory frame, and η is the asymmetry parameter described as $\eta = (\delta_{22} - \delta_{11})/\delta_{33}$.

The CSA interactions dominate for nuclei of spin $I = \frac{1}{2}$, such as ^{31}P. The CSA and asymmetry values give an indication of the orientation and dynamics of re-orientation and consequently serve as a probe to study the perturbation induced by AMP on lipid membranes.

15.2.2 Dipole–Dipole Interactions

Two nuclei of spin, I, separated by a distance, r, will interact *via* the through-space dipole–dipole interactions:[9]

$$D_{i,j}(\theta,\phi)=\frac{\mu_0\hbar\gamma_i\gamma_j}{8\pi^2r^3}(A+B+\ldots) \tag{15.2}$$

where μ_0 is the vacuum permittivity, $\hbar$ is the reduced Planck constant, $\gamma_{i,j}$ is the gyromagnetic ratio of each individual spin, r is the distance between the two nuclei, $A=-I_{1z}I_{2z}(3\cos^2\theta-1)$ describes the interaction between heteronuclear spins, and $B=\frac{1}{4}[I_{1+}I_{2-}+I_{1-}I_{2+}](3\cos^2\theta-1)$ is for homonuclear dipolar interactions. Additional terms (C–F) do not contribute significantly to the dipolar interactions, especially at high $\boldsymbol{B}_0$ field.

The pre-factor term is called the dipolar coupling constant and is independent of the molecular orientation but is coupled to the laboratory frame by the terms A and B. This leads to a direct effect on NMR spectra seen as an additional broadening of the observed lineshape.[9,18] Dipolar couplings can be used to measure the distance between two nuclei[19,20] and, therefore, assess intramolecular (*e.g.* AMP conformation) or intermolecular (*e.g.* AMP–lipid) distances.[21,22]

15.2.3 Quadrupolar Interactions (for Spin $I>\frac{1}{2}$)

Nuclei of spin $I>\frac{1}{2}$ possess an electric quadrupole moment (eQ), and the associated electric field gradient tensor interacts strongly with the magnetic field to produce the typical quadrupolar splitting pattern. The quadrupolar interactions dominate other interactions, except the Zeeman interaction with $\boldsymbol{B}_0$. For spin $I=1$, such as ^{2}H or ^{14}N nuclei, the spectrum of an isolated nucleus at a particular orientation displays a doublet separated by the quadrupolar splitting, according to:[15]

$$\Delta\nu_Q=\frac{3}{4}\frac{e^2qQ}{h}[(3\cos^2\theta-1)+\eta\cos^2(2\alpha)\sin^2\theta] \tag{15.3}$$

where e^2qQ/h is the quadrupolar constant (*ca.* 167 kHz for ^{2}H), θ and α are the Euler angles used to describe the tensor in the laboratory frame, and η is the asymmetry parameter.

From the measured quadrupolar splitting, the order parameter of the nucleus can be obtained by rewriting Eqn (15.3) as:

$$\Delta\nu_Q=\frac{3}{4}\frac{e^2qQ}{h}\left[S_z+\frac{1}{3}\eta(S_x-S_y)\right] \tag{15.4}$$

with $S_{x,y,z}$ representing the degree of re-orientational order induced by local molecular motions. In simple cases where fast reorientations average the quadrupolar tensor, $S_x=S_y$, and the quadrupolar splitting observed is directly proportional to the order parameter of the segmental bond of the observed nucleus relative to $\boldsymbol{B}_0$.

15.2.4 Magic Angle Spinning Techniques

Interactions with the magnetic field that are angle dependent can be mechanically averaged out by using magic angle spinning (MAS). Sample rotation (0.2 to 70 kHz) similar to NMR timescales modulates the anisotropic spin interactions such as CSA, dipole–dipole and quadrupole. By tilting the sample at the so-called magic angle, $\theta = \arccos\sqrt{1/3} \approx 54.74°$, and using a spinning frequency, ν_r, larger than the dipolar couplings (Eqn 15.2), the spectrum is reduced to a single resonance at the isotropic chemical shift, δ_0.[23]

If the speed is not greater than the CSA, spinning sidebands will be observed at frequencies spaced at multiples of the spinning frequency ($\delta_0 \pm N\nu_r$). The sideband intensities follow the relation:[24,25]

$$I(N, \nu_r) = \frac{1}{4\pi^2}\int_0^{2\pi} d\alpha \int_0^{\pi} P_2(\theta) d\theta \frac{1}{4\pi^2}\left\{\left[\int_0^{2\pi} d\gamma \cos(\Phi')\right]^2 + \left[\int_0^{2\pi} d\gamma \sin(\Phi')\right]^2\right\} \tag{15.5}$$

where $P_2(\theta) \propto \sin(\theta)$ for a spherical distribution of the nuclear tensor, and Φ' gives the contribution to the sideband, N, at a spinning speed of ν_r by a crystallite with chemical shift parameters δ and η, at an angle (α,θ,γ) with respect to $\boldsymbol{B}_0$:

$$\Phi' = N\gamma + \frac{\delta\eta}{\nu_r}\Phi_A - \frac{\delta}{\nu_r}\Phi_B, \text{ and}$$

$$\begin{aligned}\Phi_A = {} & \frac{1}{24}\cos 2\alpha(3 + \cos 2\theta)\sin 2\gamma + \frac{\sqrt{2}}{6}\cos 2\alpha \sin 2\theta \sin\gamma \\ & + \frac{\sqrt{2}}{3}\sin 2\alpha \sin\theta \cos\gamma + \frac{1}{6}\sin 2\alpha \cos\theta \cos 2\gamma\end{aligned} \tag{15.6}$$

$$\Phi_B = \frac{1}{8}(1 - \cos 2\theta)\sin 2\gamma - \frac{1}{\sqrt{2}}\sin 2\theta \sin\gamma$$

Gehman *et al.* have used slow spinning techniques to extract accurate information on AMP interacting with multilamellar vesicles (MLV) mimicking *Escherichia coli* membranes (see Section 15.3), where the chemical shift parameters δ and η were obtained with high accuracy.[25]

15.2.5 The Timescale of Lipid Molecular Motions in Bilayer Structures

Solid-state NMR can provide information on lipid motions within a bilayer in the range from seconds to nanoseconds (Figure 15.1).[11,13,15,26,27] The longitudinal (along $\boldsymbol{B}_0$) relaxation time, T_1, is sensitive to the density of local fluctuations on a timescale of $(\gamma_i B_0)^{-1} \approx 10^{-9}$ s. Therefore, T_1 measurements can report on perturbation of individual lipid dynamics upon introduction of AMPs. The transverse (perpendicular to $\boldsymbol{B}_0$) relaxation rate, T_2, is sensitive

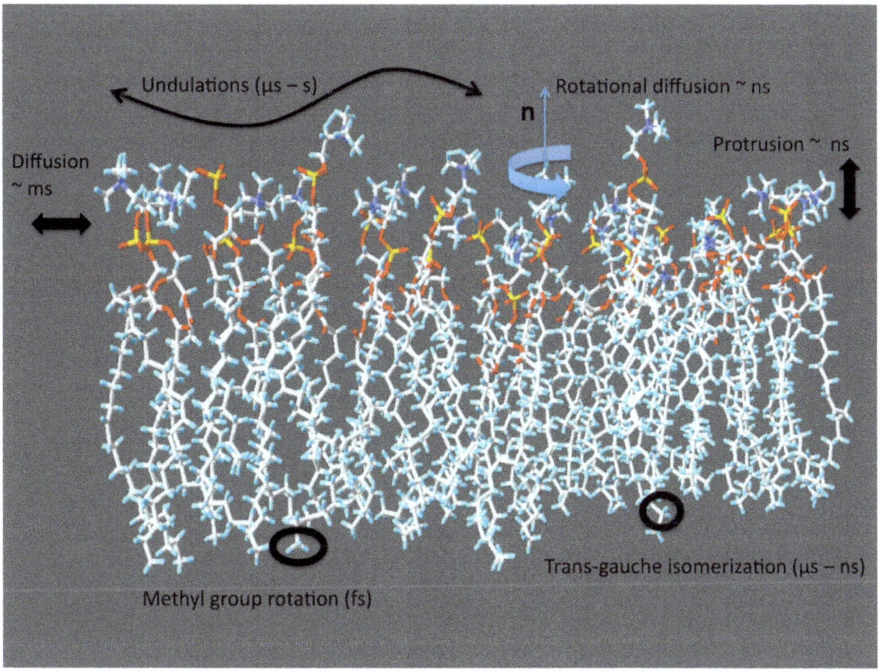

Figure 15.1 Timescales of phospholipid dynamics experienced in a membrane bilayer.

to random fluctuations in the local magnetic field, which for lipids corresponds to collective motions on a timescale of $\sim 10^{-3}$ s.

15.3 Lipid Membrane Systems Relevant for NMR Studies of AMPs

15.3.1 Liposomes

Multilamellar vesicles are frequently used as model membrane systems, where lipids are packed within concentric bilayers that form a vesicular macrostructure of size ranging from about 300 nm to over 1000 nm diameter. NMR of phospholipids in a vesicular structure yield a typical powder pattern spectrum where the spectral density at the chemical shift $\delta(\theta)$ is determined by:[10,18]

$$\delta(\theta) = \frac{\sin\theta}{\partial\delta(\theta)\,/\,\partial\theta} \tag{15.7}$$

This model system has advantages and inconveniences: the pros are that almost any lipid composition can be used; and MLVs are very stable in time within a broad range of temperatures, hydration levels and buffer compositions. The cons are that the lipid to peptide ratio is difficult to regulate due to indeterminate peptide distribution through the multilayers;

the inter-bilayer space can limit the free diffusion of the peptides; and the membrane curvature increases towards the centre of the MLVs, which could modulate interactions with some AMPs. In terms of NMR conditions, the lower hydration levels often nicely improve the signal-to-noise with a high quantity of sample centred in the coil of the probe. The quantity of material to obtain a good signal-to-noise (≥5 mg of lipids) in a short time-frame (~5 h) is reasonable. If needed, MAS experiments can be performed with little risk of shearing and thermal degradation.

Digression 1: Rotor packing

Packing an NMR rotor might look like a black art, but is a critical step in the experiment preparation since spinning stability and signal distortion problems can easily be encountered if not done properly. The usual low viscosity [$0.5 < \text{mass}_{\text{buffer}}/(\text{mass}_{\text{buffer}} + \text{mass}_{\text{lipids}}) < 0.8$] of an MLV sample makes it tricky to deliver into narrow rotors (especially 3.2 mm and 4 mm o.d. rotors). Fortunately, there exists an easy and rapid way to pack an MLV sample without losing a significant amount of material in the process. MLV samples usually necessitate several freeze–thaw cycles to improve the size homogeneity of the vesicles and diffuse external molecules through the multilayer structure. Before the last thawing step, the frozen sample can be easily transferred into a micropipette tip (1 mL tips work very well), which can be inserted into the pre-assembled rotor, fixed into an Eppendorf. The sample can then be spun down the rotor by using a benchtop centrifuge at around 2500 rpm. Ideal packing should be obtained.

Owing to their size, MLVs tumble slowly on the NMR timescale ($\tau_c > 10^{-5}$ s). This broadens the spectral features due to a strong angular dependence of the interactions between the nuclei and the static magnetic field, as described above. For instance, ^{31}P NMR experiments are often used to study the molecular organization of the phospholipids. The typical lineshapes obtained for possible lipid arrangements are displayed in Figure 15.2. When AMPs interact with the lipid membrane, they often induce severe changes in the lipid organization, which are reflected in the spectral lineshape.

Digression 2: Temperature control under MAS

Notably, depending on the rotor diameter and spinning frequency, the temperature can rise significantly under MAS. The temperature difference obtained for our samples packed into a 3.2 mm rotor was assessed by monitoring the chemical shift of lead nitrate at various MAS frequencies. We found that the temperature inside the rotor followed the empirical formula:

$$T_{\text{inside}}(^{\circ}\text{C}) = T_{\text{set}}(^{\circ}\text{C}) - 2\,^{\circ}\text{C} + 0.046\,^{\circ}\text{C} \times \text{speed}(\text{kHz})^2$$

Of course, this formula is specific to the probe, the pneumatic apparatus that controls the bearing and drive pressures, and the type of rotor used.

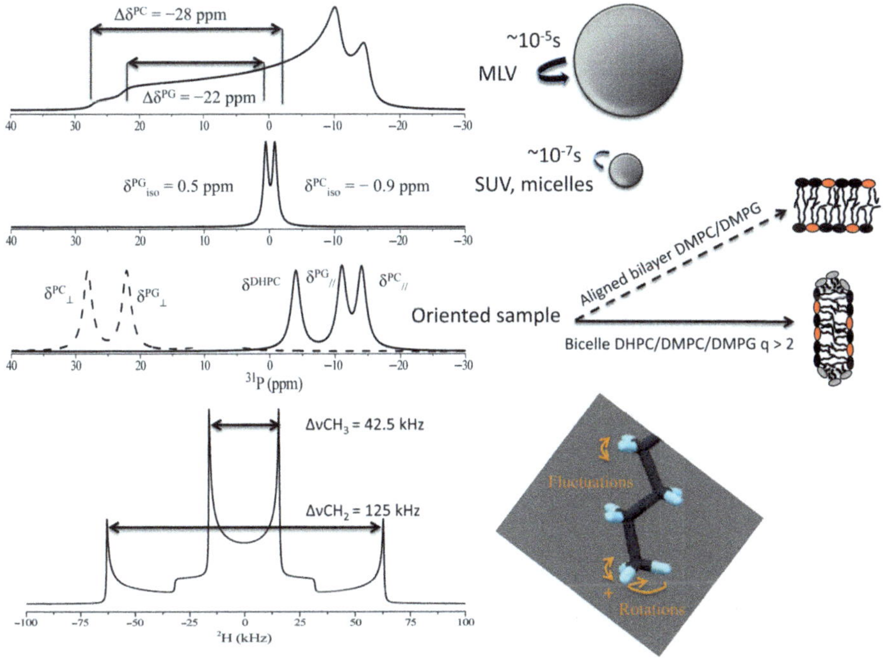

Figure 15.2 Typical orientation and motion modulated NMR ^{31}P and ^{2}H lineshapes. *Top spectrum*: slow tumbling MLVs of DMPC/DMPG (1 : 1 molar ratio) in the fluid phase; *middle spectrum*: fast tumbling SUVs of DMPC/DMPG (1 : 1 molar ratio); and *bottom spectra*: aligned bilayers with a parallel (*solid line*, bicelles with *black* DMPC, *red* DMPG and *grey* DHPC, 1 : 1 : 1 molar ratio) or perpendicular (*dashed line*, glass plate type, DMPC/DMPG 1 : 1 molar ratio) orientation. *Bottom panel*: ^{2}H quadrupolar lineshapes of a CH_2 and a CH_3 lipid segment. The ^{31}P and ^{2}H NMR spectra were simulated using an Hahn echo and a quadrupolar echo, respectively, with the SIMPSON program.[49,50] The parameters for ^{31}P and ^{2}H experiments were: $\tau_{hecho} = \tau_{qecho} = 40$ μs and line broadening (LB) = 200 Hz.

Unilamellar vesicles, either small (SUVs, diameter < 50 nm) or large (LUVs, diameter ~100–400 nm), are impractical for studies on the impact of AMPs on lipid membranes. SUVs tumble rapidly and, therefore, average out interactions to result in the isotropic chemical shift. However, SUVs are suitable for NMR experiments that study the peptide conformation or to determine distance to the lipid surface, although the high curvature may modulate the lipid–peptide interactions. Also, SUVs tend to have a high aggregation/fusion rate, which can be an issue for longer experiments. LUVs present similar features to SUVs excepting that for a diameter of 400 nm the tumbling rate drops and a powder pattern is recovered. Unfortunately, they are not structurally stable and break down into lipid bilayer fragments or fuse to form an MLV structure with time.

15.3.2 Mechanically and Magnetically Oriented Bilayers

Another method of reducing the orientation dependence of the CSA, dipole–dipole and quadrupolar interactions is to impose a single orientation of the lipid bilayers with respect to $\boldsymbol{B}_0$. Two techniques used to obtain aligned bilayers are:

(i) Lipids can be deposited onto glass plates, inducing stacked and aligned bilayers parallel to the glass plate.[28] The lipid mixtures that align well are limited and a flat coil is often used. A goniometric probe to tilt the sample at specific angles may be required. Note that the coil angle with respect to $\boldsymbol{B}_0$ modulates the pulse transfer efficiency, with a perpendicular orientation being optimum. Sample preparation may take several days, with the hydrating process often being tedious and generating inhomogeneities during long experimental times. The temperature range can be limited and a significant quantity of material is required (>10 mg lipid).

(ii) Lipid bicelles may spontaneously align with the lipid long axis either perpendicular or parallel (by addition of lanthanide ions or the use of biphenyl lipids) to $\boldsymbol{B}_0$.[29–31] The lipid compositions forming a bicelle structure are very restricted, and must contain a short chain lipid, such as a six-carbon acyl chain phosphatidylcholine (DHPC), to form the edge of a disk-like bilayer structure. Also, bicelles orient only in restricted temperature and hydration ranges, which depend on the lipid composition.[32] The long-to-short chain ratio, q, dictates the size of the bicelles and the temperature/hydration level at which they align.

Few lipid mixtures appear to form a bicelle structure, mainly DHPC lipids mixed with either dimyristoylphosphatidylcholine (DMPC) or sphingomyelin, DMPC with up to 50 mol% dimyristoylphosphatidylglycerol (DMPG), DMPC with up to 20 mol% cardiolipin, or DMPC with up to 10 mol% dimyristoylphosphatidylethanolamine (DMPE).[33] Longer chains, unsaturated lipids, ceramides and gangliosides can also produce aligned bicelles, but with narrow composition and temperature ranges.[34] Other short-chain lipids or detergents are also being investigated to replace DHPC as well as the use of ether instead of ester lipids to extend the conditions and stability of aligned bicelles.[35–38]

Interestingly, the addition of lanthanide trivalent cations induces a flip by 90° of the lipid long axis.[29,39] A similar effect is produced by replacing DMPC by a biphenylphosphatidylcholine (TBBPC), which extends significantly the temperature range but with the caveat of requiring a minimum of 95% of TBBPC as a long-chain lipid. Having the lipid long axis parallel to $\boldsymbol{B}_0$ extends the chemical shift range or quadrupolar splitting and higher resolution is obtained for aligned molecules.

15.4 Perturbation of Phospholipid Membranes by Frog AMPs

Antimicrobial peptides derived from the skin secretions of Australian tree frogs may provide valuable leads as a new class of antibiotics based upon their selectivity for prokaryotic *versus* mammalian cell membranes.[40,41] These AMPs are short membrane-active cationic peptides, typically ranging from 13 to 25 amino acids, that are unstructured in aqueous solution but adopt helical conformations in the presence of lipids.[6] Understanding the relationship between AMP activity and the composition of bacterial lipid membranes is important. Although a similar conformational change is observed for these frog AMPs, their activity against bacterial strains are very different.[42,43] For instance, the 13-residue peptide aurein 1.2 is very potent against Gram-negative bacteria, while the 21-residue maculatin 1.1 is very potent against Gram-positive bacteria. Upon binding to the lipid surface, they disrupt the lipid packing and dynamics of model membranes.[6,25,40,42,44–47] Solid-state NMR is an excellent tool to study the specific perturbations of lipid membranes induced by AMPs, with ^{31}P NMR to study the phospholipid headgroup region and ^{2}H NMR to probe perturbations of the hydrophobic region of the membrane, thus providing a means to study AMP activity at a molecular level and establish which membrane parameters modulate their specific activities.

15.4.1 The Effect of Lipid Composition on AMP Activity

Bacterial membranes contain high amounts of anionic phospholipids in comparison to eukaryotic cells. The net negatively charged surface attracts the cationic peptides, and most likely is responsible for a higher affinity for bacterial over host cells.[2] Solid-state NMR studies of aurein 1.2 and maculatin 1.1 with neutral DMPC MLVs *versus* the negatively charged DMPC/DMPG system have provided valuable insights into the different molecular mechanisms by which frog AMPs lyse lipid membranes.

Aurein 1.2 decreased by less than 8% the quadrupolar splittings of deuterated acyl chains, indicating little penetration of the peptide into the hydrophobic core of DMPC MLVs.[44,47] However, the peptide decreased the ^{31}P CSA by 15% and the T_1 and T_2 values by more than 50%.[41] This, together with the peptide being only 13 residues in length, indicates that it associated with the zwitterionic lipid headgroups in a manner that propagates part of the way down the hydrophobic acyl chains but not significantly beyond the hydrophilic/hydrophobic interface. In the anionic lipid system, aurein 1.2 induced similar but lesser changes in the quadrupolar splittings, ^{31}P CSA, T_1 and T_2. However, using the MAS technique to resolve DMPC and DMPG isotropic chemical shifts, T_1 and T_2 values could be obtained for each lipid. Aurein 1.2 disturbed more the DMPG lipids and may have segregated the anionic lipids from the zwitterionic DMPC. Unfortunately, deuterated DMPG was not used in this study to monitor if DMPG acyl chains were disturbed more significantly. Only a single ^{31}P CSA was observed, although peptide-free and bound lipids could provide two different CSAs. Overall, these NMR studies demonstrate that aurein 1.2 seems to be a surface-active peptide with a small preference for anionic lipids. Similarly for the related peptides, aurein 2.2 and 2.3, oriented circular dichroism and ^{31}P solid-state NMR at lower hydration levels showed little discrimination between anionic and neutral bilayers.[48]

The longer peptide, maculatin 1.1, significantly disturbed the headgroup region of DMPC MLVs. Indeed, the ^{31}P CSA was reduced by ~21% and, similar to aurein, the peptide triggered changes in dynamics with reduced T_1 and T_2 values but had even less effect on the deuterated acyl chains. Interestingly, maculatin 1.1 showed marked differences when DMPG was included in the bilayers.[41] The deuterated DMPC, which was disordered by the presence of DMPG, regained the characteristic order of pure DMPC in the presence of the peptide and ^{31}P CSA, T_1 and T_2 values were only slightly perturbed. Consequently, maculatin 1.1 appears to associate with zwitterionic lipid bilayers only at the surface and is not particularly effective at disrupting membrane integrity. However, in the mixed lipid bilayers, maculatin 1.1 inserts into the hydrophobic core and has greater interaction with the anionic lipid, such that the zwitterionic lipid is excluded by default into significantly enriched domains.[6,45]

15.4.2 The Effect of Lipid Phase on AMP Activity

Thermotropism is a critical property of lipid membranes. By controlling the fluidity of the lipid acyl chains, different cellular processes are regulated, such as cell division or protein activity. As well as non-lamellar structures, changes in lipid membrane fluidity, such as transitions from liquid-crystalline to ripple to gel phases, can influence the activity of AMP.

The effect of maculatin 1.1 and aurein 1.2 binding on bilayer structure through the fluid-to-gel lipid phase transition was observed by ^{31}P and ^{2}H NMR measurements at 30, 20 and 15 °C, and reheating to 30 °C.[45,47] In DMPC MLVs, lipid fluidity decreased the effect of the peptide. At 20 °C, DMPC MLVs are in the ripple phase, where coexistence of rigid and fluid lipid domains creates defects at the boundary of each phase. Aurein 1.2 and maculatin 1.1 induced significant lipid reorganization, with the appearance of an isotropic peak in the ^{31}P static NMR spectrum contributing to 5% and 30%, respectively, with maculatin 1.1 also resulting in a 50% reduction in CSA. A further decrease in temperature to 15 °C led to complete destabilization of the DMPC bilayer by both peptides, with the ^{31}P spectra showing a single isotropic line that indicated that small, rapidly diffusing lipid structures were formed. Re-heating the sample back to the fluid phase partially recovered the typical CSA powder pattern, but with an isotropic peak contributing to roughly half of the intensity and a slightly reduced CSA compared to the value originally obtained at 30 °C.

For the mixed DMPC/DMPG lipid bilayers, aurein 1.2 and maculatin 1.1 did not induce an isotropic peak in the ripple phase (20 °C) but did contribute to 52% and 15%, respectively, in the gel phase (15 °C). Upon reheating, the membrane bilayer structure was significantly restored, but aurein 1.2 preserved an isotropic peak contributing 15% intensity, while maculatin 1.2 produced only a severe reduction in the ^{31}P CSA by 40%.

Overall, the gel phase was more susceptible to the peptides with the appearance of an isotropic peak, which partially recovered into a lamellar organization when maculatin 1.1 was present. This is consistent with highly curved regions created inside a bilayer structure, such as encountered in a toroidal pore rather than disrupting the bilayer organization into small vesicles or aggregates, which is likely the mode of action of aurein 1.2. Furthermore, the electrostatic interaction seemed to reduce peptide insertion in the gel phase, with a higher amount of peptide on the surface decreasing the quantity inserted into the hydrophobic core and disrupting less the bilayer organization.

15.4.3 Tackling the Complexity of Lipid Mixtures

The difficulty in extracting ^{31}P CSA information for lipid mixtures was recently overcome by using slow MAS NMR and a Boltzmann maximum entropy analysis.[25] The approach uses 1D MAS spectra at slow spin rates, which also provides a sufficient signal/noise ratio in less time than a static spectrum. Furthermore, MAS reduces the complexities of lifetime broadening and magnetic field-induced lipid alignment that complicate interpretation of static spectra. The method is free of user-based initializations for deconvolution of static spectra to fit several CSA (δ), asymmetry (η) and spherical deformation factors to reproduce the experimental lineshape, and employs a Boltzmann-type maximum entropy strategy that gives a pseudo-continuous distribution of CSA and asymmetry parameters.

This approach was used to study the effect of 2 mol% maculatin 1.1 on a binary lipid mixture of 70% DMPE and 30% DMPG, which better approximates *E. coli* outer membranes[2,34] than the usual DMPC/DMPG systems (Figure 15.3). The study was conducted at 37 °C, between the gel-to-fluid phase transition for DMPE (50 °C) and DMPG (24 °C), which can create mixed lipid phases and lipid exchange between these domains. The static ^{31}P NMR spectra of these systems featured a number of overlapped broad line spectral components. Lipids in the fluid phase undergo rapid motions that produce a reduced symmetry static lineshape, whereas spectra of gel phase lipids are not averaged by long axis lipid rotation and produce a broad asymmetric line. With additional parameters needed to account for the broad linewidths inherent in the short lifetime signal, all attempts to deconvolute the static spectra into a small number of components proved unsatisfactory. MAS at 1.2, 1.5, and 1.8 kHz was used in a 14.1 T magnetic field to extract the CSA and asymmetry information contained in the three sets of spinning sidebands. In the peptide-free system, δ was centred

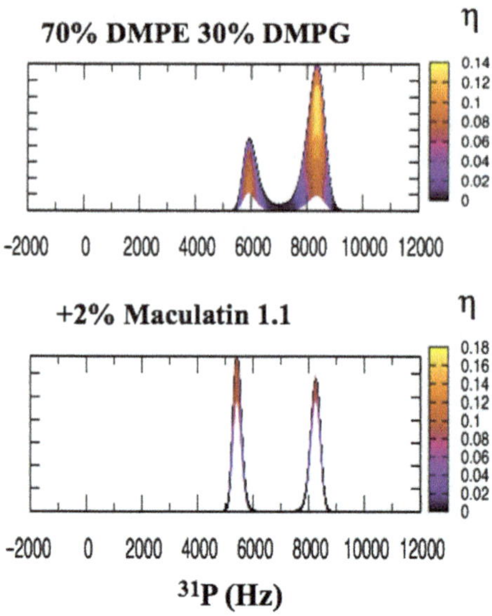

Figure 15.3 ^{31}P CSA and η distributions obtained from the Boltzmann maximum entropy analysis of MAS spectra of MLVs made of DMPE/DMPG (7 : 3) lipids (*upper panel*) and with the addition of 2% molar of maculatin 1.1. The peptide increased the component with a CSA of 22.3 ppm (5.4 kHz) and reduced the asymmetry of the component at 34.3 ppm (8.3 kHz). The experiments were performed at 37 °C on a 600 MHz spectrometer.

at ~34.3 ppm (or 8.3 kHz) and span about 8 ppm, and slightly asymmetric with $\eta \approx 0.1$–0.2. The addition of 2% maculatin 1.1 produced an increase in the isotropic component and the more disordered population with a CSA of $\Delta\delta \approx 23.3$ ppm (or 5.4 kHz), and a decreased asymmetry ($\eta \leq 0.1$) and intensity of the $\Delta\delta \approx 34.3$ ppm component. These changes are consistent with a peptide-induced lowering of the bulk lipid-phase transition temperature, or a change in lipid mixing characteristics.

The slow spinning technique coupled with the maximum entropy analysis revealed subtle potential lipid environment differences in lipid mixtures perturbed by a small amount of AMP. Another advantage of such an approach is fast acquisition, processing and analysis with minimal user-induced bias.

15.4.4 NMR Studies of Maculatin with Lipid Bicelles

The higher resolution obtained using oriented samples can lead to a more accurate description of specific interactions. For instance, the preferential interaction between DMPG and maculatin 1.1 was investigated in a mixed lipid bicelle system of DMPC/DMPG/DHPC (4 : 1 : 1.25). Figure 15.4 displays typical ^{31}P and ^{2}H NMR spectra of bicelles at 31°C. The ^{31}P chemical shift is sensitive to the orientation of the phospholipid headgroup with respect to

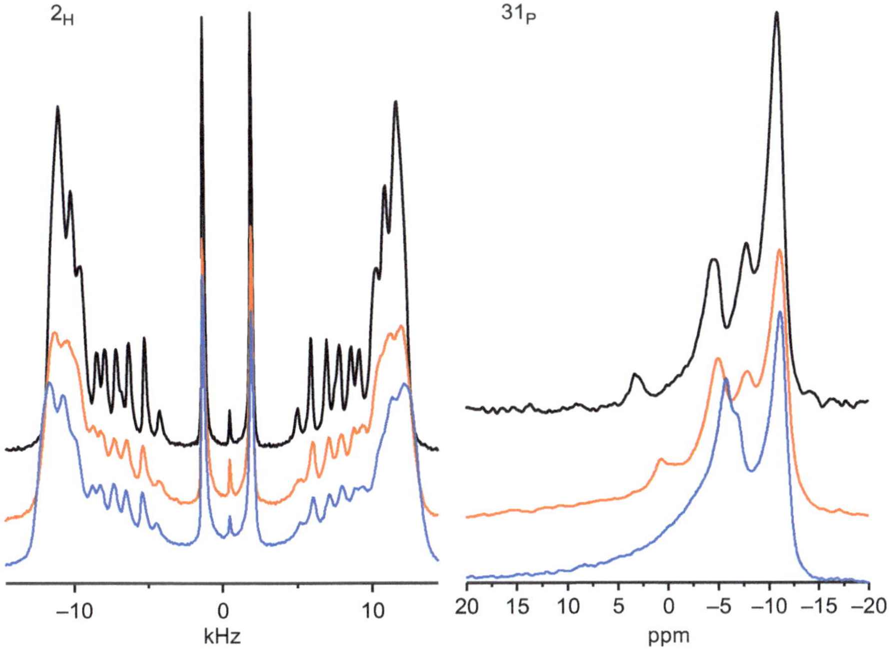

Figure 15.4 (*Left*) ^{2}H NMR spectra and (*right*) corresponding ^{31}P NMR spectra of d_{54}-DMPC/DMPC/DMPG/DHPC (2 : 2 : 1 : 1.25) bicelles (*black line*), after exogenous maculatin 1.1 addition (*red line*), and after maculatin 1.1 incorporation (*blue line*). Experiments were performed at 31 °C.

the magnetic field and is used to assess the degree of bicelle alignment. Similarly, aligned samples give rise to well-resolved ^{2}H spectra with discrete resonances able to discern the quadrupolar splittings.

In Figure 15.4, the ^{31}P static spectra of d_{54}-DMPC/DMPC/DMPG/DHPC (2 : 2 : 1 : 1.25) with lipids only (black line), maculatin 1.1 at a lipid: peptide molar ratio 30 : 1 added exogenously (red line) and incorporated during the lipid co-solubilization step (blue line) are compared. The ^{31}P resonances in the lipid-only spectrum were attributed to, from right to left, DMPC (~ -12 ppm), DMPG (~ -7 ppm) and DHPC (~ -4 ppm). The small peak centred at ~ 3 ppm arose from phosphate buffer, some of which may have coordinated to the headgroup region and caused the apparent broadening. The resolution of these peaks clearly indicates bicelle alignment (Figure 15.4). The corresponding ^{2}H spectrum also is consistent with sample alignment, evidenced by high resolution of single orientation acyl chain splittings compared to unoriented bilayers. With exogenous peptide addition, the intensities of the ^{31}P DMPC and DMPG peaks were attenuated slightly, indicating that the peptide interacts with the phospholipid headgroup region of both DMPC and DMPG. Importantly, however, the sample alignment appeared to be largely preserved. This was confirmed by the corresponding ^{2}H spectrum, which remained relatively well resolved.

Notably, the quadrupolar splitting, which is indicative of acyl chain order, was not significantly changed, indicating that the peptide did not perturb the order of the acyl chain region of DMPC. Incorporation of maculatin 1.1 resulted in a coalescence and broadening of the ^{31}P DMPG and DHPC peaks, which is indicative of stronger headgroup perturbations and disruption of bicelle alignment. In accord with the ^{2}H spectrum obtained after peptide addition, however, the quadrupolar splittings of DMPC were not significantly altered by peptide incorporation, indicating that incorporated peptide did not change the order of DMPC chains (although the change in lineshape indicates that the peptide appears to have changed lipid dynamics or T_2).

By incorporating deuterated DMPG instead of deuterated DMPC, differences in acyl chain order could be observed between the anionic and zwitterionic lipids after peptide addition and incorporation. In Figure 15.5, the ^{2}H spectra of the d_{54}-DMPG component of bicelles only (black line) are compared to bicelles after exogenous addition (red line) and incorporation (blue line) of maculatin 1.1. While exogenous addition of maculatin 1.1 decreased slightly the quadrupolar splittings of DMPG, and thus the order of the acyl chains, peptide incorporation had a similar but more pronounced effect. This could be due to maculatin 1.1 being distributed more evenly between the bilayers when "incorporated", or that peptide–lipid interactions may have reached a more advanced stage, with increased DMPG segregation. The peptide decreased the order of the acyl chain region, particularly near the glycerol backbone. Notably, these perturbations were not observed in the

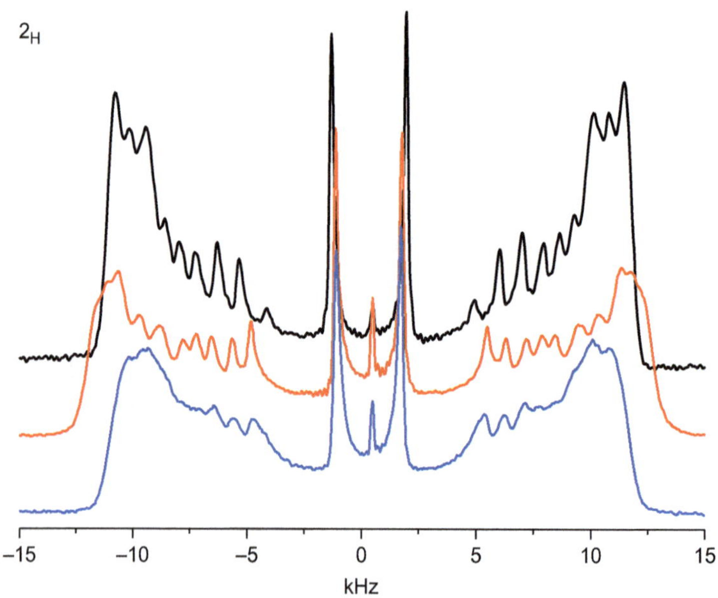

Figure 15.5 ^{2}H NMR spectra of DMPC/d_{54}-DMPG/DHPC (4 : 1 : 1.25) (*black line*), after exogenous maculatin 1.1 addition (*red line*), and after maculatin 1.1 incorporation (*blue line*). Experiments were performed at 31°C.

^{2}H spectra of d_{54}-DMPC, which suggests that the peptide selectively interacts with the DMPG component. Incorporation of the peptide significantly reduces DMPG order and may be associated with toroidal pore formation.

15.5 Conclusions and Perspectives

Both ^{31}P and ^{2}H solid-state NMR show the effect of AMP on phospholipid bilayers and give insight into their mechanism of action and also interactions with specific phospholipids. Particularly, a large array of membrane properties – curvature, lipid composition and thermotropism – and sample preparation, where incorporated *versus* added peptide can shed light on different equilibrium states, bring valuable insights into the mechanism of membrane-active peptides. In combination with other techniques, solid-state NMR can lead to design of more effective peptides with better discrimination between eukaryotic and bacterial membranes. Further work will concentrate on the peptide structure in model membranes, with a perspective focusing on peptide structure in bacterial cells and studies of peptide–lipid interactions in the membranes of live cells.

Acknowledgments

This work is supported in part by grants from the Australian Research Council and the National Health & Medical Research Council.

References

1. K. A. Brogden, *Nat. Rev.*, 2005, **3**, 238–250.
2. R. M. Epand, S. Rotem, A. Mor, B. Berno and R. F. Epand, *J. Am. Chem. Soc.*, 2008, **130**, 14346–14352.
3. R. E. W. Hancock and G. Diamond, *Trends Microbiol.*, 2000, **8**, 402–410.
4. H. Jenssen, P. Hamill and R. E. Hancock, *Clin. Microbiol. Rev.*, 2006, **19**, 491–511.
5. Y. Shai, *Biochim. Biophys. Acta*, 1999, **1462**, 55–70.
6. M. A. Sani, T. C. Whitwell and F. Separovic, *Biochim. Biophys. Acta, Biomembr.*, 2012, **1818**, 205–211.
7. M. Auger, *Curr. Issues Mol. Biol.*, 2000, **2**, 119–124.
8. A. Grelard, C. Loudet, A. Diller and E. J. Dufourc, *Methods Mol. Biol.*, 2010, **654**, 341–359.
9. A. Abragam, *The Principles of Nuclear Magnetism*, Clarendon Press, Oxford, 1961.
10. B. Blümich, R. Born and H. W. Spiess, *Solid State NMR*, Springer, Berlin, 1994.
11. E. J. Dufourc, C. Mayer, J. Stohrer, G. Althoff and G. Kothe, *Biophys. J.*, 1992, **61**, 42–57.
12. J. M. Pope, L. Walker, B. A. Cornell and F. Separovic, *Mol. Cryst. Liq. Cryst.*, 1982, **89**, 137–150.

13. F. Separovic, B. Cornell and R. Pace, *Chem. Phys. Lipids*, 2000, **107**, 159–167.
14. F. Separovic, R. Pax and B. Cornell, *Mol. Phys.*, 1993, **78**, 357–369.
15. J. H. Davis, *Biochim. Biophys. Acta*, 1983, **737**, 117–171.
16. J. Keeler, *Understanding NMR spectroscopy*, Wiley, Chichester, 2nd edn, 2010.
17. M. H. Levitt, *Spin Dynamics: Basics of Nuclear Magnetic Resonance*, Wiley, Chichester, 2nd edn, 2008.
18. M. Mehring, *Principles of High-Resolution NMR in Solids*, Springer, Berlin, 2nd edn, 1983.
19. T. Gullion and J. Schaefer, *Adv. Magn. Reson.*, 1989, **13**, 57–83.
20. T. Gullion and J. Schaefer, *J. Magn. Reson.*, 1989, **81**, 196–200.
21. A. Drechsler and F. Separovic, *IUBMB Life*, 2003, **55**, 515–523.
22. M. Hong and K. Schmidt-Rohr, *Acc. Chem. Res.*, 2013.
23. E. R. Andrew, A. Bradbury and R. G. Eades, *Nature*, 1959, **183**, 1802–1803.
24. J. Herzfeld and A. E. Berger, *J. Chem. Phys.*, 1980, **73**, 6021–6030.
25. M. A. Sani, F. Separovic and J. D. Gehman, *Biophys. J.*, 2011, **100**, L40–L42.
26. J. Frye, A. D. Albert, B. S. Selinsky and P. L. Yeagle, *Biophys. J.*, 1985, **48**, 547–552.
27. M. P. Milburn and K. R. Jeffrey, *Biophys. J.*, 1987, **52**, 791–799.
28. B. Bechinger, Y. Kim, L. E. Chirlian, J. Gesell, J. M. Neumann, M. Montal, J. Tomich, M. Zasloff and S. J. Opella, *J. Biomol. NMR*, 1991, **1**, 167–173.
29. A. Diller, C. Loudet, F. Aussenac, G. Raffard, S. Fournier, M. Laguerre, A. Grelard, S. J. Opella, F. M. Marassi and E. J. Dufourc, *Biochimie*, 2009, **91**, 744–751.
30. C. Loudet, S. Manet, S. Gineste, R. Oda, M. F. Achard and E. J. Dufourc, *Biophys. J.*, 2007, **92**, 3949–3959.
31. C. R. Sanders, 2nd and J. P. Schwonek, *Biochemistry*, 1992, **31**, 8898–8905.
32. M. N. Triba, D. E. Warschawski and P. F. Devaux, *Biophys. J.*, 2005, **88**, 1887–1901.
33. I. Marcotte, E. J. Dufourc, M. Ouellet and M. Auger, *Biophys. J.*, 2003, **85**, 328–339.
34. D. E. Warschawski, A. A. Arnold, M. Beaugrand, A. Gravel, E. Chartrand and I. Marcotte, *Biochim. Biophys. Acta.*, 2011, **1808**, 1957–1974.
35. A. Andersson and L. Maler, *Langmuir*, 2005, **21**, 7702–7709.
36. O. V. Nolandt, T. H. Walther, S. L. Grage and A. S. Ulrich, *Biochim. Biophys. Acta*, 2012, **1818**, 1142–1147.
37. C. R. Sanders and R. S. Prosser, *Structure*, 1998, **6**, 1227–1234.
38. M. N. Triba, P. F. Devaux and D. E. Warschawski, *Biophys. J.*, 2006, **91**, 1357–1367.
39. R. S. Prosser, H. Bryant, R. G. Bryant and R. R. Vold, *J. Magn. Reson.*, 1999, **141**, 256–260.
40. D. I. Fernandez, J. D. Gehman and F. Separovic, *Biochim. Biophys. Acta*, 2009, **1788**, 1630–1638.

41. J. D. Gehman, F. Luc, K. Hall, T. H. Lee, M. P. Boland, T. L. Pukala, J. H. Bowie, M. I. Aguilar and F. Separovic, *Biochemistry*, 2008, **47**, 8557–8565.
42. J. H. Bowie, F. Separovic and M. J. Tyler, *Peptides*, 2012, **37**, 174–188.
43. C. S. Chia, Y. Gong, J. H. Bowie, J. Zuegg and M. A. Cooper, *Biopolymers*, 2011, **96**, 147–157.
44. D. I. Fernandez, A. P. Le Brun, T. C. Whitwell, M. A. Sani, M. James and F. Separovic, *Phys. Chem. Chem. Phys.*, 2012, **14**, 15739–15751.
45. D. I. Fernandez, T. H. Lee, M. A. Sani, M. I. Aguilar and F. Separovic, *Biophys. J.*, 2013, **104**, 1495–1507.
46. D. I. Fernandez, M. A. Sani, J. D. Gehman, K. S. Hahm and F. Separovic, *Eur. Biophys. J.*, 2011, **40**, 471–480.
47. D. I. Fernandez, M. A. Sani, A. J. Miles, B. A. Wallace and F. Separovic, *Biochim. Biophys. Acta*, 2013, **1828**, 1863–1872.
48. J. T. Cheng, J. D. Hale, M. Elliott, R. E. Hancock and S. K. Straus, *Biochim. Biophys. Acta*, 2011, **1808**, 622–633.
49. M. Bak, J. T. Rasmussen and N. C. Nielsen, *J. Magn. Reson.*, 2000, **147**, 296–330.
50. M. Bak, J. T. Rasmussen and N. C. Nielsen, *J. Magn. Reson.*, 2011, **213**, 366–400.

CHAPTER 16

Dynamic Structure Analysis of Peptides in Membranes by Solid-State NMR

ERIK STRANDBERG[a] AND ANNE S. ULRICH*[a,b]

[a] Karlsruhe Institute for Technology (KIT), Institute for Biological Interfaces (IBG-2), P.O.B. 3640, 76021 Karlsruhe, Germany; [b] KIT, Institute of Organic Chemistry, Fritz-Haber-Weg 6, 76131 Karlsruhe, Germany
*Email: anne.ulrich@kit.edu

16.1 Introduction

Biological membranes are complex systems composed of a lipid bilayer and a multitude of proteins, involved in numerous biological functions. To better understand such systems, much effort has been made to study membrane-bound peptides of typically 10–30 amino acids in length. These peptides can either have a biological function of their own (such as antimicrobial,[1,2] cell-penetrating[3,4] or fusogenic peptides[5,6]), represent a functional segment of a larger protein (for example, the transmembrane helix of an integral membrane protein),[7] or serve as model peptides that are designed to investigate a specific aspect of peptide–membrane interactions (like hydrophobic mismatch).[8,9] In all these cases, it is of interest to obtain information on the orientation and dynamics of the peptide molecule in a lipid bilayer with a certain preferred composition, and how the orientation changes under different conditions.[10] Solid-state NMR structure analysis in macroscopically oriented membrane samples is the method of choice for such

New Developments in NMR No. 3
Advances in Biological Solid-State NMR: Proteins and Membrane-Active Peptides
Edited by Frances Separovic and Akira Naito

Published by the Royal Society of Chemistry, www.rsc.org

studies under quasi-natural conditions, which are usually carried out in the liquid crystalline phase.[11–15]

There have been some recent advances in the analysis of solid-state NMR data of peptides in membranes as improved methods have been developed to extract the entangled information about the molecular orientation and dynamics.[16,17] In this chapter we explain these new ideas and give examples of studies where the choice of a suitable dynamic model in the analysis of NMR data was critical for a better understanding of the system.

16.2 Determination of Peptide Orientation

This section gives a short explanation on how to determine the orientation of a peptide in a membrane from solid-state NMR data. In the next section we introduce the effect of dynamics in the analysis.

16.2.1 Solid-State NMR Constraints

To calculate the orientation of the peptide in the membrane, the orientational dependence of anisotropic NMR parameters is utilized. In solution-state NMR systems, molecules move fast and isotropically on the NMR timescale such that the orientational dependence is averaged out. In the case of solid-state NMR, the dynamics are much slower and the membrane systems are highly anisotropic, hence orientational information can be obtained from the measured parameters.

For solid-state NMR analysis in membranes, small peptides are usually synthesized with selective isotope labels like ^{2}H, ^{13}C, ^{15}N or ^{19}F, either in the peptide backbone or in the side chains. For side-chain labeling, the NMR label should be rigidly attached to the peptide backbone in order to provide information about the orientation and dynamics of the entire folded unit. In particular, deuterated alanine with a CD_3 methyl group (Ala-d_3) has been frequently and routinely employed.[9,18] More recently, we have established the analogous use of highly sensitive CF_3-labeled amino acids,[19] like CF_3-Phg (4-trifluoromethyl-phenylglycine),[4,20–22] CF_3-Bpg [3-(trifluoromethyl)-bicyclopent-[1.1.1]-1-ylglycine],[23–27] CF_3-Ala,[28] and CF_3-Pro,[29–31] as well as CF_3 analogues for Ser/Thr and Phe.[32,33]

The anisotropic NMR parameters used are chemical shifts, dipolar couplings (for example in CF_3 groups) and quadrupolar couplings (for example in CD_3 groups). In the case of an axially symmetric tensor (which is a good approximation for CD_3 and CF_3 groups) the splitting has a $(3\cos^2\theta-1)/2$ dependence on the angle θ between the direction of the magnetic field and the relevant tensor axis. The chemical shift has a more complex orientational dependence, but can be described by a similar formula including two polar angles.[34] In the following, we will focus on dipolar and quadrupolar couplings which are analyzed from conformationally restricted side-chain labels, but a similar analysis can also be performed using chemical shifts[35] and/or backbone labels. In Section 16.4 we will describe such studies of peptides labeled with ^{15}N in the backbone amide position, where ^{15}N chemical shifts are used to obtain orientational information, often in combination with ^{1}H–^{15}N dipolar couplings.

In the case of dipolar couplings, θ is defined as the angle between the internuclear vector and the magnetic field, while for quadrupolar couplings, θ is the angle between the electric field gradient of the chemical bond surrounding the quadrupolar nucleus and the magnetic field. For CF_3 and CD_3 groups, the fast rotation around the C–CF_3 or C–CD_3 bond leads to an averaging around this bond; hence the relevant θ angle is the angle between this C–C bond and the magnetic field direction, as illustrated in Figure 16.1A.

We can measure the experimental NMR splitting, Δ_i^{exp}, for any label that has been placed at residue i in the peptide sequence, using a simple one-pulse ^{19}F NMR or quadrupolar ^{2}H NMR experiment, to obtain a set of local orientational constraints θ_i. The spectral splitting depends on the local angle θ_i of the respective label and on the relevant maximum splitting Δ^{max} (which depends on the observed nucleus), according to Eqn (16.1):

$$\Delta_i^{exp} = \Delta^{max}(3\cos^2\theta_i - 1)/2 \qquad (16.1)$$

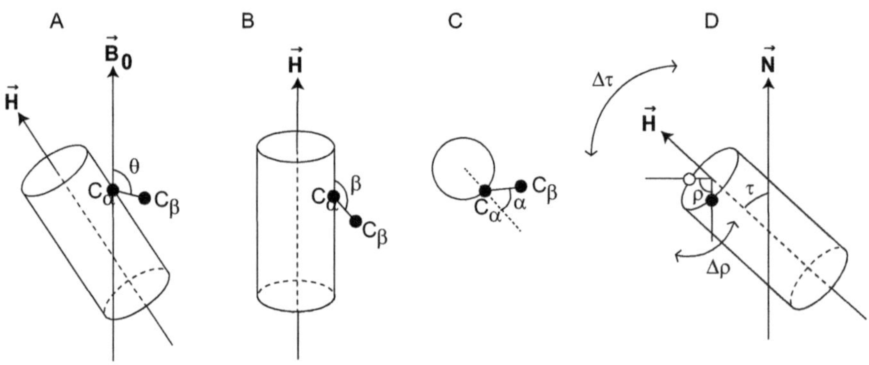

Figure 16.1 Definition of angles describing the peptide orientation (τ, ρ), dynamics ($\Delta\tau$, $\Delta\rho$), and molecular conformation (α, β). (A) The NMR spectral splitting measured for a CD_3 group depends on the angle θ between the C–CD_3 bond vector and the magnetic field direction $\boldsymbol{B}_0$. (B) For a regular backbone conformation (*e.g.* an α-helix), the orientation of the C–CD_3 bond in the peptide can be defined by the angle β between the labeled bond vector and the helix axis $\boldsymbol{H}$; and (C) by the angle α between the C–CD_3 bond and a line (perpendicular to the helix axis) from the helix axis to the C_α atom. The same is valid for a CF_3 group. (D) The peptide orientation in the membrane is defined by the tilt angle τ between the helix axis $\boldsymbol{H}$ and the membrane normal vector $\boldsymbol{N}$, and by the azimuthal angle ρ describing the rotation around the helix axis of a reference group (*filled circle*) away from a starting point (*empty circle*) at which $\rho = 0$.

16.2.2 Peptide Structure

To determine the orientation of a folded peptide in the membrane from a limited set of solid-state NMR data, the peptide structure must first be known or at least assumed. One possibility is to use a PDB structure, defining the coordinates of each atom in the peptide, which may be available from X-ray crystallography or solution-state NMR in detergent micelles.[36] Simple 3D structures can also be generated using a modeling program like SYBYL (Tripos, St Louis, USA), assuming a regular peptide conformation such as an α-helix.[9,20] Another possibility is to assume an ideal regular peptide structure that is described by characteristic backbone torsion angles, which is often a useful approximation for α-helical peptides and β-sheets. When the isotope labels are placed in the amino acid side chain and are conformationally restricted relative to the peptide backbone, it is more convenient to define the NMR tensor using two other angles, α and β. In the case of Ala-d_3 or CF_3-Bpg, these angles give the orientation of the C_α–C_β vector relative to the peptide backbone,[37,38] as shown in Figures 16.1B and 16.1C.

Once the peptide structure is known, the orientation of the relevant NMR tensors in the membrane-embedded molecule can be determined, and these constraints can then be related to the orientation and dynamics of the peptide.

16.2.3 Definition of Orientational Angles

The orientation of a folded peptide can be defined by two angles: the tilt angle τ, defined as the angle between the peptide long axis and the membrane normal, and the azimuthal rotation angle ρ, defining the rotation around the peptide long axis from a specified starting point. These angles are illustrated for a rod-like α-helical peptide in Figure 16.1D.

There is no consensus about the detailed definitions of these angles in the literature. Each group (or even individual publication) tends to use its own definitions and, therefore, in each case the definitions need to be checked to be able to visualize the membrane-bound peptide from the published τ and ρ angles. The tilt angle is fairly well defined and can usually be understood intuitively, but some groups define the peptide long axis as going from the N- to the C-terminus, while other groups use the opposite definition.[9,20] This has, for example, consequences on the sense of rotation around the long axis, leading to different definitions of ρ. The starting point at which $\rho = 0°$ can also be defined in different ways.[9,17] However, the final orientation obtained for the peptide in the membrane should eventually be the same, independent of any angular definitions. The results may only vary slightly with the backbone torsion angles, or side-chains angles α and β, used in the particular study. Far greater deviations and errors, however, will result from inadequate consideration of peptide dynamics, which is the main point covered in this review.

16.2.4 Calculation of Peptide Orientation

The general method to determine the orientation of a peptide in a membrane from NMR data is quite straightforward: several NMR measurements of specifically labeled analogues are performed to obtain a number of independent orientational constraints for the different labeled residues. These experiments are preferentially performed on macroscopically oriented samples in which the membranes are aligned on a glass support in the magnetic field, but unoriented vesicle samples can also be used provided that the peptides undergo fast rotation around the bilayer normal.[9,18,38] The data analysis is based on the known 3D structure of the peptide, the known local orientation of NMR tensors within the molecule, and the known orientation of the membrane in the magnetic field. The theoretically expected NMR splittings can be calculated for different values of τ and ρ, which describe the orientation of the peptide in the membrane.[9,20,38] A formula for these splittings as a function of τ, ρ, α and β has been derived, which is valid for the specific definitions of τ and ρ used in the study, and can be modified for other definitions.[38]

For a specific peptide structure, the best-fit orientation is taken to correspond to those τ and ρ values giving the smallest difference between the calculated and experimental splittings, as usually defined by the root mean square deviation (RMSD) (Eqn 16.2):

$$\text{RMSD} = \sqrt{\sum_{i=1}^{N} \frac{[\Delta_i^{\text{exp}} - \Delta_i^{\text{calc}}(\tau, \rho)]^2}{N}} \tag{16.2}$$

Here, Δ_i^{exp} is the experimental splitting for residue i, $\Delta_i^{\text{calc}}(\tau,\rho)$ is the splitting calculated for residue i for a specific pair of angles τ and ρ, and the sum is taken over all labeled residues, with N being the number of labels.

16.3 Peptide Dynamics

The method described in Section 16.2 only takes into account the orientation in the membrane of a static peptide, and should only work well if there is no motion of peptides in the membrane that leads to a change in peptide orientation (*i.e.*, nothing but translational motion and rotational diffusion around the bilayer normal). However, membranes are very dynamic systems under biologically relevant conditions as the lipids are usually in a liquid crystalline state, and the peptides are small and highly mobile molecules. Thus a suitable and reliable fit between experimental and calculated splittings can only be obtained when dynamics is included in the calculations. This can be done in several ways, using different models to describe the molecular motions.

16.3.1 S_{mol} (Order Parameter) Model

A simple way of including dynamics is to assume some unspecified wobble that leads to a uniform scaling of the NMR splittings, which can be described by a molecular order parameter or scaling factor, S_{mol}. If there is no motion of the peptide, $S_{mol} = 1$, while a fast isotropic motion (like in solution) gives $S_{mol} = 0$ and a complete collapse of the splittings. In this simplified approach, the best-fit orientation and dynamics can be determined by finding appropriate values of τ, ρ and S_{mol} that produce a minimal RMSD, in this case defined according to Eqn (16.3):

$$\text{RMSD} = \sqrt{\sum_{i=1}^{N} \frac{[\Delta_i^{\text{exp}} - S_{\text{mol}}\Delta_i^{\text{calc}}(\tau, \rho)]^2}{N}} \tag{16.3}$$

This approach has been used in many studies of transmembrane[39–43] and antimicrobial peptides in membranes,[4,21,24,44] with much method development centered around the antimicrobial α-helical peptide PGLa substituted with Ala-d_3 or with CF_3-labeled amino acids.[2,18,20,25,45–49] These studies typically found $S_{mol} = 0.5$–0.7 for such small monomeric or dimeric peptides, with larger values obtained for bigger complexes. While this S_{mol} approximation is easy to implement and often gives decent fits to the data, detailed information about the dynamics is not revealed. In some cases, erroneous results for highly dynamic peptides were obtained. It was, therefore, necessary to develop a dynamical model to address these issues.

16.3.2 Distribution of Angles Model

An early approach that did not include any peptide dynamics was used to analyze ^{2}H NMR data of transmembrane model peptides of the WALP family, which were labeled with Ala-d_3 at different positions. When these peptides were reconstituted in lipid bilayers of different acyl chain lengths, the analysis suggested that the helix tilt angle (τ) always remained small, around 3–8°.[9,38] This finding was unexpected, because transmembrane helices were predicted to become tilted when the hydrophobic length of the peptide exceeded the hydrophobic thickness of the membrane. Under such positive hydrophobic mismatch, tilting should minimize the exposure of hydrophobic parts of the peptide to the polar environment outside the membrane, hence the tilt angle was expected to increase with mismatch.[9,38] Molecular dynamics (MD) simulations of analogous systems indeed confirmed the expected mismatch dependence of tilt and suggested larger tilt angles than those obtained from the early ^{2}H NMR analysis.[50–53] A solid-state ^{15}N NMR analysis of the transmembrane helix of Vpu also showed a clear mismatch dependence of the tilt angle.[54]

A first indication that this discrepancy could be due to the static model used in the ^{2}H NMR data analysis was observed when the ^{2}H quadrupolar splittings for the Ala-d_3 labeled positions were back-calculated from the simulation and compared with the experimental values. The simulated NMR parameters, averaged over the simulation trajectories, were found to be not very different from the experimental data. However, when the MD data, obtained from mobile peptides with large helix tilt angles, were analyzed using the same method as in the original ^{2}H NMR study, inadequate best-fit values were obtained with too small tilt angles describing the peptides.[50,51]

When the influence of dynamics on the analysis of ^{2}H NMR data was investigated using several different dynamics models, the data from mobile peptides could be fitted using static or semi-static models, but this would usually lead to large RMSD values.[16] When the S_{mol} model was used, better RMSD values were obtained, but the helix tilt angle was still strongly underestimated in some cases.[16]

When a more complex but also more descriptive dynamical model was used, larger tilt angles were obtained for the WALP peptides, in much better agreement with the values found in the MD simulations.[16] In these models, peptide motion is described by fluctuating τ and ρ angles. Instead of assuming a fixed orientation of the peptide, some variation of these angles is allowed around their mean values τ_0 and ρ_0 (see Figure 16.1D). The distribution of angles for these random fluctuations is described by Gaussian distribution functions with respective widths (standard deviation) σ_τ and σ_ρ. The impact of whole-body fluctuations of the peptide is thus represented by the values of σ_τ and σ_ρ. In the extreme case of $\sigma_\tau = 0$, there is no variation of the τ angle, and when σ_τ and σ_ρ are both zero this corresponds to an immobile peptide. Larger values of σ_τ and σ_ρ describe a more dynamic situation.

It is assumed that peptide orientations fluctuate fast on the timescale of the NMR experiment (several ms) and that the measured splitting is the average value, which can be calculated by integrating over the distribution functions.[16,55] Figure 16.2 illustrates the angular probability distributions of τ and ρ. The static case would correspond to a delta function at τ_0 and ρ_0. To assess a mobile situation, for each (τ, ρ) point the expected splittings are calculated, and after multiplication with their respective probabilities they are integrated over the whole (τ, ρ) space, and then normalized by the integral of the probability function.

Thus, in this dynamical model the experimental NMR data are compared with splittings calculated with a more complex formula that includes averaging over Gaussian distributions, according to Eqn (16.4):

$$\mathrm{RMSD} = \sqrt{\sum_{i=1}^{N} \frac{[\Delta_i^{\mathrm{exp}} - \Delta_i^{\mathrm{calc}}(\tau_0, \sigma_\tau, \rho_0, \sigma_\rho)]^2}{N}} \tag{16.4}$$

The splittings are calculated with Eqn (16.5):

$$\Delta_i^{\mathrm{calc}}(\tau_0,\sigma_\tau,\rho_0,\sigma_\rho) = \frac{\int_{\tau_0-3\sigma_\tau}^{\tau_0+3\sigma_\tau}\int_{\rho_0-3\sigma_\rho}^{\rho_0+3\sigma_\rho} G(\tau-\tau_0,\sigma_\tau)G(\rho-\rho_0,\sigma_\rho)\sin\tau\,\Delta_i^{\mathrm{calc}}(\tau,\rho)\mathrm{d}\rho\mathrm{d}\tau}{\int_{\tau_0-3\sigma_\tau}^{\tau_0+3\sigma_\tau}\int_{\rho_0-3\sigma_\rho}^{\rho_0+3\sigma_\rho} G(\tau-\tau_0,\sigma_\tau)G(\rho-\rho_0,\sigma_\rho)\sin\tau\,\mathrm{d}\rho\mathrm{d}\tau} \tag{16.5}$$

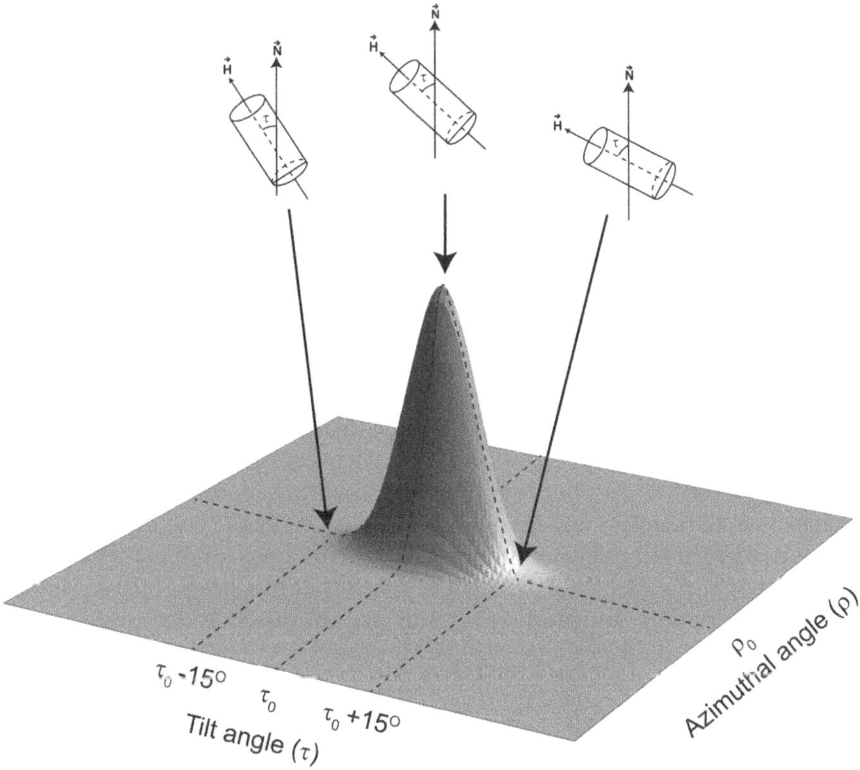

Figure 16.2 The dynamical distribution analysis assumes that there is a probability distribution of peptide orientations described by Gaussian distributions of the τ and ρ angles. Each (τ,ρ) point corresponds to one specific peptide orientation, and the highest probability is found for the mean values τ_0 and ρ_0. Other orientations have lower probabilities. The widths of the Gaussian distributions are given by the standard deviations σ_τ and σ_ρ. Fast exchange between all orientations is assumed, so that the measured NMR splitting corresponds to the time average over the entire distribution of angles. In this example, $\sigma_\tau = 5°$, and essentially the integral over $\pm 3\sigma_\tau$ covers the full range of fluctuations.

Here $\Delta_i^{calc}(\tau,\rho)$ is the same as in Eqns (16.2) and (16.3), the Gaussian distribution functions $G(x,\sigma)$ are defined by Eqn (16.6), and the integration runs in principle over the whole possible range of the τ and ρ angles. In practice, however, a range of $\pm 3\sigma_\rho$ and $\pm 3\sigma_\tau$ is sufficient as this covers 99.74% of the Gaussian distribution. The division is applied to normalize the distribution.

$$G(x,\sigma) = \frac{1}{\sigma\sqrt{2\pi}} \exp\left(-\frac{x^2}{2\sigma^2}\right) \tag{16.6}$$

It can be noted that S_{mol} is not included as an extra parameter in this case, because dynamics is fully taken into account by σ_τ and σ_ρ. For a comparative evaluation, both the S_{mol} and the dynamical distribution methods were used to analyze the data from WALP23 in several lipid systems and from PGLa in different environments. For the amphipathic surface-bound PGLa peptide, the two methods yielded essentially the same helix orientation. Even with a static model without any dynamics, almost the same orientation was found, but with a very large RMSD.[16] The dynamical distribution model revealed only small values of σ_τ and σ_ρ, up to 20°, meaning that the peptide is not so mobile in the membrane. In this situation with relatively modest mobility, there was not much difference between the different models of dynamics used in the NMR data analysis. On the other hand, in the case of the transmembrane peptide WALP23, a dramatic difference between the different models was observed. Using static models, only small tilt angles of 3–7° were obtained, while the S_{mol} model showed a larger mismatch dependence in the range of 7–15°. Finally, the dynamical distribution model gave very high tilt angles τ_0 of up to 35°, with strong azimuthal fluctuations in σ_ρ up to 100°.[16] In this case of vigorous peptide mobility, clearly only the dynamical distribution model gives the correct helix tilt angles, which are in agreement with the expected mismatch dependence. Notably for both PGLa and WALP23, the best-fit ρ_0 angle was not influenced when using a static or dynamical model, indicating that the azimuthal rotation angle is more robustly determined by solid-state NMR data.[16] Figure 16.3 (below) gives an overview of the orientation and dynamics of PGLa and WALP23 according to ^2H NMR data analyzed with the dynamical distribution model.

A more thorough analysis of ^2H NMR data from many different model transmembrane peptides was then performed using WALP, KALP and XWALP peptides of varying length in different lipid systems.[56] A dynamical

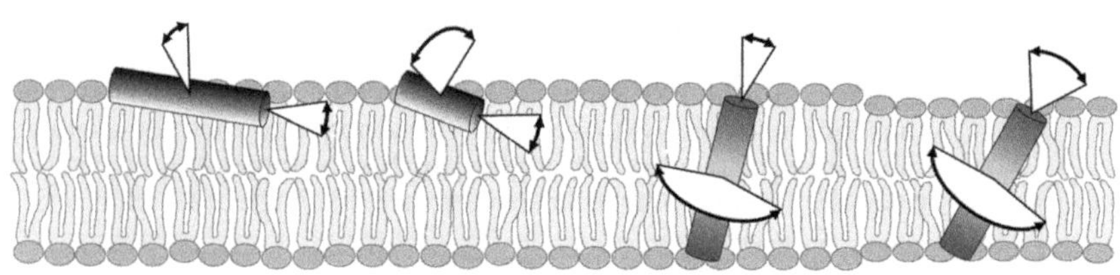

Figure 16.3 Orientation and mobility of some amphiphilic α-helical peptides in membranes according to analysis of solid-state NMR data using the dynamical distribution model. The helix tilt angle τ is given with respect to the membrane normal, together with the widths of the tilt and azimuthal angle distributions, shown as $\pm\sigma_\tau$ and $\pm\sigma_\rho$, respectively. For BP100, the analysis gave larger error bars than for the other cases so the values given are not strictly best-fit results but are estimated from a combined analysis of ^{19}F and ^{15}N solid-state NMR and OCD data.

distribution model was used, in which ρ_0, τ_0 and σ_ρ were fit while $\sigma_\tau = 0$ was kept fixed. In the dynamical distribution model there are four free parameters, but in some cases only four experimental NMR data points were available for some of the peptides. Since the full analysis of WALP23 showed that the main effect was in the azimuthal fluctuations, represented by the σ_ρ parameter (tilt fluctuations σ_τ were usually small), all data were analyzed using this simplified dynamical distribution model. It was thus found that the helix tilt angle of these peptides was indeed dependent on hydrophobic mismatch, as originally expected. Especially for WALP peptides, with flanking tryptophan residues, the tilt angle followed almost exactly the expected cosine dependence of τ on mismatch. KALP peptides, with flanking lysine residues, showed smaller tilt angles than expected from the simple theory, probably due to the snorkeling ability of the long and charged lysine side chains.[56]

In several subsequent studies, similar transmembrane model peptides of the WALP and XWALP families were studied with ^{2}H NMR, and the data analyzed using both the S_{mol} and dynamical distribution models. In general, for these transmembrane peptides the S_{mol} model yields a smaller tilt angle than the distribution model, while the ρ angle does not change much.[39–42] This effect depends on the intrinsic mobility of the peptide, as the more mobile peptides have a larger impact on the adequacy of the model. The more mobile the peptide, the more important that dynamics be taken into account in the NMR data analysis. GWALP peptides were found to be less mobile than WALP peptides and, in general, peptides with two flanking aromatic residues on each terminus were more mobile than peptides with only one tryptophan at each end.[42,56] The introduction of an arginine residue in the GWALP23 sequence also reduced the dynamics.[41] Other transmembrane peptides have been studied with ^{2}H NMR and data analyzed with distribution models.[57,58]

Interestingly, the ^{2}H NMR structure analysis of the antimicrobial peptide MSI-103, which has a similar sequence and amphipathic helical structure as PGLa, showed almost no difference in orientation when the data were analyzed with the S_{mol} or the dynamical distribution models.[44] The distribution model gave small values of σ_τ (0–5°) and σ_ρ (up to 20°) in various different lipid environments, and the corresponding τ_0 and ρ_0 values were very close to within a few degrees of the values obtained with the S_{mol} model. It thus appeared that the dynamical distribution model was only necessary for the analysis of highly mobile transmembrane peptides, whereas amphipathic peptides lying on the membrane surface are less dynamic and, therefore, also less sensitive to the choice of a dynamical model.

Both PGLa and MSI-103 have a length of 21 amino acids, similar to the transmembrane WALP-type peptides. However, when we studied a much smaller amphipathic peptide, BP100, with only 11 amino acids in the sequence, in this case dynamics played a much more prominent role.[59] When this peptide, labeled with CF_3-Bpg, was analyzed with the S_{mol} model, a tilt angle of $\tau = 155°$ was found, suggesting an almost transmembrane orientation of the helix. However, complementary data from oriented circular dichroism spectroscopy (OCD) indicated a tilt angle of around 90°, *i.e.* that the helix lies flat on the membrane surface. On the other hand, in the case of PGLa and MSI-103, NMR and OCD data were in good agreement.[60] This apparent contradiction could be resolved when the ^{19}F NMR data were analyzed using the dynamical distribution model. This model showed that a large range of τ_0 values between 110° and 160° gave similarly low RMSD values, while the ρ angle was well defined with only a small uncertainty. Thus, a combined analysis of the ^{19}F NMR and OCD data could give the correct peptide orientation, but only when the dynamical distribution model was used in the ^{19}F NMR analysis.[59] The orientation and dynamics of BP100 according to this analysis are shown in Figure 16.3.

16.4 Complementary ^{15}N NMR Studies

Having focused until now on specifically labeled side chains, however, the orientation of peptides can also be determined with solid-state NMR from isotope labels in the backbone. The most common method is to label the amide nitrogen with ^{15}N. From the chemical shift of a single such label, it is possible to obtain an approximate tilt angle of the peptide in the membrane.[61] When using several labels at once, usually as a uniformly ^{15}N-labeled peptide, a 2D ^{15}N–^{1}H PISEMA (polarization inversion spin exchange at the magic angle)[62] experiment can give more accurate tilt angles of helical peptides. In the 2D PISEMA spectrum, the peaks from a regular α-helix will fall on a so-called PISA (polarity index slant angle) wheel whose size and position directly reflect on the helix tilt angle.[63–65] By incorporating selective labels at specific positions in the sequence, the azimuthal ρ angle can also be determined.[54,66]

In early ^{15}N NMR studies, dynamics was usually not taken into account in the analysis of PISEMA spectra.[67,68] When dynamical models were included in the data analysis, the S_{mol} model was mostly used,[66,69–72] until the dynamical distribution methods mentioned above were introduced. We have performed a systematic study of how the distributions of orientational angles affect the appearance of PISEMA spectra and influence the data analysis.[17] An earlier study had only considered small values of σ_τ and σ_ρ (up to 20°) and reported negligible effects,[73] but we found clear effects when larger values of σ_τ and σ_ρ were taken into account. We also showed that there is a fundamental difference in the sensitivity of σ_τ and σ_ρ to dynamics when using side-chain ^{2}H/^{19}F labels compared to backbone ^{15}N labels.[17,74] As argued above, high peptide mobility (*i.e.*, large values of σ_ρ) can lead to an underestimation of the tilt angle when the S_{mol} model is used to analyze ^{2}H NMR data. For PISEMA, the effect of a large σ_ρ is to make the PISA wheel smaller, without moving the position of the center of the wheel, which contains the critical information about the tilt angle. Therefore, the correct tilt angle from a PISA wheel can be obtained by using the simple S_{mol} model, even in the presence of vigorous azimuthal fluctuations. The effect of σ_τ is both to move the center of the wheel and to reduce its size, which may then lead to an overestimation of the tilt angle in the case of transmembrane peptides.[17,74]

The dynamical models were used to compare the ^{2}H and ^{15}N NMR data of two model transmembrane peptides, WLP23 and GWALP23, in oriented lipid bilayers.[75] For both sets of data from both peptides, a good fit with consistent tilt angles was obtained when the dynamical distribution model was used, but the S_{mol} model gave very different tilt angles for the ^{2}H and ^{15}N NMR data sets, and worse RMSDs.[75] The σ_ρ values found in the dynamical distribution fits were similar for the ^{2}H and ^{15}N NMR data, being quite large. It was noted that GWALP23 is less dynamic than WLP23 (*i.e.*, has a smaller σ_ρ value) and hence the difference between the S_{mol} and dynamical distribution analyses was more pronounced for WLP23.[75] In yet another case, ^{15}N NMR PISEMA data of a multimeric transmembrane peptide were fitted with the S_{mol} and dynamical distribution model, but as this peptide was rather immobile the choice did not make much of a difference,[76] as also seen for PGLa and MSI-103 when studied with ^{2}H NMR.[16,44]

16.5 Conclusions

The orientation and dynamics of peptides in lipid membranes can be determined comprehensively using solid-state NMR. However, orientation and dynamics are intrinsically linked, in the sense that both parameters influence the spectral splittings and chemical shifts measured in the NMR experiment. Therefore both orientation and dynamics need to be taken into account in the analysis of such data. In some cases, peptide dynamics are moderate and then the choice of a dynamical model for the analysis is not so important. However, in cases where the molecules are highly mobile, such as monomeric transmembrane helices or short amphiphilic peptides, a simple S_{mol} model can give erroneous results. In these systems, dynamical models that include distributions of the τ and ρ angles in the analysis are more appropriate and have been shown to perform better.

Acknowledgments

We thank Jesús Salgado, Santi Esteban-Martín and Stephan Grage for fruitful discussions about the dynamical models, and Parvesh Wadhwani for synthesizing the peptides studied in our group. Financial support is acknowledged from CFN TP E1.2.

References

1. K. A. Brogden, *Nat. Rev. Microbiol.*, 2005, **3**, 238–250.
2. E. Strandberg, P. Tremouilhac, P. Wadhwani and A. S. Ulrich, *Biochim. Biophys. Acta*, 2009, **1788**, 1667–1679.

3. Ü. Langel, *Cell-Penetrating Peptides: Processes and Applications*, CRC Press, Boca Raton, FL, 2002.
4. P. Wadhwani, J. Bürck, E. Strandberg, C. Mink, S. Afonin and A. S. Ulrich, *J. Am. Chem. Soc.*, 2008, **130**, 16515–16517.
5. J. Reichert, D. Grasnick, S. Afonin, J. Bürck, P. Wadhwani and A. S. Ulrich, *Eur. Biophys. J.*, 2007, **36**, 405–413.
6. D. Grasnick, U. Sternberg, E. Strandberg, P. Wadhwani and A. S. Ulrich, *Eur. Biophys. J.*, 2011, **40**, 529–543.
7. V. Wray, R. Kinder, T. Federau, P. Henklein, B. Bechinger and U. Schubert, *Biochemistry*, 1999, **38**, 5272–5282.
8. J. A. Killian, I. Salemink, M. R. de Planque, G. Lindblom, R. E. Koeppe II and D. V. Greathouse, *Biochemistry*, 1996, **35**, 1037–1045.
9. P. C. A. Van der Wel, E. Strandberg, J. A. Killian and R. E. Koeppe II, *Biophys. J.*, 2002, **83**, 1479–1488.
10. S. L. Grage, S. Afonin and A. S. Ulrich, in *Antimicrobial Peptides. Methods and Protocols*, ed. A. Giuliani and A. C. Rinaldi, Springer/Humana Press, New York, 2010, vol. 618, pp. 183–209.
11. E. Strandberg and A. S. Ulrich, *Concepts Magn. Reson. A*, 2004, **23A**, 89–120.
12. P. Wadhwani, P. Tremouilhac, E. Strandberg, S. Afonin, S. L. Grage, M. Ieronimo, M. Berditsch and A. S. Ulrich, in *Current Fluoroorganic Chemistry: New Synthetic Directions, Technologies, Materials, and Biological Applications*, ed. V. A. Soloshonok, K. Mikami, T. Yamazaki, J. T. Welch and J. F. Honek, American Chemical Society, Washington, DC, 2007, pp. 431–446.
13. A. S. Ulrich, *Progr. Nucl. Magn. Reson. Spectrosc.*, 2005, **46**, 1–21.
14. K. Koch, S. Afonin, M. Ieronimo, M. Berditsch and A. S. Ulrich, *Top. Curr. Chem.*, 2012, **306**, 89–118.
15. P. Wadhwani and E. Strandberg, in *Fluorine in Medicinal Chemistry and Chemical Biology*, ed. I. Ojima, Blackwell, London, 2009, pp. 463–493.
16. E. Strandberg, S. Esteban-Martín, J. Salgado and A. S. Ulrich, *Biophys. J.*, 2009, **96**, 3223–3232.
17. S. Esteban-Martín, E. Strandberg, G. Fuertes, A. S. Ulrich and J. Salgado, *Biophys. J.*, 2009, **96**, 3233–3241.
18. E. Strandberg, P. Wadhwani, P. Tremouilhac, U. H. N. Dürr and A. S. Ulrich, *Biophys. J.*, 2006, **90**, 1676–1686.
19. V. S. Kubyshkin, I. V. Komarov, S. Afonin, P. K. Mykhailiuk, S. L. Grage and A. S. Ulrich, in *Fluorine in Pharmaceutical and Medicinal Chemistry: From Biophysical Aspects to Clinical Applications*, ed. V. Gouverneur and K. Müller, Imperial College Press, London, 2012, pp. 91–138.
20. R. W. Glaser, C. Sachse, U. H. N. Dürr, P. Wadhwani and A. S. Ulrich, *J. Magn. Reson.*, 2004, **168**, 153–163.
21. E. Strandberg, N. Kanithasen, J. Bürck, P. Wadhwani, D. Tiltak, O. Zwernemann and A. S. Ulrich, *Biochemistry*, 2008, **47**, 2601–2616.
22. E. Strandberg, D. Tiltak, M. Ieronimo, N. Kanithasen, P. Wadhwani and A. S. Ulrich, *Pure Appl. Chem.*, 2007, **79**, 717–728.

23. P. K. Mikhailiuk, S. Afonin, A. N. Chernega, E. B. Rusanov, M. O. Platonov, G. G. Dubinina, M. Berditsch, A. S. Ulrich and I. V. Komarov, *Angew. Chem. Int. Ed.*, 2006, **45**, 5659–5661.
24. P. Wadhwani, E. Strandberg, N. Heidenreich, J. Bürck, S. Fanghänel and A. S. Ulrich, *J. Am. Chem. Soc.*, 2012, **134**, 6512–6515.
25. S. Afonin, P. K. Mikhailiuk, I. V. Komarov and A. S. Ulrich, *J. Pept. Sci.*, 2007, **13**, 614–623.
26. M. Salwiczek, P. K. Mikhailiuk, S. Afonin, I. V. Komarov, A. S. Ulrich and B. Koksch, *Amino Acids*, 2010, **39**, 1589–1593.
27. P. K. Mykhailiuk, N. M. Voievoda, S. Afonin, A. S. Ulrich and I. V. Komarov, *J. Fluorine Chem.*, 2010, **131**, 217–220.
28. D. Maisch, P. Wadhwani, S. Afonin, C. Böttcher, B. Koksch and A. S. Ulrich, *J. Am. Chem. Soc.*, 2009, **131**, 15596–15597.
29. P. K. Mykhailiuk, S. Afonin, G. V. Palamarchuk, O. V. Shishkin, A. S. Ulrich and I. V. Komarov, *Angew. Chem. Int. Ed.*, 2008, **47**, 5765–5767.
30. V. S. Kubyshkin, P. K. Mykhailiuk, S. Afonin, A. S. Ulrich and I. V. Komarov, *Org. Lett.*, 2012, **14**, 5254–5257.
31. V. S. Kubyshkin, P. K. Mykhailiuk, S. Afonin, S. L. Grage, I. V. Komarov and A. S. Ulrich, *J. Fluorine Chem.*, 2013, **152**, 136–143.
32. A. N. Tkachenko, P. K. Mykhailiuk, S. Afonin, D. S. Radchenko, V. S. Kubyshkin, A. S. Ulrich and I. V. Komarov, *Angew. Chem. Int. Ed.*, 2013, **52**, 1486–1489.
33. A. N. Tkachenko, D. S. Radchenko, P. K. Mykhailiuk, S. Afonin, A. S. Ulrich and I. V. Komarov, *Angew. Chem. Int. Ed.*, 2013, **52**, 6504–6507.
34. M. J. Duer, *Solid-State NMR Spectroscopy: Principles and Applications*, Blackwell Science, Malden, MA, 2002.
35. S. Afonin, U. H. N. Dürr, P. Wadhwani, J. B. Salgado and A. S. Ulrich, *Top. Curr. Chem.*, 2008, **273**, 139–154.
36. M. Michalek, E. S. Salnikov, S. Werten and B. Bechinger, *Biochemistry*, 2013, **52**, 847–858.
37. A. S. Ulrich, P. Wadhwani, U. H. N. Dürr, S. Afonin, R. W. Glaser, E. Strandberg, P. Tremouilhac, C. Sachse, M. Berditchevskaia and S. L. Grage, in *NMR Spectroscopy of Biological Solids*, ed. A. Ramamoorthy, CRC Press, Boca Raton, FL, 2006, pp. 215–236.
38. E. Strandberg, S. Özdirekcan, D. T. S. Rijkers, P. C. A. Van der Wel, R. E. Koeppe II, R. M. J. Liskamp and J. A. Killian, *Biophys. J.*, 2004, **86**, 3709–3721.
39. J. M. Rankenberg, V. V. Vostrikov, D. V. Greathouse, C. V. Grant, S. J. Opella and R. E. Koeppe II, *Biochemistry*, 2012, **51**, 10066–10074.
40. J. M. Rankenberg, V. V. Vostrikov, C. D. DuVall, D. V. Greathouse, R. E. Koeppe II, C. V. Grant and S. J. Opella, *Biochemistry*, 2012, **51**, 3554–3564.
41. V. V. Vostrikov, B. A. Hall, M. S. P. Sansom and R. E. Koeppe II, *J. Phys. Chem. B*, 2012, **116**, 12980–12990.

42. N. J. Gleason, V. V. Vostrikov, D. V. Greathouse, C. V. Grant, S. J. Opella and R. E. Koeppe II, *Biochemistry*, 2012, **51**, 2044–2053.
43. V. V. Vostrikov, B. A. Hall, D. V. Greathouse, R. E. Koeppe II and M. S. Sansom, *J. Am. Chem. Soc.*, 2010, **132**, 5803–5811.
44. E. Strandberg, D. Tiltak, S. Ehni, P. Wadhwani and A. S. Ulrich, *Biochim. Biophys. Acta*, 2012, **1818**, 1764–1776.
45. M. Ieronimo, S. Afonin, K. Koch, M. Berditsch, P. Wadhwani and A. S. Ulrich, *J. Am. Chem. Soc.*, 2010, **132**, 8822–8824.
46. S. Afonin, S. L. Grage, M. Ieronimo, P. Wadhwani and A. S. Ulrich, *J. Am. Chem. Soc.*, 2008, **130**, 16512–16514.
47. P. Tremouilhac, E. Strandberg, P. Wadhwani and A. S. Ulrich, *Biochim. Biophys. Acta*, 2006, **1758**, 1330–1342.
48. P. Tremouilhac, E. Strandberg, P. Wadhwani and A. S. Ulrich, *J. Biol. Chem.*, 2006, **281**, 32089–32094.
49. R. W. Glaser, C. Sachse, U. H. N. Dürr, S. Afonin, P. Wadhwani, E. Strandberg and A. S. Ulrich, *Biophys. J.*, 2005, **88**, 3392–3397.
50. S. Esteban-Martin and J. Salgado, *Biophys. J.*, 2007, **93**, 4278–4288.
51. S. Özdirekcan, C. Etchebest, J. A. Killian and P. F. Fuchs, *J. Am. Chem. Soc.*, 2007, **129**, 15174–15181.
52. S. K. Kandasamy and R. G. Larson, *Biophys. J.*, 2006, **90**, 2326–2343.
53. W. Im and C. L. Brooks 3rd, *Proc. Natl. Acad. Sci. U.S.A.*, 2005, **102**, 6771–6776.
54. S. H. Park and S. J. Opella, *J. Mol. Biol.*, 2005, **350**, 310–318.
55. A. Holt, L. Rougier, V. Reat, F. Jolibois, O. Saurel, J. Czaplicki, J. A. Killian and A. Milon, *Biophys. J.*, 2010, **98**, 1864–1872.
56. E. Strandberg, S. Esteban-Martin, A. S. Ulrich and J. Salgado, *Biochim. Biophys. Acta*, 2012, **1818**, 1242–1249.
57. M. C. McDonald, V. Booth and M. R. Morrow, *Biophys. J.*, 2011, **100**, 656–664.
58. K. Bertelsen, B. Paaske, L. Thogersen, E. Tajkhorshid, B. Schiott, T. Skrydstrup, N. C. Nielsen and T. Vosegaard, *J. Am. Chem. Soc.*, 2009, **131**, 18335–18342.
59. P. Wadhwani, E. Strandberg, J. van den Berg, C. Mink, J. Bürck, R. Ciriello and A. S. Ulrich, *Biochim. Biophys. Acta.*, 2013, DOI: 10.1016/j.bbamem.2013.11.001, in press.
60. J. Bürck, S. Roth, P. Wadhwani, S. Afonin, N. Kanithasen, E. Strandberg and A. S. Ulrich, *Biophys. J.*, 2008, **95**, 3872–3881.
61. B. Bechinger, L. M. Gierasch, M. Montal, M. Zasloff and S. J. Opella, *Solid State Nucl. Magn. Reson.*, 1996, **7**, 185–191.
62. C. H. Wu, A. Ramamoorthy and S. J. Opella, *J. Magn. Reson., Ser. A*, 1994, **109**, 270–272.
63. J. K. Denny, J. F. Wang, T. A. Cross and J. R. Quine, *J. Magn. Reson.*, 2001, **152**, 217–226.
64. J. Wang, J. Denny, C. Tian, S. Kim, Y. Mo, F. Kovacs, Z. Song, K. Nishimura, Z. Gan, R. Fu, J. R. Quine and T. A. Cross, *J. Magn. Reson.*, 2000, **144**, 162–167.

65. F. M. Marassi and S. J. Opella, *J. Magn. Reson.*, 2000, **144**, 150–155.
66. T. H. Walther, S. L. Grage, N. Roth and A. S. Ulrich, *J. Am. Chem. Soc.*, 2010, **132**, 15945–15956.
67. A. A. Nevzorov and S. J. Opella, *J. Magn. Reson.*, 2003, **160**, 33–39.
68. C. Tian, P. F. Gao, L. H. Pinto, R. A. Lamb and T. A. Cross, *Protein Sci.*, 2003, **12**, 2597–2605.
69. C. Muhle-Goll, S. Hoffmann, S. Afonin, S. L. Grage, A. A. Polyansky, D. Windisch, M. Zeitler, J. Bürck and A. S. Ulrich, *J. Biol. Chem.*, 2012, **287**, 26178–26186.
70. S. H. Park, A. A. De Angelis, A. A. Nevzorov, C. H. Wu and S. J. Opella, *Biophys. J.*, 2006, **91**, 3032–3042.
71. S. D. Müller, A. A. De Angelis, T. H. Walther, S. L. Grage, C. Lange, S. J. Opella and A. S. Ulrich, *Biochim. Biophys. Acta*, 2007, **1768**, 3071–3079.
72. R. Heinzmann, S. L. Grage, C. Schalck, J. Bürck, Z. Banoczi, O. Toke and A. S. Ulrich, *Eur. Biophys. J.*, 2011, **40**, 463–470.
73. S. K. Straus, W. R. P. Scott and A. Watts, *J. Biomol. NMR*, 2003, **26**, 283–295.
74. S. Esteban-Martín, E. Strandberg, J. Salgado and A. S. Ulrich, *Biochim. Biophys. Acta*, 2010, **1798**, 252–257.
75. S. L. Grage, E. Strandberg, P. Wadhwani, S. Esteban-Martin, J. Salgado and A. S. Ulrich, *Eur. Biophys. J.*, 2012, **41**, 475–482.
76. R. C. Page, S. Kim and T. A. Cross, *Structure*, 2008, **16**, 787–797.

CHAPTER 17

Structural Dynamics of Retinal in Rhodopsin Activation Viewed by Solid-State ^{2}H NMR Spectroscopy

MICHAEL F. BROWN*[a,b] AND ANDREY V. STRUTS[a,c]

[a] Department of Chemistry and Biochemistry, University of Arizona, Tucson, AZ 85721, USA; [b] Department of Physics, University of Arizona, Tucson, AZ 85721, USA; [c] Laboratory of Biomolecular NMR, St Petersburg State University, St Petersburg 199034, Russia
*Email: mfbrown@u.arizona.edu

17.1 Introduction

G-protein-coupled receptors (GPCRs) are responsible for signal transduction through cellular membranes[1,2] and have long been a focus of structural studies employing site-directed spin-labelling,[3–6] crystallographic analysis,[7–12] and NMR spectroscopy.[13–22] However, the structures of most GPCRs are currently unknown due to difficulties in their crystallization for X-ray studies and their relatively high molecular weight for solution NMR spectroscopy. In the last few years, X-ray structures of several GPCRs, including rhodopsin,[7–12] the β_1- and β_2-adrenergic receptors,[23–25] the adenosine A_{2A} receptor,[26] and opioid receptors,[27–30] have made possible significant advances in the understanding of their activation mechanisms.[1,2,31,32] Currently, rhodopsin is one of the few GPCRs for which the 3D structures in various functional states have

New Developments in NMR No. 3
Advances in Biological Solid-State NMR: Proteins and Membrane-Active Peptides
Edited by Frances Separovic and Akira Naito

Published by the Royal Society of Chemistry, www.rsc.org

been recently established.[11,12,33–35] Nevertheless, many important details of its structure and activation mechanism remain speculative, and require further investigation by different methods; in particular, the orientation of retinal in the active metarhodopsin II (Meta II) state is not unambiguously determined.[11,12,35]

To understand the activation mechanism of the visual receptor rhodopsin, first, one must characterize the atomistic motions of the ligand and the protein. Next, one must explain how the local changes of the chromophore structure lead to the activating conformational changes of the receptor. A further aspect is that one must establish the role of the membrane lipid bilayer, because it is well known that the lipid environment affects the activation of rhodopsin.[36–39] In the visual process, rhodopsin activation involves the 11-*cis* to 11-*trans* isomerization of retinal due to photon absorption, which leads to movements of the transmembrane helices[3,6] that expose recognition sites on the cytoplasmic loops for the G-protein (transducin). Major local changes include deprotonation of the retinylidene Schiff base that triggers large-scale helical rearrangements in the protein structure, accompanied by disruption of two ionic locks and hydrogen-bonding networks around Glu122 and Asp83 that stabilize the dark-state conformation. Apart from X-ray analysis[7–12] and spin-label ESR studies,[3–6] rhodopsin structure and dynamics have also been studied by vibrational Raman spectroscopy,[40] which gives information about torsional twisting of the retinal chromophore during light activation.

Notably, biophysical and bioorganic investigations of rhodopsin mutants and rhodopsin recombined with various retinal analogues have been extremely useful in identifying key interactions that are implicated in its biological activity.[41–43] Fourier transform infrared studies of desmethyl analogues of retinal bound to rhodopsin have revealed that deletion of the C5-, C9-, or C13-methyl groups of retinal shifts the metarhodopsin equilibrium towards the inactive metarhodopsin I (Meta I) state, thus implicating them in the receptor functioning. Another valuable tool that is complementary to X-ray crystallography is solid-state ^{2}H NMR spectroscopy. One of the advantages of NMR is that membrane proteins are studied in a native-like membrane environment[13,18,44–50] where protein function is preserved.[37,38] Moreover, the structural information obtained from ^{2}H NMR, for example bond orientations, may be more accurate than from X-ray studies. Last, solid-state NMR relaxation experiments[21] reveal dynamical information that cannot be readily obtained with X-ray crystallography or other biophysical methods.[51]

17.2 Order Parameters and Bond Orientations for Membrane Proteins are Determined from Solid-State ^{2}H NMR Spectroscopy

For decades, solid-state ^{2}H NMR spectroscopy has been an irreplaceable tool for studying the structure and molecular motions in polymers and liquid crystals (including lyotropic liquid crystals such as lipid bilayers). Progress in NMR spectroscopy has made it possible to extend the application of solid-state ^{2}H NMR to the investigation of membrane proteins.[13,44–46,52] Two major difficulties in this area are related to the large linewidth in solid-state ^{2}H NMR spectroscopy: first, this affects the signal intensity for relatively high molecular weight membrane proteins and, second, only a single or at most a few selectively labelled groups can be studied in one experiment. Similar obstacles apply for other solid-state NMR methods that involve nuclei other than deuterium, such as polarization-inversion spin-exchange at the magic angle (PISEMA) and dipolar-assisted rotational resonance (DARR) spectroscopy.[18,53–60] For reviews of solid-state NMR spectroscopy as applied to membrane proteins and peptides, see the Hong *et al.*[56] and McDermott[58] articles.

Figure 17.1 shows theoretical solid-state ^{2}H NMR spectra calculated for a single deuterated group in (a) unoriented or (b, c) aligned systems. The solid-state ^{2}H NMR spectrum is described by the distribution of the spectral intensity as a function of the quadrupolar frequencies, $\nu_Q^{\pm}$, or equivalently the reduced frequencies, $\xi_{\pm}$, for the two single-quantum transitions of the spin $I=1$ ^{2}H nucleus. For spin $I=1$ nuclei, the two spectral branches correspond to transitions between energy levels with different projection quantum numbers, $m=-1, 0, +1$, for the spin angular momentum. For an axially symmetric coupling tensor $\xi_{\pm} \equiv \nu_Q^{\pm}/(3/4)\chi$, where χ is the quadrupolar coupling constant (static $\chi=\chi_Q=170$ kHz; or residual $\chi=\langle\chi_Q\rangle$). Reduced frequencies depend on the angle $\tilde{\theta}$ between the EFG tensor principal z-axis (assuming axial symmetry) and the static external magnetic field $\boldsymbol{B}_0$ according to $\xi_{\pm}=\pm\frac{1}{2}(3\cos^2\tilde{\theta}-1)$.[61] In the case of random isotropic orientations of molecules in unoriented (non-aligned) samples as in Figure 17.1(a), the angular probability distribution of the nuclear coupling tensor $p(\xi_{\pm})$ and spectral intensities can be expressed as $1/|\cos\tilde{\theta}|$, yielding a so-called powder-type spectrum (or Pake doublet). In the case of a semi-random uniaxial distribution of the coupling tensor, Figure 17.1(b,c), the ^{2}H NMR lineshapes depend on the angle θ of the alignment symmetry axis with respect to the main magnetic field $\boldsymbol{B}_0$, as well as the angle θ_B between the coupling tensor principal symmetry axis and the alignment axis. Consequently, useful structural information, namely bond orientations, is available from ^{2}H NMR spectroscopy of aligned membrane samples.[13,44–46]

Next, in Figure 17.2, we show representative experimental ^{2}H NMR spectra of selectively deuterated methyl groups for retinal bound to the dark-state rhodopsin in non-oriented POPC membranes. The spectral splitting indicates fast spinning of the C5-, C9-, or C13-methyl groups about their C_3 axes with correlation times $<10^{-5}$ s even at low temperatures (down to −150 °C, as shown for the C9-methyl group; Figure 17.2a). The motion-averaged ^{2}H NMR spectral lineshapes involve residual quadrupolar couplings (RQCs) designated by $\langle\chi_Q\rangle$ that correspond directly to the segmental order parameters (Figure 17.2b). The order parameter characterizes the amplitude of the

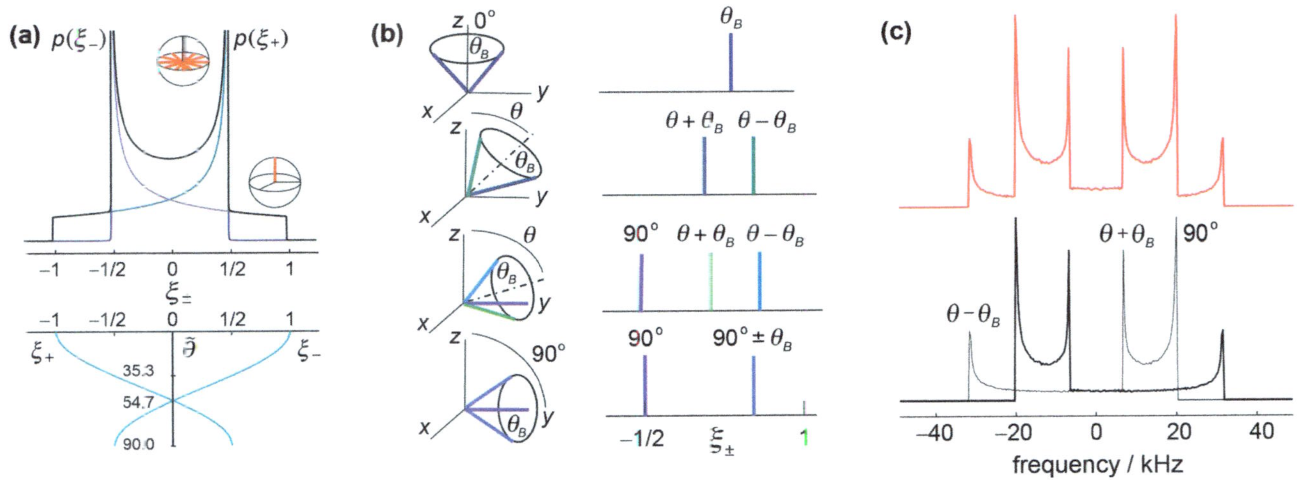

Figure 17.1 In solid-state ^{2}H NMR the spectral frequency distribution is a map of the angular distribution of the electric field gradient (EFG) tensor. (a) For unoriented membrane systems, the principal symmetry axis of the coupling tensor is randomly distributed *versus* the static external magnetic field $\boldsymbol{B}_0$, giving a powder-type spectrum (so-called Pake doublet). The ^{2}H NMR spectrum for the two spectral branches of an $I=1$ spin system is graphed as a function of the reduced frequency $\xi_\pm$ and corresponds to the principal *z*-axis of the EFG tensor randomly distributed over the surface of a sphere. Note that the probability distribution $p(\xi_\pm)$ scales as $1/|\cos\tilde{\theta}|$, where $\tilde{\theta}$ is the overall angle between the EFG tensor *z*-axis (assumed axially symmetric) and the $\boldsymbol{B}_0$ field. Weak singularities are evident at $\tilde{\theta}=90°$ (equator) or 35°, with the intensity extending to a minimum at $\tilde{\theta}=0°$ (poles). The spectral centre is where $\tilde{\theta}=54.7°$ (so-called magic angle). Angular dependence of the reduced frequencies is graphed at the *bottom*. (b) Example of semi-random angular distribution of coupling (EFG) tensor principal axis, where θ is the tilt angle of the alignment axis to the main magnetic field $\boldsymbol{B}_0$ and θ_B is the coupling tensor orientation to the alignment axis. For rotating methyl groups, as in the case of retinal bound to rhodopsin, the principal direction of the EFG is the three-fold axis. At *left*, the various orientations giving the intensity maxima for an individual spectral branch of the $I=1$ nucleus are depicted. A static uniaxial distribution is considered where the coupling tensor *z*-axes are confined to the rim of a cone. Owing to inversion and reflection symmetry, only the distribution for a quarter-sphere is needed. The adjacent schematic (*stick*) spectra are for a single spectral branch. For zero tilt (*top spectrum*), an axially symmetric spectrum is obtained; with increasing tilt angle the spectral features represent the sum and difference of θ and θ_B. At large enough tilt angle, a further weak singularity appears at $\theta=90°$ as for a random distribution. (c) Calculated ^{2}H NMR spectrum for a static uniaxial distribution of methyl groups having three-fold rotation, *e.g.* the case of an aligned membrane protein. The two spectral branches of the $I=1$ ^{2}H nucleus are shown in different colours at the *bottom*, with the resultant spectrum at the *top*. To correspond to the stick spectra in panel (b), the line broadening is intentionally reduced. (Reproduced from Brown *et al.*,[116] with permission from Elsevier.)

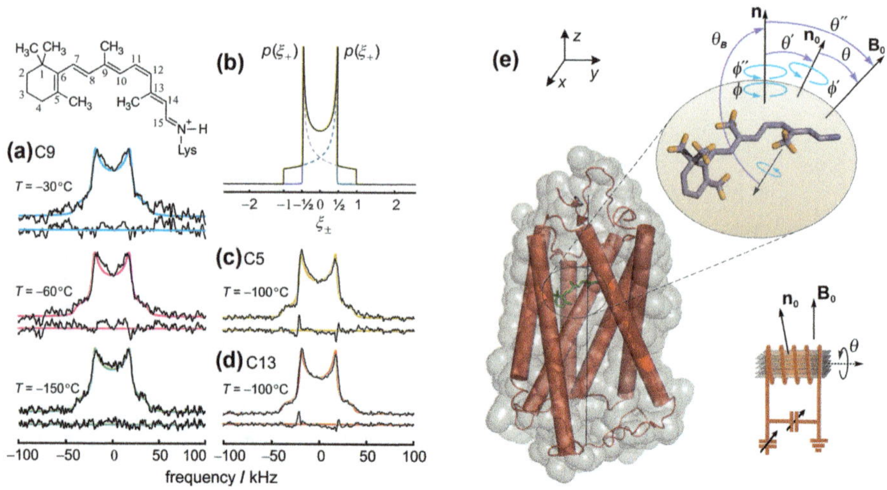

Figure 17.2 Powder-type ^{2}H NMR spectra for membrane-bound rhodopsin with ^{2}H-labelled retinal in the dark state indicate that the methyl groups undergo rapid rotation with large order parameters. (a) Experimental ^{2}H NMR spectra for 11-*Z*-[9-C^{2}H$_3$]-retinylidene rhodopsin, *i.e.* having 11-*cis*-retinal deuterated at the C9-methyl group, in gel-phase POPC membranes (1:50 molar ratio) at different temperatures. (b) Theoretical ^{2}H NMR spectrum for randomly oriented C–C^{2}H$_3$ groups (Pake doublet) undergoing rapid three-fold rotation on the ^{2}H NMR timescale ($<(3\chi_Q/8)^{-1} \approx 10$ μs) plotted on a reduced frequency scale. (c,d) Representative ^{2}H NMR spectra for dark-state 11-*Z*-[5-C^{2}H$_3$]-retinylidene rhodopsin and 11-*Z*-[13-C^{2}H$_3$]-retinylidene rhodopsin, *i.e.* with 11-*cis*-retinal deuterated at the C5-methyl or C13-methyl group, respectively, in POPC membranes (1:50). Theoretical ^{2}H NMR spectra for C–C^{2}H$_3$ groups undergoing axial rotation (*continuous colour lines*) are superimposed on the experimental spectra, with residuals below. (e) Structure and orientation of retinal ligand within the rhodopsin binding cavity are established by solid-state ^{2}H NMR spectroscopy of aligned membranes. For a given methyl group, θ_B is the angle of the C–C^{2}H$_3$ bond axis to the local membrane normal $\boldsymbol{n}$, with static rotational symmetry given by the azimuthal angle ϕ. Alignment disorder is described by the angle θ' of $\boldsymbol{n}$ to the average membrane normal $\boldsymbol{n}_0$ and is also uniaxially distributed as characterized by ϕ'. Next, θ is the tilt angle from $\boldsymbol{n}_0$ to the main magnetic field $\boldsymbol{B}_0$, about which there is cylindrical symmetry. Lastly, θ'' is the angle from $\boldsymbol{n}$ to $\boldsymbol{B}_0$, where ϕ'' is the azimuth. The van der Waals surface of rhodopsin is depicted in *light grey* and the seven transmembrane helices are indicated by *rods*, with the N-terminus at the *top* (extracellular side) and the C-terminus at *bottom* (cytoplasmic side). The schematic depiction of a stack of aligned membranes containing rhodopsin within the radiofrequency coil of the NMR spectrometer shows the geometry relative to the magnetic field.
(Adapted from Struts *et al.*,[20] with permission from Elsevier.)

fluctuations of the coupling tensor, and is defined by $S_{C_3} \equiv \frac{1}{2}\langle 3\cos^2\beta_{IM} - 1\rangle$, where β_{IM} is the angle between the instantaneous and average orientations of the methyl coupling tensor principal symmetry axes. For a rapidly spinning methyl group, rotation about the three-fold symmetry axis averages the static coupling constant to $\chi_Q^{eff} = \chi_Q(3\cos^2 109.47^\circ - 1)/2 = -\frac{1}{3}\chi_Q = -56.67$ kHz. Off-axial fluctuations additionally reduce the quadrupolar coupling constant, and can be characterized by $S_{C_3} = \langle\chi_Q\rangle/\chi_Q^{eff}$, which is the order parameter of the methyl three-fold (C_3) axis, where the brackets denote a time average. The order parameter for off-axial motion of methyl groups is more accurately determined for a powder-type sample, because fewer fitting parameters are needed than for an oriented sample.

Interestingly, the RQCs and corresponding order parameters for the spinning methyl groups of the retinylidene ligand change little over a broad range of temperature (Figure 17.2c,d). The order parameter S_{C_3} for the off-axial fluctuations of all the methyl groups studied (C5, C9, C13) is approximately the same ($\sim$0.9). The order parameters depend on the mean orientation of the nuclear interaction tensor and are related to the static as well as dynamical disorder. In addition, the intrinsic line broadening is proportional to the transverse (R_2) relaxation rate, which in turn is connected with molecular mobility, whereby narrower NMR lines correspond to faster motion. Evidently, neither the order parameters nor the line broadening show any noticeable changes over the wide temperature range studied (−30 to −150 °C). Further information about dynamics can be obtained from longitudinal (R_{1Z}) and quadrupolar (R_{1Q}) relaxation experiments, as discussed subsequently.

In Figure 17.2(e) we show how the theoretical ^{2}H NMR spectra for the oriented samples can be calculated by assuming a static uniaxial distribution about the average membrane normal.[62] Effectively, we introduce the Jacobian for transforming from the geometric space of the NMR sample to the frequency space of the corresponding NMR spectrum. The theoretical lineshapes depend on the methyl bond orientation θ_B with respect to the membrane normal and on the alignment disorder (mosaic spread), which is assumed to be Gaussian with a standard deviation of σ. The ^{2}H NMR spectra also depend on the tilt angle (orientation of the average membrane normal $\boldsymbol{n}_0$ with respect to the main magnetic field $\boldsymbol{B}_0$). In the gel phase of a lipid bilayer, the membrane proteins do not undergo significant rotational diffusion[63] on the ^{2}H NMR timescale ($\sim$10 μs). The quadrupolar frequencies of the two spectral branches can be expressed as:[61]

$$\nu_Q^{\pm} = \pm\,\frac{3}{4}\chi D_{00}^{(2)}(\Omega_{XL}) \tag{17.1}$$

where χ denotes the static or residual coupling constant. The symbol $\boldsymbol{D}^{(j)}$ designates the Wigner rotation matrix of rank j, and $\Omega_{XL} = (\tilde{\phi}, \tilde{\theta}, 0)$ are the Euler angles corresponding to overall transformation of the irreducible components of the coupling tensor from its principal axes system (PAS; molecule fixed) to the laboratory coordinate frame, defined by the main magnetic field $\boldsymbol{B}_0$. We assume an axially symmetric coupling tensor, *i.e.* asymmetry parameter $\langle\eta_Q\rangle = 0$ for the case of rotating methyl groups.

Using closure,[61,64] the overall rotational transformation from the PAS to the laboratory frame may then be represented by a series of intermediate frames (Figure 17.2e). The elements of the Wigner rotation matrix can be expressed according to:[61]

$$D_{00}^{(2)}(\Omega_{XL}) = \sum_{m'=-2}^{2} \sum_{m=-2}^{2} D_{0m'}^{(2)}(\Omega_{XN}) D_{m'm}^{(2)}(\Omega_{ND}) D_{m0}^{(2)}(\Omega_{DL}) \tag{17.2}$$

The first set of Euler angles, $\Omega_{XN} \equiv (0, \theta_B, \phi)$, characterizes the orientation of the PAS of the coupling tensor (static or residual; X frame in general) relative to the local membrane normal $\boldsymbol{n}$ where θ_B is the bond orientation, and ϕ is the random azimuthal rotation about the local normal (N frame); *cf.* Figure 17.2(e). The second transformation $\Omega_{ND} \equiv (0, \theta', \phi')$ treats the disorder of the local membrane normal *versus* the average normal $\boldsymbol{n}_0$ to the membrane surface (D frame) in terms of θ' as well as the random azimuth ϕ'. The last transformation describes the tilt of the average membrane normal $\boldsymbol{n}_0$ to the laboratory (L) frame (static magnetic field $\boldsymbol{B}_0$). Next, the analytical calculation can be simplified by introducing rank-1 rotation matrix elements (so-called Poor Man's closure),[62] giving the following results:

$$\cos\tilde{\theta} = \cos\theta_B \cos\theta'' - \sin\theta_B \sin\theta'' \cos(\phi + \phi'') \tag{17.3}$$

$$\cos\theta'' = \cos\theta' \cos\theta - \sin\theta' \sin\theta \cos\phi' \tag{17.4}$$

Here $\tilde{\theta}$ is the overall angle of the C–C^{2}H$_3$ bond axis to the magnetic field $\boldsymbol{B}_0$. The angles $\Omega_{NL} \equiv (\phi'', \theta'', 0)$ describe the overall rotation of the local frame to the laboratory frame, where θ'' is the tilt of the local normal, ϕ'' is a phase factor for the two combined transformations,[62] and θ is the tilt angle of the alignment axis to the main magnetic field $\boldsymbol{B}_0$.

After some lengthy calculations,[62] we obtain the following expressions for the two spectral branches of the ^{2}H NMR lineshape:

(i) For $\alpha > \gamma > \delta > \beta$ or $\gamma > \alpha > \beta > \delta$ then:

$$|p(\xi_\pm)| \propto \frac{1}{|\cos\tilde{\theta}|} \int_0^{\pi} \frac{1}{y} \boldsymbol{K}\left(\frac{x}{y}\right) \exp\left(\frac{-\theta'^2}{2\sigma^2}\right) \sin\theta' d\theta' \tag{17.5}$$

(ii) For $\gamma > \alpha > \delta > \beta$ or $\alpha > \gamma > \beta > \delta$ then:

$$|p(\xi_\pm)| \propto \frac{1}{|\cos\tilde{\theta}|} \int_0^{\pi} \frac{1}{x} \boldsymbol{K}\left(\frac{y}{x}\right) \exp\left(\frac{-\theta'^2}{2\sigma^2}\right) \sin\theta' d\theta' \tag{17.6}$$

Note that $x \equiv \sqrt{(\gamma - \delta)(\alpha - \beta)}$ and $y \equiv \sqrt{(\alpha - \delta)(\gamma - \beta)}$ in which the cosines of the sum and difference angles are defined as follows: $\alpha \equiv \cos(\tilde{\theta} - \theta_B)$; $\beta \equiv \cos(\tilde{\theta} + \theta_B)$; $\gamma \equiv \cos(\theta - \theta')$; and $\delta \equiv \cos(\theta + \theta')$. In our results, the kernel $\boldsymbol{K}(k) = F(\pi/2, k)$ represents a complete elliptic integral of the first kind in the normal trigonometric form:

$$\boldsymbol{K}(k) \equiv \int_0^{\pi/2} \frac{dx}{\sqrt{1 - k^2 \sin^2 x}} \tag{17.7}$$

Here the 3D alignment disorder (mosaic spread) is formulated by a Gaussian distribution of the local membrane normal *versus* the average bilayer normal, $p(\theta') = (1/\sigma\sqrt{2\pi})\exp(-\theta'^2/2\sigma^2)$, where σ is the standard deviation about the mean of $\langle\theta'\rangle = 0$.

The reader can also verify that the above results yield the appropriate limits for large and small alignment disorder. In the limit of very small mosaic spread, the probability distribution $p(\xi_\pm)$ scales as:[45]

$$|p(\xi_\pm)| = 1/|\cos\tilde{\theta}|\,[\cos(\theta + \theta_B) - \cos\tilde{\theta}]^{1/2}[\cos\tilde{\theta} - \cos(\theta - \theta_B)]^{1/2} \tag{17.8}$$

On the other hand, for the limit of very large mosaic spread the results amount to a random (spherical) distribution. The lineshape formula then gives us the well-known Pake doublet:

$$|p(\xi_\pm)| \propto \frac{1}{|\cos\tilde{\theta}|} \propto \frac{1}{\sqrt{1 \pm 2\xi_\pm}} \tag{17.9}$$

Thus the expected results[65,66] are readily obtained.

17.3 Deuterium NMR Spectral Lineshapes Reveal Changes in Structure of Retinal Due to Light Activation of Rhodopsin

Notable applications of site-directed solid-state ^{2}H NMR spectroscopy include membranes containing bacteriorhodopsin,[44–46,62] DNA fibers,[67] and the structure of retinal bound to rhodopsin in the dark and pre-active Meta I state.[20] In our first example, bovine rhodopsin was regenerated with retinal containing selectively deuterated methyl groups at the C5-, C9-, and C13-methyl positions, and recombined with phospholipid bilayers.[13] The membranes were aligned on ultra-thin glass slides, and ^{2}H NMR spectra were measured at different orientations with respect to the main magnetic field $\boldsymbol{B}_0$, as illustrated in Figure 17.2(e). A tilt series of representative ^{2}H NMR spectra for 11-*cis*-retinal bound to dark-state rhodopsin are shown in Figure 17.3(a), together with the root-mean-square deviation (RMSD) of experimental *versus* simulated spectra (Figure 17.3b,c).

Global (simultaneous) fitting of all ^{2}H NMR spectra at different orientations gives well-defined minima, allowing accurate determination of the bond orientation θ_B and mosaic spread σ of the aligned membranes (Figure 17.3b,c). Theoretical ^{2}H NMR spectra are calculated using the above

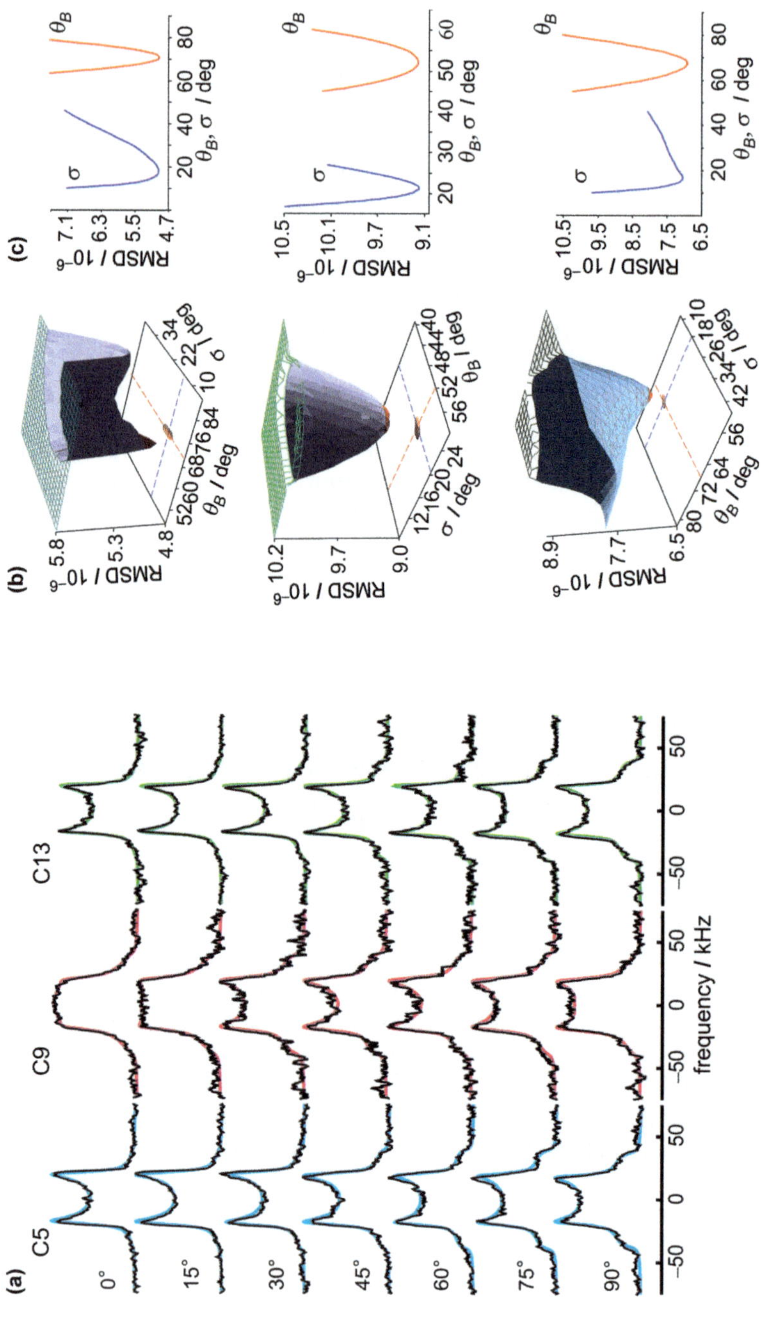
(a)
C5
C9
C13
0°
15°
30°
45°
60°
75°
90°
frequency / kHz
(b)
RMSD / 10⁻⁶
θ_B / deg
σ / deg
(c)
RMSD / 10⁻⁶
θ_B, σ / deg
σ
θ_B

formulas (17.5)–(17.7) for a static uniaxial distribution.[62] Alternatively, a Monte Carlo simulation can be employed, where the lineshapes are accumulated numerically by randomly generating the angular variables according to their distribution functions as in Eqns (17.1)–(17.4), together with a Gaussian distribution for the mosaic spread.[62] Both mathematical techniques – the closed-form analysis and Monte Carlo simulation – give essentially identical spectra, thus validating our results. Notably the fitting parameters include the C–C^2H_3 bond orientation, the mosaic spread, the residual coupling $\langle\chi_Q\rangle$, and the intrinsic line broadening. By comparing the residual coupling $\langle\chi_Q\rangle$ to the effective coupling χ_Q^{eff} we obtain the order parameter for the fluctuations of the C–C^2H_3 bond axis directly.

In this way, for 11-*cis*-retinal bound to the dark state of rhodopsin, average C–C^2H_3 bond orientations are obtained of $70 \pm 3°$, $52 \pm 3°$, and $68 \pm 2°$ for the C5-, C9-, and C13-methyl groups, respectively, as shown in Figure 17.3(b). Values for the mosaic spread ($\sigma = 18$–$21°$) are larger than in previous studies of aligned purple membranes containing bacteriorhodopsin.[45] This finding may be due to the larger molar mass of rhodopsin ($M_r = 40$ kDa) compared to bacteriorhodopsin ($M_r = 26$ kDa), and to the presence of substantial extramembranous domains.[68] It is interesting to note that the ^{2}H NMR bond orientations are in good agreement with independent results from X-ray crystallography.[8,10]

For our next example, experimental solid-state ^{2}H NMR spectra for the retinylidene C5-, C9-, and C13-methyl groups of rhodopsin in bilayers aligned at $\theta = 0°$ are shown in Figures 17.4(a) and 17.4(b) for the dark state and Meta I state, respectively.[47] Further analysis of the retinal conformation entails a simple three-plane model, as illustrated by Figures 17.4(c) and 17.4(d). Deviation of the torsion angles from the ideal *cis* or *trans* configuration is neglected, except for the C6–C7 and C12–C13 bonds, shared by the A–B and B–C pairs of planes, respectively, and the C11=C12 bond.

Figure 17.3 Tilt series of ^{2}H NMR spectra for the 11-*cis*-retinal ligand of rhodopsin in the dark state are explicable by methyl bond orientations and mosaic spread of aligned membranes. (a) ^{2}H NMR spectra for 11-*Z*-[5-C^2H_3]-retinylidene rhodopsin (*blue*), 11-*Z*-[9-C^2H_3]-retinylidene rhodopsin (*magenta*), and 11-*Z*-[13-C^2H_3]-retinylidene rhodopsin (*green*), respectively. Experimental ^{2}H NMR spectra are shown as a function of tilt angle θ between the normal to membrane stack $\boldsymbol{n}_0$ and the magnetic field $\boldsymbol{B}_0$ for rhodopsin/POPC bilayers (1:50) at pH 7 and $T = 150$ °C. Theoretical lineshapes (*continuous colour lines*) assume a static uniaxial distribution of rhodopsin molecules with rotating retinylidene C–C^2H_3 groups. Global fitting of ^{2}H NMR spectra for 11-*cis*-retinal gives accurate values for methyl bond orientations for the dark state of rhodopsin in aligned membranes. (b) RMSD of calculated *versus* experimental ^{2}H NMR spectra for rhodopsin with retinal deuterated at C5-, C9-, or C13-methyl groups, respectively. (c) Corresponding cross-sections through surfaces at left showing well defined minima in bond orientation θ_B and mosaic spread σ of aligned membranes.
(Adapted from Struts *et al.*,[20] with permission from Elsevier.)

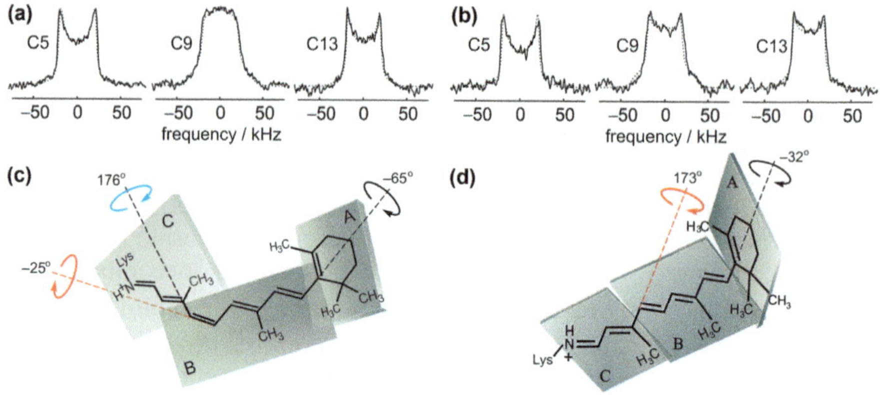

Figure 17.4 Solid-state ^{2}H NMR spectroscopy characterizes the structure of the retinal ligand bound to rhodopsin in aligned membranes. Examples of solid-state ^{2}H NMR spectra are shown for (a) dark and (b) Meta I states of rhodopsin regenerated with 11-*cis*-retinal specifically ^{2}H-labelled at C5-, C9-, or C13-methyl positions in aligned POPC membranes (1:50 molar ratio). The spectra are measured for the parallel orientation ($\theta = 0°$) of the membrane normal to the magnetic field $\boldsymbol{B}_0$ at temperatures of −150 and −100 °C for rhodopsin and Meta I, respectively. (c, d) The solid-state NMR structures proposed for retinal for (c) the dark state of rhodopsin and (d) the Meta I photo-intermediate.

Plane A includes the C5=C6 bond of the β-ionone ring, and planes B and C represent the polyene chain to either side of the C12–C13 bond (see Figure 17.4). Note that each plane is defined by two vectors; thus two angular constraints (degrees of freedom) are needed to specify its spatial orientation. Since the three molecular planes have two common bonds, a total of four independent parameters (degrees of freedom) are needed to define the retinal structure and orientation ($3 \times 2 - 2 = 4$). Together with the C5-, C9-, and C13-methyl group orientations, the electronic transition dipole moment of the retinylidene chromophore from linear dichroism measurements can be introduced as a fourth orientational restraint.[69–72]

According to Figure 17.4, we see that the three ^{2}H-labelled methyl groups lie in the three different planes (A, B, C). This allows the values of the torsion angles $\chi_{i,k}$ between the pairs of planes to be calculated from the individual C–C^{2}H$_3$ bond orientations by the formula:[13]

$$\chi_{i,k} = \cos^{-1}\left(\frac{\cos\theta_i \cos\theta_B^{(i,k)} - \cos\theta_B^{(i)}}{\sin\theta_i \sin\theta_B^{(i,k)}}\right) - \cos^{-1}\left(\frac{\cos\theta_k \cos\theta_B^{(i,k)} - \cos\theta_B^{(k)}}{\sin\theta_k \sin\theta_B^{(i,k)}}\right) \quad (17.10)$$

where $\theta_B^{(i)}$ and $\theta_B^{(k)}$ are the orientations of methyl groups i and k to the local membrane normal $\boldsymbol{n}$, and θ_i and θ_k are angles of the two methyl axes to the common bond between the planes. The angle between the common bond and the local membrane normal is defined as $\theta_B^{(i,k)}$, and is calculated from the C9-methyl orientation, together with the transition dipole moment.

Multiple solutions for the C6–C7 and C12–C13 torsion angles between the A, B, or C planes (and multiple retinal geometries) are obtained for a single set of experimental methyl bond orientations. A total of 64 combinations for the relative orientations of the A, B, and C planes is possible, yielding 128 possible retinal configurations. However, circular dichroism data[73,74] and carbon–carbon distances obtained from solid-state rotational resonance ^{13}C NMR studies[14,17] can be introduced as additional geometrical constraints to find a single physical solution, *e.g.* in analogy to the case of solution NMR spectroscopy.[75]

By this means, Eqn (17.10) is used to obtain the torsion angle of the C12–C13 bond (*i.e.*, the dihedral angle between planes B and C), which is found to be $\chi_{9,13} = +150 \pm 4^\circ$ $(+147 \pm 4^\circ)$. The two values correspond to the C11=C12–C13 bond angle equal to 120° or 130°, respectively. Distortion of the C11=C12–C13 bond angle from ideal orbital geometry (120°) can arise due to interaction of the C13-methyl group with the retinal H10 hydrogen, or it may be induced by the specific configuration of the ligand within the rhodopsin binding pocket. For the β-ionone ring, using positional restraints for the C8-to-C18 and C8-to-C16/C17 distances from solid-state ^{13}C NMR,[14] together with the chirality from circular dichroism studies,[73] the physical solution for the C6–C7 dihedral angle is found to be $\chi_{5,9} = -65 \pm 6^\circ$. The negatively twisted 6-*s-cis* conformation (Figure 17.4c) is most probably due to the non-bonded interactions between the C5-methyl of the β-ionone ring and the H8 hydrogen of the polyene chain.[74,76] The distorted 6-*s-cis* conformation of retinal in rhodopsin agrees with solid-state ^{13}C NMR chemical shift data[77] and differs from the planar 6-*s-trans* conformation seen in bacteriorhodopsin.[45,78] Clearly, the protein environment can modulate the β-ionone ring conformation substantially, which in turn can influence the electronic structure of the polyene[79] and protonation of the retinylidene Schiff base.[80] Notably, the β-ionone ring may undergo torsional fluctuations about the C6–C7 bond, and may be less restricted than the polyene chain, as indicated by molecular dynamics simulations.[81] Mobility of the β-ionone ring within its hydrophobic binding pocket[82] might also explain the fact that it is more difficult to establish a unique conformation for this group using X-ray crystallography.[8,13,14,81,83,84]

Next, to further improve the 11-*cis*-retinylidene ^{2}H NMR structure, it can be inserted into the binding pocket of the X-ray structure of rhodopsin in the dark state (PDB accession code 1U19).[10] However, a simple three-plane model does not fit into the binding pocket, due to multiple steric clashes of retinal with Tyr178 and Cys187 of extracellular loop E2, plus Met207, Tyr268, and Ala292 on helices H5, H6, and H7, respectively. Detailed structural analysis shows that the answer is to allow further twisting of the C11=C12 bond of the polyene chain.[20] Twisting about the C11=C12 bond locates the β-ionone ring within its hydrophobic cavity and is energetically optimal, while maintaining the retinylidene Schiff base near its complex counterion, as in the dark-state rhodopsin structure.[10] The value of the C11=C12 torsion angle can be determined by fitting the C10-to-C20 and C11-to-C20 distances to ^{13}C NMR rotational resonance data.[17] For the extended three-plane model, the optimum torsion angles for the C11=C12 and C12–C13 bonds are $-25\pm10°$ and $176\pm6°$, respectively. Thus, the twist is predominantly localized to the C11=C12 bond. The solid-state NMR structure for retinal is shown in Figure 17.4(c), and it fits perfectly into the retinal binding pocket. Note that the C10-to-C20 and C11-to-C20 distances are relatively insensitive to the C11=C12 torsion angle, giving somewhat large errors. Our proposed torsional twisting of retinal about the C11=C12 double bond agrees with high-resolution X-ray data,[10] and is key to understanding its ultra-fast photoreaction dynamics.[40,85–87] Another interesting aspect is that pre-twisting about the 11-*cis* double bond can govern the chromophore trajectory upon photon absorption.[88] More specifically, we can propose that the binding pocket induces distortion: it prepares the retinal for the isomerization. A negative direction of isomerization about the C11=C12 bond is consistent with hybrid quantum mechanics/molecular mechanics simulations, which indicate that steric interaction of the C13-methyl group with Ala117 hinders rotation toward positive angles.[89] Moreover, the positive helical twist about the C12–C13 bond is in agreement with circular dichroism and bioorganic studies of locked retinoids by Nakanishi and co-workers[74] and with our previous results.[13] Analogous conclusions regarding local twisting about the C11=C12 and C12–C13 bonds have been drawn from combined quantum mechanics/molecular mechanics simulations.[90,91]

The Meta I state has also been cryo-trapped to investigate changes in the retinal conformation and its orientation induced by photon absorption.[20,47] Figure 17.4(b) shows examples of ^{2}H NMR spectra acquired at −100 °C for POPC membranes containing Meta I macroscopically aligned at the $\theta=0°$ tilt angle. The largest differences in the ^{2}H NMR spectra *versus* the dark state are seen for the C9–C^{2}H$_3$ group, with smaller variations for the C5–C^{2}H$_3$ and C13–C^{2}H$_3$ groups. These differences may be pertinent to changes of the bond orientation θ_B and also the mosaic spread σ. The bond orientations of the C5-, C9-, and C13-methyl groups obtained from the ^{2}H NMR spectra in the Meta I state are $\theta_B=72\pm4°$, $53\pm3°$, and $59\pm3°$, respectively. Owing to the greater alignment disorder (mosaic spread) of Meta I ($\sigma=22$–$25°$) compared to the dark state ($\sigma=18$–$21°$), the ^{2}H NMR spectra for the C9-methyl are different, whereas the bond orientations are similar. On the other hand, for the C13-methyl group a smaller spectral difference is evident, with a larger change in bond orientation. For Meta I, the ^{2}H NMR spectra indicate that the retinylidene methyl substituents undergo rapid rotation as in the dark state, with order parameters describing off-axial fluctuations of ~0.9 within the ligand binding pocket.[20]

Allowable solutions for the C6–C7 torsion angle in Meta I are calculated from the orientational restraints of the methyl groups and the transition dipole moment, together with the ^{13}C NMR distance constraints, yielding $\pm32\pm7°$and $\pm57\pm7°$ as the results. In addition, two possible values for the C12–C13 dihedral angle are obtained: $\pm173\pm4°$. It follows that eight combinations of the dihedral angles corresponding to the retinal structure are allowed. Because electron crystallography does not detect any rearrangements of the seven transmembrane helices,[92] for retinal the calculated Meta I structures can be further restrained by inserting them into the dark-state rhodopsin structure.[10] Most of the possible retinal structures are eliminated due to the different position of the β-ionone ring in the binding pocket, except for the structure shown in Figure 17.4(d) with $\chi_{5/9} = -32°$ and $\chi_{9/13} = 173°$, and the structure obtained by reflection (mirror symmetry transformation) in a vertical plane with $\chi_{5/9} = 32°$ and $\chi_{9/13} = -173°$. The first structure fits best within the binding pocket, having its only close contact with the functionally important Trp265 side chain. On the other hand, the mirror structure makes close contacts with the side chains of Glu122, Trp265, Tyr268, and Ala292, and thus is less likely.

Figure 17.5 compares the ^{2}H NMR structure of 11-*cis*-retinal in the pre-activated Meta I state (red) *versus* the dark state (green). Evidently, the main changes in the retinal conformation in Meta I encompass rotation of the polyene chain adjacent to the C13-methyl towards the C9-methyl group, accompanied by straightening and elongation of the ligand. The resulting displacement of the β-ionone ring towards helix H5 is also observed by X-ray crystallography and solid-state ^{13}C NMR[11,12,33,93] in different rhodopsin photo-intermediates, and the active Meta II state. Indeed, it plays a crucial role

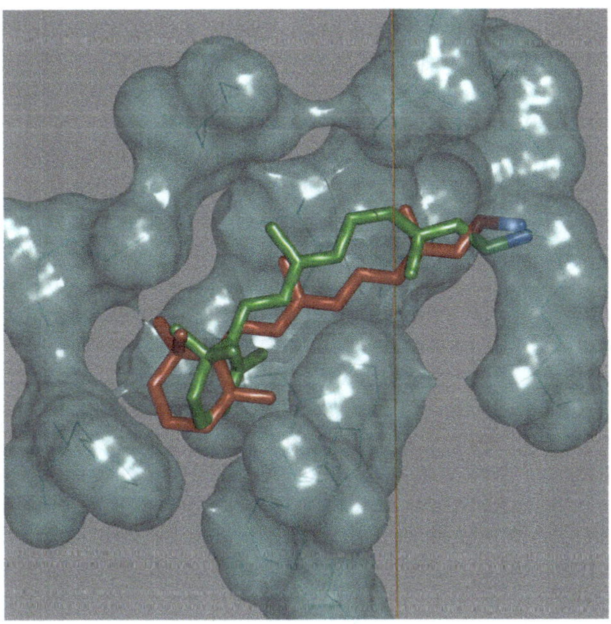

Figure 17.5 Solid-state NMR provides structure of retinal ligand bound to rhodopsin in a natural membrane lipid environment. The vertical direction corresponds to the membrane normal, where the extracellular side is up and cytoplasmic side is down. The ^{2}H NMR structure of 11-*cis*-retinal in the pre-activated Meta I state (*red*) is compared to retinal inserted into the dark state (*green*) X-ray structure (access code 1U19). (Reproduced from Brown *et al.*,[116] with permission from Elsevier.)

in the receptor activation.[94] Rotation of the (proximal) part of the polyene chain containing the C13-methyl group may destabilize the ionic lock involving the protonated Schiff base, and thus facilitate its deprotonation. An apparent steric clash between Trp265 and the retinal may displace the β-ionone ring towards Glu122, whereas Trp265 moves towards helix H5 and the cytoplasmic side of the membrane. These motions are pivotal to the activation process.[95] The β-ionone ring is responsible for disruption of the hydrogen-bonding network around Glu122, which connects helices H3 and H5, and stabilizes the inactive rhodopsin conformation,[94] whereas the displacement of Trp265 underlies the outward activating rotation of the helix H6.[16,18,19,93]

Two additional examples are provided to further highlight the possibilities of solid-state ^{2}H NMR spectroscopy in structural studies of membrane proteins. The next case illustrates the accuracy and sensitivity of the method. In Figure 17.6(a,b) we show experimental and simulated ^{2}H NMR spectra of rhodopsin/POPC recombinant membranes containing 11-*cis*-retinal deuterated at the C5-methyl group of the β-ionone ring. Deuterium NMR spectra for aligned membranes were recorded for a temperature of −60 °C at tilt angles of $\theta = 0°$ or 90° relative to the main magnetic field $\boldsymbol{B}_0$. In Figure 17.6(a) the experimental ^{2}H NMR spectra are superimposed with theoretical fits for a static uniaxial distribution with a bond axis orientation of $\theta_B = 70 \pm 3°$ and a mosaic spread of $\sigma = 21 \pm 2°$. Values for the C5-methyl group were determined from simultaneous non-linear regression fitting of the angular-dependent spectral data.[13] For comparison, Figure 17.6(b) shows theoretical simulations of the same experimental ^{2}H NMR spectra, assuming a C–C^{2}H$_3$ bond orientation of 21° as proposed earlier,[83] together with the above-mentioned value of the mosaic spread. In this case, the experimental line shapes for the C5-methyl group are not reproduced, and thus we conclude that the value of $70 \pm 3°$ is more consistent with the experimental data. Even for relatively large values of the mosaic spread (21° in this example), accurate bond orientations can be determined from the ^{2}H NMR spectra of aligned samples.

The last example in Figure 17.6 illustrates how the experimental ^{2}H NMR spectra of aligned bilayers containing rhodopsin can distinguish between different counterion-switch mechanisms for the protonated Schiff base (PSB) of retinal in the dark state-to-Meta I transition.[96] Figure 17.6(c) shows results for the counterion-switch model, where Glu113 is deprotonated and acts as the counterion to the PSB in the dark state. Proton transfer occurs to Glu113 from Glu181 (tautomerization), which is the new counterion to the PSB upon light absorption. By contrast, in the complex-counterion model shown in Figure 17.6(d), both Glu113 and Glu181 are deprotonated, and the main counterion changes from Glu113 to Glu181 after a photon is absorbed. To decide between these two alternatives, theoretical spectra for the retinal ligand of rhodopsin were calculated numerically for the C5-, C9-, and C13-methyl groups using atomistic molecular dynamics (MD) simulations.[82,96] The continuous colour lines represent theoretical ^{2}H NMR spectra calculated from the methyl group orientational distributions[96] for either the counterion-switch model in Figure 17.6(c) or a complex-counterion

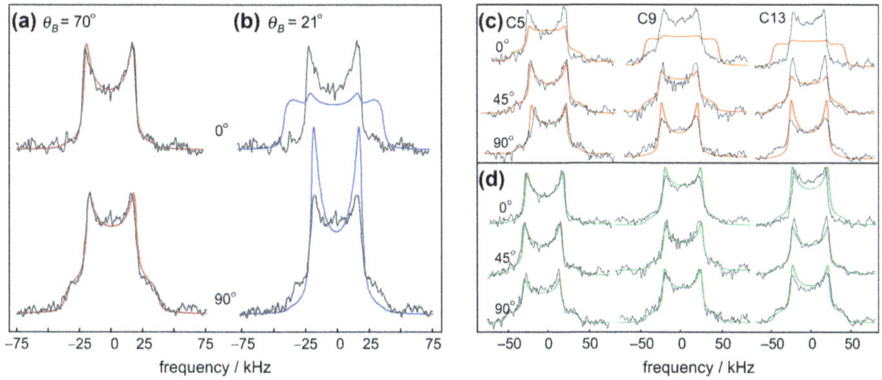

Figure 17.6 Solid-state ^{2}H NMR spectra distinguish various proposed structures for the retinal ligand of dark-state rhodopsin. Experimental ^{2}H NMR spectra are compared to theoretical lineshape simulations.[62] (a) Examples of a C–C^{2}H$_3$ bond orientation[13] of $\theta_B = 70 \pm 3°$ or (b) an alternative[83] of $\theta_B = 21°$ for aligned membranes containing rhodopsin labelled with ^{2}H at the retinal C5-methyl group. Evidently, the former angle is in better agreement with experiment. Results are shown for the $\theta = 0°$ and 90° orientations of the membrane normal to the static $\boldsymbol{B}_0$ magnetic field at −60 °C. Experimental ^{2}H NMR spectra are indicated in *grey* and the *continuous colour lines* depict theoretical spectra for the two proposed[13,83] retinal structures. (c, d) Solid-state ^{2}H NMR spectra for Meta I with retinal ^{2}H-labelled at the C5-, C9-, or C13-methyl groups for $\theta = 0°$, 45°, and 90° tilt angles in aligned membranes at −100 °C.[20] Bond orientation distributions from numerical molecular dynamics (MD) simulations were used to weight the theoretical lineshapes computed for a static uniaxial distribution with a mosaic spread.[62] *Continuous colour lines* based on the MD simulations are shown.[96] Examples are given for (c) counterion-switch model and (d) complex-counterion model for the ionic lock of the protonated Schiff base in the dark-to-Meta I transition. Agreement of the simulated spectra with the experimental data indicates the complex-counterion model applies to rhodopsin.
(Adapted from Salgado *et al.*,[13] and Martínez-Mayorga *et al.*,[96] with permission from American Chemical Society.)

model in Figure 17.6(d). For each methyl group, three representative orientations of $\theta = 0°$, 45°, and 90° for the bilayer normal relative to the magnetic field $\boldsymbol{B}_0$ are shown. Comparing the experimental ^{2}H NMR results to those of the atomistic MD simulations in Figure 17.6 unambiguously favors the complex-counterion switch model.[96]

17.4 Retinal Dynamics within the Binding Cavity of Rhodopsin are Investigated by Solid-State ^{2}H NMR Relaxation Methods

In the case of rhodopsin, ^{2}H NMR relaxation gives a unique, site-directed approach to study the local dynamics of the retinylidene methyl groups within the ligand-binding pocket.[95] Deuterium spin–lattice relaxation time (T_{1Z}) measurements (decay of Zeeman order) for the C5-, C9-, and C13-methyl groups of retinal have been made for rhodopsin in the dark state, and for the Meta I and Meta II states.[21] In the dark state, the experiments have entailed both non-oriented (powder-type) and oriented membrane samples. Additional experiments in the dark state have determined ^{2}H NMR relaxation times (T_{1Q}) for the decay of quadrupolar order. Measurements were conducted as a function of temperature over a broad interval from −160 to −30 °C. Use of aligned membranes on ultra-thin glass slides with relatively low optical density allowed trapping and characterization of the Meta I and Meta II states of rhodopsin using electronic (optical) spectroscopy.[36,38] By conducting measurements below the melting temperature of the POPC lipid bilayer (−4 °C), the rotational diffusion of the rhodopsin molecules within the membrane was suppressed to reveal the internal dynamics of the receptor-bound ligand.[20] Because the studies were also carried out below the freezing point of water, the interfering signal from the residual ^{2}H nuclei of the water was eliminated.

For the dark state, in Figure 17.7 the experimental temperature dependencies of the deuterium relaxation times for Zeeman (T_{1Z}) and quadrupolar (T_{1Q}) order of the retinal ligand are summarized. First, we note that in Figures 17.7(a,b) the T_{1Z} and T_{1Q} curves with minima at approximately −120 °C are observed for the C5-methyl group. By contrast, the ^{2}H NMR relaxation data indicate that the T_{1Z} and T_{1Q} relaxation times for the C9- and C13-methyl groups of retinal increase monotonically with the temperature (Figures 17.7c,d).[21] The observation of a minimum is important because it corresponds to an optimal matching of the spectral density for the molecular fluctuations to the nuclear Larmor frequency ω_0. It enables an effective correlation time $\tau_c \approx 1/\omega_0$ to be identified for the motions at the corresponding temperature, in accord with NMR relaxation theory.[64] The different T_{1Z} and T_{1Q} minima allow essentially model-free conclusions to be reached about the mobility of the retinylidene methyl groups within the binding cavity of rhodopsin in terms of differences in their correlation times. Shifting the minimum to lower temperature corresponds to faster motions if other conditions (*e.g.* activation energies) are the same. For the C5-methyl group (Figure 17.7a,b), the experimental rotational correlation time exceeds 2 ns at temperatures below approximately −120 °C. However, for the C9-methyl in Figure 17.7(c) and the C13-methyl group in Figure 17.7(d), all motional correlation times are less than $1/\omega_0 \approx 2$ ns down to −160 °C, meaning that the corresponding minima must occur at lower temperatures. Consequently, the ^{2}H NMR relaxation data show significant site-specific differences in the internal mobility of the retinylidene methyl groups within the binding pocket of rhodopsin.

Application of Redfield theory[66] then allows us to establish a connection of experimental relaxation data (Figure 17.7) to the fluctuations of the retinylidene methyl groups within the binding pocket of rhodopsin. Detailed

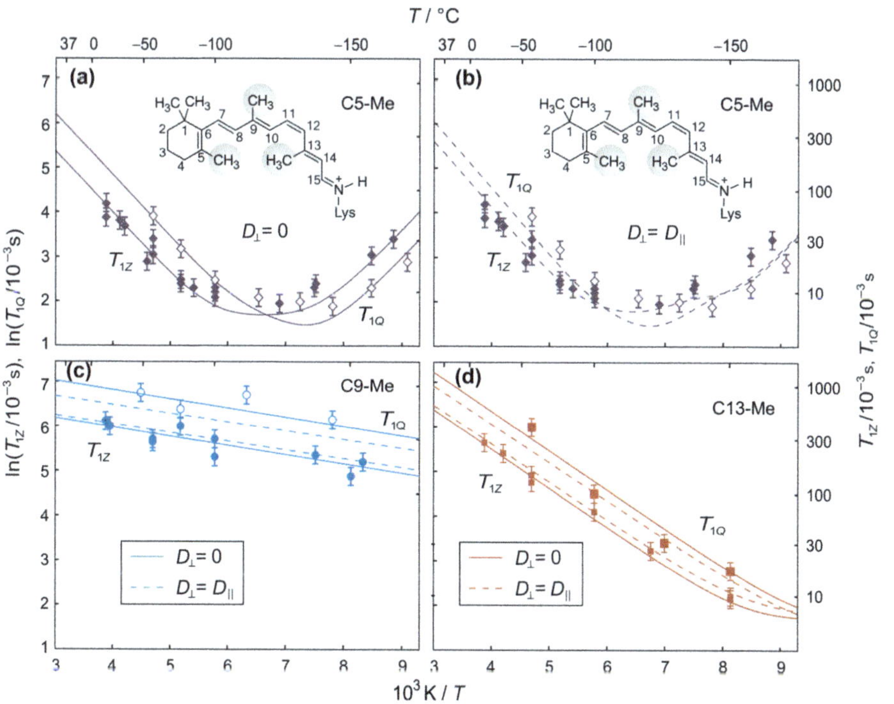

Figure 17.7 Solid-state ^{2}H NMR relaxation uncovers site-specific differences in the mobility of retinal ligand bound to rhodopsin in a membrane environment. Spin–lattice (T_{1Z}) and quadrupolar-order (T_{1Q}) relaxation times are plotted against inverse temperature for rhodopsin in POPC membranes (1:50 molar ratio) in the dark state with retinal ^{2}H-labelled at (a, b) the C5-methyl, (c) the C9-methyl, and (d) the C13-methyl groups. The experimentally determined T_{1Z} and T_{1Q} times were fit using analytical models for molecular dynamics[97] (see text): axial three-fold jump model (rate constant k, *solid lines*) and continuous rotational diffusion model (coefficients $D_\parallel$ and $D_\perp$ with *solid lines* for $D_\perp = 0$ and *dashed lines* for $D_\perp = D_\parallel$).

information about the ligand dynamics is provided, as shown Figure 17.8. The spin–lattice relaxation rates are related to the spectral density (power spectrum) of the thermal fluctuations of the irreducible components of the quadrupolar coupling tensor near the resonance frequency by:[97]

$$R_{1Z} = 1 / T_{1Z} = \tfrac{3}{4}\pi^2 \chi_Q^2 [J_1(\omega_0) + 4J_2(2\omega_0)] \tag{17.11}$$

and

$$R_{1Q} = 1 / T_{1Q} = \tfrac{9}{4}\pi^2 \chi_Q^2 J_1(\omega_0) \tag{17.12}$$

Here χ_Q is the static quadrupolar coupling constant, ω_0 is the resonance (Larmor) frequency, and $J_m(m\omega_0)$ denotes the spectral densities ($m = 1,2$) in an irreducible representation. In terms of generalized model-free analysis as introduced by Brown,[97] the spectral densities are written as:

$$J_m(\omega) = \sum_{r,q} |D^{(2)}_{0r}(\Omega_{PI})|^2 \Big[\Big\langle |D^{(2)}_{rq}(\Omega_{IM})|^2\Big\rangle - \Big|\Big\langle D^{(2)}_{rq}(\Omega_{IM})\Big\rangle\Big|^2 \delta_{r0}\delta_{q0}\Big] j^{(2)}_{rq}(\omega)\, |D^{(2)}_{qm}(\Omega_{ML})|^2 \quad (17.13)$$

where $\omega = m\omega_0$ and $m = 1, 2$. The quantities in square brackets are the mean-square amplitudes of the fluctuations. Assuming single-exponential relaxation, the corresponding reduced spectral densities for the fluctuations are given by:

$$j^{(2)}_{rq}(\omega) = 2\tau_{rq} / (1 + \omega^2\tau^2_{rq}) \quad (17.14)$$

In Eqn (17.13), $D^{(2)}_{m'm}(\Omega_{ij})$ indicates the Wigner rotation matrix elements with indices m', $m = -2, -1, 0, 1, 2$, and $\Omega_{ij} \equiv (\alpha_{ij}, \beta_{ij}, \gamma_{ij})$ are the Euler angles for rotational transformations between the coordinate systems i and j, where $i,j = P,I,M,D,L$; see Figure 17.8. In our designation, P denotes the principal axis system of the static electric field gradient tensor for a deuterium atom (ordinarily the principal z-axis is parallel to the C–^{2}H bond), I specifies the coordinate system related to the instantaneous orientation of the labelled methyl group, M indicates the system for the average methyl orientation, and L signifies the laboratory axis system (due to the external

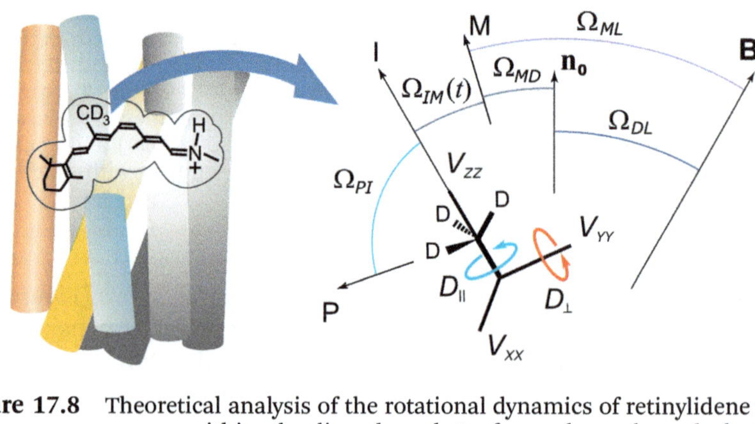

Figure 17.8 Theoretical analysis of the rotational dynamics of retinylidene methyl groups within the ligand pocket of membrane-bound rhodopsin. Euler angles $\Omega_{ij} \equiv (\alpha_{ij}, \beta_{ij}, \gamma_{ij})$ transform among the various coordinate systems $i,j \equiv P,I,M,D,L$. The principal axis system (PAS) of the static EFG tensor of the ^{2}H nucleus is denoted by P (z-axis parallel to C–^{2}H bond), I is the intermediate frame for methyl rotation (z-axis is instantaneous C–C^{2}H$_3$ bond direction), M is the methyl coordinate system (z-axis is average C–C^{2}H$_3$ bond direction), D is the membrane system (z-axis is surface normal $\boldsymbol{n}_0$), and L is the laboratory system (z-axis along external magnetic field $\boldsymbol{B}_0$).[109]

magnetic field $\boldsymbol{B}_0$). According to this notation, for the retinylidene methyl groups $\Omega_{PI} = (0°, 70.5°, 0°)$ and $\Omega_{ML} = (0°, \beta_{ML}, 0°)$. The final coordinate transformation can be further expanded to include the dependence on the membrane orientation relative to the main magnetic field $\boldsymbol{B}_0$ according to the formula:

$$D_{qm}^{(2)}(\Omega_{ML}) = \sum_{n} D_{qn}^{(2)}(\Omega_{MD}) D_{nm}^{(2)}(\Omega_{DL}) \tag{17.15}$$

where D corresponds to the axis system of the membrane normal. For unoriented (powder-type) samples, averaging over all membrane tilt angles is performed.[97]

Further analysis of the rotational dynamics of the retinylidene methyl groups involves introduction of a specific model for molecular motion (*e.g.* discrete jump, continuous diffusion).[97–102] The choice of a specific motional model enters into the correlation times which appear in the reduced spectral densities (Eqn 17.14). For discussion of motional models, see the work of Meirovitch and Freed.[100–102] In general, NMR relaxation is relatively insensitive to the functional form of the potential,[97,103] although differences between jump and continuous diffusion models are observed for methyl groups in solid amino acids.[104] In our work on rhodopsin, the relaxation was observed to be exponential within experimental error for all temperatures employed.[95] Because deviations from exponential relaxation are not observed, one can adopt the azimuthal average for an N-site jump model, *e.g.* as described by Torchia and Szabo.[99] Alternatively, a continuous diffusion model[105,106] may be employed.[97] If rotation about a single axis is considered, *i.e.* off-axial motion is neglected as in the case of a methyl group, then the result of generalized model-free analysis[97] (Eqn 17.13) simplifies to:[107]

$$J_m(\omega) = \sum_{r \neq 0} |D_{0r}^{(2)}(\Omega_{PI})|^2 j_r^{(2)}(\omega) |D_{rm}^{(2)}(\Omega_{IL})|^2 \tag{17.16}$$

where[97,99]

$$1/\tau_{rq} \rightarrow 1/\tau_r = \begin{cases} 4k\sin^2(\pi r/N) & (N\text{-site jump with azimuthal averaging}) \\ r^2 D_{\|} & (\text{axial diffusion}) \end{cases} \tag{17.17}$$

Here k is the rate constant for discrete N-fold axial nearest-neighbour jumps, and $D_{\|}$ is the diffusion constant for continuous axial rotation.

The above results can also encompass both axial and off-axial motions, as included in Eqn (17.13) for the generalized model free approach.[97] For simplicity, here we adopt[98] a model of continuous rotation diffusion in a potential of mean torque,[97,108] as originally introduced by Pier Luigi Nordio.[105,106] The rotational correlation times are then given by:

$$\frac{1}{\tau_{rq}} = \frac{\mu_{rq}}{\left\langle |D_{rq}^{(2)}(\Omega_{IM})|^2 \right\rangle - \left|\left\langle D_{rq}^{(2)}(\Omega_{IM}) \right\rangle\right|^2 \delta_{r0}\delta_{q0}} D_{\perp} + (D_{\|} - D_{\perp})r^2 \tag{17.18}$$

where $D_{\|}$ and $D_{\perp}$ are the rotational diffusion coefficients for rotation of the labelled methyl group about its three-fold axis and for the off-axial fluctuations, respectively. In addition, μ_{qn} and $\langle |D_{qn}^{(2)}(\Omega_{MD})|^2\rangle$ are the moments and mean-squared moduli of the Wigner rotation matrix elements for a generalized potential of mean torque.[98,109] Their values are tabulated in terms of the order parameters in Trouard *et al.*[98] In the limit of a strong-collisional approximation, the methyl orientation is assumed to change randomly by any amount, due to rapid variations in the torque acting upon the segment. The orientation after a collision is independent of that before a collision, and the time taken for a transition is negligible. Hence, the orientation after a collision is given by the orientational probability distribution, leading to:[97]

$$1/\tau_{rq} \rightarrow 1/\tau_r = 6D_{\perp} + (D_{\|} - D_{\perp})r^2 \tag{17.19}$$

Lastly, if we assume that the correlation times follow an Arrhenius activation law, then the temperature dependence for either the *N*-site jump[99] or continuous diffusion[97] model is a simple exponential. For the $N = 3$-site jump model, $1/\tau_{rq} \rightarrow 1/\tau_r = 3k$, where:

$$k = k_0 \exp(-E_a/RT) \tag{17.20}$$

in which k_0 is the jump rate in the absence of the potential and E_a is the potential barrier height or activation energy. For a continuous rotational diffusion model, $1/\tau_r$ is related to the principal values of the rotational diffusion tensor (Eqn 17.18) by:

$$D_{\|} = D_{0\|} \exp(-E_{a\|}/RT) \tag{17.21}$$

and

$$D_{\perp} = D_{0\perp} \exp(-E_{a\perp}/RT) \tag{17.22}$$

The pre-exponential factors ($D_{0\|}$, $D_{0\perp}$) represent the maximum diffusion constants when $E_a \ll RT$, *i.e.* at infinite temperature. Here the values of k_0 (or D_0) and E_a describe the local protein packing influences on retinal within the rhodopsin binding cavity.

In our work, both the three-fold jump model of axial methyl rotations[99] and the continuous diffusion model[97] have been employed to fit the T_{1Z} and T_{1Q} temperature dependencies for retinal bound to rhodopsin[21] in the dark state. In principle, T_{1Z} measurements at different orientations for aligned samples, or at different frequencies for the powder-type samples, allow one to distinguish between the different types of motion. The axial jump model predicts a 20% difference in the methyl relaxation depending on orientation, whereas for an axial free-diffusion model the methyl relaxation is angle independent.[99] For powder-type samples, due to the relatively low signal-to-noise ratio, T_{1Z} is typically measured for the $\beta_{ML} = 90°$ orientation of the methyl axis to the main magnetic field $\boldsymbol{B}_0$ (close to the $\theta = 90°$ peaks in the

^{2}H NMR spectra). By contrast, aligned samples are often studied with the average membrane normal parallel to the magnetic field ($\beta_{DL} = 0°$). Here, the β_{ML} angle is determined by the orientation of the methyl group in the sample with respect to the membrane normal (director). Differences in the relaxation times for different orientations were found to be undetectable within the error of the measurements, consistent with a diffusion model for the methyl rotational dynamics.

On the other hand, the error of the T_{1Z} measurements (10–20%) is also comparable with the predicted relaxation anisotropy for a jump model, so that firm conclusions cannot be reached regarding the motional mechanism. For the diffusion model, the $\eta_D = D_{\|}/D_{\perp}$ ratio could not be determined unambiguously based only on the T_{1Z} relaxation times. The T_{1Z} temperature dependencies could be fitted with η_D values of either 1 or ∞ (axial diffusion). The differences in E_a do not exceed 20%, depending on whether only axial diffusion ($D_{\perp} = 0$) is considered or off-axial motions ($D_{\perp} = D_{\|}$) are also considered. However, simultaneous fitting of the T_{1Z} and T_{1Q} temperature dependencies for two continuous diffusion models shows that $\eta_D = \infty$ gives good results (Figure 17.7; solid lines), while the fit with $\eta_D = 1$ (Figure 17.7; dashed lines) cannot be regarded as satisfactory, indicating that $\eta_D \gg 1$. This result is probably to be expected, considering the restricted off-axial fluctuations of the retinal methyl groups. The value of $\sim$0.9 for the S_{C_3} order parameter corresponds to an amplitude of $\beta_{IM} \approx 15°$ for the off-axial fluctuations of the methyl groups of retinal within the binding pocket of rhodopsin. This motion includes libration of the methyl groups with respect to the unsaturated polyene, as well as any reorientations of the polyene chain and β-ionone ring within the binding cavity. Evidently, the inertial tensors for off-axial fluctuations are larger than for methyl spinning. Correlation times are obtained in the range of 2–12 ps at 30°C and 3–45 ps at −60°C, depending on the methyl position, which are in the range expected for methyl rotation.[21]

At temperatures above the T_{1Z} minimum, the slopes of the T_{1Z} and T_{1Q} dependencies (Figure 17.7) correspond to the activation energy for methyl rotation, and are related to the heights of the rotational potential barriers.[21] The ^{2}H NMR relaxation studies show large site-specific differences in the dynamics, as manifested by both the pre-exponential factor and activation barriers for the retinal motions. To a certain degree, this is surprising considering that the S_{C_3} order parameters and the amplitudes of the off-axial fluctuations for the C5-, C9-, and C13-methyl groups are approximately the same. Values of the experimentally determined activation energies range from 2 to 15 kJ mol^{-1}, depending on the methyl position, and are essentially independent of whether a three-fold jump or a continuous diffusion model is considered. The highest activation energy is for the C5-methyl, with $E_a \approx 15$ kJ mol^{-1}, as is typical of methyl groups in organic solids. The high activation energy for the C5-methyl group is consistent with a 6-*s*-*cis* conformation of the β-ionone ring, which is in agreement with solid-state ^{13}C NMR chemical shift data[77] as discussed above. The surprisingly low activation energy for the C9-methyl group in the dark state can be explained by compensation of intra-retinal 1–6 interactions with hydrogens H7 and H11 (Figure 17.7c), because as shown by Struts *et al.*[21] the corresponding rotational potentials are displaced *versus* one another by about 180°. The higher activation energy for the C13-methyl groups is evidently due to intra-retinal 1–7 interactions with the H10 hydrogen atom. Contributions due to interactions with the side chains in the binding pocket of rhodopsin in the dark state are secondary, on account of the large distances from the retinylidene methyl groups (>4 Å), *e.g.* see Figure 17.5. Interestingly, the ^{2}H NMR relaxation data imply that the C5-methyl is the least mobile at low temperatures, despite the fact that the β-ionone ring may have multiple conformations that interconvert within the hydrophobic binding pocket on a longer timescale.[81]

After isomerization, notable changes in the molecular dynamics of retinal are evident in the Meta I and Meta II states, as shown in Figure 17.9. The largest light-induced difference involves the C9-methyl group, which is known to be crucial to the process of rhodopsin activation.[110] The E_a barrier for the C9-methyl group is increased approximately two-fold upon isomerization. At the same time, the E_a value for the C13-methyl group is reduced in the Meta I state, and the values of T_{1Z} for the C9- and C13-methyl groups become nearly the same. This seems logical, considering the all-*trans* conformation of the retinal after isomerization, where the C13- and C9-methyl groups may have similar rotational potentials (Figure 17.9). A reduction of the C13-methyl activation energy in Meta I can be explained, analogously to the C9-methyl group in the dark state, in terms of anti-phase potentials for the intra-retinal C13-to-H11 and C13-to-H15 1–6 interactions.[21] The more restricted mobility for the C9-methyl group upon isomerization may indicate a more crowded environment, *e.g.* smaller C9-to-H7 and C9-to-H11 distances in all-*trans*-retinal with respect to 11-*cis*-retinal, or non-bonded interactions with surrounding side chains in the binding pocket (Tyr268 or Thr118). For the C5-methyl group in Meta I, other differences are seen in the temperature dependence of the T_{1Z} relaxation times (Figure 17.9), with possible evidence for two individual minima. In this case, fitting of the experimental relaxation data requires two components with different activation energies E_a and pre-exponential factors (k_0 and D_0). The two components may correspond to different types of motion of the methyl groups (rotation and reorientation), or alternatively two 6-*s*-*cis* conformers of the retinal may be present.[81]

In the transition from the Meta I to the Meta II state, further changes are observed in the dynamics of the retinal. Theoretical fits in Figure 17.9 indicate the activation energy E_a and diffusion constants D_0 increase slightly for the C9-methyl group but decrease for the C13-methyl group.[21] The Meta II state is of particular interest because it is the activated receptor conformation.[111] Minor differences in the rotational dynamics and the activation energies of the C9- and C13-methyl groups in Meta II most likely can be attributed to interactions with amino acid side chains forming the

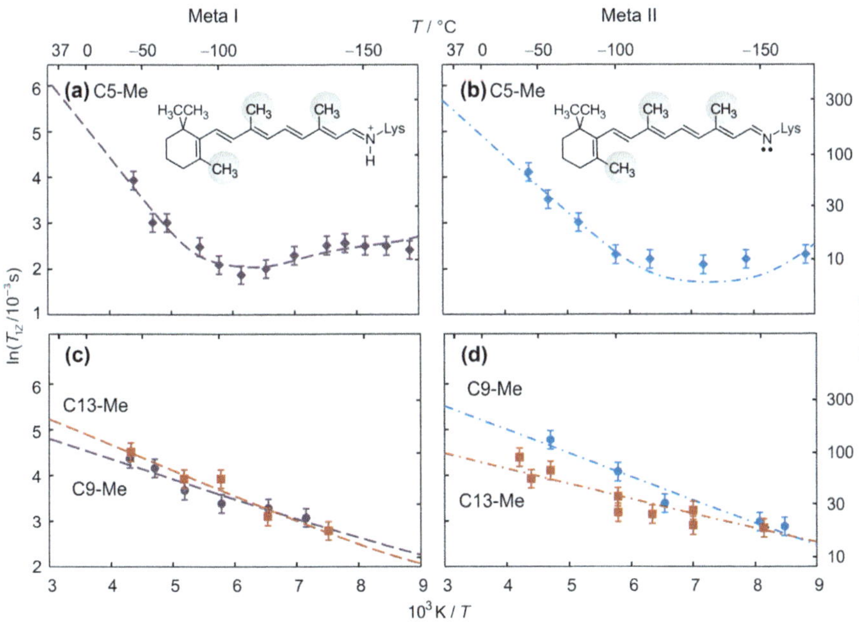

Figure 17.9 Solid-state ^{2}H NMR relaxation illuminates how the dynamics of the retinylidene methyl groups vary upon light activation of rhodopsin in membranes. Spin–lattice relaxation times (T_{1Z}) are plotted against inverse temperature for rhodopsin in POPC membranes (1:50 molar ratio) with retinal ^{2}H-labelled at the C5-, C9-, or C13-methyl groups in (a, c) the Meta I and (b, d) the Meta II states, respectively. The relaxation data are fit with an axial three-fold jump model (rate constant *k*) as well as with a diffusion model (axial coefficient $D_{\|}$ with $D_{\perp}=0$). For the C5-methyl group in Meta I, either different rotational diffusion constants were assumed ($D_{\|} \neq D_{\perp}$, not shown) or different conformers.[95] Following 11-*cis* to 11-*trans* isomerization, a reduction occurs from three distinct methyl environments in the dark state to roughly two environments in Meta I and Meta II, suggesting a more planar retinal conformation.
(Adapted from Brown and Struts,[117] with permission from Springer.)

chromophore binding pocket. Yet the relatively low activation energies of the C9- and C13-methyl groups in the Meta II state indicate the absence of steric clashes of these groups with residues in the binding cleft. Surprisingly, the C13-methyl has a lower activation energy E_a for axial rotation than the C9-methyl group, although rotation of the C13-methyl group due to retinal isomerization should bring it very close to the β4 strand of extracellular loop E2. A possible explanation for the absence of interactions between the C13-methyl and Cys187 on the β4 strand is that a displacement of the retinal polyene chain away from the E2 loop occurs, due to the polyene chain straightening. This is in agreement with ^{13}C NMR distance measurements[54] and the X-ray structures reported for the putative Meta II state.[11,12,35] According to our relaxation data, such a displacement already takes place in the Meta I state. Thus, if a steric clash occurs between the C13-methyl and the β4 strand, it is transient. Interestingly, the overall behavior of the T_{1Z} relaxation times for the C5-methyl group in the dark, Meta I, and Meta II states is quite similar (the differences in E_a hardly exceed 20% for a given model of the rotational dynamics, *e.g.* see Figures 17.7 and 17.9). Hence, one can conclude that the conformation and the environment of the β-ionone ring remain similar after rhodopsin activation.[21] A similar conclusion has been reached based on ^{13}C NMR chemical shift data.[16]

17.5 Solid-State ^{2}H NMR Spectroscopy Illuminates the Activation Mechanism of Rhodopsin in a Membrane Environment

Current knowledge of the rhodopsin activation process entails a sequence of transformations among various states and substates (including photorhodopsin, bathorhodopsin, lumirhodopsin, Meta I, Meta II_a, Meta II_b, and Meta II_bH^+), where Meta II_b and Meta II_bH^+ are considered as the active conformational substates. The substates from Meta I to Meta II_bH^+ are in a quasi-thermodynamic equilibrium after the last stage of the activation is reached. It has been suggested that an ensemble of substates occurs due to the entropy gain produced by release of the inhibitory retinal lock.[95] The picture of rhodopsin activation[21,95] shown in Figure 17.10 is revealed by combining data for the ^{2}H NMR retinal structure and dynamics in different states with the results of other structural and biophysical studies.[6,11,12,18,33,74,93,94,112]

First, photo-isomerization of retinal leads to the rotation of the C13-methyl group and the adjacent part of the retinal polyene chain towards the β4 strand of the extracellular loop E2.[20,33] In addition, the polyene chain of the retinal straightens and elongates due to the 11-*cis* to 11-*trans* isomerization.[11,20] Straightening of the polyene chain leads to a steric clash with Trp265, accompanied by its displacement towards the cytoplasmic end of the protein, and towards helix H5 (see Figure 17.10).[11,18–20] The displacement of Trp265 of helix H6 may enable outward rotation of helix H6.[18–20] The retinal elongation also results in a translation of the β-ionone ring towards the H5 helix. Rotation of the retinal polyene chain at the Schiff base end of the ligand destabilizes the hydrogen bonding networks connecting the E2 loop, helices H1–H3, H6, H7, and retinal in the inactive state.[8,10,11,112] Additionally, the ionic lock of the protonated Schiff base with its counterion may be destabilized by the 11-*cis*- to all-*trans*-retinal isomerization, which changes the pK_a of the protonated Schiff base.[112] As a result, the hydrogen-bonding network rearranges[11] and the Schiff base deprotonates, breaking the ionic lock and enabling a rearrangement of the helical bundle. Displacement of the β-ionone ring (whose C5-methyl group is in the vicinity of Glu122) towards helix H5 is apparently coupled to rearrangement of another hydrogen-bonding network between Glu122 on H3 and His211 on H5,[11,20,94] which initially stabilizes the inactive conformation. The rearrangement of a

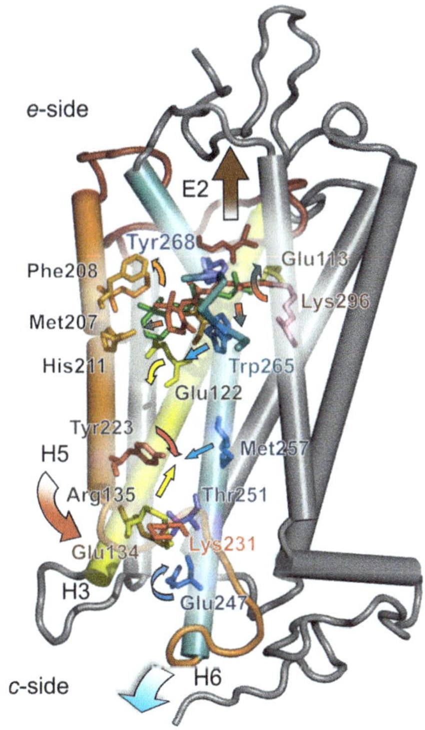

Figure 17.10 Local retinal dynamics underlie large-scale protein rearrangements leading to activation of the membrane-bound photo-receptor. Data from solid-state ^2H NMR spectroscopy[21] and other biophysical measurements indicate the C13-methyl and the $C{=}NH^+$– moieties of the protonated Schiff base change their orientation upon ligand isomerization. The β-ionone ring with the C5-methyl group remains in its hydrophobic pocket, but its displacement due to elongation of retinal triggers allosteric movements of helices H5 and H6 that lead to rhodopsin activation (see text).
Figure produced with PyMOL [http://pymol.sourceforge.net/].

hydrogen bond between Glu122 and His211 may be responsible for rotation of the cytoplasmic end of the helix H5 towards to H6.[93] This motion destabilizes the ionic lock between helices H3 and H6 involving Glu134, Arg135, and Glu247 by bringing Tyr223 into close contact with the Arg135 residue, as indicated by Figure 17.10.

Consequently, the activating rotation of the helix H6 as first shown by Hubbell and co-workers[3,6] provides for a transient opening of the transducin (G_t) binding site.[11,12,21,95,113] New interactions between Tyr223–Arg135 and Glu247–Lys231–Thr251 are formed that stabilize the active Meta II conformation. Rhodopsin activation involves collective fluctuations of transmembrane helices (H5 and H6) in the Meta I–Meta II equilibrium. The reversible helix movements have large activation barriers; they occur infrequently due to their collective nature, because they involve displacements of many atoms within the protein on the ms timescale. They match the catalytic turnover rate for G_t binding and activation by rhodopsin. An important role of collective helix fluctuations in activation of transducin in the photo-transduction cascade is suggested,[95] where a single photo-activated rhodopsin catalyzes the activation of hundreds of transducin molecules. Receptor activation is thus due to a fluctuating equilibrium among states and substates, which may be characteristic for other GPCRs in a membrane lipid environment.

17.6 Conclusions

Solid-state ^{2}H NMR spectroscopy gives a unique avenue for investigating G-protein-coupled receptors and other membrane proteins that is complementary to other biophysical approaches, such as X-ray crystallography. A distinct advantage of solid-state ^{2}H NMR spectroscopy is that rhodopsin is studied in a native-like membrane lipid environment.[114,115] Using ^{2}H NMR methods, the structure and dynamics of retinal have been investigated in the rhodopsin dark state, as well as in the Meta I and Meta II photointermediates. Despite its phenomenal stability in the dark state, the retinal cofactor of rhodopsin has significant dynamics within the binding pocket of the receptor. The ^{2}H NMR data indicate that the methyl groups are rapidly rotating. However, the off-axial motions are highly restricted in all the states of rhodopsin investigated (dark, Meta I, Meta II), with order parameters of $\sim$0.9 due to interactions within the binding pocket. Solid-state ^{2}H NMR data combined with the results of other structural and biophysical studies suggest that displacement of the retinal β-ionone ring and rotation of the polyene chain occur upon photo-isomerization. Disruption of hydrogen bonding networks and ionic locks that stabilize the inactive conformation then lead to triggering of the activating helical fluctuations of the receptor. The Meta I–Meta II equilibrium of rhodopsin entails a complex energy landscape, with multiple substates rather than a simple on-off switch, with broad implications for biological signalling in general.

Acknowledgments

Research was supported in part by grants from the U.S. National Institutes of Health. Discussions with K. P. Hofmann, W. L. Hubbell, J. W. Lewis, G. F. J. Salgado, R. Vogel, and X. Xu are gratefully acknowledged.

References

1. D. M. Rosenbaum, S. G. F. Rasmussen and B. K. Kobilka, *Nature*, 2009, **459**, 356–363.
2. K. P. Hofmann, P. Scheerer, P. W. Hildebrand, H.-W. Choe, J. H. Park, M. Heck and O. P. Ernst, *Trends Biochem. Sci.*, 2009, **34**, 540–552.
3. D. L. Farrens, C. Altenbach, K. Yang, W. L. Hubbell and H. G. Khorana, *Science*, 1996, **274**, 768–770.

4. W. L. Hubbell, C. Altenbach, C. M. Hubbell and H. G. Khorana, *Adv. Protein Chem.*, 2003, **63**, 243–290.
5. B. Knierim, K. P. Hofmann, O. P. Ernst and W. L. Hubbell, *Proc. Natl. Acad. Sci. U. S. A.*, 2007, **104**, 20290–20295.
6. C. Altenbach, A. K. Kusnetzow, O. P. Ernst, K. P. Hofmann and W. L. Hubbell, *Proc. Natl. Acad. Sci. U. S. A.*, 2008, **105**, 7439–7444.
7. K. Palczewski, T. Kumasaka, T. Hori, C. A. Behnke, H. Motoshima, B. A. Fox, I. Le Trong, D. C. Teller, T. Okada, R. E. Stenkamp, M. Yamamoto and M. Miyano, *Science*, 2000, **289**, 739–745.
8. D. C. Teller, T. Okada, C. A. Behnke, K. Palczewski and R. E. Stenkamp, *Biochemistry*, 2001, **40**, 7761–7772.
9. T. Okada, Y. Fujiyoshi, M. Silow, J. Navarro, E. M. Landau and Y. Shichida, *Proc. Natl. Acad. Sci. U.S.A.*, 2002, **99**, 5982–5987.
10. T. Okada, M. Sugihara, A.-N. Bondar, M. Elstner, P. Entel and V. Buss, *J. Mol. Biol.*, 2004, **342**, 571–583.
11. H.-W. Choe, Y. J. Kim, J. H. Park, T. Morizumi, E. F. Pai, N. Krauß, K. P. Hofmann, P. Scheerer and O. P. Ernst, *Nature*, 2011, **471**, 651–656.
12. J. Standfuss, P. C. Edwards, A. D'Antona, M. Fransen, G. Xie, D. D. Oprian and G. F. X. Schertler, *Nature*, 2011, **471**, 656–660.
13. G. F. J. Salgado, A. V. Struts, K. Tanaka, N. Fujioka, K. Nakanishi and M. F. Brown, *Biochemistry*, 2004, **43**, 12819–12828.
14. P. J. R. Spooner, J. M. Sharples, M. A. Verhoeven, J. Lugtenberg, C. Glaubitz and A. Watts, *Biochemistry*, 2002, **41**, 7549–7555.
15. P. J. R. Spooner, J. M. Sharples, S. C. Goodall, H. Seedorf, M. A. Verhoeven, J. Lugtenburg, P. H. M. Bovee-Geurts, W. J. DeGrip and A. Watts, *Biochemistry*, 2003, **42**, 13371–13378.
16. P. J. R. Spooner, J. M. Sharples, S. C. Goodall, P. H. M. Bovee-Geurts, M. A. Verhoeven, J. Lugtenburg, A. M. A. Pistorius, W. J. DeGrip and A. Watts, *J. Mol. Biol.*, 2004, **343**, 719–730.
17. P. J. E. Verdegem, P. H. M. Bovee-Geurts, W. J. de Grip, J. Lugtenburg and H. J. M. de Groot, *Biochemistry*, 1999, **38**, 11316–11324.
18. A. B. Patel, E. Crocker, M. Eilers, A. Hirshfeld, M. Sheves and S. O. Smith, *Proc. Natl. Acad. Sci. U.S.A.*, 2004, **101**, 10048–10053.
19. E. Crocker, M. Eilers, S. Ahuja, V. Hornak, A. Hirshfeld, M. Sheves and S. O. Smith, *J. Mol. Biol.*, 2006, **357**, 163–172.
20. A. V. Struts, G. F. J. Salgado, K. Tanaka, S. Krane, K. Nakanishi and M. F. Brown, *J. Mol. Biol.*, 2007, **372**, 50–66.
21. A. V. Struts, G. F. J. Salgado and M. F. Brown, *Proc. Natl. Acad. Sci. U. S. A.*, 2011, **108**, 8263–8268.
22. S. H. Park, B. B. Das, F. Casagrande, Y. Tian, H. J. Nothnagel, M. Chu, H. Kiefer, K. Maier, A. A. De Angelis, F. M. Marassi and S. J. Opella, *Nature*, 2012, **491**, 779–783.
23. V. Cherezov, D. M. Rosenbaum, M. A. Hanson, S. G. F. Rasmussen, F. S. Thian, T. S. Kobilka, H.-J. Choi, P. Kuhn, W. I. Weis, B. K. Kobilka and R. C. Stevens, *Science*, 2007, **318**, 1258–1265.

24. S. G. F. Rasmussen, H.-J. Choi, D. M. Rosenbaum, T. S. Kobilka, F. S. Thian, P. C. Edwards, M. Burghammer, V. R. P. Ratnala, R. Sanishvili, R. F. Fischetti, G. F. X. Schertler, W. I. Weis and B. K. Kobilka, *Nature*, 2007, **450**, 383–387.
25. T. Warne, M. J. Serrano-Vega, J. G. Baker, R. Moukhametzianov, P. C. Edwards, R. Henderson, A. G. W. Leslie, C. G. Tate and G. F. X. Schertler, *Nature*, 2008, **454**, 486–491.
26. V.-P. Jaakola, M. T. Griffith, M. A. Hanson, V. Cherezov, E. Y. T. Chien, J. R. Lane, A. P. Ijzerman and R. C. Stevens, *Science*, 2008, **322**, 1211–1217.
27. A. Manglik, A. C. Kruse, T. S. Kobilka, F. S. Thian, J. M. Mathiesen, R. K. Sunahara, L. Pardo, B. K. Kobilka and S. Granier, *Nature*, 2012, **485**, 321–326.
28. S. Granier, A. Manglik, A. C. Kruse, T. S. Kobilka, F. S. Thian, W. I. Weis and B. K. Kobilka, *Nature*, 2012, **485**, 400–404.
29. H. Wu, D. Wacker, M. Mileni, V. Katritch, G. W. Han, E. Vardy, W. Liu, A. A. Thompson, X.-P. Huang, F. I. Carroll, S. W. Mascarella, R. B. Westkaemper, P. D. Mosier, B. L. Roth, V. Cherezov and R. C. Stevens, *Nature*, 2012, **485**, 327–332.
30. A. A. Thompson, W. Liu, E. Chung, V. Katritch, H. Wu, E. Vardy, X.-P. Huang, C. Trapella, R. Guerrini, G. Calo, B. L. Roth, V. Cherezov and R. C. Stevens, *Nature*, 2012, **485**, 395–399.
31. R. Nygaard, T. M. Frimurer, B. Holst, M. M. Rosenkilde and T. W. Schwartz, *Trends Pharmacol. Sci.*, 2009, **30**, 249–259.
32. S. Ahuja and S. O. Smith, *Trends Pharmacol. Sci.*, 2009, **30**, 494–502.
33. H. Nakamichi and T. Okada, *Angew. Chem. Int. Ed.*, 2006, **45**, 4270–4273.
34. D. Salom, D. T. Lodowski, R. E. Stenkamp, I. Le Trong, M. Golczak, B. Jastrzebska, T. Harris, J. A. Ballesteros and K. Palczewski, *Proc. Natl. Acad. Sci. U. S. A.*, 2006, **103**, 16123–16128.
35. X. Deupi, P. Edwards, A. Singhal, B. Nickle, D. Oprian, G. Schertler and J. Standfuss, *Proc. Natl. Acad. Sci. U. S. A.*, 2012, **109**, 119–124.
36. M. F. Brown, *Chem. Phys. Lipids*, 1994, **73**, 159–180.
37. A. V. Botelho, N. J. Gibson, Y. Wang, R. L. Thurmond and M. F. Brown, *Biochemistry*, 2002, **41**, 6354–6368.
38. A. V. Botelho, T. Huber, T. P. Sakmar and M. F. Brown, *Biophys. J.*, 2006, **91**, 4464–4477.
39. M. F. Brown, *Biochemistry*, 2012, **51**, 9782–9795.
40. P. Kukura, D. W. McCamant, S. Yoon, D. B. Wandschneider and R. A. Mathies, *Science*, 2005, **310**, 1006–1009.
41. T. P. Sakmar, S. T. Menon, E. P. Marin and E. S. Awad, *Annu. Rev. Biophys. Biomol. Struct.*, 2002, **31**, 443–484.
42. S. Lüdeke, R. Beck, E. C. Y. Yan, T. P. Sakmar, F. Siebert and R. Vogel, *J. Mol. Biol.*, 2005, **353**, 345–356.
43. K. Nakanishi and R. Crouch, *Isr. J. Chem.*, 1995, **35**, 253–272.
44. A. S. Ulrich, I. Wallat, M. P. Heyn and A. Watts, *Nat. Struct. Biol*, 1995, **2**, 190–192.

45. S. Moltke, A. A. Nevzorov, N. Sakai, I. Wallat, C. Job, K. Nakanishi, M. P. Heyn and M. F. Brown, *Biochemistry*, 1998, **37**, 11821–11835.
46. S. Moltke, I. Wallat, N. Sakai, K. Nakanishi, M. F. Brown and M. P. Heyn, *Biochemistry*, 1999, **38**, 11762–11772.
47. G. F. J. Salgado, A. V. Struts, K. Tanaka, S. Krane, K. Nakanishi and M. F. Brown, *J. Am. Chem. Soc.*, 2006, **128**, 11067–11071.
48. M. Etzkorn, S. Martell, O. C. Andronesi, K. Seidel, M. Engelhard and M. Baldus, *Angew. Chem. Int. Ed.*, 2007, **46**, 459–462.
49. K. Varga, L. Tian and A. E. McDermott, *Biochim. Biophys. Acta*, 2007, **1774**, 1604–1613.
50. A. Watts, *Nat. Rev. Drug Discovery*, 2005, **4**, 555–568.
51. M. F. Brown, K. Martínez-Mayorga, K. Nakanishi, G. F. J. Salgado and A. V. Struts, *Photochem. Photobiol.*, 2009, **85**, 442–453.
52. V. Copié, A. E. McDermott, K. Beshah, J. C. Williams, M. Spyker-Assink, R. T. Gebhard, J. Lugtenberg, J. Herzfeld and R. G. Griffin, *Biochemistry*, 1994, **33**, 3280–3286.
53. A. A. Nevzorov and S. J. Opella, *J. Magn. Reson.*, 2003, **160**, 33–39.
54. S. Ahuja, V. Hornak, E. C. Y. Yan, N. Syrett, J. A. Goncalves, A. Hirshfeld, M. Ziliox, T. P. Sakmar, M. Sheves, P. J. Reeves, S. O. Smith and M. Eilers, *Nat. Struct. Mol. Biol.*, 2009, **16**, 168–175.
55. F. Hu, W. Luo and M. Hong, *Science*, 2010, **330**, 505–508.
56. M. Hong, Y. Zhang and F. Hu, *Annu. Rev. Phys. Chem.*, 2012, **63**, 1–24.
57. A. E. McDermott, *Curr. Opin. Struct. Biol.*, 2004, **14**, 554–561.
58. A. E. McDermott, *Annu. Rev. Biophys.*, 2009, **38**, 385–403.
59. T. A. Cross, M. Sharma, M. Yi and H.-X. Zhou, *Trends Biochem. Sci.*, 2011, **36**, 117–125.
60. M. Sharma, M. Yi, H. Dong, H. Qin, E. Peterson, D. D. Busath, H.-X. Zhou and T. A. Cross, *Science*, 2010, **330**, 509–512.
61. M. F. Brown, in *Biological Membranes. A Molecular Perspective from Computation and Experiment*, ed. K. Merz Jr. and B. Roux, Birkhäuser, Basel, 1996, pp. 175–252.
62. A. A. Nevzorov, S. Moltke, M. P. Heyn and M. F. Brown, *J. Am. Chem. Soc.*, 1999, **121**, 7636–7643.
63. A. Kusumi and J. S. Hyde, *Biochemistry*, 1982, **21**, 5978–5983.
64. M. F. Brown and S. I. Chan, in *Encyclopedia of Nuclear Magnetic Resonance*, ed. D. M. Grant and R. K. Harris, Wiley, New York, 1996, vol. 2 , pp. 871–885.
65. K. Schmidt-Rohr and H. W. Spiess, *Multidimensional Solid State NMR and Polymers*, Academic Press, San Diego, 1994.
66. C. P. Slichter, *Principles of Magnetic Resonance*, Springer, Heidelberg, 3rd edn, 1990.
67. A. A. Nevzorov, S. Moltke and M. F. Brown, *J. Am. Chem. Soc.*, 1998, **120**, 4798–4805.
68. T. Huber, A. V. Botelho, K. Beyer and M. F. Brown, *Biophys. J.*, 2004, **86**, 2078–2100.
69. M. Chabre and J. Breton, *Vision Res.*, 1979, **19**, 1005–1018.

70. M. Michel-Villaz, C. Roche and M. Chabre, *Biophys. J.*, 1982, **37**, 603–616.
71. J. W. Lewis, C. M. Einterz, S. J. Hug and D. S. Kliger, *Biophys. J.*, 1989, **56**, 1101–1111.
72. S. Jäger, J. W. Lewis, T. A. Zvyaga, I. Szundi, T. P. Sakmar and D. S. Kliger, *Proc. Natl. Acad. Sci. U. S. A.*, 1997, **94**, 8557–8562.
73. Y. Fujimoto, J. Ishihara, S. Maki, N. Fujioka, T. Wang, T. Furuta, N. Fishkin, B. Borhan, N. Berova and K. Nakanishi, *Chem. Eur. J.*, 2001, 7, 4198–4204.
74. Y. Fujimoto, N. Fishkin, G. Pescitelli, J. Decatur, N. Berova and K. Nakanishi, *J. Am. Chem. Soc.*, 2002, **124**, 7294–7302.
75. J. Cavanaugh, W. J. Fairbrother, A. G. Palmer III, N. J. Skelton and M. Rance, *Protein NMR Spectroscopy: Principles and Practice*, Academic Press, New York, 2nd edn, 2006.
76. N. Fishkin, N. Berova and K. Nakanishi, *Chem. Rec.*, 2004, **4**, 120–135.
77. S. O. Smith, I. Palings, M. E. Miley, J. Courtin, H. de Groot, J. Lugtenberg, R. A. Mathies and R. G. Griffin, *Biochemistry*, 1990, **29**, 8158–8164.
78. G. S. Harbison, S. O. Smith, J. A. Pardoen, J. M. L. Courtin, J. Lugtenburg, J. Herzfeld, R. A. Mathies and R. G. Griffin, *Biochemistry*, 1985, **24**, 6955–6962.
79. B. Mertz, M. Lu, M. F. Brown and S. E. Feller, *Biophys. J.*, 2011, **101**, L18–L19.
80. S. Zhu, M. F. Brown and S. E. Feller, *J. Am. Chem. Soc.*, 2013, **135**, 9391–9398.
81. P.-W. Lau, A. Grossfield, S. E. Feller, M. C. Pitman and M. F. Brown, *J. Mol. Biol.*, 2007, **372**, 906–917.
82. N. Leioatts, B. Mertz, K. Martínez-Mayorga, T. D. Romo, M. C. Pitman, S. E. Feller, A. Grossfield and M. F. Brown, *Biochemistry*, 2014 in press.
83. G. Gröbner, I. J. Burnett, C. Glaubitz, G. Choi, A. J. Mason and A. Watts, *Nature*, 2000, **405**, 810–813.
84. G.-F. Jang, V. Kuksa, S. Filipek, F. Bartl, E. Ritter, M. H. Gelb, K. P. Hofmann and K. Palczewski, *J. Biol. Chem.*, 2001, **276**, 26148–26153.
85. Q. Wang, R. W. Schoenlein, L. A. Peteanu, R. A. Mathies and C. V. Shank, *Science*, 1994, **266**, 422–424.
86. S. W. Lin, M. Groesbeek, I. van der Hoef, P. Verdegem, J. Lugtenburg and R. A. Mathies, *J. Phys. Chem. B*, 1998, **102**, 2787–2806.
87. T. Andruniów, N. Ferré and M. Olivucci, *Proc. Natl. Acad. Sci. U. S. A.*, 2004, **101**, 17908–17913.
88. D. Polli, P. Altoè, O. Weingart, K. M. Spillane, C. Manzoni, D. Brida, G. Tomasello, G. Orlandi, P. Kukura, R. A. Mathies, M. Garavelli and G. Cerullo, *Nature*, 2010, **467**, 440–443.
89. J. A. Gascon and V. S. Batista, *Biophys. J.*, 2004, **87**, 2931–2941.

90. M. Sugihara, J. Hufen and V. Buss, *Biochemistry*, 2006, **45**, 801–810.
91. S. Hayashi, E. Tajkhorshid and K. Schulten, *Biophys. J.*, 2009, **96**, 403–416.
92. J. J. Ruprecht, T. Mielke, R. Vogel, C. Villa and G. F. X. Schertler, *EMBO J.*, 2004, **23**, 3609–3620.
93. S. Ahuja, E. Crocker, M. Eilers, V. Hornak, A. Hirshfeld, M. Ziliox, N. Syrett, P. J. Reeves, H. G. Khorana, M. Sheves and S. O. Smith, *J. Biol. Chem.*, 2009, **284**, 10190–10201.
94. R. Vogel, F. Siebert, S. Lüdeke, A. Hirshfeld and M. Sheves, *Biochemistry*, 2005, **44**, 11684–11699.
95. A. V. Struts, G. F. J. Salgado, K. Martínez-Mayorga and M. F. Brown, *Nat. Struct. Mol. Biol.*, 2011, **18**, 392–394.
96. K. Martínez-Mayorga, M. C. Pitman, A. Grossfield, S. E. Feller and M. F. Brown, *J. Am. Chem. Soc.*, 2006, **128**, 16502–16503.
97. M. F. Brown, *J. Chem. Phys.*, 1982, 77, 1576–1599.
98. T. P. Trouard, T. M. Alam and M. F. Brown, *J. Chem. Phys.*, 1994, **101**, 5229–5261.
99. D. A. Torchia and A. Szabo, *J. Magn. Reson.*, 1982, **49**, 107–121.
100. E. Meirovitch, Y. E. Shapiro, A. Polimeno and J. H. Freed, *Prog. Nucl. Magn. Reson. Spectrosc.*, 2010, **56**, 360–405.
101. E. Meirovitch, A. Polimeno and J. H. Freed, *J. Chem. Phys.*, 2010, **132**, 207101-1–207101-2.
102. E. Meirovitch, A. Polimeno and J. H. Freed, *J. Phys. Chem. B*, 2006, **110**, 20615–20628.
103. B. Halle, *J. Phys. Chem.*, 1991, **95**, 6724–6733.
104. L. S. Batchelder, C. H. Niu and D. A. Torchia, *J. Am. Chem. Soc.*, 1983, **105**, 2228–2231.
105. P. L. Nordio and P. Busolin, *J. Chem. Phys.*, 1971, **55**, 5485–5490.
106. P. L. Nordio, G. Rigatti and U. Segre, *J. Chem. Phys.*, 1972, **56**, 2117–2123.
107. A. A. Nevzorov and M. F. Brown, *J. Chem. Phys.*, 1997, **107**, 10288–10310.
108. A. Szabo, *J. Chem. Phys.*, 1984, **81**, 150–167.
109. A. A. Nevzorov, T. P. Trouard and M. F. Brown, *Phys. Rev. E*, 1998, **58**, 2259–2281.
110. R. Vogel, S. Lüdeke, F. Siebert, T. P. Sakmar, A. Hirshfeld and M. Sheves, *Biochemistry*, 2006, **45**, 1640–1652.
111. T. Okada, O. P. Ernst, K. Palczewski and K. P. Hofmann, *Trends Biochem. Sci.*, 2001, **26**, 318–324.
112. M. Mahalingam, K. Martínez-Mayorga, M. F. Brown and R. Vogel, *Proc. Natl. Acad. Sci. U. S. A.*, 2008, **105**, 17795–17800.
113. P. Scheerer, J. H. Park, P. W. Hildebrand, Y. J. Kim, N. Krauß, H.-W. Choe, K. P. Hofmann and O. P. Ernst, *Nature*, 2008, **455**, 497–502.
114. M. F. Brown, G. F. J. Salgado and A. V. Struts, *Biochim. Biophys. Acta*, 2010, **1798**, 177–193.

115. B. Mertz, A. V. Struts, S. E. Feller and M. F. Brown, *Biochim. Biophys. Acta*, 2012, **1818**, 241–251.
116. M. F. Brown, M. P. Heyn, C. Job, S. Kim, S. Moltke, K. Nakanishi, A. A. Nevzorov, A. V. Struts, G. F. J. Salgado and I. Wallat, *Biochim. Biophys. Acta*, 2007, **1768**, 2979–3000.
117. M. F. Brown and A. V. Struts, in *Encyclopedia of Biophysics*, ed. G. C. K. Roberts, Springer, Berlin, 2013, pp. 2231–2243.

CHAPTER 18

Helical Membrane Protein Structure: Strategy for Success

NABANITA DAS,[a,b] DYLAN T. MURRAY,[a,b] YIMIN MIAO[a,c] AND TIMOTHY A. CROSS*[a,b,c]

[a] National High Magnetic Field Laboratory, Florida State University, Tallahassee, FL 32310, USA; [b] Institute of Molecular Biophysics, Florida State University, Tallahassee, FL 32306, USA; [c] Department of Chemistry and Biochemistry, Florida State University, Tallahassee, FL 32306, USA
*Email: cross@magnet.fsu.edu

18.1 Introduction

Membrane protein structural characterization represents a unique opportunity for the solid-state NMR (ssNMR) community. It is now clear that inadequate membrane mimetic environments easily distort native helical membrane protein structure. No other technology can characterize membrane protein structure in liquid crystalline lipid bilayers or potentially even in *E. coli* membranes that represent good models for the native environment. This chapter discusses the unique properties of membrane environments and, while membrane protein structural characterization by ssNMR has been challenging, a strategy for such characterizations based on the current literature is outlined. This chapter also redefines what is needed for characterizing a "good" or "high resolution" membrane protein structure, since definitions from solution NMR and X-ray crystallography are inappropriate for ssNMR of membrane proteins.

New Developments in NMR No. 3
Advances in Biological Solid-State NMR: Proteins and Membrane-Active Peptides
Edited by Frances Separovic and Akira Naito

Published by the Royal Society of Chemistry, www.rsc.org

18.2 Defining the Lipid Bilayer Environment for Native Structures

18.2.1 Membrane Environment

Natural membranes are a complex environment that is home to helical membrane proteins. The membrane is composed of a multitude of molecules including lipids, proteins, and other biological molecules. Together these molecules form a highly impermeable barrier to hydrophilic compounds that delineate the extracellular from intracellular environment and extraorganelle from intraorganelle environs. For the study of membrane proteins that facilitate transport or communication across membranes the characteristics of these membranes must be appreciated. Water concentration, dielectric, hydrophobicity, and lateral pressure all take the form of gradients that define the anisotropic nature of the membrane.[1] The native tertiary and quaternary structure of helical membrane proteins is stabilized by a unique balance of inter- and intramolecular interactions, resulting not only in a stable conformation but one that is uniquely oriented with respect to its environment, a key feature for the functionality of these proteins.

The molecular forces that hold membrane proteins together are no different than those that stabilize soluble proteins. Hydrogen bonding, van der Waals contacts, additional electrostatic interactions, the hydrophobic effect, and pressure all influence protein structure. However, within the confines of a membrane these forces have altered relative strengths depending on the location within the membrane, because of the substantial gradients across it. The dielectric and water concentration gradients are very large across the membrane. In the hydrophobic interior the dielectric is ~ 2 increasing to ~ 4 in the glycerol backbone region of the lipid molecules. It further increases by up to two orders of magnitude in the polar headgroup region before returning to the value experienced by soluble proteins of ~ 80 in the bulk aqueous environment.[2] Water concentration follows a similar pattern over many orders of magnitude from the bilayer center to 55 M in the aqueous environment.

For electrostatic interactions, these forces are strengthened in the hydrophobic core of the membrane becoming weaker in the headgroup region before returning to the strengths we often think of in the aqueous environment. As a result, the secondary structure of helical membrane proteins and the tertiary structure of β-barrel membrane proteins are stabilized by strong hydrogen bonds in the membrane interior, whereas in the interfacial region, water competes for the amide hydrogen bonds and electrostatic interactions are greatly weakened. Within the hydrophobic core of the membrane, the protein/lipid interface is dominated by van der Waals contacts because of scant polar and charged residues in the transmembrane (TM) domain of helical membrane proteins.[3] Since the hydrophobic effect is driven by the presence of water it is, therefore, not important in the membrane interior but is very significant in the interfacial region where hydrophobic moieties can be buried in the fatty acyl environment and hydrophilic groups in the interfacial or aqueous environs.

The lateral pressure profile is unique to the membrane environment. Within the hydrocarbon core there is a positive pressure due to entropic disorder and there is also a significant positive pressure due to charge repulsion in the lipid head groups.[4] Conversely, the glycerol backbone region exhibits a considerable negative pressure. This anisotropic profile adds to the heterogeneity of the membrane environment and to the importance of modeling this environment accurately for achieving native-like membrane protein structures. As noted above, the collection of interactions that stabilize helical membrane proteins vary much more significantly across a helical TM protein than across globular proteins. Therefore, great care must be employed when considering the stabilization of helical TM proteins in membrane mimetic environments.

18.2.2 Synthetic Lipid/Membrane Protein Samples

Despite the intricate nature of the membrane environment, several appropriate mimetic environments exist for the study of native membrane protein structures. The reconstitution of membrane proteins in synthetic lipids generate samples of isolated and purified protein in a model membrane in which functional assays can be performed to confirm the native-like functionality of the protein, whether it is the wild-type or a mutated sequence. Typically, a wide range of pH, temperature, and salt conditions can be achieved in such samples. Lipid vesicles or liposomes can be used for magic angle spinning (MAS) ssNMR studies of such binary and ternary lipid mixtures, including the incorporation of cholesterol.[5–9] Similarly, uniformly aligned lipid bilayers on glass slides prepared from these proteoliposomes are used for oriented sample (OS) ssNMR, representing a good membrane mimetic for membrane proteins.[10,11] These lipid preparations can mimic the gradients in water concentration, dielectric, hydrophobicity, and potentially the lateral pressure profile may also be well modeled. However, such environments lack the leaflet asymmetry of the native environment, but otherwise represent an excellent approximation to the natural environs.[1] In Figure 18.1, the structure of the M2 protein is shown in the synthetic lipid bilayer in which it was characterized.[11] Today, many membrane protein samples have been prepared in these membrane mimetic environments for ssNMR characterization.

A variety of other membrane protein sample preparation strategies have been described in the literature, focused more on achieving high-resolution spectra than on forming a native-like environment for membrane proteins. Nano- or microcrystalline preparations have the same problem that X-ray quality crystals have with a high protein concentration

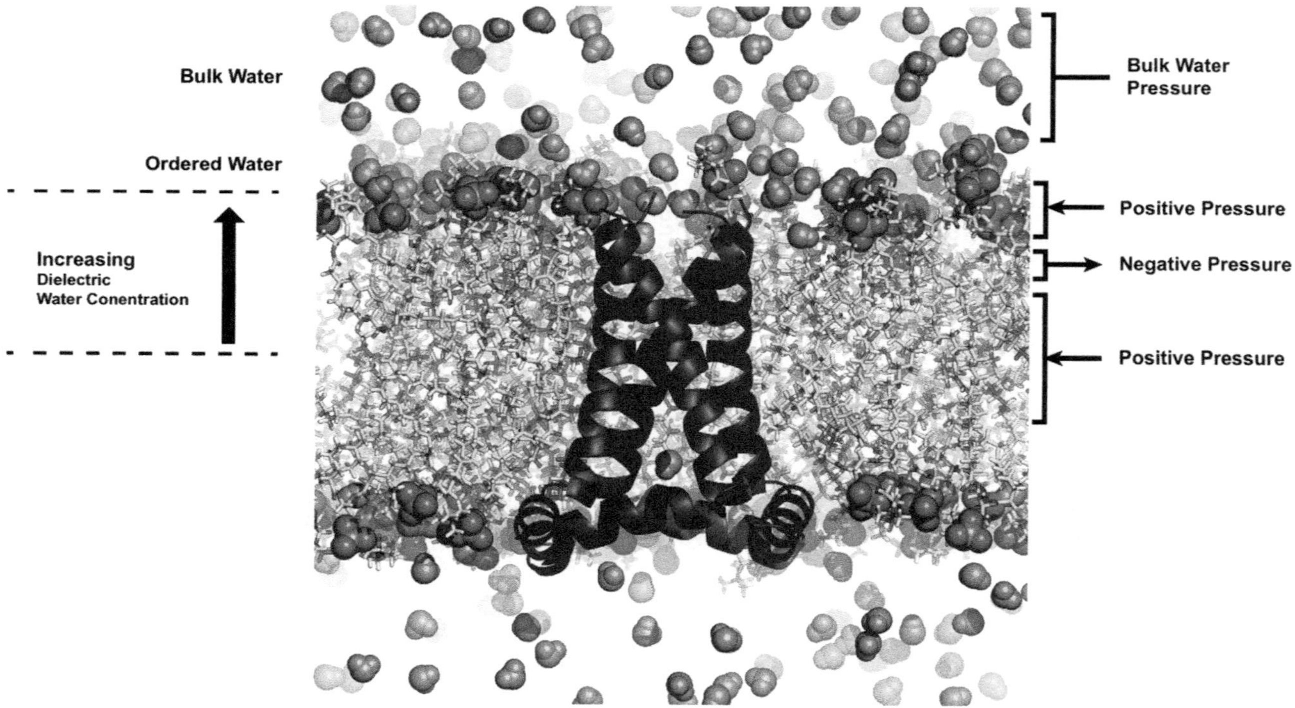

Figure 18.1 Restrained molecular dynamics simulation of a membrane protein (M2 from influenza A) in a bilayer environment of DOPC and DOPE lipids.[11] The gradients for electrostatic interactions are shown on the *left side* of the figure while the profile for lateral pressure is indicated on the *right side* of the figure.

and crystal lattice contacts. Precipitated proteins have similar problems due to heterogeneity and protein–protein interactions. Even bicelles for ssNMR that nominally have a lipid bilayer environment for the membrane protein may contain a considerable fraction of detergents and consequently have a substantial concentration of monomeric detergent molecules that can interfere with water-soluble domains. Clearly, if a proteoliposome or a planar bilayer preparation of the membrane protein can be prepared, it presents an excellent opportunity for characterizing the native structure of the protein. High-quality spectra are also needed, but this is more a reflection of the quality of the sample preparation, *i.e.* the complete removal of detergents, quality temperature control and, potentially, the choice of lipid. It is possible that some proteins may display heterogeneities that could be minimized by binding a ligand to the protein.[12] In other words, if the protein structure is native and homogeneous the spectral lines ought to be reasonably narrow for both OS and MAS spectroscopy.

18.2.3 Cellular Membranes

Native cellular membranes are exceptionally complex. Bacterial membranes have hundreds of different lipids in an asymmetric distribution between bilayer leaflets and they contain hundreds, if not thousands, of different membrane proteins. The proteins are often modified by glycosylation, acylation, and phosphorylation. Eukaryotic membranes are even more complex and contain two additional classes of lipids, including cholesterol and sphingomyelin. Since *E. coli* is often used for protein expression and since membrane proteins are often expressed into the cellular membranes, there is an opportunity to isolate the *E. coli* membranes and to observe the expressed protein as it is inserted into the membranes by the cellular machinery without ever exposing the protein to the denaturing influence of detergents. While *E. coli* membranes may not be the native membrane of the membrane protein being studied, it clearly represents an enhanced membrane mimetic by including not only bilayer asymmetry, but enhanced heterogeneity of the environment through the hundreds to thousands of other membrane proteins and lipids surrounding the expressed protein. While such heterogeneity might be expected to induce structural heterogeneity in the cloned protein and as a result resonance broadening, this has not been apparent or at least not very significant, as shown in Figure 18.2. This figure shows MAS spectra of the full-length M2 protein from influenza A virus that has been purified and reconstituted into synthetic lipid bilayers compared with spectra of the protein in isolated *E. coli* membranes without protein purification.[13] These results show a virtual overlap of the ^{13}C Val and ^{13}C Ala resonances, thereby validating the purification and bilayer reconstitution protocols.

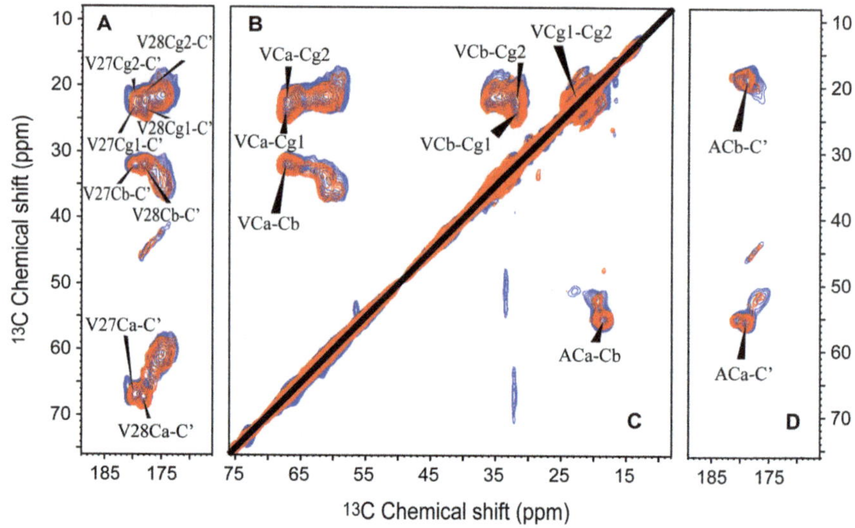

Figure 18.2 Comparison of ^{13}C–^{13}C correlation spectra using 50 ms DARR mixing time at 243 K of (A, B) uniformly [^{13}C,^{15}N]valine-labeled M2FL and (C, D) uniformly [^{13}C,^{15}N]alanine-labeled M2FL, in reconstituted liposomes at pH 7.5 (*blue*) and in *E. coli* membranes at pH 8.0 (*red*). (A, D) Cα, Cβ, Cγ1, and Cγ2 correlations with C' for valine and Cα, Cβ correlations with C' for alanine-labeled samples. (B, C) Correlations between Cα, Cβ, Cγ1, and Cγ2 for valine and Cα–Cβ correlations for alanine-labeled samples.[13]

18.3 Defining a "Good" Structure

It is important to consider what defines a "good" structure for a membrane protein. For years, X-ray crystallographers have defined a "good" structure as one that diffracts to high resolution and for which there is functional validation. NMR spectroscopists have often defined a "good" structure by a large number of qualitative restraints per residue. For membrane proteins and ssNMR, these definitions need to be rethought.

All structural methods for membrane proteins use membrane mimetic environments. For solution NMR, detergent micelles are typically used, although isotropic bicelles that include a small fraction of lipid are sometimes used. Nanodiscs that are small lipid bilayer fragments bounded by polymers hold significant promise, but while this latter method has been used for the structure of a β-barrel membrane protein, no such success has been reported for helical membrane proteins.[14] The other membrane mimetics for solution NMR are dominated by detergents and have often led to misleading structural characterizations of helical membrane proteins in which helix packing is non-native-like, despite the requisite number of restraints for a "good" structure.[15] For crystallography, detergents are used almost exclusively in place of lipids, and while many of the high molecular weight helical membrane protein structures appear to have an appropriate packing of the TM helices, there are quite a few others that have distorted helix packing, again despite having the crystallographic badge of "good" structure based on high-resolution diffraction. Clearly, the traditional definitions of a good structure have broken down.

In many of these cases of "structural distortion", functional assays have been conducted and used for structural validation. However, these assays for membrane proteins are typically performed in a lipid bilayer environment, as opposed to the environment required for the structural methodology. This is not to say that the assay is not important, it is, but for the purposes of validating the construct, rather than validating the tertiary or quaternary structure observed in the environment dictated by the structural technique.

For the most part, there is little doubt that the structure characterized in these membrane mimetics is the actual structure present in the sample environment; the problem is that the native structure of the protein results from the sum of the interactions within the protein coupled with the interactions between the protein and its environment, as discussed above. Indeed, protein structure is not dictated by the amino acid sequence alone, but by the sum of all interactions between the protein and its environment as well as those interactions within the protein.[16] A clear example of this structural sensitivity to the protein's environment is demonstrated by the X-ray and NMR (both solution and solid-state) structures of the M2 protein from influenza A virus obtained in multiple environments (Figure 18.3). In DHPC micelles the helix tilt is 16° smaller than in lipid bilayers and consequently the anti-viral drugs do not bind in the pore, while in liquid crystalline bilayers with the greater helix tilt the drugs bind in the pore, as occurs in the native viral coat of influenza A.[3,17] The solution NMR structure is a well-characterized structure, yet the tetrameric structure is not "native-like". The early X-ray crystal structures of the M2 protein both with and

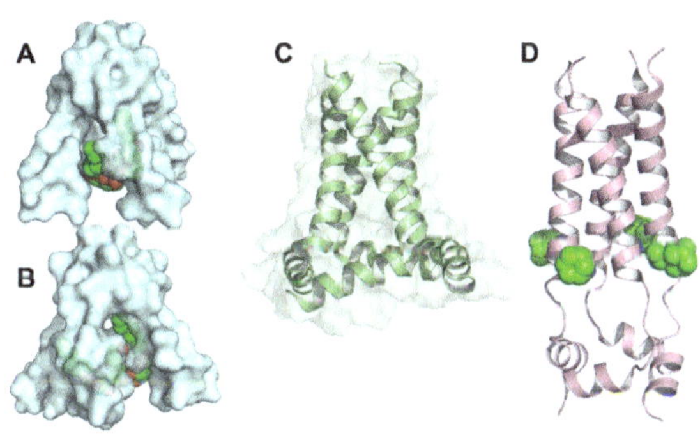

Figure 18.3 M2 proton channel structures as determined by: (A, B) X-ray crystallography in an octylglucoside environment (PDB 3BKD);[18] (C) ssNMR in a synthetic bilayer of DOPC and DOPE lipids (PDB 2L0J);[11] and (D) solution NMR in a DHPC micelle (PDB 2RLF).[17]

without bound drug showed the C-terminus of TM helices so splayed apart that large fenestrations in the helix bundle were open to the fatty acyl chain environment (Figure 18.3).[15,18] In other words, the aqueous pore was exposed to the very hydrophobic fatty acyl environment. In addition, the structure without the drug was highly asymmetric for this homo-tetrameric structure. Not only were detergents bound in three of the four helix–helix interfaces, but a significant tetramer–tetramer salt bridge was formed in the crystal lattice, suggesting that the influence of high monomeric detergent concentrations plus crystal lattice interactions in the high protein concentration in the crystal could result in non-"native-like" structure. Significantly, two years later, another X-ray structure was published for a thermally stabilized mutant that did not have the splayed helices and had a Cα RMSD of 0.6 Å (P_{25}–W_{41}), with the ssNMR structure characterized in a liquid crystalline lipid bilayer.[3,19] The structural differences between the crystal structures were not the result of the relatively minor differences in crystallographic resolution. Multiple examples of such "structural distortions" in both solution NMR and X-ray diffraction structures have been documented.[15]

It is important, therefore, to reconsider a definition for "good" or "native-like" helical membrane protein structures. Clearly, the number of restraints or the crystallographic diffraction limits is not adequate. To address this issue we will consider the quality and properties of the structural restraints that we can obtain from ssNMR and also consider what is most important from a membrane protein structural perspective.

18.3.1 From a Structural Perspective

For helical membrane proteins, the packing of the TM helices is fundamental to the structure or structures of a given protein. If the helical packing is distorted, the positions of the side chains will not be accurate and consequently the first priority is to obtain the native helix packing for the protein. Furthermore, the numerous side chains that face the dynamic lipid environment are most likely highly dynamic and identifying an averaged or most highly occupied conformation is of little import. Side chains in the helix–helix interfacial space are likely to be more rigid and certainly more important to characterize. In addition, other side chains in the interior of the structure, especially those that are hydrophilic and may be associated with functionality, will be more important. However, even for these side chains one can anticipate today that with our knowledge of protein structure and helix packing that computational efforts will give rise to an excellent model of the side chains and, therefore, the emphasis must be upon obtaining an accurate structure of the backbone conformation. Obviously, the more experimental restraints that can be obtained for the side chains the better, but without an accurately characterized backbone structure the side chain conformations are not meaningful.

18.3.2 From a Restraint Perspective

The structural restraints from ssNMR can be divided into two classes: orientational and distance or torsional restraints. The fundamental difference between these types of structural restraints is whether or not they are measured with respect to an external frame of reference. Orientational restraints, whether from OS or MAS ssNMR, are measured with respect to a frame external to the protein, namely the normal to the lipid bilayer.[20–24] Conversely, distance and torsion angle restraints are measured with respect to the molecular frame of the protein under study. The orientational restraints are classified as absolute, as are residual dipolar couplings used in solution NMR spectroscopy, and the distance and torsion angles as relative according to the frames used in the interpretation of experimental data.[25]

The difference between these structural restraints is profound. As a result of the different frames of reference for which the restraints are measured, error is propagated through the structure in a different way. For each absolute restraint, the error in each measurement is independent of other absolute measurements, because they are each determined with respect to a fixed external frame of reference. The uncertainty in one site has nothing to do with the uncertainty in the other sites and, therefore, the errors are independent when calculating a structure. The uncertainty for each site is simply the standard deviation for that site. However, when structural restraints are measured within the same frame, such as relative restraints within a helix, the errors add in a substantially different way. To illustrate this point we provide a simple example of adding distances. In Figure 18.4A a set of points in the molecular frame are depicted as black circles. If two interatomic distances, x_1 and x_2, are measured for these sites the distance x_3 can be calculated. Often the sites x_1 and x_3 are too far apart from each other

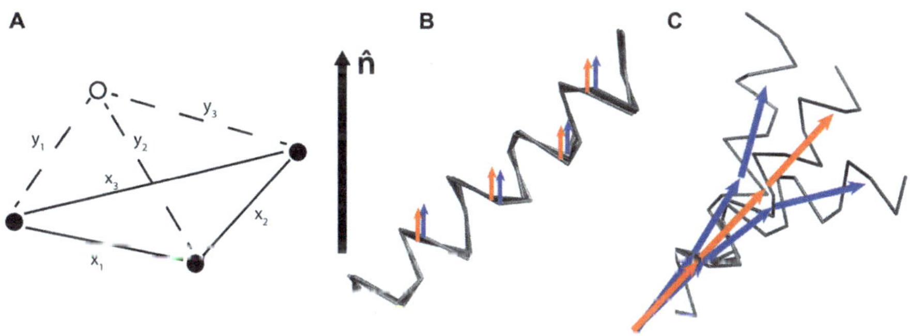

Figure 18.4 Absolute *vs.* relative restraints: (A) relative distances are between *solid circles* while absolute distances are between the *solid circles* and the *open circle*; (B) illustrates the range of helical structure consistent with a set of absolute orientational restraints between individual peptide planes and the bilayer normal fixed in the laboratory frame of reference; and (C) illustrates the range of helical structure consistent with a set of relative restraints within the helix.

for the distance to be experimentally determined and the distance must, therefore, be calculated. The distance is simply the vector sum of x_1 and x_2. Unfortunately, the error in this third distance is larger than the error in either x_1 or x_2. Using simple error propagation for the sum operation[26] the error of the third distance is given by:

$$\Delta x_3 = \sqrt{(\Delta x_1^2 + \Delta x_2^2)} \tag{18.1}$$

and, if the errors are equivalent, then it reduces to:

$$\Delta x_3 = \sqrt{(2\Delta x^2)} = \sqrt{2} \cdot \Delta x \tag{18.2}$$

The total error for the system would be the sum:

$$\Delta X = \sqrt{(\Delta x_1^2 + \Delta x_2^2 + \Delta x_3^2)} = \sqrt{(2\Delta x^2 + 2\Delta x^2)} = 2 \cdot \Delta x \tag{18.3}$$

To illustrate how error propagation is different for absolute distances, consider the open circle in Figure 18.4A, which represents a point in an external frame. If three distances were measured to a site in an external frame, y_1, y_2, and y_3, then the calculated distance between y_1 and y_3 would have a different uncertainty. When used to calculate a structure, the errors do not add in the same way. In this case we would obtain for the total error:

$$\Delta Y = \sqrt{(\Delta y_1^2 + \Delta y_2^2 + \Delta y_3^2)} = \sqrt{(3\Delta y^2)} = \sqrt{3} \cdot \Delta y \tag{18.4}$$

which is less than ΔX, the error for relative distances, assuming both measurements have the same intrinsic error. The error in combined absolute restraints is always less than that for combined relative restraints.

The absolute restraints that are observed by ssNMR are orientational restraints to the bilayer normal. These restraints are highly complementary to the relative restraints (distances and torsion angles) that can also be obtained from ssNMR. However, limitations in using these restraints independently are very different. For absolute restraints, the orientation of individual peptide planes are determined with respect to a common fixed frame of reference, whereas the relative distances are between sites within the helix and each of these sites has a positional uncertainty (Figure 18.4B and 18.4C). For a helix that is shown by relative restraints to have uniform torsion angles and uniform intrahelical Cα–Cα distances, the errors permit this helix to be significantly curved or bent, whereas if the set of peptide plane orientations in the first and last turns of the helix indicate the same tilt, the helix will be straight, assuming that the rotational positions of the resonances about the PISA wheel (discussed below) are consistent with a continuous helix. In other words, when the structure is determined using relative restraints, the positions between distal sites in the helix contains much more error than the positions of nearby sites in the helix. As a result, small sections of the helix are well defined but the overall uniformity of the helix can vary substantially and still remain within experimental error. Consequently, it can be anticipated that helices characterized by relative restraints may display more bent structures than those from orientational restraints. Indeed, it is anticipated that in the low dielectric environment of the membrane, helices do have a more uniform structure induced by stronger hydrogen bonding. Of course, the presence of Pro and potentially the presence of Gly residues can perturb the uniformity and consequently we do not mean to imply that all TM helices have a linear axis.

18.3.3 Restraints per Residue

It is common for the quality of NMR structures to be evaluated based on a number of restraints per residue ("good" is sometimes thought to require six restraints per residue). This comes from a history of qualitative relative restraints and such a number is not relevant when considering the quantitative and absolute nature of orientational restraints obtained from OS and, more recently, MAS ssNMR.[27] Orientational restraints for the backbone of a helix lead to a pattern of resonances, known as a PISA (polar index slant angle) wheel in PISEMA or SAMPI4 spectra of uniformly aligned samples.[28,29] The shape of the resonance pattern is very sensitive to the torsion angles of the helix and even to the covalent geometry of the polypeptide backbone. As shown in Figure 18.5A, changes in the torsion angles

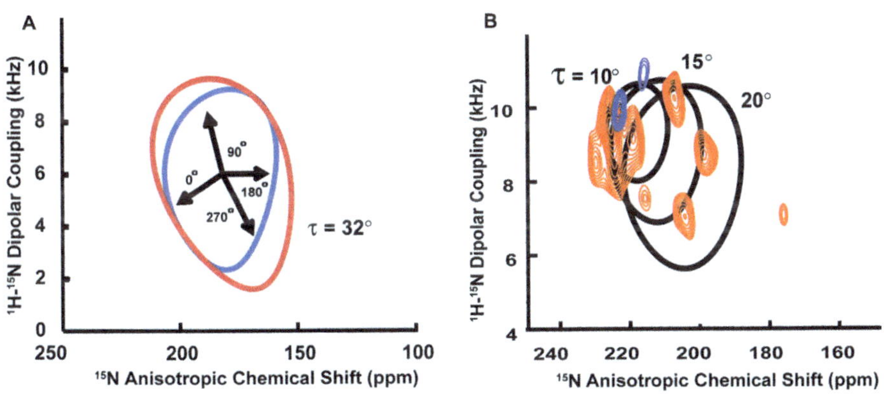

Figure 18.5 Interpretation of PISEMA data. (A) Two PISA wheels at 32° tilt angle (τ) are calculated for typical membrane protein α-helical torsion angles (*blue*, ϕ = –60°, ψ = –45°) and typical water-soluble protein α-helical torsion angles (red, ϕ = –65°, ψ = –40°). In addition, a coordinate system is imposed on the pattern, with 0° indicating amide nitrogen sites at the top of the helix and 180° indicating the bottom of the helix. (B) PISEMA spectra of [^{15}N]leucine (*red*) and [^{15}N]threonine (*blue*) labeled CrgA membrane protein in POPC/POPG (1:1 mol/mol) at 286 K are superimposed and helical wheels for tilt angles of 10, 15, and 20° are superimposed on the data.

of just a few degrees can alter the shape of the pattern. Despite significant variation in the chemical shift tensors, TM helices have been characterized and the average torsion angles for these helices have been shown to be significantly different from those of water-soluble helices where the backbone hydrogen bonds compete with water. The result is that the carbonyl groups are consistently tilted in toward the helix axis by approximately 4° more than in water-soluble protein helices and the torsion angles are typically −60°/−45° instead of −65°/−40° as they are for water-soluble helices.

In addition to the high-resolution structural details for helices derived from the PISA wheel, the size, shape, and position of the pattern of resonances is exquisitely sensitive to the tilt of the TM helices with respect to the bilayer normal. In Figure 18.5B we show a superimposed set of PISEMA spectra for amino acid specific labeled samples showing a clear PISA wheel pattern. Since we have shown that helix tilt can be critically important for the functionality of membrane proteins, as exemplified by the M2 protein structures, and that even high-resolution "good" membrane protein structural characterization can get this parameter wrong, this represents a great advantage for OS ssNMR. Indeed, such an experiment could be used to validate solution NMR and X-ray crystal structures without trying to complete the structure by ssNMR. The membrane protein samples would not have to be uniformly labeled, but could be labeled with a specific amino acid type, as in Figure 18.5B, so as to minimize the spectral complexity. While the tilt of the helix or helices can often be characterized without resonance assignments, a few resonance assignments can identify the PISA wheel pattern as belonging to a specific helix in the protein and also define the rotational orientation of the helix or helices. Since the resonance pattern is closely allied with a helical wheel having 3.6 residues per turn, the PISA wheel has 3.6 resonances per turn and also has a well-defined helix rotational orientation in the bilayer. In other words, the plane through the bilayer normal and helix axis defines the "top" and "bottom" for a helix having corresponding resonance positions (0° and 180°) on opposite sides of the PISA wheel, as shown in Figure 18.5A.

With the helix backbone characterized both with regard to the torsion angles and with respect to the helix tilt and rotational angles, the backbone structure out to the Cβ atoms is defined. In other words, $\chi 1$ and $\chi 2$ torsion angles for those residues that have such angles are not defined by the experiments discussed so far. Indeed, there are many Ile, Leu, and Phe residues in TM helices that have both $\chi 1$ and $\chi 2$ angles and yet a recent analysis of residues in TM proteins shows that 2/3rds of these bulky residues face the lipid environs.[3] The $\chi 1$ and $\chi 2$ angles for these residues have only minor importance unless one is studying heteroprotein complexes. More importantly, the residues facing the interior of the protein are often small, such as Gly, Ala, Ser, and Thr. Only the latter two have $\chi 1$ angles, while Gly and Ala residues do not have $\chi 1$ or $\chi 2$ angles to be characterized and these residues account for more than a quarter of the residues in the interior of TM domains. Consequently, there are relatively few $\chi 1$ and $\chi 2$ angles that are potentially important. At a time when computation methods are making significant progress in predicting the secondary and tertiary structures of proteins from the amino acid sequence in a given environment, based on much improved methodology and force fields, it is reasonable to expect that computational methods can predict a few side-chain torsion angles with significant accuracy given a backbone structure.[30] Of course, when possible it is always better to experimentally characterize such angles as long as the characterization is performed in a lipid bilayer environment. In addition, such characterizations are often fraught with complicating dynamics, a factor that can be characterized but is often ignored by characterizing samples in a frozen state.

The bottom line is that the number of restraints per residue does not have to be large if absolute restraints are obtained: the backbone structure is far more important to experimentally characterize than the side chains. However, for residues that are critical for functional activity, experimental characterization of both the side-chain structure and dynamics will be important. The quality of the structure is largely dictated by the characterization of the helix packing that is greatly facilitated by a high-resolution characterization of helix tilt and rotation angles, but also requires distance restraints between each interacting helix pair. Because of the quality of the orientational restraints there is no need for a large number of these distance restraints between each helix pair, but sparse distance restraints are needed, indeed they are essential for the characterization of helix packing. This leads us to the concept of combining absolute and relative restraints.

18.4 Combining Absolute and Relative Restraints

Since the loops between helices are not uniquely characterized by orientational restraints and because these loops may also be dynamic, orientational restraints alone cannot define helix packing, although it effectively limits the packing possibilities by defining the backbone structure and orientation of each helix. Because of the demand for many relative restraints to define a complete 3D structure without orientational restraints, this approach has yet to produce a high-resolution structure for a helical membrane protein in a lipid bilayer.[31] The combination of absolute and relative restraints has been used to achieve atomic resolution structures of membrane proteins.[6,11,32–34] However, there is much that can be done to optimize the experiments and labeling strategies to facilitate the structural characterization of this important class of membrane proteins.

In addition to numbers of restraints per residue, another unfortunate concept has been suggested from time to time: that selective labeling is not efficient enough to be a viable strategy for protein structural characterization. However, the important concept is that a native structure needs to be characterized and, as we have discussed, this may require the structural characterization of helical membrane proteins in liquid crystalline lipid bilayers. Spectral dispersion for hydrophobic side chains in the TM domain is limited while linewidths are substantial. This does not preclude assigning resonance frequencies to the residues, but does preclude unique assignments for many sites and consequently it is very difficult to define distance restraints involving ambiguous assignments. Fortunately, what is needed is only sparse distance restraints for a strategy that combines orientational and distance restraints and, to achieve sparse distances, a limited labeling strategy is helpful.

18.4.1 Labeling ^{13}C and ^{15}N Sites

Uniform ^{15}N labeling provides initial spectra for OS NMR and uniform ^{13}C and ^{15}N labeling provides the same for MAS NMR, but sequence-specific assignments are very difficult to achieve. Even in 3D spectra for MAS NMR the spectral dispersion is not adequate and residual scalar and dipolar couplings result in broad resonances. Moreover, significant dipolar truncation effects limit opportunities to correlate resonances over anything more than minimal distances.[35,36] Improvement in these spectra can be achieved by calculating difference spectra between long and short DARR mixing times.[37] Limiting ^{13}C–^{13}C couplings through [1,3-^{13}C]- or [2-^{13}C]glycerol labeling or through specifically labeled glucose enhances spectral linewidths and reduces dipolar truncation and spectral congestion.[38,39] The use of NCACX, NCOCX, and CANCO experiments can lead to numerous assignments. However, one of the major drawbacks to glycerol labeling is the partial labeling of numerous sites generating numerous weak signals in the spectra.

18.4.2 Selective Labeling

Two approaches for amino acid specific labeling have been developed. The direct addition of isotopically labeled amino acids to the culture and the addition of unlabeled amino acids to cultures with uniform ^{15}N and/or ^{13}C isotope sources. This latter approach is referred to as reverse labeling.[40] For OS ssNMR, both labeling strategies have been effective for proteins up to several TM helices and many orientational restraints have been obtained for a recent seven TM helix GPCR structure.[41] Here, "effective" means that it has been possible to resolve the resonances and make the sequence specific assignments. Assignments in OS ssNMR can be made spectroscopically by incorporating a ^{15}N–^{15}N mixing element into the pulse sequence to encode i, $i+1$ backbone nitrogen correlations in a PISEMA-type experiment with either 2D or 3D evolution. Mismatched Hartmann–Hahn mixing has been employed in a 2D version to assign resonances for Pfu1 in aligned bicelles and proton-driven spin diffusion mixing has been used in a 3D experiment to assign resonances for sarcolipin in aligned bicelles.[42,43] Alternatively, it may be useful to add not only a ^{15}N labeled amino acid to the expression culture, but also a ^{13}C carbonyl labeled amino acid to split one or more ^{15}N resonances and to facilitate the assignment of particularly challenging sites.

Assigning the resonances at the center of the TM helix and working the assignments out toward the first and last turns of the helix, where PISA wheel distortions may be more prevalent, is a useful strategy. Even without full assignments, the PISA wheels provide the tilt of the TM helices or helical segments and a couple of assignments in each wheel are enough to achieve the rotational orientation of each helical segment. Reverse labeling is useful for OS NMR for those residues present in low abundance in the TM domain and for those residues where the ^{15}N may be scrambled during protein expression. In MAS, reverse labeling is more commonly used, because $^{13}C/^{15}N$ labeled amino acids are more expensive and because the goal is to achieve inter-residue connectivity. In the search for distance restraints, eliminating many of the common amino acid residues that are not assigned is very useful. This is especially true in the later stages of the structural characterization when an initial model of the structure has been achieved and distances can be predicted and, therefore, a targeted search can be conducted with limited labeling.

18.5 Computational Refinement of Membrane Protein Structure

In a multi-step process, OS and MAS ssNMR restraints are initially used to develop a structural model and later on these restraints are combined with an all-atom model of the lipid bilayer for restrained molecular dynamics to achieve a refined structure of the protein in the same environment from which the experimental restraints were obtained. Since the structure of helical membrane proteins is, in part, dictated by its environment, it is essential to refine the protein structure in the same environment in which it was characterized.

In the first step, an initial monomeric model is developed, taking advantage of the known helical segments of the sequence from the OS and MAS NMR data. In Xplor-NIH, simulated annealing OS orientational and MAS distance restraints are included along with helix specific restraints, such as torsion angles and hydrogen bond distances.[44] In the second step, an oligomeric state is generated if such a state is implied by native gels, SDS-PAGE, or gel filtration chromatography. Additional experimental restraints specific to the oligomeric state would be utilized in this step of the simulated annealing. The third step is to refine the low-resolution membrane protein structure (monomer or oligomer) by inserting the simulated annealing structure into a lipid bilayer environment for molecular dynamics (MD) simulations. These simulations are restrained by all of the structural restraints.[11,45] In this step an equilibrated lipid bilayer, such as DOPC/DOPE (4:1 mol/mol) or POPC/POPG (4:1 mol/mol), is generated with solvated water molecules from CHARMM GUI so that restrained MD simulations can be performed in the same bilayer environment as was used to obtain the experimental restraints.[11,46] The final step includes restraining the orientation of experimentally determined side chains. In this step, an MD force field can be used to achieve a realistic model for the other side chains, leading to an optimized structure of the membrane protein in lipid bilayers.

18.6 Conclusions

The goal here has been to lay out the strategies for characterizing the native structures of helical membrane proteins. The ssNMR strategy for accomplishing these structures is rapidly evolving as many more research groups enter this challenging arena. Fast MAS and uniform deuteration have great potential for reducing ^{13}C linewidths. ^{1}H detection also has great potential for enhancing sensitivity. All of this, coupled with the next generation of high-field magnets using high-temperature superconductors, should make routine the structural characterization of helical membrane proteins in liquid crystalline lipid bilayers or in cellular membranes. Through this structural biology approach, ssNMR, which has always demonstrated great promise, will fulfill its promise by having a major impact on basic biomedical research and on drug development.

Acknowledgements

This work was in part supported by NIH grants AI 074805, AI 073891, and AI 023007 and the National Science Foundation through Cooperative Agreement 1157490 between the Division of Materials Research and the State of Florida.

References

1. T. A. Cross, D. T. Murray and A. Watts, *Eur. Biophys. J.*, 2013, **42**, 731–755.
2. H. Nymeyer and H. X. Zhou, *Biophys. J.*, 2008, **94**, 1185–1193.
3. H. X. Zhou and T. A. Cross, *Protein Sci.*, 2013, **22**, 381–394.
4. R. S. Cantor, *Biophys. J.*, 1999, **76**, 2625–2639.
5. M. P. Bhate, B. J. Wylie, L. Tian and A. E. McDermott, *J. Mol. Biol.*, 2010, **401**, 155–166.
6. T. V. Can, M. Sharma, I. Hung, P. L. Gor'kov, W. W. Brey and T. A. Cross, *J. Am. Chem. Soc.*, 2012, **134**, 9022–9025.
7. L. B. Andreas, M. T. Eddy, R. M. Pielak, J. Chou and R. G. Griffin, *J. Am. Chem. Soc.*, 2010, **132**, 10958–10960.
8. M. Renault, A. Cukkemane and M. Baldus, *Angew. Chem. Int. Ed.*, 2010, **49**, 8346–8357.
9. T. Wang, S. D. Cady and M. Hong, *Biophys. J.*, 2012, **102**, 787–794.
10. R. Verardi, L. Shi, N. J. Traaseth, N. Walsh and G. Veglia, *Proc. Natl. Acad. Sci. U. S. A.*, 2011, **108**, 9101–9106.
11. M. Sharma, M. Yi, H. Dong, H. Qin, E. Peterson, D. D. Busath, H. X. Zhou and T. A. Cross, *Science*, 2010, **330**, 509–512.

12. C. Li, H. Qin, F. P. Gao and T. A. Cross, *Biochim. Biophys. Acta*, 2007, **1768**, 3162–3170.
13. Y. M. Miao, H. J. Qin, R. Q. Fu, M. Sharma, T. V. Can, I. Hung, S. Luca, P. L. Gor'kov, W. W. Brey and T. A. Cross, *Angew. Chem. Int. Ed.*, 2012, **51**, 8383–8386.
14. T. Y. Yu, T. Raschle, S. Hiller and G. Wagner, *Biochim. Biophys. Acta*, 2012, **1818**, 1562–1569.
15. H. X. Zhou and T. A. Cross, *Annu. Rev. Biophys.*, 2013, **42**, 361–392.
16. C. B. Anfinsen, *Science*, 1973, **181**, 223–230.
17. J. R. Schnell and J. J. Chou, *Nature*, 2008, **451**, 591–595.
18. A. L. Stouffer, R. Acharya, D. Salom, A. S. Levine, L. Di Costanzo, C. S. Soto, V. Tereshko, V. Nanda, S. Stayrook and W. F. DeGrado, *Nature*, 2008, **451**, 596–599.
19. R. Acharya, V. Carnevale, G. Fiorin, B. G. Levine, A. L. Polishchuk, V. Balannik, I. Samish, R. A. Lamb, L. H. Pinto, W. F. DeGrado and M. L. Klein, *Proc. Natl. Acad. Sci. U. S. A.*, 2010, **107**, 15075–15080.
20. T. A. Cross and S. J. Opella, *J. Am. Chem. Soc.*, 1983, **105**, 306–308.
21. C. H. Wu, A. Ramamoorthy and S. J. Opella, *J. Magn. Reson., Ser. A*, 1994, **109**, 270–272.
22. B. B. Das, H. J. Nothnagel, G. J. Lu, W. S. Son, Y. Tian, F. M. Marassi and S. J. Opella, *J. Am. Chem. Soc.*, 2012, **134**, 2047–2056.
23. B. J. Wylie, L. J. Sperling, A. J. Nieuwkoop, W. T. Franks, E. Oldfield and C. M. Rienstra, *Proc. Natl. Acad. Sci. U. S. A.*, 2011, **108**, 16974–16979.
24. G. J. Hou, I. J. L. Byeon, J. Ahn, A. M. Gronenborn and T. Polenova, *J. Chem. Phys.*, 2012, 137.
25. N. Tjandra, S. Grzesiek and A. Bax, *J. Am. Chem. Soc.*, 1996, **118**, 6264–6272.
26. J. R. Taylor, *An Introduction to Error Analysis: The Study of Uncertainties in Physical Measurements*, University Science Books, Sausalito, CA, 2nd edn, 1997.
27. B. B. Das, H. J. Nothnagel, G. J. Lu, W. S. Son, Y. Tian, F. M. Marassi and S. J. Opella, *J. Am. Chem. Soc.*, 2012, **134**, 2047–2056.
28. J. Wang, J. Denny, C. Tian, S. Kim, Y. Mo, F. Kovacs, Z. Song, K. Nishimura, Z. Gan, R. Fu, J. R. Quine and T. A. Cross, *J. Magn. Reson.*, 2000, **144**, 162–167.
29. F. M. Marassi and S. J. Opella, *J. Magn. Reson.*, 2000, **144**, 150–155.
30. Y. Shen, R. Vernon, D. Baker and A. Bax, *J. Biomol. NMR*, 2009, **43**, 63–78.
31. M. Tang, A. E. Nesbitt, L. J. Sperling, D. A. Berthold, C. D. Schwieters, R. B. Gennis and C. M. Rienstra, *J. Mol. Biol.*, 2013.
32. K. Nishimura, S. G. Kim, L. Zhang and T. A. Cross, *Biochemistry*, 2002, **41**, 13170–13177.
33. S. D. Cady, K. Schmidt-Rohr, J. Wang, C. S. Soto, W. F. Degrado and M. Hong, *Nature*, 2010, **463**, 689–692.
34. D. T. Murray, N. Das and T. A. Cross, *Acc. Chem. Res.*, 2013.
35. P. Hodgkinson and L. Emsley, *J. Magn. Reson.*, 1999, **139**, 46–59.

36. S. Kiihne, M. A. Mehta, J. A. Stringer, D. M. Gregory, J. C. Shiels and G. P. Drobny, *J. Phys. Chem. A*, 1998, **102**, 2274–2282.
37. Y. Miao, T. A. Cross and R. Fu, *J. Biomol. NMR*, 2013, **56**, 265–273.
38. F. Castellani, B. van Rossum, A. Diehl, M. Schubert, K. Rehbein and H. Oschkinat, *Nature*, 2002, **420**, 98–102.
39. B. Habenstein, A. Loquet, K. Giller, S. Becker and A. Lange, *J. Biomol. NMR*, 2013, **55**, 1–9.
40. G. W. Vuister, S. J. Kim, C. Wu and A. Bax, *J. Am. Chem. Soc.*, 1994, **116**, 9206–9210.
41. S. H. Park, B. B. Das, F. Casagrande, Y. Tian, H. J. Nothnagel, M. Chu, H. Kiefer, K. Maier, A. A. De Angelis, F. M. Marassi and S. J. Opella, *Nature*, 2012, **491**, 779–783.
42. W. X. Tang, R. W. Knox and A. A. Nevzorov, *J. Biomol. NMR*, 2012, **54**, 307–316.
43. K. R. Mote, T. Gopinath, N. J. Traaseth, J. Kitchen, P. L. Gor'kov, W. W. Brey and G. Veglia, *J. Biomol. NMR*, 2011, **51**, 339–346.
44. C. D. Schwieters, J. J. Kuszewski, N. Tjandra and G. M. Clore, *J. Magn. Reson.*, 2003, **160**, 65–73.
45. X. Cheng and W. Im, *Biophys. J.*, 2012, **102**, L27–L29.
46. S. Jo, T. Kim and W. Im, *PLoS One*, 2007, **2**, e880.

CHAPTER 19

Chemistry and Structure via Solid-State NMR

JUDITH HERZFELD

Department of Chemistry, Brandeis University, Waltham, MA 02454-9110, USA
Email: herzfeld@brandeis.edu

19.1 Introduction

At one level, NMR makes absolutely no sense: the signal is weak and needs to be averaged for long periods on expensive equipment. Indeed, obituaries were declared for NMR several decades ago. Yet the technique has persisted, become more powerful, and extended its range of application, even as other methods of interrogating macromolecules have provided stiff competition. The answer to this riddle is that NMR, *sotto voce*, is more versatile than any other probe. Whereas diffraction offers only structural information (*i.e.*, probes of coordinates), and electronic and vibrational spectroscopies offer only chemical information (*i.e.*, probes of bonding and interactions), NMR offers both. Solid-state NMR, which has still greater sensitivity problems than solution NMR, has the important advantage of being applicable to systems that are neither soluble nor crystalline. This chapter illustrates the versatility of solid-state NMR with studies of two long-standing problems that have been refractory to other approaches.

New Developments in NMR No. 3
Advances in Biological Solid-State NMR: Proteins and Membrane-Active Peptides
Edited by Frances Separovic and Akira Naito

Published by the Royal Society of Chemistry, www.rsc.org

19.2 Chemistry: Enforcing Vectoriality in the Ion-motive Photocycle of Bacteriorhodopsin

Bacteriorhodopsin (bR), found in the cell membranes of *Halobacterium salinarum*, is either an outward H^+ pump or an inward OH^- pump (the two being so far impossible to distinguish). The retinylidene chromophore, which absorbs green photons, is formed by a Schiff base between retinal and a lysine sidechain buried deep inside a bundle of seven transmembrane helices (Figure 19.1). Not surprisingly, photoisomerization of the polyene of the chromophore from all-*trans* in the resting state to 13-*cis* (see Figure 19.2) perturbs the system sufficiently to cause p*K* changes in the active site that result in cascading protonation changes. The question regarding energy transduction is how the protein causes discharged ions to be replaced from the opposite side of the membrane, so as to enforce the vectoriality of the pump. A hint at what is going on is that the Schiff base continues to be connected to the extracellular side of the membrane after photoisomerization, until proton transfer neutralizes the charges of the Schiff base and its counterion in the L→M transition (see Figure 19.3). This implies significant, electrostatically-induced distortion in the interim.

As bR spontaneously self-aggregates into crystalline patches in the cell membrane, it has been relatively straightforward for crystallography to provide a high-resolution picture of the resting state.[1] However, the results

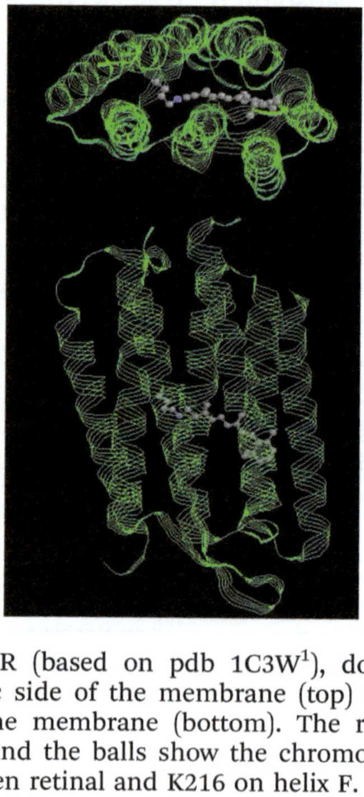

Figure 19.1 Views of bR (based on pdb 1C3W[1]), down the lumen from the cytoplasmic side of the membrane (top) and from the side in the plane of the membrane (bottom). The ribbon shows the peptide backbone and the balls show the chromophore formed by a Schiff base between retinal and K216 on helix F.

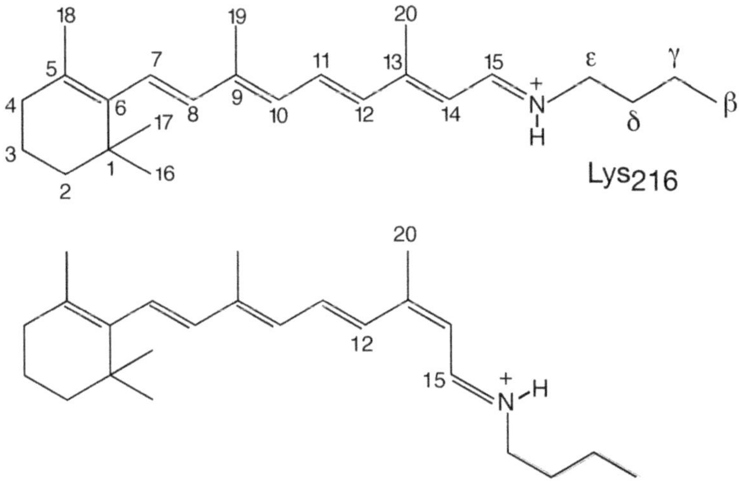

Figure 19.2 The chromophore of bR, in the resting state and in the photoisomerized state in their relaxed, planar forms. Because the ionone ring end of the molecule is immobilized by the protein, isomerization of the C13=C14 bond reorients the Schiff base proton.

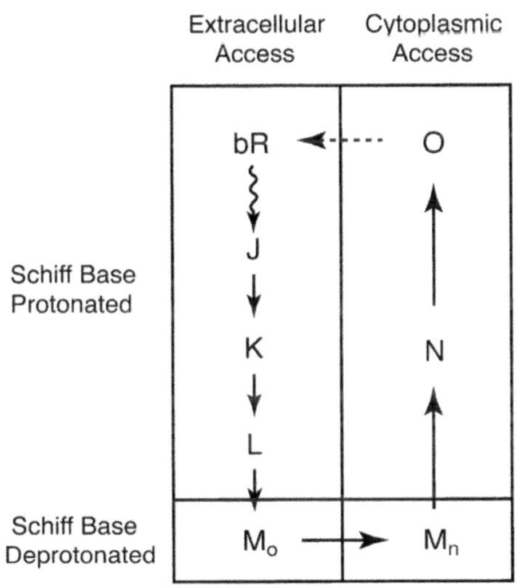

Figure 19.3 The ion-motive photocycle of bR. The *letters* represent photocycle intermediates that have been detected by various forms of spectroscopy and the *wavy arrow* represents photoisomerization. All other steps occur by thermal relaxation.

of crystallography of cryo-trapped photocycle intermediates have been controversial. The problem is that cryogenic temperatures generally trap mixtures of intermediates that all contribute to the diffraction. This is not a problem for NMR because intermediates contribute distinct, identifiable

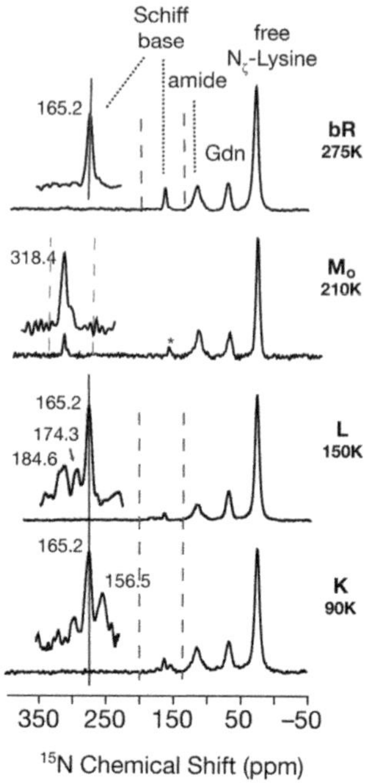

Figure 19.4 ^{15}N spectra of [ζ-^{15}N]lysine-labeled bR that has been irradiated at different temperatures to accumulate different photocycle intermediates (as indicated on the *right*).[6] Following irradiation, spectra are obtained at 90K in the dark, using dynamic nuclear polarization for signal enhancement. The *insets* enlarge the Schiff base portion of each spectrum (between the *vertical dashed lines*) and the *vertical solid line* marks the chemical shift of the resting state which remains present in K and L preparations. (The signal with an *asterisk* in the M spectrum is a spinning sideband of the highly deshielded Schiff base.)

signals to the spectra. As illustrated in Figure 19.4, irradiating at successively lower temperatures traps successively earlier photocycle intermediates, with variations in ^{15}N chemical shielding that show changes in the bonding and/or interactions of the Schiff base. Concomitant variations in the ^{13}C chemical shifts along the chromophore reflect the effects of new interactions on the distribution of electron density.

Changes are also seen in the protein. The ^{13}C difference spectrum in Figure 19.5 shows that an aspartic acid residue that is ionized in the resting state is protonated in the M_o state. This is known from FTIR studies to be Asp85. Asp212, the other aspartate near the Schiff base, is also perturbed. The ^{15}N difference spectrum in Figure 19.6 shows that an arginine residue with ordinary hydrogen bonding in the resting state finds itself in an extraordinarily asymmetric environment in the M_o state. This is presumably Arg82, since it is the only buried arginine residue and a similarly unusual perturbation is seen at rest in the D85N mutant. Diffraction studies indicate

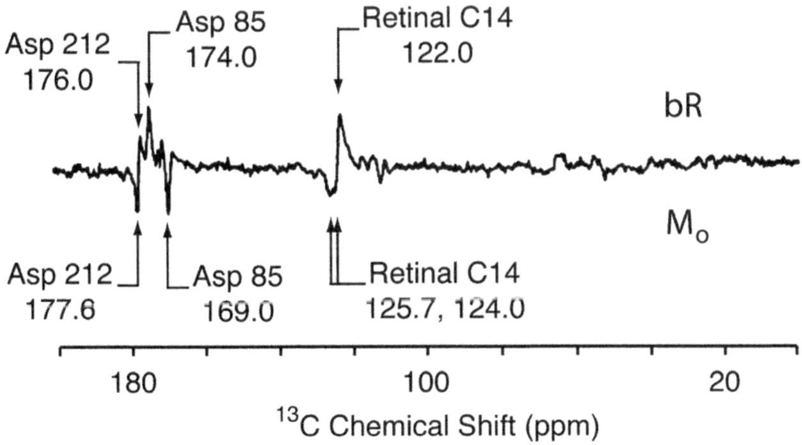

Figure 19.5 ^{13}C bR–M_o difference spectrum of [14-^{13}C]retinal,[4-^{13}C]aspartate-labeled purple membranes.[15] The [14-^{13}C]retinal signal is a marker for the change in state and the [4-^{13}C]aspartate signals show changes near the Schiff base, including protonation of Asp85.

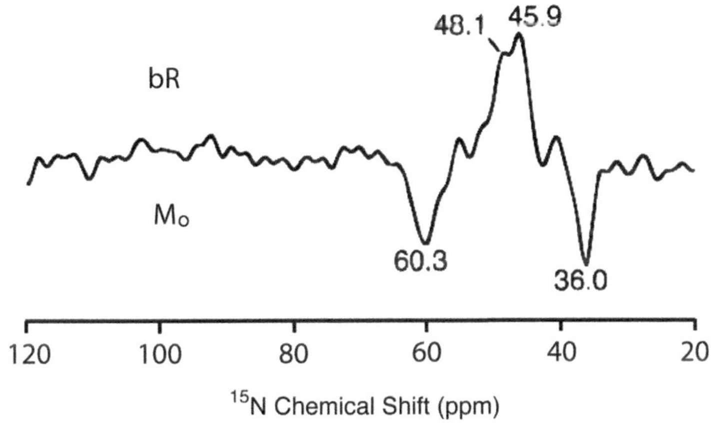

Figure 19.6 ^{15}N bR–M_o difference spectra of [^{15}N]guanidyl-labeled purple membranes.[31]

that the sidechain of Arg82 changes orientation during the photocycle. The NMR tells us that this new environment is very unusual.

Important insight into the photocycle comes from thermal relaxation experiments which demonstrate continuity in the photocycle. As shown in Figure 19.7, although only a single K state is observed, thermal relaxation of that K state leads to four "L" substates (which are better resolved in multidimensional spectra made possible by signal enhancement using

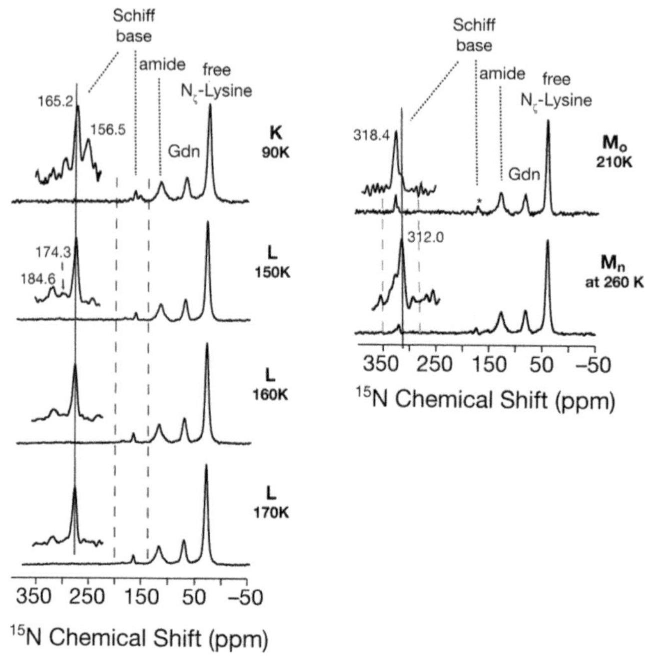

Figure 19.7 ^{15}N spectra of [ζ-^{15}N]lysine-labeled bR in states obtained by thermal relaxation of previously accumulated K (*top left*) and M_o (*top right*).[6] The appearance of L's from K, and M_n from M_o, demonstrates that cryo-trapping preserves continuity in the photocycle. The significance of the *insets* and the *vertical dashed lines* is as in Figure 19.4. The *vertical solid line* on the *left* marks the chemical shift of the resting state. The *vertical solid line* on the *right* marks the chemical shift of the M_n state.

dynamic nuclear polarization). The multiplicity of L explains some of the controversy that has surrounded it. Raising the temperature in small increments, we find that three of these substates relax back to the resting state and one is functional in that it persists until it relaxes to the M state.[2] This indicates a rugged and convoluted energy surface in this part of the photocycle.

How do we interpret the chemical shifts? While the different intermediates provide reference points for one another, it is important to also have simpler, outside reference points. This involves the examination of model compounds. From spectra of retinylidenes with various partners,[3,4] it is evident that the ^{15}N chemical shift of the Schiff base is very sensitive to the availability of a proton, varying (as shown in Figure 19.8) between the upfield extreme of a fully protonated Schiff base (*i.e.*, missing a counterion with which to share the proton) and the downfield extreme of a fully deprotonated Schiff base (*i.e.*, missing a hydrogen bonding partner with which to share a proton).

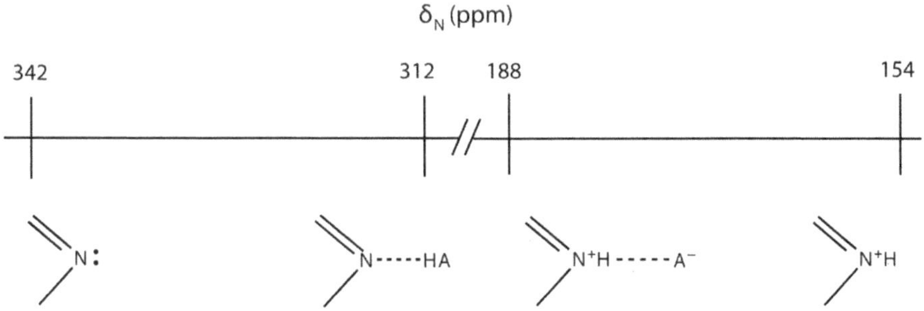

Figure 19.8 The 200 ppm range of the ^{15}N chemical shift of the Schiff base in relaxed retinylidenes.

From these data, we have been able to make several detailed inferences:

(1) In the resting state the counterion of the protonated Schiff base is too weak to be a carboxylate and, therefore, must be a hydrogen bonded complex that delocalizes its charge by including at least one water molecule.[5] To avoid Schiff base hydrolysis, this water must be immobilized in the complex. A decade later, crystallography found three immobilized water molecules forming an interhelical complex with zero net charge by hydrogen bonding to the protonated Schiff base at Lys216, Asp212 (one turn away on helix F), and Asp85 and Arg82 (one turn apart on helix C).[1]
(2) In the K photocycle intermediate, the earliest intermediate that has been cryotrapped, the protonated Schiff base has lost contact with its counterion.[6]
(3) In the functional L intermediate, the still protonated Schiff base has regained a counterion and this counterion is much stronger than the one in the resting state. Since the Schiff base in L is poised to transfer a proton into the transport pathway, this strong counterion is most likely the immediate proton acceptor.[6,7]
(4) The failed L intermediates all have signals further upfield than that of the persistent L, indicating that they have failed to find a strong new counterion. This suggests that the new counterion at this point in the photocycle has an important functional role.[6,7]
(5) In the M_o intermediate, the deprotonated Schiff base has a hydrogen bonding partner, most likely the group that has just accepted the proton that had been on the Schiff base.[8]
(6) In the M_n intermediate, the deprotonated Schiff base is more strongly hydrogen bonded, most likely by the group poised to reprotonate the Schiff base.[8]

If bR is an inward OH^- pump, then it seems likely that the hydrogen bonding partner in M is a water molecule, formed by combination of the proton on the Schiff base in L with an OH^- that is abstracted from the

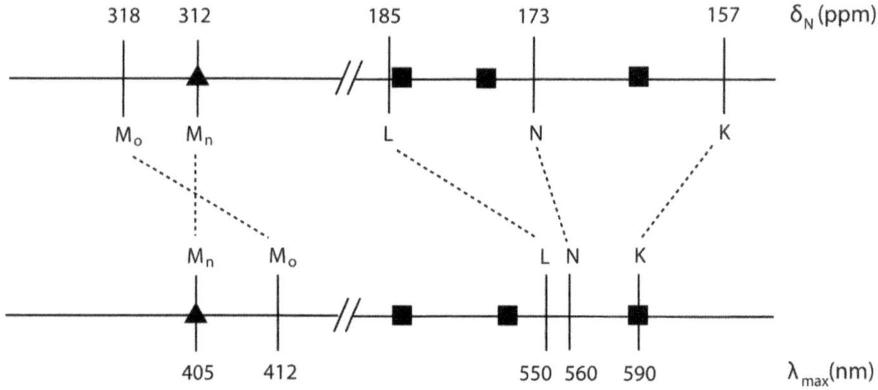

Figure 19.9 Correlation of the ^{15}N chemical shifts and wavelengths of maximum visible absorption in 13-*cis* bR intermediates compared to relaxed 13-*cis* retinylidenes.[4] The relaxed model for the deprotonated state (▲) is the resting state of the D85N mutant at high pH (electrically neutral at the Schiff base and at residue 85) and the relaxed models for the protonated states (■) are the chloride, bromide, and iodide salts (in order of decreasing chemical shifts and increasing absorption wavelengths).

active site, leaving an H^+ behind that is, directly or indirectly, taken up by Asp85.

Proton availability is also known to affect the visible spectrum of retinylidenes and color tracks closely with ^{15}N chemical shift in relaxed retinylidenes (Figure 19.9). However, color is also affected by twisting in the polyene chain: twisting around nominal single bonds, which stabilizes the ground state relative to excited states, causes a blue shift, and twisting around nominal double bonds has the opposite effect. These more distant effects can be isolated by comparing the color with the ^{15}N chemical shift. The comparison (Figure 19.9) indicates single bond torsion in K, double bond torsion in L and M_o, and no torsion in M_n and N. Thus, energy in K is stored primarily in the electrostatic mode by separation of the Schiff base from its counterion, enabled by a twist around a relatively floppy single bond. Subsequently, energy is transferred to torsion of a double bond, which is presumably driven by the electrostatic attraction between the still protonated Schiff base and its new, strong counterion. The chromophore is thus ready to unwind as soon as the Schiff base and the counterion charges are neutralized by proton transfer. This unwinding will change the connectivity of the Schiff base between the two sides of the membrane and prevent ion backflow. By the M_n and N states, much of the energy has been dissipated and slow relaxation to the resting state follows.

A clue as to the location of the single bond torsion in the K state and the double bond torsion in the L state comes from the ^{13}C chemical shift of C12 (Figure 19.10). Model compound studies in polyenes have found that when isomerization causes the protons on two carbon atoms to approach one

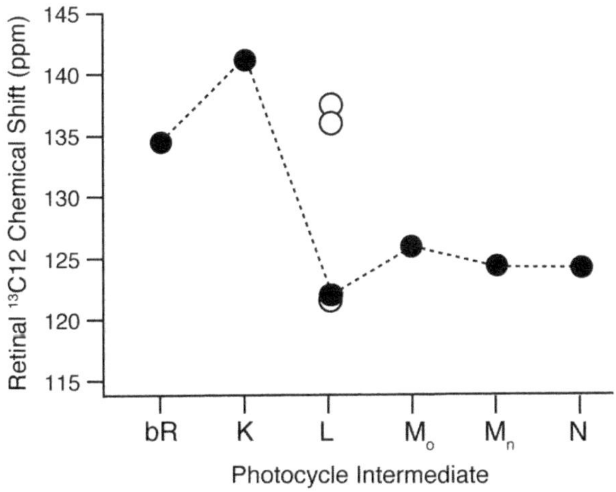

Figure 19.10 ^{13}C chemical shifts of C12 in various photocycle intermediates of bR.[2] The γ-effect (upfield shift at carbons three bonds apart due to steric interactions) is not seen until L, well after photoisomerization.

another, electron density is shifted toward the carbon atoms, as manifest in increased carbon shielding. In retinylidenes, this steric interaction occurs between the protons of C20 and C15 in the fully planar all-*trans* conformation and between the protons of C12 and C15 in the fully planar 13-*cis* conformation (Figure 19.2). Accordingly, the ^{13}C chemical shift of C12 is further upfield in the M state than in the resting state. However, this upfield shift is not seen yet in K, indicating the proton of C12 or C15 is pointing out-of-plane, due to torsion around one or both of the single bonds flanking the isomerized C13=C14 bond, most likely C14–C15 because it is closer to the electrostatically constrained Schiff base. On the other hand, the expected upfield shift in C12 does appear in the functional L state. This indicates that the C13=C14 conformation has fully relaxed and suggests that the double bond torsion in L is located closer to the Schiff base, *i.e.* around the C15=N bond. Of the three failed L's, two do not reach the relaxed C13=C14 state. The one that does presumably has problems in other degrees of freedom in the active site that cause it to fail.

The overall picture is one in which the absorbed electromagnetic energy is converted first into charge separation (in K) and then into electrostatically induced double bond torsion (in L) that is released when proton transfer causes the Schiff base and its counterion to be neutralized. This timing of a decisive change in the access of the Schiff base from the extracellular to the intracellular side of the pathway is ideal for enforcing vectoriality.[9,10] The proton transfer from the Schiff base at Lys216 to the carboxyl at Asp85 also weakens the association between helices C and F such that helix F is seen to shift outward at this point in the photocycle.[11–13] Since the ionone ring is immobilized, the effect of the outward movement of helix F is probably to elongate the Lys216 sidechain. However, this is not likely to be functionally significant since retinal alkylamine Schiff bases bound to the K216G mutant of bR are able to pump ions even though there is no covalent connection between the Schiff base and helix F.[14]

Internuclear distance measurements in the photocycle intermediates of bR can add details to the above story and some distances have been measured.[15,16] However, we illustrate the application of distance measurements in a different system below.

19.3 Structure: Building Buoyancy in Aquatic Microorganisms

The assembly and disassembly of gas vesicles allows aquatic microorganisms to move up and down the water column without the use of flagellar motors. The vesicles are air-filled capsules with thin, rigid walls made entirely of protein (Figure 19.11). Grown from small biconicals, they are bipolar with a low pitch helix of GvpA molecules forming ribs on each side of a putative insertion seam. FTIR[17] and X-ray diffraction[18] indicate a substantial presence of β-strands, the former suggesting an antiparallel configuration and the latter extended sheets.

In nature, the GvpA shell is reinforced by GvpC adhering to the outside, but the shell remains intact and rigid even after GvpC is removed with detergent. None of the other gas vesicle gene products are incorporated into

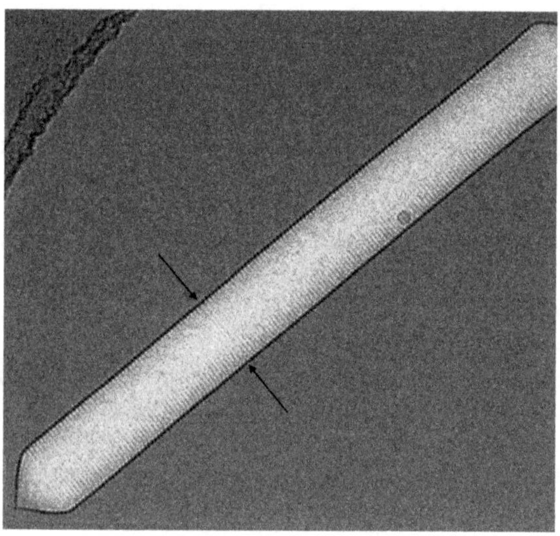

Figure 19.11 Electron micrograph of a 52 nm wide gas vesicle from *Anabaena flos-aquae* in water. Notice the very low electron density inside the vesicle, the ribs running across the vesicle, and the bipolar structure of the vesicle indicated by the orientations of the rib projections on either side of the seam marked by *arrows*.
(Courtesy of Chen Xu and Marina Belenky, Brandeis University.)

the vesicles in significant amounts. Some presumably act as chaperones and others are thought to be associated with the vesicle tips and the putative insertion seam.

The proportions and strengths of the vesicles vary among species. As expected, narrower vesicles, with higher ratios of surface protein to gas volume, have higher collapse pressures. This investment in greater strength is found in organisms that live in deep water, while organisms that live in shallow water economize. *Anabaene flos-aquaea* (*A.f-a.*), an inhabitant of cold fresh water lakes, is an example of the former, while *Halobacterium salinarum* (*H.s.*), an inhabitant of salt flats, is an example of the latter.

An important feature of the vesicles is their permeability to gas molecules,[19] including the 200 Da C_4F_8.[20] As a result of this permeability, changes in pressure only cause collapse if they are abrupt. In general, the contents of the vesicles are at equilibrium with dissolved gases in the surroundings and, by extension, the contents of the atmosphere. Given the penetration of large gas molecules and the ability of water molecules to traverse hydrophobic cavities in nanotubes[21–23] and lipid bilayers,[24] it seems likely that water molecules are able to enter gas vesicles. The question then is why this water vapor does not condense. Homogeneous nucleation of droplets would be a very rare event in such small volumes. However, GvpA must provide an interior surface that is highly inhospitable to heterogeneous nucleation. This suggests an aliphatic surface, *i.e.* a surface covered with hydrogen atoms that have small dispersive attractions due to their paucity of electrons.

Because the vesicle shells are not soluble in detergent, and because the very strong contrast in electron density between the gaseous interior and the aqueous exterior defeats high-resolution electron microscopy of the thin intervening layer of protein, only solid-state NMR has the prospect of providing high-resolution structural information. Figure 19.12 shows the sequences and predicted secondary structures for the ~7 kDa GvpA's of *A.f-a.* and *H.s.* The similarity in predicted structure is not surprising, given the strong sequence homology (despite the huge evolutionary distance and

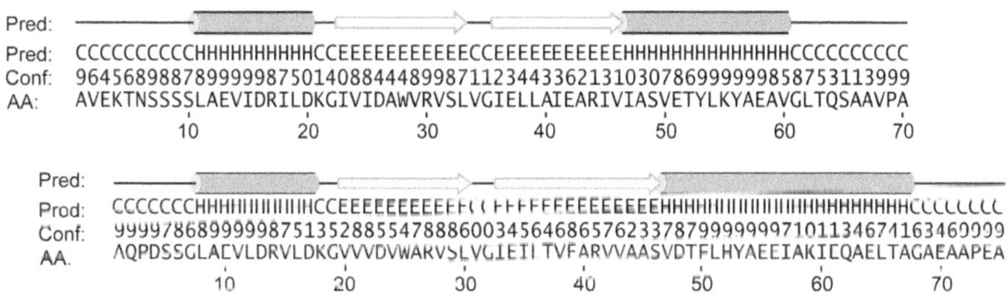

Figure 19.12 GvpA sequences and PSIPRED secondary structure predictions[32] for *A.f-a.* (*top*) and *H.s.* (*bottom*). The sequences (AA) are aligned to emphasize the very strong homology in the central portions of these proteins. The predictions (Pred) are shown graphically and by letters (C = coil, E = strand, H = helix), with accompanying confidence (Conf) estimates (0 = low and 9 = high).

habitat difference between the organisms). The first step in NMR evaluation of this prediction is signal assignment through a variety of backbone and sidechain correlation experiments, including a powerful pulse sequence that captures intra-residue and inter-residue correlations in one 3D spectrum.[25] Analysis of the chemical shifts of the assigned signals suggests that, in both *A.f-a.*[26] and in *H.s.*,[27] the secondary structure is close to the

coil (c1) – helix (α1) – strand (β1) – strand (β2) – helix (α2) – coil (c2)

predicted from the amino acid sequence, an interesting result given that the predictions were based on structural databases of soluble proteins.

The assignment process uncovered an interesting difference between vesicles from *A.f-a.* and *H.s.*: whereas assignments were unique in *H.s.*, some duplicates of similar intensity were found in *A.f-a.* (*e.g.*, in the glycine correlations shown in Figure 19.13). Thus, while *H.s.* uses a single,

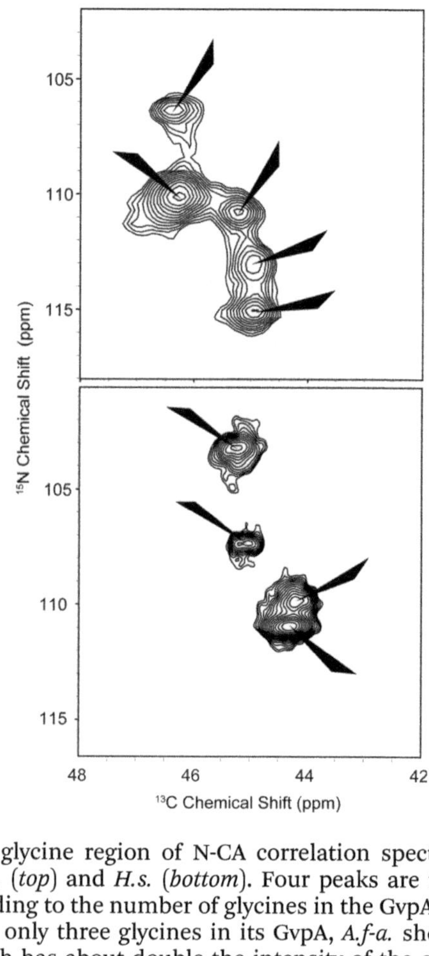

Figure 19.13 The glycine region of N-CA correlation spectra for gas vesicles of *A.f-a.* (*top*) and *H.s.* (*bottom*). Four peaks are found for *H.s.*, corresponding to the number of glycines in the GvpA molecule.[27] However, with only three glycines in its GvpA, *A.f-a.* shows five peaks, one of which has about double the intensity of the others.[29]

presumably lowest energy, GvpA fold, *A.f-a.* also makes use of a second, presumably somewhat higher energy, fold. The duplications in *A.f-a.* are particularly found at the transitions between different secondary structures, suggesting that the polymorphism is primarily in the location of these transitions.

Given assignments, dipolar recoupling with varying mixing times allows interactions between nuclei to be detected over various distances, calibrated in compounds of known structure. So far, this has been done only for *A.f-a.* vesicles.[28] Confirming portions of the secondary structure, correlations are observed between residues on successive turns in each of the helices, α1 and α2. The two stretches of β-strand are confirmed at the level of tertiary structure by correlations between residues in β1 and β2 that are as expected for a β-hairpin. However, consistent with the observations of duplicate chemical shifts, there are two such sets of correlations, one corresponding to an odd-numbered β-turn centered on Val34 and the other corresponding to an odd-numbered β-turn centered on Gly35. The odd-numbered turns are functionally significant because they form a sheet that is entirely aliphatic on one side.

Indications of quaternary structure are supplied by intermolecular correlations. One set shows that β1 strands in neighboring molecules associate in antiparallel fashion and another set shows that β2 strands in neighboring molecules also associate in antiparallel fashion. In both cases, the alignment and strand orientation occurs with electrostatically favorable pairing of acidic and basic residues on one face of the sheet and only aliphatic residues exposed on the other face. The latter presumably presents to the lumen of the vesicle, where it can inhibit heterogeneous nucleation of water condensation.

The presence of intermolecular β1–β1 and β2–β2 contacts, rather than intermolecular β1–β2 contacts, indicates that the repeat unit in the vesicle wall is a GvpA dimer, rather than the GvpA monomer. The overall picture (Figure 19.14) is of an extended, anti-parallel β-sheet formed by intramolecular β1–β2 contacts and intermolecular β1–β1 and β2–β2 contacts. Since extended cross-β structures are hallmarks of amyloid fibrils, the ribs of gas vesicles may be regarded as a functional amyloid.[28] This is consistent with the insolubility of gas vesicles in detergents.

The polymorphism of GvpA in *A.f-a.* suggests that the gas vesicle building block is an asymmetric dimer, rather than a symmetric dimer. In fact, modeling[29] shows that an asymmetric dimer, formed by a GvpA with an odd-numbered turn on V34 and a GvpA with an odd-numbered turn on G35, allows the hydrogen bonds in the extended β-sheet to be oriented the closest to the "magic angle" which distributes stress most uniformly.[30] Thus, the polymorphism of GvpA in *A.f-a.* seems to be an additional means by which the organism invests in vesicle strength, while the monomorphism of GvpA employed by *H.s.* reflects the greater economy allowed by its less rigorous habitat.

What remains is to understand how the vesicle ribs are held together. Since the helix and coil segments of GvpA extend from the edges of the β-sheet (Figure 19.14), they are expected to provide ample inter-rib contacts.

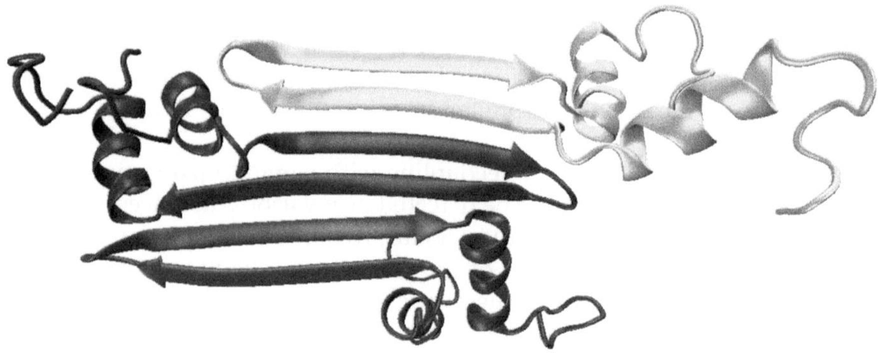

Figure 19.14 Model of a three-monomer segment of the amyloid ribbon of *A.f-a.* gas vesicles. The two dark monomers form an asymmetric dimer. (Choosing different cell boundaries, the light monomer and the adjacent dark one also form an asymmetric dimer). The model was generated by including the secondary structure and internuclear constraints described in the text. The α-helical and coil portions of GvpA are included without any tertiary constraints. Located at the edges of the amyloid ribbon, they are well situated to be involved in rib–rib interactions.
(Courtesy of Eugenio Daviso, Brandeis University.)

The challenge is to discern the corresponding correlations in multidimensional NMR spectra.

19.4 Summary

As described above for gas vesicles, chemical shifts in structural studies serve primarily to resolve and assign signals that can then be interrogated for spatial information by recoupling methods. The observed strengths of the internuclear correlations provide a measure of internuclear distances that interface well with molecular modeling programs. As also illustrated by the studies of gas vesicles, the sensitivity of the chemical shift to bonding can be interpreted in terms of secondary structure based on chemical shift data for proteins of known structure. However, reliable determinations of secondary structure, as well as the higher level tertiary and quaternary structures, are based on internuclear distance measurements.

As demonstrated above for bacteriorhodopsin, chemical shifts can play a dominant role in studies of function where subtle spatial changes correspond to substantial electronic changes. The chemical shift perturbations observed in these cases can potentially be interpreted in terms of structural perturbations *via* chemical shift calculations using high-level quantum mechanics. However, the more reliable route remains spectroscopy of compounds chosen to model the degrees of freedom relevant to hypothesized functional mechanisms. In the case of bacteriorhodopsin, this includes retinylidenes in different conformations and protonation states, with different counterions or hydrogen bonding partners.

References

1. H. Luecke, B. Schobert, H. T. Richter, J. P. Cartailler and J. K. Lanyi, *J. Mol. Biol.*, 1999, **291**, 899–911.
2. V. S. Bajaj, M. L. Mak-Jurkauskas, M. Belenky, J. Herzfeld and R. G. Griffin, *Proc. Natl. Acad. Sci. U. S. A.*, 2009, in press.
3. H. J. M. Degroot, G. S. Harbison, J. Herzfeld and R. G. Griffin, *Biochemistry*, 1989, **28**, 3346–3353.
4. J. G. G. Hu, R. G. Griffin and J. Herzfeld, *J. Am. Chem. Soc.*, 1997, **119**, 9495–9498.
5. H. J. M. Degroot, S. O. Smith, J. Courtin, E. Vandenberg, C. Winkel, J. Lugtenburg, R. G. Griffin and J. Herzfeld, *Biochemistry*, 1990, **29**, 6873–6883.
6. M. L. Mak-Jurkauskas, V. S. Bajaj, M. K. Hornstein, M. Belenky, R. G. Griffin and J. Herzfeld, *Proc. Natl. Acad. Sci. U. S. A.*, 2008, **105**, 883–888.
7. J. G. Hu, B. Q. Sun, A. T. Petkova, R. G. Griffin and J. Herzfeld, *Biochemistry*, 1997, **36**, 9316–9322.
8. J. G. Hu, B. Q. Sun, M. Bizounok, M. E. Hatcher, J. C. Lansing, J. Raap, P. J. E. Verdegem, J. Lugtenburg, R. G. Griffin and J. Herzfeld, *Biochemistry*, 1998, **37**, 8088–8096.
9. J. Herzfeld and J. C. Lansing, *Annu. Rev. Biophys. Biomol. Struct.*, 2002, **31**, 73–95.
10. J. Herzfeld and B. Tounge, *Biochim. Biophys. Acta, Bioenerg.*, 2000, **1460**, 95–105.
11. N. Radzwill, K. Gerwert and H. Steinhoff, *Biophys. J.*, 2001, **80**, 2856–2866.
12. S. Subramaniam, M. Gerstein, D. Oesterhelt and R. Henderson, *EMBO J.*, 1993, **12**, 1–8.
13. F. Hendrickson, F. Burkard and R. Glaeser, *Biophys. J.*, 1998, **75**, 1446–1454.
14. N. Friedman, S. Druckmann, J. Lanyi, R. Needleman, A. Lewis, M. Ottolenghi and M. Sheves, *Biochemistry*, 1994, **33**, 1971–1976.
15. J. M. Griffiths, A. E. Bennett, M. Engelhard, F. Siebert, J. Raap, J. Lugtenburg, J. Herzfeld and R. G. Griffin, *Biochemistry*, 2000, **39**, 362–371.
16. L. K. Thompson, A. E. McDermott, J. Raap, C. M. Vanderwielen, J. Lugtenburg, J. Herzfeld and R. G. Griffin, *Biochemistry*, 1992, **31**, 7931–7938.
17. T. Strunk, K. Hamacher, F. Hoffgaard, H. Engelhardt, M. D. Zillig, K. Faist, W. Wenzel and F. Pfeifer, *Mol. Microbiol.*, 2011, **81**, 56–68.
18. A. E. Blaurock and A. E. Walsby, *J. Mol. Biol.*, 1976, **105**, 183–199.
19. A. E. Walsby, *Proc. R. Soc. London, Ser. B*, 1971, **178**, 301–326.
20. A. E. Walsby, *J. Gen. Microbiol.*, 1982, **128**, 1679–1684.
21. G. Hummer, J. C. Rasaiah and J. P. Noworyta, *Nature*, 2001, **414**, 188–190.

22. H. Kyakuno, K. Matsuda, H. Yahiro, Y. Inami, T. Fukuoka, Y. Miyata, K. Yanagi, Y. Maniwa, H. Kataura, T. Saito, M. Yumura and S. Iijima, *J. Chem. Phys.*, 2011, **134**, 244501.
23. Y. Maniwa, H. Kataura, M. Abe, S. Suzuki, Y. Achiba, H. Kira and K. Matsuda, *J. Phys. Soc. Jpn.*, 2002, **71**, 2863–2866.
24. A. Finkelstein, *J. Gen. Physiol.*, 1976, **68**, 127–135.
25. E. Daviso, M. T. Eddy, L. B. Andreas, R. G. Griffin and J. Herzfeld, *J. Biomol. NMR*, 2013, **55**, 257–265.
26. A. C. Sivertsen, M. J. Bayro, M. Belenky, R. G. Griffin and J. Herzfeld, *Biophys. J.*, 2010, **99**, 1932–1939.
27. E. Daviso, M. Belenky, R. Griffin and J. Herzfeld, *J. Mol. Microbiol. Biotechnol.*, 2013, **23**, 281–289.
28. M. J. Bayro, E. Daviso, M. Belenky, R. G. Griffin and J. Herzfeld, *J. Biol. Chem.*, 2012, **287**, 3479–3484.
29. A. C. Sivertsen, M. J. Bayro, M. Belenky, R. G. Griffin and J. Herzfeld, *J. Mol. Biol.*, 2009, **387**, 1032–1039.
30. A. E. Walsby, *Microbiol. Rev.*, 1994, **58**, 94–144.
31. A. T. Petkova, J. G. G. Hu, M. Bizounok, M. Simpson, R. G. Griffin and J. Herzfeld, *Biochemistry*, 1999, **38**, 1562–1572.
32. L. J. McGuffin, K. Bryson and D. T. Jones, *Bioinformatics*, 2000, **16**, 404–405.

CHAPTER 20

Photoactivated Structural Changes in Photoreceptor Membrane Proteins as Revealed by in situ Photoirradiation Solid-State NMR Spectroscopy

AKIRA NAITO* AND IZURU KAWAMURA

Graduate School of Engineering, Yokohama National University, 79-5 Tokiwadai, Hodogaya-ku, Yokohama 240-8501, Japan
*Email: naito@ynu.ac.jp

20.1 Introduction

20.1.1 Photoirradiation Solid-State NMR Spectroscopy

Photoirradiation solid-state NMR spectroscopy is particularly useful for studying photoreceptor membrane proteins and revealing the photoreaction cycle and the photoactivated structural changes of the systems. However, such experiments are not easy because photoillumination systems for solid-state NMR are quite difficult to build due to the complicated MAS probe assemblies. Photoirradiation NMR experiments have been reported by several groups. Most of the experiments involved photoillumination of the

New Developments in NMR No. 3
Advances in Biological Solid-State NMR: Proteins and Membrane-Active Peptides
Edited by Frances Separovic and Akira Naito

Published by the Royal Society of Chemistry, www.rsc.org

outside of the probe and freezing of the sample followed by transfer to the NMR probe to measure the signals of the trapped photoactivated intermediates.[1–8] Using photoirradiation systems, the light-adapted state and M- and N-intermediates of bacteriorhodopsin (bR) have been characterized. The torsion angles of the retinylidene chromophore in bovine rhodopsin and metharhodopsin-I and metharhodopsin-II were determined simultaneously in the photoactivated membrane-bound state. An *in situ* photoirradiation solid-state NMR spectrometer involving passage of light through a photofibre to the probe and illumination from outside the rotor tube has been described.[9–11] Using this system, the L-intermediate of bR was characterized.[9] By combining with the DNP method, the heterogeneity of dark-adapted bR, distortion in the K-state, and four discrete L-substates were detected.[10] Furthermore, a photointermediate designated bathorhodopsin was generated by illuminating the rhodopsin sample *in situ*.[11] By contrast, in the system described in this chapter, light illumination is carried out from the inside of the rotor tube though a glass rod inserted into the rotor.[12] This illumination system allows us to illuminate the sample with extremely high efficiency. This experimental system opens the door to studies of the photoreaction processes of photoreceptor membrane proteins.

20.1.2 Photoreaction Cycle of *Pharaonis* Phoborhodopsin

Pharaonis phoborhodopsin (*p*pR or sensory rhodopsin II) is a negative phototaxis receptor of *Natronomonas pharaonis*, which forms a 2:2 complex with the cognate transducer *p*HtrII that transmits the photosignal into the cytoplasm.[13,14] Light absorption by *p*pR initiates *trans–cis* photoisomerization of the retinal chromophore, followed by a cyclic chemical reaction consisting of several intermediates (K, L, M and O),[15] as shown in Figure 20.1A.

The K(540)-intermediate has a half-life of 1 μs and has a 13-*cis*,15-*anti* retinal configuration. The L(488)-intermediate can be transformed from the K-intermediate and has a half-life of 30 μs and a 13-*cis*,15-*anti* retinal configuration. Subsequent to the L-intermediate, a proton is removed from the Schiff base (SB) group to form the M(390)-intermediate, which has a long half-life of 1.7 s. The M-intermediate has a 13-*cis*,15-*anti* retinal configuration with a deprotonated SB. Interestingly, a very long half-life of 2 h for the M-intermediate has been reported in a membrane mimetic containing *n*-octyl-β-D-glucoside (OG).[16] By taking protons, the M-intermediate transforms into the O(560)-intermediate, which has a half-life of 770 ms and has an all-*trans* retinal configuration.[17] Here, 540, 488, 390 and 560 denote the wavelengths (in nm) of the maximum absorption, λ_{max}, for the K-, L-, M- and O-intermediates, respectively. The M- and O-intermediates are thought to be active states for signal transduction.

The crystal structure of the *p*pR/*p*HtrII complex suggests the formation of two specific hydrogen bonds between Tyr199^{ppR} and Asn74^{pHtrII}, and between Thr189^{ppR} and Glu43^{pHtrII}/Ser62^{pHtrII}.[18] Thr204 is an important residue for color tuning and the photocycle kinetics of *p*pR,[19] and further observations provide additional information about the important roles of Thr204 in *p*pR for the negative phototaxis function of the complex.[20] Steric hindrance between C14–H of retinal and Thr204 occurs upon formation of the K-intermediate.[21] At the same time, a specific hydrogen bonding alteration occurs between Thr204 and Tyr174 in a *p*HtrII-dependent manner.[20,22] Helix movement of *p*pR, and outward tilting of helix F, during the photocycle was suggested by various groups,[23–25] and is thought to be an essential step for the activation of *p*HtrII. However, no helix tilting was observed in the crystal structure of the M-intermediate of the *p*pR/*p*HtrII complex.[26] Subsequently, TM2 helical rotation occurs, which transfers the signal to the phosphorization cascade to initiate the rotation of the bacterial flagellar motor, resulting in negative phototaxis (Figure 20.1C). Thus, the structural changes that occur upon formation of the active M-intermediate continue to be an exciting topic of discussion, although high-resolution structural studies have been a major challenge. In this regard, solid-state NMR techniques can be applied to elucidate site-specific positions for these membrane-embedded systems.[27,28]

In this chapter, we report that the M-intermediate was successfully trapped using newly developed *in situ* photoirradiation solid-state NMR

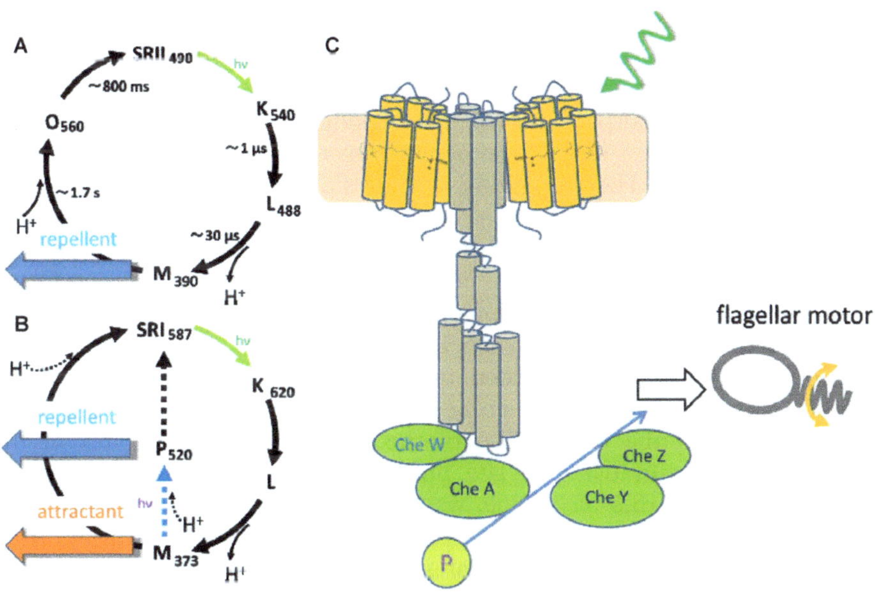

Figure 20.1 (A) Photochemical reaction cycle of *pharaonis* phoborhodopsin (*p*pR). *p*pR absorbs blue light and forms K(540)-, L(488)-, M(390)- and O(560)-intermediates. The M-intermediate has a long half-life of 1.7 s in a membrane, but it has a much longer half-life of 2 h in a membrane mimetic containing *n*-octyl-β-D-glucoside. (B) Photochemical reaction cycle of HrSRI. (C) Schematic of the 2 : 2 *p*pR/*p*HtrII complex. Photoactivated retinal isomerization leads to transition to *p*pR, which then relays the signal to the transducer protein and finally to the flagellar motor.

spectroscopy,[12,29] which was used to obtain NMR signals from the retinal of *pp*R, to gain significant insights into the mechanism of signal transduction in the *pp*R/*p*HtrII complex.

20.1.3 Photoreaction Cycle of Sensory Rhodopsin I

Sensory rhodopsin I (SRI) functions as a color-discriminating receptor in halobacterial phototaxis.[30] SRI has seven transmembrane α-helices with an all-*trans* retinal chromophore.[31,32] SRI forms a 2:2 complex with its cognate transducer protein, the halobacterial transducer protein for SRI, HtrI.[33] The essential photocycle of the SRI from *Halobacterium salinarum* (HsSRI) proceeds as follows: $SRI_G(587) \rightarrow (h\nu) \rightarrow SRI_K(625) \rightarrow SRI_L(540) \rightarrow SRI_M(373) \rightarrow SRI_G(587)$ (Figure 20.1B). Here, the $SRI_P(520)$-intermediate is produced from the M-intermediate by a second photon absorption by near-UV light as a back reaction: $SRI_M(373) \rightarrow (h\nu) \rightarrow SRI_P(520) \rightarrow SRI_G(587)$.[34–36] During the photocycle, the M- and P-intermediates are thought to be essential for positive and negative phototaxis, respectively (Figure 20.1B). Little is known about the structural changes or the signal relay mechanism of this type of phototaxis because HsSRI is unstable under various conditions and its inherent instability hampers the elucidation of its molecular mechanism. Recently, a novel SRI protein from the eubacterium *Salinibacter ruber* (SrSRI) was cloned and characterized. SrSRI was the first eubacterial SRI identified as a functional protein.[37] SrSRI has an all-*trans* retinal configuration as a chromophore in the dark and an absorption maximum at a longer wavelength (557 nm) than SRII (500 nm), and a slower photocycle than the light-driven ion-pumping rhodopsins (bR and HR), indicating that it is similar to HsSRI.[37,38]

Thus, the structural changes of both the protein moiety and the retinal chromophore upon formation of the active M- and P-intermediates continue to be stimulating topics of discussion, although conducting high-resolution structural studies has been a major challenge. However, solid-state NMR techniques can be applied to elucidate with atomic resolution the conformations of the photoactivated intermediates and the pathway of the photocycle for such membrane-embedded systems under natural conditions.

Here, we describe observation of the individual intermediates of the photocycle of SrSRI using *in situ* photoirradiaton solid-state NMR spectroscopy.[39] Using this technique, we obtained ^{13}C NMR signals from the [20-^{13}C]retinal of the long-lived M- and P-intermediates and pathways from the G-state to the photointermediates, thereby providing insights into the mechanism of the photocycle of SrSRI.

20.2 Experimental

20.2.1 Materials

[15,20-^{13}C]Retinal-*pp*R and SrSRI with a His-Tag (6×His) at the C-terminus were expressed in *Escherichia coli* strain BL21(DE3) in M9 medium by induction with 1 mM isopropyl-1-thio-β-D-galactoside (IPTG) and 10 μM [15,20-^{13}C]retinal. The protein was solubilized using *n*-dodecyl-β-D-maltoside (DDM) and purified using an Ni-NTA column (QIAGEN, Hilden, Germany), as described previously.[27] The truncated transducer, *p*HtrII(1–159), was prepared using the method described above. Purified proteins in DDM micelles were incorporated into a lipid film of egg PC and egg PG (*p*pR:PC and SrSRI:PG molar ratio of 1:30), and gently stirred at 4°C overnight. DDM was removed using Bio-Beads (Bio-Rad, Hercules, USA) to yield egg PC and PG bilayers. Reconstituted suspensions were concentrated by centrifugation and then resuspended in a buffer solution of 5 mM 4-(2-hydroxyethyl)-1-piperazineethanesulfonic acid (HEPES) and 10 mM NaCl (pH 7). Pelleted samples of *p*pR or a 1:1 molar ratio of *p*pR:*p*HtrII complex embedded in egg PC bilayers and pelleted samples of SrSRI embedded in PG bilayers were placed in a 5 mm outer diameter (o.d.) zirconia pencil-type rotor for MAS solid-state NMR experiments. To prepare a sample of long-lived M-intermediate, solubilized *p*pR in DDM was concentrated using an Amicon Ultra device and exchanged into OG buffer (50 mM KCl, 10 mM citrate, 1% nondetergent sulfo-betaine [NDSB]-195 and 1% OG).[16]

20.2.2 Photoirradiation System for Solid-State NMR

We developed an *in situ* photoirradiation system for solid-state NMR measurements under MAS. *In situ* continuous photoirradiation was carried out using an optical fibre passed from outside the magnet through a tightly sealed piece of cap made of glass rod glued to a zirconia rotor (Figure 20.2). The tip of the glass rod was grained so as to provide illumination perpendicular to the rotor wall, to which the fluid membrane proteins were attached as a thin film. Therefore, light is directed perpendicular from inside of the spinner to the rotor wall. If the irradiation is not perpendicular to the rotor wall, the light will be completely reflected off the surface of the film sample and will not penetrate into the sample. Another group has mixed 100 μm diameter glass beads with frozen 200 μm diameter rhodopsin samples to improve light penetration.[11] However, the light was directed from outside the rotor, and the irradiation efficiency is improved when the light is provided from the inside. Using a CMX-400 Infinity NMR spectrometer equipped with this photoirradiation system, we could efficiently irradiate samples in the rotor with 50 mW of green (520 nm), orange (590 nm) and blue (365 nm) LED lights.

20.2.3 NMR Measurements

In the *in situ* solid state NMR measurements, 100 μL (50 mg) each of *p*pR-PC and *p*pR/*p*HtrII-PC were placed into the zirconia rotor and sealed with the glass rod as a cap. A CP pulse sequence was used with a contact time of 1ms, followed by acquisition with 50 kHz TPPM proton decoupling pulses.[40] It was noted that the ^{13}C NMR signals of [20-^{13}C]retinal were only observed

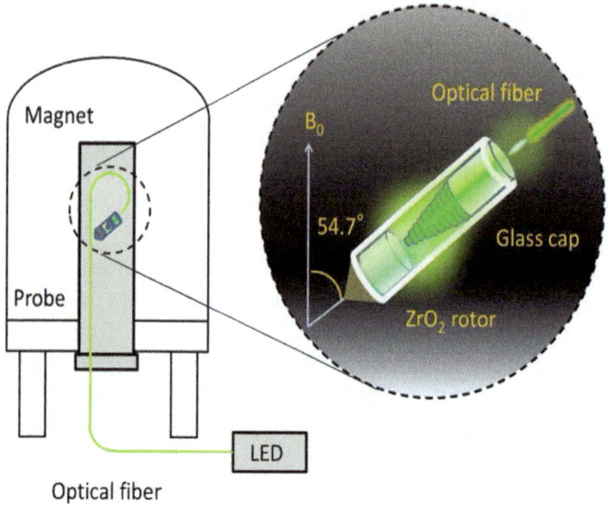

Figure 20.2 *In situ* photoirradiation solid-state NMR spectrometer. An optical fibre is guided from the outside of the magnet and connected to an LED light (520, 595 and 365 nm, 50 mW). An optical fibre is guided from the bottom to the top of the probe head and accurately aligned to the top of the MAS rotor. A glass rod in which the tip is ground to direct the light perpendicular to the spinner axis, so that the strong light irradiates the sample from inside the sample tube, caps the top part of a zirconia rotor. Specifically, the light penetrates from the top part of the spinner through an aligned optical fibre into the zirconia sample tube without any contact with the optical fibre, so that rapid spinning is established.

with the CP pulse sequence. The MAS frequency was set to 4 kHz and the temperature was set to 0, −20 and −40 °C using a gas flow temperature control system. Typically, 20 000 transients were accumulated for the dark and light experiments. Samples could be continuously irradiated with three different LED lights (365, 560 and 595 nm) during NMR measurements.

20.2.4 Trapping of Photointermediates Using *in situ* Photoirradiation Solid-State NMR

In situ photoirradiation is particularly useful for studying the photocycle of retinal proteins, as shown in Figure 20.1A. The half-life of the *p*pR M-intermediate is ~1.7 s, which is much longer than that of the other intermediates. Thus, when a sample is continuously irradiated with green light (520 nm), the M-intermediates are trapped in a stationary fashion. This makes it possible to detect relatively short-lived intermediates in a stationary state when their half-life differs from those of other intermediates. The trapping efficiency can be increased by decreasing the temperature, because the half-life of the M-intermediate can be extended. Also, the M-intermediate can be eliminated by illuminating with 365 nm LED light since the M-intermediate has a maximum absorbance frequency of 390 nm, which differs from those of the other intermediates. Therefore, the M-intermediate can be selectively eliminated by directly applying 365 nm blue light. Activation of consecutive photoreaction processes, such as the double-photon process by illuminating with multiple wavelengths, can be detected. Thus it is possible to select a particular intermediate by adjusting either temperature or wavelength *in situ.*

20.3 *In situ* Photoirradiated Solid-State NMR Study of *p*pR and the *p*pR/*p*HtrII Complex

20.3.1 An Active Intermediate of the *p*pR Photoreceptor Dissolved in OG

To investigate the trapping of an active intermediate of a photoreceptor membrane protein, photoirradiation-induced isomerization of *p*pR from the ground-state (G-state) to the M-intermediate was performed on [20-^{13}C]retinal–*p*pR dissolved in OG at 20 °C. This is possible because the half-life of the M-intermediate (2 h) is much longer than those of the other intermediates. The sample was irradiated outside the magnet and quickly inserted into the magnet to observe the ^{13}C directly detected (DD)-MAS NMR spectrum of the M-intermediate (Figure 20.3). The ^{13}C DD-MAS NMR spectrum obtained from the G-state of *p*pR in OG showed a peak at 13.4 ppm. A new peak from the M-intermediate appeared at 21.5 ppm and the peak at 13.4 ppm completely disappeared. The observed chemical shift value is consistent with that of the M-intermediate of bacteriorhodopsin.[10] This observation clearly

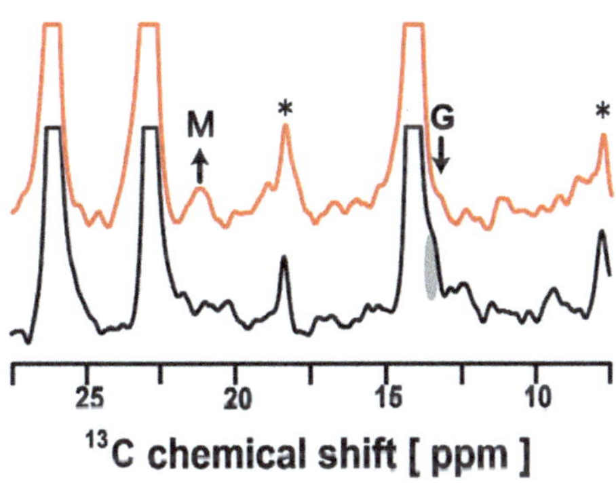

Figure 20.3 Directly detected ^{13}C MAS NMR spectra of [20-^{13}C]retinal–*p*pR in OG. Photoirradiated state (*top*, *red*); ground state (*bottom*, *black*). The ^{13}C NMR signals from NDSB are marked by an asterisk (*). The peaks labeled M and G indicate the ^{13}C NMR signals from [20-^{13}C]retinal in the M-intermediate and the G-state, respectively. (Adapted from Tomonaga *et al.*,[12] and reproduced with permission from Cell Press.)

showed that all-*trans* retinal in the G-state was transformed into the 13-*cis*,15-*anti* configuration of retinal in the M-intermediate. Thus, most of the G-state was converted to the M-intermediate, which remained for hours in our experiments and enabled observation of the M-intermediate in very high yield.

20.3.2 Photocycle of *pp*R

Second, *in situ* photoirradiation cross-polarization (CP)-MAS solid-state NMR experiments were performed on a sample of [20-^{13}C]retinal–*pp*R in egg PC under near physiological conditions but at 0 °C and −20 °C (Figure 20.4). In addition to the long half-life of the M(390)-intermediate (1.7 s), the relatively

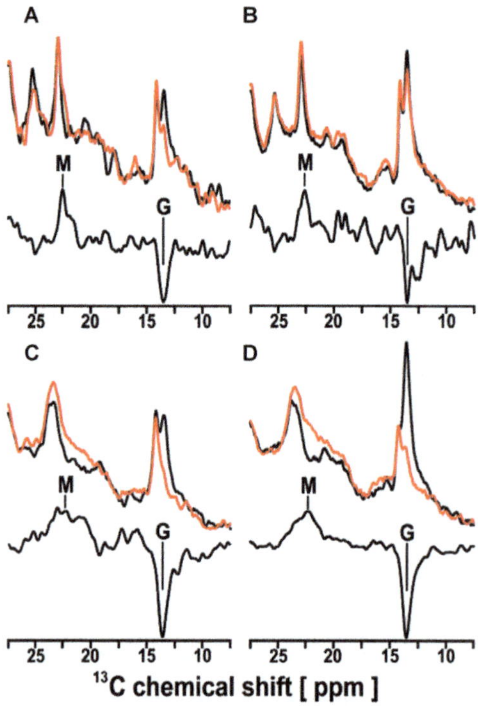

Figure 20.4 ^{13}C CP-MAS NMR spectra of [20-^{13}C]retinal–*pp*R (A and C) and the *pp*R/*p*HtrII complex (B and D) taken at 0 °C (A and B) and −20 °C (C and D). The *top* spectra were obtained from the light (*red*) and dark (*black*) states of *pp*R (A and C) and the *pp*R/*p*HtrII complex (B and D). The *bottom* spectra show the differences between the light (*red*) and dark (*black*) spectra. The peaks labeled M and G indicate the ^{13}C NMR signals of [20-^{13}C]retinal in the M-intermediate and the G-states, respectively.
(Adapted from Tomonaga *et al.*,[12] and reproduced with permission from Cell Press.)

long-lived O(560)-intermediate (0.77 s) can be irradiated with 520 nm LED light and, therefore, only the M-intermediate would be expected to be trapped under continuous photoirradiation. In the G-state, a peak for 20-C in retinal appeared at 13.5 ppm, and the signal shifted to 22.4 ppm for the M-intermediate at 0 °C (Figure 20.4A). At least three distinct M-intermediates appeared at 24.1 (M_3), 22.5 (M_2) and 21.7 ppm (M_1) in photoirradiated *p*pR at −20 °C (Figure 20.4C, Table 20.1). The overall yields of the M-intermediates were measured as 40% and 80% at 0 and −20 °C, respectively. This difference in yields can be attributed to a longer half-life of the M-intermediate at low temperature compared with high temperature. It is important to point out that a single peak with a shoulder was observed for the M-intermediate at 0 °C, whereas multiplet lines were observed at −20 °C. These multiplet signals can be attributed to the coexistence of several different interactions between the retinal and protein.

20.3.3 Photocycle of the *p*pR/*p*HtrII Complex

The M-intermediates were also trapped for the [20-^{13}C]retinal–*p*pR/*p*HtrII complex, which has a signal transduction function, as revealed by the appearance of a CP-MAS ^{13}C NMR peak at 22.6 ppm at 0 °C (Figure 20.4B) and three distinct peaks at 23.5, 22.3 and 21.3 ppm at −20 °C (Table 20.1). Thus, multiple M-intermediates were also observed in the *p*pR/*p*HtrII complex at − 20 °C (Figure 20.4D). Note that the signal from the M-intermediate for *p*pR was slightly different from that of the *p*pR/*p*HtrII complex. The chemical shift values of [20-^{13}C]retinal in the M-intermediate of *p*pR were more dispersed and at lower field than those of *p*pR/*p*HtrII complex. This observation reveals that the interaction of retinal with only *p*pR is different from that with the *p*pR/*p*HtrII complex.

Table 20.1 Chemical shift values of [20-^{13}C]retinal proteins.

Retinal proteins	*G-state (ppm)*[a]	*M-intermediate (ppm)*[a]	*P-intermediate (ppm)*[a]
Retinylidenepropylamine[b]	14.3 (all-*trans*)	22.2 (13-*cis*)	–
BR[c]	13.3 (all-*trans*) 22.0 (13-*cis*,15-*syn*)	21.5 (13-*cis*,15-*anti*)	–
*p*pR (0 °C)	13.3 (all-*trans*)	22.3 (13-*cis*,15-*anti*)	–
*p*pR (−20 °C)	13.5 (all-*trans*)	24.1, 22.5, 21.7 (13-*cis*,15-*anti*)	–
*p*pR/*p*HtrII (0 °C)	13.6 (all-*trans*)	22.7 (13-*cis*,15-*anti*)	–
*p*pR/*p*HtrII (−20 °C)	13.5 (all-*trans*)	23.5, 22.3, 21.3 (13-*cis*,15-*anti*)	–
*p*pR in OG	13.4 (all-*trans*)	21.6 (13-*cis*,15-*anti*)	–
SrSRI	13.8 (all-*trans*)	19.8 (13-*cis*)	24.9 (13-*cis*)

[a]As referenced to tetramethylsilane (TMS).
[b]Harbison *et al.*[43]
[c]Bajaj *et al.*[10]

20.3.4 Trapping of the N-Intermediate of *pp*R

After the M-intermediates of *pp*R were trapped in a stationary manner by illumination with green light, the 520 nm LED light was switched to 365 nm LED light (Figure 20.5). The intensities of the M_1 and M_2 intermediates decayed for 2.25 h after the light was switched from green to blue (Figure 20.5B), compared to the case where the green light was turned off (Figure 20.5A). In contrast, the M_3 intermediate did not decay over the same period. The M-intermediate is reported to be directly transformed back to the G-state by irradiation with blue light (365 nm),[41] since the M-intermediate has a maximum absorbance of 390 nm. Thus, the results indicate that M_1 and M_2 are M-intermediates although they are of slightly different configurations. On the other hand, M_3 is not an M-intermediate but most likely an N-intermediate since it was not activated by illumination with blue light (365 nm). Spectroscopic evidence for the formation of an N-intermediate was recently characterized in a transient absorption study,[42]

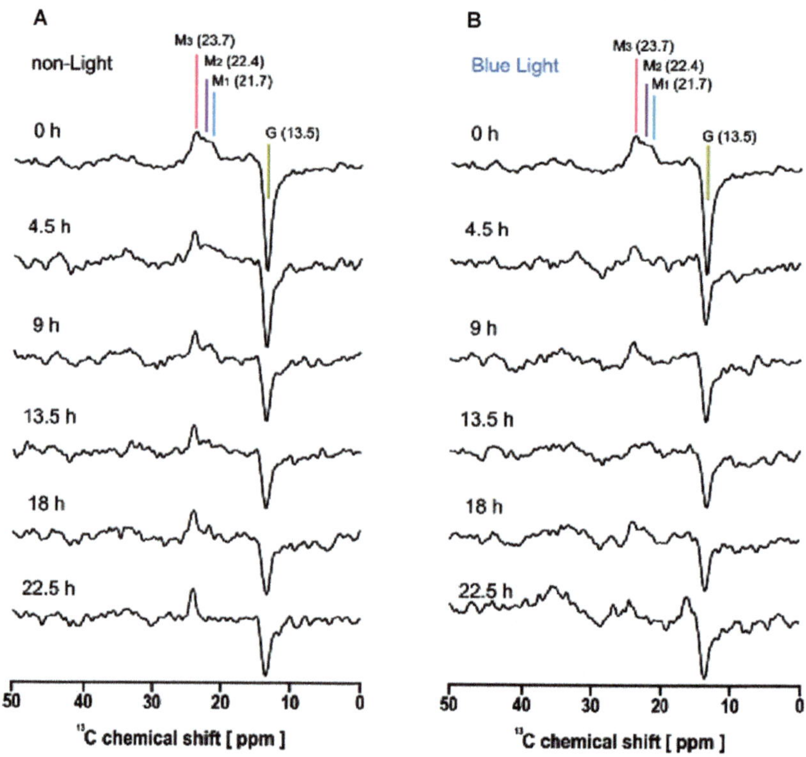

Figure 20.5 ^{13}C CP-MAS NMR spectra of [20-^{13}C]retinal-*pp*R taken at −40 °C. The spectra at 0 h were obtained by taking difference spectra between the light and dark states. (A) Thermal decay process after the M-intermediates were trapped by irradiation with 520 nm LED light. (B) Photoactivated decay process with 365 nm LED light after trapping the M-intermediates by illumination with 520 nm LED light.

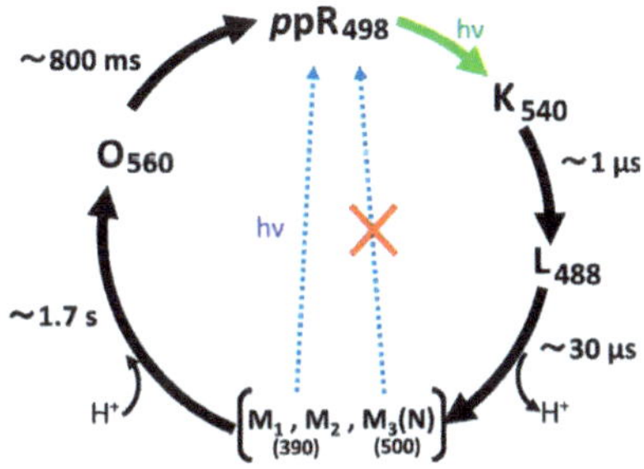

Figure 20.6 Estimated photoreaction cycle of the *p*pR/*p*HtrII complex and retinal configurations. Multiple M-intermediates (M_1, M_2, M_3) were observed under continuous illumination with 520 nm LED light. The M_3 intermediate is an N-intermediate since it is not activated by 365 nm light.

which reported that decay of the M-intermediate does not directly produce an N-intermediate but rather produces an O-intermediate that is in equilibrium with the N-intermediate. However, the pathway from the M- to the N-intermediate could not be detected in this experiment. Thus, as shown in Figure 20.6, the N-intermediate is located in the M-intermediate group. This photocycle involves the M_1- and M_2-intermediates as well as the N-intermediate. Further experiments, however, are necessary to clarify the pathway from the M- to the O-intermediate.

20.4 *In situ* Photoirradiation Solid-State NMR Study of SrSRI

20.4.1 Photocycle of SrSRI

First, *in situ* photoirradiated ^{13}C solid-state NMR spectra, as shown in Figure 20.7A and B, were collected by irradiating the G-state of [20-^{13}C]retinal–SrSRI with 520 or 595 nm LED light at −40 °C.

Since the signals of lipids are heavily overlapped, difference spectra were obtained to resolve the light-induced changes. The difference spectrum between the dark and light states clearly indicated that the ^{13}C NMR signal of the G-state at 13.8 ppm decreased and that of the M-intermediate at 19.8 ppm increased (Figure 20.7A and B). The signals marked with an asterisk (*) could be due to lipids that changed intensity during the photoillumination. This result indicated that the G-state with all-*trans* retinal transformed into the M-intermediate with 13-*cis*,15-*anti* retinal and deprotonated Schiff base (SB) bonding. These configurations of the retinals of SrSRI could be clearly determined by comparing the ^{13}C chemical shift values of [20-^{13}C]retinal with those of *p*pR[12] and bR,[1,10,43] as summarized in Table 20.1. The same results were obtained by irradiation with 520 nm LED light (Figure 20.7A) and 595 nm LED light (Figure 20.7B). The M-intermediate was trapped

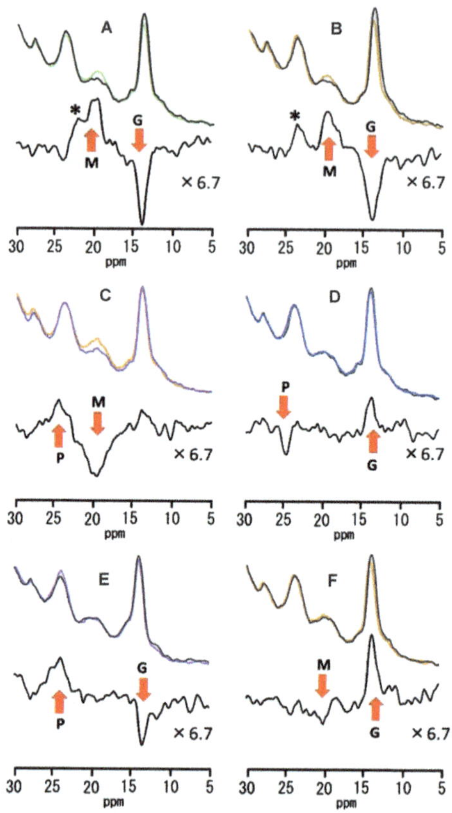

Figure 20.7 ^{13}C NMR signals of the [20-^{13}C]retinal–SrSRI–PG system in the dark (*black*), under illumination with (A) 520 nm (*green*) and (B) 595 nm (*orange*) LED light (*top*); and difference spectra between the light and dark states (*bottom*). ^{13}C NMR signals of the [20-^{13}C]retinal–SrSRI–PG system under illumination of the M-intermediate with (C) 595 nm (*orange*) and 365 nm (*blue*) LED light (*top*), and difference spectra between the M- and P-intermediates (*bottom*); and (D) relaxation process from the P-intermediate (*blue*) to the G-state (*black*). (E) ^{13}C NMR signals of the [20-^{13}C]retinal–SrSRI–PG system in the dark (*black*) and under illumination of the P-intermediate with 365 nm LED illumination (*top*), difference spectra between the P-intermediate and G-state (*bottom*), and (F) the relaxation process from the M-intermediate to the G-state.

stationary upon photoillumination because the half-life of the M-intermediate is relatively long compared with those of the K- and L-intermediates, allowing a high population to remain in the photocycle, whereas the K- and L-intermediates could not be trapped through the photocycle to the M-intermediate, as seen in Figure 20.7A and B.

As the M-intermediate was trapped stationary by illumination with 520 nm LED light, it was subsequently irradiated with 365 nm LED light at −40 °C to examine the double-photon process. As shown in Figure 20.7C, the signal of the M-intermediate at 19.8 ppm decreased as that of the P-intermediate at 24.8 ppm simultaneously increased. This result indicated that the M-intermediate transformed into the P-intermediate by absorbing the second irradiation with 365 nm LED light (blue). It was noted that the P-intermediate could be trapped stationary. This result indicated that the half-life of the P-intermediate should be long enough to enable a high population to remain at −40 °C. Since the ^{13}C chemical shift values of [20-^{13}C]retinal of the P-intermediate was 24.8 ppm, the configuration of retinal should be 13-*cis*,15-*anti*, based upon the chemical shift values of *pp*R[12] and bR[1,10,43] (Table 20.1), although the possibility of a 13-*cis*,15-*syn* configuration cannot be completely ruled out. This result is in contrast to a report tentatively assigning the configuration of the P-intermediate as all-*trans* in SRI.[35] This result clearly shows the formation of a repellent phototaxis intermediate that retains the 13-*cis* configuration of the retinal chromophore (Figure 20.8). As the absorption maximum of the P-intermediate is located at a higher wavelength (~525 nm),[37] the P-intermediate is likely to have a protonated retinal SB (Figure 20.8). Additionally, because the chemical shift value of the P-intermediate differs significantly from that of the M-intermediate, a rather large conformational change occurred, which may cause the significant changes in protein interaction that lead to the functional switch between positive and negative phototaxis.[44]

After the P-intermediate was trapped stationary by irradiating with 365 nm LED light, the illumination was stopped and the intensity change of the P-intermediate observed, as shown in Figure 20.7D. The signal intensity of the P-intermediate at 24.8 ppm decreased and that of the G-state at 13.8 ppm increased, indicating that the P-intermediate relaxed to reach the G-state. Therefore, the P-intermediate was concluded to return directly back to the G-state under dark conditions through a thermal process. Although the P-intermediate relaxed within a half day at −40 °C, it was trapped stationary under illumination with 365 nm LED light (Figure 20.7C). This indicates that the P-intermediate and G-state should be in equilibration under conditions of illumination with 365 nm LED light. Thus, to examine the direct pathway of transformation from the G-state to the P-intermediate, the G-state was irradiated with 365 nm LED light (Figure 20.7E). The result indicated a decrease in the G-state and simultaneous increase in the P-intermediate. This also clearly revealed that the P-intermediate is directly generated from the G-state upon irradiation with 365 nm LED light. The details are discussed in the following section. Briefly, this can be explained by considering two pathways from absorption of the UV light by the G-state (blue lines in Figure 20.8).

In contrast, the M-intermediate transformed into the G-state thermally under dark conditions, as shown in Figure 20.7F. Also, the half-life of the M-intermediate was as long as one day at −40 °C. This result confirmed the normal pathway of the single-photon photocycle under illumination with green light rather than UV light.

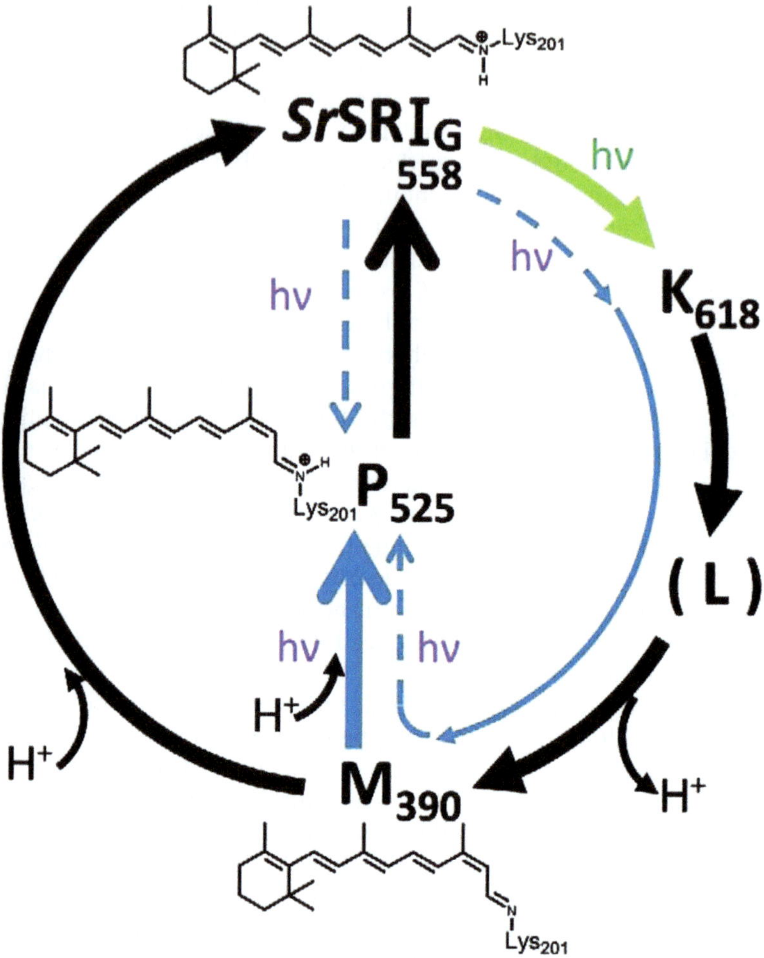

Figure 20.8 Estimated photoreaction cycle of SrSRI and retinal configurations. The two photoactivated pathways from the G-state to the P-intermediate under blue light illumination were revealed to be single- and double-photon processes by these experiments. The retinal configuration of the P-intermediate was determined to be 13-*cis*.

20.4.2 Mechanism of the Photocycle of SrSRI

In summary, the photoreaction cycle and possible configurations of retinals in the intermediates of SrSRI were revealed (Figure 20.8). In this photoreaction cycle, the M-intermediate was trapped stationary in an attractant state under green light (520 nm) illumination, and subsequently the P-intermediate was trapped stationary by a second illumination with UV light (365 nm), acting in a repellent state. The configurations of the M- and P-intermediates were determined to be 13-*cis*,15-*anti* with deprotonation of the SB and 13-*cis* with protonation of the SB, respectively.

Interestingly, the M-to-P transition occurred while the 13-*cis* configuration was retained. The pathway from the G-state to the P-intermediate was also observed by illumination with UV light at 365 nm. SrSRI has been shown to function in both positive and negative phototaxis in contrast to SRII, which functions only in negative phototaxis. SrSRI exerts a positive phototaxis function under green light illumination and a negative phototaxis function when the light has a UV component, presumably to effect escape from harmful UV light.

To determine whether the G-state absorbs in the UV region, we measured the reaction of SrSRI with hydroxylamine. The UV absorption spectrum (data not shown) of SrSRI differed at several times after the addition of hydroxylamine. A decrease in the absorbance at 556 nm for SrSRI and concomitant increase in the absorbance at 361 nm was observed. Interestingly, there was a small negative peak at 429 nm with an isosbestic point around 400 nm. The change of SrSRI absorbance was plotted against the time. The bleaching rate of the peak at 429 nm (0.00043 min^{-1}) was similar to that of the peak at 556 nm (0.00057 min^{-1}), indicating that the G-state had two peaks at 429 and 556 nm. The absorption at 429 nm can be explained by allowing for P←G and/or K←G transitions (Figure 20.8). That is, the G-state can be transformed into the P-intermediate through the K-, L- and M-intermediates in a double-photon process. It is also possible that the G-state can be directly transformed into the P intermediate in a single-photon process. These two pathways can explain why the P-intermediate can be trapped stationary under blue light illumination.

20.5 Conclusion

In conclusion, we successfully observed the M-intermediate from *p*pR and the *p*pR/*p*HtrII complex embedded in a model membrane using *in situ* photoirradiation solid-state MAS NMR spectroscopy. Our results show that multiple M-intermediates coexist at −20 °C due to different retinal–protein interactions in both *p*pR and the *p*pR/*p*HtrII complex. Since only the M-intermediate can be trapped, it is now possible to observe the structural changes of the protein in the active state by *in situ* photoirradiation solid-state NMR experiments.

The photocycle of SrSRI was examined using *in situ* photoirradiation solid-state NMR spectroscopy. The M-intermediate was trapped stationary by illumination with 520 or 595 nm LED light at −40 °C to show positive phototaxis. Subsequently, the M-intermediate was transformed into the P-intermediate, and the P-intermediate was trapped stationary by irradiation with 365 nm LED light to show negative phototaxis. The configuration of retinal in the P-intermediate was determined to be the 13-*cis*,15-*anti* form, retaining the configuration of the intermediate with 13-*cis*. Proton transfer is likely induced by formation of the P-intermediate. The G-state was also transformed into the P-intermediate upon illumination with 365 nm light. This process indicates that bacteria can discriminate colors and thus increase the population of the P-intermediate, through a direct single-photon process in addition to a double-photon process from the G-state to the P-intermediate *via* the M-intermediate, in order to escape harmful UV light.

Acknowledgements

This work was supported by a Grant-in-Aid for Scientific Research on Priority Area (21017002), a Grant-in-Aid for Young Scientists (B) (22770101) from the Ministry of Culture, Sports, Science and Technology of Japan; and SUNBOR GRANT from the Suntory Institute for Bioorganic Research and Naito Foundation. The authors thank Mr Y. Makino, Mr Y. Tomonaga, Mr T. Hidaka and Mr H. Yomoda for help with NMR measurements.

References

1. S. O. Smith, H. J. M. de Groot, R. Gehard, J. M. L. Courtin, J. Lugtenburg, J. Herzfeld and R. G. Griffin, *Biochemistry*, 1989, **28**, 8897.
2. A. E. McDermott, L. K. Thompson, C. Winkel, M. R. Farrar, S. Pelletier, J. Lugtenburg, J. Herzfeld and R. G. Griffin, *Biochemistry*, 1991, **30**, 8366.
3. M. R. Farrar, K. V. Lakshmi, S. O. Smith, R. S. Brown, J. Raap, J. Lugtenburg, R. G. Griffin and J. Herzfeld, *Biophys. J.*, 1993, **65**, 310.
4. K. V. Lakshmi, M. R. Farrar, J. Raap, J. Lugtenburg, R. G. Griffin and J. Herzfeld, *Biochemistry*, 1994, **33**, 8853.
5. J. G. Hu, B. Q. Sun, M. Bizounok, M. F. Matcher, J. G. Lansing, J. Raap, P. J. E. Yerdegem, J. Lugtenburg, R. G. Griffin and J. Herzfeld, *Biochemistry*, 1998, **37**, 8088.
6. A. T. Petkova, M. Hatanaka, C. P. Jaroniec, I. G. Hu, M. Belenky, M. Verhoeven, J. Lugtenburg, R. G. Griffin and J. Herzfeld, *Biochemistry*, 2002, **41**, 2429.
7. X. Feng, P. T. E. Verdegeor, M. Eden, D. Sandstrom, Y. K. Lee, P. H. M. BovoeGrouts, W. J. De Geip, J. Lugtenburg, H. J. M. de Groot and M. H. Levitt, *J. Biomol. NMR*, 2000, **16**, 1.
8. E. Crocker, M. Eilers, S. Ahuja, V. Hornak, A. Hirshfeld, M. Sheves and S. O. Smith, *J. Mol. Biol.*, 2006, **357**, 163.
9. J. G. Hu, B. Q. Sun, A. T. Petkova, R. G. Griffin and J. Herzfeld, *Biochemistry*, 1997, **36**, 9316.
10. V. S. Bajaj, M. L. Mark-Jurkauskas, M. Belenky, J. Herzfeld and R. G. Griffin, *Proc. Natl. Acad. Sci. U. S. A.*, 2009, **106**, 9244.
11. M. Conciste, A. Gansmuller, N. McLean, O. G. Johannessen, I. M. Montesines, P. H. M. Bovee-Genrts, P. Verdegen, J. Lugtsunburg, R. C. D. Brown, W. J. DeGrip and M. H. Levitt, *J. Am. Chem. Soc.*, 2008, **130**, 10490.
12. Y. Tomonaga, T. Hidaka, I. Kawamura, T. Nishio, K. Osawa, T. Okitsu, A. Wada, Y. Sudo, N. Kamo, A. Ramamoorthy and A. Naito, *Biophys. J.*, 2011, **101**, L50.
13. J. L. Spudich and H. Luecke, *Curr. Opin. Struct. Biol.*, 2002, **12**, 540.

14. D. Suzuki, H. Irieda, M. Homma, I. Kawagishi and Y. Sudo, *Sensors*, 2010, **10**, 4010.
15. N. Kamo, K. Shimono, M. Iwamoto and Y. Sudo, *Biochemistry (Moscow)*, 2001, **66**, 1277.
16. Y. Sudo, T. Nishihori, M. Iwamoto, K. Shimono, C. Kojima and N. Kamo., *Biophys. J.*, 2008, **95**, 753.
17. Y. Imamoto, Y. Shichida, J. Hirayama, H. Tomioka, N. Kamo and T. Yoshizawa, *Biochemistry*, 1992, **31**, 2523.
18. V. I. Gordeliy, J. Labahn, R. Moukhametzinov, R. Etremov, T. Granzin, R. Sohlesinger, G. Buldt, T. Savopol, A. J. Scheidig, J. P. Klare and M. Engelhard, *Nature*, 2002, **419**, 484.
19. K. Shimono, T. Hayashi, Y. Ikeura, Y. Sudo and N. Kamo, *J. Biol. Chem.*, 2003, **278**, 23882.
20. Y. Sudo, Y. Furutani, H. Kandori and J. L. Spudich, *J. Biol. Chem.*, 2006, **281**, 34239.
21. Y. Sudo, Y. Furutani, A. Wada, M. Ito, N. Kamo and H. Kandori, *J. Am. Chem. Soc.*, 2005, **127**, 16036.
22. Y. Furutani, K. Kamada, Y. Sudo, K. Shimono, N. Kamo and H. Kandori, *Biochemistry*, 2005, **44**, 2909.
23. A. A. Wegener, I. Chizhov, M. Engelhard and H. J. Steinhoff, *J. Mol. Biol.*, 2000, **30**, 881.
24. J. L. Spudich, *Mol. Microbiol.*, 1998, **28**, 1051.
25. H. Yoshida, Y. Sudo, K. Shimono, M. Iwamoto and N. Kamo, *Photochem. Photobiol. Sci.*, 2004, **3**, 537.
26. R. Moukhametzianov, J. P. Klare, R. Efremov, C. Baeken, A. Goppner, J. Labahn, M. Engelhard, G. Buldt and V. I. Gordeliy, *Nature*, 2006, **440**, 115.
27. I. Kawamura, H. Yoshida, Y. Ikeda, S. Yamaguchi, S. Tuzi, H. Saito, N. Kamo and A. Naito, *Photochem. Photobiol.*, 2008, **84**, 921.
28. M. Etzkorn, K. Seidel, L. Li, S. Martin, M. Geyer, M. Engelhard and M. Baldus, *Structure*, 2010, **18**, 293.
29. I. Kawamura, N. Kihara, M. Ohmine, K. Nishimura, S. Tuzi, H. Saito and A. Naito, *J. Am. Chem. Soc.*, 2007, **129**, 1016.
30. J. L. Spudich and R. A. Bogomolni, *Nature*, 1984, **312**, 509.
31. W. D. Hoff, K.-H. Jung and J. L. Spudich, *Annu. Rev. Biophys. Biomol. Struct.*, 1997, **26**, 223.
32. D. Suzuki, H. Irieda, M. Homma, I. Kawagishi and Y. Sudo, *Sensors*, 2010, **10**, 4010.
33. X. Chen and J. L. Spudich, *Biochemistry*, 2002, **41**, 3891.
34. R. A. Bogomolni and J. L. Spudich, *Biophys. J.*, 1987, **52**, 1071.
35. T. E. Swartz, I. Szundi, J. L. Spudich and R. A. Bogomoli, *Biochemistry*, 2000, **38**, 15101.
36. I. Szundi, T. E. Swartz and R. A. Bogomolni, *Biophys. J.*, 2001, **80**, 469.
37. Y. Kitajima-Ihara, Y. Furutani, K. Suzuki, K. Ihara, H. Kandori, M. Homma and Y. Sudo, *J. Biol. Chem.*, 2008, **283**, 23533.

38. D. Suzuki, Y. Sudo, Y. Furutani, H. Takahashi, M. Homma and H. Kandori, *Biochemistry*, 2008, **47**, 12750.
39. H. Yomoda, Y. Makino, Y. Tomonaga, T. Hidaka, I. Kawamura, T. Okitsu, A. Wada, Y. Sudo and A. Naito, unpublished data.
40. E. Bennett, C. M. Rienstra, M. Auger, K. V. Lakshmi and R. G. Griffin., *J. Chem. Phys.*, 1995, **103**, 6951.
41. S. Ray, T. Kikukawa, P. Sharma and N. Kamo, *IEEE Trans. Nanobiosci*, 2006, **5**, 178.
42. Y. Tateishi, T. Abe, J. Tamogami, Y. Nakao, T. Kikukawa, N. Kamo and M. Unno, *Biochemistry*, 2011, **50**, 2136.
43. G. S. Harbison, S. O. Smith, J. A. Pardoen, P. P. J. Mulder, J. L. Lugtenburg, J. Herzfeld, R. Mathies and R. G. Griffin, *Biochemistry*, 1984, **23**, 2662.
44. O. A. Sineschchekov, J. Sasahi, B. J. Phillip and J. L. Spudich, *Proc. Natl. Acad. Sci. U. S. A.*, 2008, **105**, 16159.

CHAPTER 21

A Promising Prognosis for Solid-State NMR of Functional Membrane Protein Complexes

MICHAEL J. HARRIS[a] AND LYNMARIE K. THOMPSON*[a,b]

[a] Department of Chemistry and [b] Program in Molecular and Cellular Biology, University of Massachusetts, Amherst, MA 01003, USA
*Email: thompson@chem.umass.edu

21.1 Advances in Methodology

Solid-state NMR (SS NMR) spectroscopy of biological macromolecules has seen steady progress over the last decade and a half, demonstrating numerous "firsts" in the application of the technique to increasingly complex systems. The foundation for these achievements are methodological advances that span the range from NMR probe engineering and methods development to clever labeling strategies and novel sample types. These methods and achievements suggest to us an exciting convergence that will poise SS NMR spectroscopy to successfully tackle functional membrane protein complexes. These systems are critical to many cellular processes, notably transmembrane signaling in prokaryotes[1] and eukaryotes.[2] Structure/function analysis of membrane proteins lags behind their soluble counterparts, and studies of their complexes present even greater challenges. SS NMR has made important contributions to understanding individual membrane proteins, such as bacteriorhodopsin,[3] the influenza

New Developments in NMR No. 3
Advances in Biological Solid-State NMR: Proteins and Membrane-Active Peptides
Edited by Frances Separovic and Akira Naito

Published by the Royal Society of Chemistry, www.rsc.org

virus M2 proton channel,[4,5] and the KcsA potassium channel,[6–9] and is starting to provide insights into large membrane protein complexes.

21.1.1 Structure Determination of Proteins by Solid-State NMR Spectroscopy

Structure determination by SS NMR has advanced rapidly from small model systems to proteins in the 20–35 kDa range. Progress in achieving isotropic chemical shift assignments[10] in the solid state with magic angle spinning (MAS), the primary focus of this chapter, SS NMR was followed by structures of microcrystalline samples of small $^{13}C/^{15}N$-labeled soluble proteins, *e.g.* the SH3 domain of α-spectrin (~7.2 kDa)[11] and ubiquitin (~8.5 kDa)[12] shortly thereafter. For α-spectrin SH3, metabolic ^{13}C-labeling schemes that minimize ^{13}C pairs, along with uniform ^{15}N labeling of protein samples, made it possible to determine multiple homonuclear distances (~300 restraints). In the case of ubiquitin, 122 dihedral angle restraints obtained by chemical shift analysis were included with 336 ^{13}C–^{13}C distance restraints. In both studies, labeled proteins were diluted into an excess of natural abundance protein to identify intermolecular contacts in the 2D correlation spectra. For BPTI,[10] about 0.2–0.4 μmol of labeled protein was used (800 MHz spectrometer), while for SH3 domain and ubiquitin the experiments were conducted with ~0.9 μmol of material. A *de novo* structure determination of a chemotactic tripeptide[13] was also obtained during this time period, utilizing dihedral angle and internuclear distance measurements. Dihedral angles were determined using multiple 3D dipolar-chemical shift experiments, while $^{15}N/^{13}C$ distances were obtained by frequency-selective rotational echo double resonance (REDOR). The 6 kDa β-1 immunoglobulin binding domain of protein G (GB1) has served as an excellent test system for structural methods development, being the first to be solved using proton-detected triple resonance 3D experiments under very fast MAS (VFMAS at 39 kHz) on uniform-$^{13}C,^{15}N,^{2}H$ samples back-exchanged in $^{1}H_2O$.[14] Rienstra and co-workers also used GB1 to demonstrate the value of angular restraints for improving the precision of the structure calculation from a backbone RMSD of 1.01 Å (distance restraints only) to 0.31 Å (with angular restraints from both chemical shifts and dipole tensor-based measurements).[15]

Incorporation of additional types of restraints has aided the expansion of SS NMR structure determination to larger proteins. Paramagnetic restraints were important for solving the structures of human oxidized Cu/Zn superoxide dismutase (SOD, ~16 kDa)[16] and matrix metalloprotease 12 (MMP12, 17.6 kDa).[17] The MMP12 structure incorporated pseudocontact shifts, measured as the difference in chemical shift between the diamagnetic Zn form and the paramagnetic Co form of the protein. The structure of Cu/Zn SOD was solved using proton-detected triple resonance 3D experiments under ultrafast MAS at 60 kHz on $^{13}C,^{15}N$ perdeuterated protein (as above) to obtain proton–proton distance restraints. Paramagnetic relaxation enhancements obtained by comparing relaxation measurements on diamagnetic (reduced) and paramagnetic (oxidized) states of the protein improved the calculated structure from a backbone RMSD of 2.90 Å to 1.66 Å. As will be discussed below, the introduction of paramagnetic tags allows paramagnetic restraints to be incorporated into the structural analysis of any protein.[18]

Recently, "rotationally aligned" SS NMR, which combines aspects of MAS and oriented sample (OS) experiments, was used to solve the structure of a ~6 kDa mercury transporter[19] and a ~34 kDa G-coupled protein receptor (GPCR), CXCR1, in lipid bilayers.[20] This method relies upon the rotational diffusion of proteins about the bilayer normal, where chemical shift assignments are first made using multidimensional MAS SS NMR experiments, and then the frequency edges of the motionally averaged powder pattern are measured. In this approach, which has been validated against results obtained from mechanically and magnetically aligned oriented samples,[21] both dihedral angle constraints (from the isotropic chemical shifts) and orientation constraints (heteronuclear dipolar coupling and anisotropic chemical shift values from the powder pattern edges) are used for structure calculations. This novel method appears powerful for determining structures of single membrane proteins such as a GPCR containing seven transmembrane helices, but very large multiprotein complexes are unlikely to satisfy the required rapid ($>10^5$ Hz) axial rotation in the membrane. These examples illustrate the development of methodologies for measurement of additional restraints to aid in structure determination of increasingly large proteins by SS NMR.

21.1.2 Expanding the Toolkit for Large Functional Biomolecules

To enable SS NMR to successfully tackle larger and more complex systems, recent developments that preserve biochemical functionality and enhance spectroscopic accessibility of these systems are critical.

A key advance for preserving native states of proteins and protein complexes is the development of NMR probes that minimize sample heating by radiofrequency pulses, making SS NMR compatible with samples at physiological ionic strengths. The cell is a complex environment, with total ion concentrations in excess of 0.2 M,[22] and some proteins or protein assemblies require these high ionic strength conditions to retain native structure. In an NMR sample, these ions absorb the electric field component of the pulses, resulting in significant sample heating: for example, a sample at 160 mM ionic strength can heat by 55 °C.[23] The development of scroll coil[23] and "low-E"[24] NMR probes have been of tremendous value for reducing sample heating, improving both spectral properties and sample preservation. A scroll coil probe was used in the proton-detected VFMAS SS NMR structure determination of GB1,[14] as this coil design is optimized for proton channel performance. A commercial "low-E" probe was employed to study a membrane protein sample in lipid vesicles at >0.1 M ionic strength.[25] Such probes have also made it possible to extend SS NMR experiments to unfrozen whole cell preparations.[26]

Considering the range of NMR and sample parameters that influence heating, it is prudent to use a "sacrificial sample" containing the intended buffer components and unlabeled protein to assess the extent of sample heating as a function of buffer composition (in particular ionic strength) and instrumental parameters such as decoupling strength, MAS frequency, cooling gas flow rate, and recycle delay.[27] Careful probe detuning measurements will reveal whether the NMR experiment causes sample thawing. An internal chemical shift thermometer such as TmDOTP or samarium stannate ($Sm_2Sn_2O_7$) provides the most rigorous check on sample heating;[27] however, an external KBr chemical shift thermometer can also provide a more convenient check of the sample temperature.[28] These strategies and new NMR probe designs enable SS NMR measurements of proteins under conditions that preserve their native structure and function.

For studies of membrane proteins, there are now a number of sample conditions that mimic native lipid bilayers. One option is incorporation of the membrane protein into nanodiscs (Figure 21.1A), which are lipid bilayers of about 10 nm diameter encircled by two molecules of membrane scaffold protein (MSP1) that stabilize the hydrophobic edge of the disc.[29] A study of the membrane-associated metalloenzyme CYP3A4 demonstrated that reconstitution of the protein into nanodiscs maintained the expected biochemical activity.[30] Characterization of the MSP1 component of nanodiscs with SS NMR yielded chemical shifts most consistent with each protein adopting one long helical "belt" around the nanodisc.[31] A larger-scale "macrodisc" suitable for OS SS NMR studies was developed[32] based on an amphipathic helix derived from plasma lipoproteins.[33] These peptides form small diameter particles ($\sim$10 nm) comparable in size to nanodiscs at low lipid:peptide ratios, but can form larger diameter particles ($\sim$30 nm) at higher lipid:peptide ratios. This larger assembly was shown to be magnetically alignable akin to bicelles.[32] Bicelles ('bi'layer mi'celles') are another bilayer medium for membrane proteins, with applications for both solution and solid-state NMR.[34] Bicelles (Figure 21.1B) are mixtures of long-chain lipids with short-chain lipids or detergents that form bilayer-like structures such as discs or perforated sheets, where the hydrophobic edges are "shielded" from the aqueous phase by the short-chain lipids/detergents. These lipid mixtures demonstrate a wide variety of morphologies as a function of composition and temperature, some of which align in a magnetic field. This makes them viable sample environments for NMR studies of membrane proteins, particularly for oriented sample experiments.

Bicelle and nanodisc preparations provide two alternatives to incorporation of membrane proteins into vesicles (Figure 21.1C). Another option, if the target of interest is an oligomer or protein complex involving interactions with the external domain of a membrane protein, is to assemble the

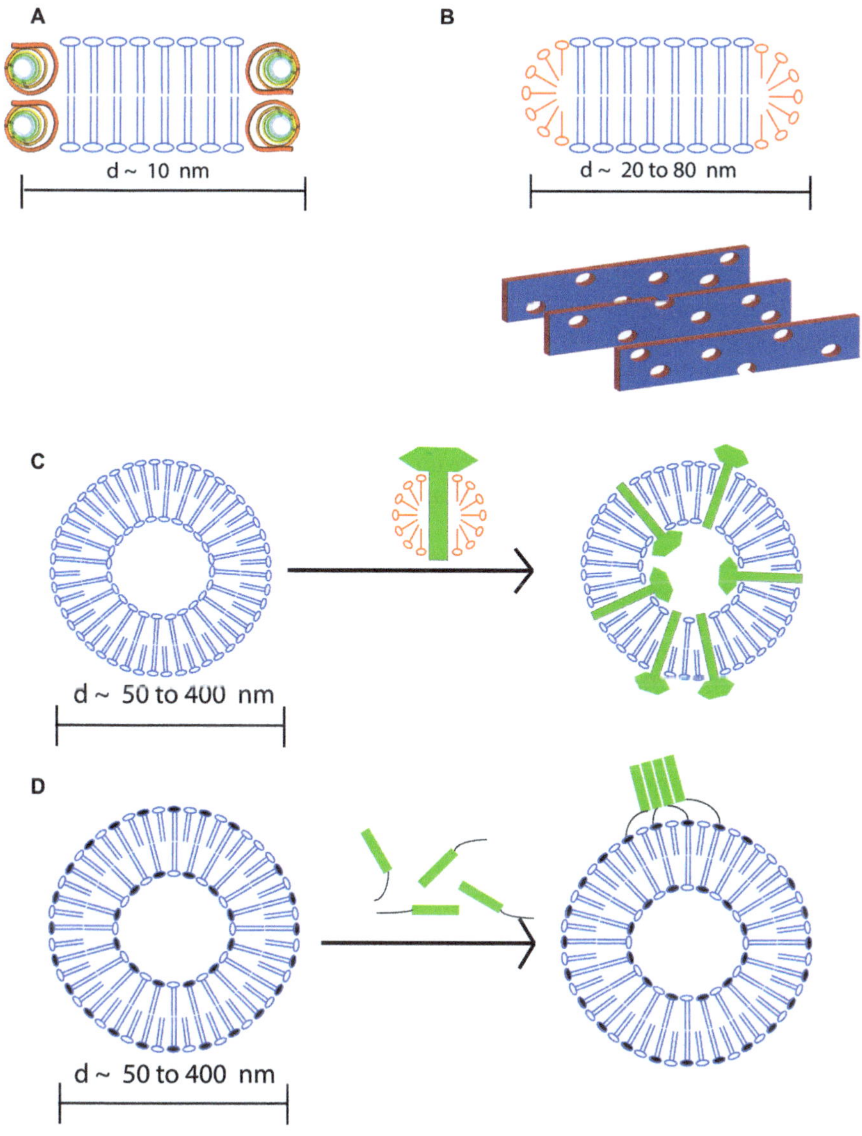

Figure 21.1 Sample environments for membrane proteins and membrane protein complexes for SS NMR. (A) Nanodiscs are monodisperse ~10 nm diameter disks of a bilayer of lipids (*blue*) encompassed by a helical protein belt of MSP1 (*rainbow*). (B) Bicelles are mixtures of long-chain lipids (*blue*) and detergents (*red*) forming a bilayer structure with detergent rims (*top*), but are known to adopt a variety of morphologies (such as perforated lamellae, *bottom*) depending on a number of parameters. (C) Unilamellar vesicles can be prepared between ~50 to 400 nm in diameter. When combined with detergent solubilized membrane proteins, the proteins insert in both orientations to form proteoliposomes. (D) Templating vesicles include a nickel-chelating headgroup (*black*) that binds a His-tagged membrane protein fragment to facilitate assembly of oligomers and complexes on the vesicle surface.

complex on a vesicle surface (Figure 21.1D). This "template-directed assembly" method was originally developed by Weis and co-workers for biochemical and biophysical studies of bacterial chemotaxis protein complexes,[35] and has also been used for studies of eukaryotic receptor fragments.[36,37] Templating vesicles are prepared from a lipid mixture that includes a lipid with a metal-chelating headgroup, which is used to bind nickel and a soluble histidine-tagged fragment of the membrane protein of interest. Binding of this fragment to the vesicle enforces the native geometry at the membrane surface and enhances its oligomerization, which can facilitate binding of additional proteins of the complex. Application of this methodology to prepare functional signaling complexes of bacterial chemotaxis proteins, which occur in the cell membrane in ~200 nm arrays,[38] overcomes some limitations of nano/macrodiscs (much smaller sizes) and vesicles (random protein orientation). These different sample types (Figure 21.1), which each provide an excellent mimic of lipid bilayers with different advantages, will facilitate SS NMR studies of active membrane proteins and their complexes.

Membrane proteins are also being studied in native membrane bilayers and whole cells. SS NMR studies of proteins in native membrane vesicles has expanded beyond studies of naturally abundant membrane proteins such as bacteriorhodopsin, to overexpressed membrane proteins and peptides, such as site-directed distance measurements in bacterial chemotaxis receptors[39,40] and secondary structure characterization of a transmembrane peptide of LR11.[41] Recent contributions to the area of "*in vivo*" protein NMR include a comparative study of whole cells and cell envelopes[26] that demonstrated the ability to identify and characterize a recombinantly expressed membrane protein as well as endogenous membrane-associated macromolecules using both $^{15}N/^{13}C$ and $^{13}C/^{13}C$ correlation spectroscopy. With dynamic nuclear polarization (DNP) of such samples,[42] significant signal enhancements on membrane proteins, membrane-associated components, and cytosolic molecules were detected in 1D and 2D spectra. The average ^{13}C linewidth in the 2D spectra was ~0.8–0.9 ppm, even at 100 K.

In addition to the improvements in NMR probes and sample types that enable expansion of SS NMR into increasingly physiological states of proteins, the NMR toolkit now includes several labeling approaches that increase the spectroscopic accessibility of very large complexes. The importance of spin dilution – usually attained *via* sparse labeling patterns available from selectively labeled carbon sources such as [1,3-^{13}C]glycerol – becomes ever more critical in tackling large proteins and complexes to limit the number of resonances in the spectra. First applied to solution NMR relaxation measurements of thioredoxin,[43] a known distribution of isotope incorporation in the 20 standard amino acids results. This method was first extended to structure determination in the solid state for α-spectrin SH3[11] and ubiquitin,[12] and has become something of a standard technique in the SS NMR spectroscopist's toolbox. The selectively labeled precursors are commercially available and reduce the number of directly bonded ^{13}C pairs, with their strong dipolar couplings that result in dipolar truncation, enabling measurement of long-range dipolar couplings.

Differential labeling, first demonstrated in SS NMR studies of thioredoxin,[44] provides a powerful strategy for probing biological assemblies, where components of the assembly/complex are prepared separately using different isotope labeling schemes and then combined to form the protein complex. A combination of these methods was applied to obtain a highly-resolved model of the type 3 secretion system needle from *Salmonella typhimurium*.[45] Spin dilution was used to obtain high-quality spectra that were fully assigned, and differential subunit labeling was used to determine which restraints were inter- *vs.* intra-subunit distances in this helical filament assembly of 80-residue subunits.

The applicability of paramagnetic SS NMR for long-distance measurements within proteins and across subunit interfaces is another promising development for the study of functional (membrane) protein complexes. In addition to the study of naturally occurring paramagnetic proteins highlighted above, the introduction of a paramagnetic center *via* either organic free radicals or metal-binding tags permits structural information to be obtained at a length scale of 1–2 nm.[46] This method has recently shown promise for characterizing inter-protein interfaces of sensory rhodopsin in proteoliposomes,[47] which were found to differ from the interfaces observed in crystals. With an assigned 2D spectrum and differential subunit labelling (spin-labeled subunit plus isotopically labeled subunit), this method should provide a powerful general approach to obtain multiple long-distance constraints to dock two proteins into a complex.

21.1.3 Sensitivity Enhancement: A Critical Advance for More Complex Systems

Because there is no need for rapid molecular tumbling, SS NMR can be applied to systems of arbitrarily large size (*e.g.* many copies of an 80-residue protein forming micron-long needles), but applications to systems of greater complexity (*e.g.* a complex of several different large proteins) remain limited by sensitivity. Spectroscopic feasibility of such systems, in which the concentration of the protein or interface of interest may be quite low, is increased by recent developments that significantly improve sensitivity for SS NMR experiments, as summarized in Table 21.1. These include methods for improving the time efficiency of the NMR experiment, by increasing the information content of each spectrum using pulse sequences for non-uniform sampling or dual acquisition from multiple nuclei, and by decreasing the time between spectra using fast MAS with paramagnetic relaxation enhancement. Furthermore, strengthening the NMR signal itself is being achieved *via* detection of protons or DNP enhancement of resonances of various nuclei.

Paramagnetic SS NMR can not only provide long-distance restraints, as described above, but also enhance sensitivity per unit time of the experiment. The reduction in ^1H T_1 times due to the presence of a paramagnetic

Table 21.1 Methods for sensitivity enhancement.

Method	Sample	*Results* Reduction in experiment time	Ref.
Paramagnetic relaxation	GB1	~4- to 8-fold[a]	48
	DsbB	5-fold	49
Dual acquisition MAS (DUMAS)	Ubiquitin; PLN	~2-fold	50
Non-uniform sampling (NUS) & paramagnetic relaxation	Dynein LC8	16-fold	51
		S/N increase	
NUS	f-MLF, thioredoxin	1.5- to 2-fold[b]	52
^{1}H detection	GB1	18-fold[c]	14
Dynamic nuclear polarization	Mistic	20- to 30-fold	53
	bR	90-fold	3

[a]Also smaller sample (1/6) because fast MAS is performed in smaller rotors.
[b]Per indirect dimension.
[c]For a ^{1}H-detected ^{1}H–^{15}N spectrum relative to a ^{15}N-detected ^{1}H–^{15}N spectrum.

center (CuEDTA in solution or bound to the protein) allows for more frequent acquisitions.[48,54] A high acquisition rate with high-power pulses might damage the probe or sample, so very fast MAS (~40 kHz) is used to attenuate proton dipolar couplings so that low-power decoupling (*e.g.* 7 kHz)[54] is sufficient. Although higher concentrations are needed for the larger amyloid fibrils than for ubiquitin, CuEDTA in solution can reduce the experiment time by at least 7-fold.[54] The combination of paramagnetic tagging and fast MAS was recently applied to a ~20 kDa membrane protein, DsbB,[49] where a five-fold reduction was observed in the amount of time to collect a 2D ^{13}C–^{13}C correlation spectrum that shows the same 2D resonances as the spectrum of the non-paramagnetic sample. Interestingly, the paramagnetic (Cu^{2+}) doping was accomplished in two separate ways: (i) by attaching a chelator tag to the protein and (ii) by including a chelating lipid that exchanged with endogenous lipid bound to DsbB. Enhancement was comparable for both cases and yielded the same resonances as seen in studies of diamagnetic DsbB.

The introduction of higher MAS frequencies, ranging from 20 to ~70 kHz, also enhances sensitivity by permitting proton detection for organic and biological solids with good resolution, as noted earlier for both GB1[14] and Cu/Zn SOD.[16] At these increased MAS frequencies, both ^{1}H–^{1}H and ^{1}H–^{13}C dipolar couplings are significantly attenuated so as to permit reduced ^{1}H decoupling or even its omission.[54,55] Samples must be fully deuterated, and are then back-exchanged in H_2O to various protonation levels at the exchangeable sites. With ultrafast MAS (60 kHz) the linewidths are equivalent for 10% and 100% protonation at the exchangeable sites, so 100% re-protonated samples can be used for maximum sensitivity.[56]

Recently developed pulse sequences also permit more time-efficient acquisition of NMR spectra. Gopinath and Veglia introduced "dual acquisition" MAS (DUMAS) NMR,[50] which permits the simultaneous acquisition of multiple MAS SS NMR spectra in less experiment time with no need for additional hardware. Here, simultaneous cross polarization (SIM-CP) from ^{1}H to both ^{13}C and ^{15}N is used, and the long-lived ^{15}N magnetization is stored along the *z*-axis before the first, ^{13}C-edited, experiment. The ^{15}N magnetization is then brought back into the transverse plane for the second, ^{15}N-filtered, experiment. DUMAS reduced the experiment time approximately two-fold for 2D spectra of microcrystalline [U-^{13}C/^{15}N]ubiquitin and [U-^{13}C/^{15}N]phospholamban in vesicles. Non-uniform sampling (NUS) is another approach that can enhance sensitivity by 1.5- to 2-fold in each indirect dimension of the SS NMR experiment, as demonstrated for the chemotactic peptide MLF and reassembled differentially labeled thioredoxin.[52]

Further sensitivity gains can be achieved by implementing these methods in parallel. For example, a combination of NUS and paramagnetic relaxation enhancements (with 40 kHz MAS) has recently been demonstrated on the dynein light chain LC8 to accelerate the acquisition of 3D correlation spectra by a factor of 16.[51] Combining these methods, none of which require any specialized equipment outside of commercially available fast MAS capabilities, should make it possible to expand SS NMR to the characterization of more complex protein samples.

DNP also offers tremendous opportunity for the enhancement of SS NMR signals, utilizing polarization transfer from electrons to nuclei. The theoretical potential signal enhancement (ε) is $\sim$660 (γ_e/γ_N), and significant signal enhancements for model compounds have been observed.[57] For MAS SS NMR, this typically requires a high-power microwave source and low-temperature capabilities, as well as polarization agents that serve as the electron spin reservoir. This technology has been commercialized [http://www.bruker.com/products/mr/nmr/dnp-nmr/overview.html], which bodes well for broad implementation to a wide range of systems.

Application of DNP SS NMR has yielded significant signal enhancements for membrane proteins[3,53] and other complex biological samples such as intact viruses,[58] bacterial cell walls,[59] and whole cells.[26] Relative to spectra acquired without microwave irradiation, signal enhancements spanning the range from 10 to nearly 100 have been observed in these types of studies, often permitting the investigation of challenging samples and/or functional states. For example, in a study of bacteriorhodopsin,[3] resolution and identification of previously indistinguishable coexisting states in the photocycle, each present at very low levels, was possible due to the signal enhancement provided by DNP. DNP also facilitated the study of a low-abundance membrane-associated protein ($\sim$16 nmol of Mistic) in an unpurified bacterial membrane extract.[53] In a study of bacterial cell walls,[59] it was found that the affinity of polarizing agent for the cell wall polymers allowed for a difference method that could selectively examine signals from either the cell wall or other cellular components.

In addition to the aforementioned technical requirements for DNP-enhanced MAS SS NMR, certain general sample requirements seem to be vital for successful implementation. The formation of a glassy sample seems to ensure homogenous dispersion of a polarization agent, permitting uniform polarization enhancement and minimizing sample line-broadening. Glass-forming solvent mixtures using glycerol or dimethyl sulfoxide have been established as suitable sample environments,[57] although formation of a sedimented protein film[60] or the high natural lipid content of bacterial membranes[53] also seems to fulfill this criterion. It was noted that usage of 15–40% glycerol also improved the stability and viability of whole cell samples for DNP.[26] The concentration of a polarizing agent is also a factor, as a balance between signal enhancement and line broadening must be found.[61] The introduction of biradicals such as TOTAPOL[62] were shown to be effective in the low millimolar range, although lower concentrations have worked well on complex systems, *e.g.* 20 μM TOTAPOL facilitated the $^{15}N/^{13}C$ chemical shift assignments of the intact Pf1 virion,[58] with signal enhancements for both protein and DNA at ~17. Both transition metals[63] and endogenous radicals[64] are intriguing alternative polarization agents, especially given their presence in biological systems.

21.1.4 Complementary Approaches Enable Structural Studies

Application of complementary methods synergistically with SS NMR is another key approach enabling studies of larger proteins and protein complexes. For example, the structure of the 20 kDa alpha-B-crystallin and its oligomer was determined with MAS SS NMR and small-angle X-ray scattering (SAXS).[65] MAS SS NMR was used to determine a curved structure for the homodimer, as well as to identify intermolecular contacts within oligomers. The homodimer motif was used as a building block for analysis of the SAXS data on the oligomeric state at pH 7.5. A 24-dimer structure was found to best fit the available data and proposed models of the oligomer. Both NMR and SAXS were able to provide information on structural changes in the protein at pH 6.5: NMR revealed local changes in a substrate binding site *via* chemical shift perturbations and SAXS revealed global changes *via* a change in the radius of gyration.

The SS NMR study of the Type 3 secretion system needle, discussed above as an example of the utility of differential and metabolic labeling, also benefitted from electron microscopy and computational methods.[45] Structural parameters from previous electron microscopy studies were utilized along with SS NMR restraints in the Rosetta modeling of the overall needle architecture. The resulting model placed the N-terminal domain of the needle protomer on the exterior surface of the needle, in contrast to its interior location in the structure of a homologous needle. This novel structural feature of the SS NMR-derived model was confirmed by immunoelectron microscopy.

In addition to cases like these in which complementary methods are important in developing and testing SS NMR-derived structural models, SS NMR restraints can also improve available protein structures. This is likely to be especially important for membrane proteins since their crystal structures are often at lower resolution than structures of soluble proteins. A recent study[66] demonstrated the utility of SS NMR restraints in refining the 3.4 Å X-ray structure of a 41-kDa membrane protein complex, DsbB/DsbA. The inclusion of ~400 dihedral angle restraints and ~800 ^{13}C–^{13}C distance restraints with the X-ray data improved the calculated overall backbone RMSD from 1.70 Å to 1.03 Å. In particular, the transmembrane helices showed a reduction in backbone RMSD from 2.20 Å to 0.92 Å. Interestingly, protein geometry validation on the transmembrane helices of DsbB after joint refinement showed an additional 22% of residues now in the most favored regions of Ramachandran space *versus* the baseline X-ray structure.

21.2 Functional Membrane Protein Complexes

The developments described above have laid a strong foundation for the application of SS NMR to large membrane protein complexes to gain insight into mechanisms of the many key biological processes that they mediate. For many systems of interest there is some structure/function information available from crystallography and biochemical approaches. The goal is to use SS NMR to address the remaining questions, by obtaining high-resolution structural information on components that are unresolved by other methods and on functional complexes in different states to define the molecular mechanism of an important process.

Crystal structures of three such membrane protein complexes are depicted in Figure 21.2. We briefly summarize the evolution of studies of these systems to illustrate how SS NMR methods are being used to advance beyond these static structures to gain insight into the structural changes involved in the mechanism. Traditional approaches involving complete assignment and structure determination are complemented for larger systems by approaches that yield selected distance restraints without complete assignments.

21.2.1 Disulfide Bond Generating Complex, DsbA/DsbB

The DsbA/DsbB complex (Figure 21.2A) catalyzes disulfide bond formation in periplasmic proteins in *E. coli*. This complex between the 21 kDa periplasmic protein DsbA (green) and the 20 kDa transmembrane protein DsbB (cyan) is responsible for re-oxidation of the active site disulfide in DsbA. Rienstra and co-workers have employed SS NMR in a systematic progression, beginning with resonance assignments of DsbA[67] and DsbB[68,69] individually. Furthermore, assignment of resonances of ubiquinone bound to DsbB and observation of its chemical shift changes in mutants of DsbB and the DsbA/B complex provide insight into its redox chemistry.[70] For DsbB this work has culminated in a demonstration that SS NMR restraints (distance restraints

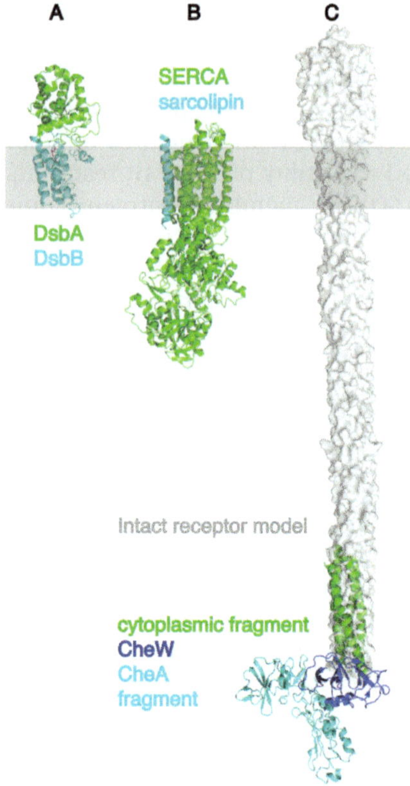

Figure 21.2 Three membrane protein complexes under study by solid-state NMR. Crystal structures for (A) the DsbA/DsbB complex (3e9j), (B) the sarcolipin/SERCA complex (4h1w), and (C) the chemoreceptor fragment/CheA/CheW (3ur1) overlaid on a model of the intact receptor.

and dihedral angles based on chemical shifts) can be combined with X-ray structure restraints to improve on the 3.4 Å crystal structure and reveal new details about the protein, such as a membrane-imbedded hydrogen bond that was confirmed by REDOR.[71]

This work has set the stage for the recent mapping of site-specific chemical shift changes in DsbA upon formation of the complex with DsbB.[72] The observed chemical shift changes are not fully accounted for by the differences in structure and solvent exposure between crystals of DsbA and DsbA/DsbB. The additional changes may indicate that the structure being measured by SS NMR in native lipids differs from the structure measured by crystallography in detergent micelles. These NMR approaches have mapped changes that occur in DsbA in two intermediates of the disulfide bond transfer pathway and will likely yield a structure of the DsbA/DsbB complex in native lipids, to provide new insights into the mechanism of this membrane protein complex.

21.2.2 Complexes of the Calcium Pump SERCA

The ~20 kDa components of the DsbA/DsbB complex are of a similar size to the largest proteins for which complete structure determination with SS NMR has been possible. For systems that contain protein subunits of larger size (Figure 21.2B and 21.2C), targeted studies on sites of interest, rather than complete structure determination, offer a means to obtain mechanistically important information.

Solid-state NMR studies of the complexes between the very large sarcoplasmic reticulum calcium ATPase (110 kDa SERCA, green in Figure 21.2B) and its inhibitors, the small membrane proteins sarcolipin (31 residues, cyan in complex with SERCA in Figure 21.2B) or phospholamban (52 residues, not shown), are now emerging. Veglia and co-workers have done detailed work on this system using multiple methods. Solution NMR in DPC micelles and oriented sample SS NMR in lipid vesicles were used to determine structures of phospholamban monomers[73] and pentamers.[74] The conformational equilibria of phospholamban that is thought to be important in the control of SERCA has been characterized in DPC micelles by solution NMR[75] and in lipid bilayers by MAS NMR.[76] Sarcolipin also exhibits multiple conformations in lipid bilayers, as characterized in oriented samples by SS NMR.[77]

Studies of SERCA complexes with these small proteins will benefit from the detailed structural information available for SERCA (crystal structures in multiple states of its pumping cycle) and for phospholamban and sarcolipin (NMR measurements of the conformational equilibria). Since SERCA retains activity in DPC micelles, its interactions with phospholamban[78] and with sarcolipin[79] were first probed in DPC micelles, using solution NMR measurements of chemical shift perturbation of phospholamban and sarcolipin resonances. This work has set the stage for current MAS SS NMR studies of the SERCA/phospholamban complex, using differential isotopic labeling of the two proteins and spin labeling for paramagnetic relaxation enhancement measurements.[80] These approaches will likely yield a structural model for the complex and insights into how the phosphorylation-controlled conformational equilibrium of phospholamban exerts its influence on SERCA function, which will inform the design of therapeutics for cardiac disease.

The complexes described thus far illustrate two approaches: (1) studies of the DsbA/DsbB complex and it subunits that employ primarily MAS SS NMR, as its components are within reach for complete structure determination by SS NMR; and (2) multiple methods that are combined for studies of phospholamban, sarcolipin, and their complexes with SERCA, to yield complete structures of the small components (phospholamban and sarcolipin) and structural restraints to define how these proteins form complexes with SERCA.

21.2.3 Signaling Complexes of Chemotaxis Receptors

A third type of system, represented by bacterial chemotaxis receptor signaling complexes (Figure 21.2C), involves subunits that are all too large for complete structure determination by SS NMR (with the exception of CheW, for which crystal and NMR structures are already available). The approach we have used for this system has been to collect structural restraints, initially on the 60 kDa chemoreceptor in native lipid vesicles, and then on active complexes of a 30 kDa cytoplasmic fragment of the receptor (a smaller fragment is shown in green in Figure 21.2C), the 18 kDa CheW (blue), and the 70 kDa kinase CheA (a smaller fragment is shown in cyan). These protein complexes function in extended (~200 nm) arrays in the cell and thus are not functional in detergent micelles. We developed a site-directed distance measurement approach and applied it to the intact chemoreceptor to begin to map the ligand-induced conformational change that transmits the initial signal through the periplasmic[39] and transmembrane[40] regions of the receptor. Complementary biochemical and mass spectrometry studies have investigated the role of changes in clustering[81] and dynamics (unpublished data) in the signal. Solid-state NMR experiments on reconstituted functional ternary complexes have become possible because assembly of active complexes of a smaller 30 kDa cytoplasmic fragment can be mediated by vesicle binding[35] or by poly(ethyleneglycol).[82]

Chemoreceptor signaling complexes formed with receptor cytoplasmic fragments are being characterized by X-ray crystallography, solution NMR, and SS NMR. A crystal structure of the complex (Figure 21.2C) reveals interactions between the tip of the dimeric receptor and a ring of CheA and CheW proteins, but lacks information on contacts between receptor dimers that mediate formation of the extended array.[83] Our site-directed $^{13}C\{^{19}F\}$ REDOR distance measurement near the receptor tip provides a dimer–dimer distance restraint of ^{13}C to $^{19}F \geq 7.9$ Å in the kinase-on state of functional receptor complexes.[82] This measurement ruled out the receptor packing first proposed based on an early crystal structure of receptor cytoplasmic fragments alone (1qu7, Figure 21.3B). In the kinase-activating state of the complex, the receptor tips must be farther apart than in the crystals to move the fluorine (cyan) outside of the dotted sphere indicating the distance restraint. More recently, modeling of the crystal structure of the ternary complex (Figure 21.2C) into cryoelectron tomograms[83] has yielded a model for the array, which is shown in Figure 21.3A, as a lateral view along the plane of the membrane (left) and also looking up from below toward the membrane (right and expansion). Although this is a low-resolution (~20 Å) model, it is satisfying that the 8.1 Å distance in the model is consistent with the REDOR constraint (Figure 21.3A expanded view). Further SS NMR measurements are needed to provide high-resolution information on the receptor–receptor packing in different functional states of the signaling array to map structural changes that may propagate the signal.

Solution NMR measurements by Dahlquist and co-workers on similar complexes have measured chemical shift perturbations of CheW, CheA, and receptor resonances to identify the interaction sites of these proteins.[85,86] Solid-state NMR experiments are in progress to obtain further information about each of the protein–protein interfaces in the active complexes, and how they change during signaling. Because SS NMR can provide selected

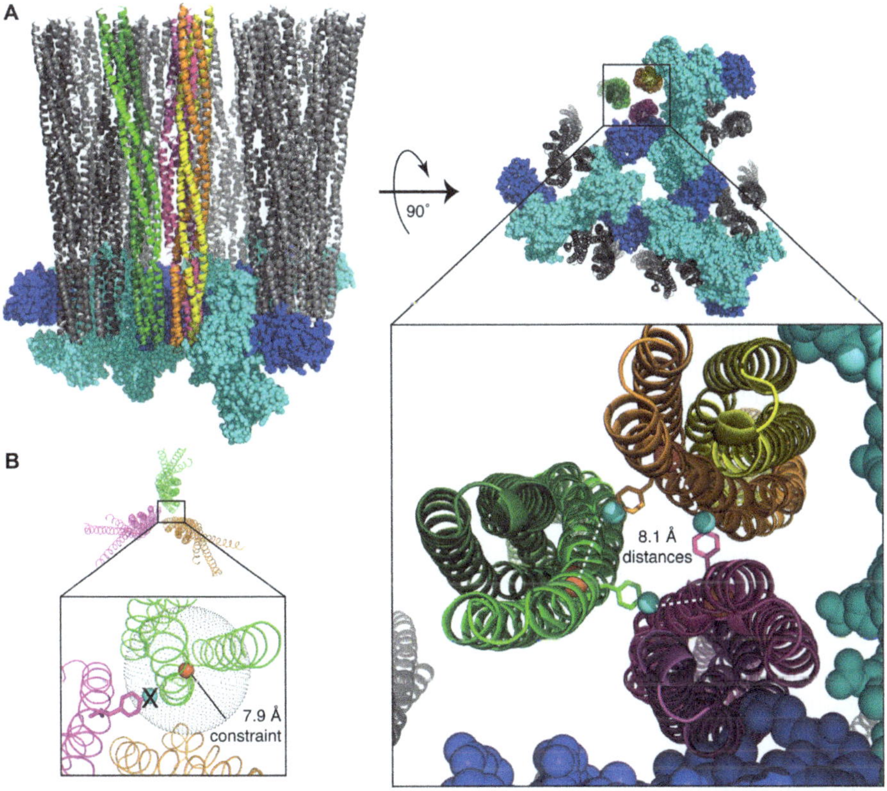

Figure 21.3 Models for receptor–receptor packing. (A) Model for the chemoreceptor array (coordinates kindly provided by Brian Crane), based on docking a crystal structure of complexes (3ur1, Figure 21.2C) into electron cryotomograms of arrays.[83] The array is viewed both laterally (*left*) as in Figure 21.2C and looking up from below the array towards the membrane (*right* and expansion). The model shown is one ring of the hexagonal array, containing six trimers of receptor dimers (*gray*, with one trimer of dimers shown in color), with CheW (*blue*) and CheA (*cyan*) bound in a plane at the membrane-distal tips of the receptors. (B) An alternate model for receptor–receptor packing in the trimer of dimers based on a receptor fragment crystal structure (1qu7).[84] This view from the membrane down toward the receptor tips best shows the convergence of the tips in this model, which places the fluorine (*cyan*) in violation of the REDOR distance constraint (*dotted sphere*).[82]

high-resolution information on the functional complexes in large native-like arrays, it is a promising approach for investigating the signaling mechanism in the chemoreceptor system.

21.3 Outlook

These three examples illustrate several SS NMR approaches for obtaining structural information and mechanistic insights on a large range of types of membrane protein complexes. The smaller proteins, phospholamban, sarcolipin, and even the 20 kDa range DsbA and DsbB, are suitable for complete structure determination and detailed characterization of structural equilibria relevant to the mechanism by SS NMR. Larger systems are now also being tackled: these cases typically involve integrating information from multiple techniques, and taking advantage of structural information on components of the complex. Selected paramagnetic or REDOR restraints across protein–protein interfaces are being used in the SERCA and chemoreceptor systems to gain insight into structures of functional complexes that will ultimately show the role of structural changes in the mechanisms of transmembrane pumping and signaling processes. The many recent advances in SS NMR methodology, particularly those that increase sensitivity, have opened the door for solid-state NMR studies to contribute insight into the structures and mechanisms of very large membrane protein complexes.

Acknowledgments

This research was supported by NIH Grants GM47601 and GM085288. The authors thank all of our current and former group members for their contributions to this research, Robert Weis and co-workers for helpful discussions and collaborations on chemoreceptor studies, and Aruni Karunanayake for comments on the manuscript.

References

1. D. Bray, M. D. Levin and C. J. Morton-Firth, *Nature*, 1998, **393**, 85.
2. J. R. Cochran, D. Aivazian, T. O. Cameron and L. J. Stern, *Trends Biochem. Sci.*, 2001, **26**, 304.
3. V. S. Bajaj, M. L. Mak-Jurkauskas, M. Belenky, J. Herzfeld and R. G. Griffin, *Proc. Natl. Acad. Sci. U. S. A.*, 2009, **106**, 9244.
4. S. D. Cady, K. Schmidt-Rohr, J. Wang, C. S. Soto, W. F. DeGrado and M. Hong, *Nature*, 2010, **463**, 689.
5. J. Hu, T. Asbury, S. Achuthan, C. Li, R. Bertram, J. R. Quine, R. Fu and T. A. Cross, *Biophys. J.*, 2007, **92**, 4335.
6. C. Ader, R. Schneider, S. Hornig, P. Velisetty, V. Vardanyan, K. Giller, I. Ohmert, S. Becker, O. Pongs and M. Baldus, *EMBO J*, 2009, **28**, 2825.
7. C. Ader, O. Pongs, S. Becker and M. Baldus, *Biochim. Biophys. Acta, Biomembr.*, 2010, **1798**, 286.
8. M. P. Bhate, B. J. Wylie, L. Tian and A. E. McDermott, *J. Mol. Biol.*, 2010, **401**, 155.
9. M. P. Bhate and A. E. McDermott, *Proc. Natl. Acad. Sci. U. S. A.*, 2012, **109**, 15265.
10. A. McDermott, T. Polenova, A. Bockmann, K. W. Zilm, E. K. Paulson, R. W. Martin, G. T. Montelione and E. K. Paulsen, *J. Biomol. NMR*, 2000, **16**, 209.

11. F. Castellani, B. van Rossum, A. Diehl, M. Schubert, K. Rehbein and H. Oschkinat, *Nature*, 2002, **420**, 98.
12. S. G. Zech, A. J. Wand and A. E. McDermott, *J. Am. Chem. Soc.*, 2005, **127**, 8618.
13. C. M. Rienstra, L. Tucker-Kellogg, C. P. Jaroniec, M. Hohwy, B. Reif, M. T. McMahon, B. Tidor, T. Lozano-Pérez and R. G. Griffin, *Proc. Natl. Acad. Sci. U. S. A.*, 2002, **99**, 10260.
14. D. H. Zhou, J. J. Shea, A. J. Nieuwkoop, W. T. Franks, B. J. Wylie, C. Mullen, D. Sandoz and C. M. Rienstra, *Angew. Chem. Int. Ed.*, 2007, **46**, 8380.
15. W. T. Franks, B. J. Wylie, H. L. F. Schmidt, A. J. Nieuwkoop, R.-M. Mayrhofer, G. J. Shah, D. T. Graesser and C. M. Rienstra, *Proc. Natl. Acad. Sci. U. S. A.*, 2008, **105**, 4621.
16. M. J. Knight, A. J. Pell, I. Bertini, I. C. Felli, L. Gonnelli, R. Pierattelli, T. Herrmann, L. Emsley and G. Pintacuda, *Proc. Natl. Acad. Sci. U. S. A.*, 2012, **109**, 11095.
17. I. Bertini, A. Bhaumik, G. De Paëpe, R. G. Griffin, M. Lelli, J. R. Lewandowski and C. Luchinat, *J. Am. Chem. Soc.*, 2010, **132**, 1032.
18. P. S. Nadaud, J. J. Helmus, N. Höfer and C. P. Jaroniec, *J. Am. Chem. Soc.*, 2007, **129**, 7502.
19. B. B. Das, H. J. Nothnagel, G. J. Lu, W. S. Son, Y. Tian, F. M. Marassi and S. J. Opella, *J. Am. Chem. Soc.*, 2012, **134**, 2047.
20. S. H. Park, B. B. Das, F. Casagrande, Y. Tian, H. J. Nothnagel, M. Chu, H. Kiefer, K. Maier, A. A. De Angelis, F. M. Marassi and S. J. Opella, *Nature*, 2012, **491**, 779.
21. S. H. Park, B. B. Das, A. A. De Angelis, M. Scrima and S. J. Opella, *J. Phys. Chem. B*, 2010, **114**, 13995.
22. H. Lodish and P. Matsudaira, *Molecular Cell Biology*, Freeman, New York, 2000.
23. J. A. Stringer, C. E. Bronnimann, C. G. Mullen, D. H. Zhou, S. A. Stellfox, Y. Li, E. H. Williams and C. M. Rienstra, *J. Magn. Reson.*, 2005, **173**, 40.
24. P. L. Gor'kov, E. Y. Chekmenev, C. Li, M. Cotten, J. J. Buffy, N. J. Traaseth, G. Veglia and W. W. Brey, *J. Magn. Reson*, 2007, **185**, 77.
25. M. A. M. Ahmed, V. V. Bamm, G. Harauz and V. Ladizhansky, *Biophys. J.*, 2010, **99**, 1247.
26. M. Renault, R. Tommassen-van Boxtel, M. P. Bos, J. A. Post, J. Tommassen and M. Baldus, *Proc. Natl. Acad. Sci. U. S. A.*, 2012, **109**, 4863.
27. D. J. Fowler, M. J. Harris and L. K. Thompson, *J. Magn. Reson.*, 2012, **222**, 112.
28. K. R. Thurber and R. Tycko, *J. Magn. Reson.*, 2009, **196**, 84.
29. T. K. Ritchie, Y. V. Grinkova, T. H. Bayburt, I. G. Denisov, J. K. Zolnerciks, W. M. Atkins and S. G. Sligar, *Methods Enzymol.*, 2009, **464**, 211.
30. A. Z. Kijac, Y. Li, S. G. Sligar and C. M. Rienstra, *Biochemistry*, 2007, **46**, 13696.
31. Y. Li, A. Z. Kijac, S. G. Sligar and C. M. Rienstra, *Biophys. J.*, 2006, **91**, 3819.

32. S. H. Park, S. Berkamp, G. A. Cook, M. K. Chan, H. Viadiu and S. J. Opella, *Biochemistry*, 2011, **50**, 8983.
33. G. M. Anantharamaiah, J. L. Jones, C. G. Brouillette, C. F. Schmidt, B. H. Chung, T. A. Hughes, A. S. Bhown and J. P. Segrest, *J. Biol. Chem.*, 1985, **260**, 10248.
34. U. H. N. Dürr, M. Gildenberg and A. Ramamoorthy, *Chem. Rev.*, 2012, **112**, 6054.
35. A. L. Shrout, D. J. Montefusco and R. M. Weis, *Biochemistry*, 2003, **42**, 13379.
36. X. Zhang, J. Gureasko, K. Shen, P. A. Cole and J. Kuriyan, *Cell*, 2006, **125**, 1137.
37. F. Shi, S. E. Telesco, Y. Liu, R. Radhakrishnan and M. A. Lemmon, *Proc. Natl. Acad. Sci. U. S. A.*, 2010, **107**, 7692.
38. P. Zhang, C. M. Khursigara, L. M. Hartnell and S. Subramaniam, *Proc. Natl. Acad. Sci. U.S.A.*, 2007, **104**, 3777.
39. O. J. Murphy, F. A. Kovacs, E. L. Sicard and L. K. Thompson, *Biochemistry*, 2001, **40**, 1358.
40. B. Isaac, G. J. Gallagher, Y. S. Balazs and L. K. Thompson, *Biochemistry*, 2002, **41**, 3025.
41. R. Fu, X. Wang, C. Li, A. N. Santiago-Miranda, G. J. Pielak and F. Tian, *J. Am. Chem. Soc.*, 2011, **133**, 12370.
42. M. Renault, S. Pawsey, M. P. Bos, E. J. Koers, D. Nand, R. Tommassen-van Boxtel, M. Rosay, J. Tommassen, W. E. Maas and M. Baldus, *Angew. Chem. Int. Ed.*, 2012, **51**, 2998.
43. D. M. LeMaster and D. M. Kushlan, *J. Am. Chem. Soc.*, 1996, **118**, 9255.
44. D. Marulanda, M. L. Tasayco, A. McDermott, M. Cataldi, V. Arriaran and T. Polenova, *J. Am. Chem. Soc.*, 2004, **126**, 16608.
45. A. Loquet, N. G. Sgourakis, R. Gupta, K. Giller, D. Riedel, C. Goosmann, C. Griesinger, M. Kolbe, D. Baker, S. Becker and A. Lange, *Nature*, 2012, **486**, 276.
46. I. Sengupta, P. S. Nadaud and C. P. Jaroniec, *Acc. Chem. Res.*, 2013, **46**, 2117.
47. S. Wang, R. A. Munro, S. Y. Kim, K.-H Jung, L. S. Brown and V. Ladizhansky, *J. Am. Chem. Soc.*, 2012, **134**, 16995.
48. P. S. Nadaud, J. J. Helmus, I. Sengupta and C. P. Jaroniec, *J. Am. Chem. Soc.*, 2010, **132**, 9561.
49. M. Tang, D. A. Berthold and C. M. Rienstra, *J. Phys. Chem. Lett.*, 2011, **2**, 1836.
50. T. Gopinath and G. Veglia, *Angew.Chem. Int. Ed.*, 2012, **51**, 2731.
51. S. Sun, S. Yan, C. Guo, M. Li, J. C. Hoch, J. C. Williams and T. Polenova, *J. Phys. Chem. B*, 2012, **116**, 13585.
52. S. Paramasivam, C. L. Suiter, G. Hou, S. Sun, M. Palmer, J. C. Hoch, D. Rovnyak and T. Polenova, *J. Phys. Chem. B*, 2012, **116**, 7416.
53. T. Jacso, W. T. Franks, H. Rose, U. Fink, J. Broecker, S. Keller, H. Oschkinat and B. Reif, *Angew. Chem. Int. Ed.*, 2012, **51**, 432.

54. N. P. Wickramasinghe, S. Parthasarathy, C. R. Jones, C. Bhardwaj, F. Long, M. Kotecha, S. Mehboob, L. W.-M. Fung, J. Past, A. Samoson and Y. Ishii, *Nat. Chem. Biol.*, 2009, **6**, 215.
55. N. P. Wickramasinghe, M. A. Shaibat, C. R. Jones, L. B. Casabianca, A. C. de Dios, J. S. Harwood and Y. Ishii, *J. Chem. Phys.*, 2008, **128**, 052210.
56. J. R. Lewandowski, J.-N. Dumez, U. Akbey, S. Lange, L. Emsley and H. Oschkinat, *J. Phys. Chem. Lett.*, 2011, **2**, 2205.
57. A. B. Barnes, G. D. Paëpe, P. C. A. van der Wel, K.-N Hu, C.-G Joo, V. S. Bajaj, M. L. Mak-Jurkauskas, J. R. Sirigiri, J. Herzfeld, R. J. Temkin and R. G. Griffin, *Appl. Magn. Reson.*, 2008, **34**, 237.
58. I. V. Sergeyev, L. A. Day, A. Goldbourt and A. E. McDermott, *J. Am. Chem. Soc.*, 2011, **133**, 20208.
59. H. Takahashi, I. Ayala, M. Bardet, G. De Paëpe, J.-P. Simorre and S. Hediger, *J. Am. Chem. Soc.*, 2013, **135**, 5105.
60. E. Ravera, B. Corzilius, V. K. Michaelis, C. Rosa, R. G. Griffin, C. Luchinat and I. Bertini, *J. Am. Chem. Soc.*, 2013, **135**, 1641.
61. S. Lange, A. H. Linden, Ü. Akbey, W. T. Franks, N. M. Loening, B.-J. vanRossum and H. Oschkinat, *J. Magn. Reson*, 2012, **216**, 209.
62. C. Song, K.-N. Hu, C.-G Joo, T. M. Swager and R. G. Griffin, *J. Am. Chem. Soc.*, 2006, **128**, 11385.
63. B. Corzilius, A. A. Smith, A. B. Barnes, C. Luchinat, I. Bertini and R. G. Griffin, *J. Am. Chem. Soc.*, 2011, **133**, 5648.
64. T. Maly, D. Cui, R. G. Griffin and A.-F. Miller, *J. Phys. Chem. B*, 2012, **116**, 7055.
65. S. Jehle, P. Rajagopal, B. Bardiaux, S. Markovic, R. Kühne, J. R. Stout, V. A. Higman, R. E. Klevit, B.-J vanRossum and H. Oschkinat, *Nat. Struct. Mol. Biol.*, 2010, **17**, 1037.
66. M. Tang, L. J. Sperling, D. A. Berthold, C. D. Schwieters, A. E. Nesbitt, A. J. Nieuwkoop, R. B. Gennis and C. M. Rienstra, *J. Biomol. NMR*, 2011, **51**, 227.
67. L. J. Sperling, D. A. Berthold, T. L. Sasser, V. Jeisy-Scott and C. M. Rienstra, *J. Mol. Biol.*, 2010, **399**, 268.
68. Y. Li, D. A. Berthold, H. L. Frericks, R. B. Gennis and C. M. Rienstra, *ChemBioChem*, 2007, **8**, 434.
69. Y. Li, D. A. Berthold, R. B. Gennis and C. M. Rienstra, *Protein Sci.*, 2008, **17**, 199.
70. M. Tang, L. J. Sperling, D. A. Berthold, A. E. Nesbitt, R. B. Gennis and C. M. Rienstra, *J. Am. Chem. Soc.*, 2011, **133**, 4359.
71. M. Tang, A. E. Nesbitt, L. J. Sperling, D. A. Berthold, C. D. Schwieters, R. B. Gennis and C. M. Rienstra, *J. Mol. Biol.*, 2013, **425**, 1670.
72. L. J. Sperling, M. Tang, D. A. Berthold, A. E. Nesbitt, R. B. Gennis and C. M. Rienstra, *J. Phys. Chem. B*, 2013, **117**, 6052.
73. N. J. Traaseth, L. Shi, R. Verardi, D. G. Mullen, G. Barany and G. Veglia, *Proc. Natl. Acad. Sci. U. S. A.*, 2009, **106**, 10165.
74. R. Verardi, L. Shi, N. J. Traaseth, N. Walsh and G. Veglia, *Proc. Natl. Acad. Sci. U. S. A.*, 2011, **108**, 9101.

75. M. Gustavsson, N. J. Traaseth and G. Veglia, *Biochemistry*, 2011, **50**, 10367.
76. M. Gustavsson, N. J. Traaseth and G. Veglia, *Biochim. Biophys.Acta*, 2012, **1818**, 146.
77. J. J. Buffy, N. J. Traaseth, A. Mascioni, P. L. Gor'kov, E. Y. Chekmenev, W. W. Brey and G. Veglia, *Biochemistry*, 2006, **45**, 10939.
78. J. Zamoon, F. Nitu, C. Karim, D. D. Thomas and G. Veglia, *Proc. Natl. Acad. Sci. U.S.A.*, 2005, **102**, 4747.
79. J. J. Buffy, B. A. Buck-Koehntop, F. Porcelli, N. J. Traaseth, D. D. Thomas and G. Veglia, *J. Mol. Biol.*, 2006, **358**, 420.
80. G. Veglia, M. Gustavsson, V. V. Vostrikov and R. Verardi, *Biophys. J.*, 2013, **104**, 539a.
81. F. C. Sferdean, R. M. Weis and L. K. Thompson, *Biochemistry*, 2012, **51**, 6920.
82. D. J. Fowler, R. M. Weis and L. K. Thompson, *Biochemistry*, 2010, **49**, 1425.
83. A. Briegel, X. Li, A. M. Bilwes, K. T. Hughes, G. J. Jensen and B. R. Crane, *Proc. Natl. Acad. Sci. U. S. A.*, 2012, **109**, 3766.
84. K. K. Kim, H. Yokota and S. H. Kim, *Nature*, 1999, **400**, 787.
85. X. Wang, A. Vu, K. Lee and F. W. Dahlquist, *J. Mol. Biol.*, 2012, **422**, 282.
86. A. Vu, X. Wang, H. Zhou and F. W. Dahlquist, *J. Mol. Biol.*, 2012, **415**, 759.

CHAPTER 22

Structural Topologies of Phosphorylated and Non-phosphorylated Oligomeric Phospholamban in Lipid Membranes by a Hybrid NMR Approach

VITALY VOSTRIKOV[a] AND GIANLUIGI VEGLIA*[a,b]

[a] Department of Biochemistry, Molecular Biology and Biophysics; [b] Department of Chemistry, University of Minnesota, Minneapolis, MN 55455, USA
*Email: vegli001@umn.edu

22.1 Introduction

Phospholamban (PLN) is a 52 amino acid membrane protein found in the sarcoplasmic reticulum (SR) of cardiac muscle.[1] PLN is a key protein in excitation-contraction coupling, regulating the apparent Ca^{2+} affinity of SERCA, the sarco(endo)plasmic reticulum Ca^{2+}-ATPase.[2] In humans, the SERCA/PLN complex transports approximately 70% of Ca^{2+} ions, playing a central role in heart muscle diastole.[3,4] In its non-phosphorylated form, PLN binds SERCA, reducing the ATP-ase's apparent Ca^{2+} affinity and decreasing Ca^{2+} ion uptake in the SR. β-Adrenergic stimulation unleashes protein kinase A that subsequently phosphorylates PLN at Ser16, an event that

New Developments in NMR No. 3
Advances in Biological Solid-State NMR: Proteins and Membrane-Active Peptides
Edited by Frances Separovic and Akira Naito

Published by the Royal Society of Chemistry, www.rsc.org

reverses the PLN inhibitory effect and augments Ca^{2+} uptake.[5] The PLN primary sequence is highly conserved among different species (Figure 22.1A). It comprises a hydrophilic stretch (domain Ia, residues 2–17, residues 18–21), harboring the phosphorylation sites (regulatory region), a short loop, and a transmembrane (TM) inhibitory region, divided into a more hydrophilic Gln-rich juxtamembrane domain Ib (residues 22–30) and a highly hydrophobic stretch of approximately 20 amino acids (domain II). Naturally occurring mutations in human PLN degenerate into heart failure, causing either dilated cardiomyopathy or hypertrophy (Figure 22.1B).[6–10]

PLN's TM domain is necessary and sufficient for SERCA inhibition; however, the cytoplasmic domain alone is insufficient to inhibit SERCA either by itself[11] or when fused with an unrelated TM domain, such as that from cytochrome b_5.[12] The cytoplasmic domain instead has a regulatory function,

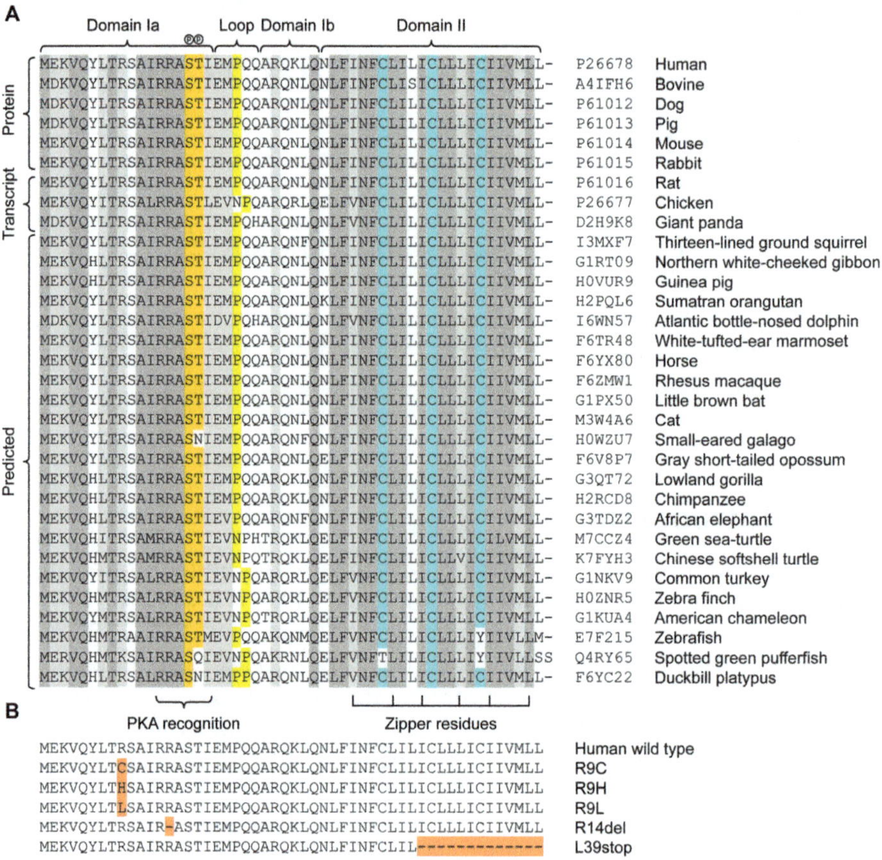

Figure 22.1 (A) Sequence alignment of PLN variants across different species. Absolutely conserved residues are highlighted in *dark grey* and conservative substitutions in *light grey*. Structural features discussed in the text (two phosphorylation sites, proline and cysteine residues) are highlighted. (B) Naturally occurring PLN mutations in humans leading to a progression of heart disease.

and in addition to Ser16 can be phosphorylated at Thr17 by calmodulin-dependent kinase (CamKII). Mutagenesis, functional assays, and cross-linking studies have shown that PLN de-oligomerizes prior to binding SERCA, the stoichiometry of the complex being 1:1.[5,13] SERCA is a 110 kDa membrane-embedded P-type ATPase that transfers two Ca^{2+} ions per enzymatic turn-over.[14] SERCA 2a (found in the heart) and SERCA 1a (found in skeletal muscle) comprise 10 TM domains, which include calcium binding sites and three major cytoplasmic regions: the N-domain, harboring the ATP binding site, the A-domain, the actuator domain, and the P-domain, which includes the autophosphorylation site Asp351.[15,16] SERCA has been crystallized in many different forms that mimic the main intermediate states along the enzymatic cycle.[17] Recently, crystals of SERCA1a, the skeletal muscle isoform of SERCA, were obtained in the presence of sarcolipin (SLN), a single-pass membrane protein homologous to PLN that binds and regulates SERCA in both skeletal and cardiac muscles.[18,19] These structures clearly revealed that the hydrophobic domain of SLN binds in a defined groove between TM2 and TM9 of SERCA, suggesting that PLN and SLN adopt a similar binding mode. A recent structure of the SERCA/PLN complex support the position of the TM domain of PLN within the SERCA binding groove.[89] However, the cytoplasmic domain of PLN can be visualized only using solid-state NMR spectroscopy.[90]

In addition to its regulatory role in SERCA function, it has been hypothesized that PLN alone might have a role in maintaining Ca^{2+} homeostatic balance.[20] In fact, both in reconstituted bilayers and cell preparations, PLN has a strong propensity to oligomerize into relatively stable homopentamers that are in equilibrium with small populations of monomers.[21] While the monomers have been considered the active species,[22] there has been an active debate regarding the role of PLN pentamerization. Two possible biological functions have been proposed for the homopentamer: (a) selective ion channels for Ca^{2+} or Cl^{-} ions, contributing to ion homeostasis in the SR;[20] or (b) a reservoir of active monomers, which exert direct control on the extent of SERCA inhibition *via* de-oligomerization.[23]

Our group has been studying the structural topology of the PLN pentamer to understand its role in Ca^{2+} ion transport.[24] We use a combination of several NMR techniques (*i.e.*, a hybrid NMR approach),[25–27] including solution NMR in micelles, magic angle spinning (MAS) NMR in vesicle preparations, and oriented solid-state NMR (O-ssNMR) with either mechanically or magnetically aligned lipid bilayer membrane systems. Solution NMR data are used to restrain the secondary structure elements, using distances (internuclear NOEs) and backbone torsion angles, as well as to identify long-range distances (*i.e.*, inter-helical NOEs). On the other hand, solid-state data in lipid vesicles are used to confirm the secondary structure elements (chemical shifts) in the native environment of the membrane and to determine inter-helical distances using dipolar-assisted rotational resonance (DARR). Finally, O-ssNMR data are used to define the orientation of the secondary, tertiary, and quaternary assembly of these proteins with respect to the lipid membrane. We believe that the convergence of the results from multiple structural biology techniques is a powerful approach to define the structural biology of complex systems, particularly when supported by biophysical, biochemical, as well as functional assays.

In this chapter, we describe our integrative NMR approach to determine the structure of the PLN pentamer, both in its phosphorylated and non-phosphorylated forms. We stress the importance of a sufficient number of restraints to obtain high-resolution structural models and determine their orientation with respect to the membrane. From our structural data and biophysical as well as biochemical/functional assays, we conclude that PLN oligomerization is a mechanism to control the concentration of PLN monomers available to bind SERCA, keeping the activity of the ATPase within a physiological window of function. Therefore, PLN oligomerization plays a key role in the regulatory cycle of the cardiac muscle.

22.2 Hybrid NMR Approach

Emerging approaches to the structural biology of large biomacromolecular machineries involve the combination of structural and functional data originating from multiple, varying sources, such as X-ray, SAXS, NMR, EPR, and cryo-EM.[28] Based on this philosophy, we have recently implemented new protocols to include data from different NMR techniques carried out with diverse membrane-mimicking systems: micelles, vesicles, and mechanically and magnetically aligned membrane systems.[25,26] Detergent micelles are a versatile membrane mimetic system also utilized extensively by X-ray crystallographers to obtain high-resolution structures of membrane proteins in a so-called "native" state. Micelles form large complexes with membrane proteins, coating the hydrophobic TM domains of these proteins with their hydrocarbon chains. These large complexes are still tractable using TROSY-based NMR techniques and have enabled the structure determination of medium and large molecular weight membrane proteins.[29] It has been shown that detergent micelles typically do not affect the more rigid TM domains, preserving the fold of multispan membrane proteins.[30] In contrast, more malleable domains, such as amphipathic helices and loops, are often affected by the geometry of micelles.[31] In particular, amphipathic helices have a tendency to adapt to the micelle surface, following its curvature. Nonetheless, structures of membrane proteins in detergent micelles still provide a first view of their secondary and tertiary structure.

The most faithful mimics of the membrane environment remain lipid bilayers. There are two major approaches to "image" membrane proteins at the atomic resolution that utilize lipid bilayers: magic angle spinning (MAS)[32] and oriented solid-state NMR (O-ssNMR).[33] Sample preparations for MAS NMR spectroscopy are relatively easy.[91,92] Once the membrane composition is chosen, the samples are reconstituted into lipid bilayers either by dissolving the proteins in organic solvents or by reconstituting membrane proteins in detergent micelles and mixing the preparations with lipid films prepared by organic solvent evaporation. The former method is fairly harsh and care must be taken to ensure that the protein is not aversely affected.

The detergent-mediated reconstitution is a more laborious procedure but offers milder reconstitution conditions.[34] Vesicles are then prepared and extensive dialysis in aqueous buffer ensures both the elimination of the detergent molecules and the formation of proteoliposomes. These preparations are then concentrated by ultracentrifugation to eliminate bulk water and inserted into rotors for MAS spectroscopy. Sample preparations for O-ssNMR are significantly more laborious. Membrane proteins need to be uniformly aligned in either mechanically (glass plates) or magnetically (bicelles) aligned systems. The latter requires several different steps. PLN preparations for both MAS and O-ssNMR studies have been described by Veglia *et al.*[34] In all cases, however, it is imperative to maintain high lipid:protein ratios (100–200 to 1 for PLN and SERCA). Lipid depletion techniques fail to preserve the function of the ATPase.

In our hybrid approach, we used MAS techniques to assign the chemical shifts of the backbone and side chains in lipid membranes. In addition, short- and medium-range distances were obtained from DARR experiments.[35] When combined with asymmetric labeling techniques,[36,37] DARR experiments provided quaternary distance restraints (see below). The O-ssNMR experiments were utilized to obtain orientational restraints and define the topology of the membrane proteins. Specifically, anisotropic chemical shifts (CSA) and dipolar couplings (DC) in oriented membrane systems enable one to obtain topological parameters, such as tilt and azimuthal angles, using separated local field experiments (SLF).[38,39] If determined for several amide sites along the secondary structure elements, CSA and DC can be used in the calculation protocols in a manner similar to that defined for residual dipolar coupling by Bax and Tjandra.[40,41] The hybrid calculation protocol we implemented includes a hybrid energy function that involves both chemical information and experimental restraints, in which the latter are obtained from both solution and solid-state NMR experiments.[26] We also carried out a validation of the structural restraints utilizing an *R*-factor described by Cross and co-workers.[42]

22.3 Structural Topology of PLN

22.3.1 Secondary Structure in Micelles and Lipid Membranes

To obtain well-behaving NMR spectra (*i.e.*, well-resolved with narrow lines), we reconstituted recombinant wild-type PLN in perdeuterated dodecylphosphocholine micelles.[43] Under these conditions, PLN forms stable symmetric pentamers.[25] The latter characteristic enables one to observe only one set of resonances out of the five protomers. A representative [^{1}H,^{15}N]-HSQC spectrum is shown in Figure 22.2A. While multidimensional NOESY-type experiments are very sensitive, COSY-type experiments are insensitive. Perdeuteration and both ^{15}N and ^{13}C labeling were necessary to assign both backbone and side-chain resonances. Chemical shifts as well as short-range distances and torsion angles obtained from these datasets were used to

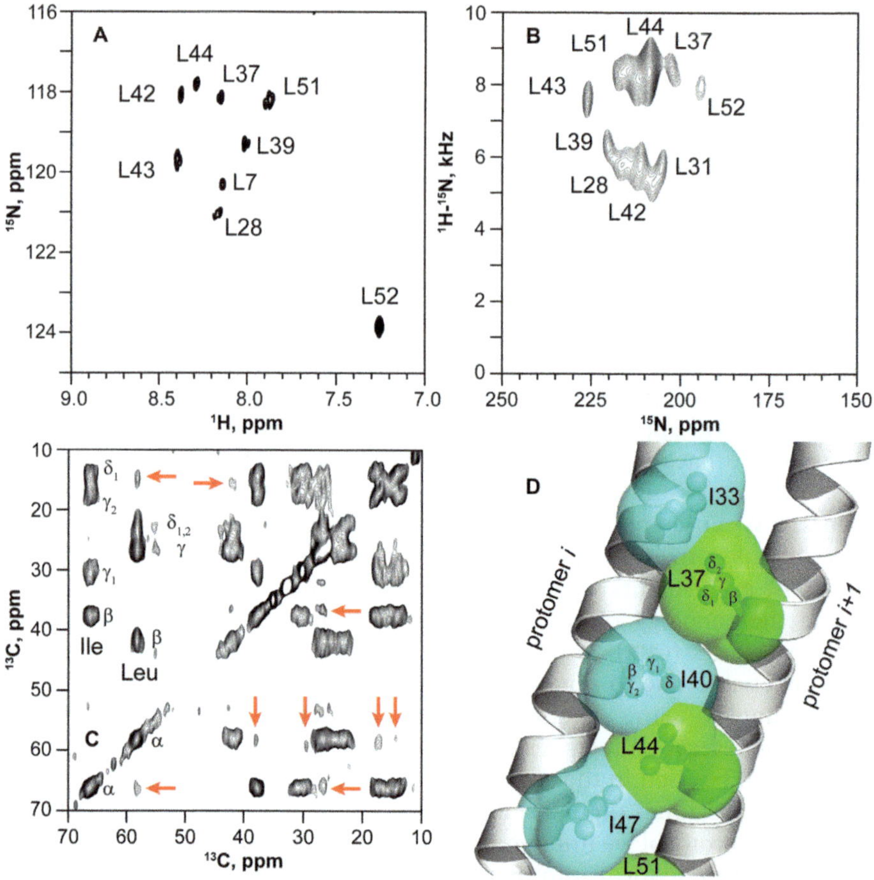

Figure 22.2 Representative NMR data obtained for the membrane proteins. (A) Solution NMR of ^{15}N-Leu PLN in DPC micelles. (B) Separated local field spectrum of ^{15}N-Leu PLN aligned in DOPC:DOPE 4:1 glass slides. (C) MAS DARR spectrum of asymmetrically labeled (^{13}C/^{15}N-Leu + ^{13}C/^{15}N-Ile) pSer16 PLN in DOPC:DOPE 4:1 multilamellar vesicles. Cross peaks between the Leu and Ile spin systems are indicated with *arrows*. (D) Leu-Ile zipper motif in the TM domain of PLN, giving rise to the spectrum in C.

restrain the secondary structures of both the TM and cytoplasmic domains. As in the solution NMR spectra, MAS spectra present one set of peaks, with ^{13}C chemical shift values for backbone and side chains in close agreement with those measured in detergent micelles.[25] The chemical shifts define the canonical helix-turn-helix motif of PLN, with a higher helical content in the TM and a lower helical content in the cytoplasmic domain. The absence of long-range NOEs in micelle preparations excludes the presence of tertiary structure arrangements. In fact, paramagnetic quenching experiments[44,45] carried out in the monomeric mutant of PLN (PLN^{AFA}) suggest that the relative arrangements of the two helical domains are approximately perpendicular, with the TM domain embedded into the core of the micelle and the cytoplasmic domain absorbed on the surface of the micelles.

Residual dipolar couplings (RDCs) measured in stretched polyacrylamide gels display an interesting pattern (data not shown): the TM domain shows a typical RDC helical pattern, with alternating values along the residues and attenuated values starting from domain Ib (the juxtamembrane region) and ending at the cytoplasmic domain. Fitting of the RDC values using classical simulated annealing and energy minimization protocols generates distorted structures, with "banana-shape" PLN TM domains. In fact, RDCs for mobile regions are dynamically averaged and can be over-fitted, ignoring dynamical scaling.[44] Owing to the dynamic nature of the RDCs, we decided not to include these data in the original hybrid structure determination protocol.

22.3.2 Inter-Protomer Distances in Micelles and Lipid Membranes

The quaternary arrangement of PLN is a feature that has generated significant interest in the design of new membrane protein folds.[46–48] To restrain PLN protomers, we needed unambiguous inter-protomer restraints. Usually, to measure long-range protein–protein interactions in oligomers, a combination of half-filtered experiments and asymmetric labeling schemes is necessary.[49] Unfortunately, these methods did not perform well with PLN, where the dispersion of the chemical shifts of the different chemical groups is limited and the sensitivity of the samples is substantially lower than soluble proteins. To overcome this hurdle, we implemented a new asymmetric labeling scheme using methyl group labeling.[36] Since several hydrophobic residues form a Leu/Ile zipper in the TM domain, we prepared a 1:1 mixture of PLN, individually labeling either Ile/Leu or Val residues in the C-terminal end of the TM region. Using this strategy, we were able to unambiguously assign NOE distances between the adjacent protomers and reconstruct the quaternary arrangement of the pentamer. Notably, the same information can be obtained for lipid preparations using DARR experiments for MAS NMR. In this case, we prepared PLN pentamers in membranes composed of DOPC/DOPE phospholipids and measured the dipolar contacts between ^{13}C resonances of adjacent protomers. Using the asymmetric labeling strategy, the contacts found between the protomers in detergent micelles were confirmed and supported by the data obtained in lipid membranes.[25] A portion of the [^{13}C,^{13}C]-DARR experiment is shown in Figure 22.2C, while the packing of the Leu/Ile zipper is shown in Figure 22.2D.

22.3.3 Orientation of PLN in Lipid Bilayers

The most unambiguous way to define the orientation of membrane proteins in lipid bilayers is to use O-ssNMR. While the oriented preparations with membrane proteins in lipid bilayers aligned mechanically onto glass plates or magnetically in bicelles are rather challenging, once alignment is achieved CSA and DC can be measured relatively easily using new SLF experiments that have improved sensitivity and resolution.[50–54] A significant advantage of CSA and DC with respect to RCSA and RDC is that the orientation of the alignment tensor is known, which greatly reduces the degeneracy and ambiguity of the system.[26,55] For both the PLN monomer (*i.e.*, PLN^{AFA} mutant) and pentamer, we first utilized mechanically aligned lipid membranes.[56–58] These systems are highly reliable and, when fully hydrated, can give rise to spectra at high resolution. Nonetheless, the linewidths obtained with these preparations still prevent the routine implementation of multidimensional NMR techniques. Therefore, to assign PLN resonances, we utilized an array of selectively labeled samples prepared either by recombinant protein preparation and selective labeling techniques or by chemical synthesis of the entire protein with site-specific isotopic labeling. The [^{1}H,^{15}N]-PISEMA spectrum of a uniformly labeled leucine sample is shown in Figure 22.2B, while the corresponding solution NMR sample in dodecylphosphocholine (DPC) micelles is shown in Figure 22.2A.

22.3.4 Structure Calculation Protocol

The structure determination of the PLN pentamers was carried out using the simulated annealing protocols as described in the XPLOR-NIH software package.[59] The hybrid protocol is summarized in Figure 22.3. The structures of the oligomers were calculated starting from the fully extended conformation of the monomers to avoid any bias. The chemical shifts derived from solution NMR and MAS experiments were included in the calculations of the secondary structure elements. In a following stage, orientational restraints, namely DC and CSA from O-ssNMR measurements, were implemented as harmonic potentials.[26,60,61] These restraints are fully compatible with the distance and torsion angle restraints derived from chemical shifts, as they do not cause any distortion in the protein secondary structure.[26]

To generate the symmetric pentamers, we created five identical copies of the proteins and assembled the quaternary structure using the inter-helical contacts (NOEs and DARR). The symmetry in the pentamer during the calculations was maintained using a non-crystallographic symmetry function built in XPLOR-NIH as well as pseudo-NOE restraints.[62] Once energy functions for chemical shift, NOEs, and orientational restraints were minimized, we included a depth of insertion potential to embed the protein into a virtual membrane.[63] Figure 22.4 visualizes the type of restraints utilized to define the secondary structure (Figure 22.4A), the orientation of the peptide planes and the helical domains with respect to the lipid membrane (Figure 22.4B), the inter-protomer distances obtained from NOESY and DARR experiments (Figure 22.4C), and the restraints to maintain the symmetry of the pentamer (Figure 22.4D); the modeling procedure was followed by molecular dynamics calculations with the PLN structural model embedded into a fully atomistic

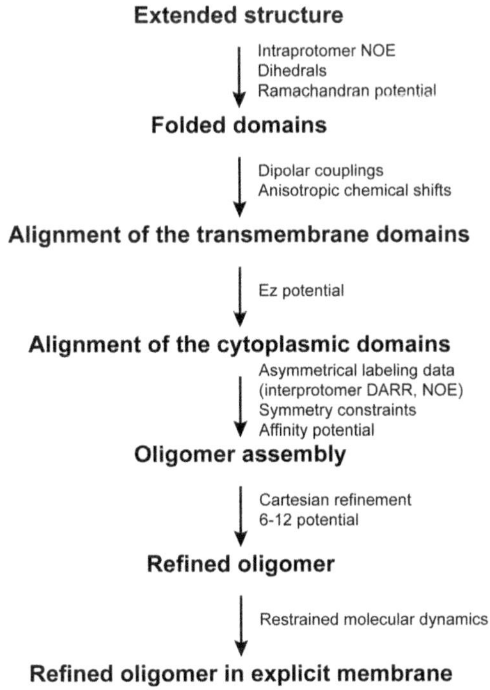

Figure 22.3 Flow chart of the hybrid structure calculation method. All steps with the exception of the final one are performed with XPLOR-NIH.

membrane. The latter step ensures more realistic electrostatic interactions between proteins and lipids and a better packing of the residues thanks to force fields used from the MD simulations, improved with respect to the more essential force fields available for modeling.[64]

22.4 Structural Topology of Non-phosphorylated and Phosphorylated Pentamers

Chemical shifts and NMR dynamic parameters define four different domains in PLN: a cytoplasmic amphipathic helix (domain Ia), a mobile linker (loop), a juxtamembrane region (domain Ib), and a hydrophobic helix (domain II). Within the membrane, PLN adopts a "pinwheel" architecture, with domain Ia absorbed on the membrane surface with its hydrophobic residues pointing toward the hydrocarbon region of the membrane, while the hydrophilic residues face the bulk solvent (Figure 22.5). The loop, which shows a preference toward a β-turn in micelles, is very dynamic; in contrast, the juxtamembrane region and domain II are more rigid and remain helical. The TM domain crosses the lipid membrane at a tilt angle of approximately 11°. In contrast, we previously found that the tilt angle of monomeric PLN^{AFA} is approximately 23°, suggesting that in the absence of the protein–protein

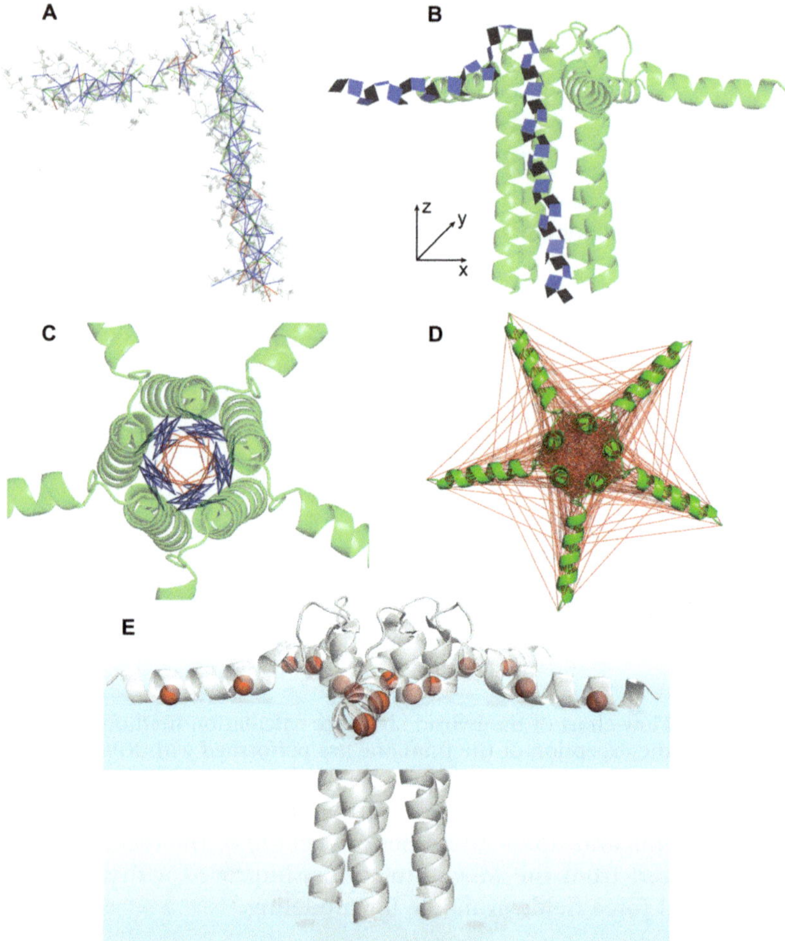

Figure 22.4 Restraints imposed on the structure at different steps of the refinement. (A) Intraprotomer NOE restraints, colored as short range (*red*), medium range (*green*), and long range (*blue*). (B) Dihedral angles, calculated *via* TALOS+[88] from the isotropic chemical shifts and SLF data, defining the topology of the helices in lipid bilayer membranes. (C) Interprotomer constraints, derived from asymmetrical labeling: DARR (*blue*) and NOE (*red*). (D) Restraints, imposed to constraint the C_{5v} symmetry of the pentamer. (E) Planar restraints, defining the relative position of the cytoplasmic helix relative to the TM domain.

interactions present in the quaternary assembly, the hydrophobic mismatch causes PLN to adopt a larger tilt angle.[57,65] The interactions between domain II of the different monomers are hydrophobic in nature and involve a Ile/Leu zipper, located in the C-terminal region of the TM domain, that holds the pentamer together in a knobs-and-holes motif. These interactions contribute to the thermostability of the quaternary structure (Figure 22.2D).

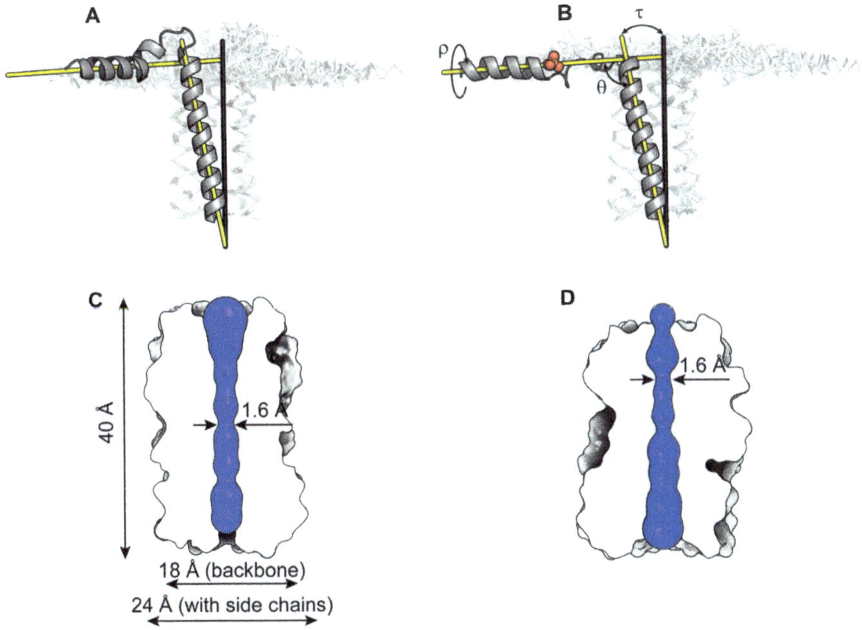

Figure 22.5 Structural features of the pentameric PLN bundles. (A) NMR ensemble of the non-phosphorylated PLN (2KYV). One protomer is shown in the cartoon representation. (B) NMR ensemble of the pSer16 phosphorylated PLN (2M3B).[27] (C) Cross-section of the TM domain of non-phosphorylated PLN, indicating the central cavity. Note that the pore constriction is inconsistent with the proposed channel function. (D) Cross-section of the TM domain of pSer16 PLN.

In addition to the Leu/Ile zipper, the pentamer is held together by a relatively stable hydrogen bond network in the juxtamembrane region, which is in agreement with solid-state measurements by Smith *et al.*[66] The cysteine residues 36 and 41 are not involved in disulfide bridges; rather they are positioned at the interface between the different protomers, contributing to the hydrophobic interactions that hold the pentamer together. In fact, cysteine mutations to phenylalanine, alanine, and/or serine perturb the interdigitation of the side chains, promoting de-oligomerization as for the PLN^{AFA}, PLN^{SSS}, or PLN^{C41F} mutants.[67,68] The contact area between the adjacent protomers in the pentamer bundles is very similar between the phosphorylated and non-phosphorylated forms, and is approximately 1250 $Å^2$ (calculated with 4 Å distance cutoff). Mutations that affect oligomerization are located both in the inner core of the pentamer and at the interface between the protomers. Alanine scanning mutagenesis of domains Ib and II by the Jones group revealed that oligomerization is entirely obliterated by substituting either of the Leu/Ile zipper residues[69] (Figure 22.6). The same study using Phe-scanning mutagenesis illustrated that it is not only the hydrophobicity of these residues that is crucial for the formation of

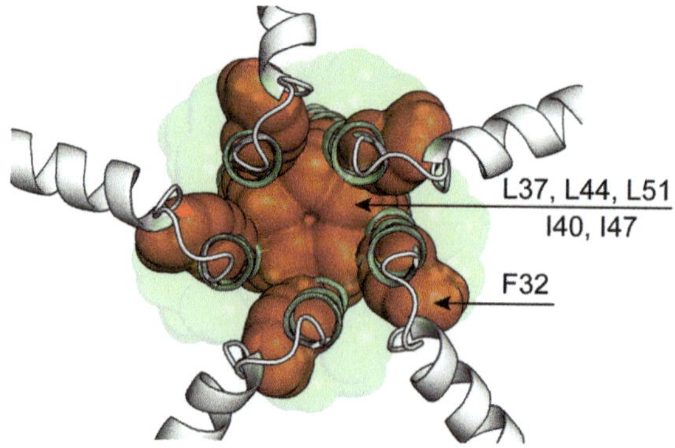

Figure 22.6 Residues in the domains Ib and II of PLN that promote destabilization of the pentameric assembly.

the pentamer, but also their surface complementarity, as expected for the interdigitized Leu/Ile side chains. The pore defined by the helices in the TM domain is lined exclusively with hydrophobic residues and measures approximately 25 Å in length with an average diameter of 2 Å (Figure 22.5C). A combined theoretical and electrochemical study showed that the pentamer, reconstituted in synthetic lipid membranes, is not able to conduct ions through its hydrophobic pore. The free energy barriers for the passage of ions through the pore was found to be ~19 kcal mol^{-1} and 40 kcal mol^{-1} for Cl^- and Ca^{2+}, respectively, precluding the spontaneous translocation of ions in the absence of another energy source.[70]

The structural topology of the phosphorylated pentamer is very similar to that of the non-phosphorylated form. The phosphorylated pentamer adopts a pinwheel topology with tilt and azimuthal angles that are essentially the same as in the non-phosphorylated bundle (Figures 22.5B and 22.5D). The inter-helical contacts determined by DARR experiments define the same Leu/Ile zipper with distances that are identical to the non-phosphorylated form, showing that domain II is not perturbed by phosphorylation in the cytoplasmic region. Additionally, the juxtamembrane domain Ib shows chemical shifts virtually identical to those in the non-phosphorylated form. The structural changes induced by phosphorylation are confined to the residues of the loop and domain Ia. In particular, the amphipathic helix of domain Ia undergoes slight unwinding upon phosphoryl transfer. This effect is even more prominent in the spin relaxation parameters measured in detergent micelles.

The membrane architecture of pentameric PLN as determined by the hybrid NMR method is consistent with a plethora of biophysical and biochemical data. First, our data agree with several structural studies carried out using fluorescence spectroscopy, which predicted the pinwheel topology both *in vitro* and in cell.[71–73] In addition, EPR studies revealed that in the major conformation of PLN, the TM domain crosses the lipid membrane, and domain Ia is absorbed onto the surface of the lipid membrane.[74–76]

Finally, Lorigan and co-workers carried out ssNMR studies using both MAS and O-ssNMR supporting the structural arrangement of the pentamer.[77]

22.5 Conformational Dynamics of Phosphorylated and Non-phosphorylated PLN

The backbone conformational dynamics of PLN were characterized by solution NMR spectroscopy measuring the nuclear spin relaxation rates of amide groups (R_1 and R_2) and heteronuclear [^{1}H,^{15}N] NOEs.[78] These parameters clearly identify the different regions of PLN, with a rigid TM domain II, a more mobile juxtamembrane domain Ib, a highly dynamic loop, and a relatively mobile domain Ia. The spin relaxation data are plotted on the PDB coordinates of the two pentamers in Figure 22.7. As expected for the TM segments lacking helix breaking residues (such as proline or glycine), domain II of PLN is characterized by slow R_1 and fast R_2 relaxation rates, as well as high heteronuclear NOE values. At 600 MHz, the spin relaxation rates fall in a range of 1.0–1.1 s^{-1} for R_1 and 18–20 s^{-1} for R_2, respectively, while the heteronuclear NOE values cluster around $\sim$0.8. These parameters are indicative of the well-defined secondary structure elements, lacking conformational mobility in the backbone region.[79] The rigidity of the TM domain is explained by the strong network of hydrogen bonds in the apolar environment of the hydrophobic region, and the Ile-Leu zipper, which locks together the five identical transmembrane domains of the PLN pentamer in the left-handed coiled-coil conformation.

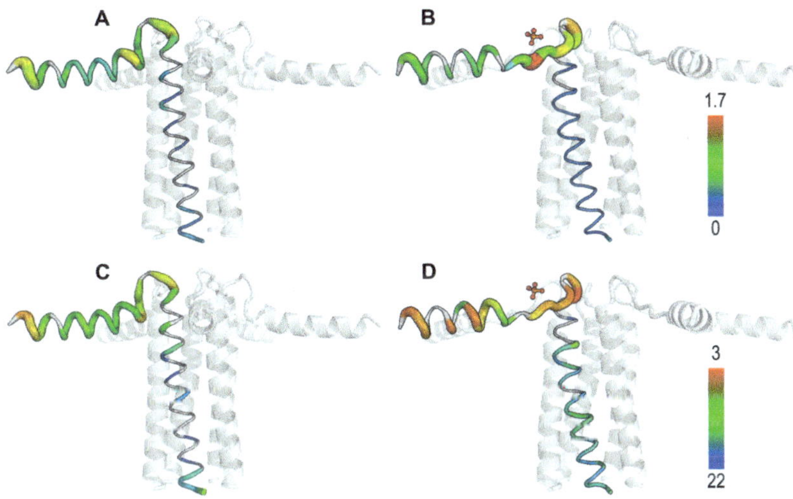

Figure 22.7 Solution NMR dynamics of the pentameric PLN. (A) $R_1 \times (1 - \text{NOE})$ product, characteristic of fast dynamics, for the non-phosphorylated protein. (B) Same product, plotted for pSer16 PLN. (C) R_2/R_1 ratio, describing slow dynamics, plotted for the non-phosphorylated protein. (D) Same ratio, plotted for pSer16 PLN pentamer.

The juxtamembrane region comprises several hydrophilic residues, including highly conserved positive residues (Lys or Arg) at position 25, that are able to establish polar interactions with the lipid headgroups or – in functional terms – with PLN binding partners.[80] These residues are also more accessible to solvent, and a gradual change of the relaxation parameters is observed for residues 26–30 (domain Ib). Notably, these amino acids adopt a helical conformation and extend the TM domain toward the cytoplasm. The helical conformation is interrupted by a proline residue at position 21 of the PLN sequence. Disruption of the hydrogen bonding pattern in the hydrophilic media is apparent from the heteronuclear NOE data, with values as low as 0.2–0.4 (as measured at 600 MHz). Finally, for the N-terminal domain Ia of PLN, a membrane-attached amphipathic helix, we observe an increase of R_1 in comparison with the TM domain (1.4–1.5 s^{-1}), and a concomitant decrease of both R_2 (15–17 s^{-1}) and heteronuclear NOEs (0.6–0.7). These relaxation parameters are characteristic of segments with moderate-to-low mobility, compared with those of the rigid TM helix and flexible loop. Similar to the juxtamembrane domain, there is a gradual increase in mobility near the N-terminus of domain Ia. The latter is attributable to the lack of hydrogen bonds in the first three residues at the N-terminus. Notably, the cytoplasmic domain of PLN is polymorphic, *i.e.* it exists in the membrane-attached T-state, which is the predominant conformation detected in micelles and in the membrane-detached unwound R-state.[74,81] While these states were clearly identified in the monomeric mutant PLN^{AFA}, the nuclear spin relaxation rates measured for the pentameric species suggest that a similar equilibrium is present in the oligomeric states.

22.6 Functional Implication of PLN Oligomerization

The homeostatic balance of Ca^{2+} in the heart muscle is critical to maintaining physiological contractility parameters. Imbalances in Ca^{2+} handling by SERCA cause muscle dysfunction and disease.[82] Therefore, SERCA's apparent affinity for Ca^{2+} needs to be kept within a well-defined physiological window. Exceeding this narrow window causes aberrant concentrations of Ca^{2+} ions in the cytoplasm or lumen that degenerate heart muscle function.[82] As PLN regulates SERCA's apparent affinity for Ca^{2+} ions, it is directly responsible for maintaining the ATP-ase's function within the physiological window. Overexpression,[83] ablation,[6] or silencing the PLN gene using RNAi[84] force SERCA activity outside this range, degenerating into cardiac disease. Additionally, PLN inhibitory function can be modulated by mutations that promote de-oligomerization.[83] The latter suggests that oligomerization is another way to regulate SERCA's function, by reducing the concentration of "active" monomers that are able to interact with SERCA.[22] *In vivo* studies show that by promoting the monomeric state of PLN with pentamer-destabilizing mutants such as PLN^{C41F} it is possible to affect Ca^{2+} transients in mice hearts.[85] On the other hand, promoting oligomerization of PLN may also result in Ca^{2+} mishandling. In fact, we found that a lethal mutation of PLN discovered in humans and located in the cytoplasmic domain Ia (PLN^{R9C}) is able to stabilize the formation of pentamers both *in vitro* and in cell, hindering SERCA regulation.[86]

It has also been reported that oligomerization may be affected by the phosphorylation state of PLN. This point remains still obscure. Our results, however, show that there are no major changes in PLN structural topology; rather the differences are confined to cytoplasmic domain Ia, influencing the stability of the amphipathic helix. Taken together, these studies reveal an active role of the pentamer in the regulatory cycle of the SERCA/PLN complex. In other words, the equilibrium between monomer and pentamer represents another level of control of SERCA function, regulating the concentration of monomeric PLN (phosphorylated or non-phosphorylated) that is available to bind and down-regulate (in the non-phosphorylated form) or up-regulate (in the phosphorylated form) SERCA function.

22.7 Concluding Remarks

A combination of solution and solid-state NMR methods ("hybrid NMR method") enabled the structural analysis of the oligomeric state of PLN at the atomic level. The convergence of the data obtained in different membrane-mimicking systems and the native lipid bilayers proved crucial to obtain a complete view of these complex membrane protein systems. Using this approach we were able to elucidate the pinwheel topology of PLN, which is consistent with several different biological and biochemical investigations as well as with the interpretation of oligomerization as a mechanism to control monomeric PLN's ability to interact with and inhibit SERCA. Our structural and biological activity data do not support the interpretation of PLN oligomers as selective Ca^{2+} or Cl^{-} channels. Approximately 35% of the proteins in a cell form oligomers,[87] which include a large number of membrane proteins. Therefore, the combination of asymmetric labeling techniques combined with the hybrid NMR method applied to PLN is likely to be beneficial for many other large membrane proteins. Naturally, obtaining high-resolution structures of the individual proteins provides only a limited understanding of the physiological phenomena. The intrinsic power of solution and solid-state (MAS and O-ss) NMR is the ability to vary the conditions and surroundings of a protein, which encompasses not only membrane and aqueous environments, but also its binding partners. We envision that the approach described herein will prove essential for the elucidation of protein complexes between membrane proteins or membrane proteins and their globular interaction partners.

Acknowledgements

This work is supported by the National Institute of Health (R01 GM 64742) and American Heart Association (13POST14670054).

References

1. H. K. Simmerman, J. H. Collins, J. L. Theibert, A. D. Wegener and L. R. Jones, *J. Biol. Chem.*, 1986, **261**, 13333.
2. D. H. MacLennan and E. G. Kranias, *Nat. Rev. Mol. Cell. Biol.*, 2003, **4**, 566.
3. D. M. Bers, *Nature*, 2002, **415**, 198.
4. D. M. Bers, *Annu. Rev. Physiol.*, 2008, **70**, 23.
5. P. James, M. Inui, M. Tada, M. Chiesi and E. Carafoli, *Nature*, 1989, **342**, 90.
6. K. Haghighi, F. Kolokathis, L. Pater, R. A. Lynch, M. Asahi, A. O. Gramolini, G. C. Fan, D. Tsiapras, H. S. Hahn, S. Adamopoulos, S. B. Liggett, G. W. Dorn 2nd, D. H. MacLennan, D. T. Kremastinos and E. G. Kranias, *J. Clin. Invest.*, 2003, **111**, 869.
7. J. P. Schmitt, M. Kamisago, M. Asahi, G. H. Li, F. Ahmad, U. Mende, E. G. Kranias, D. H. MacLennan, J. G. Seidman and C. E. Seidman, *Science*, 2003, **299**, 1410.
8. K. Haghighi, F. Kolokathis, A. O. Gramolini, J. R. Waggoner, L. Pater, R. A. Lynch, G. C. Fan, D. Tsiapras, R. R. Parekh, G. W. Dorn 2nd, D. H. MacLennan, D. T. Kremastinos and E. G. Kranias, *Proc. Natl. Acad. Sci. U. S. A.*, 2006, **103**, 1388.
9. J. P. Schmitt, F. Ahmad, K. Lorenz, L. Hein, S. Schulz, M. Asahi, D. H. MacLennan, C. E. Seidman, J. G. Seidman and M. J. Lohse, *Circulation*, 2009, **119**, 463.
10. A. Medeiros, D. G. Biagi, T. J. Sobreira, P. S. de Oliveira, C. E. Negrao, A. J. Mansur, J. E. Krieger, P. C. Brum and A. C. Pereira, *Am. Heart J.*, 2011, **162**, 1088.
11. L. R. Jones and L. J. Field, *J. Biol. Chem.*, 1993, **268**, 11486.
12. Y. Kimura, M. Asahi, K. Kurzydlowski, M. Tada and D. H. MacLennan, *FEBS Lett.*, 1998, **425**, 509.
13. Y. Kimura, K. Kurzydlowski, M. Tada and D. H. MacLennan, *J. Biol. Chem.*, 1996, **271**, 21726.
14. C. Toyoshima and G. Inesi, *Annu. Rev. Biochem.*, 2004, **73**, 269.
15. C. Toyoshima, M. Nakasako, H. Nomura and H. Ogawa, *Nature*, 2000, **405**, 647.
16. M. Bublitz, J. P. Morth and P. Nissen, *J. Cell Sci.*, 2011, **124**, 2515.
17. J. V. Moller, C. Olesen, A. M. Winther and P. Nissen, *Q. Rev. Biophys.*, 2010, **43**, 501.
18. A. M. Winther, M. Bublitz, J. L. Karlsen, J. V. Moller, J. B. Hansen, P. Nissen and M. J. Buch-Pedersen, *Nature*, 2013, **495**, 265.
19. C. Toyoshima, S. Iwasawa, H. Ogawa, A. Hirata, J. Tsueda and G. Inesi, *Nature*, 2013, **495**, 260.
20. R. J. Kovacs, M. T. Nelson, H. K. Simmerman and L. R. Jones, *J. Biol. Chem.*, 1988, **263**, 18364.
21. H. K. Simmerman and L. R. Jones, *Physiol. Rev.*, 1998, **78**, 921.

22. Y. Kimura, K. Kurzydlowski, M. Tada and D. H. MacLennan, *J. Biol. Chem.*, 1997, **272**, 15061.
23. S. L. Robia, N. C. Flohr and D. D. Thomas, *Biochemistry*, 2005, **44**, 4302.
24. N. J. Traaseth, K. N. Ha, R. Verardi, L. Shi, J. J. Buffy, L. R. Masterson and G. Veglia, *Biochemistry*, 2008, **47**, 3.
25. R. Verardi, L. Shi, N. J. Traaseth, N. Walsh and G. Veglia, *Proc. Natl. Acad. Sci. U. S. A.*, 2011, **108**, 9101.
26. L. Shi, N. J. Traaseth, R. Verardi, A. Cembran, J. Gao and G. Veglia, *J. Biomol. NMR*, 2009, **44**, 195.
27. V. V. Vostrikov, K. R. Mote, R. Verardi and G. Veglia, *Structure*, 2013, **21**, 2119–2130.
28. F. Alber, F. Forster, D. Korkin, M. Topf and A. Sali, *Annu. Rev. Biochem.*, 2008, **77**, 443.
29. C. Hilty, G. Wider, C. Fernandez and K. Wuthrich, *ChemBioChem*, 2004, **5**, 467.
30. Y. Zhou, T. Cierpicki, R. H. Jimenez, S. M. Lukasik, J. F. Ellena, D. S. Cafiso, H. Kadokura, J. Beckwith and J. H. Bushweller, *Mol. Cell*, 2008, **31**, 896.
31. J. J. Chou, J. D. Kaufman, S. J. Stahl, P. T. Wingfield and A. Bax, *J. Am. Chem. Soc.*, 2002, **124**, 2450.
32. M. Hong, Y. Zhang and F. Hu, *Annu. Rev. Phys. Chem.*, 2012, **63**, 1.
33. T. A. Cross and S. J. Opella, *Curr. Opin. Struct. Biol.*, 1994, **4**, 574.
34. G. Veglia, K. N. Ha, L. Shi, R. Verardi and N. J. Traaseth, *Methods Mol. Biol.*, 2010, **654**, 303.
35. K. Takegoshi, S. Nakamura and T. Terao, *Chem. Phys. Lett.*, 2001, **344**, 631.
36. N. J. Traaseth, R. Verardi and G. Veglia, *J. Am. Chem. Soc.*, 2008, **130**, 2400.
37. R. Verardi, N. J. Traaseth, L. R. Masterson, V. V. Vostrikov and G. Veglia, *Adv. Exp. Med. Biol.*, 2012, **992**, 35.
38. J. S. Waugh, *Proc. Natl. Acad. Sci. U. S. A.*, 1976, **73**, 1394.
39. A. Ramamoorthy, Y. F. Wei and D. K. Lee, *Annu. Rep. NMR Spectrosc.*, 2004, **52**, 1.
40. N. Tjandra and A. Bax, *Science*, 1997, **278**, 1111.
41. S. K. Straus, W. R. Scott, C. D. Schwieters and D. A. Marvin, *Eur. Biophys. J.*, 2011, **40**, 221.
42. S. Kim, J. R. Quine and T. A. Cross, *J. Am. Chem. Soc.*, 2001, **123**, 7292.
43. J. Zamoon, A. Mascioni, D. D. Thomas and G. Veglia, *Biophys. J.*, 2003, **85**, 2589.
44. L. Shi, N. J. Traaseth, R. Verardi, M. Gustavsson, J. Gao and G. Veglia, *J. Am. Chem. Soc.*, 2011, **133**, 2232.
45. M. S. Al-Abdul-Wahid, R. Verardi, G. Veglia and R. S. Prosser, *J. Biomol. NMR*, 2011, **51**, 173.
46. A. M. Slovic, C. M. Summa, J. D. Lear and W. F. DeGrado, *Protein Sci.*, 2003, **12**, 337.

47. A. M. Slovic, S. E. Stayrook, B. North and W. F. Degrado, *J. Mol. Biol.*, 2005, **348**, 777.
48. A. M. Slovic, J. D. Lear and W. F. DeGrado, *J. Pept. Res.*, 2005, **65**, 312.
49. K. J. Walters, A. E. Ferentz, B. J. Hare, P. Hidalgo, A. Jasanoff, H. Matsuo and G. Wagner, *Methods Enzymol.*, 2001, **339**, 238.
50. T. Gopinath and G. Veglia, *J. Am. Chem. Soc.*, 2009, **131**, 5754.
51. T. Gopinath and G. Veglia, *Chem. Phys. Lett.*, 2010, **494**, 104.
52. T. Gopinath, N. J. Traaseth, K. Mote and G. Veglia, *J. Am. Chem. Soc.*, 2010, **132**, 5357.
53. T. Gopinath, R. Verardi, N. J. Traaseth and G. Veglia, *J. Phys. Chem. B*, 2010, **114**, 5089.
54. T. Gopinath, K. R. Mote and G. Veglia, *J. Chem. Phys.*, 2011, **135**, 074503.
55. J. C. Hus, L. Salmon, G. Bouvignies, J. Lotze, M. Blackledge and R. Bruschweiler, *J. Am. Chem. Soc.*, 2008, **130**, 15927.
56. N. J. Traaseth, J. J. Buffy, J. Zamoon and G. Veglia, *Biochemistry*, 2006, **45**, 13827.
57. N. J. Traaseth, L. Shi, R. Verardi, D. G. Mullen, G. Barany and G. Veglia, *Proc. Natl. Acad. Sci. U. S. A.*, 2009, **106**, 10165.
58. N. J. Traaseth, R. Verardi, K. D. Torgersen, C. B. Karim, D. D. Thomas and G. Veglia, *Proc. Natl. Acad. Sci. U. S. A.*, 2007, **104**, 14676.
59. C. D. Schwieters, J. J. Kuszewski and G. M. Clore, *Prog. Nucl. Magn. Reson. Spectrosc.*, 2006, **48**, 47.
60. R. Bertram, J. R. Quine, M. S. Chapman and T. A. Cross, *J. Magn. Reson.*, 2000, **147**, 9.
61. R. Bertram, T. Asbury, F. Fabiola, J. R. Quine, T. A. Cross and M. S. Chapman, *J. Magn. Reson.*, 2003, **163**, 300.
62. S. I. O'Donoghue and M. Nilges, in *Biological Magnetic Resonance*, ed. N. Rama Krishna and L. J. Berliner, Springer, New York, 2002, vol. 17, p. 131.
63. A. Senes, D. C. Chadi, P. B. Law, R. F. Walters, V. Nanda and W. F. DeGrado, *J. Mol. Biol.*, 2007, **366**, 436.
64. R. R. Ketchem, B. Roux and T. A. Cross, *Structure*, 1997, **5**, 1655.
65. V. V. Vostrikov, A. E. Daily, D. V. Greathouse and R. E. Koeppe 2nd, *J. Biol. Chem.*, 2010, **285**, 31723.
66. W. Liu, J. Z. Fei, T. Kawakami and S. O. Smith, *Biochim. Biophys. Acta*, 2007, **1768**, 2971.
67. C. B. Karim, J. D. Stamm, J. Karim, L. R. Jones and D. D. Thomas, *Biochemistry*, 1998, **37**, 12074.
68. D. K. Ceholski, C. A. Trieber, C. F. Holmes and H. S. Young, *J. Biol. Chem.*, 2012, **287**, 26596.
69. H. K. Simmerman, Y. M. Kobayashi, J. M. Autry and L. R. Jones, *J. Biol. Chem.*, 1996, **271**, 5941.
70. L. Becucci, A. Cembran, C. B. Karim, D. D. Thomas, R. Guidelli, J. Gao and G. Veglia, *Biophys. J.*, 2009, **96**, L60.
71. S. L. Robia, N. C. Flohr and D. D. Thomas, *Biochemistry*, 2005, **44**, 4302.

72. J. M. Autry, J. E. Rubin, S. D. Pietrini, D. L. Winters, S. L. Robia and D. D. Thomas, *J. Biol. Chem.*, 2011, **286**, 31697.
73. S. L. Robia, K. S. Campbell, E. M. Kelly, Z. Hou, D. L. Winters and D. D. Thomas, *Circ. Res.*, 2007, **101**, 1123.
74. C. B. Karim, T. L. Kirby, Z. Zhang, Y. Nesmelov and D. D. Thomas, *Proc. Natl. Acad. Sci. U. S. A.*, 2004, **101**, 14437.
75. Y. E. Nesmelov, C. B. Karim, L. Song, P. G. Fajer and D. D. Thomas, *Biophys. J.*, 2007, **93**, 2805.
76. H. Ghimire, S. Abu-Baker, I. D. Sahu, A. Zhou, D. J. Mayo, R. T. Lee and G. A. Lorigan, *Biochim. Biophys. Acta*, 2012, **1818**, 645.
77. S. Abu-Baker, J. X. Lu, S. Chu, K. K. Shetty, P. L. Gor'kov and G. A. Lorigan, *Protein Sci.*, 2007, **16**, 2345.
78. A. G. Palmer III, *Annu. Rev. Biophys. Biomol. Struct.*, 2001, **30**, 129.
79. A. G. Palmer III and F. Massi, *Chem. Rev.*, 2006, **106**, 1700.
80. Y. Kimura, M. Asahi, K. Kurzydlowski, M. Tada and D. H. MacLennan, *J. Biol. Chem.*, 1998, **273**, 14238.
81. M. Gustavsson, N. J. Traaseth and G. Veglia, *Biochim. Biophys. Acta*, 2012, **1818**, 146.
82. D. H. MacLennan, *Eur. J. Biochem.*, 2000, **267**, 5291.
83. E. Zvaritch, P. H. Backx, F. Jirik, Y. Kimura, S. de Leon, A. G. Schmidt, B. D. Hoit, J. W. Lester, E. G. Kranias and D. H. MacLennan, *J. Biol. Chem.*, 2000, **275**, 14985.
84. L. T. Bish, M. M. Sleeper, C. Reynolds, J. Gazzara, E. Withnall, G. E. Singletary, G. Buchlis, D. Hui, K. A. High, G. Gao, J. M. Wilson and H. L. Sweeney, *Hum. Gene Ther.*, 2011, **22**, 969.
85. G. Chu, L. Li, Y. Sato, J. M. Harrer, V. J. Kadambi, B. D. Hoit, D. M. Bers and E. G. Kranias, *J. Biol. Chem.*, 1998, **273**, 33674.
86. K. N. Ha, L. R. Masterson, Z. Hou, R. Verardi, N. Walsh, G. Veglia and S. L. Robia, *Proc. Natl. Acad. Sci. U. S. A.*, 2011, **108**, 2735.
87. M. H. Ali and B. Imperiali, *Bioorg. Med. Chem.*, 2005, **13**, 5013.
88. Y. Shen, F. Delaglio, G. Cornilescu and A. Bax, *J. Biomol. NMR*, 2009, **44**, 213.
89. B. L Akin, T. D. Hurley, Z. Chen and L. R. Jones, *J. Biol. Chem.*, 2013, **288**(42), 30181–30191.
90. M. Gustavsson, R. Verardi, D. G. Mullen, K. R. Mote, N. J. Traaseth, T. Gopinath and G. Veglia, *Proc. Natl. Acad. Sci. USA*, 2013, **110**(43), 17338–17343.
91. N. Das, D. T. Murray and T. A. Cross, *Nat. Protoc.*, 2013, **8**(11), 2256–2570.
92. G. Veglia, N. J. Traaseth, L. Shi, R. Verardi, T. Gopinath and M. Gustavsson, *Comprehensive Biophysics*, 2012, **1**, 182–198.

CHAPTER 23

Structural Insights from Solid-State NMR into the Function of the Bacteriorhodopsin Photoreceptor Protein

PETER J. JUDGE, GARRICK F. TAYLOR, LOUIC S. VERMEER AND ANTHONY WATTS*

Biomembrane Structure Unit, Biochemistry Department, South Parks Road, Oxford, OX1 3QU, UK
*Email: anthony.watts@bioch.ox.ac.uk

23.1 Introduction

Membrane-embedded photoreceptors are a diverse group of polytopic proteins that enable organisms to detect and harvest energy from sunlight. Photoreceptors are suitable subjects for detailed biophysical and structural studies, since they are typically found in high abundance in specialized membranes within cells.[1,2] One major family of photoreceptors, the retinal-containing proteins commonly known as opsins, are found in all domains of life and play key roles in a variety of biochemical pathways, including phototaxis and photosynthesis.[3,4] Although the basic architecture of the proteins in this family is consistent (all have seven transmembrane helices), there is considerable diversity in their response to light, from the transport of

New Developments in NMR No. 3
Advances in Biological Solid-State NMR: Proteins and Membrane-Active Peptides
Edited by Frances Separovic and Akira Naito

Published by the Royal Society of Chemistry, www.rsc.org

Table 23.1 Selected retinal-containing opsin proteins for which NMR structural data and spectra have been published.

Protein	*Organism*	*Function*
Rhodopsin	Mammals	GPCR receptor for vision
Bacteriorhodopsin (bR)	*H. salinarum*	H^+ ion transport
Halorhodopsin (HR)	*H. salinarum*	Cl^- ion transport
Sensory rhodopsin II (*pp*R)[a]	*N. pharaonis*	Phototaxis receptor
Proteorhodopsin (PR)	γ-Proteobacteria	H^+ ion transport
Anabaena sensory rhodopsin (ASR)	*Anabaena* PCC 7120	Light-driven molecular rotor

[a]Sensory rhodopsin II is also known as *pharonis* phoborhodopsin.

charged ions in archaebacteria (*e.g.* bacteriorhodopsin) to the initiation of a signalling cascade in the mammalian eye (*e.g.* rhodopsin) (see Table 23.1). Although detailed structural information is now available for many members of the opsin family, crystallographic structures of intermediates are not able on their own to provide a complete description of the mechanisms of light absorption and energy transduction. Solid-state NMR is a powerful method for providing structural information for integral membrane proteins and is of particular importance for opsin photoreceptors, since it is able to resolve the interactions between the protein and retinal to sub-atomic resolution.[5]

Bacteriorhodopsin (bR) is a retinal containing transmembrane H^+ pump, found in the purple membrane of *Halobacterium salinarum* archaea, which enables the organism to generate a proton motive force across the cytoplasmic membrane for ATP synthesis.[6] The 26 kDa protein contains seven transmembrane α helices and a retinal group (in the form of retinylidine), covalently bound to the ε-amino group of Lys216 on helix 7.[7] The energy from the light absorbed by the protein causes an isomerization of the retinal (from *trans* to 13-*cis*), which in turn induces conformational changes within the protein, including a change in the tilt angle of the E and F helices.[8,9] Changes in the packing of the helices alter the exposure of a series of ionizable residues within the protein (including Asp85, Asp96, Glu194 and Glu294) causing stepwise changes in the sidechain pK_a values of up to 3 pH units, resulting in H^+ ion flux to the extracellular side of the protein.[10]

The ordered sequence of conformational changes which occurs in bR as a result of light absorption is known as the photocycle and consists of at least eight distinct states, characterized by changes in the wavelength of maximum absorbance (Scheme 23.1).[10] Structures for five of the states have been proposed, based on the crystal structures of illuminated bR mutants (some of which may be in mixed states and in which the photocycle has been arrested).[11] In this review we discuss the contributions from solid-state NMR (ssNMR) experiments to the understanding of bR photocycle intermediates and the mechanism of proton pumping, focusing in particular on the M and L intermediates. We also identify instances where ssNMR has highlighted key mechanistic and structural differences between the diverse members of the opsin family.

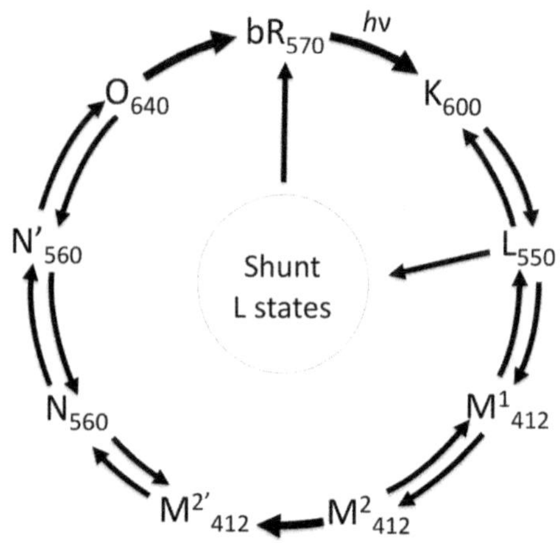

Scheme 23.1 Photocycle of bacteriorhodopsin (bR) showing intermediates identified by the wavelength of their maximum absorbance. Absorption of a photon ($h\nu$) occurs between the bR ground state and the K_{600} state. For clarity, the short-lived J state (a precursor to the K_{600} state with a lifetime of less than 1 ps) has been omitted. The shunt L states identified in Bajaj *et al.*[24] are also shown. Release of an H^+ ion to the extracellular side of the membrane occurs during the M^2 to $M^{2'}$ transition and uptake of an H^+ ion from the cytoplasmic side occurs during the N to N′ transition.

23.2 Determination of the Retinal Chromophore Structure by MAS ssNMR

A key question for early structural studies of opsin proteins was the accessibility of different conformational states of retinal in the ground or dark-adapted states. The NMR chemical shifts of ^{13}C and ^{15}N resonances are dependent upon the electronic environments of labelled atoms, and ssNMR is able to provide precise information on molecular structure and dynamics. Early crystal structures of bovine rhodopsin had predicted that only one conformation of the retinal chromophore could be accommodated within the binding pocket of the ground state; however, the existence of a second state was demonstrated by ssNMR using rhodopsin with either 11-*Z*-[8,18-$^{13}C_2$]retinal or 11-*Z*-[8,16/17-$^{13}C_2$]retinal regenerated into it. Comparison of the ^{13}C MAS chemical shift data revealed that a second conformation of retinal was present in ~26% of the sample.[12] Similarly, ^{13}C MAS spectra revealed that retinal is in equilibrium between 13-*cis*,15-*syn* and all-*trans* isomers in the dark-adapted state of bR (see Figure 23.1 for structures of retinal isomers). In contrast, the ^{13}C spectra of [$^{13}C_2$]retinal incorporated into dark-adapted proteorhodopsin (PR; a bacterial homologue of bacteriorhodopsin) show only a single resonance for each labelled atom, suggesting that only one isomer of retinylidene is present in PR.[13,14]

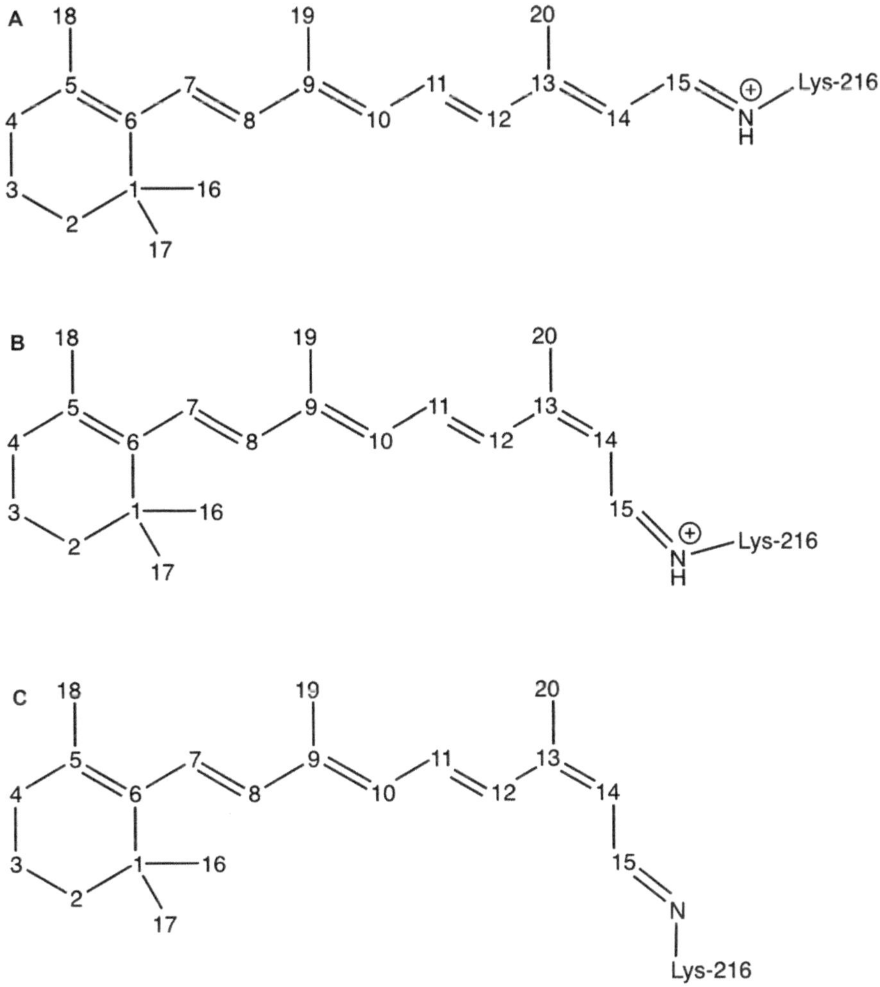

Figure 23.1 Structures of the three common retinal isomers present within bR. (A) All-*trans*,15-*anti* retinal Schiff base, bR_{568}. (B) 13-*cis*,15-*syn* retinal Schiff base, bR_{555}. (C) 13-*cis*,15-*anti* retinal Schiff base, M_{405}. The subscripted names indicate in which photocycle intermediate the isomer is commonly found and the associated wavelength at which the intermediate absorbs maximally.

The accurate structure of a chromophore at its binding site in a photoreceptor may be calculated from ssNMR-generated distance constraints to sub-Ångstrom resolution.

Rotational echo double resonance (REDOR) and rotational resonance (R^2) are two solid-state NMR experiments commonly used to provide precise measurements of chromophore structure in different conformational states of the receptor photocycle. Under the conditions of magic angle spinning (MAS), the dipolar couplings between labelled nuclei are averaged; however, the selective pulses applied during the two experiments allow distance-sensitive recoupling to be introduced, producing a series of constraints from which a structure may be accurately calculated. R^2 was used to produce distance restraints between labelled retinal atoms in the ground state of bovine rhodopsin.[15] The resulting distance constraints of C8 to C16 and C17 (4.05 ± 0.25 Å) and from C8 to C18 (2.95 ± 0.15 Å) show that the major portion of retinylidene in rhodopsin has a twisted 6-s-*cis* conformation.[15]

The characteristics of the protein environment immediately surrounding the chromophore are responsible for fine-tuning the absorbance maximum and in constraining the freedom of the retinal to undergo conformational changes. Of the eight tryptophan residues present in bR, four are in close proximity to retinal: Trp138 and Trp189 bind to the C1–C6 ring, whereas Trp86 and Trp182 lie adjacent to the C11–C15 atoms. Differences between the Trp indole-^{15}N MAS chemical shifts in the ground states and M states indicate a change in the hydrogen bonding patterns involving the four residues. In particular, the proximity of Trp182 to the C13–C14 bond suggested that the position of this residue would be particularly sensitive to the isomerization of the bond. REDOR measurements of the distance between the Trp182 indole-N and the C12 position of [20-^{13}C]retinal indicated that the two groups came closer together in the M_{412} state than in the light-adapted (LA) state.[16] A refinement of the standard REDOR experiment, simultaneous frequency and amplitude modulation (SFAM) REDOR,[17] is more tolerant of B_1 field inhomogeneity[18] and has been used to observe a 0.3 Å movement in the ring of Trp86 towards the C14 position of retinal, during the transition from the 13-*cis*-15-*syn*-retinal to the all-*trans*-15-*anti* conformer.[19] Helmle *et al.* have noted that there are clear differences in the apparent retinal conformation between the published X-ray crystal structures and highlight in particular the fact that the limitations in resolution for the crystal and electron microscopy structures of bR present an opportunity for ssNMR data to refine structural models generated by other biophysical techniques.[19]

23.3 Determination of Protein Structure by MAS NMR

Although crystallographic methods are able to provide detailed high-resolution structural information for the most rigid parts of proteins (in the case of integral membrane proteins this is generally the transmembrane secondary structure elements), they are often less successful at providing detailed structural information for regions with the highest degree of conformational flexibility. The solvent-facing loops between transmembrane domains of retinal proteins play a key role in forming inter-protein contacts required by signalling cascades and facilitate conformational changes during substrate transport and retinal isomerization. NMR can provide a useful method of determining the structures of these more mobile elements, using the dependence of the precise chemical shift of backbone Cα, Cβ and CO resonances on the torsion angles of adjacent bonds.[20] A critical stage in structure determination by this method is the *ab initio* assignment of individual resonances within a given region of the protein from complex two- and three-dimensional correlation spectra, which (in the case of membrane proteins) are acquired under MAS conditions. Once assigned, backbone dihedral angle restraints are derived from the TALOS database of chemical shifts and the resulting structures are calculated and refined using a simulated annealing protocol.[21]

Figure 23.2 shows the strip plots and resulting calculated structure determined for the AB loop of uniformly labelled ground state bR, acquired at a ^{1}H frequency of 800 MHz.[22] Some key differences are observed between the structure calculated from NMR and that obtained by crystallography. It is possible that artifacts of the crystallization process are responsible for introducing perturbations into the loop region and it should be noted that the ssNMR-derived structure is obtained from experimental spectra of bR in its native environment. A similar study of the mobility of the N- and C-termini and AB, CD and FG loops of PR has also been carried out using this method.[14]

Further photocycle intermediates of bR have been characterized by dynamic nuclear polarization-enhanced (DNP) solid-state NMR, which offers a significantly increased sensitivity over conventional ssNMR. DNP exploits the greater degree of spin polarization of unpaired electrons at Boltzmann equilibrium, by transferring the polarization to the NMR-active nuclei using microwave radiofrequency radiation. Enhancement of ^{13}C resonances in bR in the purple membrane has been recorded at a range of magnetic field strengths.[23] Signal enhancement by DNP of a series of cryo-trapped bR intermediates at 90 K has allowed acquisition of ^{15}N–^{13}C and ^{15}N–^{13}C–^{13}C correlation spectra, which revealed that the L-state (previously identified by crystallography as a well-defined single state) is in practice divided into four separate substates. The calculated orientation of the Schiff base N–H vector is in good agreement with quantum mechanics/molecular mechanics simulations which predict that this group is oriented towards the cytoplasmic side of the protein in the L-intermediate,[24] although there is some disagreement in the orientation of this bond in the existing crystal structures.[25,26] DNP experiments have also been used to compare the electron spin density distributions in chlorophyll in ^{15}N-labelled photosystems I and II from spinach chloroplasts.[27]

23.4 Determination of Chromophore Structure by Oriented Sample NMR

Many NMR interactions are orientation-dependent and can, therefore, yield information on the orientation of bonds, groups and even the entire protein molecule under investigation. However, in solution and MAS solid-state

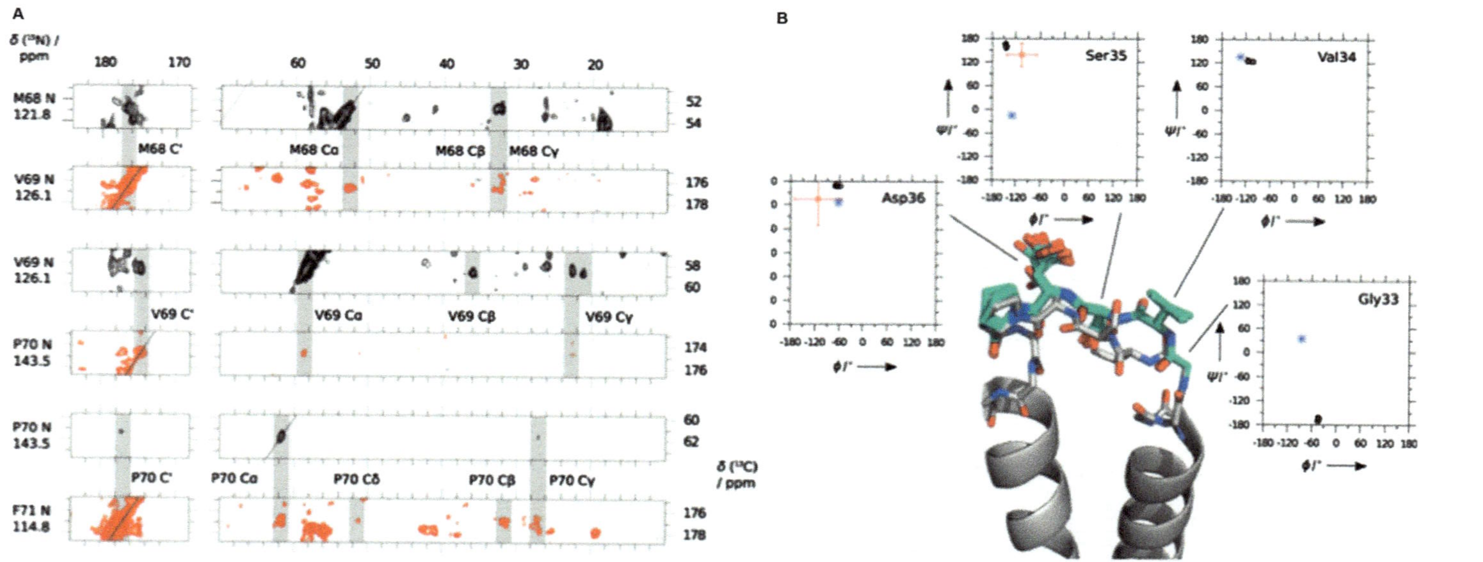

Figure 23.2 (A) Strip plots for the assignment of selected bR AB loop residues using NCACX (*black*, 20 ms mixing), NCOCX (*red*, 20 ms), CANCO (*orange*) and CAN(CO)CX (*blue*, 45 ms) spectra. (B) The AB loop of bR: the 1QHJ crystal structure is shown in *grey*, the 10 lowest-energy structures calculated using XPLOR based on the dihedral angles predicted by TALOS+ are shown in *cyan*. Side-chain conformations and Ramachandran plots are shown for Gly33–Pro37: 1QHJ crystal structure (*blue stars*), TALOS+ prediction (*red squares*), ensemble structure calculated by XPLOR-NIH (*black circles*). (Adapted from Higman *et al.*[22])

NMR these interactions are averaged out, losing much of the orientational information available. This orientational information can be extracted if the sample is manipulated so that all molecules are aligned. Mechanical orientation of membrane samples may be achieved through layering on glass plates, thus aligning the lipid bilayers and any associated proteins.[28]

The oriented sample (OS) NMR technique has been used to great success in determining the orientation of the chromophore within the retinal-containing proteins, since membranes containing these proteins orient effectively on glass plates. Deuterium NMR studies of retinal labelled at the methyl group position C19 or C20 within bovine rhodopsin in aligned dimyristoylphosphatidylcholine bilayers showed that the C–CD_3 bond vectors of the C19 and C20 segment orient at $42.5 \pm 5°$ and $30 \pm 5°$ with respect to the membrane normal.[29] The same experiments performed on labelled retinal in bR yielded angles of $40 \pm 1°$ and $32 \pm 1°$ for the C19 and C20 C–CD_3 bond vectors, respectively. These experiments also yielded an angle of $37 \pm 1°$ for C18; these angles together showed that the methyl groups are not parallel with each other and, therefore, that the retinal must be distorted with an in-plane curvature and possibly an out-of-plane twist whilst in the dark-adapted state.[30] Further deuterium studies on the cyclohexane ring of the retinal in bR found the C16 and C17 methyls oriented at $94 \pm 2°$ and $75 \pm 2°$ to the membrane normal, respectively.

23.5 Reorientation of Retinal in the M Intermediates of bR and Rhodopsin

The M_{412} intermediate is an important and well-characterized intermediate in the bR photocycle. Its importance lies in the fact that it is during this stage that the retinal isomerizes from all-*trans* to 13-*cis*,15-*anti* after absorption of a photon, inducing structural changes in the protein.[31]

In the purple membrane at room temperature, the M intermediate, which appears yellow, is short lived and is stable for only milliseconds, before reverting to the ground state.[31] However, the M intermediate may be isolated for longer (minutes) timescales using guanidine hydrochloride or temperatures below 223 K. Oriented ^2H NMR has been used to measure the C–CD_3 bond angle at C19 of the trapped M intermediate state, which was determined to be $44 \pm 2°$, $4°$ greater than in the ground state. The change in bond angle indicated that the end of the retinal covalently bound to the Schiff base swings around upon isomerization, inducing an upward tilting of the retinal long-axis, which partially compensates for this movement.[31] In a similar experiment the C16, C18 and C19 C–CD_3 bond vectors in both ground and M intermediate states increased noticeably, but the increase was greatest for the C19 bond which increased markedly, to $7 \pm 3°$ (within the margins of error of the previous study).[15] In the ground state of bovine rhodopsin, contrary to bR, the retinal is present as the 13-*cis* isomer in the ground state and isomerizes to *trans* in the light-activated M state.[32] In experiments analogous to those described for bR, aligned deuterium ssNMR studies, performed with changing tilt angles, of retinal contained within rhodopsin in the M_I state were carried out. Here the C–CD_3 bond angles for C16, C19 and C20 were found to be $72 \pm 4^\circ$, $53 \pm 3^\circ$ and $59 \pm 3^\circ$, respectively.[33] Earlier studies on the ground state found these angles to be $70 \pm 3^\circ$, $52 \pm 3^\circ$ and $68 \pm 3^\circ$ for the respective bonds.[29] These results are consistent with significant conformational distortion of retinal and, in particular, indicate that the β-ionone ring has a twisted 6-s-*cis* conformation, whereas the polyene chain is twisted 12-s-*trans*. The results from the studies of the M_I intermediate show a much less distorted conformation than for the ground state.

The results of all the above experiments are not only important for determining the mechanism of proton translocation in bR and the method of signal transduction in the rhodopsins, but also demonstrate that ssNMR is able to give highly detailed structural information for ligands at their binding sites. Ligand structures can be obtained in both activated and non-activated states, for membrane proteins in a native or near-native environment, free from the possible artifacts induced by crystallization. Similar experiments have the potential to inform a range of mechanistically driven studies, not least drug design.[34]

23.6 Determination of Protein structure by OS NMR

Crystal structures of a series of bR photocycle intermediates stabilized by site-directed mutagenesis suggest that the isomerization of the retinal group, which occurs on light absorption, in turn results in movement of the transmembrane α-helices (especially helices X and Y). The changes in tilt angles of a helix within a membrane may be calculated from changes in the ^{15}N chemical shift and the ^{15}N–^{1}H dipolar couplings arising from the backbone amide groups, since both parameters are dependent on the orientation of the N–H bond vector.

Two NMR experiments are particularly suitable for determining the orientations of α-helices in a membrane: (1) the PISEMA (polarization inversion spin exchange at the magic angle) experiment provides chemical shift information along the direct dimension and dipolar couplings in the indirect dimension;[35] and (2) the HETCOR (heteronuclear correlation) experiment correlates ^{15}N and ^{1}H chemical shifts in the direct and indirect dimensions, respectively. The key difference between these methods is that whereas PISEMA provides good resolution for transmembrane helices (aligned with the magnetic field), the resolution of HETCOR is higher for tilt angles from 60° to 90° and thus more suitable for loop regions.[36,37] The interpretation of these experiments usually involves extensive computer simulations of the lineshapes or peak positions, while taking into account the static disorder (mosaic spread) as well as the dynamics of the protein under study.[38] A combination of the SIMMOL[39] and SIMPSON[40] programs is usually used to carry out these simulations.

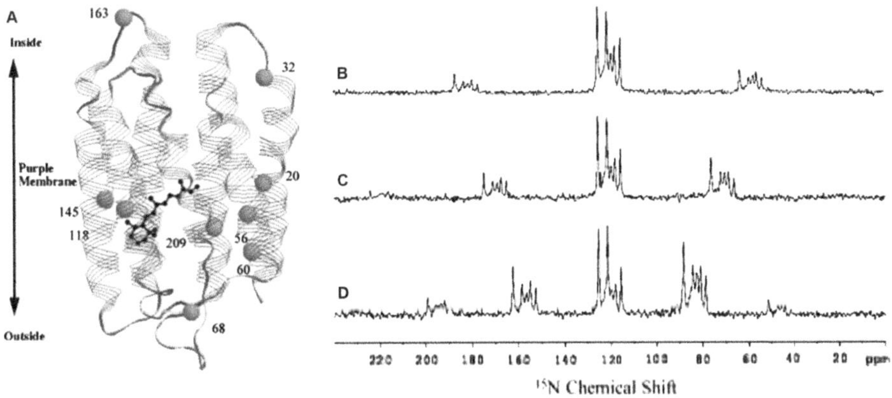

Figure 23.3 (A) Bacteriorhodopsin three-dimensional structure indicating the positions of the nine ^{15}N-labelled methionine residues, with six located in the transmembrane helices (Met20, Met56, Met60, Met118, Met145 and Met209), two in the loops (Met68 and Met163) and one (Met32) at the loop helix interface of helix A. RASMOL was used with coordinates 2BRD from the PDB, which is a structure of the protein embedded in the membrane. (B–D) ^{15}N CP MAOSS spectra for [^{15}N]methionine-labelled purple membranes prepared from buffered D_2O showing isotropic chemical shifts and spinning sideband families [9.5 mg bR, 253 K, DSX 400: (B) 2.5 kHz, (C) 2 kHz and (D) 1.5 kHz MAS frequencies; 160-, 132- and 239-k acquisitions, respectively] for the five observed residual resonances.
(Adapted from Mason *et al.*[46])

In 2005, PISEMA (in combination with the REDOR method described earlier) was used to determine orientational constraints for [^{15}N]Met-labelled bR in samples of purple membranes mechanically aligned on glass plates.[41] (The selective labelling of the protein reduces the spectral crowding and facilitates residue assignment, whilst the distribution of the nine methionine residues of the protein allows orientational information to be obtained for all seven helices; see Figure 23.3.) Pitch angles for helices A and B were determined (after spectra deconvolution using SIMPSON) and were found to be in good agreement with crystallographic structures. Interestingly, the native environment of the purple membrane lipids (present in the ssNMR samples, but absent from bR crystals) appears to influence the position of the exact boundaries between the loops and helices. Individual resolution and assignment of all methionine residues was not possible, since the methionine residues in different helices are aligned at different angles within the protein, resulting in broadened, overlapping resonances. Assignment of all the methionine resonances was possible though a subsequent series of methionine to valine mutations, D_2O to H_2O exchanges and magnetization transfer experiments, giving detailed structural information for all nine methionine residues.

Although none of the bR structures published in the Protein Data Bank (PDB) are determined by ssNMR, both PISEMA and REDOR experiments have contributed structural information in the loop regions that are often ill defined by X-ray crystallography. The orientational changes in the retinal during the photocycle have been determined by ssNMR.[31,42,43] ssNMR has also determined orientational changes in some of the side chains in the active site.[44]

23.7 Magic Angle Oriented Sample Spinning NMR of bR

Further refinement of the mechanical orientation of membrane samples for NMR was introduced with magic angle oriented sample spinning (MAOSS) NMR, which allows the line broadening induced by the mosaic spread of the sample to be averaged under MAS conditions.[45] The resulting increase in spectral resolution is sufficient to allow amide ^{15}N resonances to be assigned in selectively labelled ground-state bR and is further enhanced by ^{2}H/^{1}H exchange labelling, which reduces the dipolar couplings to protons. Structural orientational constraints may be obtained from the ^{15}N chemical shift anisotropy, which is calculated from the isotropic chemical shifts and analysis of the positions of the MAOSS spinning sidebands, although in the case of the methionine residues of bR additional mutagenesis experiments were required to assign resonances to specific amino acids.[46] Although MAOSS is generally less precise than PISEMA, its decreased linewidth resulting from the MAS averaging out the effect of mosaic spread helps to obtain a higher resolution.

The orientation dependence of the ^{2}H quadrupole interaction can also be exploited using MAOSS, since the quadrupole interaction between the electric field gradient at the ^{2}H nucleus and its electric quadrupole moment is related to the orientations of the relevant molecular and external reference frames. The orientation of C–CD_3 groups in retinal in both ground and M_{412} states has been determined by this method, by exploiting the orientational dependence of the position of ^{2}H spinning side bands.[47] Table 23.2

Table 23.2 Comparison of retinal C–C bond vector angles determined by ^{2}H ssNMR from oriented deuterated purple membranes[30] and MAOSS ssNMR.[47] Vector angles calculated from selected bR crystal structures are also given.

Study				
^{2}H ssNMR (all angles ± 10°)	*Intermediate*	*C–C_{18}*	*C–C_{19}*	*C–C_{20}*
Ulrich *et al.* (1994)[30]	bR_{568}	37°	40°	32°
Ulrich *et al.* (1994)[30]	M_{412}	–	44°	–
Glaubitz *et al.* (1999)[47]	bR_{568}	36°	–	22°
Glaubitz *et al.* (1999)[47]	M_{412}	20°	–	10°
X-ray crystal structures				
Pebay-Peroula *et al.* (1996)	–	45.2°	37.9°	9.4°
Luecke *et al.* (1999)	–	35.1°	33.8°	21.4°
Belhavi *et al.* (1999)	–	39.8°	41.2°	22.7°
Sass *et al.* (1999)	–	36.2°	39.2°	19.0°

compares the angles of retinal C–C bond vectors determined by both static and MAOSS ^{2}H NMR spectroscopy of mechanically oriented purple membranes.

23.8 Functional Insights from ssNMR into the bR Photocycle

The transport of H^{+} ions across the *H. salinarum* purple membrane is a direct result of the sequential changes in pK_a of the ionizable groups in the interior of the protein, induced by the conformation changes which follow light absorption. The retinal chromophore in bR is covalently attached to the protein moiety *via* a Schiff base linkage to Lys216. Crystal structures show that the Schiff base is protonated in the ground state and deprotonated in the M state,[48,49] and this change in protonation state is thought to be essential for proton transport from the cytoplasmic to the extracellular side of the protein. MAS ssNMR has been used to investigate the conformations of bR during the photocycle and detect photointermediates. Using [ζ-^{15}N]Lys-labelled bR, the isotropic ^{15}N chemical shifts originating from the Schiff base in the L, M^{1}, M^{2} and N intermediates are fully distinguishable.[50,51] The different intermediates are so readily distinguishable *via* the Schiff base ^{15}N chemical shift because the various conformations cover a range of almost 190 ppm. The ^{15}N chemical shift of the Schiff base in bR_{568} is strongly shielded, more so than in any of the model compounds that have been studied,[52] consistent with the presence of an extremely delocalized counter-ion, which results in the stable protonation of the Schiff base in the bR resting state. However, in the L state the ^{15}N chemical shift downshifts by approximately 130 ppm, demonstrating that the Schiff base now has a stronger interaction with its counter-ion.[50] During the M^{2} state, when the Schiff base becomes deprotonated, the ^{15}N chemical shift jumps downfield to 323 ppm and there is a large increase in chemical shift anisotropy, as demonstrated by the loss of centre-band intensity at moderate MAS frequencies.

23.9 Conclusion

Solid-state NMR provides a diversity of methodologies for obtaining functional and structural information for photocycle intermediates of retinal-containing opsin proteins. ssNMR allows this diverse family of proteins to be studied in either their native membranes or in near-native membrane environments and, unlike crystallographic methods, can study trapped intermediates without the need for mutagenesis of key functional residues. The advent of higher magnetic fields and the development of signal-enhancing technology (including DNP) offer an exciting prospect for future high-resolution characterization of biological photoreceptors.

Acknowledgements

The authors acknowledge funding from the Engineering and Physical Sciences Research Council (UK), the Medical Research Council (UK), the Biotechnology and Biological Sciences Research Council (UK) and the National Physical Laboratory (London, UK).

References

1. C. R. Worthington, *Annu. Rev. Biophys. Bioeng.*, 1974, **3**, 53–80.
2. R. Henderson, *Annu. Rev. Biophys. Bioeng.*, 1977, **6**, 87–109.
3. W. D. Hoff, K. H. Jung and J. L. Spudich, *Annu. Rev. Biophys. Biomol. Struct.*, 1997, **26**, 223–258.
4. U. Haupts, J. Tittor and D. Oesterhelt, *Annu. Rev. Biophys. Biomol. Struct.*, 1999, **28**, 367–399.
5. P. J. Judge and A. Watts, *Curr. Opin. Chem. Biol.*, 2011, **15**, 690–695.
6. M. T. Facciotti, S. Rouhani-Manshadi and R. M. Glaeser, *Trends Biochem. Sci.*, 2004, **29**, 445–451.
7. H. Luecke, B. Schobert, H. T. Richter, J. P. Cartailler and J. K. Lanyi, *J. Mol. Biol.*, 1999, **291**, 899–911.
8. H. Luecke, B. Schobert, J. P. Cartailler, H. T. Richter, A. Rosengarth, R. Needleman and J. K. Lanyi, *J. Mol. Biol.*, 2000, **300**, 1237–1255.
9. H. Luecke, B. Schobert, H. T. Richter, J. P. Cartailler and J. K. Lanyi, *Science*, 1999, **286**, 255–261.
10. J. K. Lanyi, *Biochim. Biophys. Acta*, 2006, **1757**, 1012–1018.
11. J. K. Lanyi, *Mol. Membr. Biol.*, 2004, **21**, 143–150.
12. S. Moltke, I. Wallat, N. Sakai, K. Nakanishi, M. F. Brown and M. P. Heyn, *Biochemistry*, 1999, **38**, 11762–11772.
13. L. Shi and V. Ladizhansky, *Methods Mol. Biol.*, 2012, **895**, 153–165.
14. L. Shi, E. M. Lake, M. A. Ahmed, L. S. Brown and V. Ladizhansky, *Biochim. Biophys. Acta*, 2009, **1788**, 2563–2574.
15. P. J. Spooner, J. M. Sharples, M. A. Verhoeven, J. Lugtenburg, C. Glaubitz and A. Watts, *Biochemistry*, 2002, **41**, 7549–7555.
16. A. T. Petkova, J. G. Hu, M. Bizounok, M. Simpson, R. G. Griffin and J. Herzfeld, *Biochemistry*, 1999, **38**, 1562–1572.
17. R. Fu, S. A. Smith and G. Bodenhausen, *Chem. Phys. Lett.*, 1997, **272**, 361–369.
18. L. K. Thompson, *Curr. Opin. Struct. Biol.*, 2002, **12**, 661–669.
19. M. Helmle, H. Patzelt, A. Ockenfels, W. Gartner, D. Oesterhelt and B. Bechinger, *Biochemistry*, 2000, **39**, 10066–10071.
20. S. Neal, M. Berjanskii, H. Zhang and D. S. Wishart, *Magn. Reson. Chem.*, 2006, **44**, S158–S167.
21. Y. Shen, F. Delaglio, G. Cornilescu and A. Bax, *J. Biomol. NMR*, 2009, **44**, 213–223.
22. V. A. Higman, K. Varga, L. Aslimovska, P. J. Judge, L. J. Sperling, C. M. Rienstra and A. Watts, *Angew. Chem. Int. Ed.*, 2011, **50**, 8432–8435.

23. K. J. Pike, T. F. Kemp, H. Takahashi, R. Day, A. P. Howes, E. V. Kryukov, J. F. MacDonald, A. E. Collis, D. R. Bolton, R. J. Wylde, M. Orwick, K. Kosuga, A. J. Clark, T. Idehara, A. Watts, G. M. Smith, M. E. Newton, R. Dupree and M. E. Smith, *J. Magn. Reson.*, 2012, **215**, 1–9.
24. V. S. Bajaj, M. L. Mak-Jurkauskas, M. Belenky, J. Herzfeld and R. G. Griffin, *Proc. Natl. Acad. Sci. U. S. A.*, 2009, **106**, 9244–9249.
25. A. N. Bondar, M. Elstner, S. Suhai, J. C. Smith and S. Fischer, *Structure*, 2004, **12**, 1281–1288.
26. A. N. Bondar, J. Baudry, S. Suhai, S. Fischer and J. C. Smith, *J. Phys. Chem. B*, 2008, **112**, 14729–14741.
27. A. Diller, E. Roy, P. Gast, H. J. van Gorkom, H. J. de Groot, C. Glaubitz, G. Jeschke, J. Matysik and A. Alia, *Proc. Natl. Acad. Sci. U. S. A.*, 2007, **104**, 12767–12771.
28. G. Grobner, A. Taylor, P. T. Williamson, G. Choi, C. Glaubitz, J. A. Watts, W. J. de Grip and A. Watts, *Anal. Biochem.*, 1997, **254**, 132–138.
29. G. F. J. Salgado, A. V. Struts, K. Tanaka, N. Fujioka, K. Nakanishi and M. F. Brown, *Biochemistry*, 2004, **43**, 12819–12828.
30. A. S. Ulrich, A. Watts, I. Wallat and M. P. Heyn, *Biochemistry*, 1994, **33**, 5370–5375.
31. A. S. Ulrich, I. Wallat, M. P. Heyn and A. Watts, *Nat. Struct. Biol.*, 1995, **2**, 190–192.
32. B. Aton, R. H. Callender and B. Honig, *Nature*, 1978, **273**, 784–786.
33. G. F. J. Salgado, A. V. Struts, K. Tanaka, S. Krane, K. Nakanishi and M. F. Brown, *J. Am. Chem. Soc.*, 2006, **128**, 11067–11071.
34. A. Watts, *Nat. Rev. Drug Discovery*, 2005, **4**, 555–568.
35. A. Ramamoorthy, C. H. Wu and S. J. Opella, *J. Magn. Reson.*, 1999, **140**, 131–140.
36. F. M. Marassi, C. Ma, J. J. Gesell and S. J. Opella, *J. Magn. Reson.*, 2000, **144**, 156–161.
37. T. Vosegaard and N. C. Nielsen, *J. Biomol. NMR*, 2002, **22**, 225–247.
38. S. K. Straus, W. R. Scott and A. Watts, *J. Biomol. NMR*, 2003, **26**, 283–295.
39. M. Bak, R. Schultz, T. Vosegaard and N. C. Nielsen, *J. Magn. Reson.*, 2002, **154**, 28–45.
40. M. Bak, J. T. Rasmussen and N. C. Nielsen, *J. Magn. Reson.*, 2000, **147**, 296–330.
41. M. Kamihira, T. Vosegaard, A. J. Mason, S. K. Straus, N. C. Nielsen and A. Watts, *J. Struct. Biol.*, 2005, **149**, 7–16.
42. A. S. Ulrich, M. P. Heyn and A. Watts, *Biochemistry*, 1992, **31**, 10390–10399.
43. A. S. Ulrich, A. Watts, I. Wallat and M. P. Heyn, *Biochemistry*, 1994, **33**, 5370–5375.
44. J. Herzfeld and J. C. Lansing, *Annu. Rev. Biophys. Biomol. Struct.*, 2002, **31**, 73–95.
45. C. Glaubitz and A. Watts, *J. Magn. Reson.*, 1998, **130**, 305–316.
46. A. J. Mason, S. L. Grage, S. K. Straus, C. Glaubitz and A. Watts, *Biophys. J.*, 2004, **86**, 1610–1617.

47. C. Glaubitz, I. J. Burnett, G. Grobner, A. J. Mason and A. Watts, *J. Am. Chem. Soc.*, 1999, **121**, 5787–5794.
48. H. Luecke, B. Schobert, H.-T. Richter, J.-P. Cartailler and J. K. Lanyi, *J. Mol. Biol.*, 1999, **291**, 899–911.
49. J. K. Lanyi and B. Schobert, *J. Mol. Biol.*, 2002, **321**, 727–737.
50. J. G. Hu, B. Q. Sun, A. T. Petkova, R. G. Griffin and J. Herzfeld, *Biochemistry*, 1997, **36**, 9316–9322.
51. J. G. Hu, B. Q. Sun, M. Bizounok, M. E. Hatcher, J. C. Lansing, J. Raap, P. J. E. Verdegem, J. Lugtenburg, R. G. Griffin and J. Herzfeld, *Biochemistry*, 1998, **37**, 8088–8096.
52. J. Hu, R. G. Griffin and J. Herzfeld, *Proc. Natl. Acad. Sci. U. S. A.*, 1994, **91**, 8880–8884.

CHAPTER 24

^2H Solid-State NMR Study of Peptide–Membrane Interactions in Intact Bacteria

ISABELLE MARCOTTE*[a] AND VALERIE BOOTH*[b]

[a] Department of Chemistry, Université du Québec à Montréal, GRASP/CQMF, P.O. Box 8888, Downtown Station, Montréal, Canada, H3C 3P8; [b] Department of Biochemistry and Department of Physics and Physical Oceanography, Memorial University of Newfoundland, St. John's, NL, Canada, A1B 3X9
*Email: marcotte.isabelle@uqam.ca; vbooth@mun.ca

24.1 Introduction

One of the most serious of all health challenges facing us today is the problem of pathogens that have developed resistance to conventional antibiotics.[1,2] The excessive use of antibiotics by humans and livestock is believed to be responsible for the worldwide emergence of multi-resistant bacteria. Diseases that used to be readily treated with antibiotics are nowadays considered as new threats. For instance, of the 9 million new cases yearly of tuberculosis, 5% are resistant to drugs used in standard treatment,[3] while cures for pneumonia and meningitis are also an issue.[4] Because of the variety of antibiotics to which pathogens are exposed, hospitals and healthcare facilities unwillingly participate in the transmission of bacteria with increased resistance to the most commonly used antibiotics.[5] Some of the most important are methicillin-resistant *Staphylococcus aureus*, and

New Developments in NMR No. 3
Advances in Biological Solid-State NMR: Proteins and Membrane-Active Peptides
Edited by Frances Separovic and Akira Naito

Published by the Royal Society of Chemistry, www.rsc.org

fluoroquinolone-resistant *Escherichia coli* and *Pseudomonas aeruginosa*, which are responsible for severe infections.

The bacterial cell envelope is the first barrier encountered by drugs. This biological interface is traversed by antibiotics *en route* to their active intracellular sites, and sometimes targeted by bioactive molecules such as antimicrobial peptides (AMPs) produced by higher organisms including humans, amphibians, bees and fish.[6] The study of interactions occurring in the bacterial membrane environment has thus important repercussions in drug development.

Bacterial membranes have a complex composition and organization which differ between Gram(−) and Gram(+) species. Gram(−) bacteria such as *E. coli* possess two phospholipid bilayers separated by a thin peptidoglycan layer and their outer membrane surface is covered by lipopolysaccharides (LPS) (Figure 24.1). The outer membrane is absent in Gram(+) bacteria such as *S. aureus*, which have a thicker peptidoglycan wall. The composition of bacterial membranes is different between species, as seen in Table 24.1. Moreover, bacteria regulate their membrane fluidity by adjusting the fatty acid (FA) composition of their lipids; therefore, the saturated/unsaturated phospholipid ratio will depend on growth conditions such as the temperature and chemical composition of the milieu.[7]

Solid-state nuclear magnetic resonance (ssNMR) has proven its outstanding capabilities in providing information on interactions between the membrane lipids and a wide range of molecules. Because of the complex composition of natural membranes, NMR experiments rely almost exclusively on the use of mimetics such as multilamellar vesicles and bicelles.[12] Ideally, interactions with cell membranes should be studied *in vivo*. There are a number of reasons why interactions of antimicrobial agents with whole bacteria may be modulated compared to model lipid bilayers, including the presence of the peptidoglycan and lipopolysaccharide

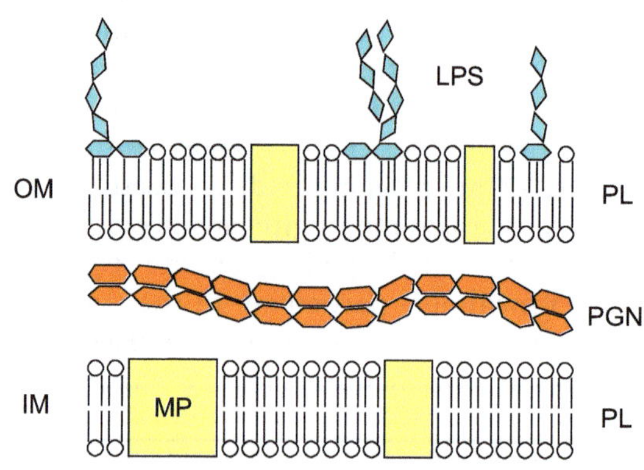

Figure 24.1 Gram(−) bacterium cell wall with outer (OM) and inner (IM) membranes comprising phospholipids (PL) and membrane proteins (MP). These membranes are separated by a peptidoglycan (PGN) layer and the OM is covered with lipopolysaccharides (LPS).

Table 24.1 Composition of selected bacterial membranes.

Bacterium	*Total lipid (%)*[a] *PE*	*PG*	*CL*
Gram(−)			
E. coli[8]	75	20	5
P. aeruginosa[9]	67	22	18
Gram(+)			
B. subtilis[10]	12	70	4
S. aureus[11]	–	58	42

[a]PE = phosphatidylethanolamine; PG = phosphatidylglycerol; CL = cardiolipin.

layers, across which AMPs have been proposed to move *via* a "self-promoted uptake" mechanism.[13] Other potential modifiers of the action of membrane-active agents on bacteria or eukaryotic cells are membrane proteins, lipid domains, bilayer asymmetry, and other complexities of membrane composition not reflected by the one- or two-component lipid systems typically used in NMR studies.[14] One demonstration of the significant impact of membrane composition complexity is a recent study showing that cholesterol exerts AMP protective effects in membrane compositions that do not form lipid rafts, but has little protective effect in membrane compositions that do form lipid rafts.[15] Another is the recent demonstration that melittin-K14 pore formation is much less prevalent in intact bacteria than in model membranes.[16] A further reason to study AMPs interacting with whole bacteria is that, in many cases, the main target may be intracellular and the AMP–bilayer interaction may be primarily required for gaining access to the inside of the cell.[17,18]

For studies of interactions of membrane-active agents with whole bacteria, NMR is advantageous in comparison to other methods such as fluorescence microscopy. The stable isotope labeling employed in NMR – when required – does not perturb the system, in contrast to the addition of reporter fluorescent probes that can change the properties of the lipids studied.[19] Moreover, *in vivo* NMR can provide information at the molecular level on the membrane interaction of molecules. Therefore, this method is particularly suited to investigate the effect of antimicrobial agents, especially by AMPs, which are known to perforate the lipid bilayer.

When studying perturbation of phospholipids in biological membranes, ^{31}P ssNMR emerges as a straightforward choice due to the high gyromagnetic ratio of this nucleus, as well as 100% natural abundance of this isotope. ^{31}P ssNMR has thus been employed to study the interaction of maculatin 1.1, caerin 1.1 and caerin 4.1 peptides from frog skin with membranes of *Bacillus cereus* and *Staphylococcus epidermidis.*[20] Because phospholipids represent only about 9% (dry mass) of the bacterial membrane,[21] these ^{31}P ssNMR experiments required large sample volumes, as well as long acquisition times, and were thus carried out using a 10-mm rotor and 6000–10 000 scans. Given the relatively long experiment times, a large portion of the bacteria may die during the experiment, complicating the interpretation of the results in relation to the *in vivo* setting of interest. ^{31}P ssNMR on such systems mainly provides information on the membrane phase and also on lysis by antimicrobial agents that is revealed by the appearance of an isotropic peak.

By contrast, introducing NMR-active labels into the entire length of the lipid acyl chains of bacteria potentially provides much more observable signals than can be derived from the single phosphate present in the head group. The lipid metabolism of microorganisms such as *E. coli* is known and can be exploited to isotopically label their membrane components.[22] This strategy has been used to deuterate bacterial phospholipids in *E. coli* and study the ^{2}H signal from deuterated acyl chains.[23–26] ^{2}H ssNMR is advantageous over ^{31}P ssNMR because the large amount of ^{2}H atoms on the lipid chains and shorter relaxation times allow spectra with a good signal-to-noise ratio to be obtained in a reasonable time (typically 2 h), ensuring a higher rate of cell survival throughout the experiment, as will be detailed in another section. The spectra, in particular the quadrupolar splitting ($\Delta\nu_Q$), can report on membrane lysis, but are also sensitive to changes in the membrane organization such as the lipid phase and ordering.[27] Two main strategies have been used to study intact *E. coli* cells using ^{2}H ssNMR. The first relies on mutant strains of bacteria which are forced to uptake exogenous deuterated fatty acids (FAs) for their phospholipids biosynthesis,[23–25] while Tardy-Laporte *et al.*[26] have used non-mutated strains. These methods will be detailed in the next section. This chapter will also present the characterization and ^{2}H ssNMR study of antimicrobial agent/membrane interactions in *E. coli.*

24.2 Production and Characterization of Deuterated *E. coli*

24.2.1 Deuteration of Phospholipid Acyl Chains in Mutants

This approach employs a doubly mutated strain of *E. coli* to enable high levels of deuteration in the lipid acyl chains of bacteria and at the same time avoid ^{2}H-labeling non-lipid bilayer molecules in the bacteria. The bacterial strain employed, called LA8,[25] cannot synthesize FAs due to a mutation in the *accB* gene, which codes for a subunit of acetyl-CoA carboxylase, the key enzyme in initiating fatty acid synthesis.[28] Absence of functional acetyl-CoA carboxylase forces bacteria to derive the FAs used in membrane biogenesis from the isotope-labeled FAs provided in the growth media. LA8 bacteria also lack the *fadE* gene, which codes for acyl-CoA dehydrogenase, a key enzyme in FA breakdown.[29] Thus, LA8 cells cannot metabolize FAs, which prevents the ^{2}H labels provided in the media from being incorporated into non-lipid components of the bacteria and generating unwanted background signals. When LA8 bacteria are grown in media containing the commercially available ^{2}H-labeled palmitic acid (PA-d_{31}), high levels of lipid acyl chain deuteration are achieved. To improve the viability of the bacteria, the growth media also contain oleic acid (undeuterated) and casamino acids.

For NMR, LA8 cells were grown to mid-log phase, harvested by centrifugation, washed with media containing 0.1% Brij-58 (without fatty acid supplements), pelleted and transferred immediately to an NMR tube. The viability of the LA8 bacteria prepared in this way has been assessed by both colony forming and MTT assay, as presented in Section 24.2.3.1.

24.2.2 Deuteration of Phospholipid Acyl Chains in Non-mutated Bacteria

In an alternative approach to employing mutated bacteria, it is also possible to exploit the molecular machinery of wild-type *E. coli* to reach high deuteration levels of phospholipid FA chains. This method is advantageous,- given the easy availability of the bacteria and the possibility of comparing different strains of bacteria. The labeling can be accomplished by growing the bacteria in a medium enriched with deuterated FAs such as PA-d_{31}. Because of the high energetic cost of FA biosynthesis, bacteria will preferably use exogenous long-chain FAs (12–18 carbon atoms) in their phospholipid production. Exogenous FAs bind to the fadL protein found within the bilayer of the outer membrane before travelling through the periplasm to reach the cell inner membrane, where they are transferred to acyl-CoA in the cytoplasm. There, they can be β-oxidized or incorporated into phosphatidic acid by the glycerol-3-phosphate acyltransferase and consequently converted into the different phospholipids present in *E. coli.*[7,30,31] Alternatively, a small proportion of exogenous FAs can also be directly incorporated into PE by an acyl carrier protein (ACP).[30] These pathways have the advantages of not permitting the accumulation of free isotopically labeled FAs inside the bacterial cell and of evenly incorporating deuterated exogenous FAs into all types of phospholipids. FA chain elongation and deuteration of LPS is not expected in the process because exogenous FAs in *E. coli* have no access to the biosynthetic acyl intermediates provided by the ACP involved in FA elongation and in lipid A biosynthesis.[30,32]

Tardy-Laporte *et al.*[26] have deuterated ampicillin-resistant *E. coli* pPD117 by growing the microorganisms in a LB medium at 37 °C supplemented with PA-d_{31} (0.25 mM) micellized using dodecylphosphocholine (DPC) to ensure optimal contact with the outer membrane of bacteria and efficient uptake by the fadL protein. Cells were harvested in the mid-log growth phase, pelleted and rinsed with a saline solution made of ^{2}H-depleted water to avoid an NMR signal from naturally occurring HDO and to remove all traces of free deuterated FAs. ^{2}H ssNMR experiments were carried out on fresh samples at a maximum of 2 hours after harvesting.

Table 24.2 Fatty acid composition of the membranes of unlabeled and labeled *E. coli* without mutating.[26]

	Fatty acid (%, w/w)				
E. coli sample	*C16:0*	*C16:0D*	*C16:1*	*C17:1*	*C18:1*
Labeled	7	76	9	1	7
Unlabeled	36	0	28	3	33

Table 24.3 Cell viability assays. Colony-forming assays for LA8 cells along with comparison of MTT reduction activities for both LA8 and unmutated (WT) cells. MTT reduction activities of labeled *E. coli* after exposure to marennine-like pigments with an estimate (%) of live population relative to that after rinsing. InP and ExP respectively refer to the intra- and extracellular forms of the marennine-like blue pigment.[25,26]

	Colony forming assay	*MTT assay MRU/(mL OD_{600})*		
Analysis	*LA8*	*LA8*	*WT-labeled*	*WT-unlabeled*
At harvest	–	–	74.8 ± 29.2	68.2 ± 17.3
After rinsing	55 ± 2	85.5 ± 1.5	58.5 ± 27.1	–
After NMR	42 ± 3	63.6 ± 5.4	24.3 (42%)	–
+InP	–	(74%)	24.1 (41%)	–
+ExP	–	–	14.0 (24%)	–

Using this protocol, 76% deuteration of the phospholipids was obtained, as shown in Table 24.2. The saturated/unsaturated (S/U) lipid ratio was increased following the reduced *de novo* synthesis of palmitoleic and oleic acid by the bacteria to counterbalance the intake of exogenous PA-d_{31}.[33,34] However, this change in lipid chain profile did not have an impact on cellular viability at 37°C (Table 24.3), or on the fluidity of the membranes, which were shown to be in a liquid crystalline phase according to the ^{2}H ssNMR spectrum. Introduction of deuterated or protonated oleic acid together with PA-d_{31} in the growth medium could help the organisms maintain an S/U FA ratio closer to that of unlabeled *E. coli*, as done by other research groups.[23–25]

24.2.3 Cell Viability and Isotopic Labeling Measurement

24.2.3.1 Cellular Viability Assays

For the plate-count assay, LA8 bacteria NMR samples were diluted in series in media and 50 μL of the diluent spread onto a solid media plate. Plates were incubated overnight at 37 °C and those that contained between 30 and 300 well-separated colonies retained for analysis. Percent viability was determined by averaging the number of colony-forming units for each set of diluted cultures (Table 24.3). A substantial number of colony-forming LA8 cells are lost during the process of preparing them for NMR; however, the loss of colony-forming units during the NMR experiment time was quite modest – less than 25%. Thus, the signal in the ^{2}H spectra of such cells derives largely from viable, living bacteria.

Cellular viability can also be assessed using an MTT [3-(4,5-dimethylthiazol-2-yl)-2,5-diphenyltetrazolium bromide] reduction assay. As detailed by Wang *et al.*,[35] this water-soluble yellow dye can be reduced by the dehydrogenase system of active cells to insoluble purple formazan crystals that are dissolved in an organic solvent and quantified using a UV-visible spectrometer. The concentration is directly proportional to the number of live bacteria in the sample. MTT assays have been performed for both LA8[36] and unmutated (WT)[26] bacterial samples. Bacteria samples were diluted in fresh medium to an absorbance of ~0.1 at 600 nm. Then 20 μL of MTT solution (5 mg mL^{-1}) heated at 37 °C was added to 200 μL of cell suspension and the sample was incubated at 37 °C for 20 min. Formazan crystals were pelleted by pulse centrifugation and dissolved in 1 mL of dimethyl sulfoxide (DMSO) at room temperature after the LB medium was removed. The absorbance was measured at 550 nm.

For LA8 cells, the MTT assays indicate a loss of about 25% of MTT activity between the start and end of the NMR experiments (Table 24.3). For the WT bacteria, the loss of MTT activity during the NMR experiments was more severe, at more than 50%, perhaps due to differences in the growth protocols. For instance, the labeled WT bacteria contain a high ratio of saturated to unsaturated fatty acid chains (see Section 24.2.3.2) compared to what would be expected from the mutated bacteria, which were grown on both palmitic and oleic acids.

24.2.3.2 Lipid Profile Analysis

A lipid extraction followed by a GC-MS (gas chromatography coupled to mass spectrometry) analysis can be performed to determine the bacteria lipid profile and isotopic labeling level of the phospholipids (Table 24.2) in the WT bacteria grown as above. The detailed procedure is described by Tardy-Laporte *et al.*[26] but will be summarized here. Lipid extraction is carried out on lyophilized bacteria sample using a 2:1 (v/v) mixture of dichloromethane (CH_2Cl_2) and methanol (MeOH) and 0.88% KCl solution in a Potter glass homogenizer. The lipid extract is then dried and the lipids are recovered with a 98:2 (v/v) CH_2Cl_2/MeOH mixture. The lipid solution is deposited in a silica gel hydrated column. Neutral lipids (triglycerides, free FAs and sterols) are first eluted with CH_2Cl_2/MeOH (98:2 v/v) and then polar lipids (phospholipids) recovered with MeOH. The phospholipid solution is evaporated and transesterified in H_2SO_4 (2% in MeOH) and toluene at 100 °C. The FA methyl esters (FAMEs) obtained from this step can then be analyzed by GC-MS using a capillary column of intermediate polarity. Peak identification and quantification are done by external calibration with a C4-C24 FAME mix and a deuterated PA methyl ester standard prepared by transesterification of PA-d_{31} as described above; C19:0 is used as an internal standard.

24.2.4 ^{2}H NMR Characterization of Bacteria

The ordering of phospholipid acyl chains in bacterial membranes has been the subject of only a few ^{2}H ssNMR studies, mainly carried out in the 1980s. Jarrell *et al.*[37] have studied the temperature dependence of the molecular order in the membranes of *Acholeplasma laidlawii* grown in a medium enriched in deuterated myristic acid and avidin, which inhibits endogenous fatty acid synthesis.[38] Pioneer work on ^{2}H ssNMR characterization of *E. coli* membranes was performed by the group of Myer Bloom on isolated outer and inner membranes, as well as intact L51 mutants grown in PA-d_{31} enriched milieu.[23,24]

Intact bacterial membranes can be characterized by ^{2}H ssNMR using a solid-echo pulse sequence[39] on fresh samples. Changes in acyl chain ordering are obtained by monitoring alterations in quadrupolar splitting ($\Delta\nu_Q$). For phospholipids in a bacterial membrane which are in a liquid-crystalline phase and undergoing fast rotation around their longitudinal axis:

$$\Delta\nu_Q(\delta) = \frac{3}{4}\frac{e^2qQ}{h}(3\cos^2\theta - 1)S_{CD} \tag{24.1}$$

where (e^2qQ/h) is the quadrupole coupling constant ($\sim$167 kHz for ^{2}H), θ is the angle between the bilayer normal and the lipid long axis, and S_{CD} is the order parameter of a deuterium bond vector.[40,41] Therefore, the quadrupolar splittings measured on the spectrum will depend on the orientation of the C–D bond with respect to the magnetic field direction as well as its position on the acyl chain, since the ordering increases when going from the more dynamic terminal CD_3 to the more rigid positions closer to the headgroup. The positions located in the first half of the phospholipid chains have a similar degree of ordering and form a plateau-like feature on the spectrum.

Specific $\Delta\nu_Q$ values are hard to measure from ^{2}H ssNMR spectra of deuterated bacteria owing to the wide distribution of quadrupolar splittings; therefore spectral moment analysis is usually carried out.[23,25,26,37] The *n*th moment can be determined as follows:[40]

$$M_n = \frac{\int_0^\infty \omega^n f(\omega)\mathrm{d}\omega}{\int_0^\infty f(\omega)\mathrm{d}\omega} \tag{24.2}$$

where ω is the angular frequency with respect to the Larmor angular frequency ω_0, and $f(\omega)$ is the intensity at that frequency, and thus $f(\omega)$ describes the spectrum lineshape. The first (M_1) and second (M_2) spectral moments report on the average orientational order parameter of the C–D bonds (the mean of the distribution of order parameters) and the mean square value of the quadrupolar splittings, respectively. Together, M_1 and M_2 allow calculation of the relative mean square deviation (Δ_2) of the distribution of order parameters (or quadrupolar splittings):

$$\Delta_2 = \frac{M2}{1.35 \times M_1^2} - 1 \tag{24.3}$$

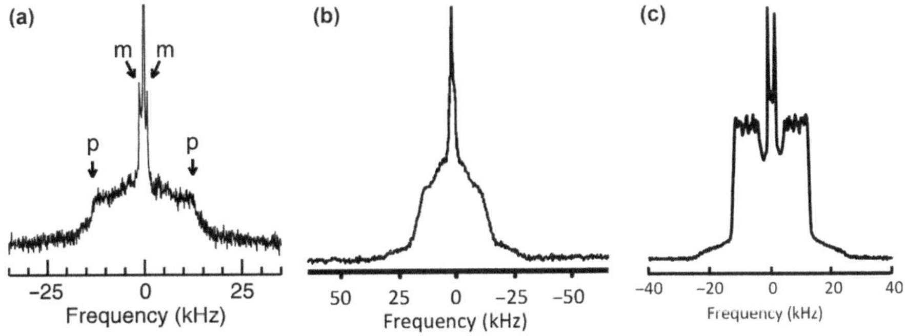

Figure 24.2 ^{2}H ssNMR spectra of (a) LA8 bacteria, (b) non-mutated *E. coli*, and (c) pure dipalmitoylphosphatidylcholine (DPPC) at 50 °C. (Adapted from Pius *et al.*[25] and Tardy-Laporte *et al.*[26])

Changes in M_1 and M_2 are sensitive to changes from a gel to a fluid phase, and the Δ_2 value gives an indication of the lipid phase inhomogeneity such as phase coexistence,[37] all parameters of prime interest in understanding the mechanisms by which antimicrobial agents affect bacterial membranes.

As displayed in Figure 24.2, in the absence of treatment with antimicrobial agents, the LA8 bacteria exhibit M_1 values of $\sim 4.6 \times 10^4$ s^{-1}, indicating the bacterial membranes are slightly less ordered than a typical model lipid membrane above its transition temperature, *e.g.* 5×10^4 s^{-1} for deuterated DPPC.[42,43] The non-mutated bacteria, grown with the protocol in Section 24.2.2, exhibit a higher M_1 value of 6.3×10^4 s^{-1}, consistent with the lower unsaturated lipid content of the membranes, and consequent higher degree of membrane order.

24.3 Membrane Interaction of Antibiotics with Deuterated *E. coli*

24.3.1 Effect of Antimicrobial Peptides

Initial studies[25] employed a potent AMP that had previously been well characterized in model lipid systems. MSI-78 is a synthetic analogue peptide that was designed based on the magainin family of AMPs from bullfrogs.[44] It is a highly charged, very potent AMP with a charge of +9 and an MIC of 4 μg mL^{-1}.[45] ssNMR studies of MSI-78 in bilayers composed of a single lipid, 1-palmitoyl-2-oleoylphosphatidylethanolamine (POPE), suggested the AMP mechanism of action includes inducing positive curvature strain in the membrane bilayer.[44] In the presence of membrane mimetic DPC micelles, MSI-78 takes on an α-helical, antiparallel, dimer structure.[46]

When ^{2}H-labeled LA8 *E. coli* are treated with MSI-78, there is an increase in the intensity at smaller splittings of the ^{2}H NMR spectra relative to that at larger splittings, indicating a decrease in average orientational order with increasing amounts of MSI-78 (Figure 24.3). The change in the shape of the

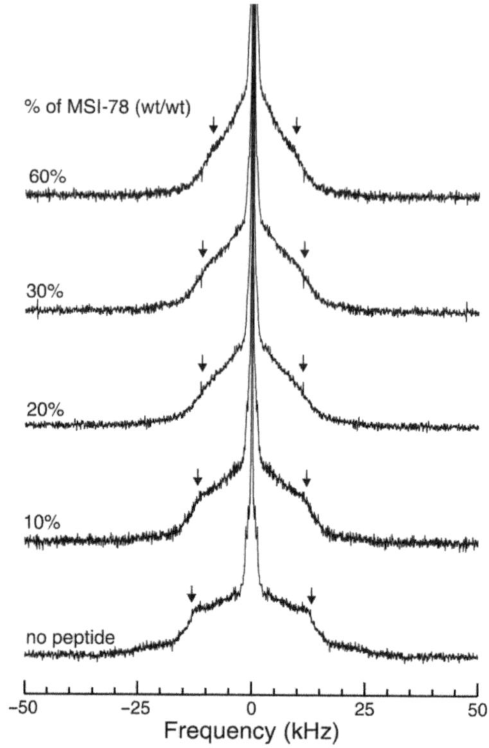

Figure 24.3 ^{2}H ssNMR spectra of membrane-deuterated LA8 bacteria with increasing concentrations of MSI-78. MSI-78 concentrations are quoted as weight MSI-78/dry weight bacteria. Each spectrum represents 32 000 scans acquired at 37 °C, on a 9.4 Tesla spectrometer.
(Reproduced from Pius *et al.*,[25] with permission from the American Chemical Society.)

spectra is reflected in the first and second spectral moments, which both decrease with MSI-78 (Table 24.4). On the other hand, Δ_2 increases with peptide concentration, indicating that MSI-78 induces a wider distribution of orientational order parameters, *i.e.* a more heterogeneous membrane bilayer. The peptide-induced changes in order parameter observed in *E. coli* agree well with earlier ^{2}H ssNMR studies of MSI-78 interacting with palmitoyloleoylphosphatidylcholine (POPC) multilamellar vesicles (MLVs), where a decrease in order parameter with addition of MSI-78 was also observed.[45] The NMR of whole bacteria thus provides an important connection between the studies of AMP–lipid interactions that employ model membrane systems, with the real functional context where AMPs carry out their biological functions.

Even more thought-provoking than this qualitative similarity between the experiments, however, is the comparison of peptide:lipid (P:L) ratios needed to observe similar effects on lipid chain order in the model membranes and

Table 24.4 Spectral moment analysis of the ^2H ssNMR spectra of ^2H-labeled LA8 *E. coli* at 37 °C in the presence of increasing concentrations of MSI-78.

Sample	M_1 (10^4 s^{-1})	M_2 (10^9 s^{-2})	Δ_2
E. coli (LA8)	4.6 ± 0.2	3.7 ± 0.4	0.27 ± 0.1
+10% MSI-78	4.0	2.8	0.30
+20% MSI-78	3.6	2.5	0.37
+30% MSI-78	3.4 ± 0.2	2.2 ± 0.2	0.37 ± 0.1
+60% MSI-78	2.8	1.5	0.45

membranes in whole bacteria. The P:L ratio required to see commensurate alterations in lipid chain order is more than 30 times greater for membranes in intact bacteria than for the POPC MLVs. One possible reason for this disparity is that although most of the peptide may reach the membrane, it is prevented by some component of the bacteria, not present in the model systems, from disrupting the bilayer chains. A more likely possibility, though, is that much of the peptide does not reach the membrane and is instead bound up in other components of the Gram(−) cell envelope, such as the LPS layer. This observation underlines the importance of examining the AMP interactions with a pathogen, and not focusing purely on the interaction with the lipid bilayer.

On the other hand, a different kind of incongruity is encountered when comparing the P:L ratio needed to observe effects on acyl chain order in the bacteria *versus* the P:L ratios needed to actually observe biological activity, *e.g.* growth inhibition. P:L ratios needed to inhibit bacterial growth are typically 100:1.[47,48] However, membrane disruption is observable in the bacteria at P:L ratios about 100 times smaller than this. Thus, substantial AMP-induced bilayer disruption occurs at levels well below those where bacterial growth inhibition takes place. One thought that follows from this is that bacteria may be able to tolerate a great deal of AMP-induced membrane disruption before actual growth inhibition occurs. Another likelihood it brings up is that, even at relatively low levels of AMP, there is enough bilayer disruption to suggest an increased permeability to the AMP that would allow it to enter the cell and interact with an intracellular target. Indeed, recent results in the Booth lab have identified a potential intracellular target for MSI-78 (unpublished).

24.3.2 Effect of Other Natural Antimicrobial Agents

In addition to providing useful information on the mechanism of action of AMPs, membrane-deuterated *E. coli* have also been employed to study antimicrobial agents such as the antibiotic polymyxin B, fullerenol nanoparticles and the detergent cetyltrimethylammonium chloride.[26] While

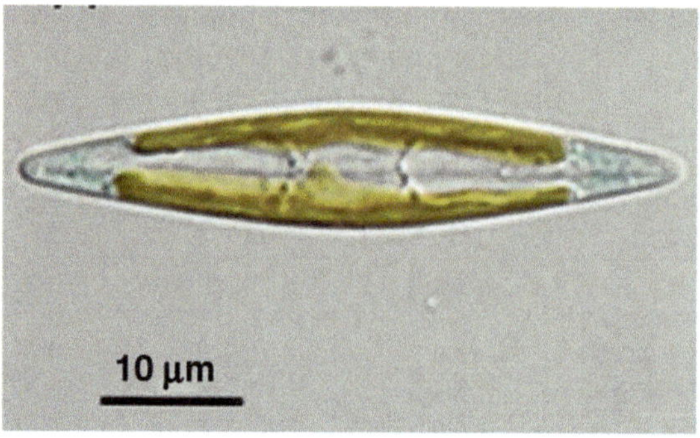

Figure 24.4 Optical microscopy image of *Haslea provincialis sp. nov.* with blue apices (intracellular form of the marennine-like pigment). (Reproduced from Tardy-Laporte *et al.*[26] with permission from Elsevier.)

AMPs are of great interest to provide a new pool of antibiotics, Nature can also provide other types of bioactive molecules potentially active against bacteria. Marine diatoms from the *Haslea* genus are known to synthesize a water-soluble blue pigment called "marennine", which has shown activity against marine bacteria and the HSV1 herpes virus cancer cells.[49] This pigment is responsible for the greening of the oyster gills cultured in Marennes-Oléron's (France) ancient salt marshes, where *H. ostrearia* is the dominant microalgal species. This pigment exists in two forms, the internal pigment (InP) found in the apical region of the cell (Figure 24.4) and the external form (ExP) excreted in the environment. Preliminary characterization of marennine from *H. ostrearia* indicates that both pigment forms are polyphenols with a molecular weight of about 10 kDa, but the two forms are believed to have different chemical structures.[50] The biocidal action mechanism of marennine has not been determined.

The membrane interaction of the marennine-like pigment produced by a newly identified species of blue diatom, *Haslea provincialis sp. nov.*,[51] has been studied by Tardy-Laporte *et al.*[26] with non-mutated *E. coli* to obtain information on its antibiotic potential. This pigment is structurally similar to marennine according to *in vivo* UV-visible spectrophotometry and Raman spectroscopy.[51] As revealed by the ^{2}H ssNMR spectra displayed in Figure 24.5, the intracellular form of the pigment, InP, has little effect on the bacteria membrane. Spectral moment analysis (Table 24.5) reveals a lower M_1 value indicative of a small decrease in phospholipid acyl chain order when InP interacts with the bacterial membranes. Concomitantly, the smaller M_2 value reflects the smaller spectrum width. The membrane is also less heterogeneous, as indicated by the decrease in Δ_2. The interaction of InP with *E. coli* membranes is not deleterious to the microorganisms since the viability is similar to that of "pure" *E. coli* after the NMR analysis (Table 24.3).

Table 24.5 Spectral moment analysis of the ^2H ssNMR spectra of labeled *E. coli* at 37 °C and in the presence of 2 mol% of the marennine-like pigments.

Sample	M_1 (10^4 s^{-1})	M_2 (10^9 s^{-2})	Δ_2
E. coli	6.3	9.0	0.68
+InP	5.5	5.1	0.25
+ExP	9.3	17.1	0.46

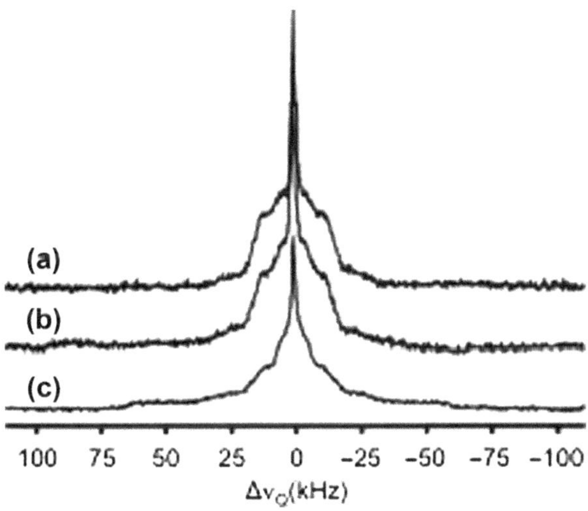

Figure 24.5 ^2H ssNMR spectra of (a) intact non-mutated *E. coli* acquired at 37 °C, and in the presence of 2 mol% of the (b) intracellular and (c) extracellular forms of the marennine-like blue pigment from *H. provincialis sp. nov.*
(Reproduced from Tardy-Laporte *et al.*,[26] with permission from Elsevier.)

Interestingly, the external form of the marennine-like pigment that is released in the water column by the diatoms, ExP, shows a different effect on *E. coli.* Figure 24.5 shows a greater contribution of the gel phase ($\Delta\nu_Q \approx 120$ kHz) to the spectrum. The quadrupolar splittings of the plateau (25 kHz) and methyl regions (2 kHz) are similar to those of unexposed cells, indicating that a proportion of the phospholipids remain in the fluid phase. The higher M_1 and M_2 values reflect the increased lipid chain order in the bacteria when exposed to ExP, and this is further illustrated by the smaller Δ_2 value, which suggests more homogeneity in the *E. coli* membranes. Altogether, the results suggest that ExP has little effect on the acyl chains of the lipids located in the fluid portion of the membrane, namely the inner membrane of the bacteria. Marennine pigments being polyhydroxylated molecules, it is very likely that ExP molecules interact with the LPS on the *E. coli* surface and restrain the phospholipid motions in the outer membrane. The antibiotic effect of ExP is more important than that of InP, as revealed in Table 24.3 with only about 24% of the bacteria still alive.

The ^{2}H ssNMR study of the interaction with intact *E. coli* of the two marennine-like pigment forms secreted by the *Haslea provincialis sp. nov.* microalgae provides evidence of different membrane interactions. A stronger membrane perturbation was observed for the more bactericidal extracellular pigment. The different membrane action and antibiotic activity support the view that the internal and released forms of the blue pigment have indeed structural differences.

24.4 Conclusion and Future Prospects

Model membranes traditionally used in ssNMR provide valuable information on the effect of antimicrobial molecules once they have reached the lipid bilayer. Such studies with purified systems have thus been instrumental in establishing models to explain how AMPs interact with the phospholipid components of bacterial cell envelopes. However, bacterial walls are complex and before reaching the lipid bilayer(s), LPS and/or peptidoglycan layers must be traversed. Additionally, there are complexities present in the inner membrane, for instance membrane proteins and subtleties of lipid composition, not reflected in the standard model membrane systems. ssNMR of intact bacteria is thus a promising avenue to improve our knowledge of the mechanism of action of AMPs, and demonstrated in this chapter with MSI-78 and marennine-like pigments.

The strategies employed for isotope labeling of bacterial lipids for ssNMR took advantage of the available wealth of knowledge regarding lipid biosynthesis pathways in bacteria. High levels of ^{2}H-labeling of membrane lipids were achieved *via* incorporation of ^{2}H-labeled palmitic acid in the growth media of both mutated and non-mutated *E. coli*. The successful acquisition of ssNMR spectra from such bacteria opens up a wealth of possibilities for probing the mechanisms and specificity of antimicrobial agents using ^{2}H ssNMR. For instance, it will be interesting to see how well the AMP-induced lipid disruption observed in intact bacteria correlates with the MIC for a variety of AMPs. For example, if there are AMPs found which show a high level of bacterial growth inhibition but only a low level of membrane disruption, this will indicate that interactions other than AMP–lipid interactions may need to be investigated as potential mechanisms of AMP action.

E. coli are routinely used for the overexpression of ^{13}C/^{15}N-labeled proteins for NMR structure determination. Recent progress in the field suggests that in-cell study of membrane interactions by ssNMR in future will most certainly rely on experiments exploiting the ^{13}C and ^{15}N labeling initially developed for protein expression. Indeed, recently the structure of cytoplasmic proteins such as TTHA1718 from *Thermus thermophilus* HB8 has been directly determined in *E. coli* using solution NMR experiments.[52] The structure and dynamics of the membrane protein PagL expressed in *E. coli* has also been studied in cell by ssNMR by Renault *et al.*,[53] who also showed that the bacterial membrane constituents such as peptidoglycans and LPS can be characterized. Also of interest, from the point-of-view of studying AMP interactions with non-lipid components of bacterial cell envelopes, are the NMR studies of intact bacteria employing specific ^{13}C- and ^{15}N-labeled moieties.[54] Therefore, more information on the interaction on antimicrobial molecules such as AMPs will be gained by studying other important bacterial wall constituents such as LPS and peptidoglycans. Already, the interaction of YajG with *E. coli* sacculi has been studied by measuring T_1 and using through-bond ^{13}C–^{13}C correlation spectra with proton-driven spin diffusion in which efficiency depends on the peptidoglycans' mobility.[55] The flexibility of the peptidoglycans was also compared in *E. coli*, *B. subtilis* and *S. aureus* by ^{13}C ssNMR using T_1, and the binding of metals to wall techoic acids was studied in Gram(+) bacteria (*B. subtilis*, *S. aureus* and *S. pneumoniae*) by ^{31}P NMR. Such experiments rely on fast magic-angle spinning and are most likely not conducted *in vivo per se*, but they will certainly improve the understanding of the bactericidal effects.

References

1. H. W. Boucher, G. H. Talbot, J. S. Bradley, J. E. Edwards, D. Gilbert, L. B. Rice, M. Scheld, B. Spellberg and J. Bartlett, *Clin. Infect. Dis.*, 2009, **48**, 1.
2. B. Spellberg, R. Guidos, D. Gilbert, J. Bradley, H. W. Boucher, W. M. Scheld, J. G. Bartlett and J. J. Edwards, *Clin. Infect. Dis.*, 2008, **46**, 155.
3. E. Marris, *Nat. News*, 2008, DOI: 10.1038/news.2008.627.
4. J. C. Wijkmans and R. P. Beckett, *Drug Discovery Today*, 2002, 7, 126.
5. K. M. Overbye and J. F. Barrett, *Drug Discovery Today*, 2005, **10**, 45.
6. M. Zasloff, *Nature*, 2002, **415**, 389.
7. Y. M. Zhang and C. O. Rock, *Nat. Rev. Microbiol.*, 2008, **6**, 222.
8. C. R. Raetz, C. M. Reynolds, M. S. Trent and R. E. Bishop, *Annu. Rev. Biochem.*, 2007, **76**, 295.
9. S. K. Ghorbal, A. Chatti, M. M. Sethom, L. Maalej, M. Mihoub, S. Kefacha, M. Feki, A. Landoulsi and A. Hassen, *Curr. Microbiol.*, 2013, **67**, 112.
10. S. Clejan, T. A. Krulwich, K. R. Mondrus and D. Seto-Young, *J. Bacteriol.*, 1986, **168**, 334.
11. R. M. Epand and R. F. Epand, *Biochim. Biophys. Acta*, 2009, **1788**, 289.
12. D. E. Warschawski, A. A. Arnold, M. Beaugrand, A. Gravel, E. Chartrand and I. Marcotte, *Biochim. Biophys. Acta*, 2011, **1808**, 1957.
13. K. L. Piers and R. E. Hancock, *Mol. Microbiol.*, 1994, **12**, 951.
14. R. F. Epand, V. K. Mishra, M. N. Palgunachari, G. M. Anantharamaiah and R. M. Epand, *Biochim. Biophys. Acta*, 2009, **1788**, 1967.

15. J. R. Brender, A. J. McHenry and A. Ramamoorthy, *Front. Immunol.*, 2012, **3**, 195.
16. M. L. Gee, M. Burton, A. Grevis-James, M. A. Hossain, S. McArthur, E. A. Palombo, J. D. Wade and A. H. Clayton, *Sci. Rep.*, 2013, **3**, 1557.
17. M. Cudic and L. J. Otvos, *Curr. Drug Targets*, 2002, **3**, 101.
18. H. Jenssen, P. Hamill and R. E. Hancock, *Clin. Microbiol. Rev.*, 2006, **19**, 491.
19. E. Strandberg and A. S. Ulrich, *Concepts Magn. Reson., Part A*, 2004, **23**, 89.
20. B. C. S. Chia, Y.-H. Lam, M. Dyall-Smith, F. Separovic and J. H. Bowie, *Lett. Pept. Sci.*, 2000, 7, 151.
21. G. C. Walker, I. I. I. R. Curtis, J. L. Ingraham, E. C. C. Lin, K. B. Low, B. Magasanick, W. S. Reznicoff, M. Riley, M. Schaechter and H. E. Umbarger, in *Escherichia Coli and Salmonella Typhimurium: Vols 1-2: Cellular and Molecular Biology*, ed. F. C. Neidhardt, American Society for Microbiology, Washington D. C., USA, 1996, vol. 2, p. 1346.
22. D. Kenanov, C. Kaleta, A. Petzold, C. Hoischen, S. Diekmann, R. A. Siddiqui and S. Schuster, *FEBSJ*, 2010, **277**, 1023.
23. J. H. Davis, C. P. Nichol, G. Weeks and M. Bloom, *Biochemistry*, 1979, **18**, 2103.
24. C. P. Nichol, J. H. Davis, G. Weeks and M. Bloom, *Biochemistry*, 1980, **19**, 451.
25. J. Pius, M. R. Morrow and V. Booth, *Biochemistry*, 2012, **51**, 118.
26. C. Tardy-Laporte, A. A. Arnold, B. Genard, R. Gastineau, M. Morancais, J. L. Mouget, R. Tremblay and I. Marcotte, *Biochim. Biophys. Acta*, 2013, **1828**, 614.
27. J. Seelig, *Q. Rev. Biophys.*, 1977, **10**, 353.
28. J. E. J. Cronan and G. L. Waldrop, *Prog. Lipid Res.*, 2002, **41**, 407.
29. J. W. Campbell and J. E. J. Cronan, *J. Bacteriol.*, 2002, **184**, 3759.
30. D. M. Byers and Z. Shen, *Can. J. Microbiol.*, 2002, **48**, 933.
31. Y. Jiang, R. M. Morgan-Kiss, J. W. Campbell, C. H. Chan and J. E. Cronan, *Biochemistry*, 2010, **49**, 718.
32. D. M. Byers, *J. Bacteriol.*, 1989, **171**, 59.
33. M. Esfahani, T. Ioneda and S. J. Wakil, *J. Biol. Chem.*, 1971, **246**, 50.
34. M. H. Weber, W. Klein, L. Muller, U. M. Niess and M. A. Marahiel, *Mol. Microbiol.*, 2001, **39**, 1321.
35. H. Wang, H. Cheng, F. Wang, D. Wei and X. Wang, *J. Microbiol. Methods*, 2010, **82**, 330.
36. J. Pius, *Biochemistry*, 2012, **51**, 118.
37. H. C. Jarrell, K. W. Butler, R. A. Byrd, R. Deslauriers, I. Ekiel and I. C. Smith, *Biochim. Biophys. Acta*, 1982, **688**, 622.
38. J. R. Silvius, N. Mak and R. N. McElhaney, *Biochim. Biophys. Acta*, 1980, **597**, 199.
39. J. H. Davis, K. R. Jeffrey, M. Bloom, M. Valic and T. P. Higgs, *Chem. Phys. Lett.*, 1976, **42**, 390.
40. J. H. Davis, *Biochim. Biophys. Acta*, 1983, **737**, 117.

41. J. Seelig and A. Seelig, *Q. Rev. Biophys.*, 1980, **13**, 19.
42. J. H. Davis, *Biophys. J.*, 1979, **27**, 339.
43. M. R. Morrow, J. P. Whitehead and D. Lu, *Biophys. J.*, 1992, **63**, 18.
44. K. J. Hallock, D. K. Lee and A. Ramamoorthy, *Biophys. J.*, 2003, **84**, 3052.
45. A. Ramamoorthy, S. Thennarasu, D. K. Lee, A. Tan and L. Maloy, *Biophys. J.*, 2006, **91**, 206.
46. F. Porcelli, B. A. Buck-Koehntop, S. Thennarasu, A. Ramamoorthy and G. Veglia, *Biochemistry*, 2006, **45**, 5793.
47. W. C. Wimley, *ACS Chem. Biol.*, 2010, **5**, 905.
48. W. C. Wimley and K. Hristova, *J. Membr. Biol.*, 2011, **239**, 27.
49. R. Gastineau, V. Leignel, B. Jacquette, Y. Hardivillier, A. Wulff, P. Gaudin, D. Bendahmane, N. A. Davidovich, I. Kaczmarska and J. L. Mouget, *Protist*, 2013, **164**, 340.
50. J.-B. Pouvreau, M. Morançais, F. Fleury, P. Rosa, L. Thion, B. Cahingt, F. Zal, J. Fleurence and P. Pondaven, *J. Appl. Phycol.*, 2006, **18**, 757.
51. R. Gastineau, PhD thesis, Université du Maine, 2011, p. 1–328.
52. D. Sakakibara, A. Sasaki, T. Ikeya, J. Hamatsu, T. Hanashima, M. Mishima, M. Yoshimasu, N. Hayashi, T. Mikawa, M. Walchli, B. O. Smith, M. Shirakawa, P. Guntert and Y. Ito, *Nature*, 2009, **458**, 102.
53. M. Renault, R. Tommassen-van Boxtel, M. P. Bos, J. A. Post, J. Tommassen and M. Baldus, *Proc. Natl. Acad. Sci. U. S. A.*, 2012, **109**, 4863.
54. S. Sharif, S. J. Kim, H. Labischinski and J. Schaefer, *Biochemistry*, 2009, **48**, 3100.
55. T. Kern, S. Hediger, P. Muller, C. Giustini, B. Joris, C. Bougault, W. Vollmer and J. P. Simorre, *J. Am. Chem. Soc.*, 2008, **130**, 5618.

CHAPTER 25

Magic Angle Spinning NMR Spectroscopy for Resolving Structure and Mechanisms of Function of Membrane Protein Assemblies Involved in Photosynthetic Energy Conversion

YULIYA MILOSLAVINA AND HUUB J. M. DE GROOT*

Leiden Institute of Chemistry, Leiden, the Netherlands
*Email: groot_h@chem.leidenuniv.nl

25.1 Introduction

In natural photosynthesis, light energy is converted and stored by proteins and protein complexes in membranes. In recent years, significant advances have been made by biological solid-state NMR techniques, mainly magic angle spinning (MAS) NMR, in resolving structure and gaining information on the mechanisms of operation of biological samples with strongly anisotropic NMR responses, immobilized to form a solid or nearly solid state.[1] Unlike soluble proteins, which rotate and move in all directions, resulting in averaged isotropic properties and narrow NMR lines, the spectra of solid

New Developments in NMR No. 3
Advances in Biological Solid-State NMR: Proteins and Membrane-Active Peptides
Edited by Frances Separovic and Akira Naito

Published by the Royal Society of Chemistry, www.rsc.org

systems are heterogeneously broadened and line narrowing techniques like MAS and multiple pulse sequences are necessary for obtaining high resolution. This chapter summarizes recent developments in using solid-state NMR for the study of structure and dynamics, energy landscapes and the link between energy conversion and signal transduction in photosynthetic systems. Important advantages of applying biological solid-state NMR for the study of membrane-bound photosynthetic protein assemblies are that crystalline or soluble samples are not required, as is the case for X-ray techniques and liquid NMR, and it gives an insight into the biological mechanisms of photosynthesis with atomic resolution. Biological solid-state NMR allows study of heterogeneous photosynthetic systems in their native environment containing all natural lipids and cofactors, which is important to avoid artifacts. When combined with specific isotope labelling or photo-CIDNP signal enhancement, large reaction centre (RC) pigment–protein complexes with molecular weights exceeding 100 kDa can be studied in great detail to obtain information about mechanisms that are otherwise inaccessible. In combination with advanced time-dependent molecular modelling, the primary processes that were selected billions of years ago for the separation of charges in membrane protein complexes can be approached. These natural processes have been highly conserved during evolution and serve today as paradigms for the construction of artificial devices for solar to fuel conversion.

25.2 Methods for Biological Solid-State NMR of Photosynthetic Assemblies

25.2.1 Magic Angle Spinning NMR

Quantum mechanically, NMR spectra are generally described by a spin Hamiltonian with five terms: the nuclear spin Zeeman interaction H_Z, the chemical shift H_{CS}, homonuclear and heteronuclear magnetic dipole–dipole interactions H_{II} and H_{IS}, and interactions with applied radio-frequency (RF) fields H_{RF}.[2] $\boldsymbol{I}$ and $\boldsymbol{S}$ are spin angular momentum operators for two different spin species. These Hamiltonians describe the energy of the spin system in the rotating frame.[3,4]

The Zeeman Hamiltonian describes the interaction of the nuclear spin with the static uniform magnetic field directed along the z-axis:

$$H_Z = \Delta\omega I_z \tag{25.1}$$

where I_z is the component of the nuclear spin along the z-axis in the direction of the magnetic field, while $\Delta\omega$ represents the difference between the nuclear Larmor frequency ω_0 and the NMR frequency ω_{RF}.

The H_{CS} represents the chemical shift, which is a frequency shift caused by the interaction of the induced electronic field with the nucleus. The electron distribution around a nucleus in a molecule is in general not spherically symmetric and will depend on the orientation of the molecule within the applied field:

$$H_{\mathrm{CS}} = \omega_0\left[\delta_{\mathrm{iso}} + \delta_{11}\sin^2\theta\cos^2\phi + \delta_{22}\sin^2\theta\sin^2\phi + \delta_{33}\cos^2\theta\right] \tag{25.2}$$

where δ_{iso}, δ_{11}, δ_{22}, δ_{33} are the isotropic chemical shift and the three principal values of the chemical shift anisotropy tensor, and θ and ϕ are polar coordinates for the direction of the magnetic field in the principal axis system. The angular terms in H_{CS} contain structural information and are dependent on the electronic structure of the system.[5] By performing accurate *ab initio* calculations on molecular systems constrained by NMR data, the information embedded in the chemical shift and its anisotropy can be extracted and put to use for structure determination or gaining biophysical understanding.

The homonuclear dipole–dipole interactions, *i.e.* between nuclear spins of the same type, are determined by the magnetogyric ratio γ_I, the distance between the two coupled spins (I_1, I_2) r_{12}, and the angle θ' between the internuclear displacement vector and the applied magnetic field directed along the *z*-axis:

$$H_{\mathrm{II}} = -\frac{\gamma_I^2\hbar}{r_{12}^3}\frac{(3\cos^2\theta' - 1)}{2}(3I_{z1}I_{z2} - I_1 \cdot I_2) \tag{25.3}$$

An analogous definition holds for the heteronuclear coupling:

$$H_{\mathrm{IS}} = -\frac{2\gamma_I\gamma_S\hbar}{r_{\mathrm{IS}}^3}\frac{(3\cos^2\theta'' - 1)}{2}I_zS_z \tag{25.4}$$

The H_{II} and H_{IS} terms contain structural information through their dependencies on internuclear distances and directions, independent of the underlying electronic structure of the molecular systems. This information is, in principle, directly accessible through NMR experiments without need for *ab initio* calculation of the electronic structure. However, in contrast to solution NMR, where the dipolar interactions are relatively weak and polarization transfer proceeds primarily *via* relaxation processes in the form of the nuclear Overhauser effect (NOE),[6,7] in the solid state the dipolar interactions are strong and coherent transfer dominates. Since the intramolecular dipolar interactions are much stronger than intermolecular couplings, relayed transfer within molecules is strongly favoured over intermolecular transfer, unless dilute labelling patterns and special pulse sequences are used to suppress competing intramolecular coherence transfer and selectively enhance the intermolecular processes.

The applied RF field is described by:

$$H_{\mathrm{RF}} = \omega_1(t)\left[I_x\cos\phi_{\mathrm{RF}}(t) + I_y\sin\phi_{\mathrm{RF}}(t)\right] \tag{25.5}$$

where, with respect to a constant-phase ω_{RF}, the $\omega_1(t)$ and $\phi_{\mathrm{RF}}(t)$ are the time-dependent RF field amplitude and phase. The H_{RF} represents the RF pulses or pulse shapes or sequences designed to excite the nuclei and manipulate the other Hamiltonian terms in order to permit the extraction of structure–function relationship information from the NMR response.

The H_{II}, H_{IS} and H_{CS} terms give rise to a strong orientation dependence of the NMR frequencies in the solid state and broadened NMR lineshapes. In order to increase the spectral resolution and sensitivity, the MAS technique can be applied, according to which a sample is mechanically spun rapidly at an angle 54.736° with respect to the magnetic field. The angle is derived from $(3\cos^2\theta - 1) = 0$ and is called the magic angle because it averages the anisotropy by rotation around a single axis, and the H_{IS} and H_{CS} terms collapse for rapid MAS.[8–10]

The most commonly studied nuclei in biological solid-state NMR of photosynthetic systems are ^{1}H, ^{13}C and ^{15}N. The most abundant is ^{1}H, which exhibits the highest gyromagnetic ratio (γ) of all stable isotope species and has a convenient relaxation time T_1, and is, thus, the most sensitive nuclei for NMR observations in solution. In the solid state, however, the proton (^{1}H) spins form a strongly coupled network, which induces strong overlap and line broadening. The observation of less abundant nuclei, such as ^{13}C and ^{15}N, suffers from low sensitivity. They have a weaker homonuclear coupling, longer relaxation times, and larger chemical shift dispersion that can still result in a well-resolved spectrum. To increase their sensitivity, cross-polarization (CP) is used, which enhances spectral resolution by transfer of the polarization from protons *via* the heteronuclear dipolar coupling, H_{IS}.[11–14]

25.2.2 Homonuclear Correlation Spectroscopy

MAS suppresses much information in solid-state NMR spectra and there are a number of ways to recover relevant information. One strategy is to rotate slower and obtain the anisotropy parameters from the side bands.[15–19] Another approach is to combine MAS with rotation-synchronized pulse sequences. Interference between the time dependency induced by MAS and by pulse sequences can prevent anisotropic spin–spin interactions from being averaged, even with high-frequency MAS.[2] This phenomenon is called "recoupling". Recoupling techniques allow anisotropic nuclear spin interactions in solids to be turned on and off at will, so that the advantages of MAS and the structural information from dipolar interactions between nuclear spins can be preserved at the same time. For better resolution of weak dipole–dipole couplings, the recoupled data are acquired in two-dimensional (2D) or three dimensional (3D) datasets. A general form of a 2D NMR pulse sequence consists of four parts: preparation, evolution, mixing and observation. An initial preparation sequence creates a non-equilibrium nuclear spin state. Then the desired coherence is allowed to evolve for a period of t_1 under the recoupled interactions. During the mixing time the selected coherence is transferred to the coherence of the observed nuclei. In the observation period, the free-induction-decay (FID) of the observed nuclear species is detected during t_2. The experiment is repeated for a number of t_1 values, which results in a data matrix $s(t_1, t_2)$. This matrix is double Fourier transformed into the 2D spectrum $S(\omega_1, \omega_2)$. The appearance of off-diagonal cross-peaks at frequencies ω_1, ω_2 in this 2D spectrum indicates that a process during the mixing period has transferred magnetization components of precession frequency ω_1 to a transition of frequency ω_2. Experimental spectra can be further compared with numerical simulations to extract the desired structural parameters.[20,21] Isotopic labeling of the samples, most commonly with ^2H, ^{13}C and ^{15}N, further facilitates the observation of the NMR signals.[22] The labeling can be done selectively, partly or uniformly.[23]

Radiofrequency driven recoupling (RFDR)[24,25] belongs to the class of homonuclear recoupling techniques for systems with a broad chemical shift range and large anisotropy. The pulse scheme is shown in Figure 25.1A. Initially, transverse polarization is prepared by a $\pi/2$ pulse, and the resulting ^1H polarization is transferred to the ^{13}C spins through CP. In the course of the evolution period of length t_1, the transverse polarizations of the ^{13}C spins precess freely at their characteristic resonance frequencies and become labelled with the evolution frequency ω_1. To eliminate the effect of the ^{13}C–^1H couplings on the observed ^{13}C spin, a continuous high-power RF-field irradiation on the ^1H spins, called continuous wave (cw) decoupling, is applied. The recoupling process takes place during the mixing period

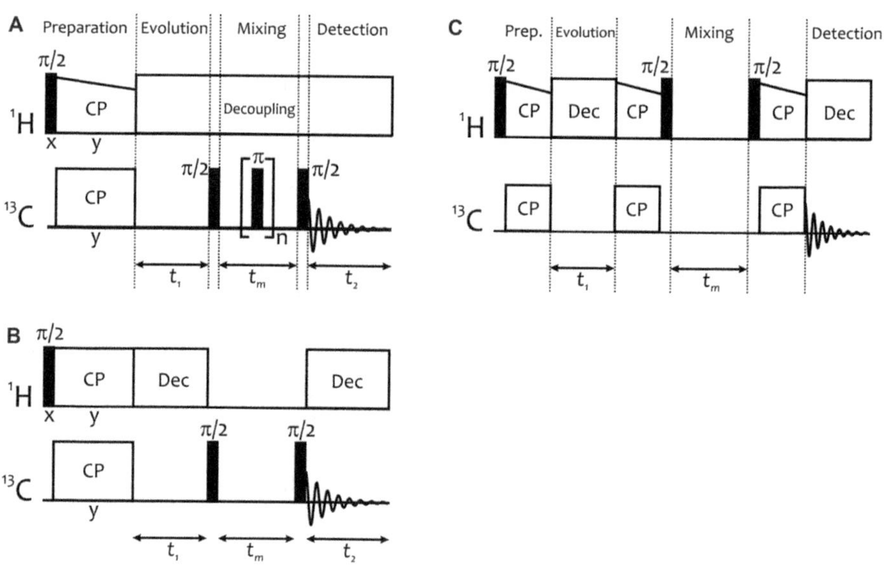

Figure 25.1 Pulse sequences for homonuclear ^{13}C–^{13}C dipolar correlation spectroscopy that are frequently employed in MAS NMR studies of proteins: (A) RFDR; (B) PDSD and DARR, using the same pulse sequence, but different ^1H decoupling, applying the rotary-resonance condition for the RF in the case of DARR; and (C) CHHC.

while the polarization is longitudinal. A third π pulse rotates the longitudinal polarization after recoupling into the *xy*-plane for detection by letting the spins precess at their ω_2 resonance frequencies and recording the induced signal as a function of the time variable t_2. During the mixing time, a train of rotor-synchronized RF pulses is applied, which introduces an additional time dependence of the spin operator term $(3I_{z1}I_{z2} - I_1I_2)$ by rotations of π around the *x*-axis in the period between the two π/2 pulses. This forces the time-dependent spin operator to be periodic with a multiple of the rotation period for recoupling of the spins. RFDR[26] has been extensively used for characterizing ^{13}C–^{13}C correlations for directly bonded carbons.[27] The drawbacks in the RF-driven polarization transfer are a fast decay of the ^{13}C polarization due to inhomogeneity of the RF irradiation, a need for strong proton decoupling and a short relaxation time under the RF field.

The other pulse sequences in Figure 25.1 are based on spin diffusion, taking into account polarization transfer through a proton bath, created by abundant protons that are strongly coupled by dipolar interactions.[28] This through-space interaction leads to correlations among large numbers of spins, allowing sampling of the longer range molecular connections as well as detecting weaker correlations than is possible with RFDR.[29] The spin diffusion schemes are distinguished from RFDR by the absence of RF pulses in the mixing period, which is responsible for "communication" between the ^{13}C spins *via* protons. This eliminates sample warming and is less sensitive to RF inhomogeneity.

In the proton-driven spin diffusion (PDSD)[30] scheme (Figure 25.1B), spectral overlap of ^{13}C resonances is caused by line broadening due to ^{13}C–^{1}H dipolar couplings. PDSD can be used with both static samples, to study anisotropic correlations in disordered solids, and with MAS for protein and peptide assignments. However, at high MAS frequencies it does not work well, since the ^{13}C–^{13}C dipolar interactions removed by MAS are not recovered. Dipolar-assisted rotational resonance (DARR) is an extension of PDSD, which improves polarization transfer at high spinning frequencies. The pulse scheme is the same as for PDSD (Figure 25.1B), but during the mixing time the ^{13}C–^{1}H dipolar interaction is recovered by ^{1}H cw irradiation with the nutation frequency corresponding to the RF intensity ω_n satisfying the rotary-resonance condition $\omega_n = n\omega_R$ ($n = 1$ or 2).[31,32] Contrary to PDSD, the ^{1}H–^{1}H dipolar interactions are not averaged and the recoupled ^{13}C–^{1}H dipolar interactions are time independent. Hence, both ^{13}C–^{1}H and ^{1}H–^{1}H dipolar couplings broaden the ^{13}C lines. The spectral ^{13}C–^{13}C overlap between two spins is realized between a spinning sideband of one ^{13}C with the ^{13}C–^{1}H dipolar broadened response of the other ^{13}C, and the same for the other ^{13}C. DARR is more effective than RFDR in recoupling long-range dipolar interactions that are weak[33,34] and is applicable for band-selective recoupling between carbonyl or aromatic carbons and aliphatic carbons.[31] Although the RF requirements are more demanding than for PDSD, the RF applied during mixing is of relatively low amplitude.[34]

The pulse sequences such as RFDR, PDSD and DARR allow us to assign the NMR responses to the chemical and electronic structure, compare solid-state chemical shifts with those from solution NMR, and study short-range ^{13}C–^{13}C or ^{13}C–^{15}N residue connectivities, distances and angles within molecules. However, it is difficult to investigate intermolecular connectivity with these pulse schemes owing to rapid relayed intramolecular spin diffusion.[35–37] An important next step in the solid-state NMR is to elucidate the 3D structural arrangements of molecules and supramolecular aggregates.[38] For these purposes the CHHC experiment can be used[39–43] to investigate intermolecular correlations for ^{13}C–^{1}H transfers up to 7Å.[21,43]

In the CHHC preparation period (Figure 25.1C), the ^{13}C transverse coherence is established by application of ramped-amplitude (RAMP) CP[44] to transfer polarization from ^{1}H to ^{13}C. RAMP CP is a modification of CP used to overcome difficulties appearing when establishing and maintaining an efficient Hartmann–Hahn matching condition at high MAS. In the RAMP CP the amplitudes of the CP field strength monotonically decrease in a ramp, resulting in stable signal intensity and wider range of matching conditions.[44]

During the evolution period t_1, the ^{13}C spins precess under the effects of chemical shift and Zeeman interaction in the presence of proton decoupling. For proton decoupling, a two-pulse phase modulation (TPPM)[11] scheme is applied, which is an improvement over cw RF decoupling for a strongly coupled homogeneous spin systems with MAS. Strongly coupled systems would need high-power cw decoupling, leading to sample degradation by overheating. TPPM implements rapid coherent modulation of RF pulses of length τ_p between two non-orthogonal phases separated by an angle ϕ. By cycling the phase of the initial ^{1}H pulse, the residual ^{13}C magnetization from the first CP step is eliminated.

In the next step, ^{1}H coherence from the ^{13}C source is established by a second RAMP CP. The subsequent proton $\pi/2$ pulse converts the transverse ^{1}H coherence into longitudinal coherence, along the z-axis.

During the mixing period t_m, the polarization migrates by spin diffusion without proton decoupling to eliminate residual ^{1}H and ^{13}C transverse coherences. The dynamics of the ^{1}H diffusion can be approximated as a diffusive process, characterized by the mean square displacement $\langle r^2 \rangle$ and an effective spin diffusion coefficient D_{eff}:

$$\langle r^2 \rangle = 6 D_{\mathrm{eff}} t_m \tag{25.6}$$

The ^{1}H spin diffusion time is a variable parameter, allowing different mean square polarization displacements to be probed.[40] For rigid organic solids, $0.6 < D_{\mathrm{eff}} < 0.8\ \mathrm{nm^2 ms^{-1}}$ is similar for many samples.[40,45]

After the mixing period, a third RAMP CP is applied to transfer ^{1}H polarization back to ^{13}C again. This ^{13}C polarization is observed in the detection period with TPPM proton decoupling.

The CHHC sequence leads to a high-resolution 2D ^{13}C–^{13}C spectrum. Every single cross-peak in the spectra potentially contains distance information. The high abundance of short ^1H–^1H distances can characterize the 3D fold of a uniformly labelled biomolecule, without chemical shift selective RF pulse schemes.[42] Comparison of multiple datasets, recorded with different mixing times, can lead to sets of distance constraints that provide information about, for example, the fold of a protein. The number of intramolecular cross-peaks increases for longer mixing times and this obscures the intermolecular transfer events. If molecular motion occurs during the mixing time, it will lead to partial averaging of the dipolar coupling, resulting in the attenuation or suppression of correlation signals.[46]

25.2.3 Photo-CIDNP MAS NMR

To increase the sensitivity of NMR signals, hyperpolarization methods are rapidly being established. These are based on enhancing the spin polarization *via* the non-Boltzmann distribution over α and β states of electron spins.[7] This can be driven by high-frequency microwave dynamic nuclear polarization (DNP),[47] chemically induced by spin sensitive reactions (CIDNP),[48] induced by photochemical reactions using light excitation (photo-CIDNP),[49–52] parahydrogen induced polarization (PHIP),[53,54] or optically induced polarization.[55] Hyperpolarization methods allow for $\sim 10^4$ times signal enhancement of NMR signals compared to the responses from nuclear systems in Boltzmann equilibrium.

Photo-CIDNP has become increasingly important in biomolecular solid-state NMR research of RCs. Originally discovered by Zysmilich *et al.*, this method gives a profound insight into the photochemical machinery of RCs with atomic selectivity for chlorophyll (Chl) and pheophytin cofactors.[56–63] Photo-CIDNP allows study of the electronic structure of the groundstate after the photocycle, *via* chemical shifts, and of the radical pair during the photocycle, *via* spectral intensities.[52]

The photo-CIDNP in photosynthetic RCs relies on the formation of a spin-correlated radical pair state by optical excitation and blocking further separation of charges.[64] The correlated electrons in the radical pair can reversibly convert their zero quantum coherence in the electronic singlet (S) state into T_0 double quantum coherence of the triplet state, forming the S–T_0 manifold spanned by electron–nuclear spin quantum states (Figure 25.2A). It can decay to the ground state *via* the S state with decay time τ_S or *via* the T_0 and the triplet of the donor (^3P) with long lifetime τ_T. The mechanism of photo-CIDNP that leads to a non-Boltzmann distribution of nuclear spin polarization in the radical pair for enhancing the NMR signal is explained by a combination of coherent and relaxation mechanisms: three-spin mixing (TSM),[65,66] differential decay (DD)[67] and differential relaxation (DR).[68]

The TSM mechanism describes the process of how coherent evolution in the S–T_0 manifold breaks the antisymmetry between the population of the nuclear states. This involves the action of the nuclear Zeeman operator, which can distinguish between opposite polarizations of nuclear spins. To

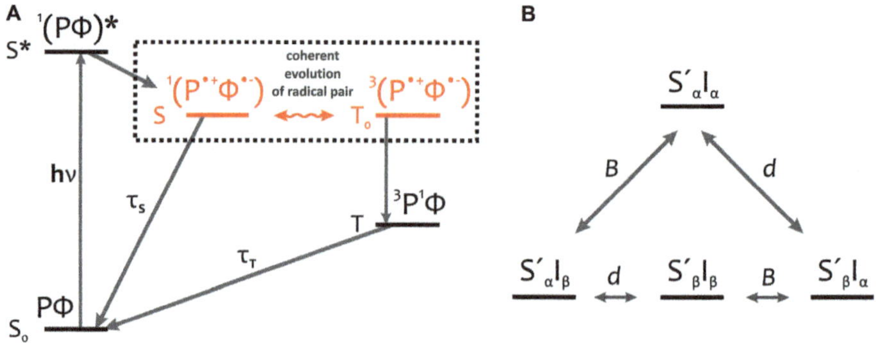

Figure 25.2 (A) Correlated radical pair mechanisms leading to photo-CIDNP in *Rb. sphaeroides*. First, the primary electron donor (P) is excited, then the electron is transferred to the primary electron acceptor (Φ), and a radical pair is created. It is formed as a singlet state that is not an eigenstate. It can either decay back to a ground state with a time τ_S or evolve coherently into a triplet state and back. The radical pair triplet can decay to a ground state with a characteristic time τ_T *via* a donor triplet state (^{3}P). (B) Three-spin mixing conditions.

first order, TSM operates *via* the interplay of electron and nuclear Zeeman operators, the hyperfine coupling between them, and the coupling between electrons.[65,69,70] TSM transfer is allowed when the energy levels of the three spins, two electrons and nucleus, are nearly degenerate under the action of the isotropic hyperfine interaction, and electronic and nuclear Zeeman interactions (*cf.* Figure 25.2B). The three-spin system is restricted to two coupled electron spins $\frac{1}{2}$ ($\boldsymbol{S_1}$, $\boldsymbol{S_2}$) and one nuclear spin $\frac{1}{2}$ ($\boldsymbol{I}$), with an anisotropic hyperfine coupling to $\boldsymbol{S_1}$.[66] It is convenient to map the S-T$_0$ manifold onto a fictitious spin $\frac{1}{2}$ ($\boldsymbol{S'}$) which undergoes a single quantum transition. The rotating frame Hamiltonian ($H_{\text{S-T0}}$) for the fictitious spin system is:

$$H_{S-T_0} = \Delta\Omega S'_z + \omega_I I_z + AS'_z I_z + BS'_z I_x - dS'_x \tag{25.7}$$

where A and B characterize the secular and pseudo-secular (diagonal and off-diagonal) parts of the hyperfine interaction, respectively, and

$$\Delta\Omega = \frac{(g_1 - g_2)\mu_e B_0}{\hbar} \tag{25.8}$$

represents the difference between the electron Zeeman frequencies of the two radical states, and

$$\omega_I = \frac{g_n \mu_n B_0}{\hbar} \tag{25.9}$$

is the nuclear Zeeman interaction. In these equations, the electron and nuclear magnetic moments are scaled according to their g-values, $g_{1,2}$ and g_n.

In addition, the combined dipolar D and exchange J couplings between the two electron spins are given by:

$$d = 2J + \frac{D}{2}\left(1 - \cos^2\theta\right) \tag{25.10}$$

where θ is the angle between the symmetry axis of the dipole–dipole interaction and the external field direction.

For RCs, $\Delta\Omega$ is larger than the hyperfine interaction or d. The TSM proceeds for the double matching condition:

$$2|\Delta\Omega| = 2|\omega_I I| = A \tag{25.11}$$

and the eigenstates of the first three commuting terms in Hamiltonian (25.7) are degenerate (Figure 25.2B). The pseudo-secular contribution B determines the mixing of the two electron spins with the nuclear spin. The TSM efficiency depends also on the magnetic field strength and performs best at moderate d values.

Hyperpolarization by the differential decay (DD) mechanism[67] can be generated by the difference in radical pair decay rates, τ_S and τ_T, which originate from a combination of coherent and incoherent reaction dynamics.[71] The matching condition for the DD is $2|\omega_I| = |A|$ and requires that the difference between τ_S and τ_T is on the order of the inverse hyperfine coupling. This mechanism works best for $d = 0$ and disappears when $\Delta\Omega$ or A is zero. Usually, positive signals in photo-CIDNP spectra from RCs are caused primarily by DD, while the negative response is predominantly from TSM mechanisms. The intensity of the lines provides information on the spin density distribution over the cofactors in the radical pair.[57,72]

For long-living triplet donors (^{3}P),[51,73] a differential relaxation (DR) mechanism is possible, which can occur when the electron spin relaxation modulates the hyperfine field of the donor in the triplet state ^{3}P, leading to stochastic transitions between the nuclear spins and resulting in different longitudinal spin relaxation rates for different nuclei. This mechanism is less frequently observed than TSM and DD and provides information about the spin density distribution over the ^{3}P.

25.3 Biological Solid-State NMR of Photosynthetic Energy Conversion Systems

25.3.1 Non-Photochemical Quenching in Light-Harvesting Antenna

Photosynthetic light-harvesting antennae serve two functions: efficient harvesting of light for the purpose of energy transfer to the RC; and dissipation of excess light energy *via* non-photochemical quenching (NPQ) in order to avoid oxidative photo-damage of the photosynthetic machinery. The exact mechanism of how the antenna switches rapidly from one mode to another is, however, not yet exactly known. Several mechanisms of NPQ in antenna were proposed: *via* formation of a radical pair,[74] formation of Chl–Chl charge transfer states,[75,76] or low excitonic carotenoid-Chl states.[77–79] Pandit *et al.*[80] used the PDSD NMR technique to study NPQ in *Chlamydomonas reinhardtii* major light-harvesting complex of photosystem II (LHCII) that was uniformly ^{13}C-labeled. The experiments were performed under two conditions: (i) unquenched trimetric LHCII solubilized in a small amount of detergent, and (ii) quenched aggregated LHCII without detergent. PDSD spectra were collected in a 17.5 T magnetic field at 244 K and were well-resolved for both LHCII preparations (Figure 25.3).

The spectra of quenched LHCII contain significantly more cross-peaks than those of unquenched LHCII. In general, the intensity of the cross-peaks in PDSD spectra is reduced by molecular motion, which weakens the dipolar couplings and suppresses spin diffusion and CP/MAS efficiency. The spectra reveal restricted dynamics for the Chl and amino acid side-chains in the protein interior, which are more mobile for the unquenched trimeric LHCII compared to quenched aggregates. The aromatic region of the PDSD spectrum was less congested with cross-peaks and allowed the assignment of

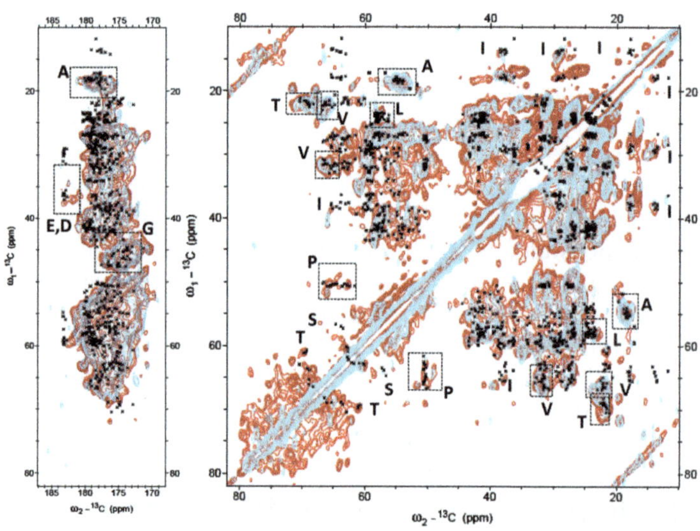

Figure 25.3 The carboxylic (*left*) and aliphatic (*right*) regions of PDSD spectra from unquenched (*blue*) and quenched (*red*) LHCII isolated from *C. reinhardtii. Black dots* represent structure-predicted NMR correlations obtained with the homologous pea LHCII crystal structure (PDB ID: 2BHW) simulated with the FANDAS[124] and SHFTX2[125] programs, which helped with amino acid assignments. Lipids and Chl aliphatic groups were not included in the simulations. Letters with rectangles indicate spin systems from aminoacids: alanine (A), threonine (T), valine (V), proline (P), serine (S), leucine (L) and isoleucine (I), glycine (G), glutamic acid (E) and aspartic acid (D).
(Reproduced from Pandit *et al.*,[80] with permission from Elsevier)

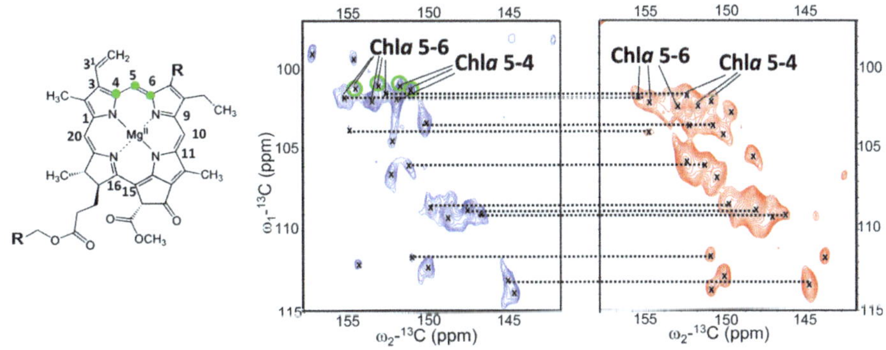

Figure 25.4 PDSD spectra for the aromatic region of LHCII samples in the quenched (*red*) and unquenched LHCII (*blue*) states, with the NMR instrument settings optimized for the detection of Chl C5-4/6 cross-signals. The conserved cross-peaks are marked by the *dotted lines.* The shifted cross-peaks for Chl *a* C4,5,6 are indicated in *green* in the structure at the *left.*
(Reproduced from Pandit *et al.*,[80] with permission from Elsevier)

carbon atoms in the Chl molecules (Figure 25.4). The largest shifts between the unquenched and quenched spectra were found in the Chl carbons 4, 5 and 6, which represent a flexing point of the molecule (left panel of Figure 25.4, marked in green). The cross-peaks from these carbons, as well as the signals from lipids,[80] are observed in the unquenched spectra only at temperatures well below 230 K. Below this temperature, the motions from solvent molecules, water, detergents and lipids in the unquenched LHCII trimers apparently freeze out; however, the cross-peaks from the quenched LHCII aggregates are visible even at 244 K. This was attributed to a low water content and a more dense LHCII packing under NPQ conditions. Hence the light-harvesting LHCII mode proceeds *via* partial mobility of the protein interior, which samples an ensemble of conformational states involving Chls, amino acid side-chains, water and detergent. This is in agreement with the non-photochemical quenching model derived by ultrafast time-resolved spectroscopy, which suggest the oligomerization and functional detachment of antenna under quenched conditions in plants[81] and diatoms.[82] In this respect, the unquenched LHCII shows essential characteristics of two classes of proteins: the energy conversion system with Chl cofactors packed in an energy potential where random noise is exploited to avoid exciton quenching due to, for example, mixing in charge transfer character; and a signal transduction system that can shift its dynamic conformational equilibrium in response to an external trigger to the quenched form where excitons are rapidly destabilized and energy is dissipated.

The MAS NMR data for the unquenched LHCII reveal conformational flexibility and absence of side-chain close-packing in its functional light-harvesting state. These characteristics are common for the molten globule, a protein state that is highly dynamic with well-defined secondary structure and fluctuating side-chain interactions. The dynamic structure of proteins in this state has been extensively investigated by NMR.[81–85] There is converging evidence that the absence of side-chain packing does not necessarily impair the capability of tight cofactor binding.[86] In addition, stable regions can be important in regulating biological function, such as protein activity or molecular recognition, and may also help to direct or facilitate the post-translational modification of substrate sites.[87]

The major free-energy barrier separating dynamic unpacked from more tightly packed protein interiors is thought to be due to the extraction or insertion of water.[88] Both dry and wet states are characterized by an expanded protein interior and a loosening of tertiary packing.[89] When loosening, the protein gains conformational entropy, the side chains unlock and the strength of the van der Waals interactions is reduced.[86,88] Factors that promote formation of these structures may not be apparent in the X-ray structure but can be sampled by biological solid-state NMR. LHCII can serve as an example of how a protein complex can switch its tertiary state under excess light.

25.3.2 BChl *a* Containing CsmA Protein from the Chlorosomal Baseplate

An important advantage of biological solid-state NMR is that it is possible to study homogeneous systems in their heterogeneous native membrane environments in cells and organelles. A recent example is found in Kulminskaya *et al.*,[90] where the structural characterization of the protein CsmA from the baseplate antenna complex of the thermophilic green sulfur bacterium *Chlorobaculum tepidum* was performed. The paracrystalline baseplate consists of many 6.2 kDa CsmA proteins forming a regular fibrillar scaffold coordinating BChl *a* molecules and binding carotenoids. The CsmA protein contains 59 residues, one of which is a single histidine (His25) that probably binds BChl *a* by coordination of the imidazole side-chain to the central Mg^{2+}.[91] CsmA accounts for more than 50% of the protein content in chlorosomes, the major light harvesting organelle of the organism. The CsmA–BChl *a* baseplate complex is important since it transfers energy from the chlorosome interior, built of self-aggregated BChl *c* rods, to the Fenna–Matthews–Olson (FMO) antenna protein, from which the energy reaches the RC.[92] Chlorosomes contain large amounts of BChl *c*, which is the most abundant chlorophyll species in the organism and the ratio of BChl *c* to BChl *a* is ~100:1.

The structure of the self-assembled BChl *c* inchlorosomes was recently determined by solid-state NMR and cryo-EM, which revealed how stacks built from alternating *syn-anti* BChl *c* can form concentric helical nanotubes.[93] To study the CsmA–BChl *a* baseplate, carotenosomes were used instead of chlorosomes. These are simplified antennae organelles derived from a *bchK* mutant lacking BChl *c* and with larger amounts of CsmA protein than for the wild-type (WT) species.[94] The structure and function of the CsmA protein are the same in chlorosomes and carotenosomes. The carotenosomes were ^{13}C,^{15}N-labeled, which allowed collection of a

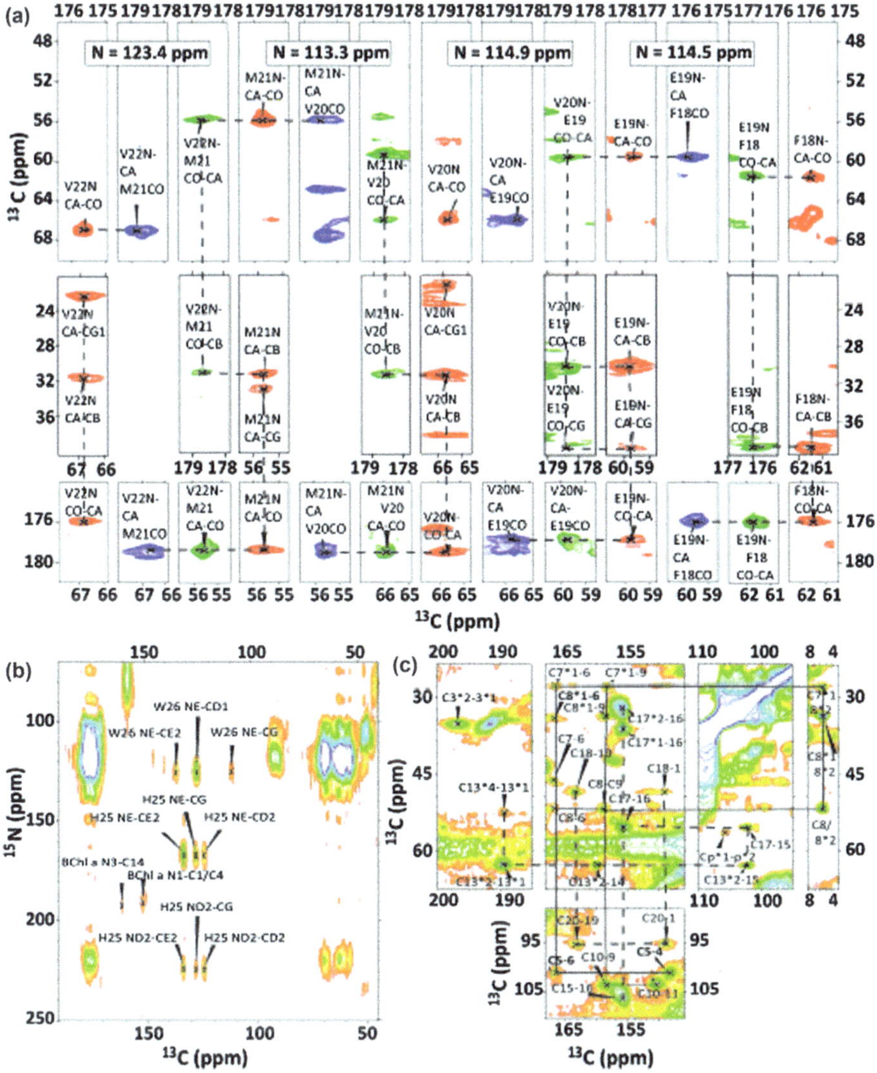

Figure 25.5 Solid-state NMR provides sequence-specific assignments for uniformly labelled $^{13}C,^{15}N$-carotenosomes. (A) Strip plots from $^{15}N,^{13}C,^{13}C$ 3D correlation spectra obtained by NCACX (*red*), NCOCX (*green*) and CONCA (*magenta*) pulse sequences show assignments for CsmA protein residues. The ^{15}N chemical shifts are indicated at the top of the strips; the top, middle and bottom rows present $^{13}C_{\alpha}$–$^{13}C'$, $^{13}C_{sidechain}$–$^{13}C'/^{13}C^{\alpha}$ and $^{13}C'$–$^{13}C^{\alpha}$ interconnections, respectively. (B) The assignments of Bchl *a*, His and Trp side-chain resonances by $^{15}N,^{13}C$ correlation spectra. (C) Correlation network for BChl *a* in a $^{13}C,^{13}C$ DARR spectrum obtained at −9°C.
(Reproduced from Kulminskaya *et al.*,[90] with permission from Wiley-VCH).

combination of 2D and 3D spectra to assign the backbone and side chains of the protein (Figure 25.5). Figure 25.5A presents the cross-peaks from the three 3D experiments required for sequential assignment of the protein: NCACX (red), NCOCX (green)[95] and CONCA (magenta).[96] In the NCACX and NCOCX pulse sequences, polarization is first transferred from ^{1}H to ^{15}N *via* CP, allowed to evolve and then is band-selectively transferred to the C^{α} carbon (CA) or C′ carbonyl (CO) using ramped double CP,[97] and allowed to evolve. This is followed by PDSD or DARR pulse sequences for polarization transfer to any other ^{13}C nearby (CX) to obtain the ^{13}C–^{13}C correlation, which gives a third dimension to the spectra. The same logic is applied for CONCA, except that the polarization is first transferred to the carbonyl, then selectively to ^{15}N and then finally to C^{α}. CONCA is a useful addition to NCACX and NCOCX, as it provides access to a sequence-specific assignment through connections between two sequential residues, *i.e.* $C^{\alpha}{}_{i}$–N_i–C'_{i-1}. Figure 25.5A shows the correlation signals from these datasets combined in strips for different ^{15}N shifts, indicated at the top. The three rows represent ^{13}C–^{13}C correlations between $^{13}C^{\alpha}$–$^{13}C'$ (top), $^{13}C_{sidechain}$–$^{13}C'/^{13}C^{\alpha}$ (middle) and $^{13}C'$–$^{13}C^{\alpha}$ (bottom); 90% of the residues were identified. The assignments of BChl *a*, His25 and tryptophan side-chain resonances were obtained from ^{15}N–^{13}C correlation data (Figure 25.5B). The ^{13}C–^{13}C DARR spectrum allowed sequential assignments of BChl *a* (Figure 25.5C). Similar to the LHCII, the interior of the CsmA baseplate antenna is subject to restricted dynamics and the BChl signals can only be detected below 264 K. The chemical shifts obtained with MAS NMR are sensitive to the protein backbone conformation.[98] Depending on their deviations from random-coil chemical shifts, obtained from unstructured peptides in solution,[99,100] the secondary structure can be determined, and the homology between the solid and solution structure can be exploited to provide insight into the interactions of the CsmA protein with BChl *a* cofactors. This deviation is also called secondary chemical shift.[101] The secondary shifts for the CsmA protein strongly support the α-helical structure of the protein that was determined by liquid-state NMR.[102]

25.3.3 The Mechanism of Charge Separation in Bacterial Reaction Centres

25.3.3.1 Photo-CIDNP MAS NMR Experiments

Photo-CIDNP MAS NMR has provided invaluable information on the mechanism of operation of photosynthetic RCs by probing the chemical shifts and the electron spin density distribution of cofactors that are active in the photochemistry. The photo-CIDNP method has been applied to RCs from green sulfur bacteria,[62] purple bacteria,[57,103,104] heliobacteria,[63] and plant photosystems I[58] and II.[59–61]

The RCs of *Rhodobacter sphaeroides* consist of: a "special pair" of strongly coupled BChls *a*, P_L and P_M, responsible for initiating charge separation (Figure 25.6C); two accessory BChls *a*, B_A and B_B; two bacteriopheophytins

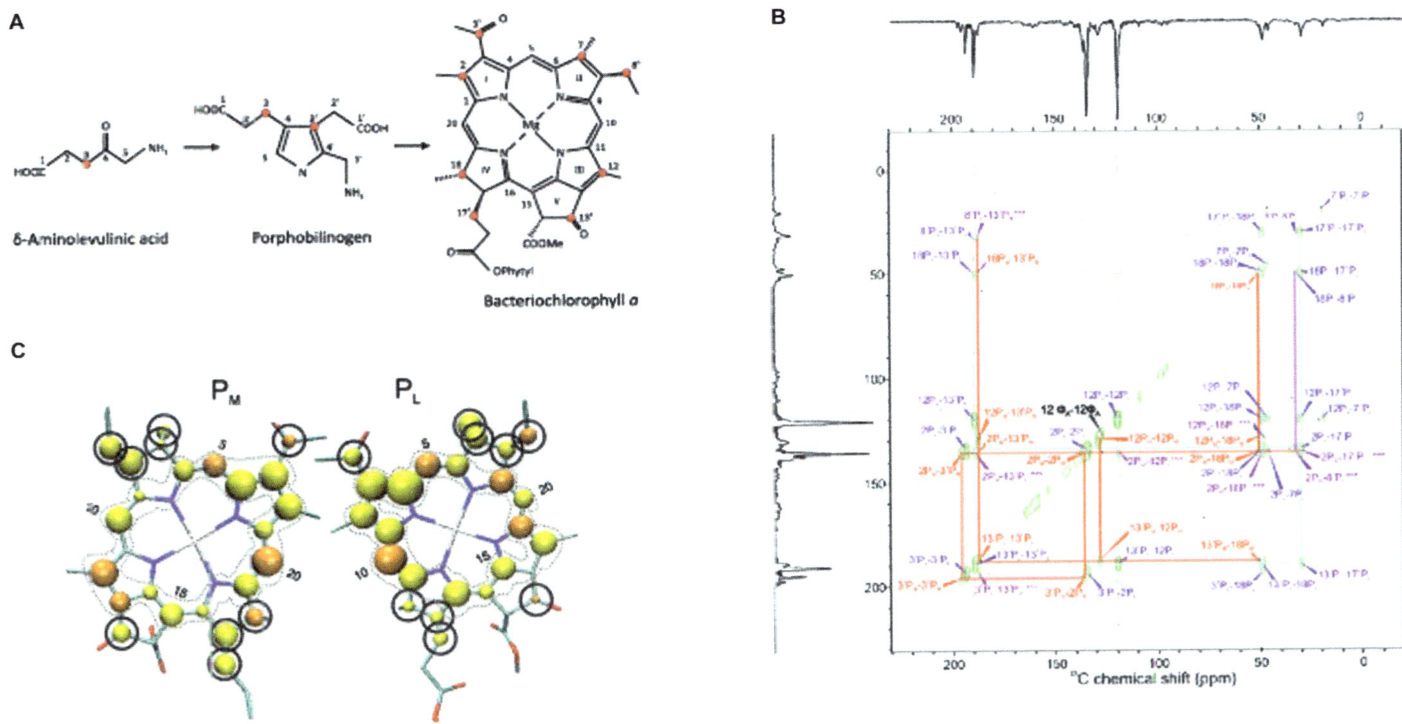

Figure 25.6 (A) Schematic representation of BChl*a* biosynthetic labeling from δ-aminolevulinic acid (ALA) precursors. The positions of labeled ^{13}C are marked with *red circles.* (B) The 1D and 2D DARR photo-CIDNP spectrum of ^{13}C-labeled RCs of *Rb. sphaeroides* WT, recorded at 4.7 T field strength, 223 K with a spinning frequency of 8 kHz; the mixing time was 2 s. (C) Electronic structures of the special pair BChls. Relative electron densities of P_L and P_M in the electronic ground state marked by *spheres* were derived from chemical shift differences with respect to monomeric BChl in acetone and corrected for ring-current effects.
(Adapted from Sai Sankar Gupta *et al.*[112])

(BPhe), Φ_A and Φ_B; and two ubiquinones, Q_A and Q_B, with a non-heme iron between them.[105] For WT RCs the symmetry in the cofactor arrangement is broken by a carotenoid cofactor that is located near B_B, while there are no carotenoids near the A-branch. However, in the mutant strain R26, for which the functional characteristics have been particularly well characterized, there is no carotenoid and the pigments are still paired in two branches with different functions, with only the A-branch "active" in electron transport starting from a special pair.[106] An electron is transferred from the donor to the accessory chlorophyll B_A, which has stronger overlap with the special pair than B_B, and then to Φ_A, which is 1.5 Å closer to B_A than Φ_B is to B_B. From Φ_A the electron is transferred to Q_A and further to Q_B. For a long time the origin of the symmetry breaking between the two BChls constituting the special pair and how this feeds into the mechanisms of function has been a mystery.

Recently, photo-CIDNP MAS NMR and time-dependent DFT modeling have provided profound insight into how a combination of static and dynamic symmetry breaking can give rise to barrier-less electron transfer along the active branch, with a sequence of proton-coupled electron transfer steps, coupled to vibrational modes that are selectively enhanced by the packing of the cofactors in the protein. One origin of static asymmetry in the special pair, which was resolved with photo-CIDNP MAS NMR and corroborated by computational studies, is in a different shaping of the two BChl macrocycles.[72] In addition, the orientations of the ester carbonyl groups of the phytyl tail and methyl ester side chain are controlled by the protein environment.[107,108] The conformations of the BChl acetyl group and the phytyl side chain can tune the energy of the π-conjugated BChl planes.[108,109]

To gain NMR and photo-CIDNP access to RC complexes with molecular weights of more than 100 kDa, specific isotope labeling was required. The labeling was performed by growing WT *Rb. sphaeroides* anaerobically in the presence of 99% labeled [3-^{13}C]-δ-aminolevulinic acid · HCl (3-ALA), which is a precursor in the biosynthesis of BChl and BPhe molecules.[110] A singly labeled 3-ALA leads to eight-fold BChl and BPhe labeled at positions C-2, 3^1, 7, 8^1, 12, 13^1, 17^1 and 18 (Figure 25.6A) with ~60% total incorporation. These are more peripheral positions in comparison to the labeling with 4-ALA[104,111] and 5-ALA.[103]

The photo-CIDNP DARR spectrum of RCs labeled with 3-ALA, hyperpolarized at 223 K by continuous illumination, is shown in Figure 25.6B. The 1D response is emissive, which is characteristic for the coherent TSM mechanism in the radical pair state, rather than DD. The signals are narrow, both in the cofactor region and for the side-chain carbons, with linewidths ~50 Hz. On the NMR scale the cofactor assembly is highly ordered, with very little static heterogeneity in the cofactor periphery and the protein-binding pocket. There are two correlation networks: one assigned to P_L (blue) with more intense signals, and the other assigned to P_M (red) (Figure 25.6B). The intermolecular crosspeaks between P_L and P_M are marked in purple. One cross peak was assigned to Φ_A.[103,104] The chemical shift differences between BChl *a* in the protein and dissolved in acetone are depicted in Figure 25.6C.

Yellow spheres represent negative shifts and reflect an increase of local electron density for the cofactors in the protein; orange spheres represent positive shifts.[72,112] The prevalence of negative shifts means that the total electron density increases for BChl in the protein compared to the monomeric BChl in solution. This reflects the partial charge transfer between BChls and their coordinating His, which has been assessed by MAS NMR and modeling.[109] This partial charge transfer may be a common property of a BChl–His motif, which was also observed in the LH2 antenna, where the B850 accumulate *ca.* −0.5 electronic equivalents of charge from coordinating His, which affects the site energy and dipole transition moments of the BChl.[113,114] However, the largest protein induced shift for the BChls in the RC is observed in the overlap region between the two BChls, especially in the first pyrrole rings, which is attributed to packing and shaping of the macrocycle.

Most of the variations between the yellow and orange spheres between P_L and P_M are found in the peripheral carbons. These shifts often complement each other, showing more positive, electron-donating character of the P_L side chains and more negative, electron-withdrawing character of the P_M side chains. This finding reflects the ability of photo-CIDNP to give structural assignment as well as characterize electron charge distribution in the special pair.

25.3.3.2 *Ab initio* Molecular Dynamics Simulations

Based on the available X-ray data and guided by the data from MAS and photo-CIDNP NMR, more details on the charge separation and electron transfer initiation in the bacterial RC were resolved with quantum-chemical modelling.[115,116] Two minimal models were simulated: for the ground state using *ab initio* molecular dynamics (AIMD),[117,118] for the excited state employing a restricted open-shell Kohn–Sham (ROKS) approach.[118] In addition during the relaxation of the excited state constrained density functional theory was applied.[119] Model 1 included the special pair with the essentials of the protein environment: axial histidines, His_L173 and His_M202, coordinated to Mg of P_L and P_M, respectively; His_L168, linked to His_L173 and the water responsible for the hydrogen bond with the 3^1-acetyl group of P_L; and phenylalanine, Phe_M197, linked to His_M202 and another water molecule.[115] Model 2 is a donor–acceptor model, which included P_M with His_M202 as the charge donor and B_A with His_L153 as the charge acceptor, while also Phe_M197, Gly_M203 and a water molecule were included for the protein environment.[116] The exclusion of P_L from the Model 2 was done to keep the system size minimal and the effect of this simplification on the energies was verified to be small.

The ground state simulations for Model 1 were done on a timescale of 1 ps and were followed by 800 fs of excited state simulations.[115] Figure 25.7A shows the time evolution of the dihedral angles for the two axial histidines, His_L173 and His_M202, which rotate with respect to the axis between the planes connecting the nitrogen atoms from the His ring to the Mg in the

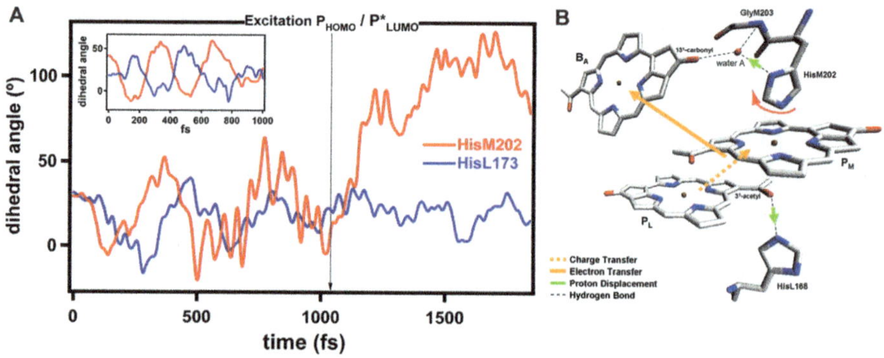

Figure 25.7 (A) Modulation of HOMO–LUMO electron density as a function of dihedral angles, determined by rotation of the axial histidines around the axis connecting the Mg from BChl plane with nitrogen from the His ring. The *red curve* represents the His_M202 dihedrals with the P_M-HOMO energy in time, while the *blue line* characterizes the His_L173 dihedrals with P_L. The *arrow* in the graph indicates the moment in time when the simulation approach was switched from the ground state to the excited state. The *inset* shows the ground state dynamics along a different trajectory. (B) The scheme of primary charge separation in the bacterial RC obtained by quantum mechanical calculations. (Adapted from Eisenmayer *et al.*[115,116])

BChl. In the ground state the electron density distribution is dynamic; it fluctuates and reversibly changes its localization from the highest occupied molecular orbital (HOMO) on P_L to the HOMO on P_M, while the electron density from the lowest unoccupied molecular orbital (LUMO) changes backwards from P_M to P_L. This fluctuation involves two collective low-frequency vibrational modes, ~50 and ~100 cm^{-1}. The ~50 cm^{-1} mode correlates with changes in the hydrogen bond between the acetyl group of P_L and His_L168, while the ~100 cm^{-1} mode depends on the changes in the angle between P_M and His_M202. Upon excitation, predominantly the ~100 cm^{-1} mode changes together with His_M202 dihedral angle rotation up to 80°, while there is no significant change for the ~50 cm^{-1} mode involving His_L168. Such behaviour is explained by formation of a $P_L{}^+P_M{}^-$ state at the end of the excitation period with strong charge transfer character,[115] which is in line with earlier investigations.[120,121]

The later work[116] extended the calculations towards the oxidized state and Model 2. In the oxidized state the amplitude of the ~50 cm^{-1} mode fluctuation is larger than for the ~100 cm^{-1} mode. This correlated with a weakening of the hydrogen bond between His_L168 and the 3^1-acetyl group of P_L. For Model 2, one more frequency was found, 30–35 cm^{-1}, in agreement with simulations based on a model Hamiltonian approach[122] and with fs absorption spectroscopy experiments.[123] This mode was assigned to the proton displacement from water A towards Gly_M203 and the 13^1-carbonyl of B_A (Figure 25.7B), with an increase of the distance between water A and His_M202 from 2.9 to 3.6 Å.

Finally, the schematic machinery of the primary charge separation mechanism in the bacterial RC is presented in Figure 25.7B. After light absorption, the process starts by the partial charge transfer from P_L to P_M in the exciton state, which initiates the conformational rearrangements in the RC. This leads to a reversible rotation of the axial His_M202, with characteristic oscillation frequency of $\sim 100\ cm^{-1}$. This rotation increases the coupling between P_M and B_A, and a proton from water A moves closer towards Gly_M203 and the 13^1-carbonyl of B_A with a frequency of 30–35 cm^{-1}. From another side, with a frequency of $\sim 50\ cm^{-1}$, the proton from the 3^1-acetyl group of P_L is displaced towards His_L168, weakening the interaction and destabilizing the $P_L^{+\cdot}$ cation radical. This step reduces the possibility for charge recombination.

25.4 Conclusions

In this chapter, techniques for biological solid-state NMR and recent applications in photosynthesis research are summarized. The focus of the first part of the chapter is on the technical description of important techniques, such as MAS, CP, homonuclear correlation, homo and hetero decoupling and photo-CIDNP, for the novice reader with a background different from NMR. The homonuclear decoupling pulse sequences, using RFDR and spin diffusion PDSD, DARR and CHHC, are explained, together with high-resolution 2D and 3D spectral assignments. The second part of the chapter gives examples of how these NMR techniques can be applied to address structure and identify the subtleties of the energy conversion mechanisms within photosynthetic light-harvesting antenna and reaction centre complexes.

Solid-state NMR was used to study NPQ by comparing the LHCII spectra isolated from the green algae *C. reinhardtii* in quenched and unquenched forms, which resulted in the assignments of Chls and amino acids. In the unquenched state the LHCII protein tertiary structure is highly dynamic and flexible, similar to the wet molten globule state, while the quenched state shows a more folded configuration with a tighter packing.

Another recent example is the CsmA protein in the chlorosome baseplate of *C. tepidum.* The baseplate is a paracrystalline entity that is inaccessible to solution NMR and diffraction methods. Recently, it was shown how a combination of different 2D and 3D protein NMR pulse sequences, such as NCACX, NCOCX and CONCA, could be applied to arrive at a 90% assignment of the NMR response from the protein in its native heterogeneous environment.

Finally, photo-CIDNP MAS NMR provides information on the electronic structure of cofactor molecules that are biosynthetically labelled with ^{13}C. The mechanism of operation of photosynthetic RCs of *Rb. sphaeroides* was studied. A detailed view on the frontier orbitals for the coupled BChls in the special pair was obtained. Subsequent *ab initio* quantum-mechanical modelling of the special pair with its close surroundings predicted pronounced formation of a charge-transfer $P_L^+P_M^-$ state upon excitation and demonstrated the protein rearrangement to assist the primary charge separation process.

References

1. A. Alia, F. Buda, H. de Groot and J. Matysik, *Annu. Rev. Biophys.*, 2013, **42**, 675–699.
2. R. Tycko, *Annu. Rev. Phys. Chem.*, 2001, **52**, 575–606.
3. A. Abragam, M. Chapelli, J. Jacquino and M. Goldman, *J. Magn. Reson.*, 1973, **10**, 322–346.
4. A. Abragam, *Phys. Rev.*, 1955, **98**, 1729–1735.
5. A. Pandit, K. Ocakoglu, F. Buda, T. van Marle, A. R. Holzwarth and H. J. M. de Groot, *J. Phys. Chem. B*, 2013, 11292–11298.
6. W. A. Anderson and R. Freeman, *J. Chem. Phys.*, 1962, **37**, 85–103.
7. A. W. Overhauser, *Phys. Rev.*, 1953, **92**, 411–415.
8. E. R. Andrew, A. Bradbury and R. G. Eades, *Nature*, 1958, **182**, 1659–1659.
9. I. J. Lowe, *Phys. Rev. Lett.*, 1959, **2**, 285–287.
10. M. M. Maricq and J. S. Waugh, *J. Chem. Phys.*, 1979, **70**, 3300–3316.
11. A. E. Bennett, C. M. Rienstra, M. Auger, K. V. Lakshmi and R. G. Griffin, *J. Chem.Phys.*, 1995, **103**, 6951–6958.
12. S. R. Hartmann and E. L. Hahn, *Phys. Rev.*, 1962, **128**, 2042–2053.
13. A. Pines, M. G. Gibby and J. S. Waugh, *J. Chem. Phys.*, 1973, **59**, 569–590.
14. J. Schaefer and E. O. Stejskal, *J. Am. Chem. Soc.*, 1976, **98**, 1031–1032.
15. M. G. Munowitz and R. G. Griffin, *J. Chem. Phys.*, 1982, **76**, 2848–2858.
16. J. Herzfeld and A. E. Berger, *J. Chem. Phys.*, 1980, **73**, 6021–6030.
17. F. J. Blanco and R. Tycko, *J. Magn. Reson.*, 2001, **149**, 131–138.
18. R. Tycko and A. E. Berger, *J. Magn. Reson.*, 1999, **141**, 141–147.
19. R. Tycko, D. P. Weliky and A. E. Berger, *J. Chem. Phys.*, 1996, **105**, 7915–7930.
20. V. E. Zorin, M. Ernst, S. P. Brown and P. Hodgkinson, *J. Magn. Reson.*, 2008, **192**, 183–196.
21. M. Aluas, C. Tripon, J. M. Griffin, X. Filip, V. Ladizhansky, R. G. Griffin, S. P. Brown and C. Filip, *J. Magn. Reson.*, 2009, **199**, 173–187.
22. K. H. Gardner and L. E. Kay, *Annu. Rev. Biophys. Biomol. Struct.*, 1998, **27**, 357–406.
23. M. Hong and K. Jakes, *J. Biomol. NMR*, 1999, **14**, 71–74.
24. A. E. Bennett, J. H. Ok, R. G. Griffin and S. Vega, *J. Chem. Phys.*, 1992, **96**, 8624–8627.
25. A. E. Bennett, C. M. Rienstra, P. T. Lansbury and R. G. Griffin, *J. Chem. Phys.*, 1996, **105**, 10289–10299.
26. A. E. Bennett, C. M. Rienstra, J. M. Griffiths, W. G. Zhen, P. T. Lansbury and R. G. Griffin, *J. Chem. Phys.*, 1998, **108**, 9463–9479.
27. D. K. Sodickson, M. H. Levitt, S. Vega and R. G. Griffin, *J. Chem. Phys.*, 1993, **98**, 6742–6748.

28. N. Bloembergen, *Physica*, 1949, **15**, 386–426.
29. J.-N. Dumez, M. C. Butler, E. Salager, B. Elena-Herrmann and L. Emsley, *Phys. Chem. Chem. Phys.*, 2010, **12**, 9172–9175.
30. N. M. Szeverenyi, M. J. Sullivan and G. E. Maciel, *J. Magn. Reson.*, 1982, **47**, 462–475.
31. K. Takegoshi, S. Nakamura and T. Terao, *Chem. Phys. Lett.*, 2001, **344**, 631–637.
32. K. Takegoshi and T. Terao, *J. Chem. Phys.*, 2002, **117**, 1700–1707.
33. K. Takegoshi, S. Nakamura and T. Terao, *J. Chem. Phys.*, 2003, **118**, 2325–2341.
34. E. Crocker, A. B. Patel, M. Eilers, S. Jayaraman, E. Getmanova, P. J. Reeves, M. Ziliox, H. G. Khorana, M. Sheves and S. O. Smith, *J. Biomol. NMR*, 2004, **29**, 11–20.
35. B. J. van Rossum, G. J. Boender, F. M. Mulder, J. Raap, T. S. Balaban, A. R. Holzwarth, K. Schaffner, S. Prytulla, H. Oschkinat and H. J. M. de Groot, *Spectrochim. Acta, Part A*, 1998, **54**, 1167–1176.
36. G. J. Boender, J. Raap, S. Prytulla, H. Oschkinat and H. J. M. de Groot, *Chem. Phys. Lett.*, 1995, **237**, 502–508.
37. F. Castellani, B. van Rossum, A. Diehl, M. Schubert, K. Rehbein and H. Oschkinat, *Nature*, 2002, **420**, 98–102.
38. S. Ganapathy, A. J. van Gammeren, F. B. Hulsbergen and H. J. M. de Groot, *J. Am. Chem. Soc.*, 2007, **129**, 1504–1505.
39. S. Zhang, B. H. Meier and R. R. Ernst, *Phys. Rev. Lett.*, 1992, **69**, 2149–2151.
40. M. Wilhelm, H. Feng, U. Tracht and H. W. Spiess, *J. Magn. Reson.*, 1998, **134**, 255–260.
41. F. M. Mulder, W. Heinen, M. van Duin, J. Lugtenburg and H. J. M. de Groot, *J. Am. Chem. Soc.*, 1998, **120**, 12891–12894.
42. A. Lange, S. Luca and M. Baldus, *J. Am. Chem. Soc.*, 2002, **124**, 9704–9705.
43. I. de Boer, L. Bosman, J. Raap, H. Oschkinat and H. J. M. de Groot, *J. Magn. Reson.*, 2002, **157**, 286–291.
44. G. Metz, X. L. Wu and S. O. Smith, *J. Magn. Reson., Ser. A*, 1994, **110**, 219–227.
45. J. Clauss, K. Schmidt-Rohr and H. W. Spiess, *Acta Polym.*, 1993, **44**, 1–17.
46. C. Gardiennet, A. Loquet, M. Etzkorn, H. Heise, M. Baldus and A. Bockmann, *J. Biomol. NMR*, 2008, **40**, 239–250.
47. R. G. Griffin and T. F. Prisner, *Phys. Chem. Chem. Phys.*, 2010, **12**, 5737–5740.
48. H. R. Ward and R. G. Lawler, *J. Am. Chem. Soc.*, 1967, **89**, 5518–5519.
49. J. Bargon, H. Fischer and Z. Naturforsch, *Teil A*, 1967, **22**, 1556–1562.
50. J. Bargon, H. Fischer, U. Johnsen and Z. Naturforsch, *Teil A*, 1967, **22**, 1551–1555.
51. P. J. Hore and R. W. Broadhurst, *Prog. Nucl. Magn. Reson. Spectrosc.*, 1993, **25**, 345–402.
52. J. Matysik, A. Diller, E. Roy and A. Alia, *Photosynth. Res.*, 2009, **102**, 427–435.

53. C. R. Bowers and D. P. Weitekamp, *Phys. Rev. Lett.*, 1986, **57**, 2645–2648.
54. C. R. Bowers and D. P. Weitekamp, *J. Am. Chem. Soc.*, 1987, **109**, 5541–5542.
55. G. Maier, U. Haeberlen, H. C. Wolf and K. H. Hausser, *Phys. Lett. A*, 1967, **A25**, 384–385.
56. M. G. Zysmilich and A. McDermott, *J. Am. Chem. Soc.*, 1994, **116**, 8362–8363.
57. S. Prakash, A. Alia, P. Gast, H. J. M. de Groot, G. Jeschke and J. Matysik, *J. Am. Chem. Soc.*, 2005, **127**, 14290–14298.
58. A. Alia, E. Roy, P. Gast, H. J. van Gorkom, H. J. M. de Groot, G. Jeschke and J. Matysik, *J. Am. Chem. Soc.*, 2004, **126**, 12819–12826.
59. J. Matysik, A. Alia, P. Gast, H. J. van Gorkom, A. J. Hoff and H. J. M. de Groot, *Proc. Natl. Acad. Sci. U. S. A.*, 2000, **97**, 9865–9870.
60. A. Diller, A. Alia, E. Roy, P. Gast, H. J. van Gorkom, J. Zaanen, H. J. M. de Groot, C. Glaubitz and J. Matysik, *Photosynth. Res.*, 2005, **84**, 303–308.
61. A. Diller, E. Roy, P. Gast, H. J. van Gorkom, H. J. M. de Groot, C. Glaubitz, G. Jeschke, J. Matysik and A. Alia, *Proc. Natl. Acad. Sci. U. S. A.*, 2007, **104**, 12767–12771.
62. E. Roy, A. Alia, P. Gast, H. van Gorkom, H. J. M. de Groot, G. Jeschke and J. Matysik, *Biochim. Biophys. Acta*, 2007, **1767**, 610–615.
63. E. Roy, T. Rohmer, P. Gast, G. Jeschke, A. Alia and J. Matysik, *Biochemistry*, 2008, **47**, 4629–4635.
64. E. Daviso, Doctoral thesis, Leiden University, 2008.
65. G. Jeschke, *J. Chem. Phys.*, 1997, **106**, 10072–10086.
66. G. Jeschke, *J. Am. Chem. Soc.*, 1998, **120**, 4425–4429.
67. T. Polenova and A. E. McDermott, *J. Phys. Chem. B*, 1999, **103**, 535–548.
68. A. McDermott, M. G. Zysmilich and T. Polenova, *Solid State Nucl. Magn. Reson.*, 1998, **11**, 21–47.
69. G. Jeschke and J. Matysik, *Chem. Phys.*, 2003, **294**, 239–255.
70. G. Jeschke, B. C. Anger, B. E. Bode and J. Matysik, *J. Phys. Chem. A*, 2011, **115**, 9919–9928.
71. S. S. Thamarath, B. E. Bode, S. Prakash, K. Gupta, A. Alia, G. Jeschke and J. Matysik, *J. Am. Chem. Soc.*, 2012, **134**, 5921–5930.
72. E. Daviso, S. Prakash, A. Alia, P. Gast, J. Neugebauer, G. Jeschke and J. Matysik, *Proc. Natl. Acad. Sci. U. S. A.*, 2009, **106**, 22281–22286.
73. S. Prakash, A. Alia, P. Gast, H. J. M. de Groot, J. Matysik and G. Jeschke, *J. Am. Chem. Soc.*, 2006, **128**, 12794–12799.
74. T. K. Ahn, T. J. Avenson, M. Ballottari, Y. C. Cheng, K. K. Niyogi, R. Bassi and G. R. Fleming, *Science*, 2008, **320**, 794–797.
75. Y. Miloslavina, A. Wehner, E. Wientjes, M. Reus, P. Lambrev, G. Garab, R. Croce and A. R. Holzwarth, *FEBS Lett.*, 2008, **582**, 3625–3631.
76. M. G. Müller, P. Lambrev, M. Reus, E. Wientjes, R. Croce and A. R. Holzwarth, *ChemPhysChem*, 2010, **11**, 1289–1296.
77. S. Bode, C. C. Quentmeier, P. N. Liao, N. Hafi, T. Barros, L. Wilk, F. Bittner and P. J. Walla, *Proc. Natl. Acad. Sci. U. S. A.*, 2009, **106**, 12311–12316.

78. A. V. Ruban, R. Berera, C. Ilioaia, I. H. M. van Stokkum, J. T. M. Kennis, A. A. Pascal, H. van Amerongen, B. Robert, P. Horton and R. van Grondelle, *Nature*, 2007, **450**, 575–578.
79. A. A. Pascal, Z. F. Liu, K. Broess, B. van Oort, H. van Amerongen, C. Wang, P. Horton, B. Robert, W. R. Chang and A. V. Ruban, *Nature*, 2005, **436**, 134–137.
80. A. Pandit, M. Reus, T. Morosinotto, R. Bassi, A. R. Holzwarth and H. J. M. de Groot, *Biochim. Biophys. Acta*, 2013, **1827**, 738–744.
81. A. R. Holzwarth, Y. Miloslavina, M. Nilkens and P. Jahns, *Chem. Phys. Let.*, 2009, **483**, 262–267.
82. Y. Miloslavina, I. Grouneva, P. H. Lambrev, B. Lepetit, R. Goss, C. Wilhelm and A. R. Holzwarth, *Biochim. Biophys. Acta*, 2009, **1787**, 1189–1197.
83. B. A. Schulman, P. S. Kim, C. M. Dobson and C. Redfield, *Nat. Struct. Biol.*, 1997, **4**, 630–634.
84. J. Baum, C. M. Dobson, P. A. Evans and C. Hanley, *Biochemistry*, 1989, **28**, 7–13.
85. P. Schanda, V. Forge and B. Brutscher, *Proc. Natl. Acad. Sci. U. S. A.*, 2007, **104**, 11257–11262.
86. R. L. Baldwin and G. D. Rose, *Curr. Opin. Struct. Biol.*, 2013, **23**, 4–10.
87. M. Tsytlonok and L. S. Itzhaki, *Arch. Biochem. Biophys.*, 2013, **531**, 14–23.
88. R. L. Baldwin, C. Frieden and G. D. Rose, *Proteins: Struct., Funct., Bioinf.*, 2010, **78**, 2725–2737.
89. S. Bhattacharyya and R. Varadarajan, *Curr.Opin. Struct. Biol.*, 2013, **23**, 11–21.
90. N. V. Kulminskaya, M. O. Pedersen, M. Bjerring, J. Underhaug, M. Miller, N. U. Frigaard, J. T. Nielsen and N. C. Nielsen, *Angew. Chem. Int. Ed.*, 2012, **51**, 6891–6895.
91. M. O. Pedersen, L. Pham, D. B. Steensgaard and M. Miller, *Biochemistry*, 2008, **47**, 1435–1441.
92. G. A. Montano, H. M. Wu, S. Lin, D. C. Brune and R. E. Blankenship, *Biochemistry*, 2003, **42**, 10246–10251.
93. S. Ganapathy, G. T. Oostergetel, P. K. Wawrzyniak, M. Reus, A. G. M. Chew, F. Buda, E. J. Boekema, D. A. Bryant, A. R. Holzwarth and H. J. M. de Groot, *Proc. Natl. Acad. Sci. U. S. A.*, 2009, **106**, 8525–8530.
94. N. U. Frigaard, G. D. Voigt and D. A. Bryant, *J. Bacteriol.*, 2002, **184**, 3368–3376.
95. J. Pauli, M. Baldus, B. van Rossum, H. de Groot and H. Oschkinat, *ChemBioChem*, 2001, **2**, 272–281.
96. Y. Li, D. A. Berthold, H. L. Frericks, R. B. Gennis and C. M. Rienstra, *ChemBioChem*, 2007, **8**, 434–442.
97. J. Schaefer, R. A. McKay and E. O. Stejskal, *J. Magn. Reson.*, 1979, **34**, 443–447.
98. D. S. Wishart and B. D. Sykes, *Methods Enzymol.*, 1994, **239**, 363–392.
99. S. Spera and A. Bax, *J. Am. Chem. Soc.*, 1991, **113**, 5490–5492.
100. O. W. Howarth and D. M. J. Lilley, *Prog. Nucl. Magn. Reson. Spectrosc.*, 1978, **12**, 1–40.

101. S. Luca, D. V. Filippov, J. H. van Boom, H. Oschkinat, H. J. M. de Groot and M. Baldus, *J. Biomol. NMR*, 2001, **20**, 325–331.
102. M. O. Pedersen, J. Underhaug, J. Dittmer, M. Miller and N. C. Nielsen, *FEBS Lett.*, 2008, **582**, 2869–2874.
103. S. Prakash, A. Alia, P. Gast, H. J. M. de Groot, G. Jeschke and J. Matysik, *Biochemistry*, 2007, **46**, 8953–8960.
104. E. A. M. Schulten, J. Matysik, A. Alia, S. Kiihne, J. Raap, J. Lugtenburg, P. Gast, A. J. Hoff and H. J. M. de Groot, *Biochemistry*, 2002, **41**, 8708–8717.
105. J. P. Allen, G. Feher, T. O. Yeates, H. Komiya and D. C. Rees, *Biophys. J.*, 1987, **51**, A377–A377.
106. A. J. Hoff and J. Deisenhofer, *Phys. Rep.*, 1997, **287**, 1–247.
107. H. Yamasaki, Y. Takano and H. Nakamura, *J. Phys. Chem. B*, 2008, **112**, 13923–13933.
108. P. K. Wawrzyniak, M. T. P. Beerepoot, H. J. M. de Groot and F. Buda, *Phys. Chem. Chem. Phys.*, 2011, **13**, 10270–10279.
109. A. Alia, P. K. Wawrzyniak, G. J. Janssen, F. Buda, J. Matysik and H. J. M. de Groot, *J. Am. Chem. Soc.*, 2009, **131**, 9626–9627.
110. *Biosynthesis of Tetrapyrroles*, ed. P. M. Jordan, Elsevier, Amsterdam, 1991.
111. E. Daviso, A. Alia, S. Prakash, A. Diller, P. Gast, J. Lugtenburg, J. Matysik and G. Jeschke, *J. Phys. Chem. C*, 2009, **113**, 10269–10278.
112. K. B. S. S. Gupta, A. Alia, H. J. M. de Groot and J. Matysik, *J. Am. Chem. Soc.*, 2013, **117**, 3287–3297.
113. A. Alia, J. Matysik, I. de Boer, P. Gast, H. J. van Gorkom and H. J. M. de Groot, *J. Biomol. NMR*, 2004, **28**, 157–164.
114. P. K. Wawrzyniak, A. Alia, R. G. Schaap, M. M. Heemskerk, H. J. M. de Groot and F. Buda, *Phys.Chem. Chem. Phys.*, 2008, **10**, 6971–6978.
115. T. J. Eisenmayer, H. J. M. de Groot, E. van de Wetering, J. Neugebauer and F. Buda, *Biophys. Chem.*, 2012, **3**, 694–697.
116. T. J. Eisenmayer, J. A. Lasave, A. Monti, H. J. de Groot and F. Buda, *J. Phys. Chem. B*, 2013, 11162–11168.
117. R. Car and M. Parrinello, *Phys. Rev. Lett.*, 1985, **55**, 2471–2474.
118. D. Marx and J. Hütter, *Ab initio Molecular Dynamics: Basic Theory and Advanced Methods*, Cambridge University Press, Cambridge, 2009.
119. Q. Wu and T. Van Voorhis, *J. Chem. Theory Comput.*, 2006, **2**, 765–774.
120. L. J. Moore, H. L. Zhou and S. G. Boxer, *Biochemistry*, 1999, **38**, 11949–11960.
121. V. A. Shuvalov and A. G. Yakovlev, *FEBS Lett.*, 2003, **540**, 26–34.
122. V. I. Novoderezhkin, A. G. Yakovlev, R. van Grondelle and V. A. Shuvalov, *J. Phys. Chem. B*, 2004, **108**, 7445–7457.
123. A. G. Yakovlev and V. A. Shuvalov, *Biochemistry*, 2003, **68**, 603–610.
124. S. Gradmann, C. Ader, I. Heinrich, D. Nand, M. Dittmann, A. Cukkemane, M. van Dijk, A. Bonvin, M. Engelhard and M. Baldus, *J. Biomol. NMR*, 2012, **54**, 377–387.
125. B. Han, Y. F. Liu, S. W. Ginzinger and D. S. Wishart, *J. Biomol. NMR*, 2011, **50**, 43–57.

CHAPTER 26

Large Protein Complexes Revealed by Solution-State NMR: G Proteins and G Protein-Activated Inwardly Rectifying Potassium Ion Channel

MASANORI OSAWA,[a] YOKO MASE,[a] MARIKO YOKOGAWA,[a,b] KOH TAKEUCHI[c] AND ICHIO SHIMADA*[a]

[a] Graduate School of Pharmaceutical Sciences, University of Tokyo, Tokyo, Japan; [b] Japan Biological Information Research Centre (JBIRC), Japan Biological Informatics Consortium (JBIC), Tokyo, Japan; [c] Molecular Profiling Research Centre for Drug Discovery, National Institute of Advanced Industrial Science and Technology (AIST), Tokyo, Japan
*Email: shimada@iw-nmr.f.u-tokyo.ac.jp

26.1 Solution-State NMR Techniques for Larger Proteins

26.1.1 Detection of NMR Signals: TROSY

Since the molecular tumbling motion of larger proteins, such as membrane proteins in the lipid bilayer, is highly restricted, their NMR spectra are mainly

New Developments in NMR No. 3
Advances in Biological Solid-State NMR: Proteins and Membrane Active Peptides
Edited by Frances Separovic and Akira Naito

Published by the Royal Society of Chemistry, www.rsc.org

affected by anisotropic magnetic interactions that depend on the orientation relative to a static magnetic field, B_0, such as chemical shift anisotropy (CSA) and dipole–dipole (DD) interactions. In order to manage these magnetic interactions, a number of solid-state NMR techniques, including magic angle spinning and cross-polarization, have been developed, enabling the NMR observation of proteins with a theoretically unlimited molecular size. Therefore, solid-state NMR is one of the best techniques to analyze membrane proteins in a lipid bilayer.[1–4] Increasing numbers of solid-state NMR studies have been reported for various membrane proteins, including G protein coupled receptors (GPCRs),[5,6] ion channels[7–13] and transporters.[14–17]

The structure and dynamics of integral membrane proteins, such as GPCRs,[18–24] voltage-dependent anion channel (VDAC),[25,26] K^+ channel (KcsA),[27–33] viral ion channels[34,35] and diacylglycerol kinase,[36] have also been successfully analyzed by solution NMR techniques. Since the high sensitivity of solution-state NMR relies on the rapid molecular tumbling motion that averages the anisotropic magnetic interactions, the membrane proteins must be reconstituted in a solubilizing medium, such as detergent micelles,[37] reversed micelles,[38,39] amphipols,[40] bicelles[41,42] or nanodiscs.[19,31,43] In addition to these sample preparation methods, transverse relaxation-optimized spectroscopy (TROSY)[44] is an essential technique for obtaining solution-state NMR signals with high sensitivity and resolution.

At high magnetic fields the dominant relaxation sources, the CSA and DD interactions, increase the overall transverse relaxation rates. TROSY suppresses the transverse relaxation by the mutual compensation of the DD and CSA interactions (Figure 26.1), which is most effective at a ^{1}H frequency

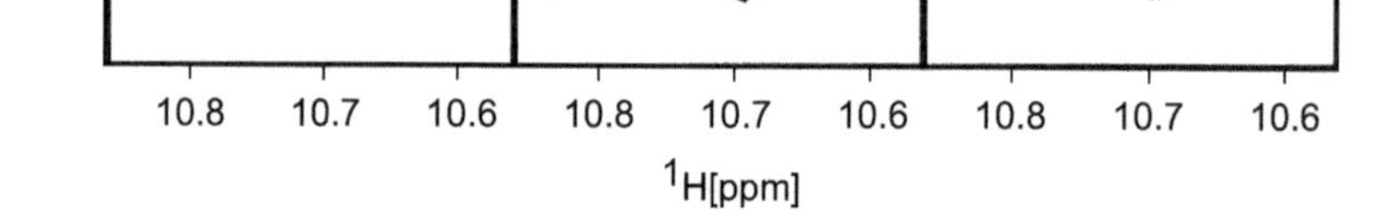

Figure 26.1 TROSY effect for a larger protein. Contour plots of ^{1}H–^{15}N correlation spectra showing the indole ^{1}H–^{15}N spin system of Trp48 recorded in a 2 mM solution of uniformly ^{15}N-labeled ftz homeodomain complexed with an unlabeled 14-bp DNA duplex in 95% H_2O/5% ^2H_2O at 4 °C, pH 6.0, measured at a ^{1}H frequency of 750 MHz.[44] (A) Conventional broad-band decoupled ^{1}H–^{15}N HSQC spectrum. (B) Conventional ^{1}H–^{15}N HSQC spectrum recorded without decoupling during t_1 and t_2. (C) TROSY-type ^{1}H–^{15}N correlation spectrum. Chemical shifts in ppm and shifts in Hz relative to the center of the multiplet are indicated in both dimensions.
(Adapted from Pervushin *et al.*[44])

of around 1 GHz for ^1H–^{15}N correlations. In combination with partial or complete deuteration of the non-exchangeable hydrogen atoms,[45] TROSY dramatically increases the spectral resolution of larger proteins. The TROSY detection of the ^1H–^{15}N correlations of amide groups has been incorporated into a number of multidimensional triple resonance experiments,[46,47] thus expanding the molecular size of proteins that can be analyzed by solution NMR techniques up to 100 kDa.[48]

For the ^1H–^{13}C correlation peaks of methyl groups, TROSY further enhances the signal sensitivity. Compared to the ^1H–^{15}N correlation of an amide group, methyl groups possess three times more ^1H atoms attached to the ^{13}C atom. The three ^1H atoms rotate more freely about a vector formed by the ^{13}C atom and the attached ^{12}C atom. Methyl TROSY enables the observation of the NMR signals of proteins with molecular masses up to 700 kDa.[49–53]

26.1.2 NMR Analyses of Protein–Protein Interactions

In addition to the transmembrane region, the hydrophilic portions in the intracellular and/or extracellular part(s) of proteins often play essential roles in their functions by binding to intra- or extracellular water-soluble ligands. For example, many ion channels possess large intracellular domains, where the binding of their intracellular ligands triggers the opening/closing (gating) of the channels. In cases where the functionally important intra- or extracellular portion of a membrane protein can be isolated without losing the function or introducing an artificial structural alteration, the interaction between the isolated portion and the ligand can be analyzed by solution-state NMR with high spectral resolution, thus contributing to elucidation of the functional mechanism of membrane proteins.

Once the NMR signals of a protein have been assigned, the interaction with its ligand can be easily detected by chemical shift perturbation (CSP) in a site-specific manner. The isotropic chemical shift observed by solution NMR reflects the electronic state of each nucleus and thus, CSP reflects the changes in the electronic state caused by the ligand binding. CSP can identify the site(s) that directly interact with the ligand and/or the conformational changes upon binding. Therefore, the residues exhibiting significant CSPs should be located at the direct binding site of the ligand and/or subjected to conformational changes upon ligand binding.

In order to distinguish the direct binding site from a site with conformational changes, distance-dependent NMR techniques, such as the intermolecular nuclear Overhauser effect (NOE), are necessary. Although direct intermolecular binding occurs mainly through side chain–side chain interactions, it is often difficult to assign the intermolecular NOE signals between the side chain protons. In the case of a low affinity interaction, a sufficient number of NOEs cannot be detected, which precludes the structure determination of a protein–protein complex.

We previously developed two NMR techniques, referred to as cross-saturation (CS) and transferred cross-saturation (TCS) (Figure 26.2).[54,55]

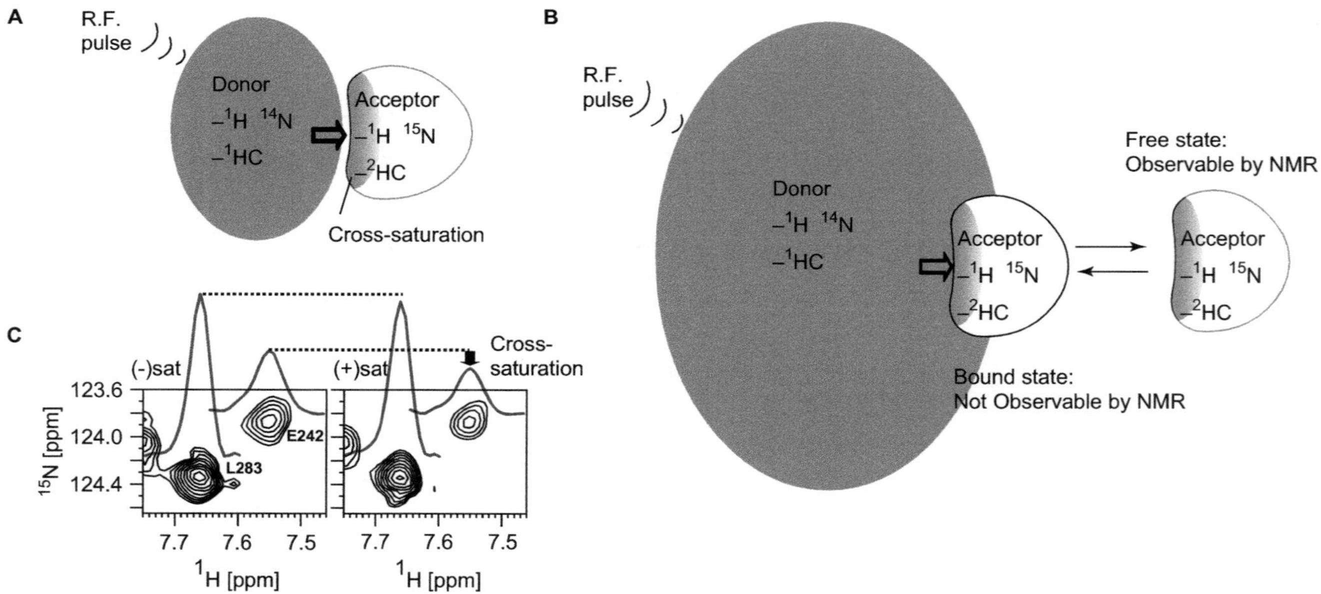

Figure 26.2 Cross-saturation (CS) and transferred cross-saturation (TCS). (A) Schematic illustration of CS.[54,55] The sample is a stoichiometric complex of two proteins, one is unlabeled (referred to as a donor) and the other is uniformly $^{2}H,^{15}N$-labeled (referred to as acceptor). Irradiation of RF pulses selectively saturates the ^{1}H resonances of the donor, which also saturates amide resonances of the acceptor (cross-saturation phenomenon). The cross-saturation phenomenon is observed as the intensity reduction of the ^{1}H–^{15}N correlation signals of the TROSY spectrum of the complex, as indicated in panel (C). (B) Schematic illustration of TCS. The sample contains excessive amounts of the accepter relative to that of the donor, under conditions that the acceptor experiences fast exchange between free and donor-bound states on the NMR chemical shift timescale. Cross-saturation is observed for the residues at the binding interface as their signal intensity reduces. (C) Resonances observed for the mixture of uniformly $^{2}H,^{15}N$-labeled GIRK$_{CP}$ and Gβγ, without (*left*) and with (*right*) RF irradiation. 1D slices of E242 and L283 are shown for comparison.

CS identifies the binding residues of a uniformly $^2H,^{15}N$-labeled protein by detecting saturation transfer from the unlabeled protein, as the reduction of the $^1H-^{15}N$ TROSY signal intensity. TCS is a modified version of CS, which is carried out under fast exchange conditions between the free and bound states on the NMR timescale. TCS generates the NMR signals of a $^2H,^{15}N$-labeled protein in the free state, which avoids the increase in the transverse relaxation rates upon binding to a large protein, while saturation transfer occurring upon binding decreases the signal intensities of the residues at the binding site. Since CS and TCS take advantage of steady-state NOEs, instead of transient NOE as observed in conventional NOESY, CS and TCS can be detected with much higher sensitivity than NOESY. Generally, CS reaches 5–8 Å, depending upon many factors, including the kinetics of association and dissociation of the complex, molecular size, duration of the RF irradiation, sample viscosity, and so on.[55]

Paramagnetic relaxation enhancement (PRE) is another sensitive technique. PRE arises from magnetic dipolar interactions between the nuclear spin and the unpaired electron spin of the paramagnetic center. These interactions enhance the relaxation of the nuclear spins, leading to line-broadening and thus intensity reduction of the NMR signals of the residues within about 20 Å of the spin-label.[56,57] In addition, PRE can detect a transient interaction because the PRE effect occurs within 250–500 μs,[58] which is a much shorter duration than that of the cross-saturation (the effective saturation times were 200–300 ms for the TCS experiments in this study).[55]

Therefore, the introduction of a paramagnetic center near the binding site of a protein results in the PRE effect at residues within 20 Å of the paramagnetic center in the complex, which is useful to determine the binding orientation of the two proteins in the complex. If the motion of the introduced paramagnetic center is sufficiently limited, then pseudo-contact shift (PCS) and RDC information is provided, which can also be used as long-range structure information.

These solution-state NMR techniques are useful to analyze the interactions in large protein complexes and can provide valuable information about the functional mechanisms of membrane proteins. In the following sections, we review the solution-state NMR analyses of interactions between the intracellular domain of an ion channel and the G-proteins that regulate the electrophysiological properties of the channel, which revealed the regulation mechanism of the ion channel.

26.2 G Protein-Activated Inwardly Rectifying Potassium Ion Channel

26.2.1 GIRK as an Effector of G Proteins

The G protein-gated inwardly rectifying potassium ion channel (GIRK)[59] is a member of the inwardly rectifying potassium ion channel (Kir) family, regulating heart rate, neuronal excitability and other physiological events.[60,61] While the opening and closing (gating) of Kir family proteins are generally regulated by phosphatidylinositol 4,5-bisphosphate, the opening of GIRK is also triggered by the direct binding of the G protein βγ subunits (Gβγ), which are released from the pertussis toxin-sensitive G protein α subunit ($G\alpha_{i/o}$).

The stimulation of G protein-coupled receptors (GPCRs) causes the exchange of GDP with GTP on the α subunit (Gα), which decreases the affinity of Gα for Gβγ. Gβγ then dissociates from Gα in the GTP-bound state [Gα(GTP)], and associates with GIRK, resulting in the opening of the GIRK gate.[60–67]

GIRK functions as a tetramer, in which a long ~88 Å K^+ permeation pathway is formed by the transmembrane (TM) and cytoplasmic pore (CP) regions (Figure 26.3).[68,69] The crystal structure of GIRK revealed two K^+-ion gates:[69] the helix bundle crossing on the cytoplasmic side of the TM region, and the loops on the membrane side of the CP region, referred to as G-loops.

26.2.2 Gβγ Binding Site on GIRK, Proposed by Mutational Analyses

Extensive mutational analyses to identify the GIRK residues that are critical for the Gβγ-induced activation revealed several critical residues, including H57, L262, L333 and G336 of GIRK1.[59,70,71] However, when these residues were mapped on the recently reported crystal structures of Kirs, they did not form a cluster on the protein surface.[72–75] Therefore, no clear consensus has been obtained regarding the region of GIRK that is essential for the Gβγ binding and/or GIRK activation. One reason might be a structural alteration introduced by the mutagenesis,[76] which could have changed the gating properties of the channel.

26.2.3 Regulation of the GIRK Properties by $G\alpha_{i/o}$

While Gβγ directly binds to and activates GIRK, Gα also plays important roles in regulating GIRK. In native tissues, GIRK is activated only through the stimulation of GPCRs that couple to the G proteins belonging to the i/o family ($G_{i/o}$).[77] The activation and the deactivation of GIRK are also accelerated when the channels are co-expressed with $G\alpha_{i/o}$.[78–80]

Efficient GIRK activation seems to be attributed to the pre-association of the heterotrimeric G protein in the i/o family [$G\alpha_{i/o}$(GDP)βγ] with GIRK, in which $G\alpha_{i/o}$ directly associates with both GPCR and GIRK, prior to GPCR activation.[81,82] On the other hand, the deactivation is accelerated by the direct interaction of $G\alpha_{i/o}$(GTP) with GIRK.[82] The interaction enables $G\alpha_{i/o}$ to rapidly bind to Gβγ upon GTP hydrolysis, which is presumably assisted by the regulator of G protein signaling (RGS),[83] resulting in the closure of the GIRK gate. Since the re-association of $G\alpha_{i/o}$(GDP) and Gβγ allows

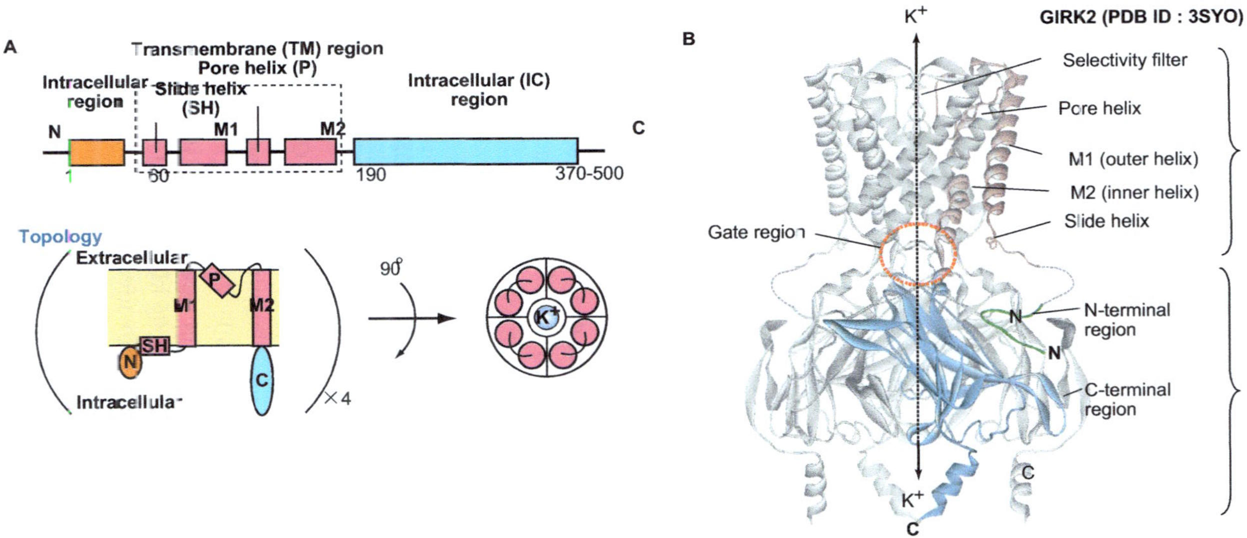

Figure 26.3 Structure of GIRK. (A) Primary structure. (B) Tertiary structure of GIRK tetramer. GIRK functions as a tetramer. One of the four subunits is colored. GIRK has a transmembrane domain (magenta) and a large intracellular domain referred to as a cytoplasmic pore (CP), which consists of N-terminal (green) and C-terminal (blue) regions of each subunit.

$G\alpha_{i/o}$(GDP)$G\beta\gamma$ to enter the next cycle of GIRK activation, the rapid re-association would also accelerate the activation of GIRK.[79] Thus, the direct interaction between GIRK and $G\alpha_{i/o}$(GTP) could contribute to the efficient activation and deactivation of the GIRK channel.

It is now evident that $G\alpha_{i/o}$ not only functions as the donor and acceptor of $G\beta\gamma$ but also modulates the gating property of GIRK. While several studies have proved the interactions between $G\alpha_{i/o}$ and the cytoplasmic region of GIRK,[78,79,84–91] how $G\alpha_{i/o}$ modulates the GIRK gating property remains unclear and, therefore, the elucidation of the molecular recognition mode of $G\alpha_{i/o}$ and GIRK is required.

26.2.4 Strategies to Investigate the GIRK Regulation Mechanism by G Proteins

In order to reveal the structural mechanism by which the G proteins regulate the gating properties of GIRK, we produced the CP regions of GIRK,[92] $G\alpha_{i3}$[93] and $G\beta\gamma$[94] and evaluated the direct binding affinities of the CP region of GIRK with G proteins by isothermal titration calorimetry (ITC) and NMR spectroscopy. Then, the $G\alpha_{i3}$ and $G\beta\gamma$ binding sites on GIRK,[94,95] as well as the GIRK binding site on $G\alpha$, were identified by TCS.[95] Furthermore, the CSP analysis of the CP region of GIRK upon the $G\beta\gamma$ binding revealed the site with the conformational rearrangements of the GIRK tetramer, which accounts for the gate opening of GIRK.[94] The binding mode of the CP region of GIRK and $G\alpha$ was further examined by PRE experiments. Based on the TCS and PRE results, a docking model of $G\alpha$ and GIRK was successfully built, suggesting the molecular basis for the efficient deactivation of the channel by $G\alpha_{i/o}$.[95]

26.3 $G\beta\gamma$ Binding and Conformational Rearrangements of GIRK1

26.3.1 Binding Affinity of the CP Region of GIRK for $G\beta\gamma$

In this study we used the cytoplasmic pore regions of mouse GIRK1 composed of residues 41–63 and 190–371 as a single polypeptide (hereafter referred to as $GIRK_{CP}$).[76,92,94] The purified $GIRK_{CP}$ protein forms a tetramer,[76] whose molecular mass is 96 kDa. The ^{1}H–^{15}N TROSY spectrum of $GIRK_{CP}$ was well dispersed and provided a number of signals somewhat consistent with the number of residues within a single subunit of $GIRK_{CP}$, indicating that the four subunits within the tetramer are in an equivalent environment in solution (Figure 26.4).[92] The purified $G\beta_1\gamma_2$ protein (46 kDa, hereafter referred to as $G\beta\gamma$) contains approximately equimolar amounts of $G\beta$ and $G\gamma$, and its binding activity to $G\alpha_{i3}$ (44 kDa) was confirmed by size exclusion chromatography.

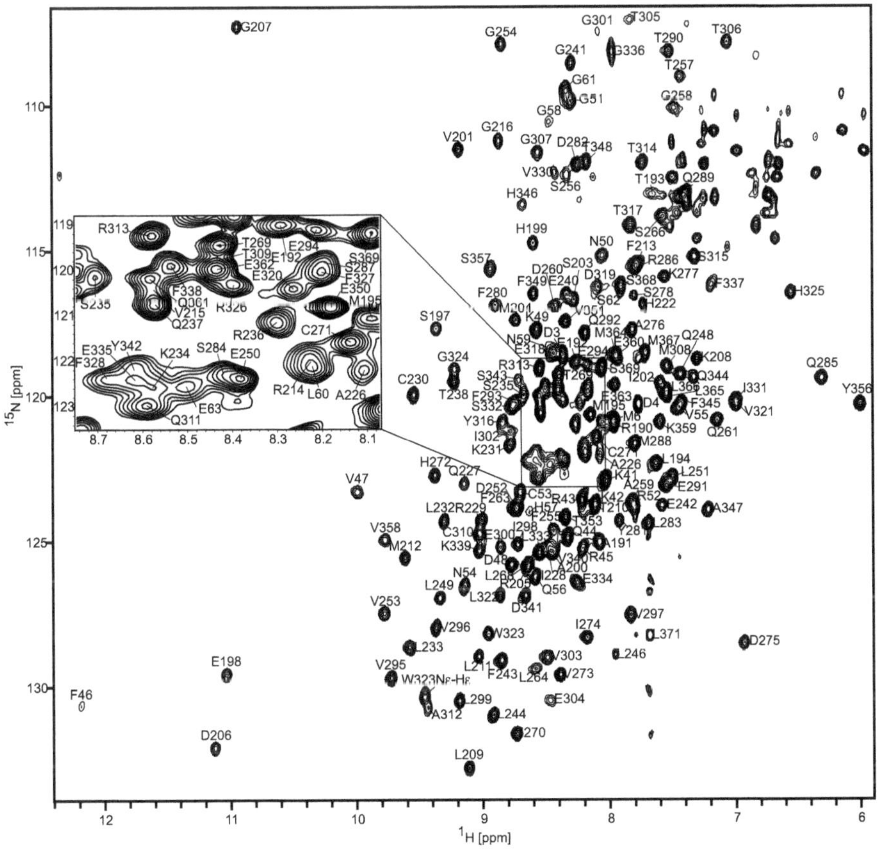

Figure 26.4 ^{1}H–^{15}N TROSY spectrum of uniformly ^{2}H,^{15}N-labeled GIRK$_{CP}$. Nearly complete backbone assignments for 186 of 200 residues (93%) have been obtained using a series of TROSY-based triple resonance experiments.[92]

ITC experiments were carried out to determine the binding affinity and stoichiometry between GIRK$_{CP}$ and Gβγ (Figure 26.5). The fitting of the isotherm resulted in an apparent dissociation constant (K_d) value of 250 μM, with a binding stoichiometry of 0.25 equivalents of GIRK$_{CP}$ tetramer per one Gβγ molecule. Since the 0.25 equivalents of GIRK$_{CP}$ tetramer correspond to the concentration of GIRK$_{CP}$ monomer, the ITC result suggests that four Gβγ molecules bind to one GIRK$_{CP}$ tetramer, where the K_d value for the binding of GIRK$_{CP}$ monomer to Gβγ is 250 μM.

The binding stoichiometry is consistent with the results of previous cross-linking[96] and electrophysiological experiments.[97] The dissociation constant between GIRK$_{CP}$ and Gβγ (K_d value of 250 μM) determined in this study is considerably larger than the reported constant for the Gβγ-induced activation of GIRK (K_{act} value of 11 nM) determined by electrophysiological experiments.[67] Theoretically, the affinity between membrane proteins could

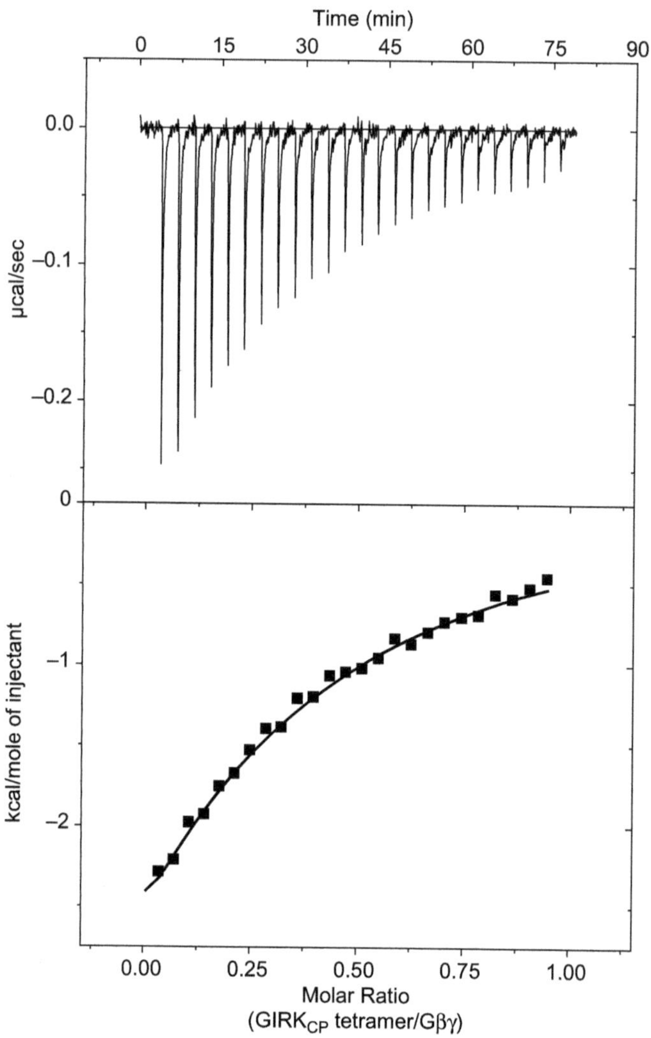

Figure 26.5 ITC results for the interaction between GIRK$_{CP}$ and Gβγ. The *upper panel* shows the isotherm upon 25 injections of a 1.5 μL aliquot of GIRK$_{CP}$ at 1.8 mM as a tetramer into the cell containing 380 μM Gβγ.

be enhanced 10^6-fold compared to that in solution, because of the effects of excluded volume, the substantially higher local concentrations and the pre-orientation of proteins in membranes.[98] Indeed, the affinities of protein–protein interactions within membranes are reported to be enhanced by 10^1- to 10^3-fold.[99,100] Since Gβγ is anchored to the membrane by the lipid moiety covalently attached to the C-terminus of Gγ, the orientational constraint of Gβγ should be smaller than those of integral membrane proteins. Therefore, the enhancement of the binding affinity in the membrane is expected to be significantly smaller than 10^6-fold, suggesting that the K_d value of the

GIRK–Gβγ interaction falls in the nM to μM range. Considering that the GIRK activation is terminated by the removal of Gβγ from GIRK through the competitive binding of the GDP-bound Gα to Gβγ (IC_{50} = 10 nM[67]), the affinity between GIRK and Gβγ should be lower than that between Gα and Gβγ (K_d = 6–9 nM[101]) to accomplish the rapid closure of GIRK by the GDP-bound Gα. Therefore, the K_d value of 250 μM, determined in this study, would be reasonable for the interaction between GIRK and Gβγ in solution.

26.3.2 Gβγ Binding Site on $GIRK_{CP}$, Revealed by TCS Experiments

In order to identify the Gβγ binding site on $GIRK_{CP}$, TCS experiments were performed. First, the backbone resonance assignments were established by a series of TROSY-type triple resonance experiments,[46,47] in combination with amino acid-selective labeling (Figure 26.4).[92] Then, the sample was prepared for TCS containing uniformly $^{2}H,^{15}N$-labeled $GIRK_{CP}$, which was 300 μM as a tetramer, and 300 μM Gβγ in the 20% H_2O/80% D_2O-based buffer of 10 mM HEPES-NaOH (pH 8.0), 50 mM KCl and 10 mM DTT. The H_2O content of 20% is intended to suppress spin diffusion between proximal amide ^{1}H atoms in $GIRK_{CP}$.[54,55]

RF irradiation of the sample resulted in selective intensity reductions of the ^{1}H–^{15}N TROSY signals of the backbone amide groups of $GIRK_{CP}$ in the Gβγ binding site (Figure 26.2C). Our relaxation matrix calculations,[102] in which on and off rates of the $GIRK_{CP}$–Gβγ interaction are also taken into account, suggest that the amide protons of $GIRK_{CP}$ within 6 Å from Gβγ exhibited significant intensity reductions. We evaluated the cross-saturation from Gβγ by subtracting the intensity reduction ratios of the control experiment in the absence of Gβγ. The differences in the reduction ratios, ΔRR, are shown in Figure 26.6A with error bars calculated based on the signal-to-noise ratios. The minimum values of ΔRR within the error ranges (ΔRR_{min}) were utilized for the evaluation.

The residues with significant intensity reductions ($\Delta RR_{min} > 0.083$) are R236, Q237, T238, E240, G241, E242, F243, L244, M308, E320, L333, E334, F337, K339, D341 and E350. These residues mostly belong to one of two regions: the region from the βD to βE strands (R236–T238 and E240–L244) and the region from the βL to βM strands (L333, E334, F337, K339 and D341). The other three residues, M308, E320 and E350, are located in the loop between the βH and βI strands, the loop between the βI and βJ strands, and in the βN strand, respectively. Although these residues are dispersed on a single subunit of $GIRK_{CP}$ (Figure 26.6B), the residues except for M308 formed two adjacent contiguous surfaces in the two neighboring subunits of the $GIRK_{CP}$ tetramer (Figure 26.6C). Note that the intensity reduction of M308 would be induced by the direct saturation of the hydrogen atoms in the hydroxyl groups of the three proximal threonine residues (T305, T306 and T309). Thus, we concluded that the Gβγ binding site on $GIRK_{CP}$ is composed of the following residues clustered between the two neighboring

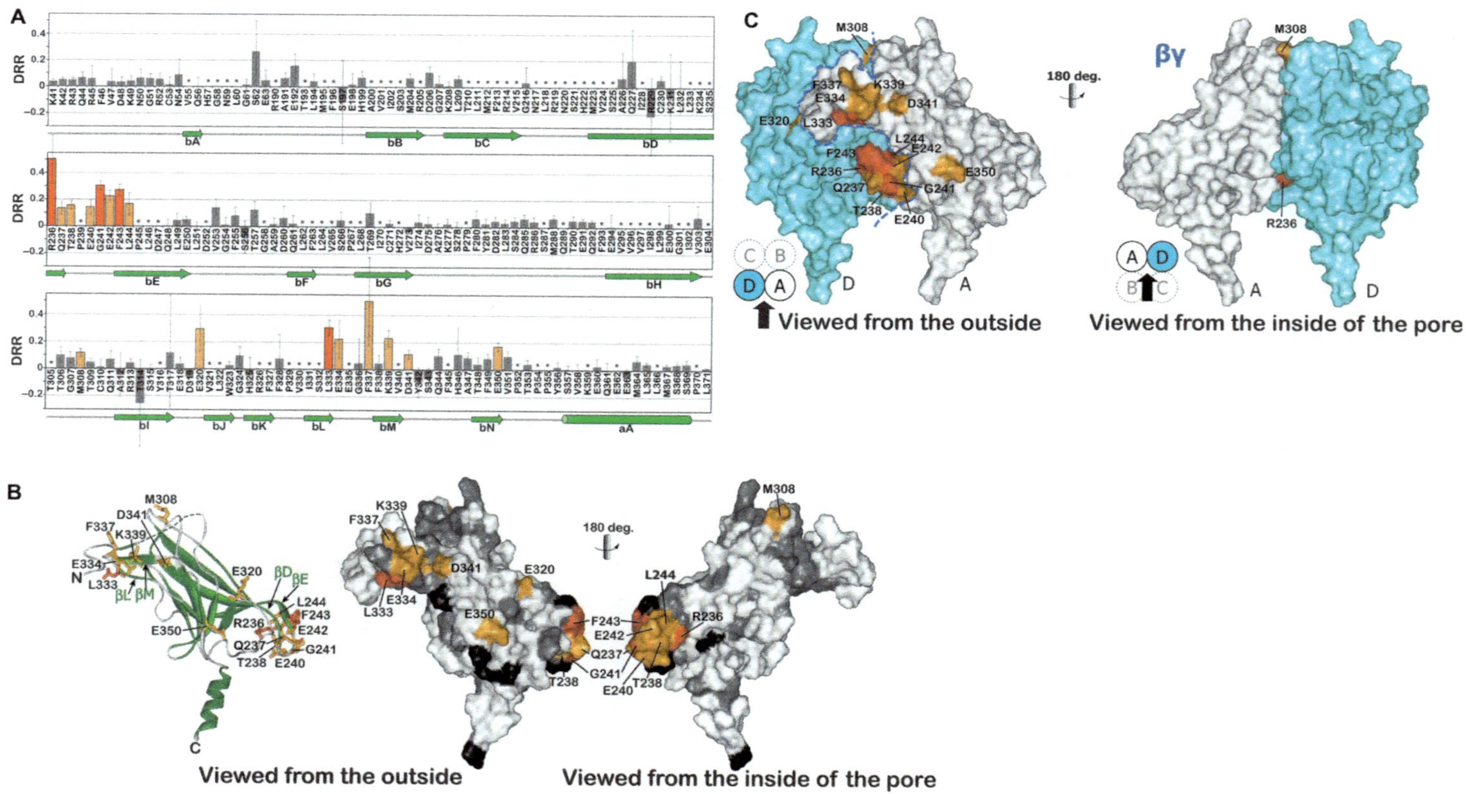

Figure 26.6 The Gβγ-binding interface on $GIRK_{CP}$ identified by TCS. (A) Signal intensity ratios with and without irradiation, normalized by the intensity ratios of the control experiment in the absence of Gβγ (ΔRR). The bars, corresponding to the residues with a ΔRR value above 0.1, are colored and mapped on the structure in panels B and C. Mapping of the affected residues on (B) a single subunit and (C) two adjacent subunits.

surfaces of the adjacent subunits: R236, Q237, T238, E240, G241, E242, F243, L244, E320, L333, E334, F337, K339, D341 and E350.

26.3.3 Chemical Shift Perturbation of $GIRK_{CP}$, upon Binding to Gβγ

The Gβγ binding effects on $GIRK_{CP}$ were also investigated by CSP experiments, which reflect the effects of the direct binding of Gβγ as well as the conformational change of $GIRK_{CP}$ induced by Gβγ binding. Upon addition of 50, 100, 200 and 400 μM of Gβγ to 50 μM of uniformly $^{2}H,^{15}N$-labeled $GIRK_{CP}$ tetramer, no signals exhibited chemical shift changes larger than 0.05 ppm but significant intensity reductions were observed for a number of signals (Figure 26.7A).

The observed intensity reductions of the NMR signals are caused by slower tumbling due to the increased average molecular mass of the complex of $GIRK_{CP}$ and Gβγ. Furthermore, the intensity reductions are induced by the chemical shift differences of $GIRK_{CP}$ between the free and bound states.[103] The differences reflect the changes in the microenvironments on the Gβγ binding site on $GIRK_{CP}$, including the disruption of the four-fold symmetry of the $GIRK_{CP}$ tetramer, as well as the concomitant conformational changes of the remote region from the binding site. The intensity reduction ratios of the TROSY signals are plotted in Figure 26.7B for the sample containing 50 μM of the $^{2}H,^{15}N$-labeled $GIRK_{CP}$ tetramer mixed with 100 μM Gβγ, in which the $GIRK_{CP}$ tetramers in the free form and 1:1 complex with Gβγ are predominantly present, assuming the K_d value of 250 μM obtained by the ITC analysis.

The residues with significantly large intensity reduction ratios (larger than 0.18) are R45, F46, V47, N50, G51, N54, A226, S235, R236, Q237, T238, E240, G241, E242, F243, L244, V253, G254, F255, S256, A259, D260, Q261, S278, T290, G307, M308, T317, E318, D319, E320, L333, E334, G336, F337, F338, Y342, S343, Q344, A347, T348 and V358 (Figure 26.7B). These residues seem to be distributed widely on the molecular surface of the single subunit of $GIRK_{CP}$ (Figure 26.7C). However, mapping on the neighboring two subunits revealed that the affected residues are clustered at the subunit–subunit interface and surrounding regions on both the inside and the outside of the pore (Figure 26.7D). The residues located on the outside of the pore reside in the Gβγ binding site (Figure 26.6C) and in proximity. The other residues are located on the inside of the pore (A226, V253, G254, F255, S256, A259, D260 and Q261), at the membrane side (G307 and M308), and at other distal sites (S278 and V358) (Figure 26.7D, right). In particular, the CSPs for the residues outside of the Gβγ binding interface (see Figure 26.6C for comparison) reflect their conformational changes as allosteric effects of the Gβγ binding.

26.3.4 Possible Binding Mode between GIRK and Gβγ

The TCS method identified the site of direct Gβγ binding, which is mainly located on the region from the βD to βE strands of one subunit and the region from the βL to βM strands of the adjacent subunit of $GIRK_{CP}$

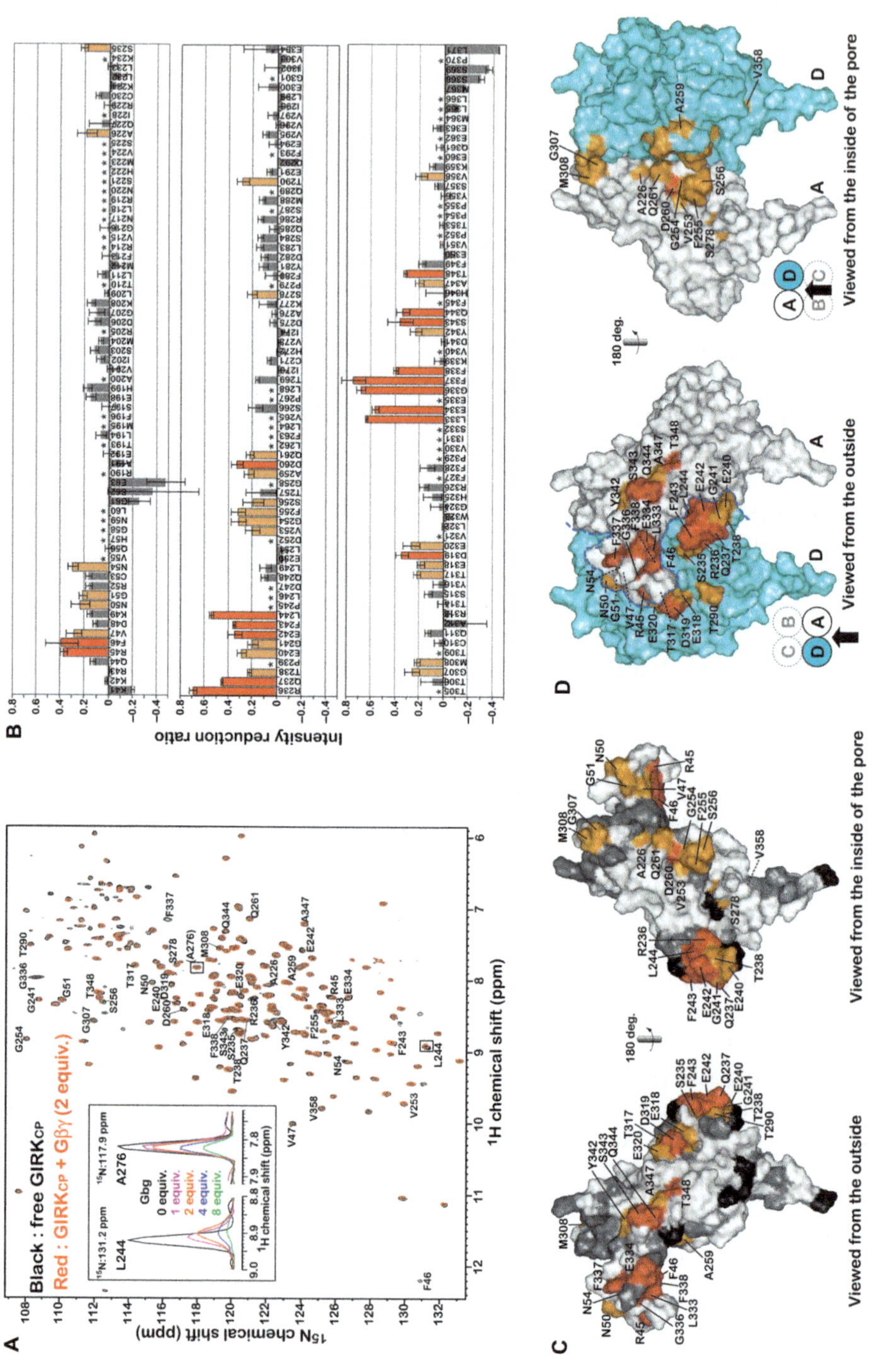
A
B
C
D
Black : free GIRKCP
Red : GIRKCP + Gβγ (2 equiv.)
15N chemical shift (ppm)
1H chemical shift (ppm)
Intensity reduction ratio
180 deg.
Viewed from the outside
Viewed from the inside of the pore
Viewed from the outside
Viewed from the inside of the pore

(Figure 26.6). The latter region contains L333 and G336, which were previously identified by mutational analyses as important residues for the Gβγ-induced activation of GIRK.[59,71,104] The other residues postulated by the mutational studies, H57 and L262, were excluded from the present analyses, unfortunately due to insufficient signal intensity and lack of backbone assignment of the NMR signal, respectively.

On the other hand, the mutational analyses have suggested several residues of Gβγ that are important for GIRK activation (residues L55, K78, I80, K89, W99, D228, D246 and W332 of Gβ).[105,106] Notably, most of these residues are located in the Gα binding site in the crystal structure of the heterotrimeric Gαβγ complex (Figure 26.8A). The four Gβ residues (L55, K78, I80 and K89) bind to the N-terminal helix of Gα, which undergoes a large conformational change upon Gβγ binding,[107] suggesting that this site on Gα is specifically formed in the complex with Gβγ. Meanwhile, the latter four residues (W99, D228, D246 and W332) of Gβ reside on the "top" face of the Gβγ torus, also known as the "hot spot" (Figure 26.8A, left) which binds to the GTPase domain of Gα (Figure 26.8A, right). The center of the binding site of the Gβγ torus on Gα contains the hydrophobic residues, I180, F195 and F211, which are located in the proximity of the W99 side chain of Gβ. These three hydrophobic residues of Gα are flanked by two acidic residues, E182 and E212, where E182 hydrogen bonds with the W99 side chain of Gβ and E212 makes an electrostatic interaction with K57 of Gβ. Indeed, the importance of W99 was indicated by the mutational analyses, and affected the activation of Gβγ effectors besides GIRK, such as phospholipase C-β2 and adenylyl cyclase 2.[108] Although Gβγ binding affects the conformation of the GTPase domain of Gα, the structural motif comprising "the hydrophobic region" and "the two flanking acidic residues" is preserved in GDP-bound Gα, in both the presence and absence of Gβγ.

A closer inspection of the Gβγ binding sites on $GIRK_{CP}$ revealed a similar structural motif, "the hydrophobic region" consisting of residues on the βE strand (F243, L244 and P245) of one subunit and the residues at the beginning of the βL strand (P329 and I331) of the neighboring subunit, and "the two flanking acidic residues" (E242 and E334) (Figure 26.8B). Thus, we conclude that the Gβγ binding sites on GIRK possess surface properties similar to those of Gα. It is tempting to speculate that, upon binding, the W99 side chain of Gβ is accommodated in "the hydrophobic region" supported by the interactions with "the two flanking acidic residues" of GIRK. The binding orientation would be possible for intact GIRK and Gβγ that

Figure 26.7 GIRK sites for direct binding of Gβγ and sites with conformational change upon Gβγ binding revealed by CSP. (A) Spectral overlay of ^1H–^{15}N TROSY in the absence (*black*) and presence (*red*) of Gβγ. The *inset* shows typical intensity reductions. (B) Signal intensity ratios in the presence and absence of Gβγ. The bars, corresponding to the residues with signal intensity reduction ratios above 0.15, are colored and mapped on the structure of (C) a single subunit and (D) two adjacent subunits of $GIRK_{CP}$.

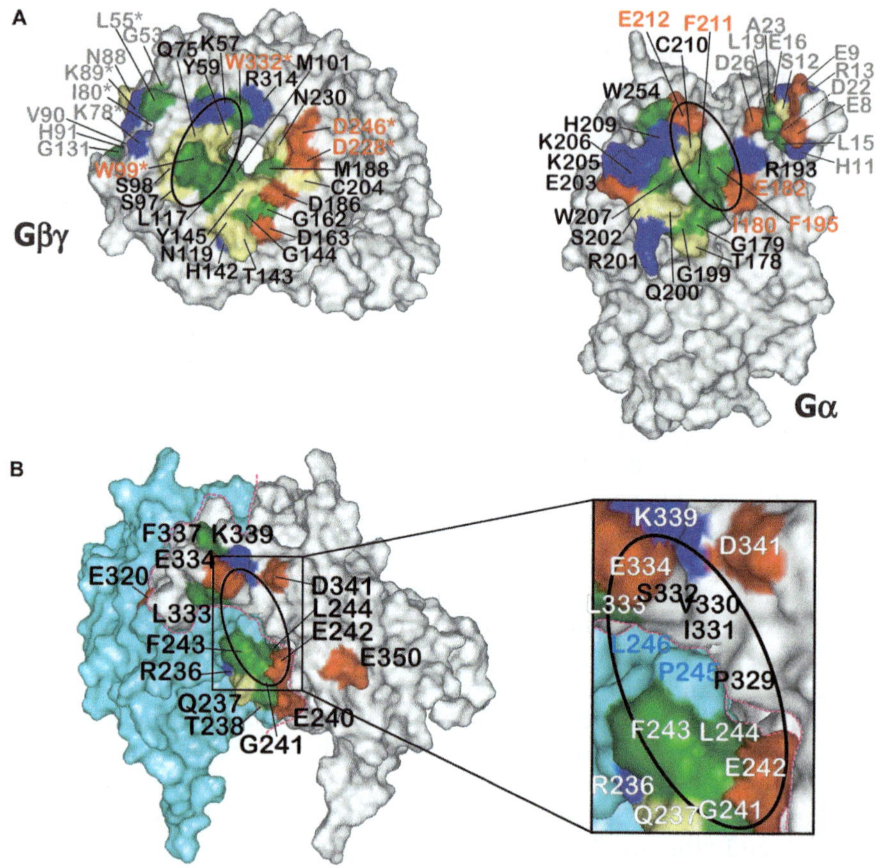

Figure 26.8 Possible binding mode between GIRK and Gβγ. (A) Surface properties of the interface between Gβγ and Gα in the Gαβγ complex. (B) Surface properties of the Gβγ binding site formed by the adjacent two subunits of $GIRK_{CP}$. Acidic, basic, hydrophobic and the rest are colored red, blue, green and yellow, respectively.

anchors to the membrane with a lipid moiety attached to the C-terminus of Gγ. Since "the hydrophobic region" spans two neighboring subunits, Gβγ binding to this site could cause the rearrangements of the CP regions of GIRK, as evidenced by the CSPs at distant sites far away from the Gβγ binding site (see below).

26.3.5 Propagation of the Gβγ-Induced Rearrangements of $GIRK_{CP}$ to the K^+- Ion Gate

The Gβγ binding induced significant intensity reductions originating from CSPs not only for the residues at the direct binding site but also for those in other regions (Figures 26.7C and 26.7D). In particular, the significant CSPs observed for the residues located at the subunit–subunit interface away from the Gβγ binding site (Figure 26.7D, right) strongly suggest that the rearrangements of the four subunits were induced by Gβγ binding. Indeed, two distinct conformations of the CP region by the rigid-body subunit rearrangements were observed in the crystal structure of a chimeric channel of KirBac1.3 and GIRK1 (PDB ID: 2QKS).[72] The recently reported structures of KirBac3.1[109] also revealed conformational changes in the orientation of the CP region relative to the TM region (PDB ID: 2WLI and 2WLJ, as the representative structures).

In the crystal structure of the chimeric channel, several residues around the K^+ iongate (F72 and T73 in the slide helix, I182 and S185 in the inner TM helix, and Q186 and P187 in the C-linker) directly interact with the residues in the loop between the βH and βI strands (H-I loop) of the CP region of GIRK1 (V303, T305, T306, M308 and C310), as summarized in Figure 26.9.

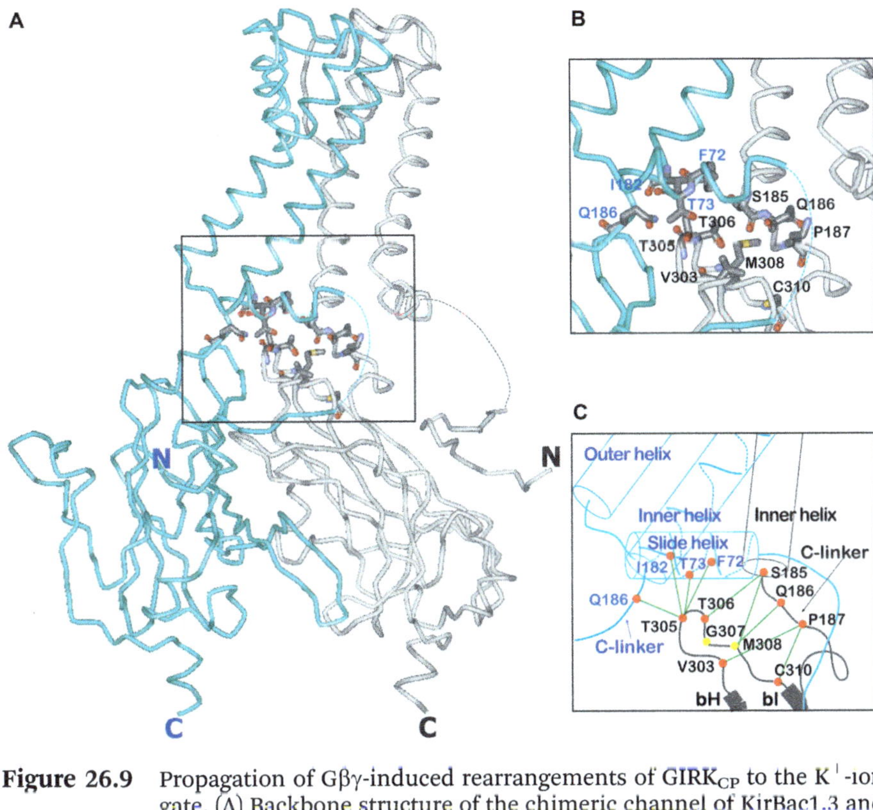

Figure 26.9 Propagation of Gβγ-induced rearrangements of $GIRK_{CP}$ to the K^+-ion gate. (A) Backbone structure of the chimeric channel of KirBac1.3 and GIRK1 (PDB ID: 2QKS). The interface between the CP and TM regions is boxed (B), and residues in the loop between the βH and βI strands of one subunit proximal to those in the TM region of the same and adjacent subunits within 4 Å are shown in *stick models*. (C) Schematic drawing of the interface between the CP and TM regions of the chimeric channel, shown in (B). Green lines connect the residue pairs within 4 Å of each other. The positions of the Cα atoms are depicted as red dots, except for the residues with significant CSPs, which are indicated as yellow dots.

In addition, T73 (slide helix) and K188 (C-linker) interact with H222 and R219, respectively, which are in the loop between the βC and βD strands (C–D loop) of the CP region. The interactions of the residues in the H-I loop with the TM region residues, including the slide helix, inner helix and C-linker, are also observed in the crystal structure of KirBac3.1, while more interactions were observed between the C–D loop and the slide helix in one form of the KirBac3.1 structures.

As shown in Figure 26.7, significant CSPs upon Gβγ binding were observed for G307 and M308 on the H–I loop, suggesting that the relative positions between the loops of the four subunits change upon Gβγ binding, which could result in the opening of the K^+-ion gate. The residue at the top of the H–I loop, T305, provided a very weak NMR signal and was excluded from the analysis and no significant CSP was observed for the adjacent residue, T306, due to the lack of the interacting TM residues in $GIRK_{CP}$ used in this study. The residues in the C–D loop were also excluded from the analysis, since their backbone NMR signals were not assigned. Further studies are awaited to reveal the mode of the Gβγ-induced structural rearrangements of the CP region of GIRK and their effects on the conformation of the K^+-ion gate.

26.4 Structural Basis for the Modulation of the Gating Property of GIRK by $G\alpha_{i/o}$

26.4.1 Binding Affinity between $G\alpha_{i3}$ and $GIRK_{CP}$ Evaluated by NMR Spectral Change

Based on previous reports indicating that the C-terminal part of GIRK1 is involved in the interaction with Gα,[78,79,84–88] we prepared a protein consisting of mouse GIRK1 residues 41–63 and 190–386 (hereafter referred to as $GIRK_{CP\text{-}L}$), whose C-terminal region is 15 residues longer than that of $GIRK_{CP}$ (residues 41–63 and 190–371). We prepared $G\alpha_{i3}$, which is a member of $G\alpha_{i/o}$ family, expressed in *E. coli*, and purified to homogeneity. We confirmed the Gβγ-binding affinity and the GDP/GTP exchange rate were comparable to reported values. The backbone NMR signal assignments of $G\alpha_{i3}$ were established by a series of triple resonance experiments.[93]

In order to investigate the direct binding between $G\alpha_{i3}$ and $GIRK_{CP\text{-}L}$, we observed a series of ^1H–^{15}N TROSY spectra of 0.28 mM uniformly ^2H,^{15}N-labeled $G\alpha_{i3}$ in the absence or presence of 0.25, 0.50, 0.75 and 1.0 mM $GIRK_{CP\text{-}L}$ as a tetramer (Figure 26.10). As the concentration of $GIRK_{CP\text{-}L}$ increased, most signals exhibited a decrease in intensity due to line-broadening without a change in their chemical shifts, while a number of signals exhibited further intensity reductions and eventually disappeared in the presence of 0.75 mM $GIRK_{CP\text{-}L}$ (*e.g.* G45). In addition, several signals exhibited small but significant chemical shift changes (*e.g.* W258). Although slowing of the overall tumbling motion upon binding to $GIRK_{CP\text{-}L}$ causes the intensity reductions, the further intensity reductions and apparent

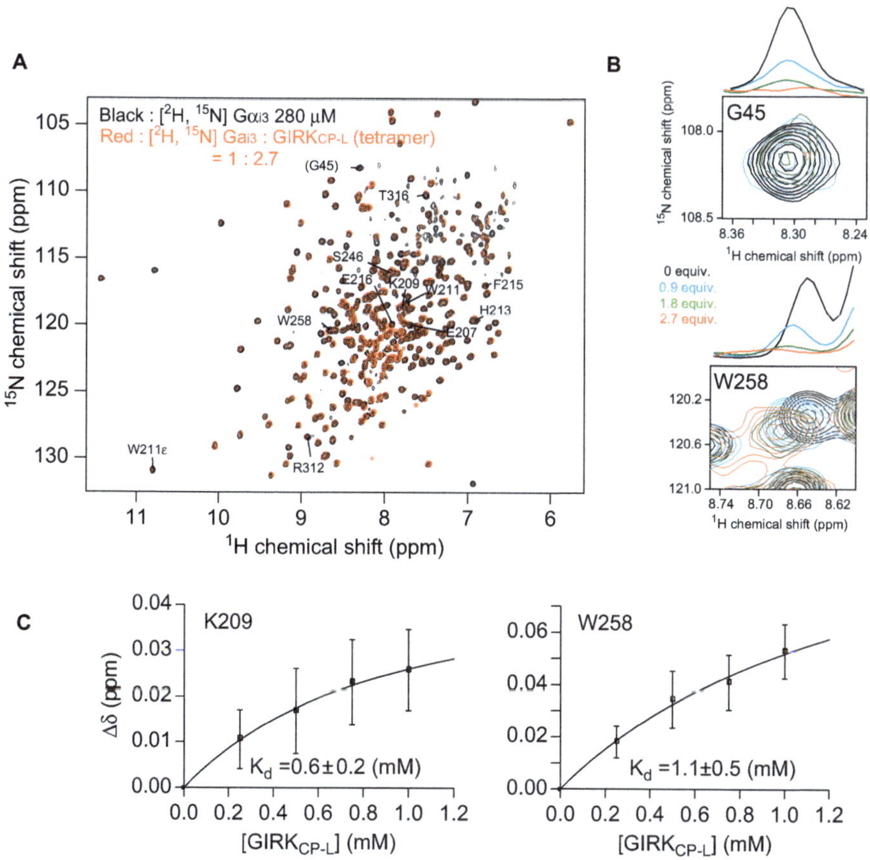

Figure 26.10 Direct binding of $G\alpha_{i3}$ and $GIRK_{CP\text{-}L}$, indicated by NMR spectral change of $G\alpha_{i3}$ upon addition of $GIRK_{CP\text{-}L}$. (A) Spectral overlay of ^1H–^{15}N TROSY of $G\alpha_{i3}$ in the absence (black) and presence (red) of $GIRK_{CP\text{-}L}$. The residues with significant chemical shift changes are labeled. (B) Changes in signal intensities. (C) Titration plots of the chemical shift of K209 and W258 signals at each concentration of added $GIRK_{CP\text{-}L}$. Fitting of the plots resulted in a K_d value of approximately 1.0 mM.

chemical shift changes reflect the direct interactions and/or the conformational change upon binding of $GIRK_{CP\text{-}L}$ to $G\alpha_{i3}$.

The weighted averages of the chemical shift differences ($\Delta\delta$) were calculated using the equation:

$$\Delta\delta = \{(\Delta\delta_{HN})^2 + (\Delta\delta_N / 6.5)^2\}^{1/2} \tag{26.1}$$

The titration curves of $\Delta\delta$ were fit to the following theoretical formula to obtain the value of the dissociation constant (K_d):

$$\Delta\delta = \Delta\delta_{sat}[[GIRK_{CP\text{-}L}]_{tot} + [G\alpha_{i3}]_{tot} + K_d - \{([GIRK_{CP\text{-}L}]_{tot} + [G\alpha_{i3}]_{tot} + K_d)^2 - 4[GIRK_{CP\text{-}L}]_{tot}[G\alpha_{i3}]_{tot}\}^{0.5}] / 2[G\alpha_{i3}]_{tot} \tag{26.2}$$

where $\Delta\delta_{sat}$ is the $\Delta\delta$ value when a saturating amount of $GIRK_{CP\text{-}L}$ is added (ppm), and the $[G\alpha_{i3}]_{tot}$ value is 0.28 (mM). The fitting of $\Delta\delta$ for the signals from K209 and W258 resulted in K_d values of 0.6 and 1.1 mM, respectively (Figure 26.10C).

The K_d value on the order of 1 mM is four times the K_d value of 250 μM as described in Section 26.3.1.[94] This is consistent with the report that the affinity of the cytoplasmic region of GIRK1 for Gβγ was 4- to 5-fold stronger than that for $G\alpha_{i3}$.[79]

26.4.2 $GIRK_{CP\text{-}L}$ Binding Site on $G\alpha_{i3}$ Revealed by TCS Experiments

In order to identify the $GIRK_{CP\text{-}L}$ binding site on $G\alpha_{i3}$, TCS experiments[54,55] were performed. The residues with large intensity reductions ($\Delta RR_{min} > 0.08$) were mapped on the $G\alpha_{i3}$ structure, which consists of a GTPase domain and a helical domain (Figure 26.11A). The affected residues are located on the helical domain (G89 in the αA helix; G112 and A114 in the αB helix) and the GTPase domain (R208, K209, W211, I212, H213, E216 and G217 in the α2 helix and the following loop; M240, K248, L249, I253, N256 and W258 in the α3 helix and flanking loops; E186 in the β2 strand). The mapping of these residues on the structure of Gα revealed that the residues identified are clustered on the GTPase domain of $G\alpha_{i3}$, indicating that this site (hereafter referred to as the "α2/α3 site") mainly contributes to $GIRK_{CP\text{-}L}$ binding (Figure 26.11A). This is also supported by two alanine mutants of the identified residues (I212A and W258A), which exhibited impaired binding affinity for $GIRK_{CP\text{-}L}$. It should be noted that E186 seems separated from the cluster by the intervening residue, F199. The amide group of F199 is buried in the protein and is more than 6 Å away from the contiguous surface for $GIRK_{CP\text{-}L}$ binding, resulting in the lack of cross-saturation for F199. Therefore, we conclude that the α2/α3 site is the major $GIRK_{CP\text{-}L}$ binding site.

The residues identified on the helical domain (G89, G112 and A114), which are distant from the α2/α3 site, do not form a contiguous surface but are located on the same side of the protein surface as the α2/α3 site, suggesting that these residues transiently contact $GIRK_{CP\text{-}L}$. We further investigated the interactions of the helical domain with $GIRK_{CP\text{-}L}$ (see below).

26.4.3 $G\alpha_{i3}$ Binding Site on $GIRK_{CP}$ Revealed by TCS Experiments

Conversely, in order to identify the $G\alpha_{i3}$ binding site on $GIRK_{CP\text{-}L}$, TCS experiments were performed. The backbone NMR signal assignments of residues 41–63 and 190–366 of $GIRK_{CP\text{-}L}$ were transferred from those of $GIRK_{CP}$.[92] The assignments of residues 367–377 were established by triple-resonance experiments (HNCACB and HNCA).[46,47] In the same manner as the TCS experiments to identify the binding site on $G\alpha_{i3}$, we evaluated the

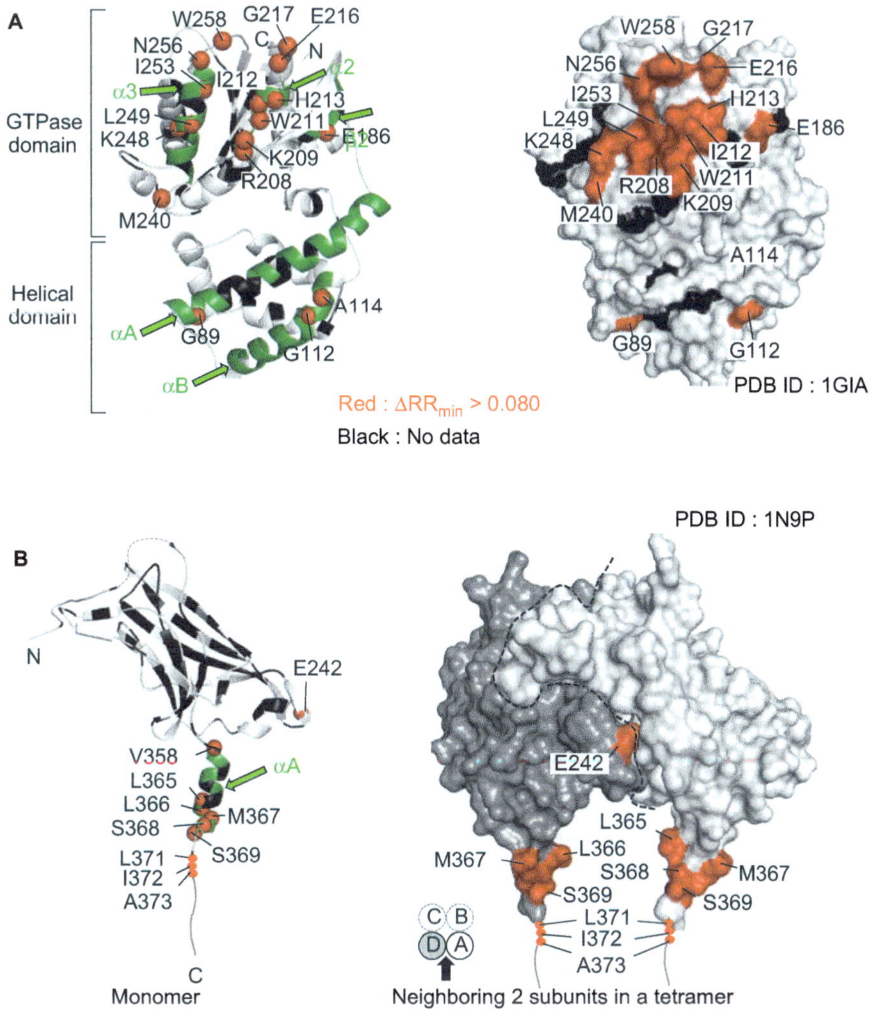

Figure 26.11 Binding sites of the $G\alpha_{i3}$–$GIRK_{CP\text{-}L}$ complex revealed by TCS experiments. (A) Residues of $G\alpha_{i3}$ with significant CS from $GIRK_{CP\text{-}L}$ are mapped on the structure of $G\alpha_{i3}$. At *left* is a ribbon drawing of Gα in which the secondary structure segments containing the affected residues are colored *red*. At *right* is a surface drawing of Gα. (B) Residues of $GIRK_{CP\text{-}L}$ with significant CS from $G\alpha_{i3}$ are mapped on the structure of $GIRK_{CP\text{-}L}$.

cross-saturation from unlabeled $G\alpha_{i3}$, by subtracting the intensity reduction ratios of the control experiment in the presence of ^{2}H-labeled $G\alpha_{i3}$.

The residues with large intensity reductions ($\Delta RR_{min} > 0.02$) are E242, V358, L365, L366, M367, S368, S369, L371, I372 and A373, as well as one of the C-terminal unassigned residues (hereafter, referred to as CT1). These residues were mapped on a single subunit (Figure 26.11B, left) and two adjacent subunits (right) of $GIRK_{CP}$. All of these residues, except for E242, are located on the C-terminal region around the αA helix of $GIRK_{CP\text{-}L}$. Therefore, we concluded that these residues are the $G\alpha_{i3}$ binding residues. It should be noted that the first two residues, I372 and A373, and CT1 were detected among the extended C-terminal 15 residues from $GIRK_{CP}$. The other C-terminal residues showed no significant intensity reductions, strongly suggesting that the extension for 15 residues is sufficient for the interaction with $G\alpha_{i3}$.

26.4.4 Contribution of the Helical Domain of $G\alpha_{i3}$ to the Interaction with $GIRK_{CP}$ as Investigated by PRE Experiments

The interaction of the helical domain of $G\alpha_{i3}$ with $GIRK_{CP\text{-}L}$ was investigated by PRE experiments. A spin-labeling reagent, *S*-(1-oxy-2,2,5,5-tetramethylpyrroline-3-methyl) methanethiosulfonate (MTSL), which can be chemically introduced to cysteine sidechains, was used to label $G\alpha_{i3}$. First, we prepared a mutant of $G\alpha_{i3}$ (C3S-C66A-C214S-C305S-C325A-C351I), in which the six natively existing cysteine residues that are exposed and reactive are substituted with other residues (hereafter, referred to as Hexa III). It should be noted that the corresponding mutant of $G\alpha_{i1}$, (C3S-C66A-C214S-C305S-C325A-C351I), named Hexa I, was folded properly and functional.[110] We also confirmed the proper folding of Hexa III by ^{1}H–^{15}N TROSY of ^{2}H,^{15}N-labeled Hexa III (data not shown). Then, either I82 or S153 was substituted by Cys and chemically modified by MTSL (Hexa III-I82C-MTSL and Hexa III-S153C-MTSL). The I82 residue is located at the centre of the three residues identified by TCS (G89, G112 and A114), and S153 exists on the opposite side from I82. Since the paramagnetism of MTSL disappears upon reduction, by addition of ascorbate, the PRE effects were evaluated by the signal intensity ratios of ^{1}H–^{15}N TROSY spectra of ^{2}H,^{15}N-labeled $GIRK_{CP\text{-}L}$ between the oxidized and reduced states in the presence of Hexa III-I82C-MTSL or Hexa III-S153C-MTSL.

Figure 26.12A shows that the PRE effects from Hexa III-I82C-MTSL were observed for 13 signals of $GIRK_{CP\text{-}L}$: T353, Y356, S357, L365, L366, M367, S368, S369, L371, I372, A373, A375 and CT1, which reside in the C-terminal region of $GIRK_{CP\text{-}L}$ (Figure 26.12B). Most of the other residues of the C-terminal region could not be analyzed, however, due to signal overlap and these residues might also be affected by MTSL. On the other hand, essentially no PRE was detected for Hexa III-S153C-MTSL (data not shown), in which S153C is located at the side opposite to I82C on the helical domain of Hexa III. Since the spin-label was introduced at I82C, which is at the center of the three residues identified by TCS, we conclude that the I82 side of the helical domain can approach within 20 Å of the C-terminal region of $GIRK_{CP\text{-}L}$.

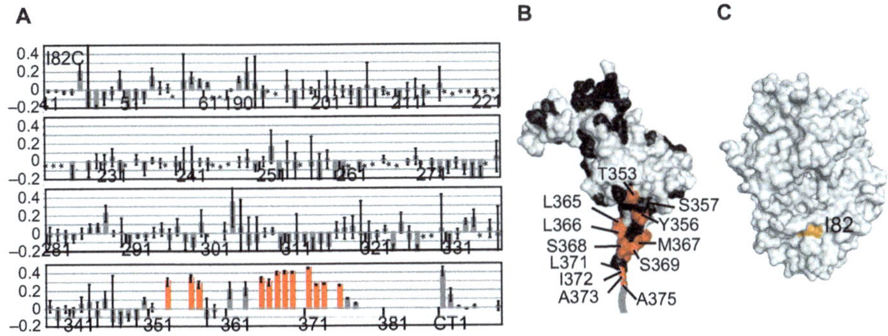

Figure 26.12 GIRK$_{CP\text{-}L}$ residues subject to PRE from I82 located on the helical domain of Gα_{i3}. (A) Plot of the signal intensity reduction ratios (RR) of paramagnetic to diamagnetic for Hexa III-I82C-MTSL. The residues indicated by asterisks are those with no data due to overlapping resonances or insufficient signal-to-noise ratio. Bars corresponding to the residues with significant intensity reductions >0.20 are colored red. (B) Mapped on the GIRK structure. The proline residues and the residues with no data are colored black. (C) The spin-labeled site of Gα_{i3}.

26.4.5 Binding Mode of Gα and the Cytoplasmic Region of GIRK

NMR analyses were performed to probe the interaction between GIRK$_{CP\text{-}L}$ and Gα_{i3}. Since cross-saturation is a phenomenon depending on intermolecular ^{1}H–^{1}H distances, the TCS results from GIRK$_{CP\text{-}L}$ to Gα_{i3} (Figure 26.11A) and *vice versa* (Figure 26.11B) indicated that the α2/α3 site of Gα_{i3} and the αA helix of GIRK$_{CP\text{-}L}$ directly interact with each other. Our relaxation matrix calculations,[55,102] in which the on and off rates of the GIRK$_{CP\text{-}L}$–Gα_{i3} interaction were also considered, suggest that under the experimental conditions the cross-saturation effect should be observed for the amide hydrogen atoms of GIRK$_{CP\text{-}L}$ within 6 Å from Gα_{i3} and for those of Gα_{i3} within 5 Å from GIRK$_{CP\text{-}L}$.

The PRE results indicated that the αA helix of GIRK$_{CP\text{-}L}$ is within 20 Å of I82 of Gα_{i3} (Figure 26.12). However, the αA helix of GIRK$_{CP\text{-}L}$, which binds to the α2/α3 site, should be more than 20 Å away from the spin-labeled site, suggesting that another αA helix in a neighboring subunit of the GIRK$_{CP\text{-}L}$ tetramer must come close to the spin-labeled site.

In order to build a model of the Gα_{i3}–GIRK$_{CP\text{-}L}$ complex satisfying the NMR-derived structural information, rigid body docking was performed using the HADDOCK program.[111] In the calculation, the residues in the α2/α3 site of Gα and in the αA helix of GIRK$_{CP\text{-}L}$, identified by the TCS experiments, were specified as the "active residues" in the program, so that they formed the interface of the complex. Although the information about the residues on the helical domain derived from the TCS and PRE experiments was not used for the calculation, the helical domain of Gα was calculated to be proximal to the αA helix in a neighboring subunit of $GIRK_{CP-L}$, probably due to the restraint of E242 on the subunit interface of $GIRK_{CP-L}$ as determined from the TCS experiments (Figure 26.13A). In this binding mode, E242 of $GIRK_{CP-L}$, at a distance from the αA helix, approaches the side chain of E186 of $G\alpha_{i3}$, which accounts for its intensity reduction in the TCS experiment.

The three residues that TCS identified on the helical domain of $G\alpha_{i3}$ (G89, G112 and A114) showed relatively weak cross-saturation and do not form a continuous binding surface, presumably because the αA helix of $GIRK_{CP-L}$

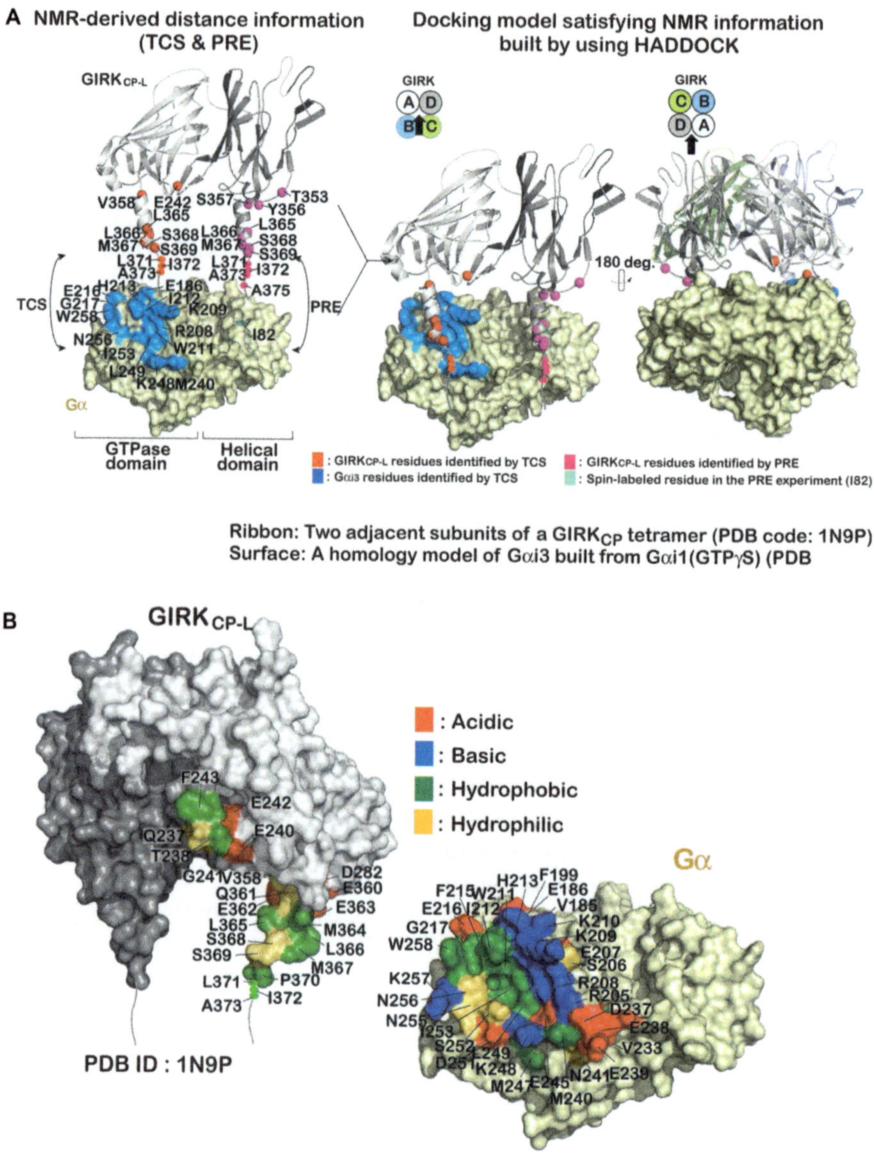

transiently accesses the $G\alpha_{i3}$ helical domain. This can be accounted for by the conformational flexibility of the αA helix. In the crystal structure, the αA helix of $GIRK_{CP}$ is stabilized by the crystal contacts with the αA helix of another tetramer. In addition, the C-terminal part of the helix (residues 367–368) exhibited very weak or no NOEs that are typical for an α helix (data not shown), and the chemical shift index calculated by the ^{13}C chemical shifts indicated that the C-terminal residues after M367 were unstructured. Altogether, we concluded that the α2/α3 site of $G\alpha_{i3}$ mainly recognizes the αA helix of $GIRK_{CP\text{-}L}$ in one subunit of the tetramer, while the helical domain of $G\alpha_{i3}$ transiently approaches the αA helix in another neighboring subunit.

Figure 26.13B shows the surface properties of the α2/α3 site of $G\alpha_{i3}$ and the αA helix of $GIRK_{CP\text{-}L}$. The α2/α3 site possesses a hydrophobic cleft at its center, which is surrounded by polar residues, while the αA helix also contains a cluster of hydrophobic residues that are exposed to the outside. This suggests that the complementarity of the shape at the direct binding sites formed by these hydrophobic residues is important for the binding. Indeed, the mutations of the hydrophobic residues of $G\alpha_{i3}$, I212A and W258A resulted in impaired affinity for $GIRK_{CP\text{-}L}$ (data not shown).

26.5 GIRK Regulation Mechanism by G Proteins

Upon ligand stimulation of the GPCR, the GDP–GTP exchange on $G\alpha_{i/o}$ in the complex with Gβγ decreases the affinity of $G\alpha_{i/o}$ and Gβγ, allowing Gβγ to activate GIRK by binding. When the GPCR stimulation ends, GTP on $G\alpha_{i/o}$ is rapidly hydrolyzed to GDP with assistance from RGS.[83] Thus, the $G\alpha_{i/o}$ in complex with the hydrolyzed GDP immediately re-associates with Gβγ and, thus, $G\alpha_{i/o}$(GTP) contributes to the rapid deactivation of GIRK.[79]

Figure 26.13 Binding mode between $GIRK_{CP\text{-}L}$ and $G\alpha_{i3}$. (A) *Left*: Two adjacent subunits of a $GIRK_{CP}$ tetramer (PDB code: 1N9P), depicted by a ribbon diagram, viewed from the inside of the tetramer, and a homology model of $G\alpha_{i3}$ built from the crystal structure of $G\alpha_{i1}$(GTPγS) (PDB code: 1GIA) shown in a surface representation. The nitrogen atoms of the $GIRK_{CP\text{-}L}$ residues identified in the TCS experiments and residues with PRE effects from MTSL, modified to I82C on the helical domain of $G\alpha_{i3}$ (green), are shown as red and magenta balls, respectively. The α2/α3 residues of $G\alpha_{i3}$ identified in TCS experiments are colored blue. *Centre*: The binding model of $G\alpha_{i3}$ in the GTP-bound state and the cytoplasmic region of GIRK1 (PDB code: 1N9P), obtained with HADDOCK software.[111] Only the two adjacent subunits from the inside of the GIRK tetramer are shown. *Right*: 180° rotation of the model (*centre*) with all subunits of the GIRK tetramer shown. (B) Properties of the binding interfaces on $G\alpha_{i3}$ and GIRK1. An "open-book" display of the Gα–GIRK complex model in (A). The residues on $GIRK_{CP\text{-}L}$ and the α2/α3 residues on $G\alpha_{i3}$ identified by TCS (bold letters) and surrounding residues with Cα atoms within 6.0 Å, are colored according to side-chain properties: acidic, basic, hydrophobic and hydrophilic residues are colored red, blue, green and yellow, respectively.

In order to address the GIRK regulation mechanism by G proteins, two types of G protein binding sites on GIRK were proposed recently ("activation sites" and "anchoring sites"), although none of these sites have been identified.[84,86] The activation sites are the Gβγ binding sites; the anchoring sites are the binding sites of $G\alpha_{i/o}$(GDP)βγ and $G\alpha_{i/o}$(GTP), where the sites for the two forms of $G\alpha_{i/o}$ are not known to be identical.

Our solution-state NMR analyses on the interactions of $G\alpha_{i3}$ and Gβγ with the CP region of GIRK revealed that the activation sites lie at the border of the two neighboring subunits of the GIRK tetramer (Figure 26.6).[94] The results of the TCS and CSP experiments indicated that the change in the orientation of the CP region of GIRK upon the Gβγ binding leads to the conformational change of one of the GIRK loops that directly interact with the K^+-ion gate (H–I loop), which is presumably the structural mechanism of the GIRK opening by Gβγ.

On the other hand, we also identified one of the "anchoring sites", *i.e.* the $G\alpha_{i/o}$(GTP) binding site on GIRK. The TCS and PRE results successfully identified the mutual binding sites on $G\alpha_{i/o}$(GTP) and $GIRK_{CP}$, and their binding orientation, which enabled building of a complex model (Figures 26.11 and 26.13).

The mapping of the Gβγ-binding residues[94] on the structure of the Gα–GIRK complex indicated that the binding sites of $G\alpha_{i3}$(GTP) and Gβγ on GIRK do not overlap with each other and, therefore, $G\alpha_{i3}$(GTP) and Gβγ can simultaneously bind to GIRK (Figure 26.14). Therefore, the $G\alpha_{i/o}$–GIRK–Gβγ ternary complex model obtained here provides the structural basis for the

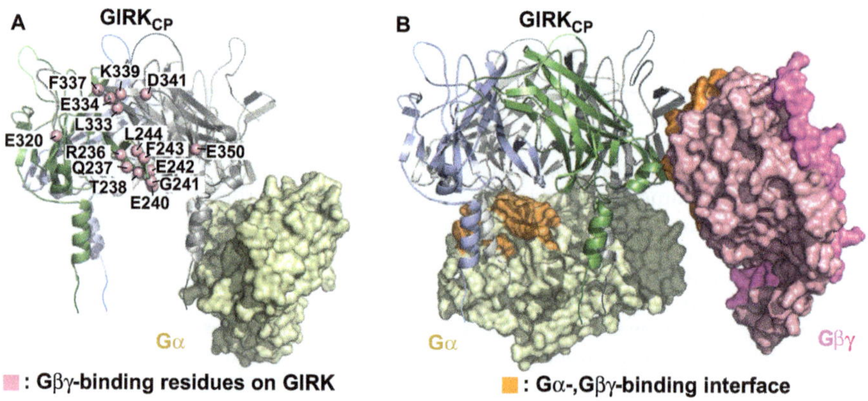

Figure 26.14 A model of the $G\alpha_{i/o}$–GIRK–Gβγ ternary complex. (A) The nitrogen atoms of the Gβγ-binding residues on GIRK1[94] are mapped as magenta balls on the complex model of Gα and the GIRK tetramer of Figure 26.12. (B) Ternary complex model of Gα–GIRK–Gβγ. First, the binding model of Gβγ–GIRK was built using HADDOCK. Gβ and Gγ are colored magenta and violet, respectively. The residues involved in the intermolecular interactions between Gα and Gβγ within a distance of 4 Å in the $G\alpha_{i1}(GDP)\beta_1\gamma_2$ structure (PDB code: 1GP2) are colored orange.

rapid closure of the GIRK channel upon signal termination, through the efficient removal of the proximal Gβγ from GIRK.

It should be noted that the α2 helix of $G\alpha_{i3}$ in the GIRK-binding site (the α2/α3 site) is located in the switch II region, which is known to alter its conformation upon GDP–GTP exchange and is buried upon binding to Gβγ. Thus, the $G\alpha_{i/o}$(GDP)βγ anchoring site to GIRK is different from that of $G\alpha_{i/o}$(GTP) revealed here. The binding of $G\alpha_{i/o}$(GDP)βγ to GIRK contributes to the efficient gate opening of GIRK by Gβγ and the specificity of GPR signalling[81,82] as well as the suppression of the basal K^+ current, referred to as "priming".[78,79,84–86,112] Therefore, structural analyses of the interaction between $G\alpha_{i/o}$(GDP)βγ and GIRK will provide a complete understanding of the modulation mechanism of the GIRK gating by $G_{i/o}$ proteins.

Acknowledgements

This work was supported in part by grants from the Japan New Energy and Industrial Technology Development Organization (NEDO) and the Ministry of Economy, Trade and Industry (METI) (to I.S.), a Grant-in-Aid for Scientific Research on Priority Areas from the Japanese Ministry of Education, Culture, Sports, Science and Technology (MEXT) (to M.O. and I.S.), Global COE Program "Medical System Innovation on Multidisciplinary Integration" from MEXT (to Y.M.), and a grant from Takeda Science Foundation (to M.O.). Y.M. is a research fellow of the Japan Society of Promotion of Science (JSPS).

References

1. M. Renault, A. Cukkemane and M. Baldus, *Angew. Chem. Int. Ed.*, 2010, **49**, 8346.
2. F. M. Marassi, B. B. Das, G. J. Lu, H. J. Nothnagel, S. H. Park, W. S. Son, Y. Tian and S. J. Opella, *Methods*, 2011, **55**, 363.
3. W. T. Franks, A. H. Linden, B. Kunert, B. J. van Rossum and H. Oschkinat, *Eur. J. Cell Biol.*, 2012, **91**, 340.
4. X. Ding, X. Zhao and A. Watts, *Biochem. J.*, 2013, **450**, 443.
5. G. Grobner, I. J. Burnett, C. Glaubitz, G. Choi, A. J. Mason and A. Watts, *Nature*, 2000, **405**, 810.
6. S. H. Park, B. B. Das, F. Casagrande, Y. Tian, H. J. Nothnagel, M. Chu, H. Kiefer, K. Maier, A. A. De Angelis, F. M. Marassi and S. J. Opella, *Nature*, 2012, **491**, 779.
7. A. Lange, K. Giller, S. Hornig, M. F. Martin-Eauclaire, O. Pongs, S. Becker and M. Baldus, *Nature*, 2006, **440**, 959.
8. C. Ader, R. Schneider, S. Hornig, P. Velisetty, E. M. Wilson, A. Lange, K. Giller, I. Ohmert, M. F. Martin-Eauclaire, D. Trauner, S. Becker, O. Pongs and M. Baldus, *Nat. Struct. Mol. Biol.*, 2008, **15**, 605.
9. C. Ader, R. Schneider, S. Hornig, P. Velisetty, V. Vardanyan, K. Giller, I. Ohmert, S. Becker, O. Pongs and M. Baldus, *EMBO J.*, 2009, **28**, 2825.

10. M. P. Bhate, B. J. Wylie, L. Tian and A. E. McDermott, *J. Mol. Biol.*, 2010, **401**, 155.
11. M. P. Bhate and A. E. McDermott, *Proc. Natl. Acad. Sci. U. S. A.*, 2012, **109**, 15265.
12. F. Hu, W. Luo and M. Hong, *Science*, 2010, **330**, 505.
13. S. D. Cady, K. Schmidt-Rohr, J. Wang, C. S. Soto, W. F. Degrado and M. Hong, *Nature*, 2010, **463**, 689.
14. S. G. Patching, G. Psakis, S. A. Baldwin, J. Baldwin, P. J. Henderson and D. A. Middleton, *Mol. Membr. Biol.*, 2008, **25**, 474.
15. V. Lange, J. Becker-Baldus, B. Kunert, B. J. van Rossum, F. Casagrande, A. Engel, Y. Roske, F. M. Scheffel, E. Schneider and H. Oschkinat, *ChemBioChem*, 2010, **11**, 547.
16. S. G. Patching, P. J. Henderson, D. J. Sharples and D. A. Middleton, *Mol. Membr. Biol.*, 2013, **30**, 129.
17. G. J. Lu, Y. Tian, N. Vora, F. M. Marassi and S. J. Opella, *J. Am. Chem. Soc.*, 2013, **135**, 9299.
18. Y. Kofuku, C. Yoshiura, T. Ueda, H. Terasawa, T. Hirai, S. Tominaga, M. Hirose, Y. Maeda, H. Takahashi, Y. Terashima, K. Matsushima and I. Shimada, *J. Biol. Chem.*, 2009, **284**, 35240.
19. C. Yoshiura, Y. Kofuku, T. Ueda, Y. Mase, M. Yokogawa, M. Osawa, Y. Terashima, K. Matsushima and I. Shimada, *J. Am. Chem. Soc.*, 2010, **132**, 6768.
20. A. Gautier, H. R. Mott, M. J. Bostock, J. P. Kirkpatrick and D. Nietlispach, *Nat. Struct. Mol. Biol.*, 2010, **17**, 768.
21. M. P. Bokoch, Y. Zou, S. G. Rasmussen, C. W. Liu, R. Nygaard, D. M. Rosenbaum, J. J. Fung, H. J. Choi, F. S. Thian, T. S. Kobilka, J. D. Puglisi, W. I. Weis, L. Pardo, R. S. Prosser, L. Mueller and B. K. Kobilka, *Nature*, 2010, **463**, 108.
22. Y. Kofuku, T. Ueda, J. Okude, Y. Shiraishi, K. Kondo, M. Maeda, H. Tsujishita and I. Shimada, *Nat. Commun.*, 2012, **3**, 1045.
23. J. J. Liu, R. Horst, V. Katritch, R. C. Stevens and K. Wuthrich, *Science*, 2012, **335**, 1106.
24. R. Nygaard, Y. Zou, R. O. Dror, T. J. Mildorf, D. H. Arlow, A. Manglik, A. C. Pan, C. W. Liu, J. J. Fung, M. P. Bokoch, F. S. Thian, T. S. Kobilka, D. E. Shaw, L. Mueller, R. S. Prosser and B. K. Kobilka, *Cell*, 2013, **152**, 532.
25. S. Hiller, R. G. Garces, T. J. Malia, V. Y. Orekhov, M. Colombini and G. Wagner, *Science*, 2008, **321**, 1206.
26. M. Bayrhuber, T. Meins, M. Habeck, S. Becker, K. Giller, S. Villinger, C. Vonrhein, C. Griesinger, M. Zweckstetter and K. Zeth, *Proc. Natl. Acad. Sci. U. S. A.*, 2008, **105**, 15370.
27. K. Takeuchi, M. Yokogawa, T. Matsuda, M. Sugai, S. Kawano, T. Kohno, H. Nakamura, H. Takahashi and I. Shimada, *Structure*, 2003, **11**, 1381.
28. K. Takeuchi, H. Takahashi, S. Kawano and I. Shimada, *J. Biol. Chem.*, 2007, **282**, 15179.

29. J. H. Chill, J. M. Louis, F. Delaglio and A. Bax, *Biochim. Biophys. Acta*, 2007, **1768**, 3260.
30. K. A. Baker, C. Tzitzilonis, W. Kwiatkowski, S. Choe and R. Riek, *Nat. Struct. Mol. Biol.*, 2007, **14**, 1089.
31. Z. O. Shenkarev, E. N. Lyukmanova, O. I. Solozhenkin, I. E. Gagnidze, O. V. Nekrasova, V. V. Chupin, A. A. Tagaev, Z. A Yakimenko, T. V. Ovchinnikova, M. P. Kirpichnikov and A. S. Arseniev, *Biochemistry (Moscow)*, 2009, **74**, 756.
32. S. Imai, M. Osawa, K. Takeuchi and I. Shimada, *Proc. Natl. Acad. Sci. U. S. A.*, 2010, **107**, 6216.
33. S. Imai, M. Osawa, K. Mita, S. Toyonaga, A. Machiyama, T. Ueda, K. Takeuchi, S. Oiki and I. Shimada, *J. Biol. Chem.*, 2012, **287**, 39634.
34. J. R. Schnell and J. J. Chou, *Nature*, 2008, **451**, 591.
35. B. OuYang, S. Xie, M. J. Berardi, X. Zhao, J. Dev, W. Yu, B. Sun and J. J. Chou, *Nature*, 2013, **498**, 521.
36. W. D. Van Horn, H. J. Kim, C. D. Ellis, A. Hadziselimovic, E. S. Sulistijo, M. D. Karra, C. Tian, F. D. Sonnichsen and C. R. Sanders, *Science*, 2009, **324**, 1726.
37. C. Fernandez and K. Wuthrich, *FEBS Lett.*, 2003, **555**, 144.
38. J. M. Kielec, K. G. Valentine, C. R. Babu and A. J. Wand, *Structure*, 2009, **17**, 345.
39. J. M. Kielec, K. G. Valentine and A. J. Wand, *Biochim. Biophys. Acta*, 2010, **1798**, 150.
40. P. Bazzacco, E. Billon-Denis, K. S. Sharma, L. J. Catoire, S. Mary, C. Le Bon, E. Point, J. L. Baneres, G. Durand, F. Zito, B. Pucci and J. L. Popot, *Biochemistry*, 2012, **51**, 1416.
41. R. S. Prosser, F. Evanics, J. L. Kitevski and M. S. Al-Abdul-Wahid, *Biochemistry*, 2006, **45**, 8453.
42. S. F. Poget and M. E. Girvin, *Biochim. Biophys. Acta*, 2007, **1768**, 3098.
43. T. Raschle, S. Hiller, M. Etzkorn and G. Wagner, *Curr. Opin. Struct. Biol.*, 2010, **20**, 471.
44. K. Pervushin, R. Riek, G. Wider and K. Wuthrich, *Proc. Natl. Acad. Sci. U. S. A.*, 1997, **94**, 12366.
45. X. Shan, K. H. Gardner, D. R. Muhandiram, N. S. Rao, C. H. Arrowsmith and L. E. Kay, *J. Am. Chem. Soc.*, 1996, **118**, 6570.
46. M. Salzmann, K. Pervushin, G. Wider, H. Senn and K. Wuthrich, *Proc. Natl. Acad. Sci. U. S. A.*, 1998, **95**, 13585.
47. M. Salzmann, G. Wider, K. Pervushin, H. Senn and K. Wuthrich, *J. Am. Chem. Soc.*, 1999, **121**, 844.
48. C. Fernandez and G. Wider, *Curr. Opin. Struct. Biol.*, 2003, **13**, 570.
49. E. Koculi, R. Horst, A. L. Horwich and K. Wuthrich, *Protein Sci.*, 2011, **20**, 1380.
50. R. Horst, E. B. Bertelsen, J. Fiaux, G. Wider, A. L. Horwich and K. Wuthrich, *Proc. Natl. Acad. Sci. U. S. A.*, 2005, **102**, 12748.
51. R. Sprangers, A. Gribun, P. M. Hwang, W. A. Houry and L. E. Kay, *Proc. Natl. Acad. Sci. U. S. A.*, 2005, **102**, 16678.

52. A. M. Ruschak, T. L. Religa, S. Breuer, S. Witt and L. E. Kay, *Nature*, 2010, **467**, 868.
53. T. L. Religa, R. Sprangers and L. E. Kay, *Science*, 2010, **328**, 98.
54. I. Shimada, T. Ueda, M. Matsumoto, M. Sakakura, M. Osawa, K. Takeuchi, N. Nishida and H. Takahashi, *Prog. Nucl. Magn. Reson. Spectrosc.*, 2009, **54**, 123.
55. M. Matsumoto, T. Ueda and I. Shimada, *J. Magn. Reson.*, 2010, **205**, 114.
56. J. L. Battiste and G. Wagner, *Biochemistry*, 2000, **39**, 5355.
57. G. M. Clore and J. Iwahara, *Chem. Rev.*, 2009, **109**, 4108.
58. G. M. Clore, *Protein Sci.*, 2011, **20**, 229.
59. C. He, H. Zhang, T. Mirshahi and D. E. Logothetis, *J. Biol. Chem.*, 1999, **274**, 12517.
60. D. Bichet, F. A. Haass and L. Y. Jan, *Nat. Rev. Neurosci.*, 2003, **4**, 957.
61. H. Hibino, A. Inanobe, K. Furutani, S. Murakami, I. Findlay and Y. Kurachi, *Physiol. Rev.*, 2010, **90**, 291.
62. D. E. Clapham and E. J. Neer, *Nature*, 1993, **365**, 403.
63. Y. Kurachi, *Am. J. Physiol.*, 1995, **269**, C821.
64. H. Ito, R. T. Tung, T. Sugimoto, I. Kobayashi, K. Takahashi, T. Katada, M. Ui and Y. Kurachi, *J. Gen. Physiol.*, 1992, **99**, 961.
65. D. E. Logothetis, Y. Kurachi, J. Galper, E. J. Neer and D. E. Clapham, *Nature*, 1987, **325**, 321.
66. E. Reuveny, P. A. Slesinger, J. Inglese, J. M. Morales, J. A. Iniguez-Lluhi, R. J. Lefkowitz, H. R. Bourne, Y. N. Jan and L. Y. Jan, *Nature*, 1994, **370**, 143.
67. K. D. Wickman, J. A. Iniguez-Lluhl, P. A. Davenport, R. Taussig, G. B. Krapivinsky, M. E. Linder, A. G. Gilman and D. E. Clapham, *Nature*, 1994, **368**, 255.
68. A. Kuo, J. M. Gulbis, J. F. Antcliff, T. Rahman, E. D. Lowe, J. Zimmer, J. Cuthbertson, F. M. Ashcroft, T. Ezaki and D. A. Doyle, *Science*, 2003, **300**, 1922.
69. M. R. Whorton and R. MacKinnon, *Cell*, 2011, **147**, 199.
70. C. He, X. Yan, H. Zhang, T. Mirshahi, T. Jin, A. Huang and D. E. Logothetis, *J. Biol. Chem.*, 2002, **277**, 6088.
71. M. Finley, C. Arrabit, C. Fowler, K. F. Suen and P. A. Slesinger, *J. Physiol.*, 2004, **555**, 643.
72. M. Nishida, M. Cadene, B. T. Chait and R. MacKinnon, *EMBO J.*, 2007, **26**, 4005.
73. M. Nishida and R. MacKinnon, *Cell*, 2002, **111**, 957.
74. S. Pegan, C. Arrabit, W. Zhou, W. Kwiatkowski, A. Collins, P. A. Slesinger and S. Choe, *Nat. Neurosci.*, 2005, **8**, 279.
75. A. Inanobe, T. Matsuura, A. Nakagawa and Y. Kurachi, *Channels (Austin)*, 2007, **1**, 39.
76. M. Osawa, M. Yokogawa, T. Muramatsu, T. Kimura, Y. Mase and I. Shimada, *J. Biol. Chem.*, 2009, **284**, 26117.
77. B. Hille, *Neuron*, 1992, **9**, 187.

78. T. Ivanina, D. Varon, S. Peleg, I. Rishal, Y. Porozov, C. W. Dessauer, T. Keren-Raifman and N. Dascal, *J. Biol. Chem.*, 2004, **279**, 17260.
79. S. Berlin, V. A. Tsemakhovich, R. Castel, T. Ivanina, C. W. Dessauer, T. Keren-Raifman and N. Dascal, *J. Biol. Chem.*, 2011, **286**, 33223.
80. A. Benians, J. L. Leaney, G. Milligan and A. Tinker, *J. Biol. Chem.*, 2003, **278**, 10851.
81. A. Kovoor and H. A. Lester, *Neuron*, 2002, **33**, 6.
82. P. A. Slesinger, E. Reuveny, Y. N. Jan and L. Y. Jan, *Neuron*, 1995, **15**, 1145.
83. C. A. Doupnik, N. Davidson, H. A. Lester and P. Kofuji, *Proc. Natl. Acad. Sci. U. S. A.*, 1997, **94**, 10461.
84. S. Berlin, T. Keren-Raifman, R. Castel, M. Rubinstein, C. W. Dessauer, T. Ivanina and N. Dascal, *J. Biol. Chem.*, 2010, **285**, 6179.
85. S. Peleg, D. Varon, T. Ivanina, C. W. Dessauer and N. Dascal, *Neuron*, 2002, **33**, 87.
86. M. Rubinstein, S. Peleg, S. Berlin, D. Brass, T. Keren-Raifman, C. W. Dessauer, T. Ivanina and N. Dascal, *J. Physiol.*, 2009, **587**, 3473.
87. S. M. Clancy, C. E. Fowler, M. Finley, K. F. Suen, C. Arrabit, F. Berton, T. Kosaza, P. J. Casey and P. A. Slesinger, *Mol. Cell. Neurosci.*, 2005, **28**, 375.
88. R. Rusinova, T. Mirshahi and D. E. Logothetis, *J. Biol. Chem.*, 2007, **282**, 34019.
89. S. R. Sprang, Z. Chen and X. Du, *Adv. Protein Chem.*, 2007, **74**, 1.
90. W. M. Oldham and H. E. Hamm, *Q. Rev. Biophys.*, 2006, **39**, 117.
91. S. G. Rasmussen, B. T. DeVree, Y. Zou, A. C. Kruse, K. Y. Chung, T. S. Kobilka, F. S. Thian, P. S. Chae, E. Pardon, D. Calinski, J. M. Mathiesen, S. T. Shah, J. A. Lyons, M. Caffrey, S. H. Gellman, J. Steyaert, G. Skiniotis, W. I. Weis, R. K. Sunahara and B. K. Kobilka, *Nature*, 2011, **477**, 549.
92. M. Yokogawa, T. Muramatsu, K. Takeuchi, M. Osawa and I. Shimada, *Biomol. NMR Assign.*, 2009, **3**, 125.
93. Y. Mase, M. Yokogawa, M. Osawa and I. Shimada, *Biomol. NMR Assign.*, 2012, **6**, 217.
94. M. Yokogawa, M. Osawa, K. Takeuchi, Y. Mase and I. Shimada, *J. Biol. Chem.*, 2011, **286**, 2215.
95. Y. Mase, M. Yokogawa, M. Osawa and I. Shimada, *J. Biol. Chem.*, 2012, **287**, 19537.
96. S. Corey and D. E. Clapham, *J. Biol. Chem.*, 2001, **276**, 11409.
97. T. T. Ivanova-Nikolova, E. N. Nikolov, C. Hansen and J. D. Robishaw, *J. Gen. Physiol.*, 1998, **112**, 199.
98. B. Grasberger, A. P. Minton, C. DeLisi and H. Metzger, *Proc. Natl. Acad. Sci. U. S. A.*, 1986, **83**, 6258.
99. F. Ozcan, P. Klein, M. A. Lemmon, I. Lax and J. Schlessinger, *Proc. Natl. Acad. Sci. U. S. A.*, 2006, **103**, 5735.
100. E. Stratikos, L. Mosyak, D. M. Zaller and D. C. Wiley, *J. Exp. Med.*, 2002, **196**, 173.

101. N. A. Sarvazyan, A. E. Remmers and R. R. Neubig, *J. Biol. Chem.*, 1998, **273**, 7934.
102. H. Takahashi, M. Miyazawa, Y. Ina, Y. Fukunishi, Y. Mizukoshi, H. Nakamura and I. Shimada, *J. Biomol. NMR*, 2006, **34**, 167.
103. H. Matsuo, J. Walters, K. Teruya, T. Tanaka, G. Gassner, S. Lippard, Y. Kyogoku and G. Wagner, *J. Am. Chem. Soc.*, 1999, **121**, 9903.
104. T. Ivanina, I. Rishal, D. Varon, C. Mullner, B. Frohnwieser-Steinecke, W. Schreibmayer, C. W. Dessauer and N. Dascal, *J. Biol. Chem.*, 2003, **278**, 29174.
105. C. E. Ford, N. P. Skiba, H. Bae, Y. Daaka, E. Reuveny, L. R. Shekter, R. Rosal, G. Weng, C. S. Yang, R. Iyengar, R. J. Miller, L. Y. Jan, R. J. Lefkowitz and H. E. Hamm, *Science*, 1998, **280**, 1271.
106. A. M. Albsoul-Younes, P. M. Sternweis, P. Zhao, H. Nakata, S. Nakajima, Y. Nakajima and T. Kozasa, *J. Biol. Chem.*, 2001, **276**, 12712.
107. D. G. Lambright, J. Sondek, A. Bohm, N. P. Skiba, H. E. Hamm and P. B. Sigler, *Nature*, 1996, **379**, 311.
108. T. L. Davis, T. M. Bonacci, S. R. Sprang and A. V. Smrcka, *Biochemistry*, 2005, **44**, 10593.
109. O. B. Clarke, A. T. Caputo, A. P. Hill, J. I. Vandenberg, B. J. Smith and J. M. Gulbis, *Cell*, 2010, **141**, 1.
110. M. Medkova, A. M. Preininger, N. J. Yu, W. L. Hubbell and H. E. Hamm, *Biochemistry*, 2002, **41**, 9962.
111. S. J. de Vries, M. van Dijk and A. M. Bonvin, *Nat. Protoc.*, 2010, **5**, 883.
112. M. Rubinstein, S. Peleg, S. Berlin, D. Brass and N. Dascal, *J. Physiol.*, 2007, **581**, 17.

CHAPTER 27

NMR Studies of Small Molecules Interacting with Amyloidogenic Proteins

ELKE PRADE,[a,b] JUAN-MIGUEL LOPEZ DEL AMO[a,b,c] AND BERND REIF*[a,b]

[a] Helmholtz-Zentrum München (HMGU), Deutsches Forschungszentrum für Gesundheit und Umwelt, Ingolstädter Landstr. 1, 85764 Neuherberg, Germany; [b] Munich Center for Integrated Protein Science (CIPS-M) at Department Chemie, Technische Universität München (TUM), Lichtenbergstr. 4, 85747 Garching, Germany; [c] Current address: CIC Energigune, Albert Einstein 48, 01510 Miñano (Álava), Spain
*Email: reif@tum.de

27.1 Introduction

In recent years, rapid progress has been made in the development of solid-state NMR as a tool to investigate the structure of non-soluble amyloid aggregates. This work has contributed to a better understanding of the fibril core, secondary structure, and provided information on the dynamic parts, as well as the topology of the amyloid fold. Prominent examples include the prion protein HET-s of *Podospora anserina*,[1,2] the 11-residue peptide fragment of the protein transthyretin, TTR(105-115),[3,4] α-synuclein fibrils,[5–8] the diabetes type II peptide hIAPP,[9,10] and fibrils formed by the Alzheimer disease Aβ(1–40) peptide which were shown to adopt at least three different polymorphous forms, featuring a two-fold symmetric structure with a

New Developments in NMR No. 3
Advances in Biological Solid-State NMR: Proteins and Membrane-Active Peptides
Edited by Frances Separovic and Akira Naito

Published by the Royal Society of Chemistry, www.rsc.org

Thioflavin T	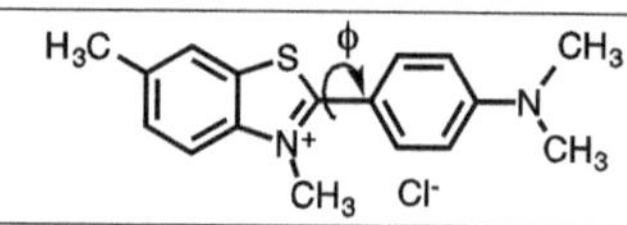
Congo Red	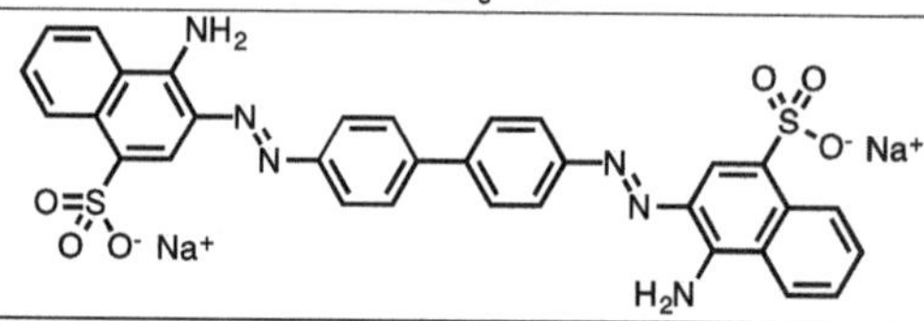
Chlorazole Black E	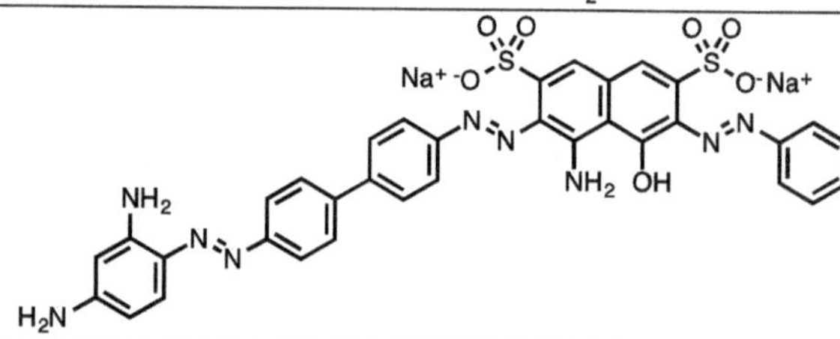
(-)-epigallocatechin -3-gallate (EGCG)	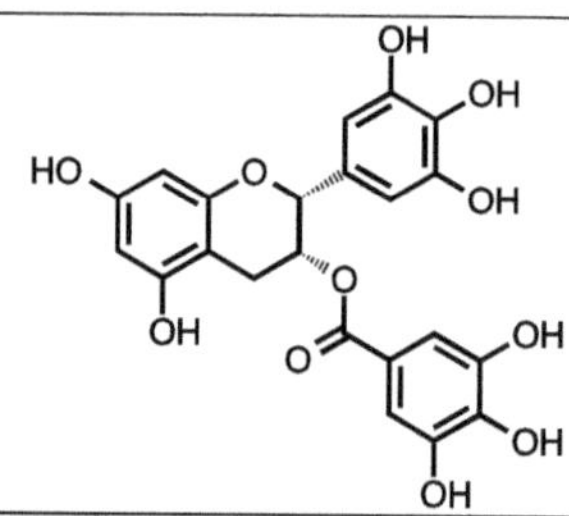
Curcumin (keto form)	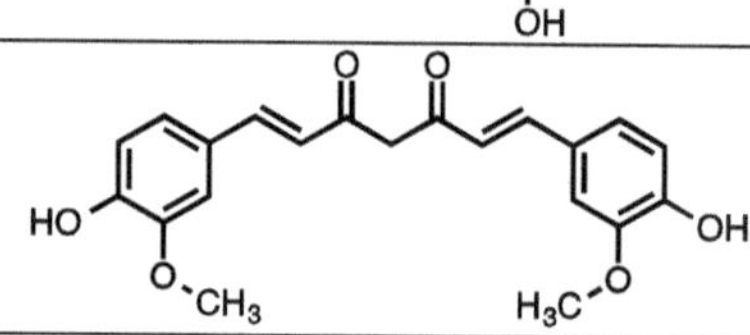
Resveratrol	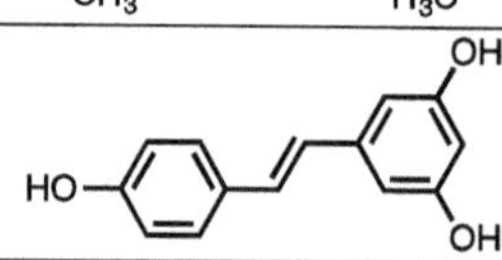
Myricetin	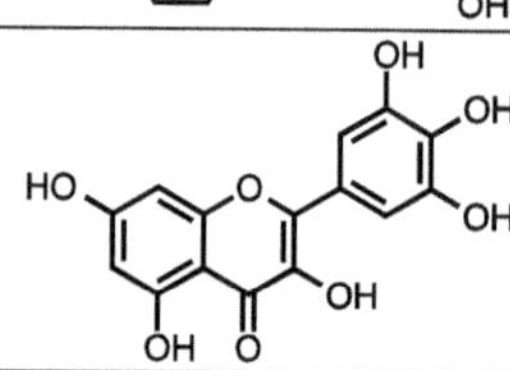

disordered N-terminus,[11] a two-fold symmetric structure with an ordered N-terminus,[12] or a three-fold symmetric structure.[13] In addition, Aβ(1–40) fibrils have been found to adopt an asymmetric dimer structure under certain conditions.[14–16] Aβ(1–42) fibrils have been reported to adopt a similar fold as Aβ(1–40) fibrils,[17] but with different C-terminal packing interactions.[18]

Nordihydro-guaiaretic Acid (NDGA)	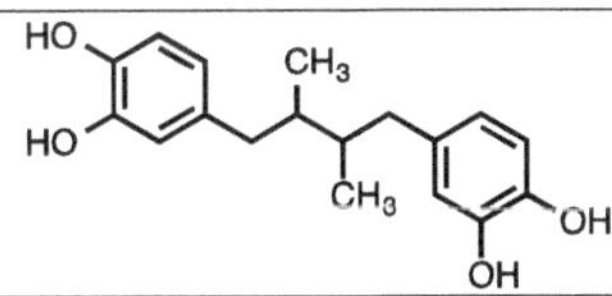
Ferulic Acid	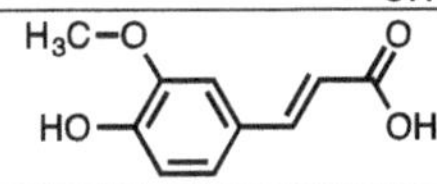
Rosmarinic Acid	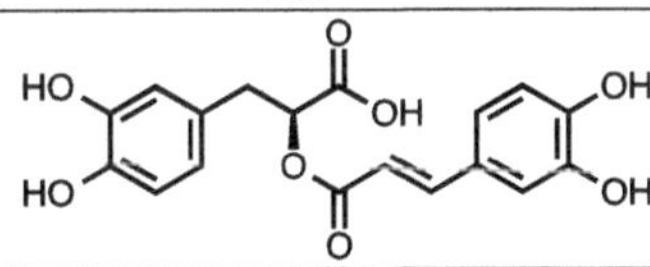
Phthalocyanine Tetrasulfonate ($PcTS\text{-}Cu^{2+}$)	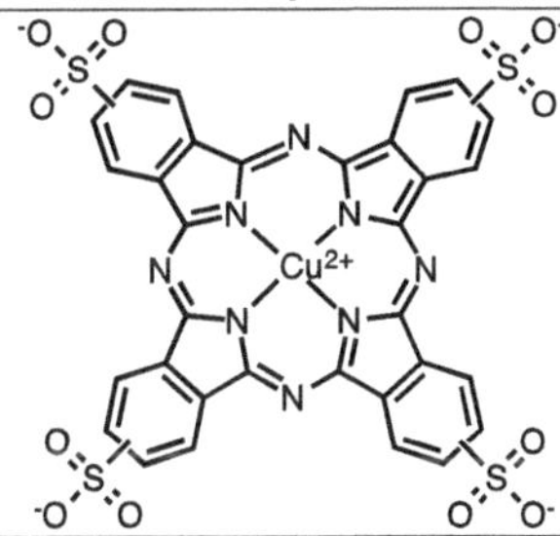
Lacmoid	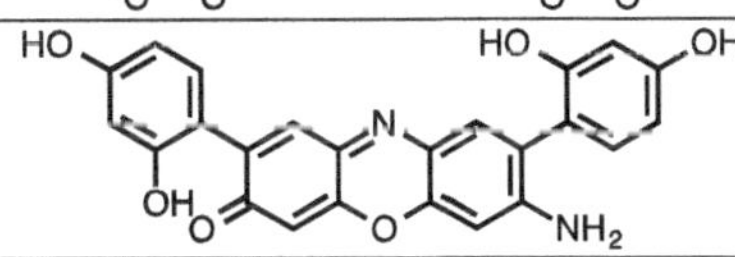
O4	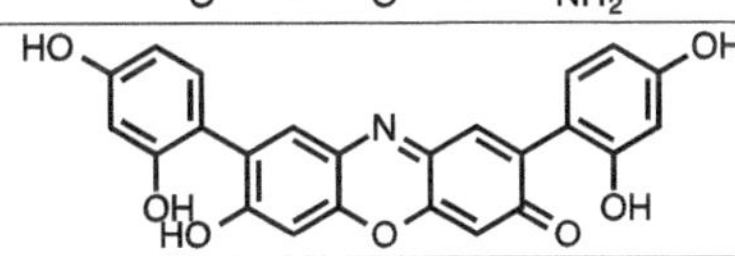

Figure 27.1 Chemical structures of amyloid dyes and aggregation inhibitors.

In addition to fibrillar structures, an increasing number of Aβ oligomer assembly models has become available.[19,20] These oligomers are considered to be the major toxic species, and are identified to be responsible for disease pathogenesis.[21–24]

For the first time, it is now possible to investigate how small molecules, peptides and proteins can potentially interact with these aggregates on a molecular level. It is still mysterious, however, how a peptide or a protein can exert a cytotoxic effect on a cell only in a particular aggregation state, such as a specific oligomeric state, whereas the monomeric or fibrillar state is void of toxicity. On the other hand, it is not understood how interference with small molecules can modulate these toxic effects.

This chapter aims to summarize the progress that has been made in this field of structural biology, and describes NMR studies involving ligand–amyloid interactions that have been published so far.

27.2 Amyloid Dyes

Amyloids are defined, amongst other properties, by their ability to bind the diazo dye Congo red (CR) or the benzothiazole dye Thioflavin T (ThT) (Figure 27.1).[25–27] The interaction between amyloids and CR yields a characteristic yellow-green birefringence. In 1922, CR was first used as an indicator for amyloidosis in patients,[28] and to date remains an important diagnostic tool to detect amyloidal aggregations in numerous neurodegenerative diseases.[31] ThT shows enhanced fluorescence upon binding to amyloids. This feature has been used since 1959 to detect amyloids in tissues[29] and is commonly used to monitor fibril formations *in vitro*.[30]

In the past, several binding modes of these dyes with respect to an amyloid structure have been suggested (Figure 27.2A, B).[31–34] These models are based on computational approaches,[32–34] confocal microscopy,[35] and polarized light microscopy,[36] among others. The first solid-state NMR studies provided experimental evidence on the mechanism of CR interaction with HET-s fibrils (Figure 27.2D).[37] However, the promiscuity of the binding to a great variety of different amyloid aggregates still poses riddles.

The structures of many small organic molecules that interact with amyloid aggregates resemble that of CR and ThT. Understanding the mechanism of the interaction is of high interest for several reasons. Radio-labeled derivatives of these amyloid dyes, such as the Pittsburgh compound B (PIB), are employed as markers in positron emission tomography (PET) imaging.[38,39] Further development of these compounds might allow differentiation between protein aggregates associated with diverse neurodegenerative diseases, and eventually facilitate the detection of an early onset disease. Complementary, knowledge of the structure of the amyloid aggregate may assist design of novel therapeutics.[40] Solid-state NMR can potentially contribute by providing the structural scaffold which in turn could be employed for rational drug design.

27.2.1 Thioflavin T

Studies involving confocal microscopy and molecular dynamics simulations have proposed that ThT binds with its long axis parallel to the fibril axis (Figure 27.2A).[33–35] Like many other compounds, ThT forms micelles with a hydrophobic core above a critical micelle concentration (CMC). AFM data suggests that ThT binds to amyloids in a micellar form.[40,41] Thioflavin T behaves as a molecular rotor in solution. Upon binding to an amyloid fibril, rotation of the dimethylaminobenzene ring with respect to the benzothiazole ring is sterically restricted, resulting in a high fluorescence quantum yield.[42–44] The relative orientation between the two ring systems is given by the torsional angle ϕ (Figure 27.1). Unbound ThT typically adopts a torsion angle of $\sim 35^\circ$.[42,43,45] Cross-peak buildup in exchange-transferred NOE experiments (trNOE) carried out for ThT interacting with insulin fibrils suggest that ϕ adopts an angle of 64° in the bound state, *i.e.* in a twisted manner.[46] Weak binding and fast exchange is a pre-requisite for the observation of a large trNOE effect.

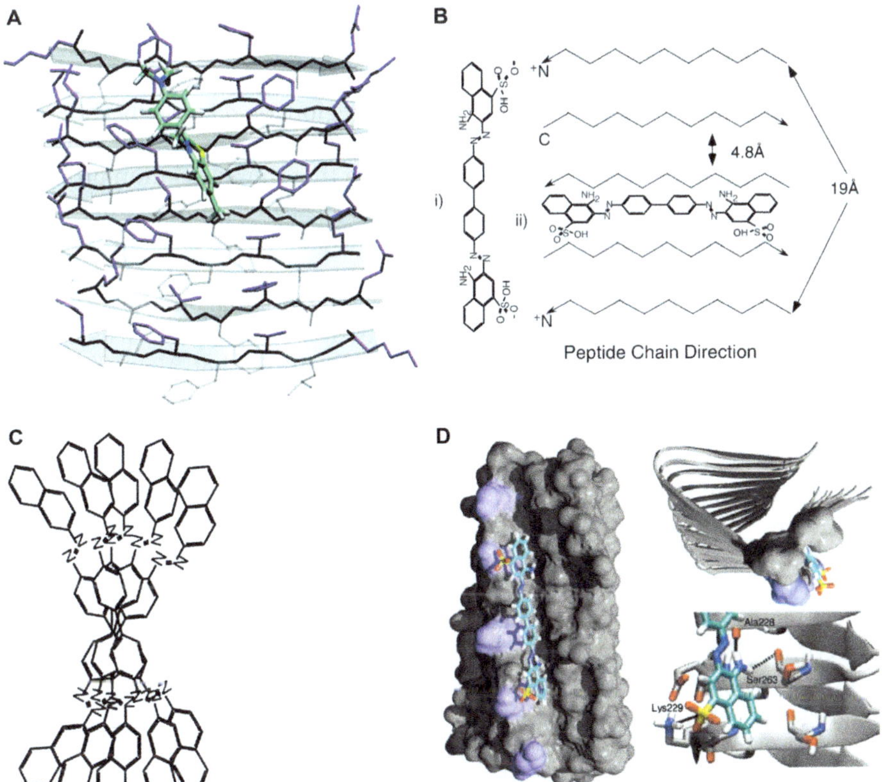

Figure 27.2 (A) Binding model of ThT to KLVFFAE protofibrils.[33] ThT binds with the long axis parallel to the fibril axis, *i.e.* perpendicular to the individual β-strands. (Reproduced with permission from the American Chemical Society.) (B) Schematic representation of CR docking to amyloid structures:[31] (i) terminal groups of CR interact with every 5th β-sheet along the fibril axis;[64,141] (ii) intercalation of CR between peptide chains.[66] (Reproduced with permission from Elsevier.) (C) Schematic model of a CR aggregate.[31] (Reproduced with permission from Elsevier.) (D) Docking model of CR interacting with HET-s(218–289) fibrils.[37] The model is based on experimental MAS solid-state NMR restraints. The mutation K229A abolishes binding of the dye. The structural model highlights specific electrostatic interactions involving the sulfonate group of CR and K229 and S263 of HET-s. (Reproduced with permission from Wiley.)

An enhancement of ThT fluorescence in twisted ThT bound to amyloid fibrils requires an interaction of ThT with aromatic residues on the fibril surface.[46]

27.2.2 Congo Red

Several studies have demonstrated that interactions between CR and monomeric amyloid peptides are typically weak.[47–51] Titration of CR or derivatives such as chlorazole black E to monomeric α-synuclein (αS) yields chemical shift changes and line broadening, in particular for the charged N-terminus of αS (residues 2–60).[47,49] Upon addition of CR at conditions of high molar excess, the hydrophobic core region of αS also becomes affected.[48,49] These effects are attenuated in samples in which the ionic strength of the buffer is increased, confirming that the association process is driven by electrostatic interactions.[49] Similarly, Aβ40 shows broadening for all resonances in the presence of CR, with the C-terminus being the least affected.[50,51]

The effect of CR on fibril formation is controversial in the literature. Originally, it was believed that CR inhibits aggregation.[52–54] This is supported by long-term NMR experiments, which show that Aβ40 incubated with the dye is maintained in solution longer in comparison to preparations which contain Aβ40 only.[55] In fact, CR acts as an inhibitor at very high concentrations, whereas it promotes aggregation at low concentrations.[56]

A more detailed analysis suggests that binding of CR to Aβ40 proceeds in two steps.[51] The first complex has a 1:1 stoichiometry with a K_D of *ca.* 5 μM. Subsequently, a second CR molecule can be incorporated into the complex with a K_D of *ca.* 300 μM. The second complex appears to be more inert, as exchange broadening is decreased and no aggregation in a long-term NMR experiment has been observed. It was, therefore, postulated that the first complex promotes fibril formation, whereas the second complex inhibits aggregation of Aβ40.

At pH 7, isothermal calorimetry experiments yield a binding model in which eight CR molecules interact with αS with an apparent binding constant of 0.4 μM.[48] In particular at low pH, CR possesses the propensity to self-associate into colloidal aggregates, a property that is common among many small molecules.[57–60] Stacking of the elongated, planar polyaromatic rings of individual molecules forms the hydrophobic core of these micelles, whereas the charged groups are solvent exposed (Figure 27.2C).[58,60] It has been postulated that colloidal aggregates are necessary for the inhibition of amyloid formations.[54]

Relaxation dispersion experiments carried out at pH 6, a condition where CR exists as colloidal aggregates, indicate that at equimolar ratio only 2% of αS is bound to CR micelles.[49] Hence, the inhibitory effect imposed by CR on αS fibrillization seems to occur due to an interaction between CR and either αS fibrils or protofibrils. TrNOEs imply that the CR bound state of αS exists in various conformations and remains largely disordered.[49] Similarly, relaxation-dispersion measurements on the Aβ–CR co-aggregates illustrate dynamic exchange between the free and the bound peptide, where the bound peptide population attributes to around 1.6% of the total peptide population.[55]

CR directs αS and Aβ40 into off-pathway aggregates which contain a large amount of β-structure.[48,50] A similar behavior has been observed for anionic detergents.[61–63] It is postulated that the two negative charges of CR promote β-sheet formation by electrostatic interactions with K16 and K28.[50] These two lysine residues flank the characteristic hairpin structure that Aβ40 peptides adopt within a fibril.

Currently, there is no general model that explains the exact binding mode between CR and insoluble amyloid species. It is generally accepted that the interaction is stabilized by hydrophobic and electrostatic contacts. The two negatively charged sulfonate groups of CR seem to be of particular importance. The two acidic groups are separated by 19.3 Å.[64] This corresponds to the common distance between five peptide chains along the fibril axis of approximately 19 Å. Therefore, CR has been suggested to bind parallel to the fibril axis and perpendicular to the individual peptides.[36,64,65] Alternatively, CR might intercalate between two anti-parallel β-strands.[32,66] Schematic representations of the two models are shown in Figure 27.2B.

To-date, the HET-s prion is the only example for which high-resolution experimental NMR data are available. In the past, the structure of the HET-s prion domain has been studied intensively by solid-state NMR. HET-s fibrils adopt a β-solenoid structure, with three β-strands forming a hydrophobic core.[1,2,67] Comparison of 2D-^{13}C,^{13}C solid-state NMR correlation spectra of HET-s incubated with CR and HET-s alone indicate that the overall fibrillar core structure is maintained in the presence of the compound, although small chemical shift perturbations are observed. Experiments using polarization transfer from the ^{1}H spins of the ligand to the ^{13}C atoms of the [U-^{2}H,^{13}C,^{15}N]-labeled protein yield a structural model of the CR-HET-s fibril complex. The results demonstrate the particular importance of the positively charged lysine repeats along the fibril surface, and indicate that CR binds with the long axis parallel to the fibril axis (Figure 27.2D).[37] In fact, mutation K229A abolishes binding of the dye. It remains to be seen how CR interacts with other fibril forming peptides and proteins.

27.3 Peptide Inhibitors

Peptide inhibitors are very convenient for structural investigations, as they are easily accessible by synthesis and can be isotopically enriched without great effort. For Alzheimer's disease Aβ, peptide inhibitors have been suggested that are either directed against the aromatic, hydrophobic core region of the peptide[68–70] or are supposedly targeting the aliphatic C-terminus.[18] In the first class of inhibitors, residues 17–20 (LVFF) of Aβ serve as a template for design. For the so-called β-sheet breaker peptides, proline residues, which are known to have a low β-sheet formation propensity, were included in the inhibitor peptide sequence to decrease its propensity to adopt β-sheet structure. In addition, one or more charged residues were added to the termini of the peptide to increase its solubility. Addition of a 20-fold excess of the inhibitor peptide iAβ5 resulted in disassembly of preformed Aβ42 fibrils and reduced their neurotoxicity *in vivo*.[69] The binding of the peptide inhibitor is very weak, as the dissociation constant is in the order of mM.[71,72]

For small molecules that weakly interact with an amyloid fibril, so-called transfer techniques such as trNOE and trRDC (residual dipolar coupling) can be employed to characterize the structure of the ligand in the bound state (Figure 27.3B). trNOEs yield the structure of the ligand. Experimentally, large

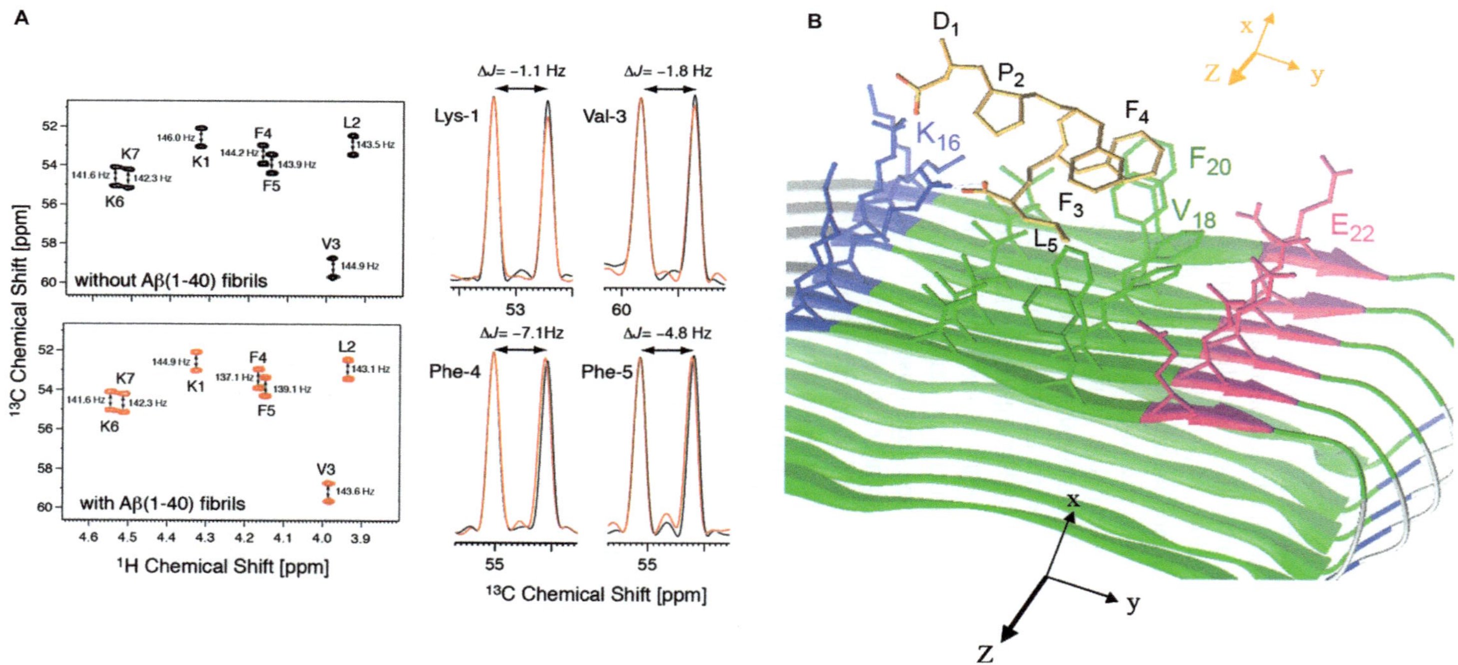

Figure 27.3 (A) trRDC measurement for a peptide inhibitor in the presence of unlabeled Aβ40 fibril. In the experiment, the peptide inhibitor KKLVFFKK suggested by Murphy and co-workers has been employed.[70] The coupled HSQC experiment on the top (on the bottom) was recorded in the absence (presence) of Aβ40 fibrils. Representative traces along the ^{13}C dimension are shown on the *right-hand side* of the figure (with and without Aβ40 fibrils in red and black, respectively). (B) Docking model of DPFFL in the presence of Aβ40 fibrils.[72] The model is based on trNOE and trRDC data, using the Aβ fibril structure suggested by Tycko and co-workers.[11]
(Figure B reproduced with permission from Elsevier.)

trNOEs are found for weakly binding peptide inhibitors.[72] trRDCs in turn allow determination of the relative orientation of the ligand in cases where the fibril can be oriented in the laboratory frame.[73] A partial orientation of an amyloid fibril is induced by the magnetic field in the NMR experiment itself. This alignment (with the fibril parallel to the external field) is based on the anisotropic diamagnetic susceptibility of the peptide bond,[74] and is employed in X-ray analysis of amyloid fibrils to obtain high-resolution diffraction patterns.[53,75] The alignment of the fibril is transferred to the transiently binding peptide inhibitor and is reflected in the size of the measured trRDCs (Figure 27.3A).

Peptide inhibitors, which target the C-terminal part of the Aβ sequence, are designed to disrupt sheet-to-sheet packing in amyloids with parallel orientation of β-strands, and with amino acids being in-register with one another.[18] The suggested peptide inhibitors are based on the framework GxFxGxF, which is complementary to the corresponding sequence IxGxMxG in Aβ.

Both types of Aβ peptide inhibitors are able to dissolve preformed fibrillar aggregates. Apparently, ligand binding induces a weakening of the fibril structure, presumably due to competition for hydrogen bond formation. Interestingly, the GxFxGxF peptide inhibitor, which is directed against the C-terminus of the Aβ sequence is able also to dissolve preformed fibrils, even though the C-terminus is not solvent accessible in all Aβ fibril structures determined so far. Depolymerization is thought to result from the dynamic nature of an amyloid fibril. In fact, several studies have demonstrated that a dynamic equilibrium exists between soluble and aggregated forms of an amyloid peptide.[76–79] A summary of peptide inhibitors employed in therapy and diagnosis of Alzheimer's disease is given by Funke and Willbold.[80]

Other Aβ targeting peptide inhibitors are derived from the human islet amyloid polypeptide (hIAPP), which has high sequence similarity to the Alzheimer's disease Aβ peptide. The double N-methylated analogue IAPP-GI has been shown to block the cytotoxic self-assembly of Aβ.[81,82] Addition of IAPP-GI yields specific chemical shift changes in the His-containing region, and results in a concentration-dependent co-aggregation of Aβ and IAPP-GI. The Aβ peptide has been suggested to be redirected into non-toxic "off-pathway" aggregates.

27.4 Polyphenolic Inhibitors

Polyphenolic small molecules have been shown to interact with many misfolding peptides and proteins involved in neurodegenerative pathologies. A prominent example of this class of compounds is (−)-epigallocatechin-3-gallate (EGCG, Figure 27.1), which is the major catechin found in the leaves of green tea. EGCG interacts with a large variety of amyloid forming proteins such as Aβ,[83] α-synuclein,[84] transthyretin,[85] and Huntingtin,[86] redirecting fibrilogenesis into the formation of sodium dodecyl sulfate stable oligomers

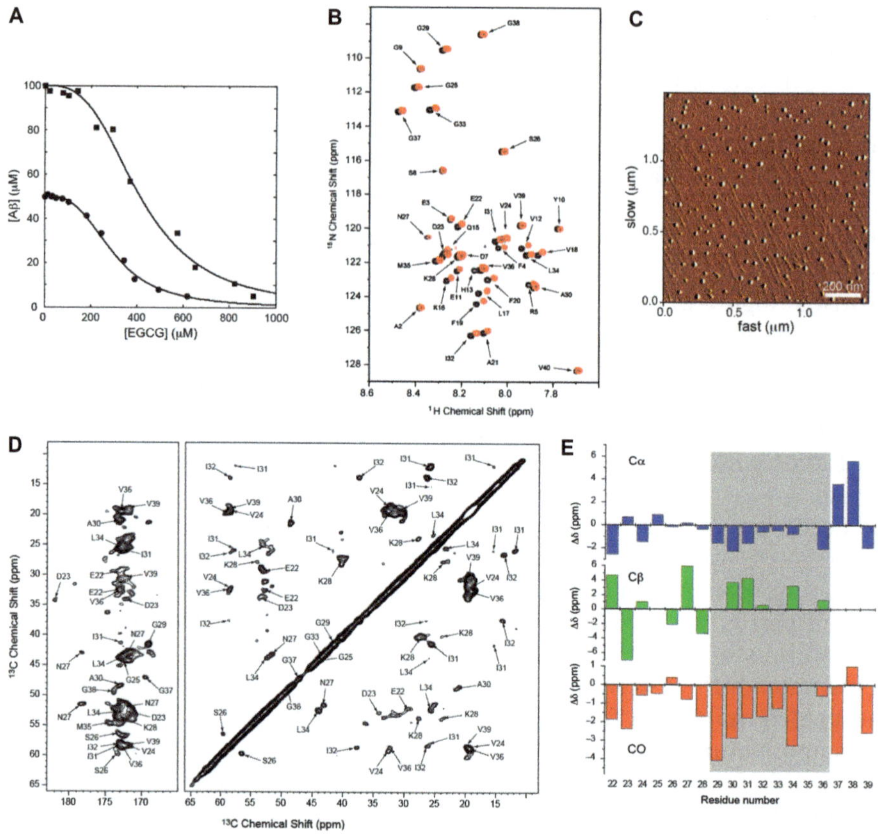

Figure 27.4 Interaction between Aβ40 and EGCG, and characterization of EGCG-induced Aβ40 oligomers.[88] (A) Fraction of soluble Aβ40 quantitated by solution-state NMR as a function of the concentration of soluble EGCG. Solution-state NMR experiments were carried out using 50 μM and 100 μM Aβ solutions (pH 7.3). (B) Solution-state NMR $^{1}H,^{15}N$ correlation spectra of Aβ40 in the presence (red) and absence (black) of a 10-fold molar excess of EGCG. (C) AFM image of the sample employed for MAS solid-state NMR studies. (D) 2D-$^{13}C,^{13}C$ PDSD (50 ms) MAS solid-state NMR correlation spectra of EGCG-induced Aβ40 oligomers; Aβ was incubated with a 10-fold molar excess of EGCG. (E) Secondary chemical shift analysis of the resonances of the EGCG-induced oligomers. Amino acids 29–36 adopt a β-sheet conformation. (Reproduced with permission from Elsevier.)

with reduced toxicity.[87] Recently, the aggregates formed in presence of EGCG have been shown to be well structured and amenable for solid-state NMR investigations (Figure 27.4).[88]

After addition of EGCG, the Aβ peptide precipitates rapidly out of solution (Figure 27.4A). Nevertheless, solution-state NMR experiments yield chemical shift changes of residues located in the hydrophobic aromatic core region of Aβ (residues 17–20) (Figure 27.4B). EGCG-induced Aβ oligomers as seen by AFM (Figure 27.4C) are relatively homogeneous. The aggregates which are formed in the presence of EGCG are not amorphous, but well structured and give rise to MAS solid-state NMR spectra with well-defined chemical shifts (Figure 27.4D). NCACX- and NCOCX-type experiments yield the assignments of residues 22–39 in the EGCG-induced Aβ oligomer. Residues 29–36 adopt a β-sheet conformation (Figure 27.4E). The aromatic hydrophobic core region, which seems to be the region where EGCG interacts initially with monomeric Aβ, is not observable in the solid state. Scalar coupling-based experiments do not yield further chemical shift information in addition to experiments that rely on dipolar interactions for magnetization transfer. The exact mode of interaction of EGCG with Aβ is, therefore, elusive and the exact structure of EGCG-induced Aβ oligomers remains to be seen.

In addition to EGCG, other plant polyphenolic compounds, such as curcumin[89,90] and resveratrol[91,92] (Figure 27.1), have been investigated by NMR. These compounds have been shown to have anti-oxidant properties and to be beneficial in cancer, inflammatory, cardiovascular and neurodegenerative diseases, among others.[93] In neurodegenerative diseases, polyphenols were recognized to modulate the amyloid fibrillization pathway.[94–98]

The interaction of curcumin with soluble monomeric α-synuclein and low molecular weight assemblies (LMW) of 100 kD was probed by ^{1}H–^{15}N HSQC experiments.[89] Monomeric α-synuclein seems inert to the presence of curcumin, whereas the resonances of the LMW αS species seemed affected upon addition of the compound. Similar to EGCG, curcumin redirects Aβ into off-pathway aggregates and is able to remodel preformed fibrils.[96,99] Other polyphenolic compounds such as myricetin, nordihydroguaiaretic acid (NDGA) and ferulic acid (Figure 27.1) seem to possess even higher inhibitory activities in comparison to curcumin. In the NMR study, myricetin appeared to be the strongest inhibitor, whereas rosmarinic acid did not interact with Aβ. ^{1}H and ^{15}N chemical shift changes were observed for similar Aβ residues, as in the case of EGCG.[88] However, interactions seem to be less specific, as a higher excess of the inhibitor is required to probe the interaction. These findings contradict reports that curcumin does not bind monomeric Aβ, but only protofibrils and fibrils.[96]

In addition, interactions between curcumin and Aβ42 fibrils were also analyzed by MAS solid-state NMR.[100] In the reported work, Aβ residues L17–A21 were ^{15}N and ^{13}C labeled. In addition, labels were incorporated at selected residues at the N-terminus. Curcumin was ^{13}C isotopically enriched at the aromatic rings. In the experiment, curcumin was titrated to preformed Aβ fibrils. In contrast to previous studies,[96] no remodeling of the fibril morphology was observed. To obtain distance restraints for the interaction, 2D-^{13}C,^{13}C DARR spectra were recorded. The labeled residues were found to be in close proximity to curcumin.

Curcumin and resveratrol were reported to have an impact on IAPP aggregation.[90,92] NMR spectra of IAPP as a function of concentration yield chemical shift changes for H18, S19 and Y37, indicating hot spots for IAPP oligomerization.[92] Resveratrol seems to interfere with the aggregation process by interacting with the side chains of residues K1 and H18. Similarly, addition of insulin results in amide chemical shift changes of the very same residues H18, S19 and Y37 in IAPP, suggesting that insulin and resveratrol might share the same binding motif and interact competitively with IAPP.[92]

27.5 Lacmoid

Lacmoid has been shown to inhibit Aβ and αS filament assembly.[52,98] Whereas binding of CR is restricted to the hydrophilic residues 1–60 of αS, lacmoid also seems to affect the hydrophobic core residues 60–100.[47,48] The C-terminus, like in the case of chlorazole black E, rosmarinic acid and phthalocyanine tetrasulfonate (PcTS-Cu^{2+}) (Figure 27.1), is largely unaffected. Lacmoid self-associates into colloidal species, similarly to CR. PFG-NMR measurements suggest that αS remains monomeric in the presence of lacmoid, and binds to small oligomeric lacmoid aggregates. Small oligomers of the compound seem to be the main effector for interactions with αS.[48]

Similarly, very small chemical shift changes are observed for Aβ upon titration of the lacmoid-like compound O4 (Figure 27.5C).[101,102] At higher concentrations of O4, a rapid decay of signal intensities is observed. At the same time, backbone amides experience differential chemical exchange broadening, yielding large fluctuations in cross-peak intensities (Figure 27.5B). As in the case of αS,[49] broadening and chemical shift changes seem to be correlated. Chemical exchange was quantified by CPMG relaxation dispersion experiments:[55] 1.5% of the total peptide population was found to populate the bound state. The non-toxic co-aggregates, which are formed upon lacmoid binding, were assumed to be unstructured according to CD data.[101] In contrast to the study by Abelein *et al.*, in which a random coil character of the off-pathway aggregates was observed, we find at least some degree of β-sheet propensity for the core of the lacmoid-induced Aβ aggregates (Figure 27.5D). At this point it is not clear if the additional amino group in the phenoxazine ring system of O4 causes the difference in the observation.

On the one hand, lacmoid seems to shift the equilibrium towards an insoluble fraction of Aβ (Figure 27.5A). On the other hand, the compound prevents further aggregation into Aβ fibrillar structures.[55] Apparently, recruitment of Aβ into a soluble O4 complex is kinetically favored in comparison to Aβ nucleation and fibril formation (Figure 27.5E). O4 colloid appears to provide a hydrophobic core which facilitates Aβ binding. The presence of a soluble, low molecular weight complex containing Aβ and O4 is supported by the observed chemical shifts changes in the solution-state NMR experiments. After formation of this soluble Aβ/O4 complex, aggregation into a non-soluble O4-containing Aβ precipitate proceeds rapidly. β-Sheet secondary structure is supported by MAS solid-state NMR experiments. Thereby, the amount of precipitate is dictated by the availability of O4.

27.6 Bifunctional Metal-Chelating Compounds

An increasing amount of data suggests a link between metal ions and the pathogenesis of numerous diseases caused by fibril-forming proteins. For instance, dyshomeostasis of metal ions[103,104] and their interactions with Aβ and the amyloid precursor protein (APP) have been reported in Alzheimer's disease (AD).[105–107] Binding of Cu^{2+} to Aβ, subsequent reduction to Cu^{+}, and production of H_2O_2 enhances oxidative stress, a hallmark of AD pathogenesis.[108,109] In the presence of various metal ions, in particular Cu^{2+} and Zn^{2+}, precipitation into amyloid aggregates is enhanced.[110–113] These aggregates have been shown to be amorphous and of non-fibrillar character.[110,114,115]

Solution-state NMR and EPR spectroscopy have shown that the imidazole rings of all histidines (His6, His13 and His14), as well as other N-terminal residues, are of particular importance for stabilizing the interaction of Aβ and metal ions.[116–123] Solid-state NMR studies of Aβ42 fibrillized in the presence of Zn^{2+} show that the D23–K28 salt-bridge is disrupted in the presence of Zn^{2+}.[124] Zn^{2+} binding causes major structural changes in the N-terminus of Aβ42. However, the overall cross-β structure appears to be maintained. Selective quenching observed in ^{13}C solid-state NMR of Cu^{2+} bound to preformed Aβ40 fibrils suggests that the primary Cu^{2+} binding sites in Aβ40 fibrils include Nε in His-13 and His-14, and carboxyl groups in Val-40, as well as the glutamate side chains (Glu3, Glu11 and/or Glu22). Similar to the Zn^{2+}-induced Aβ aggregates, Cu^{2+} binding to preformed Aβ amyloid fibrils shows by contrast that amyloid fibril structure is maintained.[125]

Structural insights of metal binding does not only contribute to the understanding of amyloid aggregate toxicity, but might also assist in drug development of metal-binding aggregation inhibitors.[126–128] To target metal ions associated with Aβ, bifunctional molecules have been designed that combine metal chelation and Aβ interaction within the same molecule.[129–131] Solution-state NMR experiments show that these molecules in fact interact with histidine residues in Aβ in the absence of metals.[128] Addition of metals results in a dissociation of amyloid fibrils and modulation of neurotoxicity. Given that the metal binding site and the hydrophobic aromatic region in Aβ are relatively far apart, the structure of a ternary complex of the bifunctional compound, the metal and the amyloid peptide remains as yet unknown.

27.7 Antibody Inhibitors

A very promising strategy for AD treatment involves immunotherapeutic approaches, which target the production, aggregation and clearance of Aβ.[132,133] Amyloid-targeting antibodies may have different modes of action involving dissolution as a consequence of direct binding to oligomeric/fibrillar Aβ in the brain, phagocytosis by microglial cells, and increased efflux of Aβ from the brain as a result of binding of monomeric Aβ in the plasma.

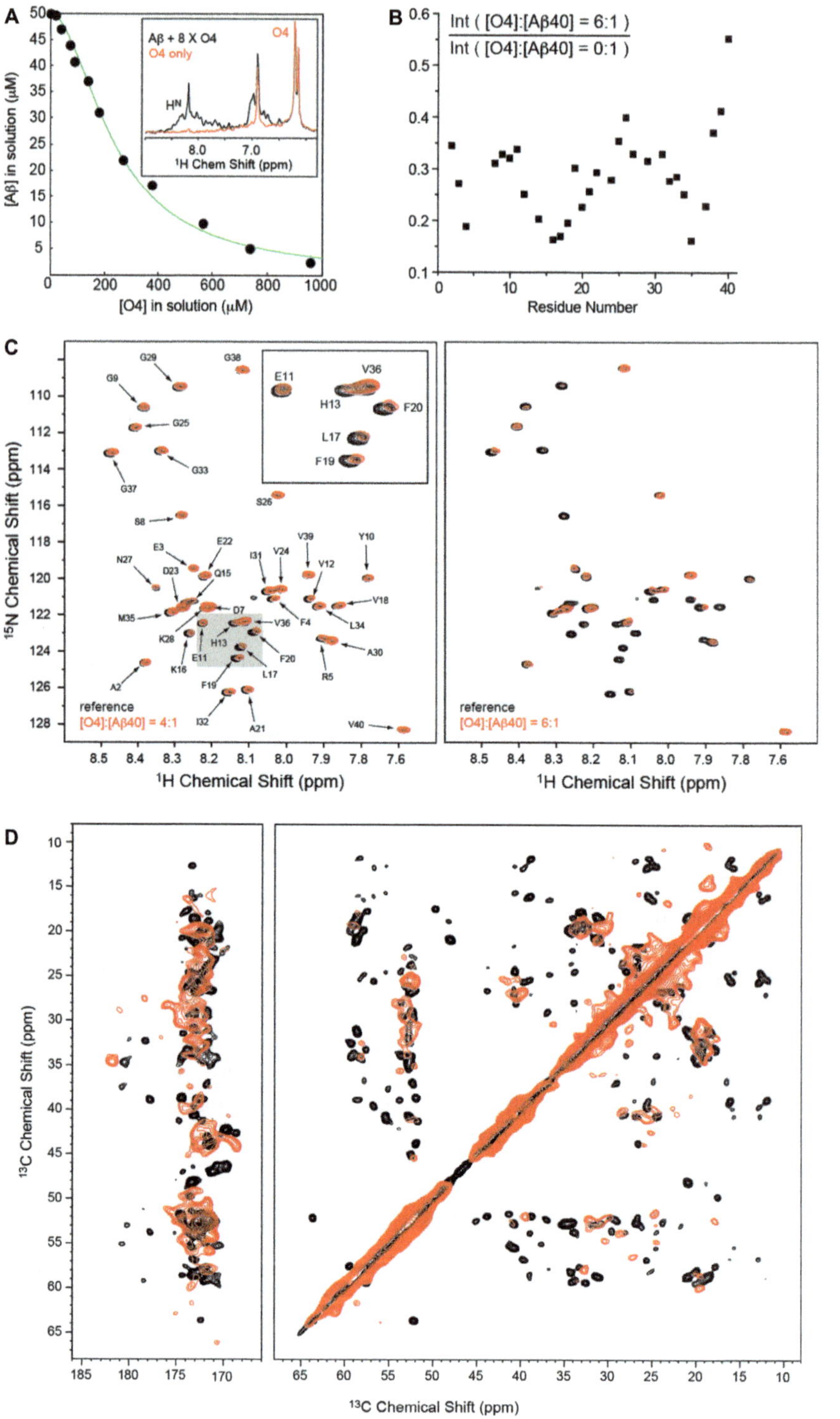
A
[Aβ] in solution (μM)
[O4] in solution (μM)
Aβ + 8 X O4
O4 only
O4
H^N
1H Chem Shift (ppm)
B
Int ([O4]:[Aβ40] = 6:1)
Int ([O4]:[Aβ40] = 0:1)
Residue Number
C
15N Chemical Shift (ppm)
1H Chemical Shift (ppm)
reference
[O4]:[Aβ40] = 4:1
[O4]:[Aβ40] = 6:1
E11
V36
H13
F20
L17
F19
D
13C Chemical Shift (ppm)

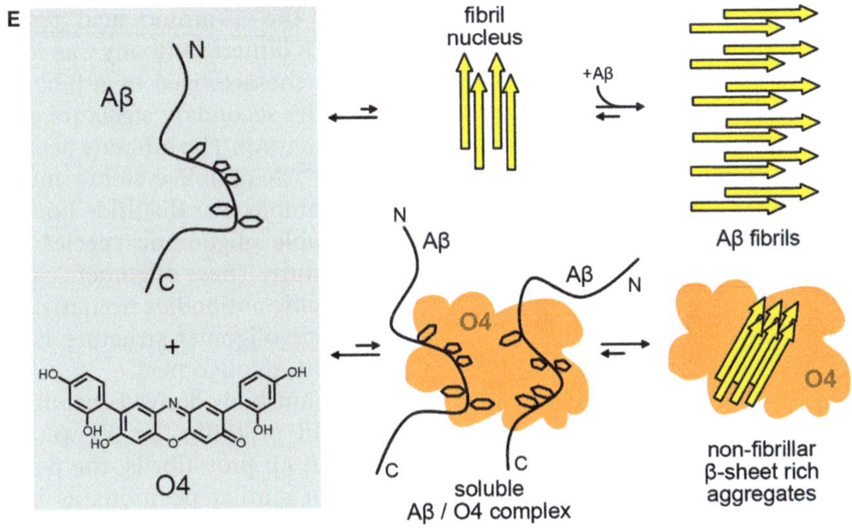

Figure 27.5 (A) Aβ40 solution-state peak intensities upon titration of the lacmoid O4. The *inset* shows representative 1D spectra in the absence of the compound and in the presence of an eight-fold molar excess. (B) Relative HSQC cross-peak intensities of Aβ40 in the presence and absence of a six-fold molar excess of O4. (C) $^{1}H,^{15}N$ correlation spectrum of Aβ40 in the presence (red) and absence (black) of a four-fold (*left*) and a six-fold (*right*) molar excess of O4.[102] (Reproduced with permission from Macmillan Science.) (D) MAS solid-state NMR 2D-$^{13}C,^{13}C$ correlation spectrum of Aβ40 fibrils (black),[14] and of O4-induced Aβ40 aggregates (red). The resolution of the O4-induced Aβ aggregates is much lower in comparison to EGCG-induced Aβ aggregates and Aβ fibrils. Secondary chemical shifts for lacmoid-induced Aβ aggregates are indicative for a certain degree of β-sheet structure. (E) Aggregation scheme of Aβ in the presence of the lacmoid molecule O4. O4 recruits Aβ molecules into a soluble colloidal state of O4 which stabilizes Aβ in a random-coil-like structure. The assumption of the population of a soluble, low molecular weight complex is in agreement with the observed chemical shift changes of Aβ in the presence of O4. The soluble state is in equilibrium with an insoluble state in which Aβ peptides are precipitated and contain a certain degree of β-sheet structure.

Interestingly, conformation-selective antibodies such as A11 can recognize oligomers of different aggregation prone peptides and proteins.[134] The mode of interaction and its intrinsic promiscuity so far is not understood. However, the approach holds the potential to structurally characterize an oligomeric intermediate state which is metastable and difficult to capture otherwise.

Subsequently, the structure of the affibody in complex with the monomeric substrate Aβ has been solved by solution-state NMR spectroscopy.[135,136] The affibody targeting the monomeric Aβ was obtained from a phage display library which is based on the 58-amino acid residue staphylococcal protein A-derived Z domain.[137] A dimeric affibody was found to bind to monomeric Aβ40. The structure is characterized by a β-hairpin comprising residues 17–36. The positions of the secondary structure elements strongly resemble those observed for fibrillar Aβ. The affibody acts as a stoichiometric inhibitor of Aβ fibrillization.[135] A double-cysteine mutant allows stabilization of the β-hairpin by an intramolecular disulfide bond.[136] The mutant peptide spontaneously forms stable oligomeric species with distinct molecular weights and secondary structure. These oligomers cannot convert into amyloid fibrils. Conformation-specific antibodies used to detect Aβ aggregates *in vivo* indicate that the wild-type oligomer structure is preserved and stabilized in double-cysteine mutant Aβ oligomers.

Protofibrillar Aβ can be stabilized with the antibody-derived fusion protein B10AP.[138] Complexes of the affibody and protofibrillar Aβ were investigated using MAS solid-state NMR.[139,140] In Aβ protofibrils, the β-sheet secondary structure elements are observed for similar positions as in Aβ fibrils. Tertiary contacts are found between residues Glu22 and Ile31. These contacts are not present in mature fibrils, and indicate that Aβ protofibrils must undergo a major structural reorientation in development into mature Aβ fibrils. It is speculated that the intramolecular hydrogen-bonded structure of the oligomer/protofibril is converted into a structure that is stabilized by intermolecular hydrogen bonds. A detailed structural picture, however, has so far not been proposed and how protofibrils are assembled remains to be seen.

Acknowledgements

This work was performed in the framework of SFB 1035 (German Research Foundation DFG, Sonderforschungsbereich 1035, Projekt B07). Furthermore, we acknowledge support from the Helmholtz-Gemeinschaft and the DFG (Re1435).

References

1. C. Ritter, M. L. Maddelein, A. B. Siemer, T. Luhrs, M. Ernst, B. H. Meier, S. J. Saupe and R. Riek, *Nature*, 2005, **435**, 844.
2. C. Wasmer, A. Lange, H. Van Melckebeke, A. B. Siemer, R. Riek and B. H. Meier, *Science*, 2008, **319**, 1523.
3. C. P. Jaroniec, C. E. MacPhee, V. S. Bajaj, M. T. McMahon, C. M. Dobson and R. G. Griffin, *Proc. Natl. Acad. Sci. U. S. A.*, 2004, **101**, 711.
4. A. W. Fitzpatrick, G. T. Debelouchina, M. J. Bayro, D. K. Clare, M. A. Caporini, V. S. Bajaj, C. P. Jaroniec, L. Wang, V. Ladizhansky, S. A. Muller, C. E. Macphee, C. A. Waudby, H. R. Mott, A. De Simone, T. P. Knowles, H. R. Saibil, M. Vendruscolo, E. V. Orlova, R. G. Griffin and C. M. Dobson, *Proc. Natl. Acad. Sci. U. S. A.*, 2013, **110**, 5468.

5. H. Heise, W. Hoyer, S. Becker, O. C. Andronesi, D. Riedel and M. Baldus, *Proc. Natl. Acad. Sci. U. S. A.*, 2005, **102**, 15871.
6. G. Comellas, L. R. Lemkau, A. J. Nieuwkoop, K. D. Kloepper, D. T. Ladror, R. Ebisu, W. S. Woods, A. S. Lipton, J. M. George and C. M. Rienstra, *J. Mol. Biol.*, 2011, **411**, 881.
7. G. Lv, A. Kumar, K. Giller, M. L. Orcellet, D. Riedel, C. O. Fernandez, S. Becker and A. Lange, *J. Mol. Biol.*, 2012, **420**, 99.
8. J. Gath, B. Habenstein, L. Bousset, R. Melki, B. H. Meier and A. Bockmann, *Biomol. NMR Assign.*, 2012, **6**, 51.
9. S. Luca, W. M. Yau, R. Leapman and R. Tycko, *Biochemistry*, 2007, **46**, 13505.
10. J. T. Nielsen, M. Bjerring, M. D. Jeppesen, R. O. Pedersen, J. M. Pedersen, K. L. Hein, T. Vosegaard, T. Skrydstrup, D. E. Otzen and N. C. Nielsen, *Angew. Chem. Int. Ed.*, 2009, **48**, 2118.
11. A. T. Petkova, W. M. Yau and R. Tycko, *Biochemistry*, 2006, **45**, 498.
12. I. Bertini, L. Gonnelli, C. Luchinat, J. Mao and A. Nesi, *J. Am. Chem. Soc.*, 2011, **133**, 16013.
13. A. K. Paravastu, R. D. Leapman, W. M. Yau and R. Tycko, *Proc. Natl. Acad. Sci. U. S. A.*, 2008, **105**, 18349.
14. J. M. Lopez del Amo, M. Schmidt, U. Fink, M. Dasari, M. Fandrich and B. Reif, *Angew. Chem. Int. Ed.*, 2012, **51**, 6136.
15. C. Sachse, M. Fandrich and N. Grigorieff, *Proc. Natl. Acad. Sci. U. S. A.*, 2008, **105**, 7462.
16. M. Schmidt, C. Sachse, W. Richter, C. Xu, M. Fandrich and N. Grigorieff, *Proc. Natl. Acad. Sci. U. S. A.*, 2009, **106**, 19813.
17. T. Luhrs, C. Ritter, M. Adrian, D. Riek-Loher, B. Bohrmann, H. Dobeli, D. Schubert and R. Riek, *Proc. Natl. Acad. Sci. U. S. A.*, 2005, **102**, 17342.
18. T. Sato, P. Kienlen-Campard, M. Ahmed, W. Liu, H. Li, J. I. Elliott, S. Aimoto, S. N. Constantinescu, J. N. Octave and S. O. Smith, *Biochemistry*, 2006, **45**, 5503.
19. S. Chimon, M. A. Shaibat, C. R. Jones, D. C. Calero, B. Aizezi and Y. Ishii, *Nat. Struct. Mol. Biol.*, 2007, **14**, 1157.
20. M. Ahmed, J. Davis, D. Aucoin, T. Sato, S. Ahuja, S. Aimoto, J. I. Elliott, W. E. Van Nostrand and S. O. Smith, *Nat. Struct. Mol. Biol.*, 2010, **17**, 561.
21. M. P. Lambert, A. K. Barlow, B. A. Chromy, C. Edwards, R. Freed, M. Liosatos, T. E. Morgan, I. Rozovsky, B. Trommer, K. L. Viola, P. Wals, C. Zhang, C. E. Finch, G. A. Krafft and W. L. Klein, *Proc. Natl. Acad. Sci. U. S. A.*, 1998, **95**, 6448.
22. D. M. Walsh, D. M. Hartley, Y. Kusumoto, Y. Fezoui, M. M. Condron, A. Lomakin, G. B. Benedek, D. J. Selkoe and D. B. Teplow, *J. Biol. Chem.*, 1999, **274**, 25945.
23. D. M. Hartley, D. M. Walsh, C. P. Ye, T. Diehl, S. Vasquez, P. M. Vassilev, D. B. Teplow and D. J. Selkoe, *J. Neurosci.*, 1999, **19**, 8876.
24. H. A. Lashuel, D. Hartley, B. M. Petre, T. Walz and P. T. Lansbury, Jr., *Nature*, 2002, **418**, 291.

25. P. Westermark, M. D. Benson, J. N. Buxbaum, A. S. Cohen, B. Frangione, S. Ikeda, C. L. Masters, G. Merlini, M. J. Saraiva and J. D. Sipe, *Amyloid*, 2007, **14**, 179.
26. P. Westermark, M. D. Benson, J. N. Buxbaum, A. S. Cohen, B. Frangione, S. Ikeda, C. L. Masters, G. Merlini, M. J. Saraiva and J. D. Sipe, *Amyloid*, 2005, **12**, 1.
27. M. Fandrich, *Cell. Mol. Life Sci.*, 2007, **64**, 2066.
28. H. Bennhold, *Münch. Med. Wochenschr.*, 1922, **69**, 1537.
29. P. S. Vassar and C. F. Culling, *Arch. Pathol.*, 1959, **68**, 487.
30. H. Naiki, K. Higuchi, K. Nakakuki and T. Takeda, *Lab. Invest.*, 1991, **65**, 104.
31. P. Frid, S. V. Anisimov and N. Popovic, *Brain Res. Rev.*, 2007, **53**, 135.
32. D. B. Carter and K. C. Chou, *Neurobiol. Aging*, 1998, **19**, 37.
33. C. Wu, Z. Wang, H. Lei, W. Zhang and Y. Duan, *J. Am. Chem. Soc.*, 2007, **129**, 1225.
34. C. Wu, Z. Wang, H. Lei, Y. Duan, M. T. Bowers and J. E. Shea, *J. Mol. Biol.*, 2008, **384**, 718.
35. M. R. Krebs, E. H. Bromley and A. M. Donald, *J. Struct. Biol.*, 2005, **149**, 30.
36. L. W. Jin, K. A. Claborn, M. Kurimoto, M. A. Geday, I. Maezawa, F. Sohraby, M. Estrada, W. Kaminksy and B. Kahr, *Proc. Natl. Acad. Sci. U. S. A.*, 2003, **100**, 15294.
37. A. K. Schutz, A. Soragni, S. Hornemann, A. Aguzzi, M. Ernst, A. Bockmann and B. H. Meier, *Angew. Chem. Int. Ed.*, 2011, **50**, 5956.
38. C. C. Rowe, U. Ackerman, W. Browne, R. Mulligan, K. L. Pike, G. O'Keefe, H. Tochon-Danguy, G. Chan, S. U. Berlangieri, G. Jones, K. L. Dickinson-Rowe, H. P. Kung, W. Zhang, M. P. Kung, D. Skovronsky, T. Dyrks, G. Holl, S. Krause, M. Friebe, L. Lehman, S. Lindemann, L. M. Dinkelborg, C. L. Masters and V. L. Villemagne, *Lancet Neurol.*, 2008, **7**, 129.
39. V. L. Villemagne, S. Burnham, P. Bourgeat, B. Brown, K. A. Ellis, O. Salvado, C. Szoeke, S. L. Macaulay, R. Martins, P. Maruff, D. Ames, C. C. Rowe and C. L. Masters, *Lancet Neurol.*, 2013, **12**, 357.
40. M. Groenning, *J. Chem. Biol.*, 2010, **3**, 1.
41. R. Khurana, C. Coleman, C. Ionescu-Zanetti, S. A. Carter, V. Krishna, R. K. Grover, R. Roy and S. Singh, *J. Struct. Biol.*, 2005, **151**, 229.
42. E. S. Voropai, M. P. Samtsov, K. N. Kaplevskii, A. A. Maskevich, V. I. Stepuro, O. I. Povarova, I. M. Kuznetsova, K. K. Turoverov, A. L. Fink and V. N. Uverskii, *J. Appl. Spectrosc.*, 2003, **70**, 868.
43. V. I. Stsiapura, A. A. Maskevich, V. A. Kuzmitsky, K. K. Turoverov and I. M. Kuznetsova, *J. Phys. Chem. A*, 2007, **111**, 4829.
44. V. I. Stsiapura, A. A. Maskevich, V. A. Kuzmitsky, V. N. Uversky, I. M. Kuznetsova and K. K. Turoverov, *J. Phys. Chem. B*, 2008, **112**, 15893.
45. M. Groenning, L. Olsen, M. van de Weert, J. M. Flink, S. Frokjaer and F. S. Jorgensen, *J. Struct. Biol.*, 2007, **158**, 358.

46. K. J. Robbins, G. Liu, V. Selmani and N. D. Lazo, *Langmuir*, 2012, **28**, 16490.
47. J. N. Rao, V. Dua and T. S. Ulmer, *Biochemistry*, 2008, **47**, 4651.
48. C. Lendel, C. W. Bertoncini, N. Cremades, C. A. Waudby, M. Vendruscolo, C. M. Dobson, D. Schenk, J. Christodoulou and G. Toth, *Biochemistry*, 2009, **48**, 8322.
49. A. S. Maltsev, A. Grishaev and A. Bax, *Biochemistry*, 2012, **51**, 631.
50. C. Lendel, B. Bolognesi, A. Wahlstrom, C. M. Dobson and A. Graslund, *Biochemistry*, 2010, **49**, 1358.
51. M. O. Pedersen, K. Mikkelsen, M. A. Behrens, J. S. Pedersen, J. J. Enghild, T. Skrydstrup, A. Malmendal and N. C. Nielsen, *J. Phys. Chem. B*, 2010, **114**, 16003.
52. M. Masuda, N. Suzuki, S. Taniguchi, T. Oikawa, T. Nonaka, T. Iwatsubo, S. Hisanaga, M. Goedert and M. Hasegawa, *Biochemistry*, 2006, **45**, 6085.
53. P. E. Fraser, J. T. Nguyen, D. T. Chin and D. A. Kirschner, *J. Neurochem.*, 1992, **59**, 1531.
54. B. Y. Feng, B. H. Toyama, H. Wille, D. W. Colby, S. R. Collins, B. C. May, S. B. Prusiner, J. Weissman and B. K. Shoichet, *Nat. Chem. Biol.*, 2008, **4**, 197.
55. A. Abelein, L. Lang, C. Lendel, A. Graslund and J. Danielsson, *FEBS Lett.*, 2012, **586**, 3991.
56. Y. S. Kim, T. W. Randolph, M. C. Manning, F. J. Stevens and J. F. Carpenter, *J. Biol. Chem.*, 2003, **278**, 10842.
57. I. Roterman, M. Krol, M. Nowak, L. Konieczny, J. Rybarska, B. Stopa, B. Piekarska and G. Zemanek, *Med. Sci. Monit.*, 2001, **7**, 771.
58. B. Stopa, B. Piekarska, L. Konieczny, J. Rybarska, P. Spolnik, G. Zemanek, I. Roterman and M. Krol, *Acta Biochim. Pol.*, 2003, **50**, 1213.
59. B. Piekarska, J. Rybarska, B. Stopa, G. Zemanek, M. Krol, I. Roterman and L. Konieczny, *Acta Biochim. Pol.*, 1999, **46**, 841.
60. M. Skowronek, B. Stopa, L. Konieczny, J. Rybarska, B. Piekarska, E. Szneler, G. Bakalarski and I. Roterman, *Biopolymers*, 1998, **46**, 267.
61. V. Rangachari, D. K. Reed, B. D. Moore and T. L. Rosenberry, *Biochemistry*, 2006, **45**, 8639.
62. D. J. Tew, S. P. Bottomley, D. P. Smith, G. D. Ciccotosto, J. Babon, M. G. Hinds, C. L. Masters, R. Cappai and K. J. Barnham, *Biophys. J.*, 2008, **94**, 2752.
63. A. Wahlstrom, L. Hugonin, A. Peralvarez-Marin, J. Jarvet and A. Graslund, *FEBS J.*, 2008, **275**, 5117.
64. W. E. Klunk, J. W. Pettegrew and D. J. Abraham, *J. Histochem. Cytochem.*, 1989, **37**, 1273.
65. G. Romhanyi, *Virchows Arch. A: Pathol. Pathol. Anat.*, 1971, **354**, 209.
66. J. H. Cooper, *Lab. Invest.*, 1974, **31**, 232.
67. H. Van Melckebeke, C. Wasmer, A. Lange, E. Ab, A. Loquet, A. Bockmann and B. H. Meier, *J. Am. Chem. Soc.*, 2010, **132**, 13765.

68. L. O. Tjernberg, J. Naslund, F. Lindqvist, J. Johansson, A. R. Karlstrom, J. Thyberg, L. Terenius and C. Nordstedt, *J. Biol. Chem.*, 1996, **271**, 8545.
69. C. Soto, E. M. Sigurdsson, L. Morelli, R. A. Kumar, E. M. Castano and B. Frangione, *Nat. Med.*, 1998, **4**, 822.
70. T. L. Lowe, A. Strzelec, L. L. Kiessling and R. M. Murphy, *Biochemistry*, 2001, **40**, 7882.
71. C. W. Cairo, A. Strzelec, R. M. Murphy and L. L. Kiessling, *Biochemistry*, 2002, **41**, 8620.
72. Z. Chen, G. Krause and B. Reif, *J. Mol. Biol.*, 2005, **354**, 760.
73. Z. Chen and B. Reif, *J. Biomol. NMR*, 2004, **29**, 525.
74. D. L. Worcester, *Proc. Natl. Acad. Sci. U. S. A.*, 1978, **75**, 5475.
75. H. Inouye, P. E. Fraser and D. A. Kirschner, *Biophys. J.*, 1993, **64**, 502.
76. S. Narayanan and B. Reif, *Biochemistry*, 2005, **44**, 1444.
77. B. O'Nuallain, S. Shivaprasad, I. Kheterpal and R. Wetzel, *Biochemistry*, 2005, **44**, 12709.
78. N. Carulla, G. L. Caddy, D. R. Hall, J. Zurdo, M. Gairi, M. Feliz, E. Giralt, C. V. Robinson and C. M. Dobson, *Nature*, 2005, **436**, 554.
79. N. L. Fawzi, J. Ying, R. Ghirlando, D. A. Torchia and G. M. Clore, *Nature*, 2011, **480**, 268.
80. S. A. Funke and D. Willbold, *Curr. Pharm. Des.*, 2012, **18**, 755.
81. L. M. Yan, A. Velkova, M. Tatarek-Nossol, E. Andreetto and A. Kapurniotu, *Angew. Chem. Int. Ed.*, 2007, **46**, 1246.
82. N. Rezaei-Ghaleh, E. Andreetto, L. M. Yan, A. Kapurniotu and M. Zweckstetter, *PLoS One*, 2011, **6**, e20289.
83. K. Rezai-Zadeh, G. W. Arendash, H. Hou, F. Fernandez, M. Jensen, M. Runfeldt, R. D. Shytle and J. Tan, *Brain Res.*, 2008, **1214**, 177.
84. J. Bieschke, J. Russ, R. P. Friedrich, D. E. Ehrnhoefer, H. Wobst, K. Neugebauer and E. E. Wanker, *Proc. Natl. Acad. Sci. U. S. A.*, 2010, **107**, 7710.
85. N. Ferreira, I. Cardoso, M. R. Domingues, R. Vitorino, M. Bastos, G. Bai, M. J. Saraiva and M. R. Almeida, *FEBS Lett.*, 2009, **583**, 3569.
86. D. E. Ehrnhoefer, M. Duennwald, P. Markovic, J. L. Wacker, S. Engemann, M. Roark, J. Legleiter, J. L. Marsh, L. M. Thompson, S. Lindquist, P. J. Muchowski and E. E. Wanker, *Hum. Mol. Genet.*, 2006, **15**, 2743.
87. D. E. Ehrnhoefer, J. Bieschke, A. Boeddrich, M. Herbst, L. Masino, R. Lurz, S. Engemann, A. Pastore and E. E. Wanker, *Nat. Struct. Mol. Biol.*, 2008, **15**, 558.
88. J. M. Lopez del Amo, U. Fink, M. Dasari, G. Grelle, E. E. Wanker, J. Bieschke and B. Reif, *J. Mol. Biol.*, 2012, **421**, 517.
89. P. K. Singh, V. Kotia, D. Ghosh, G. M. Mohite, A. Kumar and S. K. Maji, *ACS Chem. Neurosci.*, 2013, **4**, 393.
90. S. Sparks, G. Liu, K. J. Robbins and N. D. Lazo, *Biochem. Biophys. Res. Commun.*, 2012, **422**, 551.
91. J. F. Ge, J. P. Qiao, C. C. Qi, C. W. Wang and J. N. Zhou, *Neurochem. Int.*, 2012, **61**, 1192.

92. L. Wei, P. Jiang, W. Xu, H. Li, H. Zhang, L. Yan, M. B. Chan-Park, X. W. Liu, K. Tang, Y. Mu and K. Pervushin, *J. Biol. Chem.*, 2011, **286**, 6291.
93. H. Zhou, C. S. Beevers and S. Huang, *Curr. Drug Targets*, 2011, **12**, 332.
94. K. Ono, K. Hasegawa, H. Naiki and M. Yamada, *J. Neurosci. Res.*, 2004, **75**, 742.
95. H. Kim, B. S. Park, K. G. Lee, C. Y. Choi, S. S. Jang, Y. H. Kim and S. E. Lee, *J. Agric. Food Chem.*, 2005, **53**, 8537.
96. F. Yang, G. P. Lim, A. N. Begum, O. J. Ubeda, M. R. Simmons, S. S. Ambegaokar, P. P. Chen, R. Kayed, C. G. Glabe, S. A. Frautschy and G. M. Cole, *J. Biol. Chem.*, 2005, **280**, 5892.
97. K. Ono, M. M. Condron, L. Ho, J. Wang, W. Zhao, G. M. Pasinetti and D. B. Teplow, *J. Biol. Chem.*, 2008, **283**, 32176.
98. M. Necula, R. Kayed, S. Milton and C. G. Glabe, *J. Biol. Chem.*, 2007, **282**, 10311.
99. K. Ono, L. Li, Y. Takamura, Y. Yoshiike, L. Zhu, F. Han, X. Mao, T. Ikeda, J. Takasaki, H. Nishijo, A. Takashima, D. B. Teplow, M. G. Zagorski and M. Yamada, *J. Biol. Chem.*, 2012, **287**, 14631.
100. Y. Masuda, M. Fukuchi, T. Yatagawa, M. Tada, K. Takeda, K. Irie, K. Akagi, Y. Monobe, T. Imazawa and K. Takegoshi, *Bioorg. Med. Chem.*, 2011, **19**, 5967.
101. A. Abelein, B. Bolognesi, C. M. Dobson, A. Graslund and C. Lendel, *Biochemistry*, 2012, **51**, 126.
102. J. Bieschke, M. Herbst, T. Wiglenda, R. P. Friedrich, A. Boeddrich, F. Schiele, D. Kleckers, J. M. Lopez del Amo, B. A. Gruning, Q. Wang, M. R. Schmidt, R. Lurz, R. Anwyl, S. Schnoegl, M. Fandrich, R. F. Frank, B. Reif, S. Gunther, D. M. Walsh and E. E. Wanker, *Nat. Chem. Biol.*, 2012, **8**, 93.
103. M. A. Lovell, J. D. Robertson, W. J. Teesdale, J. L. Campbell and W. R. Markesbery, *J. Neurol. Sci.*, 1998, **158**, 47.
104. G. Danscher, K. B. Jensen, C. J. Frederickson, K. Kemp, A. Andreasen, S. Juhl, M. Stoltenberg and R. Ravid, *J. Neurosci. Methods*, 1997, **76**, 53.
105. L. Hesse, D. Beher, C. L. Masters and G. Multhaup, *FEBS Lett.*, 1994, **349**, 109.
106. G. Multhaup, A. Schlicksupp, L. Hesse, D. Beher, T. Ruppert, C. L. Masters and K. Beyreuther, *Science*, 1996, **271**, 1406.
107. G. Multhaup, *Biomed. Pharmacother.*, 1997, **51**, 105.
108. X. Huang, C. S. Atwood, M. A. Hartshorn, G. Multhaup, L. E. Goldstein, R. C. Scarpa, M. P. Cuajungco, D. N. Gray, J. Lim, R. D. Moir, R. E. Tanzi and A. I. Bush, *Biochemistry*, 1999, **38**, 7609.
109. M. P. Cuajungco, K. Y. Faget, X. Huang, R. E. Tanzi and A. I. Bush, *Ann. N. Y. Acad. Sci.*, 2000, **920**, 292.
110. F. Hane, G. Tran, S. J. Attwood and Z. Leonenko, *PLoS One*, 2013, **8**, e59005.

111. C. S. Atwood, R. D. Moir, X. Huang, R. C. Scarpa, N. M. Bacarra, D. M. Romano, M. A. Hartshorn, R. E. Tanzi and A. I. Bush, *J. Biol. Chem.*, 1998, **273**, 12817.
112. A. I. Bush, W. H. Pettingell, G. Multhaup, M. d Paradis, J. P. Vonsattel, J. F. Gusella, K. Beyreuther, C. L. Masters and R. E. Tanzi, *Science*, 1994, **265**, 1464.
113. A. I. Bush, W. H. Pettingell, Jr., M. D. Paradis and R. E. Tanzi, *J. Biol. Chem.*, 1994, **269**, 12152.
114. Y. Yoshiike, K. Tanemura, O. Murayama, T. Akagi, M. Murayama, S. Sato, X. Sun, N. Tanaka and A. Takashima, *J. Biol. Chem.*, 2001, **276**, 32293.
115. A. Parbhu, H. Lin, J. Thimm and R. Lal, *Peptides*, 2002, **23**, 1265.
116. N. G. Nair, G. Perry, M. A. Smith and V. P. Reddy, *J. Alzheimers Dis.*, 2010, **20**, 57.
117. S. Zirah, S. A. Kozin, A. K. Mazur, A. Blond, M. Cheminant, I. Segalas-Milazzo, P. Debey and S. Rebuffat, *J. Biol. Chem.*, 2006, **281**, 2151.
118. J. Danielsson, R. Pierattelli, L. Banci and A. Graslund, *FEBS J.*, 2007, **274**, 46.
119. E. Gaggelli, A. Janicka-Klos, E. Jankowska, H. Kozlowski, C. Migliorini, E. Molteni, D. Valensin, G. Valensin and E. Wieczerzak, *J. Phys. Chem. B*, 2008, **112**, 100.
120. P. Faller and C. Hureau, *Dalton Trans.*, 2009, 1080.
121. L. Hou and M. G. Zagorski, *J. Am. Chem. Soc.*, 2006, **128**, 9260.
122. N. Rezaei-Ghaleh, K. Giller, S. Becker and M. Zweckstetter, *Biophys. J.*, 2011, **101**, 1202.
123. C. Hureau, *Coord. Chem. Rev.*, 2012, **256**, 2164.
124. V. S. Mithu, B. Sarkar, D. Bhowmik, M. Chandrakesan, S. Maiti and P. K. Madhu, *Biophys. J.*, 2011, **101**, 2825.
125. S. Parthasarathy, F. Long, Y. Miller, Y. Xiao, D. McElheny, K. Thurber, B. Ma, R. Nussinov and Y. Ishii, *J. Am. Chem. Soc.*, 2011, **133**, 3390.
126. A. I. Bush, *J. Alzheimers Dis.*, 2008, **15**, 223.
127. K. J. Barnham, V. B. Kenche, G. D. Ciccotosto, D. P. Smith, D. J. Tew, X. Liu, K. Perez, G. A. Cranston, T. J. Johanssen, I. Volitakis, A. I. Bush, C. L. Masters, A. R. White, J. P. Smith, R. A. Cherny and R. Cappai, *Proc. Natl. Acad. Sci. U. S. A.*, 2008, **105**, 6813.
128. J. S. Choi, J. J. Braymer, R. P. Nanga, A. Ramamoorthy and M. H. Lim, *Proc. Natl. Acad. Sci. U. S. A.*, 2010, **107**, 21990.
129. A. I. Bush and R. E. Tanzi, *Neurotherapeutics*, 2008, **5**, 421.
130. S. S. Hindo, A. M. Mancino, J. J. Braymer, Y. Liu, S. Vivekanandan, A. Ramamoorthy and M. H. Lim, *J. Am. Chem. Soc.*, 2009, **131**, 16663.
131. J. J. Braymer, A. S. Detoma, J. S. Choi, K. S. Ko and M. H. Lim, *Int. J. Alzheimers Dis.*, 2010, **2011**, 623051.
132. F. Bard, C. Cannon, R. Barbour, R. L. Burke, D. Games, H. Grajeda, T. Guido, K. Hu, J. Huang, K. Johnson-Wood, K. Khan, D. Kholodenko, M. Lee, I. Lieberburg, R. Motter, M. Nguyen, F. Soriano, N. Vasquez,

K. Weiss, B. Welch, P. Seubert, D. Schenk and T. Yednock, *Nat. Med.*, 2000, **6**, 916.
133. M. Citron, *Nat. Rev. Drug Discovery*, 2010, **9**, 387.
134. R. Kayed, E. Head, J. L. Thompson, T. M. McIntire, S. C. Milton, C. W. Cotman and C. G. Glabe, *Science*, 2003, **300**, 486.
135. W. Hoyer, C. Gronwall, A. Jonsson, S. Stahl and T. Hard, *Proc. Natl. Acad. Sci. U. S. A.*, 2008, **105**, 5099.
136. A. Sandberg, L. M. Luheshi, S. Sollvander, T. Pereira de Barros, B. Macao, T. P. Knowles, H. Biverstal, C. Lendel, F. Ekholm-Petterson, A. Dubnovitsky, L. Lannfelt, C. M. Dobson and T. Hard, *Proc. Natl. Acad. Sci. U. S. A.*, 2010, **107**, 15595.
137. C. Gronwall, A. Jonsson, S. Lindstrom, E. Gunneriusson, S. Stahl and N. Herne, *J. Biotechnol.*, 2007, **128**, 162.
138. G. Habicht, C. Haupt, R. P. Friedrich, P. Hortschansky, C. Sachse, J. Meinhardt, K. Wieligmann, G. P. Gellermann, M. Brodhun, J. Gotz, K. J. Halbhuber, C. Rocken, U. Horn and M. Fandrich, *Proc. Natl. Acad. Sci. U. S. A.*, 2007, **104**, 19232.
139. H. A. Scheidt, I. Morgado, S. Rothemund, D. Huster and M. Fandrich, *Angew. Chem. Int. Ed.*, 2011, **50**, 2837.
140. H. A. Scheidt, I. Morgado and D. Huster, *J. Biol. Chem.*, 2012, **287**, 22822.
141. W. E. Klunk, M. L. Debnath and J. W. Pettegrew, *Neurobiol. Aging*, 1995, **16**, 541.

CHAPTER 28

Solid-State NMR Studies of β-Amyloid Fibrils and Related Assemblies

WEI QIANG AND ROBERT TYCKO*

Laboratory of Chemical Physics, National Institute of Diabetes and Digestive and Kidney Diseases, National Institutes of Health, Bethesda, MD 20892-0520, USA
*Email: robertty@mail.nih.gov

28.1 Introduction

As the most prevalent aging-related disease in the United States, Alzheimer's disease (AD) affects about 30% of the population over age 65.[1] One of the crucial diagnostic hallmarks of AD is the presence of β-amyloid (Aβ) plaques in the cerebral cortex.[2] These plaques contain fibrils formed by self-aggregation of Aβ peptides[3] with 39–43 amino acids, which are segments of the transmembrane amyloid precursor protein (APP) that are released by a sequence of proteolytic cleavages involving β- and γ-secretases.[4] The most common Aβ peptides are the 40-residue and 42-residue forms ($A\beta_{1-40}$ and $A\beta_{1-42}$), which occur in the human body in a ratio of approximately 4:1 and differ only by the presence or absence of two C-terminal residues. High-resolution structures of Aβ fibrils can provide important information for understanding the process of Aβ aggregation and fibril formation. In addition, structural information can provide a crucial experimental basis for the design of diagnostic imaging agents and inhibitors of fibril formation.[5–7]

New Developments in NMR No. 3
Advances in Biological Solid-State NMR: Proteins and Membrane-Active Peptides
Edited by Frances Separovic and Akira Naito

Published by the Royal Society of Chemistry, www.rsc.org

Aβ fibrils are inherently non-crystalline and insoluble in aqueous media. Therefore, direct molecular structural characterization of *bona fide* fibrils by the standard techniques of X-ray crystallography and liquid-state NMR spectroscopy is precluded. As a result, prior to the solid-state NMR studies described below, little was known about the molecular structures of Aβ fibrils or amyloid fibrils in general. Transmission electron microscope (TEM) images show that Aβ fibrils have widths in the 5–10 nm range and lengths that can exceed several microns. X-ray fiber diffraction data indicate that Aβ fibrils contain β-sheets, aligned in the fibrils so that the β-strands within the β-sheets run approximately perpendicular to the fibril growth axis and the hydrogen bonds between β-strands run approximately parallel to the fibril growth axis.[8,9] This organization of β-sheets, depicted in Figure 28.1, is called a "cross-β" structure.[10] Prior to solid-state NMR studies, it was unclear whether the cross-β structure is comprised of parallel β-sheets or antiparallel β-sheets, which segments of the amino acid sequence form the cross-β structure, and whether Aβ fibrils contain well-ordered molecular structures

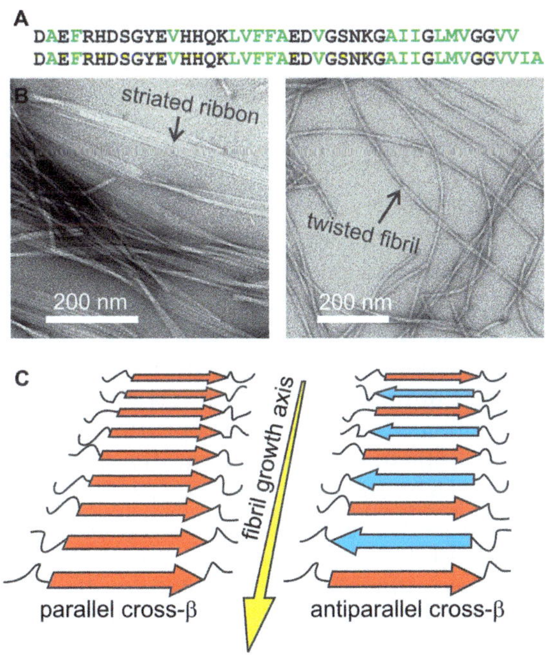

Figure 28.1 (A) Wild-type $A\beta_{1-40}$ and $A\beta_{1-42}$ sequences, with green letters indicating residues with hydrophobic sidechains. (B) TEM images of $A\beta_{1-40}$ fibrils, grown *in vitro* under conditions that produce either "striated ribbon" or "twisted" polymorphs. (C) Amyloid fibrils generally contain β-sheets with "cross-β" orientations relative to the fibril growth axis, which can be either parallel (*left*) or antiparallel (*right*) β-sheets. Solid-state NMR measurements have been used to determine the types of cross-β structures that actually exist in Aβ fibrils and to develop full molecular structural models.

with homogeneous peptide conformations and uniform intermolecular interactions. A variety of molecular structural models were proposed, none of which were constrained by substantial sets of experimental data, and none of which proved to be correct.[11–16]

Fortunately, Aβ fibrils are remarkably well-suited for a wide variety of solid-state NMR measurements. Assisted by ^{13}C and/or ^{15}N labeling, measurements of NMR chemical shifts, dipole–dipole couplings, and other parameters provide experimental restraints on specific structural features, *e.g.* site-specific peptide backbone conformations and long-range internuclear distances, from which complete molecular structural models can be developed with the assistance of computational methods. Applications of solid-state NMR methods to Aβ fibrils began in the mid-1990s. In the ensuing 20 years, these methods have made a number of significant contributions to our understanding of the structures of Aβ fibrils and related assemblies, and to our understanding of the aggregation process. In this review, we summarize results from solid-state NMR studies of a variety of Aβ species, including fibrils formed by peptide fragments, fibrils formed by full-length Aβ peptides, disease-associated mutants, protofibrils, and oligomers. We also briefly discuss studies that address possible neurotoxicity mechanisms of Aβ aggregates.

28.2 Fibrils Formed by β-Amyloid Fragments

Early solid-state NMR studies focused on short segments of Aβ peptides which contained regions that were considered important for fibril formation. In 1995, Griffin and co-workers[17] performed the first solid-state NMR studies of Aβ fibril structures, using the rotational resonance (R^2) technique[18] to estimate intramolecular and intermolecular distances in selectively ^{13}C-labeled samples of $A\beta_{34-42}$ fibrils (where $A\beta_{N-M}$ represents residues *N* through *M* of full-length Aβ). $A\beta_{34-42}$ was chosen because it represents the hydrophobic C-terminal segment of $A\beta_{1-42}$, which had been proposed to be a critically important segment that drives fibril formation.[19] In these experiments, the α-carbon (Cα) of residue *i* and the carbonyl carbon (C′) of residue $i+1$ were ^{13}C-labeled, allowing intramolecular ^{13}C–^{13}C distances that provide information about peptide backbone conformations to be measured; information about intermolecular distances was obtained from comparisons between R^2 data for fully labeled fibrils and data for fibrils in which labeled molecules were diluted in unlabeled molecules. Data from these measurements led to an antiparallel β-sheet structural model for $A\beta_{34-42}$ fibrils, consistent with a prevalent assumption (at the time of these studies) that amyloid fibrils generally contain antiparallel β-sheets.

Lynn, Meredith, Botto, and co-workers[20–22] subsequently examined fibrils formed by the longer peptide $A\beta_{10-35}$, chosen because it contains both N-terminal (residues 17–21) and C-terminal (residues 30–35) hydrophobic segments of full-length Aβ. The DRAWS solid-state NMR technique[23] was applied to a series of singly-$^{13}C'$-labeled $A\beta_{10-35}$ fibrils. Since the peptides were singly-labeled, ^{13}C–^{13}C dipole–dipole couplings could be interpreted unambiguously as measurements of intermolecular distances. Distances of approximately 5 Å were extracted from DRAWS data for all samples, irrespective of the labeled residue. This result was clearly inconsistent with an antiparallel β-sheet structure and provided the first evidence that amyloid fibrils could contain parallel β-sheets with "in-register" alignment of neighboring molecules. By now, experimental studies of amyloid fibrils formed by a wide variety of peptides and proteins have established that cross-β structures in amyloid fibrils are most often comprised of the in-register, parallel β-sheets that were first described by Lynn, Meredith, Botto, and co-workers.[24]

Soon thereafter, fibrils formed by $A\beta_{16-22}$ were studied by Balbach *et al.*[25] This peptide is particularly interesting because it is the basis for a number of inhibitors of fibril formation by full-length Aβ[5,26] and because its simplicity allows detailed computation studies.[27] A variety of solid-state NMR techniques were employed to investigate various aspects of the $A\beta_{16-22}$ fibril structure. First, two-dimensional MAS (2D-MAS) exchange[28] and constant-time double quantum-filtered dipolar recoupling (CTDQFD) experiments[29] allowed the backbone torsion angles of F19 to be determined quantitatively, confirming the formation of a β-sheet conformation centered at this residue. Second, 2D ^{13}C–^{13}C correlation spectroscopy was applied to a sample in which residues 16–21 were uniformly ^{15}N,^{13}C-labeled. This was the first example of 2D solid-state NMR spectroscopy of an amyloid fibril and the first example of measurements on a fibril sample containing uniformly-labeled residues. Relatively narrow ^{13}C NMR linewidths indicated a well-ordered structure in this fibril fragment, with chemical shifts for C′, C_{α}, and C_{β} that were fully consistent with a β-strand conformation. Finally, ^{15}N–^{13}C rotational-echo double-resonance (REDOR) experiments,[30] which detect long-range internuclear distance between selectively ^{15}N-labeled and selectively ^{13}C-labeled sites, provided direct evidence for an antiparallel β-sheet structure, with intermolecular hydrogen bonds between L17 and A21.

The registry of intermolecular hydrogen bonds in $A\beta_{16-22}$ fibrils can be symbolized by the expression "$17+k \leftrightarrow 21-k$", meaning that residue $17+k$ is hydrogen-bonded to residue $21-k$.Subsequent studies of the somewhat longer peptide $A\beta_{11-25}$ showed that $A\beta_{11-25}$ fibrils contain antiparallel β-sheets with $17+k \leftrightarrow 20-k$ hydrogen-bond registry when the fibrils are formed at pH 7.4, but $17+k \leftrightarrow 22-k$ registry when the fibrils are formed at pH 2.4.[31] Evidence for these antiparallel registries in $A\beta_{11-25}$ fibrils came from several solid-state NMR techniques, including ^{15}N–^{13}C REDOR data for selectively-labeled samples, ^{13}C–^{13}C dipolar recoupling data for selectively-labeled samples, obtained with the constant-time finite-pulse radio-frequency-driven recoupling (fpRFDR-CT) technique,[32] and proton-mediated 2D ^{13}C–^{13}C spectroscopy of samples with uniformly-labeled residues,[33] also known as 2D CHHC spectroscopy.[34] 2D CHHC spectra are particularly useful for characterizing antiparallel β-sheet structures because the short (approximately 2.1 Å) intermolecular distances between H_{α} sites of two residues that align with one another in antiparallel β-sheets lead to strong nonsequential, intermolecular $^{13}C_\alpha$–$^{13}C_\alpha$ crosspeaks between the two aligned residues.[31,33]

The dependence of hydrogen bond registries in $A\beta_{11-25}$ fibrils on pH illustrates the sensitivity of amyloid fibril structures to growth conditions. The fact that the hydrogen bond registry in $A\beta_{16-22}$ fibrils is not the same as in $A\beta_{11-25}$ fibrils, even when both are grown at pH 7.4, shows that fibril structures generally depend on the entire amino acid sequence, not on subsets thereof. The danger of drawing conclusions about an amyloid fibril structure from studies of peptide fragments is further highlighted by the fact that $A\beta_{10-35}$ fibrils contain in-register, parallel β-sheets, while $A\beta_{16-22}$ and $A\beta_{11-25}$ fibrils contain antiparallel β-sheets.

28.3 Full-Length, Wild-Type β-Amyloid Fibrils

The first solid-state NMR study of full-length Aβ fibrils was reported by Antzutkin *et al.* in 2000.[35] This work used a multiple-quantum (MQ) solid-state NMR technique[36,37] to measure the number of ^{13}C nuclei that were in close proximity within $A\beta_{1-40}$ fibrils that were prepared with ^{13}C labels at methyl sites of individual alanine residues. For samples that were labeled at either A21 or A30, up to four-quantum signals were detected with MQ excitation times of 14.4 ms, meaning at least four ^{13}C nuclei were sufficiently close in space within the fibril structure to have ^{13}C–^{13}C dipole–dipole couplings of approximately 70 Hz, corresponding to internuclear distances of approximately 5 Å. Quantitative analysis of MQ signal amplitudes demonstrated that the cross-β structure in $A\beta_{1-40}$ fibrils was composed of in-register parallel β-sheets, with the parallel intermolecular alignment extending over at least four molecules. The MQ NMR data were inconsistent with antiparallel structures and with structures in which intermolecular alignments alternated between parallel and antiparallel. Interestingly, high-order MQ signals were not detected in $A\beta_{1-40}$ fibrils in which the ^{13}C labels were placed on A2, showing that the N-terminal residues of $A\beta_{1-40}$ were not included in the β-sheets in these fibrils.

Conclusions from MQ solid-state NMR measurements were further supported by fpRFDR-CT data for a series of 11 samples with ^{13}C-labeling of single sites in residues from A2 to V39.[38] Decays of ^{13}C signals on the 30 ms timescale under the fpRFDR-CT pulse sequence, indicating intermolecular ^{13}C–^{13}C distances of approximately 5 Å, were observed for all labeled residues except A2 and F4. This result, together with ^{13}C linewidth measurements, indicated a disordered N-terminal segment and in-register parallel β-sheet structure for the rest of the $A\beta_{1-40}$ sequence in the fibrils.

An experimentally based structural model for $A\beta_{1-40}$ fibrils was proposed in 2002 by Petkova *et al.*[39] This initial model was developed from a number of NMR restraints: (1) ^{13}C and ^{15}N chemical shifts from 2D ^{13}C–^{13}C and ^{15}N–^{13}C correlation spectra of samples that were prepared with "scattered uniform labeling", meaning that subsets of the full amino acid sequence were uniformly $^{15}N,^{13}C$-labeled in each sample; (2) backbone torsion angles predicted by the TALOS program[40] using chemical shifts as input information; (3) backbone torsion angles determined from fpRFDR-CT, 2D-MAS exchange and double-quantum chemical shift anisotropy[41] (DQCSA) experiments on samples in which sequential pairs of backbone carbonyl sites were ^{13}C-labeled, especially in residues 23–30; (4) fpRFDR-CT data for singly-labeled samples described above; (5) ^{13}C NMR linewidths for backbone and sidechain carbon sites, which were 2.0 ppm or less for the most highly ordered segments and greater than 2.5 ppm for the disordered N-terminal segment in lyophilized Aβ_{1-40} fibrils; and (6) measurements of the ^{15}N–^{13}C dipole–dipole coupling between sidechain carboxylate and amino groups of D23 and K28 using frequency-selective REDOR,[42] which indicated a salt bridge interaction between these oppositely charged groups. In addition to the NMR data, the overall fibril width observed in TEM images and mass-per-length (MPL) values of individual fibrils determined from quantitative analysis of scanning transmission electron microscope (STEM) images were used as structural restraints. In this model, Aβ_{1-40} molecules adopt a U-shaped conformation, containing two β-strands, formed by residues 10–24 and 30–40, that are separated by a 180° bend in residues 25–29. The two β-strands form two separate parallel β-sheets that fold onto one another. A single layer of Aβ_{1-40} molecules, also called a single "cross-β unit", is then a double layered structure. According to the MPL data, the minimal fibril structure, also called the "protofilament", contains two cross-β units. Although experimental restraints on contacts among β-sheets were not available when this model was constructed, a possible four-layered β-sheet structure for an Aβ_{1-40} protofilament composed of two cross-β units was proposed, with contacts that maximized favorable hydrophobic interactions.

From solid-state NMR spectra and TEM images recorded in the course of the work described above, it became clear that Aβ_{1-40} fibrils are polymorphic. Fibrils with various distinctive appearances were observed in the TEM images;[43] NMR chemical shifts varied somewhat from sample to sample.[39] Eventually, Petkova *et al.* determined that both fibril morphologies and NMR chemical shifts could be controlled by careful control of the fibril growth conditions.[44] In particular, with all other conditions being identical, fibrils with "striated ribbon" morphologies (see Figure 28.1A) were formed when the Aβ_{1-40} solution was agitated during fibril growth, while fibrils with "twisted" morphologies (see Figure 28.1B) were formed when the solution was quiescent during fibril growth. Recent work by Qiang *et al.* shows that striated ribbon fibrils have slightly greater thermodynamic stability than twisted fibrils, and that the dependence on agitation arises from a greater susceptibility of striated ribbon fibrils to fragmentation in the presence of shear forces.[45]

Once morphologically homogeneous Aβ_{1-40} fibrils were available, detailed structural models for individual polymorphs were developed from additional solid-state NMR and electron microscopy measurements,[46,47] revealing molecular-level structural features that are common to different polymorphs as well as features that distinguish one polymorph from another. Figures 28.2A and 28.2B show structural models for fibrils grown under agitated and quiescent conditions, respectively. The structure in Figure 28.2A represents a single protofilament within a striated ribbon fibril.[44,46] The structure in Figure 28.2B represents a complete twisted fibril, with a specific period of twist about the fibril growth axis.[47] In both structures, $A\beta_{1-40}$ adopts the U-shaped conformation described above. 2D ^{13}C–^{13}C correlation spectra with long mixing periods indicated hydrophobic contacts between sidechains of L17 and F19 in the N-terminal β-strand, and sidechains of I32, L34, and V36 in the C-terminal β-strand. In both structures, the two β-strand segments form in-register parallel β-sheets. D23–K28 salt bridge interactions were found to exist in the striated ribbon fibrils, but not in the twisted fibrils. However, the most profound difference is in the overall symmetries of the two structures, with the protofilament in striated ribbon fibrils having two-fold symmetry with respect to rotation about the fibril growth axis and twisted fibrils having three-fold symmetry. Interactions between cross-β units are also different in the two structures, as exemplified by the presence or absence of contacts between sidechains of I31 and V39.

The two structures in Figure 28.2 are certainly not the only possible $A\beta_{1-40}$ fibril structures. Experiments by Wetzel and co-workers show that at least five distinct structures can develop, depending on growth conditions.[48]

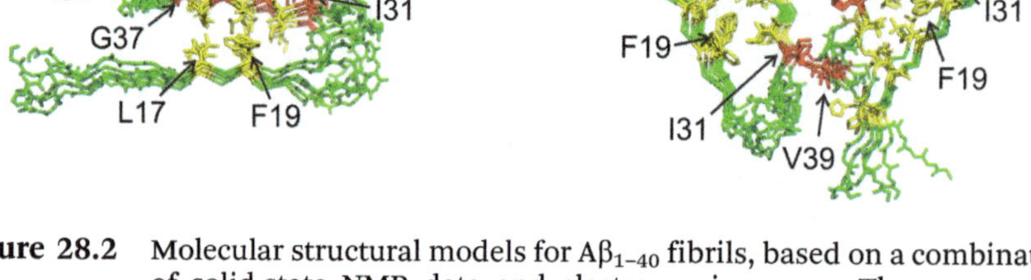

Figure 28.2 Molecular structural models for $A\beta_{1-40}$ fibrils, based on a combination of solid-state NMR data and electron microscopy. These are cross-sectional views, with the fibril growth axis perpendicular to the page. Six repeats of the (A) two-fold symmetric[46] or (B) three-fold symmetric[47] protofilament structure are shown, corresponding to the two $A\beta_{1-40}$ fibril polymorphs in the TEM images in Figure 28.1B. Each model includes residues 9–40, with residues 1–8 considered to be highly disordered. Residues that are involved in contacts within a single cross-β unit are colored yellow. Residues that are involved in contacts between cross-β units are colored red. Atomic coordinates for two-fold and three-fold symmetric protofilaments can be found in Protein Data Bank files 2LMN, 2LMO, 2LMP, and 2LMQ.

The total number of distinct $A\beta_{1-40}$ fibril structures is unknown, but is probably at least 10. Recently, a new structural model for a two-fold symmetric $A\beta_{1-40}$ fibril was reported by Bertini *et al.*[49] In this case, fibril growth conditions involved a higher pH value and more vigorous agitation. The model was generated from 2D and 3D solid-state NMR data for uniformly $^{15}N,^{13}C$-labeled $A\beta_{1-40}$ samples. Sequential assignments and specific long-range interactions were observed and utilized as structural restraints. Compared with the structure in Figure 28.2A, the structure of Bertini *et al.* exhibits differences in the details of sidechain–sidechain interactions between different β-sheet layers. While the first eight residues of $A\beta_{1-40}$ are considered to be disordered and are not shown in Figures 28.2A and 28.2B, residues 4–7 appear to form a β-strand in the fibrils prepared by Bertini *et al.* (which also include an extra N-terminal methionine due to their peptide expression system).

Given the highly polymorphic nature of Aβ fibrils, it is important to ask which fibril structures develop in human brain tissue, and whether all AD patients (as well as non-demented elderlies) develop the same fibril structures. Direct solid-state NMR measurements on Aβ fibrils from brain tissue are ruled out by the small quantities that can be extracted from the tissue and their lack of isotopic labeling. However, Paravastu *et al.* have shown that fibril structures in human brain tissue can be studied indirectly, using tissue samples from autopsies as the source of fibril "seeds" (*i.e.*, short fibril fragments, created by sonication of longer fibrils) from which isotopically labeled Aβ fibrils can be grown.[50]

In vitro studies confirm that molecular structures of Aβ fibrils can be propagated and amplified by seeded growth, provided that the quantity of seeds is sufficiently high and that the seeds themselves are structurally homogeneous.[46] When seeds are derived from polymorphic fibrils, multiple generations of seeded fibril growth can be used to purify a single structure.[50,51]

Backbone dynamics in $A\beta_{1-40}$ fibrils have also been studied by solid-state NMR, using constant time dipolar coupling and chemical shift (DIPSHIFT) experiments.[52,53] $^{13}C_{\alpha}$–^{1}H order parameters determined from the DIPSHIFT data indicate that residues 1–4 and residue 40 have greater amplitudes of motion than the rest of the sequence.[54] Interestingly, no enhanced mobility was detected in residues 23–29, consistent with structural order in the bend segment that connects N-terminal and C-terminal β-strands.

In vitro, $A\beta_{1-42}$ has been shown to aggregate more rapidly and at lower concentrations than $A\beta_{1-40}$.[55] In the human brain, antibody binding studies suggest that deposition of $A\beta_{1-42}$ in senile plaques may precede deposition of $A\beta_{1-40}$.[56] Much attention has, therefore, been directed toward $A\beta_{1-42}$ aggregates as potentially important neurotoxic species in AD. Early solid-state NMR measurements on $A\beta_{1-42}$ fibrils by Antzutkin *et al.* indicated that $A\beta_{1-42}$ fibrils and $A\beta_{1-40}$ fibrils have qualitatively similar molecular structures.[57] In particular, MPL measurements were consistent with two-fold symmetry about the fibril growth axis; fpRFDR-CT measurements on selectively

^{13}C-labeled samples indicated an in-register parallel β-sheet structure; ^{15}N–^{13}C REDOR data indicated that L34 and V40 were not in close proximity within the β-sheets, showing that the antiparallel structure identified in $A\beta_{34-42}$ fibrils by Lansbury *et al.* did not exist in $A\beta_{1-42}$ fibrils. More recently, Smith and co-workers used solid-state NMR to identify long-range inter-residue contacts for the residue pairs F19/Leu34, Q15/G37, and H13/Q15 in $A\beta_{1-42}$ fibrils formed *in vitro.*[58] These contacts indicate that the high-resolution structures of $A\beta_{1-40}$ and $A\beta_{1-42}$ fibrils can be highly similar in terms of sidechain packing patterns in the hydrophobic core and the extent of β-strands.

28.4 Fibrils Formed by Disease-Associated Mutants

Certain mutations in the amino acid sequence of Aβ are associated with early-onset familial AD. Compared with late-onset, sporadic AD, these early-onset cases typically show severe cerebral amyloid angiopathy (CAA), *i.e.* amyloid deposits within blood vessels in the brain.[59] Structural variations in Aβ fibrils induced by mutations may correlate with the differences in clinical history and pathology. Structural features of several such mutants, including the Italian mutant (E22K-$A\beta_{1-40}$), Japanese mutant (E22Δ-$A\beta_{1-40}$), and Iowa mutant (D23N-$A\beta_{1-40}$), have been investigated using solid-state NMR spectroscopy.

E22K-$A\beta_{1-40}$ aggregates more rapidly than the wild-type peptide.[60] Solid-state NMR measurements by Takegoshi *et al.* suggested that this mutation may introduce partial disruption of the N-terminal β-strand before D23, as two sets of chemical shifts for D23 were observed in 2D dipolar-assisted rotational resonance[61] (DARR) experiments, consistent with both β-strand and non-β-strand conformations. A later study by the same group, using R^2 experiments, confirmed that the mutant fibrils contained parallel in-register β-sheet structures.[62] Based on these solid-state NMR results, these authors proposed that the E22K mutation shifted the position of a non-β-strand backbone conformation from residues 25–26 to residues 22–23, although the overall β-sheet structure was not affected.

Initial studies of E22Δ-$A\beta_{1-40}$ (where E22Δ indicates deletion of E22 without replacement by another residue) suggested that this mutant might be incapable of forming fibrils, and that neurodegeneration associated with this mutant might be due to nonfibrillar aggregates.[63] Later studies showed that E22Δ-$A\beta_{1-40}$ forms fibrils instantaneously *in vitro.*[64] Measurements of intermolecular ^{13}C–^{13}C dipole–dipole couplings in selectively ^{13}C-labeled E22Δ-$A\beta_{1-40}$ fibrils, using the PITHIRDS-CT recoupling technique,[65] indicated that the labeled residues (V18 and V36) were not located in in-register parallel β-sheets. Other aspects of the E22Δ-$A\beta_{1-40}$ fibril structure have not yet been examined by solid-state NMR.

The most comprehensive solid-state NMR study of a disease-associated Aβ mutant was conducted on D23N-$A\beta_{1-40}$, a mutant that was identified in a family from the state of Iowa with a history of early-onset neurodegeneration. These patients develop both neuritic plaques and massive vascular deposition of the mutant peptides. TEM images of D23N-Aβ_{1-40} fibrils formed *in vitro* at 37°C showed polymorphism, with both twisted and non-twisted fibril morphologies. Consistent with the TEM images, solid-state NMR spectra of a selectively-labeled sample showed three resolved ^{13}C chemical shifts for each labeled site.[66] ^{13}C PITHIRDS-CT measurements revealed that only one of the three ^{13}C chemical shifts corresponded to the in-register parallel β-sheet structure observed in all previous studies of full-length, wild-type Aβ fibrils, suggesting that different D23N-Aβ_{1-40} fibril polymorphs could have qualitatively different cross-β structures. REDOR measurements on a series of selectively-labeled samples indicated intermolecular ^{15}N–^{13}C distances consistent with an antiparallel β-sheet structure in residues 17–21 for certain polymorphs, with $17+k \leftrightarrow 21-k$ registry.

Development of a high-resolution structural model for antiparallel D23N-Aβ_{1-40} fibrils required preparation of structurally homogeneous samples. For this purpose, Qiang *et al.* developed a seeding/filtration protocol that takes advantage of differences in fibril extension kinetics between parallel and antiparallel β-sheet structures.[67] In this protocol, depicted in Figure 28.3A, the heterogeneous mixture of fibrils structures prepared initially from monomeric peptide was first sonicated to produce short fibril fragments, *i.e.* fibril seeds. After addition of these seeds to a fresh solution of D23N-Aβ_{1-40} monomers, the seeds grew in length, but with more rapid extension of the parallel structures. Antiparallel structures could then be separated from parallel structures by microfiltration. Two repetitions of the seeding/filtration protocol produced sufficiently homogeneous samples.

Several types of structural restraints for antiparallel D23N-Aβ_{1-40} fibrils were obtained from solid-state NMR measurements, leading to the structural model shown in Figure 28.3B.[67] Experimental restraints included: (1) ^{13}C chemical shifts from 2D ^{13}C–^{13}C correlation spectroscopy with fpRFDR mixing; (2) intra- and intermolecular sidechain interactions from long-range crosspeaks in 2D radiofrequency-assisted diffusion (RAD) spectra[61,68] with 500 ms mixing periods; (3) antiparallel β-sheet hydrogen bond registries from 2D CHHC experiments; and (4) intermolecular C′–C′, C_β–C_β, N–C′, and N–C_β distances from ^{13}C PITHIRDS-CT and ^{15}N–^{13}C REDOR measurements. Together, these data show that the U-shaped molecular conformation of wild-type Aβ fibrils is retained in antiparallel D23N-Aβ_{1-40} fibrils, but that the location of the 180° bend in residues 24–29 alternates from one side of the fibril to the other along the length of the fibril, producing an antiparallel $17+k \leftrightarrow 21-k$ registry for β-sheets formed by residues 15–23 and an antiparallel $31+k \leftrightarrow 35-k$ registry for β-sheets formed by residues 30–36. Importantly, sidechains of L17, F19, A21, A30, I32, L34, and V36 form a hydrophobic core within the cross-β unit, just as in the wild-type Aβ_{1-40} structures shown in Figure 28.2. The similarity of molecular conformations and hydrophobic core compositions explains why a single peptide sequence can form both parallel and antiparallel cross-β structures.

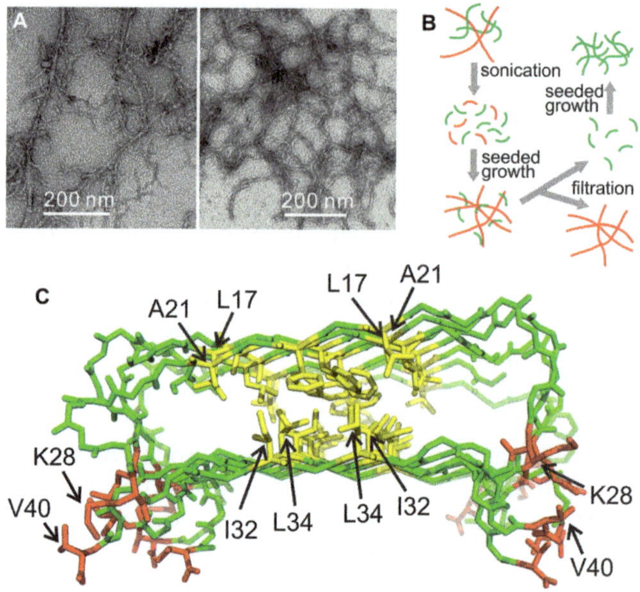

Figure 28.3 Preparation and structural modeling of antiparallel D23N-Aβ_{1-40} fibrils. (A) Negatively stained TEM images of fibrils prepared by unseeded growth from monomeric D23N-Aβ_{1-40} (*left*) and by two repetitions of the seeding/filtration protocol (*right*). This protocol results in homogeneous samples of relatively short, curved fibrils, which contain antiparallel β-sheets, resemble "protofibrils", and are metastable with respect to conversion to longer, straight fibrils that contain parallel β-sheets. (B) Schematic illustration of the seeding/filtration protocol.[67] A mixture of parallel (red) and antiparallel (green) fibrils, prepared by unseeded growth from monomeric D23N-Aβ_{1-40}, is broken into short fragments by sonication. During incubation with additional monomer for 3 h at 6°C, parallel fibrils grow more rapidly than antiparallel fibrils. The two types of structures are then separated by microfiltration. Finally, antiparallel fibrils are amplified by further incubation with monomer for 24 h. (C) Structural model for residues 15–40 in antiparallel D23N-Aβ_{1-40} fibrils, viewed in cross-section with the fibril growth axis perpendicular to the page. Residues involved in intramolecular and intermolecular interactions detected in 2D ^{13}C–^{13}C RAD spectra are shown in yellow and red, respectively. Additional structural restraints from solid-state NMR and electron microscopy are described by Qiang *et al.*[67] Atomic coordinates can be found in Protein Data Bank file 2LNQ.

According to both PITHIRDS-CT data and thioflavin T fluorescence, mixtures of parallel and antiparallel D23N-Aβ_{1-40} fibrils gradually convert to purely parallel fibrils.[67] Thus, antiparallel fibrils are an off-pathway, metastable (but quite long-lived) intermediate in the fibril formation process. Consistent with lower thermodynamic stability, ^{13}C NMR lines in spectra of pure antiparallel fibrils are broader than in spectra of morphologically pure parallel fibrils.[51] The greater linewidths appear to result from greater internal disorder, apparently including conformational disorder in residues 1–14, 37–40, and possibly 24–29. Additionally, MPL values from electron microscopy indicate that antiparallel D23N-Aβ_{1-40} fibrils do not form the two-fold or three-fold symmetric structures observed in fibrils that contain parallel β-sheets. Thus, ordered sidechain interactions for I31 and M35 are not present.

It is not yet know whether disease-associated mutants other than D23N-Aβ_{1-40} can form fibrils with antiparallel β-sheet structures similar to the structure in Figure 28.3. As discussed above, preliminary evidence exists that E22Δ-Aβ_{1-40} forms fibrils that do not have in-register parallel β-sheet structures. Antiparallel D23N-Aβ_{1-40} fibrils prepared by the seeding/filtration protocol described above are relatively short and more highly curved than parallel fibrils, resembling intermediate structures that are frequently called "protofibrils" in the amyloid literature.[69,70] The possibility exists that wild-type Aβ protofibrils may also contain antiparallel structures, as discussed further below.

28.5 β-Amyloid Oligomers and Protofibrils

Structural studies on Aβ oligomers and protofibrils have become an active area for research, primarily because these Aβ species may possess a higher level of neurotoxicity compared with monomeric Aβ peptides or mature fibrils.[71–73] These species represent either on-pathway or off-pathway intermediate structures in the process of Aβ fibrillation. By definition, on-pathway intermediates can convert to mature fibrils through internal conformational transitions, without involving the dissociation and re-association of monomers. In contrast, off-pathway intermediates may precede or coexist with mature fibrils and may also slowly convert to mature fibrils with higher thermodynamic stabilities. However, monomers have to dissociate from off-pathway intermediates and associate with the more stable mature fibrils for this conversion to occur.

Solid-state NMR studies of oligomers and protofibrils are challenging for a variety of reasons, including the facts that these intermediates are generally prepared at low concentration and low yield, are generally not morphologically homogeneous, are inherently unstable, and may be difficult to separate from mature fibrils. Three approaches have been utilized to prepare samples of Aβ oligomers and protofibrils. First, a mixture of mature fibrils, protofibrils/oligomers, and monomers may be separated, making use of the fact that these species have distinct densities and sizes. Usually, gentle centrifugation can be used first to remove the more massive fibrils, followed by ultracentrifugation to pellet the oligomers or protofibrils and separate them from monomers.[74] Second, small oligomers may be separated using size exclusion chromatography, although the experimental conditions must be chosen carefully to prevent the dissociation of oligomers during the chromatographic separation or subsequent steps.[75] Third, oligomers/protofibrils can be precipitated using conformationally specific antibodies.[76]

The first solid-state NMR studies of Aβ oligomers were reported by Ishii and co-workers.[77,78] These authors observed that relatively stable, spherical oligomers formed along the pathway of $A\beta_{1-40}$ fibrillation, particularly at low temperature (*e.g.* 4 °C). Oligomers were isolated by ultracentrifugation after an appropriate incubation time and were lyophilized to prevent further structural evolution. Surprisingly, ^{13}C chemical shifts observed in 2D solid-state NMR spectra of samples containing uniformly ^{15}N,^{13}C-labeled residues indicate that the $A\beta_{1-40}$ conformation in these spherical oligomers is nearly the same as in mature fibrils formed under the same experimental conditions. Moreover, fpRFDR-CT data for selectively ^{13}C-labeled samples indicate parallel intermolecular alignment within the oligomers.[77] Interestingly, spherical oligomers and mature fibrils were found to have distinct toxicities to cultured PC12 cells, suggesting that the molecular basis of neurotoxicity lies beyond the level of secondary structure.

Scheidt *et al.* have performed solid-state NMR measurements on $A\beta_{1-40}$ protofibrils, captured with the conformationally specific antibody B10AP.[79] TEM measurements confirmed the curved, protofibrillar morphology of $A\beta_{1-40}$ aggregates in the presence of B10AP. Antibody binding prevented structural evolution toward the mature fibrillar state. ^{13}C chemical shifts of most residues along the $A\beta_{1-40}$ sequence were determined. These data indicate the presence of β-strand segments quite similar to those in spherical oligomers and mature fibrils. Moreover, F19-L34 contacts were identified, as in the $A\beta_{1-40}$ and D23N-$A\beta_{1-40}$ fibrils discussed above.

Based on the foregoing observations, β-strand formation appears to occur at an early stage in the Aβ aggregation process. What then is the key structural transition between intermediates and mature fibrils? One possibility, suggested by results for D23N-$A\beta_{1-40}$ discussed above, is that the key transition is from an antiparallel to a parallel β-sheet structure. An alternative proposal, supported by recent crystallographic studies of oligomers formed by another amyloidogenic peptide,[80] is that certain intermediates may contain true β-hairpin conformations, which are distinct from the U-shaped conformations discussed above in that β-hairpins contain intramolecular hydrogen bonds between N- and C-terminal β-strands. The key structural transition would then involve 90° rotations of both β-strands, disruption of intramolecular hydrogen bonds, and formation of intermolecular hydrogen bonds.

Existence of a β-hairpin conformation for an Aβ monomer was first demonstrated through solution NMR studies of $A\beta_{1-40}$ when bound to an affibody protein.[81] Evidence that stabilization of a β-hairpin conformation by intramolecular disulfide crosslinking leads to stable protofibril formation was obtained subsequently.[82] Further solid-state NMR experiments by Scheidt *et al.* showed that the structural transition from protofibrils to mature fibrils includes a change in inter-residue distances between E22 and I31.[76] Evidence for this structural change came from 2D proton-driven spin diffusion (PDSD) experiments, which showed E22/I31 crosspeaks in the B10AP-stabilized $A\beta_{1-40}$ protofibrils, but not in mature $A\beta_{1-40}$ fibrils.

In addition, a recent solid-state NMR study on $A\beta_{1-40}$ protofibrils, using the R^2 technique to probe intermolecular $^{13}C-^{13}C$ dipole–dipole couplings between ^{13}C labels at carbonyl and methyl positions of A21 in samples prepared from a 1:1 mixture of molecules with labels at the two sites, found no detectable couplings, suggesting that these protofibrils did not contain the in-register parallel β-sheets found in mature fibrils.[83]

In summary, recent solid-state NMR studies on wild-type Aβ oligomers and protofibrils have begun to provide crucial structural information. Existing evidence indicates that both intermediate species have molecular conformation similar to those in mature fibrils, where residues 17–21 and 30–36 are included in β-strands. However, the supramolecular structures may be distinct. β-Hairpin conformations, with intramolecular hydrogen bonds between the two β-strand segments, may be present in some oligomers or protofibrils, but more extensive structural characterizations need to be done to verify this hypothesis. The structural model for antiparallel D23N-$A\beta_{1-40}$ fibrils in Figure 28.3 is the only complete model for a metastable Aβ aggregate that has been developed to date.

28.6 Interactions with Membranes and Inhibitors

Aβ oligomers, protofibrils, and mature fibrils are toxic to cultured neurons. For example, addition of sonicated wild-type $A\beta_{1-40}$ fibrils, parallel D23N-$A\beta_{1-40}$ fibrils, or antiparallel D23N-$A\beta_{1-40}$ fibrils to cultures of primary rat embryonic hippocampal neurons leads to the death of 50% or more of the neurons within 48 h at Aβ concentrations around 10 μM, under conditions where about 10% of the neurons die in the absence of Aβ.[44,67] One of the important motivations for structural characterizations of Aβ assemblies is to understand the cellular disruption mechanism at the molecular level. Interaction between Aβ and cell membranes during the fibrillation process has been considered as a possible mechanism.[84] *In vivo*, Aβ is produced by enzymatic cleavage of APP, which is an integral membrane protein.[4] Therefore, Aβ is naturally linked to cell membranes. Experimentally, it has been observed that growth of amyloid fibrils in the presence of synthetic lipid vesicles leads to significant vesicle leakage.[85]

Solid-state NMR spectroscopy has been used to obtain molecular-level information on Aβ/membrane interactions. Work by Ravault *et al.* provided an early example of a solid-state NMR study on membrane-bound Aβ.[86] This work focused on the C-terminal fragment $A\beta_{29-42}$ because this segment is important for fibril formation and was identified as the minimum segment of Aβ with fusogenic properties. ^{13}C chemical shift measurements showed that membrane-bound $A\beta_{29-42}$ adopted a β-strand conformation. Binding of $A\beta_{29-42}$ to lipid vesicles was detected using static ^2H NMR spectroscopy on a peptide with a single deuterated alanine methyl group. The results showed that the deuterium quadrupole splitting changed as the temperature varied across the phase transition temperature of the lipid bilayer, implying that the local dynamics of the labeled residue were affected by a phase change in the membrane bilayer. T_2 relaxation measurements indicated that, for lipid vesicles composed of phosphatidylcholine (PC) and phosphatidylethanolamine (PE), the $A\beta_{29-42}$ peptide was more likely to be associated with PE rather than PC headgroups.

Effects of full-length $A\beta_{1-42}$ on the integrity of synthetic lipid bilayers were studied by Separovic and co-workers using solid-state ^{2}H and ^{31}P NMR spectroscopy.[87] Two distinct phenomena were observed for the case of exogenous addition of $A\beta_{1-42}$ into pre-formed bilayers and the incorporation of $A\beta_{1-42}$ during the bilayer formation. For the former case, the addition of $A\beta_{1-42}$ only induced minor changes in the ^{31}P chemical shift anisotropy and ^{2}H quadrupole splitting, implying that little disruption of the membrane occurred. On the other hand, when $A\beta_{1-42}$ was incorporated into membranes during the sample preparation process, a large enhancement of isotropic signals was observed in both ^{31}P and ^{2}H NMR spectra, indicating more significant disruption of the membrane. Both *in vitro* sample preparation conditions may be biologically relevant, as the C-terminal hydrophobic segments of full-length Aβ peptides represent part of the transmembrane segment of APP. Therefore, elevated production of Aβ may induce destabilization of membrane bilayers, or formation of so-called amyloid ion channels.[88]

Another possible molecular mechanism of Aβ-induced neurotoxicity is related to the interaction between Aβ and metal ions such as Cu^{2+} and Zn^{2+}. These metal ions are enriched in amyloid plaques and such interactions was reported to affect the aggregation state of Aβ.[89,90] In addition, the metal–Aβ complex is redox-active and has been considered to relate to the generation of reactive oxygen species (ROS). Static solid-state ^{31}P NMR experiments on membrane-associated, Cu^{2+}-bound $A\beta_{1-42}$ showed that the Cu^{2+}-bound peptide, which exhibited elevated toxicity in neuronal cell cultures, was associated with the membrane surface.[91] ^{13}C NMR spectra of membrane-bound Aβ with selective isotopic labeling have shown that membrane binding causes changes in chemical shifts that indicate changes from random-coil to more extended conformations at certain sites.[92] In addition, the introduction of Cu^{2+} induced significant broadening for the NMR signal at G29 and A30, but not at L34. This experiment revealed potential binding sites of Cu^{2+} to the membrane-associated Aβ peptides.[93]

The molecular interaction between Cu^{2+} and $A\beta_{1-40}$ fibrils has been studied more recently by Ishii and co-workers.[94] Results from this work, which made use of the paramagnetic signal quenching effect upon binding of Cu^{2+}, showed that the Cu^{2+} binding sites in mature $A\beta_{1-40}$ fibrils include the Nε sites of H13 and H14, the C′ site of V40, and glutamate sidechains (most likely E3 and/or E11). Most of these sites are located in regions that are outside of the hydrophobic core of mature fibrils. This result, combined with the previous work on membrane-associated monomeric Aβ peptides, suggests a clear difference in terms of the Cu^{2+} binding sites of Aβ at different stages of fibrillation process. Therefore, the detailed molecular mechanism of the production of ROS by metal-bound Aβ must be studied with a careful consideration of the Aβ species (monomer, small oligomers, fibrils, *etc.*) that are directly involved in the process.

Prevention of Aβ fibrillation is considered to be a possible approach for the treatment of AD. The rational design of inhibitors requires high-resolution structural information of Aβ fibrils. It is possible that certain small molecules can be developed to bind to and suppress the growth of fibrils with high toxicity. Solid-state NMR spectroscopy may provide crucial information about the molecular-level details of such binding. Reif and co-workers have reported that the compound epigallocatechin-3-gallate (EGCG) can prevent the formation of $A\beta_{1-40}$ fibrils by stabilizing a non-toxic oligomeric state.[95] The binding between EGCG and $A\beta_{1-40}$ as well as the structures of non-toxic oligomers were investigated by NMR spectroscopy. Solution NMR data suggested that EGCG binds to residues 17–20, excluding these residues from the hydrophobic core region of the oligomers. Residues 10–21 were not observable by solid-state NMR spectra of the EGCG-induced oligomers.

Finally, Smith and co-workers have investigated inhibition of $A\beta_{1-40}$ fibrillation by designed peptides with sequences of the form xGxFxGxF.[6] These peptide are intended to block the interactions between cross-β units in mature fibrils that involve residues 31–37 of Aβ. For one such inhibitor, intermolecular contacts between M35 of $A\beta_{1-40}$ and the two glycines of the inhibitor were detected by solid-state NMR, providing support for the proposed mechanism of inhibition.

28.7 Concluding Remarks

Solid-state NMR has clearly made unique and fundamental contributions to the current structural understanding of Aβ fibrils and related assemblies. Given the inherently non-crystalline and insoluble nature of these species, as well as their high total molecular weights, it is difficult to imagine how the molecular-level details of peptide conformations, long-range intramolecular interactions, or supramolecular structure could have been determined without solid-state NMR data. Indeed, structural studies of Aβ fibrils and related assemblies (and analogous studies of other amyloid and prion fibrils) constitute one of the most successful examples of the application of solid-state NMR methods to problems that are of genuine biological and biomedical significance.

Future research in this area is likely to focus on the systems that are most relevant to the development, diagnosis, and treatment or prevention of AD. Currently, a clear, experimentally supported consensus does not exist regarding the nature of Aβ assemblies that play causative roles in AD. Much attention has been devoted to oligomeric and protofibrillar species, but the identities of species that exist in human brain tissue and their pathogenic mechanisms remain to be established definitively. The polymorphism of Aβ fibrils themselves raises the possibility that certain fibril structures may be more highly pathogenic than others, perhaps due to differences in the biological responses they elicit or differences in their self-propagation efficiencies. If certain fibril structures or certain oligomeric species are indeed more highly pathogenic than others, then it may be productive to investigate compounds that selectively inhibit formation of the most pathogenic Aβ assemblies or stabilize the relatively benign species. With regard to diagnosis, imaging agents that bind preferentially to the most pathogenic Aβ assemblies are likely to be most useful in predicting AD progression. Very recently, the first detailed molecular structural model for brain-derived Aβ fibrils has been reported,[96] setting the stage for such studies.

Obviously, future progress will depend heavily on advances in our understanding of the biological and biomedical aspects of AD. Nonetheless, the success of solid-state NMR studies to date indicates that solid-state NMR measurements will continue to contribute invaluable structural information, with increasing relevance to these biological and biomedical aspects.

Acknowledgements

This work was supported by the Intramural Research Program of the National Institute of Diabetes and Digestive and Kidney Diseases, National Institutes of Health.

References

1. C. Brayne, *Nat. Rev. Neurosci.*, 2007, **8**, 233–239.
2. P. Tiraboschi, L. A. Hansen, L. J. Thal and J. Corey-Bloom, *Neurology*, 2004, **62**, 1984–1989.
3. G. G. Glenner and C. W. Wong, *Biochem. Biophys. Res. Commun.*, 1984, **120**, 885–890.
4. G. Thinakaran and E. H. Koo, *J. Biol. Chem.*, 2008, **283**, 29615–29619.
5. D. J. Gordon, K. L. Sciarretta and S. C. Meredith, *Biochemistry*, 2001, **40**, 8237–8245.
6. T. Sato, P. Kienlen-Campard, M. Ahmed, W. Liu, H. L. Li, J. I. Elliott, S. Aimoto, S. N. Constantinescu, J. N. Octave and S. O. Smith, *Biochemistry*, 2006, **45**, 5503–5516.
7. S. A. Sievers, J. Karanicolas, H. W. Chang, A. Zhao, L. Jiang, O. Zirafi, J. T. Stevens, J. Munch, D. Baker and D. Eisenberg, *Nature*, 2011, **475**, 96–100.
8. S. B. Malinchik, H. Inouye, K. E. Szumowski and D. A. Kirschner, *Biophys. J.*, 1998, **74**, 537–545.
9. M. McDonald, H. Box, W. Bian, A. Kendall, R. Tycko and G. Stubbs, *J. Mol. Biol.*, 2012, **423**, 454–461.
10. M. Sunde and C. C. F. Blake, *Q. Rev. Biophys.*, 1998, **31**, 1–39.
11. M. O. Chaney, S. D. Webster, Y. M. Kuo and A. E. Roher, *Protein Eng.*, 1998, **11**, 761–767.
12. N. D. Lazo and D. T. Downing, *Biochemistry*, 1998, **37**, 1731–1735.
13. A. R. George and D. R. Howlett, *Biopolymers*, 1999, **50**, 733–741.

14. L. P. Li, T. A. Darden, L. Bartolotti, D. Kominos and L. G. Pedersen, *Biophys. J.*, 1999, **76**, 2871–2878.
15. L. O. Tjernberg, D. J. E. Callaway, A. Tjernberg, S. Hahne, C. Lilliehook, L. Terenius, J. Thyberg and C. Nordstedt, *J. Biol. Chem.*, 1999, **274**, 12619–12625.
16. M. F. Perutz, J. T. Finch, J. Berriman and A. Lesk, *Proc. Natl. Acad. Sci. U. S. A.*, 2002, **99**, 5591–5595.
17. P. T. Lansbury, P. R. Costa, J. M. Griffiths, E. J. Simon, M. Auger, K. J. Halverson, D. A. Kocisko, Z. S. Hendsch, T. T. Ashburn, R. G. S. Spencer, B. Tidor and R. G. Griffin, *Nat. Struct. Biol.*, 1995, **2**, 990–998.
18. M. H. Levitt, D. P. Raleigh, F. Creuzet and R. G. Griffin, *J. Chem. Phys.*, 1990, **92**, 6347–6364.
19. J. T. Jarrett, E. P. Berger and P. T. Lansbury, *Biochemistry*, 1993, **32**, 4693–4697.
20. T. L. S. Benzinger, D. M. Gregory, T. S. Burkoth, H. Miller-Auer, D. G. Lynn, R. E. Botto and S. C. Meredith, *Proc. Natl. Acad. Sci. U. S. A.*, 1998, **95**, 13407–13412.
21. T. L. S. Benzinger, D. M. Gregory, T. S. Burkoth, H. Miller-Auer, D. G. Lynn, R. E. Botto and S. C. Meredith, *Biochemistry*, 2000, **39**, 3491–3499.
22. T. S. Burkoth, T. L. S. Benzinger, V. Urban, D. M. Morgan, D. M. Gregory, P. Thiyagarajan, R. E. Botto, S. C. Meredith and D. G. Lynn, *J. Am. Chem. Soc.*, 2000, **122**, 7883–7889.
23. D. M. Gregory, M. A. Mehta, J. C. Shiels and G. P. Drobny, *J. Chem. Phys.*, 1997, **107**, 28–42.
24. R. Tycko and R. B. Wickner, *Acc. Chem. Res.*, 2013, **46**, 1487–1496.
25. J. J. Balbach, Y. Ishii, O. N. Antzutkin, R. D. Leapman, N. W. Rizzo, F. Dyda, J. Reed and R. Tycko, *Biochemistry*, 2000, **39**, 13748–13759.
26. L. O. Tjernberg, C. Lilliehook, D. J. E. Callaway, J. Naslund, S. Hahne, J. Thyberg, L. Terenius and C. Nordstedt, *J. Biol. Chem.*, 1997, **272**, 12601–12605.
27. D. K. Klimov and D. Thirumalai, *Structure*, 2003, **11**, 295–307.
28. D. P. Weliky and R. Tycko, *J. Am. Chem. Soc.*, 1996, **118**, 8487–8488.
29. A. E. Bennett, D. P. Weliky and R. Tycko, *J. Am. Chem. Soc.*, 1998, **120**, 4897–4898.
30. T. Gullion and J. Schaefer, *J. Magn. Reson.*, 1989, **81**, 196–200.
31. A. T. Petkova, G. Buntkowsky, F. Dyda, R. D. Leapman, W. M. Yau and R. Tycko, *J. Mol. Biol.*, 2004, **335**, 247–260.
32. Y. Ishii, J. J. Balbach and R. Tycko, *Chem. Phys.*, 2001, **266**, 231–236.
33. R. Tycko and Y. Ishii, *J. Am. Chem. Soc.*, 2003, **125**, 6606–6607.
34. A. Lange, S. Luca and M. Baldus, *J. Am. Chem. Soc.*, 2002, **124**, 9704–9705.
35. O. N. Antzutkin, J. J. Balbach, R. D. Leapman, N. W. Rizzo, J. Reed and R. Tycko, *Proc. Natl. Acad. Sci. U. S. A.*, 2000, **97**, 13045–13050.
36. Y. S. Yen and A. Pines, *J. Chem. Phys.*, 1983, **78**, 3579–3582.
37. O. N. Antzutkin and R. Tycko, *J. Chem. Phys.*, 1999, **110**, 2749–2752.

38. J. J. Balbach, A. T. Petkova, N. A. Oyler, O. N. Antzutkin, D. J. Gordon, S. C. Meredith and R. Tycko, *Biophys. J.*, 2002, **83**, 1205–1216.
39. A. T. Petkova, Y. Ishii, J. J. Balbach, O. N. Antzutkin, R. D. Leapman, F. Delaglio and R. Tycko, *Proc. Natl. Acad. Sci. U. S. A.*, 2002, **99**, 16742–16747.
40. G. Cornilescu, F. Delaglio and A. Bax, *J. Biomol. NMR*, 1999, **13**, 289–302.
41. F. J. Blanco and R. Tycko, *J. Magn. Reson.*, 2001, **149**, 131–138.
42. C. P. Jaroniec, B. A. Tounge, J. Herzfeld and R. G. Griffin, *J. Am. Chem. Soc.*, 2001, **123**, 3507–3519.
43. C. S. Goldsbury, S. Wirtz, S. A. Muller, S. Sunderji, P. Wicki, U. Aebi and P. Frey, *J. Struct. Biol.*, 2000, **130**, 217–231.
44. A. T. Petkova, R. D. Leapman, Z. H. Guo, W. M. Yau, M. P. Mattson and R. Tycko, *Science*, 2005, **307**, 262–265.
45. W. Qiang, K. Kelley and R. Tycko, *J. Am. Chem. Soc.*, 2013, **135**, 6860–6871.
46. A. T. Petkova, W. M. Yau and R. Tycko, *Biochemistry*, 2006, **45**, 498–512.
47. A. K. Paravastu, R. D. Leapman, W. M. Yau and R. Tycko, *Proc. Natl. Acad. Sci. U. S. A.*, 2008, **105**, 18349–18354.
48. R. Kodali, A. D. Williams, S. Chemuru and R. Wetzel, *J. Mol. Biol.*, 2010, **401**, 503–517.
49. I. Bertini, L. Gonnelli, C. Luchinat, J. F. Mao and A. Nesi, *J. Am. Chem. Soc.*, 2011, **133**, 16013–16022.
50. A. K. Paravastu, I. Qahwash, R. D. Leapman, S. C. Meredith and R. Tycko, *Proc. Natl. Acad. Sci. U. S. A.*, 2009, **106**, 7443–7448.
51. W. Qiang, W. M. Yau and R. Tycko, *J. Am. Chem. Soc.*, 2011, **133**, 4018–4029.
52. M. G. Munowitz, R. G. Griffin, G. Bodenhausen and T. H. Huang, *J. Am. Chem. Soc.*, 1981, **103**, 2529–2533.
53. E. R. De Azevedo, K. Saalwachter, O. Pascui, A. A. De Souza, T. J. Bonagamba and D. Reichert, *J. Chem. Phys.*, 2008, **128**, 104505.
54. H. A. Scheidt, I. Morgado, S. Rothemund and D. Huster, *J. Biol. Chem.*, 2012, **287**, 2017–2021.
55. G. Bitan, M. D. Kirkitadze, A. Lomakin, S. S. Vollers, G. B. Benedek and D. B. Teplow, *Proc. Natl. Acad. Sci. U. S. A.*, 2003, **100**, 330–335.
56. T. Iwatsubo, A. Odaka, N. Suzuki, H. Mizusawa, N. Nukina and Y. Ihara, *Neuron*, 1994, **13**, 45–53.
57. O. N. Antzutkin, R. D. Leapman, J. J. Balbach and R. Tycko, *Biochemistry*, 2002, **41**, 15436–15450.
58. M. Ahmed, J. Davis, D. Aucoin, T. Sato, S. Ahuja, S. Aimoto, J. I. Elliott, W. E. Van Nostrand and S. O. Smith, *Nat. Struct. Mol. Biol.*, 2010, **17**, 561–567.
59. L. Y. Wu, P. Rosa-Neto, G. Y. R. Hsiung, A. D. Sadovnick, M. Masellis, S. E. Black, J. P. Jia and S. Gauthier, *Can. J. Neurol. Sci.*, 2012, **39**, 436–445.
60. Y. Masuda, K. Irie, K. Murakami, H. Ohigashi, R. Ohashi, K. Takegoshi, T. Shimizu and T. Shirasawa, *Bioorg. Med. Chem.*, 2005, **13**, 6803–6809.

61. K. Takegoshi, S. Nakamura and T. Terao, *J. Chem. Phys.*, 2003, **118**, 2325–2341.
62. Y. Masuda, A. Nakanishi, R. Ohashi, K. Takegoshi, T. Shimizu, T. Shirasawa and K. Irie, *Biosci., Biotechnol., Biochem.*, 2008, **72**, 2170–2175.
63. T. Tomiyama, T. Nagata, H. Shimada, R. Teraoka, A. Fukushima, H. Kanemitsu, H. Takuma, R. Kuwano, M. Imagawa, S. Ataka, Y. Wada, E. Yoshioka, T. Nishizaki, Y. Watanabe and H. Mori, *Ann. Neurol.*, 2008, **63**, 377–387.
64. A. L. Cloe, J. Orgel, J. R. Sachleben, R. Tycko and S. C. Meredith, *Biochemistry*, 2011, **50**, 2026–2039.
65. R. Tycko, *J. Chem. Phys.*, 2007, **126**, 064506.
66. R. Tycko, K. L. Sciarretta, J. Orgel and S. C. Meredith, *Biochemistry*, 2009, **48**, 6072–6084.
67. W. Qiang, W. M. Yau, Y. Q. Luo, M. P. Mattson and R. Tycko, *Proc. Natl. Acad. Sci. U. S. A.*, 2012, **109**, 4443–4448.
68. C. R. Morcombe, V. Gaponenko, R. A. Byrd and K. W. Zilm, *J. Am. Chem. Soc.*, 2004, **126**, 7196–7197.
69. D. M. Walsh, D. M. Hartley, Y. Kusumoto, Y. Fezoui, M. M. Condron, A. Lomakin, G. B. Benedek, D. J. Selkoe and D. B. Teplow, *J. Biol. Chem.*, 1999, **274**, 25945–25952.
70. A. D. Williams, M. Sega, M. L. Chen, I. Kheterpal, M. Geva, V. Berthelier, D. T. Kaleta, K. D. Cook and R. Wetzel, *Proc. Natl. Acad. Sci. U. S. A.*, 2005, **102**, 7115–7120.
71. R. Kayed, E. Head, J. L. Thompson, T. M. McIntire, S. C. Milton, C. W. Cotman and C. G. Glabe, *Science*, 2003, **300**, 486–489.
72. I. Benilova, E. Karran and B. De Strooper, *Nat. Neurosci.*, 2012, **15**, 349–357.
73. S. E. Lesne, M. A. Sherman, M. Grant, M. Kuskowski, J. A. Schneider, D. A. Bennett and K. H. Ashe, *Brain*, 2013, **136**, 1383–1398.
74. Y. F. Mok and G. J. Howlett, *Methods Enzymol.*, 2006, **413**, 199–217.
75. A. Jan, D. M. Hartley and H. A. Lashuel, *Nat. Protoc.*, 2010, **5**, 1186–1209.
76. H. A. Scheidt, I. Morgado and D. Huster, *J. Biol. Chem.*, 2012, **287**, 22822–22826.
77. S. Chimon, M. A. Shaibat, C. R. Jones, D. C. Calero, B. Aizezi and Y. Ishii, *Nat. Struct. Mol. Biol.*, 2007, **14**, 1157–1164.
78. S. Chimon and Y. Ishii, *J. Am. Chem. Soc.*, 2005, **127**, 13472–13473.
79. H. A. Scheidt, I. Morgado, S. Rothemund, D. Huster and M. Fandrich, *Angew. Chem. Int. Ed.*, 2011, **50**, 2837–2840.
80. A. Laganowsky, C. Liu, M. R. Sawaya, J. P. Whitelegge, J. Park, M. L. Zhao, A. Pensalfini, A. B. Soriaga, M. Landau, P. K. Teng, D. Cascio, C. Glabe and D. Eisenberg, *Science*, 2012, **335**, 1228–1231.
81. W. Hoyer, C. Gronwall, A. Jonsson, S. Stahl and T. Hard, *Proc. Natl. Acad. Sci. U. S. A.*, 2008, **105**, 5099–5104.
82. A. Sandberg, L. M. Luheshi, S. Sollvander, T. P. de Barros, B. Macao, T. P. J. Knowles, H. Biverstal, C. Lendel, F. Ekholm-Petterson, A. Dubnovitsky,

L. Lannfelt, C. M. Dobson and T. Hard, *Proc. Natl. Acad. Sci. U. S. A.*, 2010, **107**, 15595–15600.
83. T. Doi, Y. Masuda, K. Irie, K. Akagi, Y. Monobe, T. Imazawa and K. Takegoshi, *Biochem. Biophys. Res. Commun.*, 2011, **428**, 458–462.
84. S. M. Butterfield and H. A. Lashuel, *Angew. Chem. Int. Ed.*, 2010, **49**, 5628–5654.
85. M. F. M. Engel, L. Khemtemourian, C. C. Kleijer, H. J. D. Meeldijk, J. Jacobs, A. J. Verkleij, B. de Kruijff, J. A. Killian and J. W. M. Hoppener, *Proc. Natl. Acad. Sci. U. S. A.*, 2008, **105**, 6033–6038.
86. S. Ravault, O. Soubias, O. Saurel, A. Thomas, R. Brasseur and A. Milon, *Protein Sci.*, 2005, **14**, 1181–1189.
87. T. L. Lau, E. E. Ambroggio, D. J. Tew, R. Cappai, C. L. Masters, G. D. Fidelio, K. J. Barnham and F. Separovic, *J. Mol. Biol.*, 2006, **356**, 759–770.
88. A. Quist, L. Doudevski, H. Lin, R. Azimova, D. Ng, B. Frangione, B. Kagan, J. Ghiso and R. Lal, *Proc. Natl. Acad. Sci. U. S. A.*, 2005, **102**, 10427–10432.
89. A. S. Pithadia and M. H. Lim, *Curr. Opin. Chem. Biol.*, 2012, **16**, 67–73.
90. V. Tougu, A. Tiiman and P. Palumaa, *Metallomics*, 2011, **3**, 250–261.
91. D. P. Smith, D. G. Smith, C. C. Curtain, J. F. Boas, J. R. Pilbrow, G. D. Ciccotosto, T. L. Lau, D. J. Tew, K. Perez, J. D. Wade, A. I. Bush, S. C. Drew, F. Separovic, C. L. Masters, R. Cappai and K. J. Barnham, *J. Biol. Chem.*, 2006, **281**, 15145–15154.
92. T. L. Lau, J. D. Gehman, J. D. Wade, K. Perez, C. L. Masters, K. J. Barnham and F. Separovic, *Biochim. Biophys. Acta, Biomembr.*, 2007, **1768**, 2400–2408.
93. J. D. Gehman, C. C. O'Brien, F. Shabanpoor, J. D. Wade and F. Separovic, *Eur. Biophys. J.*, 2008, **37**, 333–344.
94. S. Parthasarathy, F. Long, Y. Miller, Y. L. Xiao, D. McElheny, K. Thurber, B. Y. Ma, R. Nussinov and Y. Ishii, *J. Am. Chem. Soc.*, 2011, **133**, 3390–3400.
95. J. M. L. del Amo, U. Fink, M. Dasari, G. Grelle, E. E. Wanker, J. Bieschke and B. Reif, *J. Mol. Biol.*, 2012, **421**, 517–524.
96. J.-X. Lu, W. Qiang, W.-M. Yau, C. D. Schwieters, S. C. Meredith and R. Tycko, *Cell*, 2013, **154**, 1257–1268.

Subject Index

References to tables and charts are in **bold** type